WORLD HEALTH ORGANIZATION
INTERNATIONAL AGENCY FOR RESEARCH ON CANCER

RELEVANCE TO HUMAN CANCER OF

N-NITROSO COMPOUNDS,

TOBACCO AND MYCOTOXINS

Editors

**I.K. O'Neill, J. Chen
and H. Bartsch**

IARC Scientific Publications No. 105

International Agency for Research on Cancer
Lyon, 1991

Published by the International Agency for Research on Cancer
150 cours Albert Thomas, 69372 Lyon Cedex 08, France

© International Agency for Research on Cancer, 1991

Distributed by Oxford University Press, Walton Street, Oxford OX2 6DP, UK

Distributed in the USA by Oxford University Press, New York

All rights reserved. No part of this publication may be reproduced, stored in a retrieval system, or transmitted, in any form or by any means, electronic, mechanical, photocopying, recording, or otherwise, without the prior permission of the copyright holder.

The authors alone are responsible for the views expressed in the signed articles in this publication.

ISBN 92 832 2105 2

ISSN 0300 5085

Printed in the United Kingdom

CONTENTS

Foreword xvii
Organizers ixx

N-Nitroso compounds and human cancer: where do we stand?
    H. Bartsch 1

Relevance of N-nitrosamines to oesophageal cancer in China
    S.H. Lu, S.X. Chui, W.X. Yang, X.N. Hu, L.P. Guo & F.M. Li 11

Dietary practices and cancer mortality in China
    J. Chen 18

The etiology of gastric cancer
    D. Forman 22

Etiological research on gastric cancer and its precursor lesions in Shandong, China
    W.C. You, Y.S. Chang, Z.T. Yang, L. Zhang, G.W. Xu, W.J. Blot,
    R. Kneller, L.K. Keefer & J.F. Fraumeni, Jr 33

Nasopharyngeal carcinoma: epidemiology and dietary factors
    M.C. Yu 39

Aflatoxins: data on human carcinogenic risk
    F.X. Bosch & F. Peers 48

N-Nitroso compounds and tobacco-induced cancers in man
    S.S. Hecht & D. Hoffmann 54

Tobacco smoking and its effect on health in China
    Y.T. Gao, W. Zheng, R.N. Gao & F. Jin 62

DOSIMETRY METHODS FOR DETERMINING HUMAN EXPOSURE TO N-NITROSO COMPOUNDS, MYCOTOXINS AND TOBACCO

Biomonitoring of human exposure to alkylating agents by measurement of adducts to haemoglobin or DNA
    P.B. Farmer, E. Bailey, J.A. Green, C.S. Leung & M.M. Manson 71

Novel, sensitive assays for O^6-alkylguanine and its repair and their application to studies of the molecular epidemiology of this lesion in human populations
  S.A. Kyrtopoulos, V.L. Souliotis, P. Ambatzi, G. Pangalis, V. Bousiotou, G. Haritopoulos & G. Davaris 78

Comparison of urinary nitrate, N-nitrosoproline, 7-methylguanine and 3-methyladenine levels in a human population at risk for gastric cancer
  W.G. Stillwell, H.X. Xu, J. Glogowski, J.S. Wishnok, S.R. Tannenbaum & P. Correa 83

Endogenous nitrosamines and liver fluke as risk factors for cholangiocarcinoma in Thailand
  P. Srivatanakul, H. Ohshima, M. Khlat, M. Parkin, S. Sukaryodhin, I. Brouet & H. Bartsch 88

Possible effect of infection with liver fluke (*Opisthorchis viverrini*) on the monitoring of urine by enzyme-linked immunosorbent assay for human exposure to aflatoxins
  K. Makarananda, C.P. Wild, Y.Z. Jiang & G.E. Neal 96

A rapid gas chromatography–mass spectrometry method for the determination of urinary 3-methyladenine; application in human subjects
  D.E.G. Shuker, M.D. Friesen, L. Garren & V. Prévost 102

Tandem mass spectrometric approaches for determining exposure to alkylating agents
  J.R. Cushnir, J.H. Lamb, A. Parry & P.B. Farmer 107

Quantification of 4-hydroxy-1-(3-pyridyl)-1-butanone released from human haemoglobin as a dosimeter for exposure to tobacco-specific nitrosamines
  S.S. Hecht, S.S. Kagan, M. Kagan & S.G. Carmella 113

Promutagenic lesions persist in the DNA of target cells for nitrosamine-induced carcinogenesis
  C.Y. Fan, W.H. Butler & P.J. O'Connor 119

Nitrite-trapping capacity of thioproline in the human body
  M. Tsuda & Y. Kurashima 123

Practical aspects of application of dosimetry methods: summary of a workshop discussion
  D.E.G. Shuker & D. Forman 129

EXPOSURES AND ETIOLOGY

Endogenous N-nitrosation and cancer of the biliary tract
 C.P.J. Caygill, S.A. Leach, F. Fernandez, S. Duncan, M.J. Hill,
 R. Massey & R. Hall 137

Effect of ascorbic acid on the intragastric environment in patients at
 increased risk of developing gastric cancer
 P.I. Reed, B.J. Johnston, C.L. Walters & M.J. Hill 139

An approach to establishing N-nitroso compounds as the cause of gastric
 cancer
 K.X. Shi, D.J. Mao, W.F. Cheng, Y.S. Ji & L.Z. Xu 143

The N-nitrosoproline test as a measure of cancer risk in geographical
 comparison studies: results from Italy and an overall comparison
 T. Knight, D. Forman, S.A. Leach, P. Packer, G. Cocco, D. Palli &
 R. Pirastu 146

Role of nitrosamides in the high risk for gastric cancer in China
 R.F. Zhang, D.J. Deng, Y. Chen, H.Y. Wu & C.S. Chen 152

Microflora of the nasopharynx in Caucasian and Maghrebian subjects with
 and without nasopharyngeal carcinoma
 M. Charrière, S. Poirier, S. Calmels, H. De Montclos, C. Dubreuil,
 R. Poizat, M. Hamdi Cherif & G. de-Thé 158

Exposure to N-nitrosamines and other risk factors for gastric cancer in Costa
 Rican children
 R. Sierra, H. Ohshima, N. Muñoz, S. Teuchmann, A.S. Peña,
 C. Malaveille, B. Pignatelli, A. Chinnock, F. El Ghissassi, C. Chen,
 A. Hautefeuille, C. Gamboa & H. Bartsch 162

Urinary excretion of N-nitrosoproline in relation to consumption of raw and
 cooked vegetables in a Danish rural population.
 H. Møller, J. Landt, E. Pedersen, P. Jensen, H. Autrup & O.M. Jensen ... 168

N-Nitroso compounds, genotoxins and their precursors in gastric juice from
 humans with and without precancerous lesions of the stomach
 B. Pignatelli, C. Malaveille, C. Chen, A. Hautefeuille, P. Thuillier,
 N. Muñoz, B. Moulinier, F. Berger, H. De Montclos,
 H. Ohshima, R. Lambert & H. Bartsch, ... 172

Urinary nitrate, nitrite and N-nitroso compounds in bladder cancer
patients with schistosomiasis (bilharzia)
  A.R. Tricker, M.H. Mostafa, B. Spiegelhalder & R. Preussmann 178

Formation of N-trimethyl-N-nitrosourea in the gastric lumen of the
fistulated pig
  C.M. Maragos, J.H. Hotchkiss & S.L. Fubini 182

Bacterial formation of N-nitroso compounds in the rat stomach after
omeprazole-induced achlorhydria
  S. Calmels, J.-C. Béréziat, H. Ohshima & H. Bartsch 187

Epidemiological study of precursor lesions of oesophageal cancer among
young persons in Huixian, China
  J. Chang-Claude, J. Wahrendorf, S.L. Qiu, G.R. Yang, N. Muñoz,
  M. Crespi, R. Raedsch, D.I. Thurnham & P. Correa 192

Gliomas and meningiomas in men in Los Angeles county: investigation of
exposures to N-nitroso compounds
  S. Preston-Martin & W. Mack 197

Epstein–Barr virus activators, mutagens and volatile nitrosamines in
preserved food samples from high-risk areas for nasopharyngeal carcinoma
  G. Bouvier, S. Poirier, Y.M. Shao, C. Malaveille, H. Ohshima, A. Polack,
  G.W. Bornkamm, Y. Zeng, G. de-Thé & H. Bartsch 204

Dietary sources of N-nitrosamines in a high-risk area for oesophageal
cancer — Kashmir, India
  M.A. Siddiqi, A.R. Tricker, R. Kumar, Z. Fazili & R. Preussmann 210

Volatile N-nitrosamines in fish meal, with special reference to the
mechanism of formation of N-nitrosothiazolidine
  H. Tozawa & T. Kawabata 214

Regional differences in N-nitrosamine content of traditional Chinese foods
  J. Gao, J.H. Hotchkiss & J. Chen 219

Trace analysis of N-nitrosoureas by their alkylating activity
  P. Mende, B. Spiegelhalder & R. Preussmann 223

Potential occupational exposure associated with parenteral administration of N-nitrosodiethylamine to *Macaca mulatta*: evidence of post-injection release to the atmosphere
    E.B. Sansone & S.D. Keimig 226

Analysis of volatile N-nitrosamines in beer and food in China
    H.X. Xu, X.H. Shen, Z.L. Jin & X.B. Xu 230

Recent studies in Canada on the occurrence and formation of N-nitroso compounds in foods and food contact materials
    N.P. Sen 232

Occurrence of and exposure to N-nitrosamines in Sweden: a review
    B.-G. Österdahl 235

N-Nitrosoalkanolamines in cosmetics
    G. Eisenbrand, M. Blankart, H. Sommer & B. Weber 238

N-Nitrosodimethylamine content of US and Canadian beers
    R.A. Scanlan & J.F. Barbour 242

Nitrosation of tertiary aromatic amines related to sunscreen ingredients
    R.N. Loeppky, R. Hastings, J. Sandbothe, D. Heller, Y. Bao & D. Nagel .. 244

Mutagenicity of *Alternaria alternata* and *Penicillium cyclopium* isolated from grains in an area of high incidence of oesophageal cancer — Linxian, China
    Y.Z. Zhen, Y.M. Xu, G.T. Liu, J. Miao, Y.D. Xing, Q.L. Zheng, Y.F. Ma, T. Su, X.L. Wang, L.R. Ruan, J.F. Tian, G. Zhou & S.L. Yang 253

Relationships between *Alternaria alternata* and oesophageal cancer
    G.T. Liu, Y.Z. Qian, P. Zhang, Z.M. Dong, Z.Y. Shi, Y.Z. Zhen, J. Miao & Y.M. Xu 258

MECHANISMS AND BIOLOGICAL EFFECTS OF N-NITROSO COMPOUNDS AND MYCOTOXINS

Enzyme mechanisms in the metabolism of nitrosamines
    C.S. Yang, T. Smith, H. Ishizaki & J.Y. Hong 265

Modulation of mycotoxin and nitrosamine carcinogenesis by indole-3-carbinol: quantitative analysis of inhibition *versus* promotion
    G.S. Bailey, R.H. Dashwood, A.T. Fong, D.E. Williams, R.A. Scanlan & J.D. Hendricks 275

Areca-nut toxicity in cultured human buccal epithelial cells
  K. Sundqvist, Y. Liu, P. Erhardt, J. Nair, H. Bartsch & R.C. Grafström .. 281

Bioactivation of asymmetric N-dialkylnitrosamines in rat tissues derived from the ventral entoderm
  B. Ludeke, T. Meier & P. Kleihues 286

Role of oncogenes and tumour suppressor genes in human lung carcinogenesis
  C.C. Harris, R. Reddel, A. Pfeifer, D. Iman, M. McMenamin, B.F. Trump & A. Weston 294

Alkylation of DNA related to organ-specific carcinogenesis by N-nitroso compounds
  W. Lijinsky 305

Relationship between dose and risk reduction: statistical evaluation of a combination experiment with three hepatocarcinogenic N-nitrosamines in rats
  M.R. Berger, D. Schmähl & L. Edler 311

Carcinogenic activity of endogenously synthesized N-nitrosobis(2-hydroxypropyl)amine in rats
  Y. Konishi, K. Yamamoto, H. Eimoto, M. Tsutsumi, M. Sugimura, H. Nii & Y. Mori 318

Effects of long-term inhalation of N-nitrosodimethylamine in rats
  R.G. Klein, I. Janowsky, B.L. Pool, P. Schmezer, R. Hermann, F. Amelung, B. Spiegelhalder & W.J. Zeller 322

Specificity in the methylation of DNA by N-nitroso compounds
  J. Milligan, S. Skotnicki, S.J. Lu & M.C. Archer 329

Mechanism of action of the urinary bladder carcinogen N-nitrosobutyl-3-carboxypropylamine
  C. Janzowski, D. Jacob, I. Henn, H. Zankl, B.L. Pool-Zobel, M. Wiessler & G. Eisenbrand 332

Mutagenicity, DNA damage and DNA adduct formation by N-nitroso-2-hydroxyalkylamines and corresponding aldehydes
  G. Scherer, B. Ludeke, P. Kleihues, R.N. Loeppky & G. Eisenbrand 339

List of presentations

N-Nitrosobutyl(4-hydroxybutyl)amine α-hydroxylation by rat liver and urothelial cell homogenates
  L. Airoldi, C. Magagnotti, M. Bonfanti, M. Moret & R. Fanelli 343

Mechanism of DNA binding by the oesophageal carcinogen N-nitroso-N-methylaniline
  S.R. Koepke, M.B. Kroeger-Koepke & C.J. Michejda 346

Metabolic denitrosation of N-nitrosamines: mechanism and biological consequences
  K.E. Appel, S. Görsdorf, T. Scheper, B. Spiegelhalder, M. Wiessler, M. Schoepke, C. Engeholm & R. Kramer 351

Subcellular fractions of rat organs contain nitroreductases which reduce N-nitrodimethylamine to N-nitrosodimethylamine
  E. Frei, M. Hassel, B. Spiegelhalder & M. Wiessler 358

Toxicokinetic studies of N-nitrosamine carcinogenesis
  A.J. Streeter, R.W. Nims & L.K. Keefer 362

Enzyme kinetics of N-nitrosodimethylamine demethylase in rodents and humans
  J.S.H. Yoo, H. Ishizaki & C.S Yang 366

Acceleration of N-nitrosation reactions by electrophiles
  B.P. Sullivan, T.J. Meyer, M.T. Stershic & L.K. Keefer 370

Nitrosamine activation and detoxication through free radicals and their derived cations
  R.N. Loeppky & Y.E. Li 375

Alkylating potency of nitrosated amino acids and peptides
  S.E. Shephard, I. Meier & W.K. Lutz 383

Macrophages produce nitrite, nitrate and nitrosamines after addition of catalase
  M. Miwa, M. Watanabe, T. Nishida & K. Shinohara 388

Use of monoclonal antibodies to identify cytochrome P450 isozymes in rat liver microsomes that hydroxylate N-nitrosomethylamylamine at each of six positions
  S.S. Mirvish, C. Ji, Q. Huang, S. Wang, S.S. Park & H.V. Gelboin 392

Participation of phenobarbital-inducible cytochrome P450 in the mutagenic activation of *N*-nitrosopropylamines by liver and lung 9000 *g* fractions from five animal species and man
    Y. Mori & Y. Konishi 398

Biological and chemical properties of alkanediazotates as active species of *N*-nitroso compounds
    S. Ukawa & M. Mochizuki 404

Activity of O^6-alkylguanine-DNA alkyltransferase in the liver, kidney and white blood cells of rats of different ages
    A. Likhachev, N. Zhukovskaya, V. Anisimov, J. Hall & N. Napalkov 407

Excision of imidazole ring-opened *N*7-hydroxyethylguanine from chloro-ethylnitrosourea-treated DNA by *Escherichia coli* formamidopyrimidine-DNA glycosylase
    J. Laval, F. Lopès, J.C. Madelmont, D. Godenèche, G. Meyniel,
        Y. Habraken, T.R. O'Connor & S. Boiteux 412

Resistance to *N*-nitroso compounds in cells treated with various physical and chemical agents
    P. Lefebvre & F. Laval 417

Effect of long-term feeding of nivalenol on aflatoxin B_1-initiated hepato-carcinogenesis in mice
    Y. Ueno, T. Kobayashi, H. Yamamura, T. Kato, F. Tashiro, K. Nakamura
        & K. Ohtsubo 420

Distribution and excretion of 3H-sterigmatocystin in rats
    D.S. Wang, H.L. Sun, F.Y. Xiao, X.H. Ji, Y.X. Liang & F.G. Han 424

Aflatoxin B_1-DNA binding and aflatoxin B_1-glutathione conjugation with isolated hepatocytes from rats and hamsters
    L.L. Ho, E.C. Jhee, P. Gopalan, K. Tsuji & P.D. Lotlikar 427

Reliability of a short-term test for hepatocarcinogenesis induced by aflatoxin B_1
    Y. Li, R.Q. Yan, G.Z. Qin, L.L. Qin & X.X. Duan 431

Chemical transformation of human embryonic nasopharyngeal epithelial cells *in vitro*
    Z.C. Chen, S.C. Pan & K.T. Yao 434

Control over the sequence specificity of DNA alkylation: syntheses and reactions with 32P-end-labelled DNA of *N*-alkyl-*N*-nitrosoureas linked to minor groove binding lexitropsins
    B. Gold, K.M. Church, R.L. Wurdeman, Y. Zhang & F.X. Chen 439

Nitrotyrosine as a new marker for endogenous nitrosation and nitration
    H. Ohshima, I. Brouet, M. Friesen & H. Bartsch 443

TOBACCO-RELATED CANCER

Lung cancer and the changing cigarette
    D. Hoffmann, I. Hoffmann & E.L. Wynder 449

Environmental determinants of lung cancer in Shenyang, China
    Z.Y. Xu, W.J. Blot, G. Li, J.F. Fraumeni, Jr, D.Z. Zhao, B.J. Stone,
    Q. Yin, A. Wu, B.E. Henderson & B.P. Guan 460

Betel quid and oral cancer: prospects for prevention
    P.C. Gupta 466

Lung cancer: political measures
    J.M. Mackay 471

Analysis and pyrolysis of some *N*-nitrosamino acids in tobacco and tobacco smoke
    K.D. Brunnemann, M.V. Djordjevic, R. Feng & D. Hoffmann 477

Studies in tobacco carcinogenesis
    D. Hoffmann, A.A. Melikian & K.D. Brunnemann 482

Carcinogenic substances in Soviet tobacco products
    D.G. Zaridze, R.D. Safaev, G.A. Belitsky, K.D. Brunnemann &
    D. Hoffmann 485

Tobacco-specific nitrosamines in commercial cigarettes: possibilities for reducing exposure
    S. Fischer, B. Spiegelhalder & R. Preussmann 489

Occurrence of and exposure to *N*-nitroso compounds in tobacco
    A.R. Tricker & R. Preussmann 493

Localization of DNA adducts formed in the nasal cavity of the rat by the tobacco-specific nitrosamine 4-(*N*-nitrosomethylamino)-1-(3-pyridyl)-1-butanone (NNK)
J. Van Benthem, J.W.G.M. Wilmer, L. Den Engelse & E. Scherer 496

Feasibility of a prospective study of smoking and mortality in Qidong, China
J.G. Chen, R. Peto, Z.T. Sun & Y.R. Zhu 502

Lack of promoting ability of snuff in rats initiated with 4-nitroquinoline-*N*-oxide
S.L. Johansson, J.M. Hirsch, P.-A. Larsson, J. Saidi & B.-G. Österdahl ... 507

Characterization of activation and deactivation pathways of 4-(*N*-nitrosomethylamino)-1-(3-pyridyl)-1-butanone (NNK) in rat hepatocytes
L. Liu, M.A. Alaoui-Jamali, N. El Alami & A. Castonguay 510

Activation of *N'*-nitrosonornicotine by hydrogen peroxide *in vitro*
J. Nair, U.J. Nair, A.J. Amonkar & S.V. Bhide 516

Antimutagenic and anticarcinogenic effects of betel leaf extract against the tobacco-specific nitrosamine 4-(*N*-nitrosomethylamino)-1-(3-pyridyl)-1-butanone (NNK)
S.V. Bhide, P.R. Padma & A.J. Amonkar 520

Effect of vitamin A status of rats on metabolizing enzymes after exposure to tobacco extract or *N'*-nitrosonornicotine
U.J. Nair, N. Ammigan, M. Nagabhushan, A.J. Amonkar & S.V. Bhide ... 525

Inhibition of tobacco-specific nitrosamine 4-(*N*-nitrosomethylamino)-1-(3-pyridyl)-1-butanone (NNK) tumorigenesis with aromatic isothiocyanates
M.A. Morse, K.I. Eklind, S.S. Hecht & F.L. Chung 529

Modulation of genotoxic activity of tobacco smoke
R.M. Balansky, P.M. Blagoeva & Z.I. Mircheva 535

PREVENTION OF EXPOSURE TO *N*-NITROSO COMPOUNDS

Chinese tea inhibits the occurrence of oesophageal tumours induced by *N*-nitrosomethylbenzylamine and blocks its formation in rats
C. Han & Y. Xu 541

List of presentations

Inhibitory effect of Chinese tea on N-nitrosation *in vitro* and *in vivo*
  H. Wang & Y. Wu 546

Inhibitory effect of ellagic acid on genotoxicity induced by aflatoxins B_1 and G_1
  T. Górski, E. Górska, J. Odlanicki & M. Sikora 550

Some approaches to prevention of endogenous formation of N-nitrosamines in humans
  P.P. Dikun, V.B. Ermilov & I.A. Shendrikova 552

Inhibition by fatty acids of direct mutagenicity of N-nitroso compounds
  K. Takeda, S. Ukawa & M. Mochizuki 558

Potent inhibition of oesophageal metabolism of N-nitrosomethylbenzylamine, an oesophageal carcinogen, by higher alcohols present in alcoholic beverages
  V.M. Craddock & A.R. Henderson 564

Prevention of tumour production in rats fed aminopyrine plus nitrite by sea buckthorn juice
  Y. Li & H. Liu 568

Inhibition of bacterially mediated N-nitrosation by ascorbate: therapeutic and mechanistic considerations
  S.A. Leach, C.W. Mackerness, M.J. Hill & M.H. Thompson 571

Retinoids prevent epithelial carcinogenesis induced by N-nitroso compounds
  H.Y. Cai 579

Stability of mutagenic nitrosated products of indole compounds occurring in vegetables
  H.G.M. Tiedink, J.A.R. Davies, W.M.F. Jongen
    & L.W. van Broekhoven 584

Newly synthesized dithiocarbamates inhibit the metabolism and toxicity of N-nitrosodimethylamine
  N. Frank, B. Bertram, H.R. Scherf & M. Wiessler 588

New sulfenamide accelerators derived from 'safe' amines for the rubber and tyre industry
  C.-D. Wacker, B. Spiegelhalder & R. Preussmann 592

List of participants 595
Nomenclature and abbreviations for *N*-nitrosamines 603
Author index... 605
Subject index 609

FOREWORD

These proceedings are the result of the tenth meeting in a series held over the past 20 years, each time bringing together a group of researchers that over the years has become truly multidisciplinary, to discuss the relevance of N-nitroso compounds as human carcinogens. The number and (what counts more) the quality of the papers presented and published in the proceedings of these meetings (after rigorous peer review) have increased over the past two decades, indicating the continuing interest in this subject.

It all started in 1956, with the discovery by Barnes and Magee of the carcinogenicity of N-nitrosodimethylamine. Over the years, the methods for the detection of N-nitrosamines have evolved, the occurrence of these compounds in the environment has been amply demonstrated and our understanding of their mechanisms of action has advanced considerably.

During the last ten years, interest has been directed to studying their interactions with human tissues and cells and their formation in humans. This development is clearly reflected in the successive changes in emphasis of the meetings, which started primarily on analysis, formation and occurrence of N-nitroso compounds then shifted to exposure to and the effects of N-nitroso compounds in humans; closer collaboration with epidemiologists naturally evolved towards the inclusion among the subjects of the present meeting of other important human carcinogens of particular relevance to China, where this meeting had originally been scheduled to take place.

The presentations included in this volume will hopefully contribute to resolving some of the questions about the modalities and extent of exposures to these substances and, equally or perhaps even more importantly, how preventive measures can be more efficiently implemented. One of the difficulties with regard to N-nitroso compounds is that exposure may occur over a long period, in most instances to low levels, and to compounds that can be formed endogenously.

While there is little doubt that persistent infection with hepatitis B virus plays an important and fundamental role in the genesis of primary liver cancer, contrasting views exist on whether the virus or a mycotoxin is finally more important. For example, evidence has been provided that chronic infection with hepatitis B virus is probably the leading cause of hepatocellular carcinoma throughout the world, that the virus itself is a potent carcinogen *per se* and that exogenous cofactors like mycotoxins may only speed up the process of carcinogenesis in virus carriers. In contrast, it was shown in a recent publication that exposure to aflatoxins is linearly correlated with mortality from hepatocellular carcinoma, and no association was found between the prevalence of hepatitis B antigens and mortality from this cancer.

Another conspicuous example of the possible importance of a virus–chemical interaction is the role of the Epstein-Barr virus and certain food items possibly contaminated with N-nitrosamines in the origin of nasopharyngeal carcinoma. Although interactions between viruses and chemicals were an area of research that was very fashionable 20 years ago, advances in molecular biology now provide the means for studying them closely.

Tobacco smoke, and the use of tobacco in general, are included in this meeting not only because specific N-nitroso compounds are suspected to be the agents primarily responsible for tobacco-induced cancers but also (and mainly) because more efficient prevention is needed than has been available until now. The greatest concerns are the spread of smoking habits to developing countries and the increased prevalence of cigarette smoking among young women. There is no evidence that women are less susceptible than men to lung carcinogens — there is, if anything, evidence to the contrary. We may therefore expect that mortality from lung cancer will continue to increase rapidly in women too, if nothing is done to stop the habit.

A very interesting section of these proceedings concerns factors that may modulate or condition individual responses to environmental carcinogens. About 200 years ago, Dr J. Hill discovered that use of snuff resulted in oral cancer, and he warned against its immoderate use: 'no man should venture upon snuff, who is not sure that he is not so far liable to cancer: and no man can be sure of that.' In 1989, we may just be beginning to be able to identify the most susceptible individuals in a population. While it is essential to assess the extent to which genetic factors contribute to individual cancer risk due to environmental exposures, however, the most efficient prevention nevertheless lies in avoiding, whenever possible, exposures to carcinogens.

I am pleased to pay tribute here to Helmut Bartsch, who is not only a pioneer in research on N-nitroso compounds but, as secretary of the last four conferences, has gradually changed their emphasis and thus contributed substantially to bringing epidemiology and experimental oncology together, favouring in this way a closer integration of laboratory methods into epidemiological surveys.

This meeting, as mentioned before, should have been held in Beijing, where it would have provided the occasion for a wider exchange of ideas and discussions between Chinese and foreign scientists on issues of particular relevance to China. I am none the less confident that the proceedings of this meeting, with the qualified input of several Chinese scientists, will still contribute to the solution of some of the largest cancer problems in China, and that our collaboration with Chinese scientists will continue fruitfully.

Lastly, I would like to thank the persons and institutions who provided major contributions to making this meeting possible: Dr Chen of the Chinese Academy of Preventive Medicine and Dr Lu of the Chinese Academy of Medical Sciences. In addition, support from the National Cancer Institute (USA), the International Programme on Chemical Safety (WHO) and the National Institute of Environmental Health Sciences (USA) is gratefully acknowledged, as are a number of generous contributions from private donors and industries.

Lorenzo Tomatis, M.D.
Director, IARC

ORGANIZERS OF THE TENTH INTERNATIONAL MEETING ON N-NITROSO COMPOUNDS, MYCOTOXINS AND TOBACCO SMOKE: RELEVANCE TO HUMAN CANCER

Organizing Secretariat of the International Agency for Research on Cancer

H. Bartsch
I.K. O'Neill
L. Neyret
M. Wrisez
E. Heseltine

Programme Committee

J.S. Chen
B. Chen
C.C. Harris
S.S. Hecht
P. Magee
R. Peto
G.N. Wogan
C.S. Yang

Relevance to Human Cancer of *N*-Nitroso Compounds,
Tobacco Smoke and Mycotoxins.
Ed. I.K. O'Neill, J. Chen and H. Bartsch
Lyon, International Agency for Research on Cancer
© IARC, 1991

N-NITROSO COMPOUNDS AND HUMAN CANCER: WHERE DO WE STAND?

H. Bartsch

International Agency for Research on Cancer, Lyon, France

Humans are exposed not only to preformed *N*-nitroso compounds (NOC) but also to a wide range of nitrogen-containing compounds and nitrosating agents which can react *in vivo* to form NOC, a versatile class of carcinogens. Nitrosating agents and NOC can also be synthesized endogenously in reactions mediated by bacteria and activated macrophages. Thus, endogenous formation of NOC can occur at various sites in the body.

A sensitive procedure (the *N*-nitrosoproline (NPRO) test) has been developed to estimate exposure of humans to exogenous and endogenous NOC. Results of studies in human subjects with this test led to the following conclusions: (1) The process of endogenous nitrosation in humans is influenced by many factors; therefore, determination only of nitrate and nitrite in body fluids is insufficient to assess the extent of nitrosation in man in vivo. (2) In clinical studies to examine the model of gastric carcinogenesis based on bacterial colonization and nitrosation *in vivo*, progress has been made in explaining some steps, but several controversies remain. Although bacterial strains possessing enzymes that catalyse *N*-nitrosamine formation at neutrality have been isolated from the gastric juice of achlorhydric subjects, their precise role in gastric carcinogenesis remains to be clarified. (3) Formation of endogenous NOC was assessed by the NPRO test in: (i) subjects living in high- and low-incidence areas for stomach cancer in northern Japan, Costa Rica and Poland; (ii) subjects with different habits of betel-quid chewing and tobacco use; (iii) patients with urinary bladder infections; and (iv) subjects infested with liver fluke in Thailand. In all instances, greater exposure to endogenous NOC was found in high-risk subjects, but individual exposure was greatly affected by dietary modifiers of disease state: ascorbic acid efficiently lowered the body burden of intragastrically formed NOC. (4) Increased nitrosation is also observed in tobacco smokers, adding to the body burden of ingested or inhaled tobacco-related carcinogens.

These results, together with the knowledge that NOC produce tumours in 40 animal species, clearly underline the potential role of NOC (and other nitrite-reactive compounds) in human cancer etiology, particularly when exposure starts early in life and persists over a long period. The demonstrated efficacy of certain vitamins as nitrosation inhibitors also provides a plausible interpretation of epidemiological findings that have shown protective effects of fruits and vegetables (sources of vitamins and polyphenols) against various malignancies and particularly stomach cancer. Future research should include: (i) the development of additional markers of (preferably) medium- and long-term exposure to nitrosating and alkylating agents, e.g., formed by nitrosation reactions mediated by bacteria and macrophages or by nitrogen oxides in the lungs; (ii) integration of such dosimetry methods in epidemiological studies, together with the detection of (cyto)genetic damage and measures of host-susceptibility factors; and (iii) identification of high-risk subjects, followed by attempts to lower exposure to carcinogens from endogenous and exogenous sources.

The question 'Nitrosamines and human cancer: where do we stand?' is particularly pertinent since these are the proceedings of the tenth meeting in a series on *N*-nitroso compounds (NOC). More than 30 years of research have passed since the discovery of the carcinogenicity of these compounds by Magee and Barnes (1956). The progressive change in the emphasis of the meetings in this series reflects the increasingly widely held view that NOC are involved in the causation of some human cancers. This is contrary to what John Barnes believed in 1974, as documented by his statement: 'Preoccupation with the occurrence and behaviour of minute amounts of nitrosamines in the human environment will probably divert skills from more profitable studies in experimental systems.' (Barnes, 1974). That Barnes may have underestimated the etiological relevance of NOC has become clear from subsequent studies in both experimental systems and humans. Circumstantial evidence that indicates the hazard of NOC as human carcinogens includes the following: (i) We know now that many NOC are multi-species and multi-organ carcinogens, a property shared with most other human carcinogens identified in the *IARC Monographs* programme (Bartsch & Malaveille, 1989). (ii) The cellular and molecular damage produced by some well-studied nitrosamines is virtually identical in animal and human tissues (Bartsch & Montesano, 1984). (iii) There is growing evidence for widespread human exposure to exogenous and endogenous NOC (National Research Council, 1981; Shephard *et al.*, 1987). (iv) The time-dose relationships that Druckrey *et al.* (1963) observed for tumour formation in animals indicate that lifetime or chronic exposure to even low levels of certain carcinogenic NOC may represent a cancer risk to man because of his long life span. (v) Only recently, it has been found that *N*-nitroso compounds can be synthesized endogenously by multiple pathways, potentially at several organ sites in the body; for instance, nitrosation can be mediated by bacteria (Leaf *et al.*, 1989) and by macrophages (Marletta, 1988) in infected and inflamed organs. This finding opens a hitherto unexplored area for studies in human cancer etiology.

Nitrosation reactions, which can take place *in vivo* at neutral pH, can be summarized by the common scheme shown in Figure 1. Endogenous NOC synthesis can occur when bacterial enzymes catalyse nitrosation from nitrate or nitrite, probably through formation of nitric oxide. Similarly, activated macrophages use arginine as a source to produce nitric oxide. Once generated, nitric oxide can be oxidized to nitrosating agents that form NOC readily in the presence of amines. A similar reaction may occur in endothelial cells, since the endothelium-derived relaxing factor, which was once thought to be a labile protein, may be nitric oxide. Therefore, under appropriate conditions, NOC may be produced as well.

Figure 1. Proposed scheme for nitrosation reactions catalysed by bacteria, activated macrophages and activated endothelial cells through the common intermediate, nitric oxide (NO)

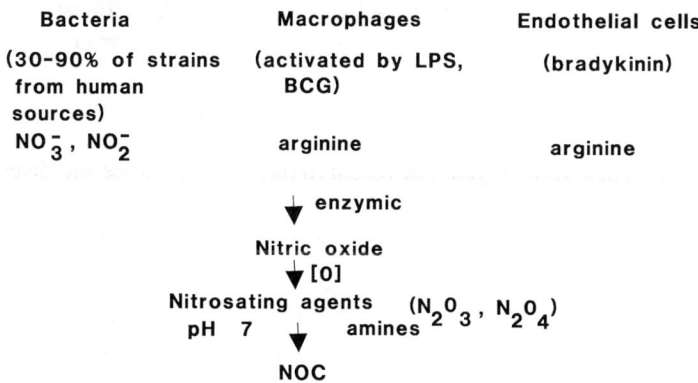

In the presence of oxygen, NO is converted to NO_x, which react with amino precursors to form N-nitroso compounds (NOC) at neutral pH; LPS, lipopolysaccharide; BCG, bacillus Calmette-Guérin. Collated from Marletta (1988) and Leaf et al. (1989)

Although the relevance of these nitrosation processes to cancer causation has not yet been demonstrated, attempts should now be made to determine whether NOC are synthesized in cancer-prone sites, as mentioned below, where the necessary NOC precursors, bacteria or parasites may be present. NOC has been shown to be synthesized in the acidic human stomach through chemical nitrosation (Crespi et al., 1987); whether bacterially mediated nitrosation also occurs in the achlorhydric stomach, which is colonized by bacteria, is still a controversial point. Recent studies have confirmed that N-nitrosamines are formed in the infected urinary bladder (Ohshima et al., 1987; Tricker et al., 1989). Parasitic infection by liver fluke leads to a strong increase in endogenous N-nitrosamine synthesis (Srivatanakul et al., this volume). Nitrosation of tobacco-specific and areca nut-specific akaloids has been demonstrated to occur in the oral cavity of betel-quid/tobacco chewers (Nair et al., 1986, 1987). Although direct experimental support is lacking, there are other sites at which NOC synthesis could occur, including the nasal cavity [bacteria normally found in the nasal cavity of subjects infected with Epstein-Barr virus could produce N-nitrosamines, and this possibility has been explored in a pilot study (Charrière et al., this volume)], infected female genital organs (e.g., cervical fluids) and (virus-)infected liver.

We have been interested in developing and applying sensitive methods for estimating nitrosation reactions in humans. These range from studies in volunteers to the application of dosimetry methods to clinical and epidemiological studies. The results that have been obtained in the last five years are summarized briefly, while quantitative data are reported in more detail elsewhere (Bartsch et al., 1989). We have concentrated mainly on endogenous formation of NOC; however, exposure to preformed NOC is well documented and is most intense and widespread in tobacco users (Hoffmann & Hecht, 1985; Hecht & Hoffmann, this volume). Exposure to NOC in certain occupational environments is also known to occur (IARC, 1978). Recent studies in molecular and metabolic epidemiology have shown significantly elevated exposures to mostly endogenously formed NOC in subjects at high risk for cancers of the oral cavity and lungs in relation to tobacco use, of the oesophagus in certain populations in northern China, of the stomach in Japan, Poland and Costa Rica, of the liver in subjects infested with liver fluke in Thailand, and of the urinary bladder in patients with urinary tract infections in the Sudan and Europe. Supporting data for these observations are given briefly, and problem areas in which more work is needed are outlined.

Results and discussion

Gastric cancer

The decline in gastric cancer incidence has been called the epidemiology of an unplanned triumph (Howson et al., 1986), although a recent survey (Parkin et al., 1988) showed that it still ranks number one worldwide, particularly in males in developing countries. If we want to implement more efficient preventive measures, we must find out whether the hypothesis of intragastric synthesis of carcinogens is valid. More importantly, we must discover which carcinogens act in the complex etiology of this disease and where they come from. Epidemiological studies have indeed supported this hypothesis, although they have given us only a vague idea of the causative factors (Howson et al., 1986). The etiology of stomach cancer has been shown to be linked to a diet rich in nitrate or nitrite, consumption of smoked foods and damage to the gastric mucosa by irritants such as salt. Gastric cancer appears to be initiated early in life. Ingestion of fruits and vegetables has been shown to reduce the risk, and the fact that both are sources of nitrosation inhibitors fits the hypothesis (Bartsch et al., 1988). One conclusion that can be drawn from the wide diversity of dietary items that have been correlated with gastric cancer risk in different parts of the world is that different carcinogen precursors must be ingested from nitrate-/nitrite-rich and smoked food items in different countries. Various nitrosatable precursors have been isolated recently in several laboratories – for example, from Chinese cabbage, from soya sauce in Japan, from fava beans in Colombia, and from smoked foods in western countries (Wakabayashi et al., 1989). These precursors of mutagenic and carcinogenic aromatic compounds are not alkylamides but are of the phenol and indole types. In addition, their nitrosation products are in many cases not NOC but aryldiazonium intermediates, which cannot be detected by the same analytical methods used for identifying nitrosamines. Attempts should be made to detect these diazonium compounds and their macromolecular adducts, using the various methods that are now available. Several members of this class are carcinogens; some produce tumours of the forestomach in mice (Toth et al., 1982); and they may also be formed in vivo, although the nitrite concentration required may be higher than for the formation of NOC.

The discovery that endogenous nitrosation may be involved in the etiology of gastric cancer was shown for the first time in humans by an ecological study of inhabitants of high- and low-risk areas for gastric cancer in Japan (Kamiyama et al., 1987); it was confirmed later in subjects in Poland (Zatonski et al., 1989) and in Costa Rica (Sierra et al., this volume).

Recently completed studies (in collaboration with local investigators) using the NPRO test (Ohshima & Bartsch, 1981, 1988), revealed greater synthesis of NOC in high-risk subjects than in low-risk subjects in all three countries; furthermore, endogenous production of NOC was efficently reduced by moderate doses of ascorbic acid. In general, poor correlations were seen between the amount of NOC synthesized and nitrate intake on a group level; however, when correlated individually, nitrosation increased significantly with total nitrate intake. These results emphasize that determination of nitrite and nitrate alone in body fluids is insufficient to assess the complex nitrosation reaction that occurs in man *in vitro*. If endogenous formation of NOC is to be associated with cancers at specific sites, individual monitoring for these compounds is required, rather than measurements of the intake of NOC precursors.

Of particular interest is the study in Costa Rican children (Sierra *et al.*, this volume), which was designed specifically to test (i) whether NOC are formed by acid-catalysed (chemical) nitrosation in the histologically normal stomach, i.e., early in life when precancerous conditions like chronic atrophic gastritis and colonization of the stomach should not yet have developed; and (ii) whether these nitrosation products play a role in gastric carcinogenesis. Remarkably, the only difference in the several parameters tested (see Sierra *et al.* for details) was a significantly increased endogenous synthesis of NPRO in children from the high-risk area, suggesting that nitrosation (probably acid-catalysed) in the stomach occurs at a higher rate prior to the onset of achlorhydria. These findings in children, reported for the first time, support results from migrant studies that indicate that gastric cancer risk emerges early in life (Howson *et al.*, 1986).

This view diverges from the classical model proposed by Correa *et al.* (1975; Correa, 1988), who suggested that gastric atrophy develops into chronic atrophic gastritis and finally into dysplasia, because the bacterial flora in the achlorhydric stomach are responsible for conversion of nitrate into nitrite and for formation of NOC, which are thought to be the gastric carcinogens. This hypothesis was tested in four studies, using the NPRO test in patients with different precancerous gastric conditions (Crespi *et al.*, 1987; Hall *et al.*, 1987a,b). According to the model, these patient groups should have produced NOC at a higher rate than controls, but none of the studies showed excess synthesis of NPRO. However, Houghton *et al.* (1989) reported a significant increase in the NPRO level in urine of gastrectomy and post-vagotomy patients and a moderate increase in the level in urine of pernicious anaemia patients after administration of proline without nitrate.

Before dismissing Correa's hypothesis, therefore, we should verify whether these negative findings could be interpreted differently. Ways of overcoming some of the limitations of the NPRO test, which could have led to these findings, include the following: (i) In comparison with many secondary amines, proline (used as a probe in NPRO test) is not a good substrate for catalysis of nitrosation by bacteria (Calmels *et al.*, 1985); therefore, better substrates should be sought. Alternatively, total NOC and their macromolecular adducts in gastric juice or gastric mucosa could be measured: Pignatelli *et al.* (1987) have developed an improved method for measuring total NOC in gastric juice, which is being used in several laboratories. (ii) The NPRO test may not reveal nitrosation reactions occurring between normal and metaplastic regions of the stomach (Charnley *et al.*, 1982). Direct measurement of DNA damage in biopsy specimens of normal and metaplastic stomach tissues should help to clarify this possibility. (iii) Exposure to NOC formed by acid-catalysed nitrosation in the normal stomach, early in life, may be as important as later bacterial nitrosation in the achlorhydric stomach in the etiology of gastric cancer. The results obtained in Costa Rican children (Sierra *et al.*, this volume) examined prior to the onset of achlorhydria lend support to this assumption. Thus, gastric cancer may involve two distinct steps: early exposure to NOC, formed intragastrically by chemical nitrosation, followed later by exposure to NOC produced by bacterial catalysis. This sequence resembles the two-stage carcinogenesis model of initiation and promotion. The role of bacteria as possible tumour promoters has not yet been investigated in depth; they could exert this effect by oxidative damage and other growth stimuli, in addition to their potential to generate NOC in the colonized stomach.

Cancer of the urinary bladder

The fact that *N*-nitrosamines are formed by microorganisms through an enzymic step at neutral pH has now been demonstrated conclusively in several laboratories (Leaf *et al.*, 1989). One of the bacterial enzymes has been isolated and purified (S. Calmels, unpublished data), but more than one enzyme seems to be involved in catalysis of nitrosation. In nondenitrifying bacteria, like *Escherichia coli*, nitrosation is induced by nitrate and is linked to expression of the nitrate reductase (*nar* GHI) gene (Calmels *et al.*, 1988; Ralt *et al.*, 1988). In denitrifying bacteria, like *Pseudomonas aeruginosa*, a different enzyme appears to be involved (Leach *et al.*, 1987). Most importantly, as many as 30% of the strains isolated from human achlorhydric gastric juice and about 90% of strains from urinary tract infections possess nitrosating enzymes (Calmels *et al.*, 1987). The obvious question is whether subjects at high risk for gastric or bladder cancer harbour more of these nitrosation-proficient microorganisms than persons at lower risk. Development of a rapid (immunological) screening assay for identifying proficient microorganisms in the clinical setting would lead to an early answer to this hypothesis. Epidemiological evidence has associated urinary tract infections, e.g., infections accompanying schistosomiasis, with an increased risk for bladder cancer of the squamous-cell carcinoma type (reviewed by Preston-Martin & Correa, 1989). Two recent studies have also supported the relationship between bacterial nitrosation and urinary tract infections; both involved analytical methods and storage conditions that excluded artefactual formation of NOC. In a pilot study in our laboratory, patients from a Lyon hospital with urinary tract infections were compared with healthy controls (Ohshima *et al.*, 1987) and levels of *N*-nitrosodimethylamine, other volatile *N*-nitroso compounds, *N*-nitrosamino acids, and particularly nitrite were found to be greatly enhanced in those with infections. Similar findings were reported in schistosomiasis patients (Tricker *et al.*, 1989). Therefore, bacterial strains that have been shown to catalyse *N*-nitrosamine formation and to reduce nitrate to nitrite appear to be responsible for the formation of NOC *in situ* in subjects with urinary tract infections. These findings support the known epidemiological association and offer mechanistic clues for the etiology of bladder cancer.

Liver cancer

Parasitic infection may also greatly increase endogenous *N*-nitrosamine synthesis in man (Srivatanakul *et al.*, this volume), as shown in subjects infested with *Opisthorchis viverrini* (liver fluke) in Thailand who are at high risk for developing cholangiocarcinoma. Based on earlier work by Srianujata *et al.* (1987), a study was conducted in Thailand using the NPRO test. About ten times more endogenous nitrosation occurred in subjects who had these parasites than in controls. The interaction between carcinogenic *N*-nitrosamines formed *in vivo* and liver fluke is thus likely to play a role in the development of cholangiocarcinoma. This notion is further supported by findings using a Syrian golden hamster model, in which a low dose of *N*-nitrosodimethylamine together with liver fluke infection produced cholangiocarcinoma in a synergistic fashion (Thamavit *et al.*, 1978). If *N*-nitrosamine synthesis in parasite-infested subjects is mediated by activated macrophages, which remains to be proven, it would be the first example of a chronic inflammatory process producing carcinogens *in vivo* and thus contributing to a human malignancy.

Oesophageal cancer

Three correlation studies were carried out in two, eight and 26 counties in China, involving almost 3000 subjects in high- and low-risk areas for oesophageal cancer (Lu *et al.*, 1986; Chen *et al.*, 1987; Lu *et al.*, 1987), to correlate indices of exposures to NOC and other cancer risk markers with mortality from oesophageal cancer, which in China shows geographic variations up to 300 fold. The results of the three studies revealed: (i) positive correlations between the amount of urinary amino acids excreted, in particular those synthesized *in vivo*, and cancer mortality; (ii) efficient reduction of endogenous NOC synthesis by moderate doses of ascorbic acid in high-risk subjects; and, in one study (Chen *et al.*, 1987), (iii) an inverse correlation between serum ascorbate levels and cancer mortality. Increased O^6-methylation of guanine was found in oesophageal DNA taken from cancer patients living in the same high-risk area as investigated by Lu *et al.* (1986), as compared to patients living in a low-risk area (Umbenhauer *et al.*, 1985).

Chinese scientists had already incriminated NOC and their precursors as etiological agents in cancer at this site in the mid-1970s (Co-ordinating Group for Research on the Etiology of Esophageal Cancer in North China, 1975). The more recent data strongly support the role of *N*-nitrosamines in the multifactorial origin of oesophageal malignancies in northern China.

Conclusions

Results obtained in the laboratory and in epidemiological studies over the past decade strengthen the role of NOC in the etiology of human cancer, although causality has not been established. Relatively inexpensive, easy measures have been identified that can be used to reduce exposure to endogenous NOC. The results provide an interpretation of earlier epidemiological findings that consistently showed a protective effect of fresh fruits and vegetables – both of which are sources of nitrosation inhibitors – against stomach cancer and other malignancies. Implementation of more effective preventive measures, however, will require further work: (i) to identify the major carcinogenic NOC and other reactive intermediates to which humans are exposed, in particular, those derived from reactions *in vivo*, and to trace their origins, possibly in the diet, using the available and new analytical tools; (ii) because of limitations of the NPRO test (as discussed above), to search for better amino substrates that are nitrosatable by nitrogen oxides (see Ohshima *et al.*, this volume), bacteria and macrophages, and to develop dosimetry methods with better molecular markers for these exposures; (iii) to conduct more studies on humans in which dosimetry, detection of (cyto)genetic damage and molecular markers of host susceptibility are integrated; and, finally, to follow up these steps by attempts to identify high-risk subjects and ways of lowering exposure to carcinogens from exogenous and endogenous sources.

Acknowledgements

I wish to thank, in particular, my colleagues, H. Ohshima, B. Pignatelli, S. Calmels, M. Friesen, D. Shuker and C. Malaveille for their outstanding research (referred to in part in this review). I also gratefully acknowledge the scientific contributions and collaborative efforts of N. Muñoz, M. Parkin and J. Kaldor (IARC, Lyon), S.V. Bhide, U. and J. Nair (Cancer Research Institute, Tata Memorial Centre, Bombay, India), J. Chen (Chinese Academy of Medical Sciences, Beijing, China), M. Crespi (The Regina Elena Institute for the Study and Therapy of Tumours, Rome, Italy), S. Kamiyama (Akita University, Akita, Japan), S.H. Lu (Academy of Medical Sciences, Cancer Institute, Beijing, China), R. Lambert, Y. Minaire and B. Moulinier (Edouard Herriot Hospital, Lyon, France), R. Peto (Imperial Cancer Research Fund Cancer Studies Unit, Oxford University, Oxford, UK), R. Sierra (University of Costa Rice, San José, Costa Rica), M. Srivatanakul (National Cancer Institute, Bangkok, Thailand), and W. Zatonski (Institute of Oncology, Warsaw, Poland). M. Wrisez is thanked for secretarial help and E. Heseltine for editorial assistance.

Part of the work described here was supported by grants from the US National Institutes of Health NO 1R01 - CA47591 and IUO1 CA 43176. One of the thermal energy analysers used in these studies was provided on loan by the US National Cancer Institute under Contract NO1 CP-50715.

References

Barnes, J.M. (1974) Nitrosamines. *Essays Toxicol.*, **5**, 1-15

Bartsch, H. & Malaveille, C. (1989) Prevalence of genotoxic chemicals among animals and human carcinogens evaluated in the IARC Monograph series. *Cell Biol. Toxicol.*, **5**, 115-127

Bartsch, H. & Montesano, R. (1984) Relevance of nitrosamines to human cancer. *Carcinogenesis*, **5**, 1381-1393

Bartsch, H., Ohshima, H. & Pignatelli, B. (1988) Inhibitors of endogenous nitrosation. Mechanisms and implications in human cancer prevention. *Mutat. Res.*, **202**, 307-324

Bartsch, H., Ohshima, H., Pignatelli, B. & Calmels, S. (1989) Human exposure to endogenous N-nitroso compounds: quantitative estimates in subjects at high risk for cancer of the oral cavity, oesophagus, stomach and urinary bladder. *Cancer Surv.*, **8**, 335-362

Calmels, S., Ohshima, H., Vincent, P., Gounot, A.-M., & Bartsch, H. (1985) Screening of microorganisms for nitrosation catalysis at pH 7 and kinetic studies on nitrosamine formation from secondary amines by *E. coli* strains. *Carcinogenesis*, **6**, 911-915

Calmels, S., Ohshima, H., Crespi, M., Leclerc, H., Cattoen, C. & Bartsch, H. (1987) N-Nitrosamine formation by microorganisms isolated from human gastric juice and urine: biochemical studies on bacteria-catalysed nitrosation. In: Bartsch, H., O'Neill, I.K. & Schulte-Hermann, R., eds, *The Relevance of N-Nitroso Compounds to Human Cancer: Exposures and Mechanisms* (IARC Scientific Publications No. 84), Lyon, IARC, pp. 391-395

Calmels, S., Ohshima, H. & Bartsch, H. (1988) Nitrosamine formation by denitrifying and non-denitrifying bacteria: implications of nitrite reductase and nitrate reductase in nitrosation catalysis. *J. Gen. Microbiol.*, **134**, 221-226

Charnley, G., Tannenbaum, S.R. & Correa, P. (1982) Gastric cancer: an etiological model. In: Magee, P.N., ed., *Nitrosamines and Human Cancer* (Banbury Rep. 12), Cold Spring Harbor, NY, CSH Press, pp. 503-522

Chen, J., Ohshima, H., Yang, H., Li, J., Campbell, T.C., Peto, R. & Bartsch, H. (1987) A correlation study on urinary excretion of N-nitroso compounds and cancer mortality in China: interim results. In: Bartsch, H., O'Neill, I.K. & Schulte-Hermann, R., eds, *The Relevance of N-Nitroso Compounds to Human Cancer: Exposures and Mechanisms* (IARC Scientific Publications No. 84), Lyon, IARC, pp. 503-506

Co-ordinating Group for Research on the Etiology of Esophageal Cancer in North China (1975) The epidemiology of esophageal cancer in north China and preliminary results in the investigation of its etiological factors. *Sci. Sin.*, 18, 131-148

Correa, P. (1988) A human model of gastric carcinogenesis. *Cancer Res.*, 48, 3354-3560

Correa, P., Haenszel, W., Cuello, C., Tannenbaum, S. & Archer, M. (1975) A model for gastric cancer epidemiology. *Lancet*, ii, 58-60

Crespi, M., Ohshima, H., Ramazzotti, V., Muñoz, N., Grassi, A., Casale, V., Leclerc, H., Calmels, S., Cattoen, C., Kaldor, J. & Bartsch, H. (1987) Intragastric nitrosation and precancerous lesions of the gastrointestinal tract: testing of an etiological hypothesis. In: Bartsch, H., O'Neill, I.K. & Schulte-Hermann, R., eds, *The Relevance of N-Nitroso Compounds to Human Cancer: Exposures and Mechanisms* (IARC Scientific Publications No. 84), Lyon, IARC, pp. 511-517

Druckrey, H., Schildbach, A., Schmähl, D., Preussmann, R. & Ivankovic, S. (1963) Quantitative analysis of the carcinogenicity of diethylnitrosamine (Ger.). *Arzneimittelforschung*, 10, 841-851

Hall, C.N., Kirkham, J.S. & Northfield, T.C. (1987a) Urinary N-nitrosoproline excretion: a further evaluation of the nitrosamine hypothesis of gastric carcinogenesis in precancerous conditions. *Gut*, 28, 216-498

Hall, C.N., Darkin, D., Viney, N., Cook, A., Kirkham, J.S. & Northfield, T.C. (1987b) Evaluation of the nitrosamine hypothesis of gastric carcinogenesis in man. In: Bartsch, H., O'Neill, I.K. & Schulte-Hermann, R., eds, *The Relevance of N-Nitroso Compounds to Human Cancer: Exposures and Mechanisms* (IARC Scientific Publications No. 84), Lyon, IARC, pp. 527-530

Hoffmann, D. & Hecht, S.S. (1985) Nicotine-derived N-nitrosamines and tobacco related cancer: current status and future directions. *Cancer Res.*, 45, 935-942

Houghton, P.W.J., Leach, S., Owen, R.W. McC., Mortensen, N.J., Hill, M.J. & Williamson, R.C.N. (1989) Use of modified N-nitrosoproline test to show intragastric nitrosation in patients at risk of gastric cancer. *Br. J. Cancer*, 60, 231-234

Howson, C.P., Hiyama, T. & Wynder, E.L. (1986) The decline in gastric cancer: epidemiology of an unplanned triumph. *Epidemiol. Rev.*, 8, 1-27

IARC (1978) *IARC Monographs on the Evaluation of the Carcinogenic Risk of Chemicals to Humans*, Vol. 17, *Some N-Nitroso Compounds*, Lyon

Kamiyama, S., Ohshima, H., Shimada, A., Saito, N., Bourgade, M.-C., Ziegler, P. & Bartsch, H. (1987) Urinary excretion of N-nitrosamino acids and nitrate by inhabitants in high- and low-risk areas for stomach cancer in northern Japan. In: Bartsch, H., O'Neill, I.K. & Schulte-Hermann, R., eds, *The Relevance of N-Nitroso Compounds to Human Cancer: Exposures and Mechanisms* (IARC Scientific Publications No. 84), Lyon, IARC, pp. 497-502

Leach, S.A., Cook, A.R., Challis, B.C., Hill, M.J. & Thompson, M.H. (1987) Bacterially mediated N-nitrosation reactions and endogenous formation of N-nitroso compounds in the gastric juice of Greek hypochlorhydric individuals. *Carcinogenesis*, 6, 1135-1140

Leaf, C.D, Wishnok, J.S. & Tannenbaum, S.R. (1989) Mechanisms of endogenous nitrosation. *Cancer Surv.*, 8, 323-334

Lu, S.H., Ohshima, H., Fu, H.-M., Tian, Y., Li, F.M., Blettner, M., Wahrendorf, J. & Bartsch, H. (1986) Urinary excretion of N-nitrosamino acids and nitrate by inhabitants of high- and low-risk areas for esophageal cancer in northern China: endogenous formation of nitrosoproline and its inhibition by vitamin C. *Cancer Res.*, 46, 1485-1491

Lu, S.H., Yang, W.X., Guo, L.P., Li, F.M., Wang, G.J., Zhang, J.S. & Li, P.Z. (1987) Determination of N-nitrosamines in gastric juice and urine and a comparison of endogenous formation of N-nitrosoproline and its inhibition in subjects from high- and low-risk areas for oesophageal cancer. In: Bartsch, H., O'Neill, I.K. & Schulte-Hermann, R., eds, *The Relevance of N-Nitroso Compounds to Human Cancer: Exposures and Mechanisms* (IARC Scientific Publications No. 84), Lyon, IARC, pp. 538-543

Magee, P.N. & Barnes, J.M. (1956) The production of malignant primary hepatic tumours in the rat by feeding dimethylnitrosamine. *Br. J. Cancer*, 10, 114-122

Marletta, M.A. (1988) Mammalian synthesis of nitrite, nitrate, nitric oxide and *N*-nitrosating agents. *Chem. Res. Toxicol.*, 1, 249-257

Nair, J., Ohshima, H., Pignatelli, B., Friesen, M., Malaveille, C., Calmels, S. & Bartsch, H. (1986) Modifiers of endogenous carcinogen formation: studies on *in vivo* nitrosation in tobacco users. In: Hoffmann, D. & Harris, C.C., eds, *New Aspects of Tobacco Carcinogenesis* (Banbury Rep. No. 23), Cold Spring Harbor, NY, CSH Press, pp. 45-61

Nair, J., Nair, U.J., Ohshima, H., Bhide, S.V. & Bartsch, H. (1987) Endogenous nitrosation in the oral cavity of chewers while chewing betel quid with or without tobacco. In: Bartsch, H., O'Neill, I.K. & Schulte-Hermann, R., eds, *The Relevance of N-Nitroso Compounds to Human Cancer: Exposures and Mechanisms* (IARC Scientific Publications No. 84), Lyon, IARC, pp. 465-469

National Research Council (1981) *The Health Effects of Nitrate, Nitrite and N-Nitroso Compounds. Part 1 of a Two-part Study*, Washington DC, National Academy Press

Ohshima, H. & Bartsch, H. (1981) Quantitative estimation of endogenous nitrosation in humans by monitoring of *N*-nitrosoproline excreted in the urine. *Cancer Res.*, 41, 3658-3662

Ohshima, H. & Bartsch, H. (1988) Urinary *N*-nitrosamino acids as an index of exposure to *N*-nitroso compounds. In: Bartsch, H., Hemminki, K. & O'Neill, I.K., eds, *Methods for Detecting DNA Damaging Agents in Humans: Application in Cancer Epidemiology and Prevention* (IARC Scientific Publications No. 89), Lyon, IARC, pp. 83-91

Ohshima, H., Calmels, S., Pignatelli, B., Vincent, P. & Bartsch, H. (1987) *N*-Nitrosamine formation in urinary tract infections. In: Bartsch, H., O'Neill, I.K. & Schulte-Hermann, R., eds, *The Relevance of N-Nitroso Compounds to Human Cancer: Exposures and Mechanisms* (IARC Scientific Publications No. 84), Lyon, IARC, pp. 384-390

Parkin, D.M., Läärä, E. & Muir, C.S. (1988) Estimates of the worldwide frequency of sixteen major cancers in 1980. *Int. J. Cancer*, 41, 184-197

Pignatelli, B., Richard, I., Bourgade, M.-C. & Bartsch, H. (1987) Improved group determination of total *N*-nitroso compounds (NOC) in human gastric juice by chemical denitrosation and thermal energy analysis. *Analyst*, 112, 945-949

Preston-Martin, S. & Correa, P. (1989) Epidemiologic evidence for the role of nitroso compounds in human cancer. *Cancer Surv.*, 8, 459-474

Ralt, D., Wishnok, J.S., Fitts, R. & Tannenbaum, S.R. (1988) Bacterial catalysis of nitrosation: involvement of the nar operon of *Escherichia coli*. *J. Bacteriol.*, 170, 359-364

Shephard, S.E., Schlatter, C. & Lutz, W.K. (1987) Assessment of the risk of formation of carcinogenic *N*-nitroso compounds from dietary precursors in the stomach. *Food Chem. Toxicol.*, 25, 591-108

Srianujata, S., Tonbuth, S., Bunyaatvej, S., Valyseva, A., Promvanit, N. & Chiavatsagul, W. (1987) High urinary excretion of nitrate and *N*-nitrosoproline in opisthorchiasis subjects. In: Bartsch, H., O'Neill, I.K. & Schulte-Hermann, R., eds, *The Relevance of N-Nitroso Compounds to Human Cancer: Exposures and Mechanisms* (IARC Scientific Publications No. 84), Lyon, IARC, pp. 544-546

Thamavit, W., Bhamarapravati, N., Sahapong, S., Vajarasthira, S. & Angsubhakorn, S. (1978) Effect of dimethylnitrosamine on induction of cholangiocarcinoma in *Opisthorchis viverrini*-infected Syrian golden hamsters. *Cancer Res.*, 38, 4634-4639

Toth, B., Nagel, D. & Ross, A. (1982) Gastric tumorigenesis by a single dose of 4-(hydroxymethyl)benzenediazonium ion of *Agaricus bisporus*. *Br. J. Cancer*, 46, 417-422

Tricker, A.R., Mostafa, M.H., Speigelhalder, B. & Preussmann, R. (1989) Urinary excretion of nitrate, nitrite and *N*-nitroso compounds in schistosomiasis and bilharzia bladder cancer patients. *Carcinogenesis*, 10, 547-552

Umbenhauer, D., Wild, C.P., Montesano, R., Saffhill, R., Boyle, J.M., Huh, N., Kirstein, U., Thomale, J., Rajewsky, M.F. & Lu, S.H. (1985) O⁶-Methyldeoxyguanosine in oesophageal DNA among individuals at high risk of oesophageal cancer. *Int. J. Cancer*, 36, 661-665

Wakabayashi, K., Nagao, M. & Sugimura, T. (1989) Mutagens and carcinogens produced by the reaction of environmental aromatic compounds with nitrite. *Cancer Surv.*, 8, 385-400

Zatonski, W., Ohshima, H., Przewozniak, K., Drosik, K., Mierzwinska, J., Krygier, M., Chmielarczyk, W. & Bartsch, H. (1989) Urinary excretion of *N*-nitrosamino acids and nitrate by inhabitants of high- and low-risk areas for stomach cancer in Poland. *Int. J. Cancer*, 44, 823-827

RELEVANCE OF N-NITROSAMINES TO OESOPHAGEAL CANCER IN CHINA

S.H. Lu[1], S.X. Chui[1], W.X. Yang[2], X.N. Hu[1], L.P. Guo[1] & F.M. Li[1]

[1]*Cancer Institute, Chinese Academy of Medical Sciences, Beijing; and*
[2]*Cancer Institute, Zhengchow, China*

Oesophageal cancer occurs at a very high frequency in certain areas of China, especially in Linxian county, Henan province. Previous studies suggested that N-nitroso compounds play a causative role. In order to study further the exposure of Linxian inhabitants, the intake of N-nitrosamines in the diet was determined. The total daily intake of volatile nitrosamines and of N-nitroso-N-methylbenzylamine (NMBzA) was higher in Linxian than in two other counties. NMBzA can induce cancer in animal and human oesophageal epithelium. Human fetal oesophageal epithelia were cultured with NMBzA for 4 h, and eight monkeys were treated with NMBzA. High-molecular-weight DNA extracted from explants and from the oesophageal epithelia of monkeys induced malignant transformation of NIH 3T3 cells. *Alu* and monkey-specific repeat sequences were present in transformed cells, and H-*ras* was found in the transforming DNA. Human fetal oesophageal epithelium cultured with NMBzA was transplanted into the mesentery of BALB/c nude mice. Tumours were found after eight months; and the *Alu* sequence was present in DNA extracted from tumours induced by NMBzA, showing that the tumours were of human origin. The results provide direct evidence that NMBzA is carcinogenic in human oesophageal epithelium and that N-nitrosamines are one of the causative factors of oesophageal cancer in Linxian county.

Oesophageal cancer is a common malignancy in certain areas of China, especially in the north. Linxian county in Henan province has the highest age-adjusted mortality rates from this cancer, at 151/100 000 in males and 115/100 000 in females. Since 1972, extensive epidemiological and clinical investigations have been conducted to identify the causative factors of oesophageal cancer in this area. N-Nitrosamines have received the most attention among the suspected etiological factors in the environment. In previous studies in Linxian county, we found (i) high levels of N-nitrosamines in foods, pickled vegetables and gastric juice; (ii) a positive correlation between the amount of N-nitrosamines in gastric juice and the severity of lesions of the oesophageal epithelium (Lu *et al.*, 1987, 1988a); and (iii) that the urine of subjects in this county contained more N-nitrosamino acids and nitrates than that of subjects in a low-incidence county (Lu *et al.*, 1986). The study reported here demonstrates that the level of N-nitrosamines in diets collected from Linxian was greater than that in diets from other counties. We have also found that N-nitrosamines activate oncogenes in oesophageal epithelia from human fetuses and monkeys, and can induce oesophageal carcinoma in human fetal tissue.

Intake of N-nitrosamines in diets collected in high- and low-risk areas for oesophageal cancer

Trace levels of volatile N-nitrosamines have been reported in certain foods (Scanlan, 1974). Low levels (0.4-1.5 µg/kg) were found in about 250 samples of various foods, including cured meat products, fried bacon, baby foods containing meat, tomato products and cooked pizza (Sen et al., 1979). Spiegelhalder et al. (1979) reported concentrations of 0.5-5.0 µg/kg in more than 3000 food samples from the Federal Republic of Germany. We found previously that foods collected in Linxian county, an area of high risk for oesophageal cancer, contained higher levels of N-nitrosamines than those from Fanxian county, which is a low-risk area (Lu & Lin, 1982). We have now estimated exposure to N-nitrosamines in areas with intermediate rates of mortality from this cancer (Table 1). N-Nitrosodimethylamine and N-nitrosodiethylamine were found in 87-100% of 103 diet samples from the three counties; the samples also contained several N-nitrosamines that can induce oesophageal cancer in animals (Druckrey et al., 1967). N-Nitrosomethylbenzylamine (NMBzA) was identified in a high proportion (78%) of diet samples collected from Linxian county but only 15% of samples collected in Yuxian.

The total daily intake of volatile N-nitrosamines and the intake of NMBzA in diets collected from the three counties was correlated with the mortality rates from oesophageal cancer in those counties (Table 1). Gough et al. (1978) found that the average intake of N-nitrosodimethylamine, N-nitrosopyrrolidine and N-nitrosopiperidine from the normal UK diet was about 1-3 µg/week, and Fine (1978) estimated the daily intake in the USA from nitrite-preserved foods to be 1 µg N-nitrosodimethylamine and 5 µg N-nitrosopyrrolidine. The daily intake of N-nitrosamines from the normal Linxian county diet is thus more than 100 times higher than those in the UK and the USA. Further, the Linxian diet contains NMBzA, which can induce oesophageal cancer in animal (S. Lu et al., unpublished data) and human fetal oesophageal epithelium (see below).

Oncogene activation in oesophageal epithelium from human fetuses and monkeys

The carcinogenic activity of N-nitroso compounds is due to their ability to generate highly reactive species that alkylate cellular DNA, and there is extensive evidence that DNA methylation is mutagenic (Doerfler, 1983). Recent studies indicate that mammary tumours induced in rats by N-methyl-N-nitrosourea contain a specific base substitution mutation in the H-ras cellular oncogene (Zarbl et al., 1985). We found previously that oesophageal cancer in Linxian county is associated with a significant frequency of amplification of the epidermal growth factor receptor gene and of the C-myc gene (Lu et al., 1988b). The transforming gene was isolated from human oesophageal carcinoma tissue collected in Linxian county, and H-ras was found in the transforming DNA using the Southern blot assay (Jiang & Lu, 1988). In order to investigate the mechanism of oncogene activation and of the carcinogenic effect of N-nitrosamines, we treated human fetal oesophageal explants

Table 1. N-Nitrosamines in diets collected from counties in China with different risks for oesophageal cancer[a]

County	Mortality rate (per 10^5)	No. of samples	NDMA		NDEA		NPYR		NPIP		NMBzA		Total (μg/day)
			%+ve	μg/day	%+ve	μg/day	%+ve	μg/day	%+ve	μg/day	%+ve	μg/day	
Linxian	151	23	87	182.6±378	91	342.7±420	17	0.44±1.0	17	0.3±0.9	78	107.0±171	633.7±845.0
Ji Yuan	98	25	100	509.8±438	88	58.5±76.0	28	2.69±10.0	32	4.0±17	40	8.2±18	576.8±504.0
Yuxian	35	55	95	219.7±451	87	38.8±77.0	44	1.12±2.9	33	1.2±4.1	15	0.2±0.5	284.5±598.0

[a] Results given as percentage of samples containing the nitrosamide (%+ve) and mean + SD of level found (μg/day); NDMA, N-nitrosodimethylamine; NDEA, N-nitrosodiethylamine; NPYR, N-nitrosopyrrolidine; NPIP, N-nitrosopiperidine; NMBzA, N-nitrosomethylbenzylamine

with NMBzA *in vitro* and determined activation of the proto-oncogene(s) at the initiation stage.

Fetal oesophagi were obtained from medium-term abortions of normal pregnancies. Each epithelial explant was divided into two for a control and an experimental sample. The explants were cultured in M199 medium for 12-24 h; then NMBzA was added to the experimental samples to a final concentration of 10 mM and dimethyl sulfoxide to the control samples; culture was continued for 4 h. The explants were then washed thoroughly in phosphate-buffered saline and preserved in liquid nitrogen. High-molecular-weight DNA was extracted from the explants and added to NIH 3T3 cells by the calcium phosphate coprecipitation method. Transformed foci were scored two to three weeks later. The transformed cells were amplified and their DNA extracted for second-round transfection. The DNA was then hybridized with a human-specific high-repeat sequence (*Alu* probe) and an H-*ras* oncogene probe to identify the oncogene type.

We found no phenotypic alteration of the explants in either the control or experimental group. DNA extracted from the experimental samples, but not that from control samples, induced malignant transformation of NIH 3T3 cells, at a transformation efficiency of 0.089 foci/µg DNA. DNA extracted from malignant transformed NIH 3T3 cells induced second-round transformation at a transformation efficiency of 0.37 foci/µg DNA. The transformed cells were anchorage-independent and formed colonies when inoculated into 0.3% soft agar. Large tumours developed two weeks after subcutaneous inoculation of transformed cells into BALB/c nude mice and were diagnosed as fibrosarcomas. Molecular hybridization showed that only the genomes of the transformed cells contained sequences that hybridized with the *Alu* probe, indicating that malignant transformation of the NIH 3T3 cells was induced by DNA transfection and not by spontaneous transformation (Figure 1). When transformed cell DNA was hybridized with H-*ras*, the BamHI-BamHI 6.6-kb sequence was found (Figure 2), proving that NMBzA activated cellular oncogene(s) in human fetal oesophagus.

Oesophageal epithelium was also obtained from eight rhesus monkeys fed NMBzA at 30 mg/kg body weight for one to five days and from two untreated monkeys. Animals were killed one, two, three or five days after treatment with NMBzA and DNA extracted. High-molecular-weight DNA was added to NIH 3T3 cells by the calcium phosphate coprecipitation method. Transforming activity was observed, secondary transformants were produced which were anchorage-independent, and large tumours developed in athymic BALB/c nude mice. No tumour developed in nude mice injected with control NIH 3T3 cells, even after two months. The transforming DNA was linked to the monkey-specific repeat sequence (0.3 kb), indicating that a common monkey DNA fragment was conserved in the tumours. H-*ras* was found in the transforming DNA using the Southern blot assay.

These data prove directly for the first time that chemical carcinogens can activate cellular oncogenes at the initiation stage of carcinogenesis in humans and monkeys. Thus, oncogene activation may be the cause and not the consequence of carcinogenesis. These results are also powerful evidence to support the hypothesis that *N*-nitrosamines are etiological factors in oesophageal cancer in China.

Figure 1. Southern blot analysis of hybridization of *Alu* probe with DNA extracted from (A) oesophageal epithelium of a human fetus, (B) normal NIH 3T3 cells, and (C) transformed NIH 3T3 cells

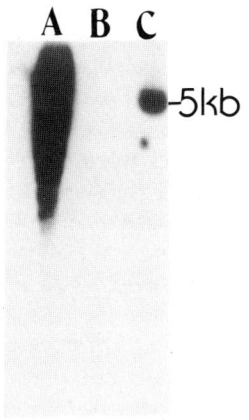

Figure 2. Southern blot analysis of hybridization of H-*ras* probe with DNA extracted from (A) human fetus, (B) normal NIH 3T3 cells, and (C) transformed NIH 3T3 cells, forming the BamHI-BamHI 6.6-kb sequence

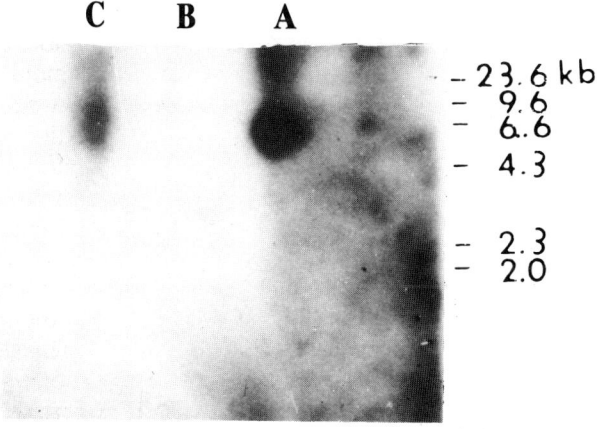

Induction of carcinomas by NMBzA in oesophageal epithelium from human fetuses

Most *N*-nitrosamines have been shown to be carcinogenic to all species of animals tested during the last 20 years, but their role in human carcinogenesis has still to be elucidated.

We therefore decided to study whether N-nitrosamines induce carcinoma in human oesophageal epithelium. When human fetal oesophagi were cultured with NMBzA in minimum essential medium for up to three months, we found considerable variation in the degree of proliferation, morphogenesis and morphological alteration. Recently, fetal epithelium was cultured with NMBzA for three weeks, and then two epithelial segments were transplanted into the mesentery of BALB/c nude mice and the mice were fed continuously with NMBzA at 5 µg/ml in the drinking–water. Tumours were found on the mesentery of two mice which were given NMBzA in drinking-water at 5 µg/ml, two months after transplantation; no tumour was found in control mice. The tumours had developed to large size by eight months. No tumour was found by macroscopic examination in the oesophagus of the mice. Morphological examination showed the mesenteric tumours to be an oesophageal squamous-cell carcinoma and a well-differentiated carcinoma (Figure 3). The *Alu* sequence was present in DNA extracted from fetal oesophageal epithelium and in tumours induced by NMBzA (Figure 4), but not in DNA from NIH 3T3 cells, indicating that the tumours induced by NMBzA in the mesentery of nude mice were of human origin.

Figure 3. Histological section of a well-differentiated squamous-cell carcinoma induced by N-nitrosomethylbenzylamine; × 40

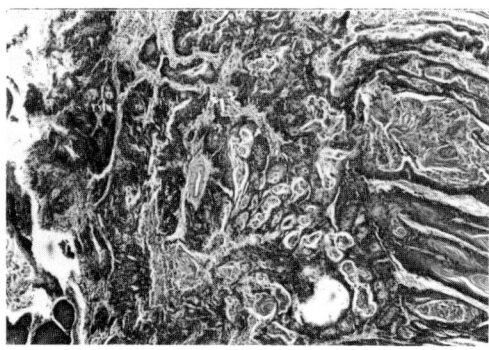

These results provide direct evidence that NMBzA is carcinogenic in human oesophageal epithelium, inducing carcinoma, and demonstrate that N-nitrosamines are a causative factor of oesophageal cancer in Linxian county.

Acknowledgements

This research was supported by the State Scientific Commission of China and by NIH grant MEP(AHR) 1RO1 ESO3646-01A1. We wish to thank Dr H. Bartsch, Dr C.S. Yang and Dr H. Ohshima for suggestions and for analysing the N-nitrosamines.

Figure 4. Southern blot analysis of hybridization of *Alu* probe with DNA extracted from (A) oesophageal epithelium of a human fetus, (B) a tumour induced by *N*-nitrosomethylbenzylamine, and (C) normal NIH 3T3 cells

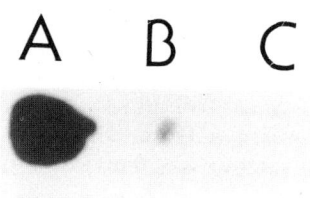

References

Doerfler, W. (1983) Methylation and gene activity. *Ann. Rev. Biochem.*, 52, 93-124

Druckrey, H., Preussmann, R., Ivankovic, S. & Schmähl, D. (1967) Organotropic carcinogenicity of 65 different *N*-nitroso compounds in BD rats (Ger.). *Z. Krebsforsch.*, 69, 103-201

Fine, D.H. (1978) An assessment of human exposure to *N*-nitroso compounds. In: Walker, E.A., Castegnaro, M., Griciute, L. & Lyle, R.E., eds, *Environmental Aspects of N-Nitroso Compounds* (IARC Scientific Publications No. 19), Lyon, IARC, pp. 267-278

Gough, T.A., Webb, K.S. & Coleman, R.F. (1978) Estimate of the volatile nitrosamine content of UK food. *Nature*, 272, 161-163

Jiang, W. & Lu, S.H. (1988) Transforming gene in human esophageal carcinoma tissue. *Chin. J. Oncol.*, 5, 330-334

Lu, S.H. & Lin, P. (1982) Recent research on the etiology of esophageal cancer in China. *Z. Gastroenterol.*, 20, 361-367

Lu, S.H., Ohshima, H., Fu, H.M., Tian, Y., Li, F.M., Blettner, M., Wahrendorf, J. & Bartsch, H. (1986) Urinary excretion of *N*-nitrosamino acids and nitrate by inhabitants of high- and low-risk areas for esophageal cancer in northern China: endogenous formation of nitrosoproline and its inhibition by vitamin C. *Cancer Res.*, 46, 1485-1491

Lu, S.H., Yang, W.X., Guo, L.P., Li, F.M., Wang, G.J., Zhang, J.S. & Lin, P. (1987) Determination of *N*-nitrosamines in gastric juice and urine and a comparison of endogenous formation of *N*-nitrosoproline and its inhibition in subjects from high- and low-risk areas of esophageal cancer. In: Bartsch, H., O'Neill, I.K. & Schulte-Hermann, R., eds, *The Relevance of N-Nitroso Compounds to Human Cancer: Exposure and Mechanisms* (IARC Scientific Publications No. 84), Lyon, IARC, pp. 538-543

Lu, S.H., Wang, G.J., Guo, L.P., Li, F.M., Lin, P.Z. & Zhang, J.S. (1988a) *N*-Nitrosamines in gastric juice of subjects from high incidence area of esophageal cancer. *Chin. J. Oncol.*, 10, 322-325

Lu, S.H., Hsieh, L.L., Luo, F.C. & Weinstein, I.B. (1988b) Amplification of the EGF receptor and C-myc gene in human esophageal cancers. *Int. J. Cancer*, 42, 502-505

Scanlan, R.A. (1974) *N*-Nitrosamines in foods. *CRC Crit. Rev. Food Sci. Technol.*, 5, 357-402

Sen, N.P., Seaman, S. & McPherson, M. (1979) Further studies on the occurrence of volatile and non-volatile *N*-nitrosamines in foods. In: Walker, E.A., Griciute, L., Castegnaro, M. & Börzsönyi, M., eds, N-*Nitroso Compounds: Analysis, Formation and Occurrence* (IARC Scientific Publications No. 31), Lyon, IARC, pp. 457-463

Spiegelhalder, B., Eisenbrand, G. & Preussmann, R. (1979) Occurrence of volatile nitrosamines in food: a survey of the West German market. In: Walker, E.A., Griciute, L., Castegnaro, M. & Börzsönyi, M., eds, N-*Nitroso Compounds: Analysis, Formation and Occurrence* (IARC Scientific Publications No. 31), Lyon, IARC, pp. 467-477

Zarbl, H., Sukumar, S., Arthur, A.V., Martin-Zanca, D. & Barbacid, M. (1985) Direct mutagenesis of Ha-ras-1 oncogenes by *N*-nitroso-*N*-methylurea during initiation of mammary carcinogenesis in rats. *Nature*, 315, 382-385

Relevance to Human Cancer of *N*-Nitroso Compounds,
Tobacco Smoke and Mycotoxins.
Ed. I.K. O'Neill, J. Chen, and H. Bartsch
Lyon, International Agency for Research on Cancer
© IARC, 1991

DIETARY PRACTICES AND CANCER MORTALITY IN CHINA

J. Chen[1]

*Institute of Nutrition and Food Hygiene,
Chinese Academy of Preventive Medicine, Beijing, China*

An ecological survey on diet, life style and cancer mortality was carried out in 65 rural counties in China, using a questionnaire comprising 285 questions on environmental factors, dietary practices and other life style characteristics; in addition, blood, urine and food were collected and analysed. Two interpretations of preliminary data are presented. One is for the finding of a positive correlation between a latent variable, namely general nutritional status, and cancer mortality rates; the other is for a positive correlation between lipid peroxidation (plasma lipid peroxidase and copper) and oesophageal and gastric cancer mortality, and a negative correlation between dietary oxidants (plasma selenium, ascorbic acid and retinol) and mortality from oesophageal and gastric cancer.

During the past 20 years, there has been renewed emphasis on the role of diet in the development of cancer; however, most epidemiological studies have focused on single nutrients or were based on questionnaire surveys. The present study was based on the assumption that cancer is a complex disease and can be affected by a considerable variety of foods. Therefore, multiple risk factors in China were studied and correlated with cancer mortality data for 1973-75 for 14 cancer sites.

The survey

Survey protocol

Sixty-five rural counties located in 24 provinces, each with a population of over 100 000, were systematically selected from a total of approximately 2000 cities and counties to represent the range of county-, sex- and organ-specific cancer mortality rates in rural populations in 1973–75 for the seven main cancers in China (nasopharynx, oesophagus, stomach, liver, lung, colon/rectum and breast). A three-stage, random, cluster sampling procedure was used to select 130 survey communes and 260 production teams. Households and 6500 individual subjects (35–64 years old) were also selected randomly.

The following was collected: (i) information on intake of food, other nutrients and other constituents in 30 households per county; (ii) blood and urine for assay of various nutrient and non-nutrient risk indicators from 100 subjects per county; (iii) approximately 600 food samples representing all plant foods consumed in the survey households; and (iv) information on life style and medical history by questionnaire from 100 subjects per county. Finally, 82 disease mortality variables (44 of which were cancer rates) and 285 risk indicator variables were obtaned. Detents of the survey method are given by Chen *et al.* (1990).

---

[1]The other three principal investigators are: Dr T.C. Campbell, Cornell University, USA; Dr J. Li, Cancer Institute, Chinese Academy of Medical Sciences,Beijing, China, and R. Peto, Oxford University, UK

In order to maximize the number of assays with the plasma and erythrocyte samples, sex- and commune-specific pools were made for each county, based on the assumption that a pool value reflects the average of the values for the individual samples comprising that pool. Several assays, including enzyme activity, lipid-soluble vitamins, nitrogen constituents and water-soluble carbohydrates, were selected to verify this assumption. All the results indicate good agreement between pooled values and averages of individual values.

Survey characteristics

A remarkable difference in cancer mortality was seen in the 65 counties under investigation (Table 1): oesophageal cancer rates for males were several hundred-fold greater in the county with the highest rate than in that with the lowest. In comparison with the highly homogeneous industrialized countries, this is an important advantage in terms of the statistical sensitivity in an ecological study.

Table 1. Cancer mortality rates in the survey populations[a]

Cancer site	Males	Females
All cancers	35–721 (21)	35–491 (14)
Nasopharynx	0–75 (75)	0–26 (26)
Oesophagus	1–435 (435)	0–286 (286)
Stomach	6–386 (64)	2–41 (70)
Liver	7–248 (35)	3–67 (22)
Colon/rectum	2–67 (34)	2–61 (30)
Lung	3–59 (20)	0–26 (26)
Breast	–	0–20 (20)
Cervix	–	4–97 (24)
Leukaemia	0–9 (9)	0–7 (7)

[a] Annual no. of cases per 100 000, truncated for ages 35-64, for 65 survey counties; numbers in parentheses are folds of the ranges

Lifetime residence is very stable: 90–94% of the survey subjects were born in the county in which they lived.

Foods consumed in each area are produced locally. There is considerable heterogeneity in the dietary patterns in different parts of China, but impressive homogeneity within the survey counties.

Data quality

An important indication of data quality is provided by the large proportion of statistically significant ($p < 0.05$) correlations in assays (85% of a total of 47 assays) in plasma and erythrocytes from two selected communes within each county; 77% of the correlations were significant at $p < 0.01$. Therefore, the averages of the values in the two communes can provide useful information about the extent to which that county differs from the other 64 counties, i.e., about the real geographic variations of that measurement.

Preliminary results

More than 100 000 correlation coefficients were generated by this study. These must, of course, be interpreted with particular caution; moreover, in a geographic survey such as this, many other variables may exist that confound any true relationships. However, if these data can be used appropriately, they should provide biologically informative associations between variables, which we hope will shed new light on concepts of nutrition and cancer. The following two examples illustrate the relative novelty of these data.

General nutritional status and cancer mortality

Five observable variables — height, weight, plasma urea-nitrogen, plasma albumin and whole blood haemoglobin — were selected from the 285 dietary, environmental and life style variables to form a latent variable, called general nutritional status based on protein intake. The correlations between this variable and mortality from the major diseases in the 65 counties were computed using the LISREL software package (Table 2). Mortality from all cancers was positively associated with general nutritional status, while total mortality from non-cancer causes was negatively correlated. Overall mortality from non-cancer causes was also negatively associated with general nutritional status, which indicates that in the 1970s overall mortality in rural China was dominated by non-cancer causes (mainly infectious and parasitic diseases).

Table 2. Correlation coefficients between general nutritional status and mortality from all cancers and from all non-cancer causes

Cause of death	Males	Females	Males and females
All causes	-0.448*	-0.454*	-0.490**
All cancers	0.463**	0.601**	0.547**
All non-cancer causes	-0.660**	-0.561**	-0.638**
Stomach cancer	0.556***	0.543***	0.581***
Oesophageal cancer	0.470***	0.392**	0.493***
Coronary heart disease	0.459***	0.289**	0.429**
Lung cancer in men	0.385**	-	-
Colon/rectal cancer	0.181	0.381**	0.385**
Cervical cancer	-	0.374**	-
Breast cancer in women	-	0.313*	-
Leukaemia	0.242	0.486***	0.304*
Liver cancer	0.089	0.252	0.172
Pneumonia	-0.423**	-0.475***	-0.461**
Chronic bronchitis, emphysema	-0.453**	-0.442**	-0.488***
Infectious diseases other than pulmonary tuberculosis	-0.590***	-0.538***	-0.577***
Pulmonary tuberculosis	-0.548***	-0.584***	-0.637***
Nasopharyngeal cancer	-0.647***	-0.419**	-0.662***

*, $p < 0.05$; **, $p < 0.01$; ***, $p < 0.001$

Antioxidant status and cancer mortality

In recent years, the possible association between antioxidant status and diseases such as cancer and heart disease has attracted much interest. Plasma ascorbic acid, selenium, β-carotene, α-tocopherol, retinol, lipid peroxides and copper, and erythrocyte catalase and superoxide dismutase were chosen as related variables to be used in a stepwise regression analysis with a number of cancer mortality rates. The results for oesophageal and gastric cancer (the two major cancer sites in the 1970s in China) are shown in Tables 3 and 4.

Although the results of simple correlation analysis showed clear correlations between lipid peroxidation and cancer mortality (data not shown), as well as between dietary antioxidants and cancer mortality, the associations become much stronger in the multiple regression models. Thus, the results implicate lipid peroxidation (plasma lipid peroxides and copper) in human carcinogenesis and show the protective effects of dietary antioxidants, especially selenium, ascorbic acid and retinol.

These are examples of the preliminary results in this study. It is the hope of the project organizers that this data base will offer important information for many years to come.

Table 3. Results of stepwise regression analysis for oesophageal cancer mortality in China

Variable	Standardized regression coefficients		
	Men	Women	Men and women
Ascorbic acid	-0.205*	-0.021	-0.154*
Selenium	-0.399***	-0.372**	-0.394***
α-Tocopherol	-0.286**	-0.151	-0.219**
Retinol	-0.206*	0.034	-0.119
Lipid peroxides	0.325***	0.250*	0.256***
Copper	0.433***	0.335**	0.379***
r^2	0.568	0.326	
No.	60	58	

*, $p < 0.05$; **, $p < 0.01$; ***, $p < 0.001$

Table 4. Results of stepwise regression analysis for gastric cancer mortality in China

Variable	Standardized regression coefficients		
	Men	Women	Men and women
Selenium	-0.383***	-0.380**	-0.392***
α-Tocopherol	-0.223*	0.192	-0.211**
Superoxide dismutase	-0.262*	-0.079	-0.182*
Lipid peroxide	0.252*	0.261*	0.244**
Copper	0.302**	0.272	0.298***
r^2	0.396	0.297	
No.	61	60	

*, $p < 0.05$; **, $p < 0.01$; ***, $p < 0.001$

Reference

Chen, J., Campbell, T.C., Li, J. & Peto, R. (1990) *Diet, Life-style and Mortality in China: A Study of the Characteristics of 65 Chinese Counties*, Oxford, Oxford University Press; Ithaca, Cornell University Press; and Beijing, People's Medical Publishing House

Relevance to Human Cancer of N-Nitroso Compounds,
Tobacco Smoke and Mycotoxins.
Ed. I.K. O'Neill, J. Chen and H. Bartsch
Lyon, International Agency for Research on Cancer
© IARC, 1991

THE ETIOLOGY OF GASTRIC CANCER

D. Forman

Imperial Cancer Research Fund, Cancer Epidemiology Unit,
Radcliffe Infirmary, Oxford, UK

We review recent evidence concerning risk factors for gastric cancer. An overview of analytical studies shows convincing evidence of a protective effect of fruit and vegetables. The specific protective constituents have not been firmly established, but micronutrients, especially ascorbic acid, are probably of importance. Other dietary factors that show a consistent pattern of effect in different studies are the moderate risks associated with high intake of preserved foods and salt. Evidence also indicates that gastric cancer is associated with tobacco consumption, although even in continuing heavy smokers the risk does not exceed two fold. Another non-dietary factor of potential importance is infection with the bacterium, *Helicobacter* (*Campylobacter*) *pylori*. The model of Correa and co-workers currently offers the best working hypothesis to explain the etiology of gastric cancer. Although the endogenous synthesis of *N*-nitroso compounds is central to the model, it is not yet clear what the rate-limiting steps are. Exposure to nitrate *per se* does not directly cause gastric cancer.

Worldwide, gastric cancer kills well over half a million people each year and, in terms of cancer mortality, is second in importance to lung cancer (Parkin *et al.*, 1988). As with all common cancers, there is much international variation in the incidence and mortality of this disease. In the period 1981–86, there was a more than six-fold range in international mortality rates, from 39.2 per 100 000 males in Japan to 5.7 per 100 000 in the USA; sub-national comparisons show ten-fold or larger differences (Muñoz, 1988). More remarkable, however, has been the consistent decline in incidence seen in nearly all countries for which statistics are available (Muñoz, 1988). In searching for the causes of gastric cancer, therefore, we can conclude that causative agents are likely to be widespread, variable in prevalence between populations and declining in importance — or that protective agents are widespread, variable and increasing in prevalence.

Dietary risk factors

Many factors have been investigated for a possible association with gastric cancer, but, although dietary factors are believed to be of major importance, causative agents have not been clearly identified (see reviews by Howson *et al.*, 1986; Mettlin, 1988). Part of the explanation for this failure lies with the problems involved in epidemiological studies of diet (Bingham, 1987).

Despite these difficulties, an overview of work in the last few years suggests that two dietary factors have appreciable protective effects – namely, vegetables and fresh fruit. Recent analytical studies show that people in the upper ranges of intake of vegetables or fruits tend to have a 30-50% reduction in risk (Tables 1 and 2). The consistency of this finding in several, occasionally very large studies in a variety of countries rules out the possibility that the results are due to chance. If the effect is truly causal, then even a 30% reduction in risk could have a substantial effect on the pattern of incidence. If, for example, the entire Italian population were in the upper intake tertile, approximately 17% of gastric cancer in that country could be avoided. If the reduction in risk were as high as 50% in the upper tertile, then 33% of the cancers could be avoided.

The mechanisms of action and the specific protective nutrients contained in fruit and vegetables have not been established. One possibility, for which there is much supporting experimental evidence, is that the associated intake of ascorbic acid inhibits the endogenous formation of carcinogenic N-nitroso compounds (Mirvish, 1986). In studies in five populations, ascorbic acid intake indices in cases and controls have been calculated (Table 3); all show an appreciable reduction in risk associated with high intake. One Italian study (Buiatti et al., 1989) also showed a significant protective effect of raw (but not of cooked) vegetables (Table 1). This suggests the involvement of a heat-labile component, such as ascorbic acid. In one small study (Stahelin et al., 1984) in which ascorbic acid levels in seven samples were measured prospectively, they were found to be lower in patients with gastric cancer than in controls.

Ascorbic acid may, of course, act by means other than inhibiting synthesis of N-nitroso compounds. It is also likely to be just one of several protective plant constituents, several of which would be highly correlated with ascorbic acid intake. β-Carotene, for example, has similar antioxidant properties to ascorbic acid and is found in many of the same foods. In three studies, total carotene or β-carotene intake has been estimated to confer as much protective effect against gastric cancer as ascorbic acid (Risch et al., 1985; La Vecchia et al., 1987; You et al., 1988), although in none of these studies were the two variables considered together in a multivariate analysis. Three prospective serological studies also indicate a protective effect of β-carotene against gastric cancer (Stahelin et al., 1984; Haenszel et al., 1985; Wald, 1987).

Planned intervention studies of vitamin supplements (Reed, 1988) should be of great value in assessing the role of different micronutrients. Present uncertainty about mechanisms should not, however, detract from the prudent public health message, which is to increase fruit and vegetable consumption. Given the range of biologically active constituents in fruit and vegetables (Ames, 1983), it is unlikely that vitamin supplementation alone will provide the optimal protective effect. Other components, such as chemicals contained in the allium family of vegetables (You et al., 1989), may also contribute; the active components in this case (if real) are more likely to be allyl salt derivatives than micronutrients.

Besides fruits and vegetables, many other dietary items and constituents have been associated with gastric cancer, either as risk factors (e.g., fresh meat, chocolate, alcohol, starch, unsaturated fats) or as protective factors (e.g., milk, dairy products, fibre, soya bean

Table 1. Vegetable intake and risk for gastric cancer[a]

Country	No. of cases	No. of controls	Variable	Highest intake level	Odds ratio[b]	Reference
China	564	1131	Fresh vegetables	Top quartile (> 156 kg/yr)	0.4*	You et al. (1988)
China	564	1131	Allium vegetables	Top quartile (> 24 kg/yr)	0.4*	You et al. (1989)
Italy	1016	1159	Raw vegetables	Top tertile (> 7.5x/week)	0.6*	Buiatti et al. (1989)
Italy	1016	1159	Cooked vegatables	Top tertile (> 8.5x/week)	1.1	Buiatti et al. (1989)
Italy	206	474	Green vegetables	Top tertile	0.3*	La Vecchia et al. (1987)
UK	95	190	Salad vegetables (summer)	>2 × per week	0.5	Coggon et al. (1989)
UK	95	190	Salad vegetables (winter)	>3 × per month	0.3*	Coggon et al. (1989)
Greece	110	110	Vegetables	6-unit increase	0.7*[c]	Trichopoulos et al. (1985)
Poland	110	110	Vegetables	>2 × per week	0.6	Jedrychowski et al. (1986)
USA (whites)	194	195	Vegetables	Above median	0.9	Correa et al. (1985)
USA (blacks)	197	195	Vegetables	Top quartile	0.5	Correa et al. (1985)
Canada	246	246	Vegetables	100 g/day increase	0.8*	Risch et al. (1985)
Canada	246	246	Pale-green vegetables	100 g/day increase	0.3	Risch et al. (1985)
Japan	5202	~250 000 (cohort)	Green-yellow vegetables	Daily	0.7*	Hirayama (1988)

[a] All known recent case-control or cohort studies of gastric cancer for which appropriate information is available about vegetable intake (either total or grouped)
[b] Reported odds ratio for highest intake level (lowest intake level = 1.0). In general, odds ratios were adjusted for potential confounding variables; however, the extent to which this was done varied between studies.
[c] Value for lettuce and onions, used as marker foods
* Odds ratio statistically significant for test utilized in study

Table 2. Fruit intake and risk for gastric cancer[a]

Country	No. of cases	No. of controls	Variable	Highest intake level	Odds ratio[b]	Reference
China	564	1131	Fresh fruit	Top quartile (>30 kg/yr)	0.6*	You et al. (1988)
Italy	1016	1159	Citrus fruit	Top tertile (>3 x /week)	0.6*	Buiatti et al. (1989)
Italy	1016	1159	Other fruit	Top tertile (>4.2 x /week)	0.4*	Buiatti et al. (1989)
Italy	206	474	Fresh fruit	Top tertile	0.7	La Vecchia et al. (1987)
Italy	206	474	Citrus fruit	Top tertile	0.6	La Vecchia et al. (1987)
UK	95	190	Fresh or frozen fruit	>5 x per week	0.6	Coggon et al. (1989)
UK	149	1934	Fruit	>4 x per week	0.4*	Burr & Holliday (1989)
Greece	110	110	Fruit	6-unit increase	0.8*[c]	Trichopoulos et al. (1985)
Poland	110	110	Fruit	>2 x per week	0.3*	Jedrychowski et al. (1986)
USA (whites)	194	195	Fruit	Top quartile	0.5*	Correa et al. (1985)
USA (blacks)	197	195	Fruit	Top quartile	0.3*	Correa et al. (1985)
Canada	246	246	Citrus fruit	100 g/day increase	0.8*	Risch et al. (1985)

[a] All known recent case-control studies of gastric cancer for which appropriate information is available about fruit intake
[b] Odds ratio for highest intake level (lowest intake level = 1.0). In general, odds ratios were adjusted for potential confounding variables; however, the extent to which this was done varied between studies.
[c] Value for oranges and lemons, used as marker foods
*Odds ratio statistically significant for test utilized in study

paste soup). Lack of consistency in the results of different studies suggests that they are isolated chance findings or that individual items vary in relevance in different cultures.

Two factors for which recent evidence is somewhat more uniform are the heterogeneous category of 'preserved foods' and salt. Included among the former are a diversity of products and preservation methods (see Table 4). Several of these foods may form N-nitroso compounds during prolonged storage, although other carcinogenic products may also be present, such as polycyclic aromatic hydrocarbons in smoked foods. A distinction should be drawn between a heavy dietary dependence on these items, which may be associated with other confounding dietary characteristics, such as low fruit and vegetable consumption, and occasional consumption in the context of a well-balanced diet. Thus, 'traditional' soup in Italy may be a marker of an overall dietary pattern rather than a specific risk factor. For very few of the food items listed in Table 4 have the odds ratios been adjusted for other dietary items; however, cured meat, smoked meat and salted fish were persistently identified in some (but not all) of the early studies of gastric cancer (see Nomura, 1982, for summary). When potent carcinogens can be identified in any of these food items, as has been the case for some preserved meats and certain types of salted fish (Forman, 1987), efforts can be made to reduce or eliminate them.

Table 3. Ascorbic acid intake and risk for gastric cancer[a]

Country	No. of cases	No. of controls	Highest intake level	Odds ratio[b]	Reference
China	564	1131	Top quartile	0.5*	You et al. (1988)
Italy	206	474	Top tertile	0.5*	La Vecchia et al. (1987)
USA (whites)	194	195	Top quartile	0.5	Correa et al. (1985)
USA (blacks)	197	195	Top quartile	0.3*	Correa et al. (1985)
Canada	246	246	1 g/day increase	0.4	Risch et al. (1985)

[a] All known recent case-control studies of gastric cancer for which appropriate information is available about ascorbic acid intake
[b] Odds ratio for highest intake level (lowest intake level = 1.0). In general, odds ratios were adjusted for potential confounding variables; however, the extent to which this was done varied between studies
*Odds ratio statistically significant for test utilized in study

Salting is used as a preserving technique, but salt itself has also sometimes been suggested as being associated with gastric cancer (Joosens & Geboers, 1981). Case-control studies in which individuals were asked about preferences for salty food, adding salt at meals, or eating foods with established high salt content have tended to show significantly positive results (see Table 5). Given the subjectivity in such questions, the consistency between the studies is quite striking, and, with the exception of the UK study, 'heavy' use of salt would be compatible with a 50% increase in risk.

In sum, a decline in dependency on preserved foodstuffs, and possibly salt, and increased availability of fruit and vegetables together offer the best available explanation for the secular decline in gastric cancer throughout the world. It is uncertain whether these factors alone are sufficient to explain the full extent of the decline; and the importance of each factor individually, independent of overall dietary pattern, is still to be determined. The

Table 4. Preserved food intake and risk of gastric cancer[a]

Country	No. of cases	No. of controls	Variable	Highest intake level	Odds ratio[b]	Reference
China	564	1131	Salted fish	> 1 kg/yr	1.4	You et al. (1988)
China	564	1131	Salted vegetables	Daily	1.1	You et al. (1988)
China	241	241	Salted and fermented soya paste	> 2 kg/yr	1.5*	Hu et al. (1988)
Japan	93	186	Dried/salted fish	> 1/week	2.0*	Tajima & Tominaga (1985)
Japan	93	186	Pickled vegetables	> 1/week	2.0*	Tajima & Tominaga (1985)
Italy	1016	1159	Dried/salted fish	Top tertile (> 0.4x/week)	1.4*	Buiatti et al. (1989)
Italy	1016	1159	Preserved meat	Top tertile (> 5.2x/week)	1.2	Buiatti et al. (1989)
Italy	1016	1159	'Traditional' soup	Top tertile (> 4.9x/week)	2.4*	Buiatti et al. (1989)
Italy	206	474	Ham	Top tertile	1.6*	La Vecchia et al. (1987)
UK	95	190	Smoked meat or fish	Sometimes eaten	1.0	Coggon et al. (1989)
USA (blacks)	197	195	Smoked meat	Top quartile	2.0*	Correa et al. (1985)
USA (blacks)	197	195	Home-cured meat	Above median	2.3*	Correa et al. (1985)
Canada	246	246	Smoked meat	100 g/day increase	2.2*	Risch et al. (1985)
Canada	246	246	Smoked fish	100 g/day increase	2.0	Risch et al. (1985)
Canada	246	246	Nitrite-preserved foods	1 mg/day increase	2.6*	Risch et al. (1985)

[a] All known recent case-control studies of gastric cancer for which appropriate information is available about preserved food intake

[b] Odds ratio for highest intake level (lowest intake level = 1.0). In general, odds ratios were adjusted for potential confounding variables; however, the extent to which this was done varied between studies.

*Odds ratio statistically significant for test utilized in study

Smoking

Another factor that has been consistently shown in case-control studies to convey a small risk for gastric cancer (usually less than two-fold) is tobacco smoke (see Table 6). The excess risk has often not been statistically significant, as many studies have not had the power to detect small increases. However, prospective cohort studies (Surgeon General, 1979: Hirayama, 1988) all show equivalent risks, and the latest data from the UK doctors study also confirms a significant trend (Table 7). Gastric cancer, it appears, should now be added to the list of tobacco-associated cancers; however, even in continuing heavy smokers, the risk does not exceed two fold.

Table 5. Salt intake and risk for gastric cancer[a]

Country	No. of cases	No. of controls	Highest exposure category	Odds ratio[b]	Reference
China	564	1131	Prefer saltier food	1.4	You et al. (1988)
Japan	93	186	Prefer salty taste	2.6*	Tajima & Tominaga (1985)
Italy	1016	1159	Prefer salty food	1.4*	Buiatti et al. (1989)
Italy	1016	1159	Often/always add salt	1.5*	Buiatti et al. (1989)
Italy	206	474	High salt use (subjective score)	1.5	La Vecchia et al. (1987)
France	163	1976	Prefer very salted food	1.8*	Tuyns (1983)
France	293	2914	Always add salt	1.8*	Tuyns (1988)
UK	95	190	High salt use (preference and addition)	6.2*	Coggon et al. (1989)
USA (whites)	~50	~60	Heavy salt user	1.2	Correa et al. (1985)
USA (whites)	~50	~60	Routinely add salt	1.4	Correa et al. (1985)
USA (blacks)	~50	~60	Heavy salt user	1.3	Correa et al. (1985)
USA (blacks)	~50	~60	Routinely add salt	1.8	Correa et al. (1985)

[a] All known recent case-control studies of gastric cancer for which appropriate information is available about exposure to salt
[b] Odds ratio for highest intake level (lowest intake level = 1.0). In general, odds ratios were adjusted for potential confounding variables; however, the extent to which this was done varied between studies.
* Odds ratio statistically significant for test utilized in study

Helicobacter pylori

One potential etiological factor, almost unheard of before 1983, is the bacterium *Helicobacter (Campylobacter) pylori*. It is found exclusively in the stomach, where it attaches to the luminal surface of the gastric epithelium (Blaser, 1987). There is now considerable evidence to associate *H. pylori* infection with development of active antral gastritis (Marshall, 1986; Blaser, 1987). Once established in the stomach, the bacteria seem to be resistant to natural host defence mechanisms and can be eradicated only with aggressive antibiotic regimens. It is not established whether untreated infections invariably progress from superficial gastritis to chronic atrophy of the gastric epithelium or whether other cofactors are needed. Either way, *H. pylori*-associated gastritis may be a precursor of the degenerative changes that predispose to gastric cancer.

Seroprevalence surveys indicate that the rate of infection with *H. pylori* can be over 50%, especially in old people, even in areas with relatively low gastric cancer rates (Jones et al., 1986). A characteristic of populations at high risk for gastric cancer may, however, be *H. pylori* infection at young ages (under 30 years; Gutierrez et al., 1988). Although this conclusion is tentative, it is consistent with other evidence relating age at onset of gastritis to gastric cancer (Correa et al., 1976). If *H. pylori* infection does predispose to gastric cancer, it is disturbing that extremely high rates of infection with *H. pylori* have been observed in young people in some populations with low levels of gastric cancer. (D. Forman, unpublished; Rouvroy et al., 1987; Wyatt et al., 1987). Pessimistically, it may be inferred that such populations (mainly African) might have an increased incidence of gastric cancer in the future. Alternatively, the infection may be harmless in the absence of some cofactor.

Table 6. Cigarette smoking and risk for gastric cancer

Country	No. of cases	No. of controls	Highest exposure category	Odds ratio[a]	Reference
China	564	1131	> 19 per day	1.5	You et al. (1988)
China	241	241	> 5 per day	1.8*	Hu et al. (1988
Japan	93	186	Current	2.1	Tajima & Tominaga (1985)
Italy	1016	1159	High	1.2	Buiatti et al. (1989)
France	40	168	> 6 per week	4.8*	Hoey et al. (1981)
UK	149	1934	> 14 per day	1.5	Burr & Holliday (1989)
Poland	110	110	Current	0.7	Jedrychowski et al. (1986)
USA (whites)	194	195	Current	1.4	Correa et al. (1985)
USA (blacks)	197	195	Current	2.7*	Correa et al. (1985)
Canada	246	246	20 pack/yr increase	1.3*	Risch et al. (1985)
Japan	5202	~250 000 (cohort)	> 24 per day	1.4*	Hirayama (1988)

[a] Odds ratio for exposure; nonsmokers = 1.0
* Odds ratio statistically significant for test utilized in study

Table 7. Cigarette smoking and risk for gastric cancer[a]

Category	No. of deaths		Relative risk
	Observed	Expected	
Nonsmokers	28	32.2	1.0
1–14 per day	27	22.1	1.4
> 15 per day	50	33.7	1.7
All current smokers	77	55.8	1.6
Ex-smokers	45	55.5	0.9
All smokers	122	111.3	1.3

[a] R. Doll, personal communication

The 'Correa model'

Our understanding of what happens to the gastric epithelium after the establishment of gastritis owes much to the work of Correa and colleagues (Correa, 1988). There is now no doubt that gastritis, when it becomes atrophic, is a risk factor for gastric cancer and results in loss of normal acidity, bacterial colonization and reduction of intragastric nitrate to nitrite. The nitrite may then nitrosate protein substrates and form carcinogenic N-nitroso compounds. This sequence of events is biologically plausible, and all of the individual steps have been demonstrated in human experimental studies (Mirvish, 1983; Bartsch & Montesano, 1984). What has yet to be established is whether such endogenously synthesized N-nitroso compounds make any material contribution to human cancer and what importance exposure to dietary nitrate has in this process.

Much epidemiological research has been focused in answering the latter question (Fraser, 1985; Forman, 1987). The complexity of assessing nitrate exposure has meant that most studies have tended to be correlational in design, i.e., estimate aggregate population exposure to nitrate and compare them with gastric cancer rates for the same populations. This method of inquiry is inevitably crude, subject to confounding and uninformative about individual risk. Despite these weaknesses, if most of these studies gave consistently positive results, it would be hard to avoid the conclusion that exposure to nitrate was of significance. The results are, however, conflicting (Fraser, 1985; Forman, 1987). Those few studies in which attempts have been made to estimate individual exposure and relate this to individual cancer risk (Risch *et al.*, 1985; Al-Dabbagh *et al.*, 1986) have failed to demonstrate a direct, positive relationship.

The lack of positive results may mean that nitrate exposure is not the rate-limiting factor in endogenous formation of N-nitroso compounds or in eventual carcinogenic insult. It could be that, in the presence of gastritis and accompanying achlorhydria and bacterial colonization, there is always sufficient nitrite in gastric juice to cause nitrosation. In such circumstances, what is then of importance is the balance of nitrosatable protein substrates, catalysts and inhibitors. A further complicating factor is the presence of nitrate that is not derived from the diet but is formed by immunostimulation (Wagner *et al.*, 1983), which may add appreciably to the total nitrate load.

The future

Adequate epidemiological testing of the Correa model in individually based studies therefore requires consideration not only of nitrate intake but also of a wide range of other dietary factors, together with details of gastric physiology and microbiology. The complexity of this task emphasizes the importance of developing new laboratory techniques that can be used in epidemiology, a precedent for which is the N-nitrosoproline test (Ohshima *et al.*, 1988). However, as shown above, questionnaire methods used appropriately in sufficiently large studies can help to unravel part of the web of cause-and-effect relationships involved in gastric cancer.

Acknowledgements

I should like to thank Sir Richard Doll, Dr T. Key and Dr H. Møller for suggesting several improvements to the manuscript, Dr D. Coggan, Dr M. Burr, Dr E. Buiatti and Sir Richard

Doll for allowing me access to unpublished data, and Miss S. Jones for typing the manuscript.

References

Al-Dabbagh, S., Forman, D., Bryson, D., Stratton, I. & Doll, R. (1986) Mortality of nitrate fertiliser workers. *Br. J. Ind. Med.*, **43**, 507-515

Ames, B.N. (1983) Dietary carcinogens and anticarcinogens. *Science*, **221**, 1256-1264

Bartsch, H. & Montesano, R. (1984) Relevance of nitrosamines to human cancer. *Carcinogenesis*, **5**, 1381-1393

Bingham, S. (1987) The dietary assessment of individuals; methods, accuracy, new techniques and recommendations. *Nutr. Abstr. Rev.*, **57**, 705-742

Blaser, M.J. (1987) Gastric, campylobacter-like organisms, gastritis, and peptic ulcer disease. *Gastroenterology*, **93**, 371-383

Buiatti, E., Palli, D., Decarli, A., Amadori, D., Avellini, C., Bianchi, S. Biserni, R., Cipriani, F., Cocco, P., Giacosa, A., Marubini, E., Puntoni, R., Vindigni, C., Fraumeni, J. & Blot, W. (1989) Italian multicenter case-control study on gastric cancer and diet. I. Frequencies of food consumption. *Int. J. Cancer*, **44**, 611-616

Burr, M.L. & Holliday, R.M. (1989) Fruit and stomach cancer. *J. Hum. Nutr. Diet*, **2**, 273-277

Coggon, D., Barker, D.J.P., Cole, R.B. & Nelson, M. (1989) Stomach cancer and food storage. *J. Natl Cancer Inst.*, **81**, 1178-1182

Correa, P. (1988) A human model of gastric carcinogenesis. *Cancer Res.*, **48**, 3554-3560

Correa, P., Cuello, C., Duque, E., Burbano, L.C., Garcia, F.T., Bolanos, O., Brown, C. & Haenszel, W. (1976) Gastric cancer in Colombia. III. Natural history of precursor lesions. *J. Natl Cancer Inst.*, **57**, 1027-1035

Correa, P., Fontham, E., Pickle, L.W., Chen, V., Lin, Y. & Haenszel, W. (1985) Dietary determinants of gastric cancer in South Louisiana inhabitants. *J. Natl Cancer Inst.*, **75**, 645-654

Forman, D. (1987) Dietary exposure to N-nitroso compounds and the risk of human cancer. *Cancer Surv.*, **6**, 719-738

Fraser, P. (1985) Nitrates – epidemiological evidence. In: Wald, N.J. & Doll, R., eds, *Interpretation of Negative Epidemiological Evidence for Carcinogenicity* (IARC Scientific Publications No. 65), Lyon, International Agency for Research on Cancer, pp. 183-194

Gutierrez, D., Sierra, F., Gomer, M.C. & Camargo, H. (1988) Campylobacter pylori in chronic environmental gastritis and duodenal ulcer patients. *Gastroenterology*, **94**, A163 (Abstract)

Haenszel, W., Correa, P., Lopez, A., Cuello, C., Zarama, G., Zavala, D. & Fontham, E. (1985) Serum micronutrient levels in relation to gastric pathology. *Int. J. Cancer*, **36**, 43-48

Hirayama, T. (1988) Actions suggested by gastric cancer epidemiological studies in Japan. In: Reed, P.I. & Hill, M.J., eds, *Gastric Carcinogenesis*, Amsterdam, Excerpta Medica, pp. 209-227

Hoey, J., Montvernay, C. & Lambert, R. (1981) Wine and tobacco: risk factors for gastric cancer in France. *Am. J. Epidemiol.*, **113**, 668-674

Howson, C.P., Hiyama, T. & Wynder, E.L. (1986) The decline in gastric cancer: epidemiology of an unplanned triumph. *Epidemiol. Rev.*, **8**, 1-27

Hu, J., Zhang, S., Jia, E., Wang, Q., Liu, S., Liu, Y., Wu, Y. & Cheng, Y. (1988) Diet and cancer of the stomach: a case-control study in China. *Int. J. Cancer*, **41**, 331-335

Jedrychowski, W., Wahrendorf, J., Popiela, T. & Rachtan, J. (1986) A case-control study of dietary factors and stomach cancer risk in Poland. *Int. J. Cancer*, **37**, 837-842

Jones, D.M., Eldridge, J., Fox, A.J., Sethi, P. & Whorwell, P.J. (1986) Antibody to the gastric campylobacter-like organism 'campylobacter pyloridis' – clinical correlations and distribution in the normal population. *J. Med. Microbiol.*, **22**, 57-62

Joosens, J.V. & Geboers, J. (1981) Nutrition and gastric cancer. *Nutr. Cancer*, **2**, 250-261

La Vecchia, C., Negri, E., Decarli, A., D'Avanzo, B. & Franceschi, S. (1987) A case-control study of diet and gastric cancer in northern Italy. *Int. J. Cancer*, **40**, 484-489

Marshall, B.J. (1986) Campylobacter pyloridis and gastritis. *J. Infect. Dis.*, **153**, 650-657

Mettlin, C. (1988) Epidemiological studies in gastric adenocarcinoma. In: Douglass, H.O., ed., *Gastric Cancer*, New York, Churchill Livingstone, pp. 1-25

Mirvish, S.S. (1983) The etiology of gastric cancer – intragastric nitrosamide formation and other theories. *J. Natl Cancer Inst.*, **71**, 629-647

Mirvish, S.S. (1986) Effects of vitamins C and E on *N*-nitroso compound formation, carcinogenesis, and cancer. *Cancer*, **58**, 1842-1850

Muñoz, N. (1988) Descriptive epidemiology of gastric cancer. In: Reed, P.I. & Hill, M.J., eds, *Gastric Carcinogenesis*, Amsterdam, Excerpta Medica, pp. 51-69

Nomura, A. (1982) Stomach. In: Schottenfeld, D. & Fraumeni, J.F., eds, *Cancer Epidemiology and Prevention*, Philadelphia, W.B. Saunders, pp. 624-637

Ohshima, H., Pignatelli, B., Malaveille, C., Friesen, M., Calmels, S., Shuker, D., Muñoz, N. & Bartsch, H. (1988) Markers for intragastric nitrosamine formation and resulting DNA damage. In: In: Reed, P.I. & Hill, M.J., eds, *Gastric Carcinogenesis*, Amsterdam, Excerpta Medica, pp. 175-185

Parkin, D.M., Läärä, E. & Muir, C.S. (1988) Estimates of the worldwide frequency of sixteen major cancers in 1980. *Int. J. Cancer*, **41**, 184-197

Reed, P.I. (1988) Chemoprevention. In: Reed, P.I. & Hill, M.J., eds, *Gastric Carcinogenesis*, Amsterdam, Excerpta Medica, pp. 253-263

Risch, H.A., Jain, M., Choi, N.W., Fodor, J.G., Pfeiffer, G., Howe, G.R., Harrison, L.W., Craib, K.J.P. & Miller, A.B. (1985) Dietary factors and the incidence of cancer of the stomach. *Am. J. Epidemiol.*, **122**, 947-959

Rouvroy, D., Bogaerts, J., Nsengiumwa, O., Omar, M., Versailles, L. & Haot, J. (1987) Campylobacter pylori, gastritis and peptic ulcer disease in central Africa. *Br. Med. J.*, **295**, 1174

Stahelin, H.B., Rosel, F., Buess, E. & Brubacher, G. (1984) Cancer, vitamins, and plasma lipids: prospective Basel study. *J. Natl Cancer Inst.*, **73**, 1463-1468

Surgeon General (1979) *Smoking and Health*, Washington, DC, US Department of Health, Education, and Welfare

Tajima, K. & Tominaga, S. (1985) Dietary habits and gastro-intestinal cancers: a comparative case-control study of stomach and large intestinal cancers in Nagoya, Japan. *Jpn. J. Cancer Res. (Gann)*, **76**, 705-716

Trichopoulos, D., Ouranos, G., Day, N.E., Tzonou, A., Manousos, O., Papadimitriou, C. & Trichopoulos, A. (1985) Diet and cancer of the stomach: a case-control study in Greece. *Int. J. Cancer*, **36**, 291-297

Tuyns, A.J. (1983) Sodium chloride and cancer of the digestive tract. *Nutr. Cancer*, **4**, 198-205

Tuyns, A.J. (1988) Salt and gastrointestinal cancer. *Nutr. Cancer*, **11**, 229-232

Wagner, D.A., Young, V.R. & Tannenbaum, S.R. (1983) Mammalian nitrate biosynthesis: incorporation of $^{15}NH_3$ into nitrate is enhanced by endotoxin treatment. *Proc. Natl Acad. Sci. USA*, **80**, 4518-4521

Wald, N.J. (1987) Retinol, beta-carotene and cancer. *Cancer Surv.*, **6**, 635-651

Wyatt, J.I., de Caestecker, J.S., Rathbone, B.J. & Heatley, R.V. (1987) Campylobacter pyloridis in tropical Africa. *Gut*, **28**, A1409 (Abstract)

You, W., Blot, W.J., Chang, Y., Ershow, A.G., Yang, Z., An, Q., Henderson, B., Xu, G., Fraumeni, J.F. & Wang, T. (1988) Diet and high risk of stomach cancer in Shandong, China. *Cancer Res.*, **48**, 3518-3523

You, W., Blot, W.J., Chang, Y., Ershow, A., Yang, Z.T., An, Q., Henderson, B.E., Fraumeni, J.F. & Wang, T. (1989) Allium vegetables and reduced risk of stomach cancer. *J. Natl Cancer Inst.*, **81**, 162-164

ETIOLOGICAL RESEARCH ON GASTRIC CANCER AND ITS PRECURSOR LESIONS IN SHANDONG, CHINA

W.C. You[1,6], Y.S. Chang[2], Z.T. Yang[3], L. Zhang[1], G.W. Xu[1],
W.J. Blot[4], R. Kneller[4], L.K. Keefer[5] & J.F. Fraumeni, Jr[4]

[1]*Beijing Institute for Cancer Research, Beijing;* [2]*Weifang Medical Institute, Weifang, Shandong;* [3]*Linqu Public Health Bureau, Linqu, Shandong, China;* [4]*National Cancer Institute, Bethesda, MD;* and [5]*National Cancer Institute, Frederick, MD, USA*

Research over the past several years in an area of Shandong, China, with one of the world's highest rates of gastric cancer, has yielded clues to the environmental determinants of this tumour. Interviews with 564 gastric cancer patients and 1131 population-based controls revealed increased risks associated with consumption of sour pancakes, a fermented staple unique to the area, in samples of which volatile N-nitrosamines have been detected. Lower risks were found among people who had a higher intake of fresh vegetables, including garlic and other *Allium* vegetables which contain constituents that can inhibit carcinogenesis by N-nitrosamines and other substances in experimental animals. A pilot study involving assays of urine and gastric juice from 60 individuals in a screening programme showed higher levels of N-nitrosoproline and of *cis-* and *trans*-N-nitroso-2-methylthiazolidine 4-carboxylic acid among persons with gastric dysplasia than in either normal controls or those with chronic atrophic gastritis. We are trying to characterize the transition and progression of precursor lesions to gastric cancer and to evaluate the role of dietary variables, nutrients, N-nitroso compounds and other factors in particular stages of the carcinogenic process.

Gastric cancer is one of the leading causes of cancer in the world, ranking first in some estimates of worldwide cancer incidence (Waterhouse *et al.*, 1982). It is the first cause of death from cancer in China (Cancer Control Office, 1980; Zhang *et al.*, 1984). There is considerable geographic variation within China, however, with elevated rates in northern and central parts of the country. One of the highest rates in China (and the world) is found in Linqu in Shandong Province: the age-adjusted (world standard) death rate from gastric cancer among males in Linqu was 70 per 100 000 per year in the 1973–75 Chinese national mortality survey — more than double the national level (Cancer Control Office, 1980) — and has remained at about this level since then.

---
[6] To whom correspondence should be addressed

The causes of gastric cancer in Linqu and elsewhere in the world are not clearly understood. Dietary factors are thought to be important, and changes in food intake, preparation and storage are considered to be responsible for the marked decline in the incidence of this cancer in the USA and other developed countries over the past several decades (Nomura, 1982). The specific causes of gastric cancer have been difficult to identify, but elevated risks have been associated with male sex, high salt intake, certain smoked and preserved foods that may contain or promote formation of N-nitroso compounds or other carcinogens, cigarette smoking, some occupational exposures, and prior chronic atrophic gastritis (Nomura, 1982; Correa et al., 1983; Tuyns, 1983; Correa et al., 1985; Montes et al., 1985; Risch et al., 1985; Trichopoulos et al., 1985; La Vecchia et al., 1987; Hu et al., 1988; You et al., 1988). Protective effects have been reported to be conferred by eating some foods (fresh fruits and vegetables) and by the nutrients they contain (e.g., carotene and ascorbic acid). The proportion of gastric cancer caused or inhibited by these factors individually or collectively has not yet been established.

To investigate the reasons for the high rates of gastric cancer in Linqu, a rural county in north-eastern China, we initiated a case-control study in 1984. During the subsequent two years, interviews were conducted with 564 gastric cancer patients and 1131 controls of the same age and sex randomly selected from the Linqu population (You et al., 1988). A structured questionnaire was used to obtain information on diet, smoking, occupation, medical history and other variables. Nearly all of the cases and controls had been born in the area, and most were farmers with little or no formal education.

Estimates of relative risk for gastric cancer according to level of consumption of certain foods are shown in Table 1. Risks were increased by 30% among people who consumed sour pancakes, a locally favourite food made of corn, wheat, dried sweet potatoes and soya bean flour that is allowed to ferment before consumption. About half of the population ate these pancakes daily, while the other half ate them hardly at all. Although the magnitude of the excess risk associated with consumption of sour pancakes seems too small to account entirely for the elevated rates of gastric cancer in Linqu, it is noteworthy that in preliminary studies sour pancake juice was shown to have mutagenic activity and to contain mycotoxins and N-nitroso compounds (H.L. Susan, unpublished data). Research to characterize further the types and concentrations of compounds in the sour pancakes is under way. In the case-control study, we also found associations between risk for gastric cancer and having a mouldy grain supply, a preference for salty food, cigarette smoking, the occurrence of gastric cancer (but not other cancers) in first-degree family members, and prior gastritis (Table 1).

The results of the case-control study also revealed strong protective effects associated with fresh vegetable intake: the relative risks for gastric cancer declined with increasing consumption, falling to half among people in the highest quartile of vegetable consumption. The mechanisms by which vegetables lower risk are not clear, but micronutrients such as ascorbic acid and tocopherol, which can block the formation of N-nitrosamines *in vivo* (Mirvish, 1983), are suspected to be involved. We found reductions in risk associated with intake of several types of fresh vegetables, with a particularly strong trend for *Allium* vegetables (You et al., 1989). Decreased risks were observed in association with consumption of each of five vegetables in this genus (garlic, garlic stalks, chives, scallions and onions) in Linqu, suggesting that a common ingredient can inhibit gastric cancer. Although this is the first such epidemiological observation, several studies have shown that extracts and oils from garlic and onions are potent inhibitors of carcinogenesis in experimental animals. Most attention has focused on allyl sulfides, which block the formation of several types of cancer induced by different chemical compounds, including N-nitrosamines (Belman, 1983; Sparnins et al., 1986; Niukian et al., 1987; Wargovich, 1987; Wargovich et al., 1988). It is also possible that garlic and other *Allium* vegetables act through their antifungal and antibacterial properties (Block, 1985). By inhibiting gastric bacterial growth, garlic may decrease the chances for bacteria-catalysed reduction of nitrate to nitrite and thus reduce the endogenous formation of N-nitroso compounds (Correa, 1987).

Table 1. Relative risks (RRs) for gastric cancer associated with selected dietary and other variables, Linqu, case-control study

Variable	Level of exposure	RR[a]	95% confidence interval
Consumption of sour pancakes	Almost never	1.0	–
	Daily	1.3	1.0–1.6
Consumption of mouldy grain	Never	1.0	–
	Yes	1.4	1.2–2.0
Consumption of *Allium* vegetables[b]	1 (low)	1.0	–
	2	0.7	0.5–0.9
	3	0.7	0.5–0.9
	4	0.5	0.4–0.8
Consumption of other fresh vegetables[c]	1 (low)	1.0	–
	2	0.7	0.5–1.0
	3	0.8	0.6–1.1
	4	0.6	0.4–0.9
Preference for salty foods	Less	1.0	–
	Average	1.1	0.8–1.6
	More	1.4	1.0–1.9
Cigarette smoking	None	1.0	–
	<20/day	1.3	0.9–1.9
	≥20/day	1.5	1.0–2.1
Cancer in family member	None	1.0	–
	Gastric cancer	1.8	1.3–2.7
	Other cancer	1.0	0.7–1.5
Prior chronic gastritis	No	1.0	–
	Yes	2.9	1.7–4.6

[a] Risks adjusted for age, sex and family income
[b] Risks adjusted for intake of other fresh vegetables
[c] Risks adjusted for intake of *Allium* vegetables

To test the feasibility of measuring N-nitroso compounds in biological specimens collected in this high-risk area, we conducted a pilot study in 1988. We collected 24-h urine specimens and samples of gastric juice after an overnight fast from 20 subjects with normal gastric mucosa, 20 subjects with chronic atrophic gastritis/intestinal metaplasia (CAG/IM) and 20 subjects with dysplasia in one Linqu village. Gastric pH was determined before the samples were shipped to the USA, where N-nitroso compounds, nitrite and nitrate were determined in all urine and gastric juice samples. Initial analyses show that median levels of N-nitrosoproline and of *cis*- and *trans-N*-nitroso-2-methylthiazolidine 4-carboxylic acid in the urine were higher in subjects with dysplasia than in those with normal mucosa, with lowest levels in people with CAG/IM. The urinary levels of the other compounds also tended to be lower in those with CAG/IM than in people with normal mucosa or with gastric dysplasia. The urinary levels of N-nitroso compounds tended to be lower in males than in females and also varied with the gastric juice pH: the urinary concentrations of nearly all of the N-nitrosamines decreased and then rose as gastric pH varied from 0–2.4 to 2.5–5.5 to 5.6 and over. The levels of N-nitroso compounds in gastric juice were lower than those in urine, which made evaluation of variation by gastric pathology difficult in these small numbers of samples.

Thus, it seems possible that N-nitroso compounds may contribute to the high risk for gastric cancer in Linqu. Our initial findings suggest that certain foods associated with high risk for this cancer contain such compounds, while the foods associated with low risk have the capacity to block their formation. Furthermore, persons with gastric dysplasia, who are exceptionally prone to gastric cancer, had elevated urinary levels of N-nitrosamines. In order to investigate further the role of N-nitroso compounds and other factors in induction of the gastric cancerous states (Correa, 1982, 1988), we are beginning a cohort study of 3000 adults in Linqu. All participants will be interviewed, given gastroscopic examinations and asked to donate blood; 20% will also be asked to donate urine and gastric juice specimens. The study will provide prospectively ascertained information on cancer incidence over a six-year follow-up period to allow direct assessment of CAG (confirmed from biopsy specimens), IM and dysplasia as risk factors for gastric cancer; body fluids (blood, urine, gastric juice) will be collected every two years, and a second gastroscopic examination will be conducted at the end of the follow-up. Although our case-control study identified prior chronic gastritis as a risk factor (the relative risk for gastric cancer associated with chronic gastritis diagnosed five or more years prior to interview was 2.9; 95% confidence interval, 1.7–4.6), information was provided only by the subjects, without confirmation from medical sources. The collection of blood will permit the first large-scale, systematic evaluation of the correlation between histological diagnoses and serological markers using pepsinogen ratios. Samloff and others (Samloff, 1982; Stemmermann *et al.*, 1987; Westerveld *et al.*, 1987) have shown that human gastric mucosa contains two immunologically distinct groups of pepsinogen — pepsinogen I (PGI) and pepsinogen II (PGII) — the ratio of which may serve as a marker of pathological change.

Low serum ferritin concentrations up to 13 years preceding diagnosis of gastric cancer have recently been reported as a marker of elevated risk (Akiba *et al.*, 1990), but this will be the first evaluation of ferritin in relation to precursor lesions. Serum nutrients (β-carotene and other fat-soluble vitamins, ascorbic acid and trace elements) will be measured in a 20% random sample. In Colombia, where the rates of gastric cancer are also high, serum levels of carotene (but not of ascorbic acid) were reported to be lower among persons with gastric dysplasia than among those with less advanced precursor lesions (Haenszel *et al.*, 1985).

In summary, continuing investigations of gastric cancer in a high-risk area of Shandong Province have provided useful etiological leads, which are being pursued in efforts to understand better the causal events and to identify preventive measures.

References

Akiba, S., Nerishii, K., Blot, W.J., Kabuto, M., Stevens, R.G., Kato, H. & Land, C.E. (1990) Serum ferritin and stomach cancer risk among Japanese. *Cancer* (in press)

Belman, S. (1983) Onion and garlic oils inhibit tumour promotion. *Carcinogenesis*, 4, 1063-1065

Block, E. (1985) The chemistry of garlic and onions. *Sci. Am.*, 252 (3), 114-119

Cancer Control Office (1980) *Data on Cancer Mortality in China*, Beijing, Ministry of Public Health, p. 197

Correa, P. (1982) Precursors of gastric and esophageal cancer. *Cancer*, 50, 2554-2565

Correa, P. (1987) Modulation of gastric carcinogenesis: updated model based on intragastric nitrosation. In: Bartsch, H., O'Neill, I.K. & Schulte-Hermann, R., eds, *The Relevance of N-Nitroso Compounds to Human Cancer: Exposures and Mechanisms* (IARC Scientific Publications No. 84), Lyon, International Agency for Research on Cancer, pp. 485-491

Correa, P. (1988) A human model of gastric carcinogenesis. Cancer Res., 48, 3554-3560

Correa, P., Cuello, C., Fajardo, L.F., Haenszel, W., Bolaños, O. & de Ramirez, B. (1983) Diet and gastric cancer: nutrition survey in a high-risk area. *J. Natl Cancer Inst.*, 70, 673-678

Correa, P., Fontham, E., Pickle, L.W., Chen, V., Lin, Y. & Haenszel, W. (1985) Dietary determinants of gastric cancer in South Louisiana inhabitants. *J. Natl Cancer Inst.*, 75, 645-654

Haenszel, W., Correa, P., López, A., Cuello, C., Zarama, G., Zavala, D. & Fontham, E. (1985) Serum micronutrient levels in relation to gastric pathology. *Int. J. Cancer*, 36, 43-48

Hu, J., Zhang, S., Jia, E., Wang, Q., Liu, S., Liu, Y., Wu, Y. & Cheng, Y. (1988) Diet and cancer of the stomach: a case-control study in China. *Int. J. Cancer*, 41, 331-335

La Vecchia, C., Negri, E., Decarli, A., D'Avanzo, B. & Franceschi, S. (1987) A case-control study of diet and gastric cancer in northern Italy. *Int. J. Cancer*, 40, 484-489

Mirvish, S.S. (1983) The etiology of gastric cancer: intragastric nitrosamide formation and other theories. *J. Natl Cancer Inst.*, 71, 629-647

Montes, G., Cuello, C., Correa, P., Zarama, G., Liuzza, G., Zavala, D., de Marin, E. & Haenszel, W. (1985) Sodium intake and gastric cancer. *J. Cancer Res. Clin. Oncol.*, 109, 42-45

Niukian, K., Schwartz, J. & Shklar, G. (1987) Effects of onion extract on the development of hamster buccal pouch carcinomas as expressed in tumor burden. *Nutr. Cancer*, 9, 171-176

Nomura, A. (1982) Stomach. In: Schottenfeld, D. & Fraumeni, J.F., eds, *Cancer Epidemiology and Prevention*, Philadelphia, W.B. Saunders, pp. 624-637

Risch, H.A., Jain, M., Choi, N.W., Fodor, J.G., Pfeiffer, C.J., Howe, G.R., Harrison, L.W., Craib, K.J.P. & Miller, A.B. (1985) Dietary factors and the incidence of cancer of the stomach. *Am. J. Epidemiol.*, 122, 947-959

Samloff, I.M. (1982) Pepsinogens I and II: purification from gastric mucosa and radioimmunoassay in serum. *Gastroenterology*, 82, 26-33

Sparnins, V.L., Mott, A.W., Barany, G. & Wattenberg, L.W. (1986) Effects of allyl methyl trisulfide on glutathione S-transferase activity and BP-induced neoplasia in the mouse. *Nutr. Cancer*, 8, 211-215

Stemmermann, G.N., Samloff, I.M., Nomura, A.M.Y. & Heilbrun, L.K. (1987) Serum pepsinogens I and II and stomach cancer. *Clin. Chim. Acta*, 163, 191-198

Trichopoulos, D., Ouranos, G., Day, N.E., Tzonou, A., Manousos, O., Papadimitriou, C. & Trichopoulos, A. (1985) Diet and cancer of the stomach: a case-control study in Greece. *Int. J. Cancer*, 36, 291-297

Tuyns, A. (1983) Sodium chloride and cancer of the digestive tract. *Nutr. Cancer*, 4, 198-205

Wargovich, M.J. (1987) Diallyl sulfide, a flavor component of garlic (*Allium sativum*), inhibits dimethylhydrazine-induced colon cancer. *Carcinogenesis*, 8, 487-489

Wargovich, M.J., Woods, C., Eng, V.W.S., Stephens, L.C. & Gray, K. (1988) Chemoprevention of N-nitrosomethylbenzylamine-induced esophageal cancer in rats by the naturally occurring thioether, diallyl sulfide. *Cancer Res.*, 48, 6872-6875

Waterhouse, J., Muir, C., Shanmugaratnam, K. & Powell, J., eds (1982) *Cancer Incidence in Five Continents*, Vol. IV (IARC Scientific Publications No. 42), Lyon, International Agency for Research on Cancer, pp. 670, 706-707

Westerveld, B.D., Pals, G., Lamers, C.B.H.W., Defize, J., Pronk, J.C., Frants, R.R., Ooms, E.C.M., Kreuning, J., Kostense, P.J., Eriksson, A.W. & Meuwissen, S.G.M. (1987) Clinical significance of pepsinogen A isozymogens, serum pepsinogen A and C levels, and serum gastrin levels. *Cancer*, **59**, 952-958

You, W.C., Blot, W.J., Chang, Y.S., Ershow, A.G., Yang, Z.T., An, Q., Henderson, B., Xu, G.W., Fraumeni, J.F. & Wang, T. (1988) Diet and high risk of stomach cancer in Shandong, China. *Cancer Res.*, **48**, 3518-3523

You, W.C., Blot, W.J., Chang, Y.S., Ershow, A., Yang, Z.T., An, Q., Henderson, B.E., Fraumeni, J.F. & Wang, T.G. (1989) Allium vegetables and reduced risk of stomach cancer. *J. Natl Cancer Inst.*, **81**, 162-164

Zhang, R.F., Sun, H.L., Jin, M.L. & Li, S.N. (1984) A comprehensive survey of etiologic factors of stomach cancer in China. *Chin. Med. J.*, **97**, 322-332

NASOPHARYNGEAL CARCINOMA: EPIDEMIOLOGY AND DIETARY FACTORS

M.C. Yu

Department of Preventive Medicine, University of Southern California School of Medicine, Los Angeles, CA, USA

Nasopharyngeal carcinoma (NPC) is a disease with a remarkable racial and geographical distribution. It is very rare (incidence of less than 1 per 100 000 person-years) in most parts of the world, and in only a handful of populations does this low-risk profile deviate; these groups include people in southern China, Eskimos and other natives of the Arctic region, natives of south-east Asia, and the mainly Arab populations of North Africa and Kuwait. Convincing evidence implicates dietary factors as the primary cause of NPC among Chinese. A series of case-control studies conducted in various Chinese populations with distinct risks of NPC, ranging from the very high-risk Cantonese populations to the relatively low-risk northern Chinese, have suggested that ingestion of salted fish and other kinds of preserved foods constitutes the most important cause of NPC among these people. Preliminary data on Malays in south-east Asia, Eskimos in Alaska and Arabs of North Africa also suggests that ingestion of preserved foods by these population groups may be responsible for their raised incidence of NPC.

Regardless of race and geography, the commonest form of nasopharyngeal cancer is that arising from the epithelial cells lining the nasopharynx. These tumours (commonly referred to as nasopharyngeal carcinoma, NPC) constitute 75-95% of nasopharyngeal cancers in low-risk populations and virtually all nasopharyngeal cancers in high-risk populations (Ho, 1971; Sugano *et al.*, 1978; Levine & Connelly, 1985).

International patterns

NPC is a rare malignancy in most parts of the world, where the age-standardized incidence rate for people of either sex is generally less than 1 per 100 000 persons per year (Muir *et al.*, 1987). Table 1 lists the handful of populations in which this pattern is known to deviate, with the age-standardized (world population) incidence rates for men and women separately. The overall incidence of NPC is elevated in China, with substantial variation between regions; in general, the incidence increases as one travels from north to south in China. Whereas rates in Chinese men in the northernmost provinces are about 2–3/100 000 person-years, the rates in those residing in the southernmost province of Guangdong are 25–40/100 000 person-years (National Cancer Control Office, 1979; Yu *et al.*, 1981; Muir *et al.*, 1987). High rates, approaching those observed in southern China, are seen in Eskimos and other natives of the Arctic region (Nielsen et al., 1977; Lanier et al., 1980). Intermediate rates of NPC (2-5/100 000 person-years) are observed among Malays in Singapore (Shanmugaratnam, 1973) and Malaysia (Armstrong et al., 1979) and among Filipinos (Muir et al., 1987). In Sabah, Malaysia, rates similar to those observed among Eskimos have been reported for the native Kadazans (Rothwell, 1979). Although data on mortality and incidence are lacking for the other indigenous people of south-east Asia, raised relative frequencies of NPC in biopsy series have been reported among natives of Thailand, Indonesia, Sarawak and Viet Nam (Muir, 1971). Intermediate rates of NPC are observed among the people of Kuwait, both local-born Arabs and immigrants from neighbouring Arab countries (Muir et al., 1987). On the basis of reviews of hospital series, it is believed that rates of NPC are also raised in the mainly Arab populations of Algeria, Tunisia, Morocco and Sudan (Muir, 1971; Cammoun et al., 1974; Hidayatalla et al., 1983). The intermediate rates of NPC among Israeli Jews born in Africa or Asia (Muir et al., 1987), many of whom were from North Africa, tend to support this view.

Table 1. Populations at increased risk for nasopharyngeal cancer

Population	Age-standardized (world) incidence[a]		Key reference
	Male	Female	
Chinese (Hong Kong)	30.0	12.9	Muir et al. (1987)
Chinese (Taipei)	8.1	3.2	Chen et al. (1988a)
Chinese (Shanghai)	4.4	2.0	Muir et al. (1987)
Chinese (Tianjin)	1.7	0.9	Muir et al. (1987)
Eskimos (Greenland)	12.3	8.5	Nielsen et al. (1977)
Eskimos, Indians, Aleuts (Alaska)	13.5	3.7	Lanier et al. (1980)
Malays (Malaysia)	2.3	0.7	Armstrong et al. (1979)
Malays (Singapore)	4.0	1.5	Muir et al. (1987)
Filipinos (Rizal)	4.7	2.6	Muir et al. (1987)
Kadazans (Sabah)	15.9	8.7	Rothwell (1979)
Kuwaitis (Kuwait)	2.1	1.3	Muir et al. (1987)
Non-Kuwaitis (Kuwait)	2.9	1.4	Muir et al. (1987)
Israeli Jews born in Africa/Asia	3.3	0.7	Muir et al. (1987)

[a] Per 100 000 person-years

Racial/ethnic variation

High risk for NPC among Chinese is confined mainly to those residing in the southern provinces of Guangdong, Guangxi, Hunan and Fujian (Figure 1; National Cancer Control Office, 1979). Several distinct racial–ethnic groups reside in this high-risk region, and these groups have different rates of NPC. The highest rates are observed among the Tankas, a sub-ethnic group of Cantonese who inhabit the Pearl River delta basin in central Guangdong. One feature that distinguishes the Tankas from the other Cantonese is that they are seafaring people (either fishermen or sea transporters) who live on houseboats moored along the banks of the many branches of the Pearl River. The rates of NPC among the Tankas are twice those of the land-dwelling Cantonese (Ho, 1978; Li et al., 1985). In turn, land-dwelling Cantonese (who comprise 98-99% of Cantonese) have a rate of NPC twice as high as those of the Hakka and Chiu Chau dialect groups, who reside in north-east Guangdong (Yu *et al.*, 1981; Li *et al.*, 1985). The people of Fukien province are culturally similar to the Chiu Chau people in Guangdong Province, and so are their rates of NPC (National Cancer Control Office, 1980). It is interesting to note that the Hakkas (who rarely intermarry with other dialect groups) originated from northern China more than 500 years ago (Ho, 1959), but their rates resemble those of their Chiu Chau neighbours instead of those of their low-risk ancestors in the north. Even after migration to south-east Asia, the Cantonese continue to exhibit a risk for NPC that is twice as high as those of the Hakkas, Chiu Chaus and Fukienese (Armstrong *et al.*, 1979; Lee *et al.*, 1988).

Figure 1. Provinces of China in which the risk for nasopharyngeal carcinoma is high

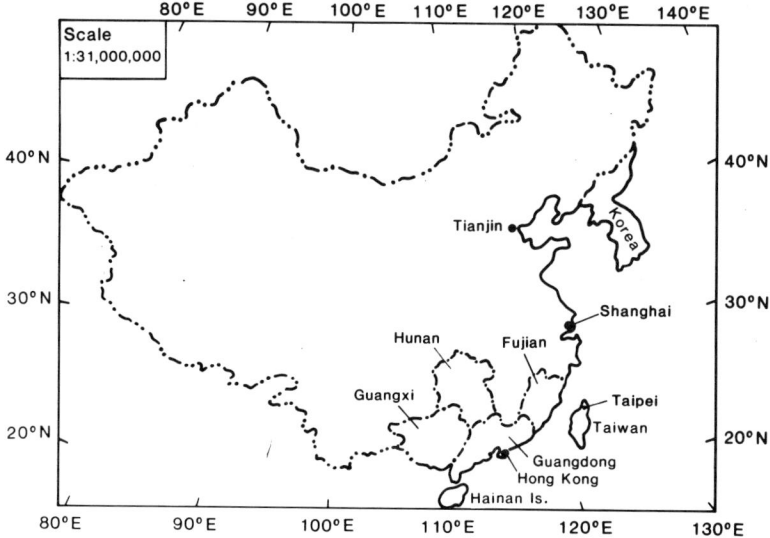

Two distinct racial groups inhabit the Autonomous Region of Guangxi. While the Zhuang people in western Guangxi have a rate of NPC that is one-fifth of that in Cantonese, the Han (the predominant race in China) people of eastern Guangxi, who are ethnically close to the Cantonese in Guangdong, show similar rates. Hunan Province borders both Guangxi and Guangdong to the north, and, not surprisingly, in areas that border Guangdong and Guangxi the rates of NPC are high. In addition, the Tujia and Miao tribes (minority races in China), who inhabit the mountainous region of western Hunan, have NPC rates approaching those in the Cantonese (National Cancer Control Office, 1980).

In summary, the geographic variation in the incidence of NPC within southern China closely parallels the distribution of the racial–ethnic groups inhabiting the region. The relatively high rates observed among the Hakkas, who originated from low-risk northern China, argues against genetic predisposition as a major cause of the different risk patterns among these population groups; however, these ethnic groups have distinct customs and food habits, and it is possible that environmental factors inherent in their traditional cultures are responsible for their differing susceptibility to NPC.

Chinese salted fish

The higher rates of NPC among the boat-dwelling Tankas as compared to the land-dwelling Cantonese was first noted by Ho (1967) of Hong Kong, who further observed that these boat people had little exposure to domestic inhalants, as they 'live all their lives in open boats and cook their food in the open air....' Ho then turned his attention to dietary factors, especially the traditional foods of southern China. He observed that salted fish is dominant in the diet of the boat people, being the principal source of supplementary food in a diet consisting mainly of rice. As they often spend long periods at sea, southern Chinese fishermen salt most of their catch as a means of preserving it, and these fishermen and their families prefer to consume salted instead of fresh fish since the latter can command much higher prices. Ho (1971) suggested that Chinese salted fish, 'a common and favourite item of food among most (Cantonese) in and outside China' be investigated as a 'possible etiological factor' for NPC.

Seven case-control studies were conducted among various Chinese populations with distinct risks for NPC to investigate this possible association (Table 2). Four were conducted in populations that are predominantly Cantonese (Chinese in California, Hong Kong and Guangzhou); while in the two earlier of these studies (Geser et al., 1978; Henderson & Louie, 1978) intake frequency at a single time point was examined, in the latter two studies (Yu et al., 1986, 1989a) we investigated intake frequencies during multiple periods throughout the subjects' lifetime. All four studies demonstrated a significant, positive association between consumption of salted fish and risk for NPC. Moreover, in the two studies in which exposures at various time points were investigated, we found that childhood exposure, especially during infancy (at weaning), was more strongly related to risk than exposure in adulthood (Yu et al., 1986, 1989a). We (Yu et al., 1989a) have pointed out that the apparent disparity in the magnitude of the relative risks observed among the Cantonese in Guangzhou compared to those in Hong Kong is an artefact related to the influence of prevalence of exposure on the diminution of the observed relative risk toward unity through random misclassification of subjects' exposure status. Indeed, a study in Hong Kong a decade earlier involving older patients (Geser et al., 1978) showed a prevalence of exposure at weaning that was almost identical to that noted in Guangzhou (Yu et al., 1989a) and, consequently, the two studies reported comparable relative risks.

Two case-control studies in which consumption of salted fish was examined in non-Cantonese populations of southern China also showed a significant association between NPC risk and ingestion of this traditional food with early age of exposure being an important codeterminant of risk. Armstrong et al. (1983) studied 100 cases of NPC in Malaysian Chinese, of whom approximately one-third were Cantonese, another one-third were Hakka, and the remaining one-third were Chiu Chau/Fukienese. Ingestion of salted fish was associated with risk for NPC in all three ethnic groups, and risk was higher for exposure during childhood than in adolescence. We (Yu et al., 1988) investigated salted fish consumption in Yulin (in eastern Guangxi), where the consumption level is relatively low, and again found that earliest exposure (at weaning) was most strongly associated with risk.

Finally, a recent case-control study in Tianjin, a coastal city in northern China where NPC is relatively rare, confirmed the association between salted fish and NPC previously observed in southern Chinese populations (Ning et al., 1990). This study clearly showed that risk increased with earlier age at first exposure. It also suggested that risk was influenced by the method of cooking; steaming the salted fish seemed to confer a higher risk for NPC than frying, grilling or boiling it. In summary, all case-control studies carried out to date have produced remarkably consistent findings to support the hypothesis that intake of Chinese salted fish is a cause of NPC. Information on Cantonese populations indicates that most cases of NPC that occur in this ethnic group are related to early ingestion of this food. There was no discernible difference in NPC risk associated with consumption of different species of salted fish.

Table 2. Case-control studies of nasopharyngeal carcinoma in which exposure to Chinese salted fish was examined

Chinese population studied	Period of diagnosis of cases (years)	Number of cases/ controls	Period of exposure to salted fish	Estimate of risk relative to no/rare consumption	Reference
California, USA	1971-74	74/109	Current	1/week, 2.1 $\geq$2/week, 3.1	Henderson & Louie (1978)
Hong Kong	1973-74	108/103	During weaning	Ever, 2.6	Geser et al. (1978)
Selangor, Malaysia	1973-80	100/100	During childhood During adolescence	Weekly, 3.1 Daily, 17.4 Weekly, 1.3 Daily, 3.5	Armstrong et al. (1983)
Hong Kong	1981-83	250/250	During weaning At age 10 Three years previously	Ever, 7.5 Weekly/daily, 37.7 Weekly, 3.2 Daily, 7.5	Yu et al. (1986)
Guangzhou, China	1983-85	306/306	During weaning At age 10 Three years previously	Ever, 2.1 Weekly, 1.1 Daily, 2.4 Weekly, 1.4 Daiiy, 1.8	Yu et al. (1989a)
Yulin, China	1984-86	128/174	During weaning At age 10	Ever, 2.6 Weekly[a], 1.5	Yu et al. (1988)
Tianjin, China	1985-86	100/300	At age 10 Three years previously	Weekly/daily, 6.7 Monthy/weekly[a], 1.2	Ning et al. (1990)

[a] No subject reported daily consumption.

The results of experiments in animals further strengthen the hypothesis that Chinese salted fish is a human nasopharyngeal carcinogen. Huang et al. (1978) successfully induced cancers of the nasal cavity in Wistar rats by feeding them cooked Chinese salted fish; this tumour otherwise rarely occurs in this animal species. We (Yu et al., 1989b) confirmed those findings in a larger study which demonstrated a dose-response relationship between level of intake and rate of tumour occurrence. Furthermore, the amounts of salted fish (in terms of percentage of total diet) that we fed to the rats were within the range of potential human (Cantonese) consumption, and the rate of nasal cavity tumours observed in the animals was in the same range as the rate of NPC occurrence in Cantonese.

Low levels (sub-parts per million) of several volatile nitrosamines, including N-nitrosodimethylamine, N-nitrosodiethylamine, N-nitrosodi-n-propylamine, N-nitrosodi-n-butylamine and N-nitrosomorpholine, have been detected in samples of Chinese salted fish (Huang *et al.*, 1981; Tannenbaum *et al.*, 1985). Most of these volatile nitrosamines can induce nasal and paranasal cavity tumours in animals (Haas *et al.*, 1973; Pour *et al.*, 1973; Althoff *et al.*, 1974; Lijinsky & Taylor, 1978). In addition, Tannenbaum *et al.* (1985) detected bacterial mutagens in Chinese salted fish that had been exposed to a nitrosating agent under simulated gastric conditions. At present, it is not clear whether the volatile nitrosamines and bacterial mutagens present in Chinese salted fish are the carcinogens that cause NPC; the food may contain other types of carcinogenic substances that have not been identified. Epstein-Barr virus-inducing substances have also been found in Chinese salted fish (Bouvier *et al.*, this volume). If confirmed, this finding would add support to a causal interpretation of the well-etablished association between Epstein-Barr virus and NPC.

Other dietary factors

Some preliminary evidence indicates that early exposure to other types of salted fish may explain at least some of the raised rates of NPC in the native peoples of south-east Asia and the Arctic region. In a small case-control study (13 cases, 50 controls) of NPC among Malays in Selangor, Malaysia, Armstrong and Armstrong (1983) found that five cases (38%) and four controls (8%) had eaten salted fish daily during childhood. In Alaska, interviews with 13 NPC patient-control pairs revealed that more cases than controls had consumed salted fish frequently during childhood (Lanier *et al.*, 1980).

Early exposure to other fermented food products has also been shown to be related to risk for NPC. In Yulin, exposure before the age of two years to *chung choi* (a salted root), salted duck eggs, salted mustard greens, dried fish and fermented soya bean paste were independent risk factors for NPC (Yu *et al.*, 1988). In Guangzhou, exposure before the age of ten to fermented fish sauce, mouldy bean curd, salted shrimp paste, and *chan pai mui* and *gar ink gee* (two kinds of preserved plum) were independently related to NPC risk (Yu *et al.*, 1989a). In Tianjin, exposure to salted shrimp paste at the age of ten was related to risk for NPC, independently of salted fish intake (Ning *et al.*, 1990). In Taiwan, exposure before the age of 20 years to fermented bean products and smoked meat increased the risk for NPC independently (Chen *et al.*, 1988b). While the levels of exposure to most of these foods during childhood were low in the case series studied, *chung choi* was a weaning food for 60% of cases in Yulin, salted shrimp paste was consumed at ten years of age by 48% of cases in Tianjin, and 63% of cases in Taiwan consumed fermented bean products before the age of 20 years. Thus, preserved foods other than salted fish may account for a substantial proportion of NPC cases in certain non-Cantonese populations.

Poirier *et al.* (1987) detected one or more volatile N-nitrosamines in samples of salted mustard greens (described by the authors as green mustard leaves fermented in brine), *chung choi* (described by the authors as radish roots and stems fermented in brine), and fermented soya bean paste. N-Nitrosodimethylamine was detected in salted mustard greens and *chung choi*, N-nitrosopiperidine in salted mustard greens, and N-nitrosopyrrolidine in salted mustard greens, *chung choi* and fermented soya bean paste. As in the case of salted fish, it is not yet clear whether any of the N-nitrosamines detected in these fermented foods are directly involved in the carcinogenic process.

The Arabs of North Africa, who are believed to have high rates of NPC, are known to consume a number of preserved foods on a regular basis. Poirier et al. (1987) examined ten preserved foods common in the Tunisian diet and found that three of them (dried mutton preserved in olive oil, turnips fermented in brine, and *touklia* stewing base) contained volatile N-nitrosamines. The preserved mutton and a Tunisian spice mixture (*harissa*) were also found to contain Epstein-Barr virus-inducing substances (Bouvier et al., this volume).

In an uncontrolled study of young (under the age of 25 years) NPC patients in Hong Kong, Anderson et al. (1978) reported that 'all families felt that vegetables and fruits were bad for babies, and the children had been fed accordingly'. These observations were later confirmed in case-control studies. In both Hong Kong and Guangzhou (Yu et al., 1986, 1989a), cases ingested significantly fewer fresh vegetables and fruit than controls, especially during childhood. Whereas in Hong Kong (Yu et al., 1986) the protective effect of fresh vegetables and fruit was no longer significant after adjustment for salted fish intake, in Guangzhou this effect was not explained by the differing consumption pattern of salted fish and other preserved foods between cases and controls (Yu et al., 1989a). In Tianjin (Ning et al., 1990), although consumption of 'fresh fruit' or 'fresh vegetables' in general was not associated with a reduction in risk for NPC, a significant protective effect was seen with intake of carrots at ten years of age; and a declining risk was seen with increasing consumption of garlic (the only other fresh vegetable mentioned on the questionnaire), although the effect was not statistically significant. Fruit and vegetables are rich in chemical compounds that have been suggested to be protective against cancer (Wattenberg, 1978). It is interesting to note that African NPC patients showed significantly lower levels of serum carotene than controls (Clifford, 1972). Unfortunately, the crudeness of the questions about consumption of fresh vegetables and fruit in these studies precludes an analysis of specific micronutrients. The hypothesis that substance(s) present in certain vegetables and fruits reduce the risk for NPC deserves further investigation.

Future research

Although a number of potentially carcinogenic substances have been identified in Chinese salted fish, it is presently unclear whether any of them are involved as nasopharyngeal carcinogens. A systematic search using laboratory methods for the compound(s) contained in Chinese salted fish and other foods that give rise to NPC in humans should receive high priority.

Although Chinese men are more than twice as likely to develop NPC as Chinese women, men have not been shown to consume salted fish or other NPC-associated foods more frequently than women. Future epidemiological studies in this population should incorporate measurements of portion size, so that the question of whether the sex difference is due to constitutional factors or to exposure level can be addressed.

It is presently unknown whether diet plays a role in the etiology of NPC in non-Chinese populations. The suggestion that the elevated rates in Malays, Alaskan natives and North Africans are related to diet requires confirmation by analytical studies. Furthermore, no information is available on the possible role of diet in the development of NPC in low-risk non-Chinese populations, such as black and white Americans.

Acknowledgements

This work was supported by Public Health Service Grant R01-CA40468 from the US National Cancer Institute.

References

Althoff, J., Mohr, U., Page, N. & Reznik, G. (1974) Carcinogenic effect of dibutylnitrosamine in European hamsters (*Cricetus cricetus*). *J. Natl Cancer Inst.*, **53**, 795-800

Anderson, E.N., Anderson, M.L. & Ho, H.C. (1978) Environmental backgrounds of young Chinese nasopharyngeal carcinoma patients. In: de-Thé, G. & Ito, Y., eds, *Nasopharyngeal Carcinoma, Etiology and Control* (IARC Scientific Publications No. 20), Lyon, IARC, pp. 231-239

Armstrong, R.W. & Armstrong, M.J. (1983) Environmental risk factors and nasopharyngeal carcinoma in Selangor, Malaysia: a cross-ethnic perspective. *Ecol. Dis.*, **2**, 185-198

Armstrong, R.W., Kannan Kutty, M., Dharmalingam, S.K. & Ponnudurai, J.R. (1979) Incidence of nasopharyngeal carcinoma in Malaysia, 1968-1977. *Br. J. Cancer*, **40**, 557-567

Armstrong, R.W., Armstrong, M.J., Yu, M.C. & Henderson, B.E. (1983) Salted fish and inhalants as risk factors for nasopharyngeal carcinoma in Malaysian Chinese. *Cancer Res.*, **43**, 2967-2970

Cammoun, M., Hoerner, G.V. & Mourali, N. (1974) Tumours of the nasopharynx in Tunisia. An anatomic and clinical study based on 143 cases. *Cancer*, **33**, 184-192

Chen, C.-J., Chen, J.-Y., Hsu, M.-M., Shieh, T., Tu, S.-M. & Yang, C.-S. (1988a) Epidemiological characteristics and early detection of nasopharyngeal carcinoma in Taiwan. In: Wolf, G.T. & Carey, T.E., eds, *Head and Neck Oncology Research*, Amsterdam-Berkeley, Kugler Publications, pp. 505-513

Chen, C.-J., Wang, Y.-F., Shieh, T., Chen, J.-Y. & Lin, M.-Y. (1988b) Multifactorial etiology of nasopharyngeal carcinoma. Epstein-Barr virus, familial tendency and environmental cofactors. In: Wolf, G.T. & Carey, T.E., eds, *Head and Neck Oncology Research*, Amsterdam-Berkeley, Kugler Publications, pp. 469-476

Clifford, P. (1972) Carcinogens in the nose and throat: nasopharyngeal carcinoma in Kenya. *Proc. R. Soc. Med.*, **65**, 682-686

Geser, A., Charnay, N., Day, N.E., Ho, H.C. & de-Thé, G. (1978) Environmental factors in the etiology of nasopharyngeal carcinoma: report on a case-control study in Hong Kong. In: de-Thé, G. & Ito, Y., eds, *Nasopharyngeal Carcinoma, Etiology and Control* (IARC Scientific Publications No. 20), Lyon, IARC, pp. 213-229

Haas, H., Mohr, U. & Kruger, F.W. (1973) Comparative studies with different doses of *N*-nitrosomorpholine, *N*-nitrosopiperidine, *N*-nitrosomethylurea, and dimethylnitrosamine in Syrian golden hamsters. *J. Natl Cancer Inst.*, **51**, 1295-1301

Henderson, B.E. & Louie, E. (1978) Discussion of risk factors in nasopharyngeal carcinoma. In: de-Thé, G. & Ito, Y., eds, *Nasopharyngeal Carcinoma, Etiology and Control* (IARC Scientific Publications No. 20), Lyon, IARC, pp. 251-260

Hidayatalla, A., Malik, M.O.A., El Hadi, A.E., Osman, A.A. & Hutt, M.S.R. (1983) Studies on nasopharyngeal carcinoma in the Sudan. I. Epidemiology and aetiology. *Eur. J. Cancer Clin. Oncol.*, **19**, 705-710

Ho, P.-T. (1959) *Studies on the Population of China, 1368-1953*, Cambridge, Harvard University Press, pp. 166-168

Ho, H.C. (1967) Nasopharyngeal carcinoma in Hong Kong. *UICC Monogr. Series*, **1**, 58-63

Ho, J.H.C. (1971) Genetic and environmental factors in nasopharyngeal carcinoma. In: Nakahara, W., Nishioka, K., Hirayama, T. & Ito, Y., eds, *Recent Advances in Human Tumor Virology and Immunology*, Tokyo, University of Tokyo Press, pp. 275-295

Ho, J.H. (1978) An epidemiologic and clinical study of nasopharyngeal carcinoma. *Int. J. Radiat. Oncol. Biol. Phys.*, **4**, 183-198

Huang, D.P., Saw, D., Teoh, T.B. & Ho, J.H.C. (1978) Carcinoma of the nasal and paranasal regions in rats fed Cantonese salted marine fish. In: de-Thé, G. & Ito, Y., eds, *Nasopharyngeal Carcinoma, Etiology and Control* (IARC Scientific Publications No. 20), Lyon, IARC, pp. 315-328

Huang, D.P., Ho, J.H.C., Webb, K.S., Wood, B.J. & Gough, T.A. (1981) Volatile nitrosamines in salt-preserved fish before and after cooking. *Food Cosmet. Toxicol.*, **19**, 167-171

Lanier, A., Bender, T., Talbot, M., Wilmeth, S., Tschopp, C., Henle, W., Henle, G., Ritter, D. & Terasaki, P. (1980) Nasopharyngeal carcinoma in Alaskan Eskimos, Indians, and Aleuts: a review of cases and study of Epstein-Barr virus, HLA, and environmental risk factors. *Cancer*, **46**, 2100-2106

Lee, H.P., Duffy, S.W., Day, N.E. & Shanmugaratnam, K. (1988) Recent trends in cancer incidence among Singapore Chinese. *Int. J. Cancer*, **42**, 159-166

Levine, P.H. & Connelly, R.R. (1985) Epidemiology of nasopharyngeal carcinoma. In: Wittes, R.E., ed., *Head and Neck Cancer*, New York, John Wiley & Sons, pp. 13-34

Li, C.-C., Yu, M.C. & Henderson, B.E. (1985) Some epidemiologic observations of nasopharyngeal carcinoma in Guangdong, People's Republic of China. *Natl Cancer Inst. Monogr.*, **69**, 49-52

Lijinsky, W. & Taylor, H.W. (1978) Relative carcinogenic effectiveness of derivatives of N-nitrosodiethylamine in rats. *Cancer Res.*, **38**, 2391-2394

Muir, C.S. (1971) Nasopharyngeal carcinoma in non-Chinese populations with special reference to south-east Asia and Africa. *Int. J. Cancer*, **8**, 351-363

Muir, C.S., Waterhouse, J., Mack, T., Powell, J. & Whelan, S., eds (1987) *Cancer Incidence in Five Continents, Vol. V* (IARC Scientific Publications No. 88), Lyon, IARC

National Cancer Control Office (1979) *Atlas of Cancer Mortality in the People's Republic of China*, Shanghai, China Map Press

National Cancer Control Office (1980) *Summary Data of the Cancer Mortality Survey in the People's Republic of China* [in Chinese], Beijing, Ministry of Public Health

Nielsen, N.H., Mikkelsen, F. & Hansen, J.P.H. (1977) Nasopharyngeal cancer in Greenland. The incidence in an Arctic Eskimo population. *Acta. Pathol. Microbiol. Scand. Sect. A*, **85**, 850-858

Ning, J.-P., Yu, M.C., Wang, Q.-S. & Henderson, B.E. (1990) Consumption of salted fish and other risk factors for nasopharyngeal carcinoma (NPC) in Tianjin, a low-risk region for NPC in the People's Republic of China. *J. Natl. Cancer Inst.*, **82**, 291-296

Poirier, S., Ohshima, H., de-Thé, G., Hubert, A., Bourgade, M.C. & Bartsch, H. (1987) Volatile nitrosamine levels in common foods from Tunisia, south China and Greenland, high-risk areas for nasopharyngeal carcinoma (NPC). *Int. J. Cancer*, **39**, 293-296

Pour, P., Kruger, F.W., Cardesa, A., Althoff, J. & Mohr, U. (1973) Carcinogenic effect of di-n-propylnitrosamine in Syrian golden hamsters. *J. Natl Cancer Inst.*, **51**, 1019-1027

Rothwell, R.I. (1979) Juvenile nasopharyngeal carcinoma in Sabah (Malaysia). *Clin. Oncol.*, **5**, 353-358

Shanmugaratnam, K. (1973) Cancer in Singapore – ethnic dialect group variations in cancer incidence. *Singapore Med. J.*, **14**, 69-81

Sugano, H., Sakamoto, G., Sawaki, S. & Hirayama, T. (1978) Histopathological types of nasopharyngeal carcinoma in a low-risk area: Japan. In: de-Thé, G. & Ito, Y., eds, *Nasopharyngeal Carcinoma: Etiology and Control* (IARC Scientific Publications No. 20), Lyon, IARC, pp. 27-39

Tannenbaum, S.R., Bishop, W., Yu, M.C. & Henderson, B.E. (1985) Attempts to isolate N-nitroso compounds from Chinese-style salted fish. *Natl Cancer Inst. Monogr.*, **69**, 209-211

Wattenberg, L.W. (1978) Inhibitors of chemical carcinogenesis. *Adv. Cancer Res.*, **26**, 197-226

Yu, M.C., Ho, J.H.C., Ross, R.K. & Henderson, B.E. (1981) Nasopharyngeal carcinoma in Chinese – salted fish or inhaled smoke? *Prev. Med.*, **10**, 15-24

Yu, M.C., Ho, J.H.C., Lai, S.-H. & Henderson, B.E. (1986) Cantonese-style salted fish as a cause of nasopharyngeal carcinoma: report of a case-control study in Hong Kong. *Cancer Res.*, **46**, 956-961

Yu, M.C., Mo, C.-C., Chong, W.-X., Yeh, F.-S. & Henderson, B.E. (1988) Preserved foods and nasopharyngeal carcinoma: a case-control study in Guangxi, China. *Cancer Res.*, **48**, 1954-1959

Yu, M.C., Huang, T.-B. & Henderson, B.E. (1989a) Diet and nasopharyngeal carcinoma: a case-control study in Guangzhou, China. *Int. J. Cancer*, **43**, 1077-1082

Yu, M.C., Nichols, P.W., Zou, X.-N., Estes, J. & Henderson, B.E. (1989b) Induction of malignant nasal cavity tumors in Wistar rats fed Chinese salted fish. *Br. J. Cancer*, **60**, 198-210

AFLATOXINS: DATA ON HUMAN CARCINOGENIC RISK

F.X. Bosch[1] & F. Peers[2]

[1]*International Agency for Research on Cancer, Lyon, France;*
and [2]*Laminga, 2 Fayle Road, PO Salt Rock, Natal 4391, Republic of South Africa*

Some of the major difficulties in assessing the role of aflatoxin (AF) in the causation of liver cancer are discussed. Firstly, exposure to AF in Africa and parts of Asia and Latin America might begin very early in life and episodically thereafter. The number of episodes and the degree of exposure to AF varies greatly by country and region, by agricultural and crop storage practices, by season and by other factors difficult to control in any questionnaire-based study. Secondly, there is a high geographical correlation between exposure to AF and to hepatitis B virus. Thirdly, long-lasting biological markers for AF are still not available. Epidemiological studies might be enhanced by the incorporation of new biological assays. It is to be hoped that better information will be generated as a result of on-going intervention projects, such as reduction of AF levels and agricultural development programmes, and by monitoring exposure to AF and the incidence of liver cancer in areas where hepatitis B virus vaccination is effectively reducing the prevalence of carriers of the viral surface antigen.

Epidemiological studies have consistently suggested that exposure to aflatoxin (AF) is involved in the causation of primary liver cancer (PLC); however, limitations to the methods available for estimating lifetime exposure to AF are still a major obstacle to an unequivocal assessment of causality. A second major difficulty is providing estimates of risk associated with exposure to AF that are independent of and larger than the strong causal association between hepatitis B virus (HBV) and PLC. A probable synergism between HBV and AF was indicated by recent experiments on woodchucks showing that infection with the woodchuck hepatitis virus enhances hepatic metabolism of chemical carcinogens, including AF (De Flora *et al.*, 1989).

High levels of AF have been found repeatedly in foodstuffs, notably corn and groundnuts, in African countries, in south-east Asia and in China, where PLC is the commonest cancer among males and a common cancer among females. Limited information is available from extensive areas of the world where the climatic conditions and crop storage facilities are appropriate for growth of the mould and production of the toxins.

Table 1 shows the proportions of AF-contaminated foodstuffs found in selected surveys in different countries and clearly indicates that sample variation is considerable (Shank *et al.*, 1972; De Campos *et al.*, 1980; Bulatao-Jayme *et al.*, 1982; Imwidthaya *et al.*, 1987; Peers *et al.*, 1987; R. Ryder, personal communication). Unpublished data from a study in Swaziland (Peers *et al.*, 1987) are presented in Tables 2 and 3, showing the effects on the average AF concentration in contaminated samples of maize and peanuts according to length of storage and to season of sampling.

Table 1. Proportion of aflatoxin-contaminated foodstuffs in different surveys

Survey	Food item					
	Cassava	Peanuts	Corn	Rice	Alcohol	Beans
Philippines, 1982 (Bulatao–Jayme et al., 1982)	100	99	84	38	47	56
Thailand, 1967, 1975, 1980, 1982, 1987 (Imwidthaya et al., 1987)	–	28–82	33–80	2–6	–	20–40
Swaziland, 1987 (Peers et al., 1987)	–	21	3.5	–	6.5	–
The Gambia, 1988 (R. Ryder, personal communication)	–	100	–	–	–	–
Hong Kong, 1972 (Shank et al., 1972)	–	0	25	0	–	13
Guatemala, 1980 (de Campos et al., 1980)	–	–	21.7	14.3	–	77.8

AF metabolites and adducts with proteins and nucleic acids can now be detected in urine (Nyathi et al., 1987; Autrup & Wakhisi, 1988), breast milk (Wild et al., 1987), sera (Denning et al., 1988; Gan et al., 1988; Wilkinson et al., 1988) and cord blood (Lamplugh et al., 1988).

Since the most commonly contaminated foodstuffs are also the staple foods in most African countries, human exposure to AF might occur very early in life (perhaps *in utero*); furthermore, adults must be exposed episodically to different amounts of AF. These characteristics of exposure to AF represent major difficulties for epidemiological studies. Correlation studies at the population level will require careful sampling schemes and large numbers in order to characterize exposure, and case–control studies using questionnaires will provide extremely inaccurate estimates of lifetime individual exposure. A further difficulty in assessing the role of AF is the strong geographical correlation that exists between exposure to AF and that to HBV (although no formal analysis is available). In spite of these limitations, several attempts have been made to assess the risk for PLC in areas of high exposure to AF, and a few have addressed the association in countries at low risk. Several reviews of the topic are available (Busby & Wogan, 1984; Groopman et al., 1988; Bosch & Muñoz, 1989).

Correlation studies at the population level have shown that average AF contamination of diets and foodstuffs is high in areas where PLC is frequent. In some of these studies, an attempt was made to adjust the estimates of the AF–PLC correlation for the prevalence rates of HBV infection in the same population, in Africa (Van Rensburg et al., 1985; Autrup et al., 1987; Peers et al., 1987) and in China (reviewed by Armstrong, 1980; Yeh et al., 1985, 1989). The conclusions of these studies seem to indicate that the variation in PLC incidence or mortality within a limited geographical area tends to correlate more closely to estimates of AF exposure than to the prevalence of certain markers of exposure to HBV, notably the hepatitis B surface antigen (HBsAg). At the international level, however, the prevalence of HBsAg is strongly correlated with PLC incidence (Szmuness, 1978). The results of correlation studies of AF and PLC are, in addition, heavily dependent on the quality of diagnosis and of cancer registration and should therefore be viewed as indicative of an association; however, given the difficulties in providing more precise estimates, these studies remain the strongest human evidence available.

Table 2. Effect of storage of maize and peanuts on the amount of total aflatoxin, Swaziland 1982–83

Length of storage	Maize				Peanuts			
	No. of samples	Contaminated		Mean AF[a] (μg/kg)	No. of samples	Contaminated		Mean AF[a] (μg/kg)
		No.	%			No.	%	
At harvest	861	71	8.2	59.8	114	23	20.2	194.8
3 months	586	47	8.0	27.6	25	5	20.0	70.2
6 months	484	31	6.4	38.7	24	5	20.8	135.2
9 months	169	18	10.7	40.0	7	2	28.6	393.5
Total	2100	167	8.0	44.7	170	35	20.6	179.8

[a] Mean value over the contaminated samples only

Table 3. Seasonal effects on total aflatoxin contamination of maize and peanuts, as taken from the plate, Swaziland 1982–83

Season[a]	Maize			Peanuts		
	% Positive	Mean[b] (μg/g)	SD	% Positive	Mean[b] (μg/g)	SD
July–September	2.7	24.2	29.1	31.5	834.1	1847.6
October–December	2.7	12.2	16.2	19.4	1427.9	1853.9
January–March	4.6	4.6	2.7	17.0	119.6	189.1
April–June	3.6	7.8	7.6	17.0	1467.6	3762.9
Total	3.4	10.9	16.2	21.2	963.9	2164.2

[a] The rainy period extends from early October to late March.
[b] Means are averages over the contaminated samples.

Two case–control studies, one in the Philippines (Bulatao-Jayme et al., 1982) and one in Hong Kong (Lam et al., 1982), have provided estimates of exposure to AF among PLC cases and controls, based on dietary questionnaires, food tables and measurements of AF in the most common foodstuffs. In the Philippines, a high relative risk was found in association with exposure to AF, and the risk was enhanced among heavy drinkers; however, the results were not adjusted for HBV status. In Hong Kong, no difference was found between PLC cases and controls with regard to the frequency of consumption of corn and beans, common sources of exposure to AF in the area. The methods used are clearly very crude, because (i), as shown in Tables 1–3, only a fraction of a given foodstuff is contaminated and therefore consumption of the food item does not accurately reflect exposure to AF, and (ii) dietary questionnaires tend to reflect 'recent' diets, which are likely to be influenced by disease status and underlying time trends in AF contamination. The same difficulties apply to case–control studies in which biological markers of recent exposure are used, such as AF in urine samples. Recent experimental work has indicated that antioxidants are potent inhibitors of the carcinogenesis of AFB_1 in rats (reviewed by Groopman et al., 1988). If the same were applicable to humans, full assessment of the components of the diet would be required in order to evaluate the role of AF properly.

One follow-up study in China indicated that the risk for developing PLC among HBsAg carriers was greater in residents of areas where AF is common in the diet, as compared to HBsAg carriers resident in areas of low exposure to AF (Yeh et al., 1985): the mortality rate ratio among HBsAg carriers in the two areas was 9.8. Among HBsAg-negative subjects, there were two cases of PLC in the high-AF area and none in the low-AF area. This type of study is certainly of great interest, and replication of the results using accurate measurements of AF exposure and HBV status would be most useful.

A few reports of cancer occurrence among workers involved in the storage and processing of agricultural foodstuffs also indicate an increased risk for PLC. Crops imported from countries where AF contamination is high are processed for animal feed production, and the presence of AF in grain and in dust has been documented (Silas et al., 1987). Table 4 summarizes the results of studies conducted in countries with low or very low background exposure to AF. In two of the studies, there was an excess of liver cancer, and in one an excess of lymphatic neoplasms. In one small study, an excess of respiratory cancer was noted.

Table 4. Studies on occupational exposures to aflatoxin

Study	Occupation	Main findings
USA (Alavanja et al., 1987a)	Grain industry	Excess of neoplasms of lymphatic and haematopoietic system ($\times$ 1.5)
Sweden (Alavanja et al., 1987b)	Grain millers	Excess of liver cancer ($\times$ 2.4)
Denmark (Olsen et al., 1988)	Livestock feed processing plants for $\geq$ 10 years	Increased risk for liver cancer and for cancer of the biliary tract ($\times$ 2.5); nonsignificant increased risk for cancer of the salivary gland and multiple myeloma
Netherlands (Hayes et al., 1984)	Peanut oil workers	Excess of respiratory cancers

Some authors have discussed the mechanisms of carcinogenesis by AF and HBV (Van Rensburg et al., 1985); however, it is now clear that accurate information on variables such as age at first exposure, age/time since exposure cessation and dose estimates are unlikely to be available for the relevant risk factors for PLC (Kaldor & Bosch, 1990). Both exposure to AF and HBV infection in high-risk areas occur very early in life (perhaps *in utero*). Chronic HBsAg carriers (the only HBV-infected individuals at risk for PLC) are likely to remain lifetime carriers, although it has been documented that HBsAg in sera decays with age, and, as described above, AF exposure occurs throughout life, probably as transient episodes with an unknown underlying trend.

Opportunities for epidemiological research to clarify the role of AF might arise in populations in whom exposure to either AF or HBV changes dramatically. Exposure to AF might be greatly reduced or eliminated through migration or as a result of intervention programmes (Bosch & Muñoz, 1989). Likewise, in countries where AF contamination is common, the HBV status of the population might change following massive HBV vaccination programmes. Follow-up studies of high-risk populations (e.g., HBsAg carriers) monitored for exposure to AF will also allow retrospective comparisons of AF exposure among PLC cases and selected controls (IARC, 1989). These studies should benefit from advances in laboratory methods to detect the presence of AF in biological specimens.

References

Alavanja, M.C.R., Rush, G.A., Stewart, P. & Blair, A. (1987a) Proportionate mortality study of workers in the grain industry. *J. Natl Cancer Inst.*, **78**, 247–252

Alavanja, M.C.R., Malker, H. & Hayes, R.B. (1987b) Occupational cancer risk associated with the storage and bulk handling of agricultural foodstuff. *J. Toxicol. Environ. Health*, **22**, 247–254

Armstrong, B. (1980) The epidemiology of cancer in the People's Republic of China. *Int. J. Epidemiol.*, **9**, 305–315

Autrup, H. & Wakhisi, J. (1988) Detection of exposure to aflatoxin in an African population. In: Bartsch, H., Hemminki, K. & O'Neill, I.K. eds, *Methods for Detecting DNA Damaging Agents in Humans: Applications in Cancer Epidemiology and Prevention* (IARC Scientific Publications No. 89), Lyon, IARC, pp. 63–66

Autrup, H., Seremet, T., Wakhisi, J. & Wasunna, A. (1987) Aflatoxin exposure measured by urinary excretion of aflatoxin B_1-guanine adduct and hepatitis B virus infection in areas with different liver cancer incidence in Kenya. *Cancer Res.*, **47**, 3430–3433

Bosch, F.X. & Muñoz, N. (1989) Epidemiology of hepatocellular carcinoma. In: Bannasch, P., Keppler, D. & Weber, G., eds, *Liver Cell Carcinoma* (Falk Symposium No. 51), Dordrecht, Kluwer Academic Publishers, pp. 3–14

Bulatao-Jayme, J., Almero, E.M., Castro, C.A., Jardeleza, T.R. & Salamat, L.A. (1982) A case–control dietary study of primary liver cancer risk from aflatoxin exposure. *Int. J. Epidemiol.*, **11**, 112–119

Busby, W.F. & Wogan, G.N. (1984) Aflatoxins. In: Searle, C.E., ed., *Chemical Carcinogens*, 2nd ed., Washington DC, American Chemical Society, pp. 945–1136

De Campos, M., Crespo Santos, J. & Olszyna-Marzys, A.E. (1980) Aflatoxin contamination in grains from the Pacific coast in Guatemala and the effect of storage upon contamination. *Bull. Environ. Contam. Toxicol.*, **24**, 789–795

De Flora, S., Hietanen, E., Bartsch, H., Camoirano, A., Izzotti, A., Bagnasco, M. & Millman, I. (1989) Enhanced metabolic activation of chemical hepatocarcinogens in woodchucks infected with hepatitis B virus. *Carcinogenesis*, **10**, 1099–1106

Denning, D.W., Onwubalili, J.K., Wilkinson, A.P. & Morgan, M.R.A. (1988) Measurement of aflatoxin in Nigerian sera by enzyme-linked immunosorbent assay. *Trans. R. Soc. Trop. Med. Hyg.*, **82**, 169–171

Gan, L.-S., Skipper, P.L., Peng, X., Groopman, J.D., Chen, J., Wogan, N. & Tannenbaum, S.R. (1988) Serum albumin adducts in the molecular epidemiology of aflatoxin carcinogenesis: correlation with aflatoxin B_1 intake and urinary excretion of aflatoxin M_1. *Carcinogenesis*, **9**, 1323–1325

Groopman, J.D., Cain, L.G. & Kensler, T.W. (1988) Aflatoxin exposure in human populations: measurements and relationship to cancer. *CRC Crit. Rev. Toxicol.*, **19**, 113–145

Hayes, R.B., Van Nieuwenhuize, J.P., Raatgever, J.W. & ten Kate F.J.W. (1984) Aflatoxin exposures in the industrial setting: an epidemiological study of mortality. *Food Chem. Toxicol.*, **22**, 39–43

IARC (1989) *International Agency for Research on Cancer Biennial Report 1988–1989*, Lyon, pp. 48–49

Imwidthaya, S., Anukarahanonta, T. & Komolpis, P. (1987) Bacterial, fungal and aflatoxin contamination of cereals and cereal products in Bangkok. *J. Med. Assoc. Thailand*, **70**, 390–396

Kaldor, J.M. & Bosch, F.X. (1990) Multistage theory of carcinogenesis: the epidemiological evidence for liver cancer. *Bull. Cancer* (in press)

Lam, K.C., Yu, M.C., Leung, J.W.C. & Henderson, B.E. (1982) Hepatitis B virus infection in southern African blacks with hepatocellular cancer. *J. Natl Cancer Inst.*, **62**, 517–520

Lamplugh, S.M., Hendrickse, R.G., Apeagyei, F. & Mwanmut, D.D. (1988) Aflatoxins in breast milk, neonatal cord blood, and serum of pregnant women. *Br. Med. J.*, **296**, 968

Nyathi, C.B., Mutiro, C.F., Hasler, J.A. & Chetsanga, C.J. (1987) A survey of urinary aflatoxin in Zimbabwe. *Int. J. Epidemiol.*, **16**, 516–519

Olsen, J.H., Dragsted, L. & Autrup, H. (1988) Cancer risk and occupational exposure to aflatoxins in Denmark. *Br. J. Cancer*, **58**, 392–396

Peers, F.G., Bosch, F.X., Kaldor, J., Linsell, A. & Pluijmen, M. (1987) Aflatoxin exposure, hepatitis B virus infection and liver cancer in Swaziland. *Int. J. Cancer*, **39**, 545–553, 1987

Shank, R.C., Wogan, G.N., Gibson, J.B. & Nondasuta, A. (1972) Dietary aflatoxins and human liver cancer. II. Aflatoxins in market foods and foodstuffs of Thailand and Hong Kong. *Food Cosmet. Toxicol.*, **10**, 61–69

Silas, J.C., Harrison, M.A., Carpenter, J.A. & Roth, I.L. (1987) Airborne aflatoxin in corn processing facilities in Georgia. *Am. Ind. Hyg. Assoc. J.*, **48**, 198–201

Szmuness, W. (1978) Hepatocellular carcinoma and the hepatitis B virus: evidence for a causal association. *Prog. Med. Virol.*, **24**, 40–69

Van Rensburg, S.J., Cook-Mozafarri, P., van Schalkwyk, D.J., van der Watt, J.J., Vincent, T.J. & Purchase, I.F. (1985) Hepatocellular carcinoma and dietary aflatoxin in Mozambique and Transkei. *Br. J. Cancer*, **51**, 713–726

Wild, C.P., Pionneau, F.A., Montesano, R., Mutiro, C.F. & Chetsanga, C.J. (1987) Aflatoxin detected in human breast milk by immunoassay. *Int. J. Cancer*, **40**, 328–333

Wilkinson, A.P., Denning, D.W. & Morgan, M.R.A. (1988) Analysis of UK sera for aflatoxin by enzyme-linked immunosorbent assay. *Human Toxicol.*, **7**, 353–356

Yeh, F.S., Mo, C.C. & Yen, R.C. (1985) Risk factors for hepatocellular carcinoma in Guangxi, People's Republic of China. *Natl Cancer Inst. Monogr.*, **69**, 47–48

Yeh, F.S., Yu, M.C., Mo, C.C., Luo, S., Tong, M.J. & Henderson, B.E. (1989) Hepatitis B virus, aflatoxins, and hepatocellular carcinoma in southern Guangxi, China. *Cancer Res.*, **49**, 2506–2509

Relevance to Human Cancer of N-Nitroso Compounds,
Tobacco Smoke and Mycotoxins.
Ed. I.K. O'Neill, J. Chen and H. Bartsch
Lyon, International Agency for Research on Cancer
© IARC, 1991

N-NITROSO COMPOUNDS AND TOBACCO-INDUCED CANCERS IN MAN

S.S. Hecht & D. Hoffmann

American Health Foundation, Valhalla, NY, USA

Human exposure to N-nitroso compounds in tobacco products is more intense and widespread than from any other known source. This paper presents evidence that two of these N-nitroso compounds, 4-(N-nitrosomethylamino)-1-(3-pyridyl)-1-butanone and N'-nitrosonornicotine are involved in causing cancers of the oral cavity, lung, oesophagus and pancreas in tobacco users. The reduction or elimination of these nitrosamines from tobacco products would probably lead to a decrease in the incidence of these cancers.

This review summarizes evidence relevant to the involvement of N-nitroso compounds, and in particular the tobacco-specific nitrosamines 4-(N-nitrosomethylamino)-1-(3-pyridyl)-1-butanone (NNK) and N'-nitrosonornicotine (NNN), in the induction of tobacco-related cancers in man. Cancers of the oral cavity, lung, oesophagus and pancreas will be discussed. Criteria for evaluating the role of N-nitroso compounds as causative agents include their abilities to induce tumours at the appropriate sites in experimental animals, their concentrations in tobacco or tobacco smoke, and biochemical data obtained from human tissues. These criteria will be considered with respect to the epidemiological evidence demonstrating that tobacco products cause these cancers. We recognize that this analysis has limitations, due principally to the complexity of tobacco and tobacco smoke. Factors such as cocarcinogenesis, tumour promotion, tumour inhibition, toxicity and other biological effects of tobacco products, as well as other factors such as diet and environmental or genetic influences on the host complicate any evaluation of the role of particular carcinogens as causative agents for cancer in humans. Nevertheless, it is useful to focus on individual compounds, as this may lead to new insights into cancer etiology and prevention.

Cancer of the oral cavity

Epidemiological studies have shown that snuff dipping causes cancer of the oral cavity; the relative risk is as high as fifty fold for cancers of the gum and buccal mucosa (Winn *et al.*, 1981; IARC, 1985; US Department of Health and Human Services, 1986). Among the carcinogenic nitrosamines present in commercial products used for snuff dipping, NNK and NNN occur in the highest concentrations, ranging from 1–100 µg/g (Hoffmann *et al.*, 1987; Hecht & Hoffmann, 1989). Other known carcinogens in snuff tobacco are benzo[*a*]pyrene, which occurs at concentrations 1000-fold less than those of NNK and NNN, aldehydes such as formaldehyde, acetaldehyde and crotonaldehyde, and ^{210}Po (Hoffmann et al., 1987). NNK and NNN are the only carcinogens in tobacco products that have been shown to induce oral tumours in experimental animals; a mixture of NNK and NNN swabbed onto the oral cavities of F344 rats induced oral tumours in eight of 30 rats (Hecht et al., 1986). As shown in Tables 1 and 2, the estimated lifetime exposure of snuff dippers to NNK and NNN is of similar magnitude to the total doses of NNK and NNN used to induce oral tumours in rats. The presence of NNK and NNN in the saliva of snuff dippers has been confirmed (Hoffmann & Adams, 1981). Metabolic studies have shown that NNK and NNN can be activated by cultured human buccal mucosa (Castonguay et al., 1983). In agreement with these observations, recent studies have demonstrated the presence of NNK- and NNN-derived haemoglobin adducts in the blood of snuff dippers (Hecht et al., 1989). Taken together, these data provide strong support for the role of NNK and NNN as causative agents of oral cancer in snuff dippers.

Table 1. Estimated lifetime exposures of smokers and snuff dippers to 4-(N-nitrosomethylamino)-1-(3-pyridyl)-1-butanone (NNK) and N'-nitrosonornicotine (NNN) [mg (mmol/kg)][a]

Nitrosamine	Smoker	Snuff dipper
NNK	180 (0.01)	450 (0.03)
NNN	285 (0.02)	9340 (0.70)

[a] Based on 488 ng/cigarette NNN and 180 ng/cigarette NNK (Adams et al., 1987) or 64 μg/g snuff NNN and 3.1 μg/g snuff NNK (Hoffmann et al., 1987), with usage of 40 cigarettes/day for 40 years or 10 g snuff/day for 40 years

Table 2. Lowest total doses of 4-(N-nitrosomethylamino)-1-(3-pyridyl)-1-butanone (NNK) and N'-nitrosonornicotine (NNN) shown to induce tumours at various sites [mg(mmol/kg)]

Nitrosamine	Respiratory tract			Rat oral cavity	Rat oesophagus	Rat pancreas
	Hamster	Rat	Mouse			
NNK	1 (0.03)[a]	6.9 (0.07)[b]	0.5 (0.13)[c]	19 ⎫ (1.6)[e]	–	6.9 (0.07)[d]
NNN	161 (6.5)[f]	–	18 (2)[g]	97 ⎭	177 (3)[h]	–

[a] Lung, tracheal and nasal cavity tumours (Hecht et al., 1983)
[b] Lung tumours (Rivenson et al., 1988)
[c] Lung adenomas in A/J mice (Hecht et al., submitted)
[d] Rivenson et al. (1988)
[e] NNK and NNN applied as a mixture (Hecht et al., 1986)
[f] Tracheal tumours (Hoffmann et al., 1981)
[g] Lung adenomas in A/J mice (Hecht et al., 1988)
[h] Castonguay et al. (1984)
–, none seen

Cigarette smoking is a cause of oral cancer, especially when combined with chronic alcohol consumption (IARC, 1986). Although NNK and NNN are the only tobacco smoke constituents known to induce oral tumours in experimental animals, their concentrations relative to those of other carcinogens are not as high in cigarette smoke as in unburnt tobacco. Bioassays in hamsters and rats have not provided evidence that NNN enhances the carcinogenicity of ethanol (McCoy et al., 1981; Castonguay et al., 1984; Griciute et al., 1986). The mechanism of the synergistic effects of smoking and alcohol in causing oral cancer is unclear at present.

Epidemiological studies have demonstrated that chewing betel quid with tobacco causes oral cancer, which is the leading cancer in Indian males. The evidence for causation is strong only when tobacco is included in the quid (IARC, 1985). NNK and NNN are the strongest carcinogens present in the tobacco used in the quids and in the saliva of betel-quid chewers (Sipahimalani et al., 1984; Wenke et al., 1984; Nair et al., 1985; Prokopczyk et al., 1987). Areca-specific nitrosamines, such as the powerful carcinogen 3-(methylnitrosamino)-propionitrile, are also present in the quids and in the saliva of betel-quid chewers (Prokopczyk et al., 1987). The carcinogenicity of this nitrosamine for the oral cavity is being evaluated.

Lung cancer

Numerous prospective and retrospective epidemiological studies have confirmed that cigarette smoking is the major cause of lung cancer (IARC, 1986). The risk for lung cancer appears to be higher in smokers of cigarettes made entirely of black tobacco, which has high levels of nitrosamines, than in smokers of blended cigarettes (Joly et al., 1983; Benhamou et al., 1985; Fischer et al., 1989; Hecht & Hoffmann, 1989).

Polynuclear aromatic hydrocarbons (PAH) and NNK appear from experimental studies to be important causes of lung cancer, for the following reasons: (i) Tobacco smoke fractions enriched with PAH are carcinogenic to mouse skin (Hoffmann & Wynder, 1971) and on intratracheal instillation (Davis et al., 1975). (ii) Tobacco smoke contains promoters and carcinogens which enhance the carcinogenicity of PAH (Hoffmann et al., 1978). (iii) PAH induce tumours of the respiratory tract after implantation in the lung (Deutsch-Wenzel et al., 1983), after intratracheal instillation (Sellakumar & Shubik, 1974; Stinson & Saffiotti, 1983) and after inhalation (Thyssen et al., 1981). Among the PAH in tobacco smoke, benzo[a]pyrene, dibenzo[a,i]pyrene and dibenzo[c,g]carbazole are powerful respiratory carcinogens (Sellakumar & Shubik, 1974; Stinson & Saffiotti, 1983). Total levels of respiratory carcinogenic PAH in mainstream cigarette smoke range from 47–178 ng/cigarette (IARC, 1986). This would lead to an estimated total dose of approximately 0.005 mmol/kg in a smoker, on the basis of the same assumptions as used in Table 1. The lowest dose of benzo[a]pyrene shown to induce lung tumours in hamsters by intratracheal instillation was 0.2 mmol/kg, or about 40-fold greater than the total dose to a smoker (Saffiotti et al., 1972).

The organospecificity for the lung of NNK is well established in rats, mice and hamsters (Hecht & Hoffmann, 1988; Rivenson et al., 1988; Anderson et al., 1989). As shown in Tables 1 and 2, the lowest doses of NNK that have been shown to produce respiratory tract tumours in hamsters and rats are only three to seven times higher than the estimated dose experienced by a lifetime smoker. Endogenous formation of NNK is also likely but is not taken into consideration in Table 1.

Human lung tissue can activate both PAH and NNK metabolically (Harris et al., 1976; Castonguay et al., 1983). ^{32}P-Postlabelling analyses of smokers' lungs have shown the presence of DNA adducts; their concentrations were related to the extent of exposure to tobacco smoke (Phillips et al., 1988; Randerath et al., 1989). The structures of these adducts are unknown, but because of the chromatographic systems used they are probably lipophilic and may result in part from PAH. NNK-DNA adducts would not be detected under these conditions. Both NNK and benzo[a]pyrene activate the K-ras oncogene in A/J mouse lung; activation occurs by a point mutation in codon 12 (Belinsky et al., 1988; You et al., 1989). Activated K-ras oncogenes with mutations in codon 12 were detected in 13 of 45 adenocarcinomas obtained from the lungs of smokers (Rodenhuis et al., 1988).

^{210}Po is a constituent of tobacco and tobacco smoke and a powerful carcinogen in Syrian golden hamster respiratory tract (Little et al., 1975). Comparisons of the doses of ^{210}Po delivered to the lungs of cigarette smokers and underground miners led to the conclusion that it is a questionable risk factor in the carcinogenicity of tobacco smoke (Harley et al., 1980).

Oesophageal cancer

Cigarette smoking is an important cause of oesophageal cancer (IARC, 1986). Nitrosamines are well-established oesophageal carcinogens in rats, and no other component of cigarette smoke is known to be organospecific for the oesophagus. Four nitrosamines that are present in cigarette smoke induce tumours of the oesophagus in rats: NNN, N'-nitrosoanabasine, N-nitrosodiethylamine and N-nitrosoethylmethylamine. The levels of the last two in mainstream cigarette smoke are typically in the range of none up to 25 ng/cigarette, while those of N'-nitrosoanabasine reach 150 ng/cigarette (IARC, 1986). In contrast, levels of NNN range from 200–3000 ng/cigarette (IARC, 1986). In Tables 1 and 2, the levels of exposure of smokers to NNN and the total doses that have induced oesophageal tumours in rats are compared. NNN produces oesophageal tumours when administered to rats in the drinking-water (Hoffmann & Hecht, 1985). The bioassays of NNN have been carried out with only relatively high doses. Oesophageal tumour incidences in these studies were also high, suggesting that NNN would be effective at lower doses, closer to those experienced by smokers.

Metabolic activation of NNN has been observed in cultured human oesophagus (Castonguay et al., 1983). The extents of activation by the various pathways of NNN metabolism appear to be different in human and rat oesophagus (Hecht et al., 1982). Although further studies are needed to define the role of nitrosamines as causative factors for oesophageal cancer in smokers, the available evidence is suggestive and more compelling than for any other tobacco smoke constituent.

Pancreatic cancer

Cohort and retrospective studies have shown that cigarette smoking is causally associated with pancreatic cancer in men and women (IARC, 1986). An increased risk for cancer of the pancreas was also indicated for snuff dippers and tobacco chewers (Heuch et al., 1983).

NNK and its major metabolite 4-(N-nitrosomethylamino)-1-(3-pyridyl)-1-butanol (NNAl) are the only constituents of tobacco smoke that have been shown to induce tumours of the exocrine pancreas in experimental animals (Rivenson et al., 1988). This observation was made in rats given 0.5 or 1.0 ppm NNK or 5 ppm NNAl in drinking-water. NNAl induced primarily ductal tumours, as seen in humans. Ductal pancreatic tumours were also observed in rats treated simultaneously with NNK and sinigrin, a constituent of cruciferous vegetables (Morse *et al.*, 1988). As shown in Tables 1 and 2, the estimated exposure levels and total doses of NNK required to produce pancreatic tumours are comparable. NNK has been shown to induce metaplasia and hyperplasia in human pancreatic explants (Parsa *et al.*, 1986).

The available data, although limited at present, suggest that NNK may be an important cause of pancreatic cancer in tobacco consumers.

Perspectives

Tobacco use causes at least 30% of all cancer in the USA and other developed countries. If tobacco were eliminated, we would see major reductions in mortality from cancers of the lung, larynx, oral cavity, oesophagus, pancreas and bladder. Although the prevalence of smoking has decreased in some groups, there are still more than 50 million smokers and more than 10 million snuff-dippers in the USA (IARC, 1986; US Department of Health and Human Services, 1989). There are hundreds of millions of tobacco users worldwide: China has approximately 250 million smokers, and a major epidemic of lung cancer has been predicted (Peto, 1986; Crofton, 1987). There is no chance that tobacco use will significantly decline in the near future; therefore, other measures must be taken to decrease the risks for cancer associated with tobacco use.

In this review, we have presented evidence supporting the role of *N*-nitroso compounds as causes of oral cancer, lung cancer, oesophageal cancer and pancreatic cancer in tobacco smokers and snuff dippers (Table 3). The reduction or elimination of *N*-nitroso compounds from tobacco products should be a major priority. Their levels should be regulated, as in other consumer products. Extensive studies of factors that influence the levels of tobacco-specific nitrosamines in tobacco and tobacco smoke have shown that curing and processing techniques, selection of appropriate tobacco types, especially with respect to nitrate concentrations, as well as other aspects of cigarette design can lead to significant reductions in the levels of *N*-nitroso compounds in tobacco and tobacco smoke (Hecht & Hoffmann, 1988, 1989).

Although the only sure way to decrease one's risk for tobacco-related cancers is to avoid tobacco, chemoprevention may be an option for those individuals whose dependence on nicotine makes this impossible. Recent studies have demonstrated that isothiocyanates are effective inhibitors of lung tumorigenesis induced by NNK (Morse *et al.*, this volume). Further evaluation of these compounds may allow their use in preliminary human trials. Their potential efficacy could be tested by monitoring their effects on levels of NNK- and NNN-induced haemoglobin adducts. Positive results in such trials could be used as a basis for larger scale intervention studies with smokers.

Acknowledgements

We thank our many colleagues at the American Health Foundation, as cited in the references, for their contributions to this work, which is supported by Grants No. 29580 and 44377 from the US National Cancer Institute.

Table 3. Probable causative agents for tobacco-related cancers[a]

Organ	Initiator or carcinogen	Possible enhancing agents
Lung, larynx	PAH NNK ^{210}Po (minor) Formaldehyde, acetaldehyde (?) Butadiene (?) Metals (Cr, Cd, Ni) (?)	Catechol tumour promoters, aldehydes
Oral cavity		
Snuff dippers/ betel-quid chewers	NNK, NNN ^{210}Po (?)	Herpes simplex, irritation, lime
Smokers	NNK, NNN, PAH (?)	Ethanol
Oesophagus	NNN	
Bladder	4-Aminobiphenyl 2-Naphthylamine	
Pancreas	NNK, NNAl	Diet (?)

[a] From Hoffmann & Hecht (1989); PAH, polycyclic aromatic hydrocarbons; NNK, 4-(N-nitrosomethylamino)-1-(3-pyridyl)-1-butanone; NNN, N'-nitrosonornicotine; NNAl, 4-(N-nitrosomethylamino)-1-(3-pyridyl)-1-butanol

References

Adams, J.D., O'Mara-Adams, K.J. & Hoffmann, D. (1987) Toxic and carcinogenic agents in undiluted mainstream smoke and sidestream smoke of different types of cigarettes. *Carcinogenesis*, 8, 729-731

Anderson, L.M., Hecht, S.S., Dixon, D.M., Dove, L.F., Kovach, R.M., Hoffmann, D. & Rice, J.M. (1989) Evaluation of the transplacental tumorigenicity of the tobacco-specific carcinogen, 4-(methylnitrosamino)-1-(3-pyridyl)-1-butanone (NNK) in mice. *Cancer Res.*, 49, 3770-3775

Belinsky, S.A., Devereux, T.R., Stoner, G.D. & Anderson, M.W. (1988) Activation of the K-ras protooncogene in lung tumours from mice treated with 4-(N-methyl-N-nitrosamino)-1-(3-pyridyl)-1-butanone (NNK) or nitrosodimethylamine (NDMA). *Proc. Am. Assoc. Cancer Res.*, 29, 139

Benhamou, S., Benhamou, E., Thimoarche, M. & Flamant, R. (1985) Lung cancer and use of cigarettes: a French case control study. *J. Natl Cancer Inst.*, 74, 1169-1175

Castonguay, A., Stoner, G.D., Schut, H.A.J. & Hecht, S.S. (1983) Metabolism of tobacco-specific N-nitrosamines by cultured human tissues. *Proc. Natl Acad. Sci. USA*, 80, 6694-6697

Castonguay, A., Rivenson, A., Trushin, N., Reinhardt, J., Stathopoulos, S., Weiss, C.J., Reiss, B. & Hecht, S.S. (1984) Effects of chronic ethanol consumption on the metabolism and carcinogenicity of N'-nitrosonornicotine in F344 rats. *Cancer Res.*, 44, 2285-2290

Crofton, J. (1987) Smoking and health in China. *Lancet*, ii, 53

Davis, B.R., Whitehead, J.K., Gill, M.E., Lee, P.N., Butterworth, A.D. & Roe, F.J.C. (1975) Response of rat lung to tobacco smoke condensate or fractions derived from it administered repeatedly by intratracheal instillation. *Br. J. Cancer*, 31, 453-461

Deutsch-Wenzel, R.P., Brune, H., Grimmer, G., Dettbarn, G. & Misfeld, J. (1983) Experimental studies in rat lungs on the carcinogenicity and dose-response relationships of eight frequently occurring environmental polycyclic aromatic hydrocarbons. *J. Natl Cancer Inst.*, 71, 539-544

Fischer, S., Spiegelhalder, B. & Preussmann, R. (1989) Tobacco-specific nitrosamines in mainstream smoke of West German cigarettes – tar alone is not a sufficient index for the carcinogenic potential of cigarette smoke. *Carcinogenesis*, 10, 169-173

Griciute, L., Castegnaro, M., Béréziat, J.C. & Cabral, J.R.P. (1986) Influence of ethyl alcohol on the carcinogenic activity of N-nitrosonornicotine. *Cancer Lett.*, 31, 267-275

Harley, N.H., Cohen, B.S. & Tso, T.C. (1980) Polonium-210: a questionable risk factor in smoking-related carcinogenesis. In: (Banbury Report No. 3), Cold Spring Harbor, NY, CSH Press, pp. 93-104

Harris, C.C., Autrup, H., Connor, R., Barrett, L.A., McDowell, E.M. & Trump, B.F. (1976) Interindividual variation in binding of benzo[a]pyrene to DNA in cultured human bronchi. *Science*, **194**, 1067-1069

Hecht, S.S. & Hoffmann, D. (1988) Tobacco-specific nitrosamines: an important group of carcinogens in tobacco and tobacco smoke. *Carcinogenesis*, **9**, 875-884

Hecht, S.S. & Hoffmann, D. (1989) The relevance of tobacco-specific nitrosamines to human cancer. *Cancer Surv.*, **8**, 273-294

Hecht, S.S., Reiss, B., Lin, D. & Williams, G.M. (1982) Metabolism of N'-nitrosonornicotine by cultured rat esophagus. *Carcinogenesis*, **3**, 453-456

Hecht, S.S., Adams, J.D., Numoto, S. & Hoffmann, D. (1983) Induction of respiratory tract tumors in Syrian golden hamsters by a single dose of 4-(methylnitrosamino)-1-(3-pyridyl)-1-butanone (NNK) and the effect of smoke inhalation. *Carcinogenesis*, **4**, 1287-1290

Hecht, S.S., Rivenson, A., Braley, J., DiBello, J., Adams, J.D. & Hoffmann, D. (1986) Induction of oral cavity tumors in F344 rats by tobacco-specific nitrosamines and snuff. *Cancer Res.*, **46**, 4162-4166

Hecht, S.S., Abbaspour, A. & Hoffmann, D. (1988) Bioassay in A/J mice of some structural analogues of tobacco-specific nitrosamines. *Cancer Lett.*, **42**, 141-145

Hecht, S.S., Morse, M.A., Amin, S.G., Stoner, G.D., Jordan, K.G., Choi, C.-I. & Chung, F.-L. (1989) Rapid single dose model for lung tumor induction in A/J mice by 4-(methylnitrosamino)-1-(3-pyridyl)-1-butanone and the effect of diet. *Carcinogenesis*, **10**, 1901-1904

Heuch, I., Kvåle, G., Jacobsen, B.K. & Bjelke, E. (1983) Use of alcohol, tobacco and coffee and risk of pancreas cancer. *Br. J. Cancer*, **48**, 637-643

Hoffmann, D. & Adams, J.D. (1981) Carcinogenic tobacco-specific N-nitrosamines in snuff and saliva of snuff-dippers. *Cancer Res.*, **41**, 4305-4308

Hoffmann, D. & Hecht, S.S. (1985) Nicotine-derived N-nitrosamines and tobacco related cancer: current status and future directions. *Cancer Res.*, **45**, 935-944

Hoffmann, D. & Hecht, S.S. (1989) Advances in tobacco carcinogenesis. In: Cooper, C.S. & Grover, P.L., eds, *Handbook of Experimental Pharmacology*, Vol. 94/I, Berlin, Springer, pp. 63-102

Hoffmann, D. & Wynder, E.L. (1971) Tumor initiators, tumor accelerators, and tumor promoting activity of condensate fractions. *Cancer*, **27**, 848-864

Hoffmann, D., Schmeltz, I., Hecht, S.S. & Wynder, E.L. (1978) Tobacco carcinogenesis. In: Gelboin, H.V. & T'so, P.O.P., eds, *Hydrocarbons and Cancer*, Vol. 1, New York, Academic Press, pp. 85-117

Hoffmann, D., Castonguay, A., Rivenson, A. & Hecht, S.S. (1981) Comparative carcinogenicity and metabolism of 4-(methylnitrosamino)-1-(3-pyridyl)-1-butanone and N'-nitrosonornicotine in Syrian golden hamsters. *Cancer Res.*, **41**, 2386-2393

Hoffmann, D., Adams, J.D., Lisk, D., Fisenne, I. & Brunnemann, K.D. (1987) Toxic and carcinogenic agents in dry and moist snuff. *J. Natl Cancer Inst.*, **79**, 1281-1286

IARC (1985) *IARC Monographs on the Evaluation of the Carcinogenic Risk of Chemicals to Humans*, Vol. 37, *Tobacco Habits Other than Smoking; Betel-quid and Areca-nut Chewing; and Some Related Nitrosamines*, Lyon

IARC (1986) *IARC Monographs on the Evaluation of the Carcinogenic Risk of Chemicals to Humans*, Vol. 38, *Tobacco Smoking*, Lyon

Joly, O.G., Lubin, J.H. & Caraballoso, M. (1983) Dark tobacco and lung cancer in Cuba. *J. Natl Cancer Inst.*, **70**, 1033-1039

Little, J.B., Kennedy, A.R. & McGandy, R.B. (1975) Lung cancer induced in hamsters by low doses of alpha radiation from polonium-210. *Science*, **188**, 737-738

McCoy, G.D., Hecht, S.S., Katayama, S. & Wynder, E.L. (1981) Differential effect of chronic ethanol consumption on the carcinogenicity of N-nitrosopyrrolidine and N'-nitrosonornicotine in male Syrian golden hamsters. *Cancer Res.*, **41**, 2849-2854

Morse, M.A., Wang, C.-X., Amin, S.G., Hecht, S.S. & Chung, F.-L. (1988) Effects of dietary sinigrin or indole-3-carbinol on O^6-methylguanine-DNA-transmethylase activity and 4-(methylnitrosamino)-1-butanone-induced DNA methylation and tumorigenicity in F344 rats. *Carcinogenesis*, **9**, 1891-1895

Nair, J., Ohshima, H., Friesen, M., Croisy, A., Bhide, S.V. & Bartsch, H. (1985) Tobacco-specific and betel nut-specific N-nitroso compounds: occurrence in saliva and urine of betel quid chewers and formation *in vitro* by nitrosation of betel quid. *Carcinogenesis*, **6**, 295-303

Parsa, I., Foye, C.A., Cleary, C.M. & Hoffmann, D. (1986) Differences in metabolism and biological effects of NNK in human target cells. In: (Banbury Report No. 23), Cold Spring Harbor, NY, CSH Press, pp. 233-244

Peto, R. (1986) Tobacco. UK and China. *Lancet*, **ii**, 1038

Phillips, D.H., Hewer, A., Martin, C.N., Garner, R.C. & King, M.M. (1988) Correlation of DNA adduct levels in human lung with cigarette smoking. *Nature*, **336**, 790-792

Prokopczyk, B., Rivenson, A., Bertinato, P., Brunnemann, K.D. & Hoffmann, D. (1987) 3-(Methylnitrosamino)propionitrile: occurrence in saliva of betel quid chewers, carcinogenicity, and DNA methylation in F344 rats. *Cancer Res.*, **47**, 467-471

Randerath, E., Miller, R.H., Mittal, D., Avitts, T.A., Dunsford, H.A. & Randerath, K. (1989) Covalent DNA damage in tissues of cigarette smokers as determined by ^{32}P-postlabelling assay. *J. Natl Cancer Inst.*, **81**, 341-347

Rivenson, A., Hoffmann, D., Prokopczyk, B., Amin, S. & Hecht, S.S. (1988) Induction of lung and exocrine pancreas tumors in F344 rats by tobacco-specific and areca-derived N-nitrosamines. *Cancer Res.*, **48**, 6912-6917

Rodenhuis, S., Sletos, R.J.C., Boot, A.J.M., Evers, S.G., Moor, W.J., Wagenaar, S.S.C., Van Bodegom, P.C. & Bos, J.L. (1988) Incidence and possible clinical significance of K-ras oncogene activation in adenocarcinoma of the human lung. *Cancer Res.*, **48**, 5738-5741

Saffiotti, V., Montesano, R., Sellakumar, A.R., Cefis, F. & Kaufman, D.G. (1972) Respiratory tract carcinogenesis in hamsters induced by different numbers of administrations of benzo[a]pyrene and ferric oxide. *Cancer Res.*, **32**, 1073-1081

Sellakumar, A. & Shubik, P. (1974) Carcinogenicity of different polycyclic hydrocarbons in the respiratory tract of hamsters. *J. Natl Cancer Inst.*, **53**, 1713-1716

Sipahimalani, A.T., Chadha, M.S., Bhide, S.V., Bratap, A.I. & Nair, Y. (1984) Detection of N-nitrosamines in the saliva of habitual chewers of tobacco. *Food Chem. Toxicol.*, **22**, 261-264

Stinson, S.F. & Saffiotti, V. (1983) Experimental respiratory carcinogenesis with polycyclic aromatic hydrocarbons. In: Reznik-Schuller, H.M., ed., *Comparative Respiratory Tract Carcinogenesis*, Boca Raton, FL, CRC Press, pp. 75-93

Thyssen, J., Althoff, J., Kimmerle, G. & Mohr, V. (1981) Inhalation studies with benzo[a]pyrene in Syrian golden hamsters. *J. Natl Cancer Inst.*, **66**, 575-577

US Department of Health and Human Services (1986) *The Health Consequences of Using Smokeless Tobacco* (NIH Publ. No. 86-2874), Bethesda, MD

US Department of Health and Human Services (1989) *Reducing the Health Consequences of Smoking. 25 Years of Progress. A Report of the Surgeon General* (DHHS Publ. No. (CDC), 89-8411), Bethesda, MD

Wenke, G., Brunnemann, K.D. & Hoffmann, D. (1984) A study of betel quid carcinogenesis. IV. Analysis of the saliva of chewers. A preliminary report. *J. Cancer Res. Clin. Oncol.*, **198**, 110-113

Winn, D.M., Blot, W.J. & Fraumeni, J.F., Jr (1981) Snuff dipping and oral cancer. *New Engl. J. Med.*, **305**, 230-231

You, M., Candrian, U., Maronpot, R.R., Stoner, G.D. & Anderson, M.W. (1989) Activation of the K-ras protooncogene in spontaneously occurring and chemically induced lung tumors of the strain A mouse. *Proc. Natl Acad. Sci. USA*, **86**, 3070-3074

Relevance to Human Cancer of N-Nitroso Compounds,
Tobacco Smoke and Mycotoxins.
Ed. I.K. O'Neill, J. Chen and H. Bartsch
Lyon, International Agency for Research on Cancer
© IARC, 1991

TOBACCO SMOKING AND ITS EFFECT ON HEALTH IN CHINA

Y.T. Gao, W. Zheng, R.N. Gao & F. Jin

*Department of Epidemiology, Shanghai Cancer Institute,
Shanghai, China*

Production and sales of manufactured cigarettes are increasing in China; furthermore, the proportion of filtered cigarettes is still low, and the tar yield thus remains high. The prevalence of smoking in the general population aged 20 and over is estimated to be 68.9% of men and 8.3% of women. Case–control studies carried out in major cities in China all showed a close relationship between lung cancer incidence and smoking. In Shanghai, the population attributable risks for lung cancer due to smoking were 0.69 in men and 0.24 in women. A limited number of case–control studies indicate that the incidences of cancers of the bladder, oesophagus and pancreas are also associated with smoking. Some studies on coronary heart disease reveal unequivocally a significant risk for this disease associated with smoking. Cancer, chronic obstructive lung diseases and coronary heart disease are now important causes of death in some large cities in China. More comprehensive investigations are needed in order to evaluate thoroughly the effect of smoking on health in China

China is the leading producer of tobacco and manufactured cigarettes in the world. It was estimated in 1982 that the total tobacco-growing area in the country was about 930 000 ha and the total quantity of tobacco produced about 2 000 000 tonnes (IARC, 1986). In 1985, the production of manufactured cigarettes in China accounted for 24% of world production (Fan, 1988). During the period 1960–80, the average annual increase in sales of manufactured cigarettes was 11.74% (China Tobacco Society, 1985). In comparison with the data for 1980, production and sales of manufactured cigarettes in 1985 increased by 53% and 50%, respectively. As a result of this increase, the financial income from the tobacco industry in the country in 1985 ranked second, just after that from the petrochemical industry (Shanghai Tobacco Society, 1986).

The proportion of filter cigarettes produced is still low — less than 20% (Zhang, 1988); the corresponding figure in Shanghai in 1985 was 28% (Shanghai Tobacco Society, 1986). The tar yield of most brands of cigarettes remains high (15–19 mg per cigarette) or very high (≥ 20 mg per cigarette). The tar yield of all brands but one produced in Shanghai and commonly available on the market exceeds 15 mg per cigarette, ranging from 19 to 32 mg per cigarette (Gao, 1986).

It is well known that excess mortality in smokers from diseases such as cancer of the lung, chronic obstructive lung diseases and ischaemic heart disease is most probably due to smoking. These diseases are now the main causes of death in several major cities of China.

Table 1 lists the five main causes of death in urban Shanghai during 1984 (Shanghai Municipal Bureau of Public Health, 1985). Table 2 lists the five main cancers in urban Shanghai during 1984 (Shanghai Cancer Registry, 1987); lung cancer can be seen to occupy an outstanding position, especially among males. Lung cancer was also reported to be the leading cancer in both males and females in Beijing and Tianjiang (Wang et al., 1988).

Table 1. Mortality from leading causes of death in urban Shanghai in 1984

Causes of death	Mortality rate	
	per 100 000	%
All causes	681.99	100.00
All cancers	182.42	26.75
Lung cancer	41.02	6.01
Cerebrovascular disease	117.94	17.29
Chronic bronchitis, emphysema and respiratory disease	86.16	12.63
All heart disease	63.69	9.34
Coronary heart disease	37.29	5.47
Accidents	40.62	5.96

Table 2. Incidence rates of five main cancers in urban Shanghai during 1984

Site (ICD-9)	Males			Site (ICD-9)	Females		
	Crude rate (per 10^5)	Age-adjusted rate^a (per 10^5)	%		Crude rate (per 10^5)	Age-adjusted rate^a (per 10^5)	%
All sites	274.8	244.4	100.0	All sites	201.1	155.7	100.0
Lung (162)	65.8	57.4	23.9	Stomach (151)	31.5	23.7	15.7
Stomach (151)	64.9	57.9	23.6	Lung (162)	27.8	20.6	13.8
Liver (155)	37.0	32.0	13.5	Breast (174)	25.5	21.1	12.7
Oesophagus (150)	18.9	17.0	6.9	Liver (155)	15.8	11.7	7.8
Rectum (154)	10.7	9.6	3.9	Colon (153)	10.5	7.9	5.2

^{a}Adjusted to world population

Prevalence of smoking in the general population

Tobacco smoking is a very popular habit among males in China. A nationwide sampling survey on smoking habits, including about 520 000 persons aged 15 and over, was carried out during 1984 (Collaborative Group for Investigating Prevalence of Smoking, 1987). The prevalence of smoking in this population was found to be 33.9%, with 61.0% in males and 7.0% in females. Among persons aged 20 and over, the prevalence was 38.9% (68.9% in males and 8.3% in females). The proportion of smokers among young women was very low.

The prevalence varied with level of education and occupation. The proportion of smokers among males with university, middle, primary and no education were 44.7%,

53.8%, 66.5% and 67.8%, respectively. Among males, the highest prevalence was observed in labourers (65.7%), then in peasants (63.7%), cadres (59.2%), physicians (56.7%), teachers (50.1%), scientific and technical workers (49.3%), high-school students (27.1%) and middle school students (5.9%). It is worth noting that more than half of male physicians and teachers are smokers.

With regard to age at beginning of smoking, 72% of male smokers started smoking at 15–24 years of age, 3.5% at 10–14 years of age and very few (0.4%) under the age of ten. Very few persons had stopped smoking: the proportion of ex-smokers was only 4.2% in males and 9.7% in females. More than two-thirds of these had stopped smoking due to illness.

In Shanghai, the largest industrial city of China, a cross-sectional survey on the prevalence of smoking among 110 000 urban adults aged 20 and over was carried out in 1982 (Deng & Gao, 1985). The proportions of smokers were found to be 55.8% among men and 6.6% among women — slightly lower than the figures reported in the national sampling survey. The prevalence was inversely related to level of education, the highest prevalences being observed among labourers and tradesmen and the lowest among teachers and medical workers, for both men and women. The median age at starting smoking was 22.6 years for men and 33.8 years for women. The proportion of male smokers aged 20–29 years who had started smoking at 15–19 years of age was much greater than among smokers aged 30 and over, implying that people are tending to start smoking earlier. The proportion of persons who had stopped smoking was again very low: 5.2% of male smokers and 9.7% of female smokers.

Analytical epidemiological studies

Lung cancer

Several case-control studies carried out in different parts of China have shown unequivocally that smoking is closely associated with cancer of the lung. A large population-based case–control study in urban Shanghai involving 733 male and 672 female cases of lung cancer and 760 male and 735 female controls (Gao *et al.*, 1988) resulted in odds ratios (ORs) for lung cancer associated with smoking of 3.9 (95% confidence interval (CI), 2.9–5.4) for males and 3.3 (95% CI, 2.5–4.2) for females. Higher ORs were observed for squamous- and oat-cell carcinomas and much lower ORs for adenocarcinoma, in both males and females (Table 3).

Table 3. Odds ratios (OR; with 95% confidence interval (CI)) for specific histological types of lung cancer associated with smoking in urban Shanghai

Type of lung cancer	Males		Females	
	OR	95% CI	OR	95% CI
All types	3.9	2.9-5.4	3.3	2.5-4.2
Squamous-cell	8.4	4.7-15.0	7.2	4.6-11.1
Adenocarcinoma	1.6	1.1-2.4	1.5	1.0-2.1
Oat-cell	7.4	2.3-24.1	7.9	3.6-17.0

The risk for lung cancer increased with the daily dose and duration of cigarette smoking; the OR for men who smoked 30 cigarettes or more per day for more than 40 years was 15.4. The trend was most impressive for squamous- and oat-cell cancers. The earlier the age at starting smoking, the higher the risk for lung cancer: in comparison with nonsmokers, the OR for male smokers who began smoking at 10–19 years of age was 5.1 (95% CI, 3.5–7.2) and for females it was 5.6 (95% CI, 3.4–9.0). In men who had stopped smoking for ten or more years, the OR decreased to 1.1, about the same risk as for nonsmokers. The population attributable risks due to smoking were estimated to be 0.69 (95% CI, 0.61–0.77) for males and 0.24 (95% CI, 0.19–0.29) for females.

Hence, cigarette smoking induces lung cancer in Shanghai to a qualitatively similar degree as in other parts of the world. However, although the incidence of female lung cancer in urban Shanghai is relatively high (see Table 2), the predominant cell type is adenocarcinoma, accounting for about 60% of all cases. This type is associated only weakly with smoking, and the prevalence of smoking among females is very low; thus, risk factors other than smoking must be responsible for the high incidence of lung cancer among women in this area.

Other studies on the association between lung cancer and smoking have been carried out in Beijing, Tianjing, Guangzhou, Harbin and Shengyang (Huang et al., 1981; Ren, 1983; Liu & Hu, 1987; Hu et al., 1988a; and Z.Y. Xu, personal communication). All have shown similar tendencies and indicate that cigarette smoking is the most important cause of male lung cancer in these areas and probably one of the main risk factors for female lung cancer, especially in the northern and north-eastern parts of China where the prevalence of smoking among women is generally two to four times higher than in other parts of China and where women usually start smoking earlier. The population attributable risks estimated from a case–control study of lung cancer in Shengyang were 0.55 for men and 0.37 for women (Z.Y. Xu, personal communication).

Cancers of the urinary and digestive tracts

Few studies have been carried out in China on associations of other cancers with smoking. In Shanghai, You et al. (1981) matched 317 male cases of bladder cancer with the same number of controls by age and residential area and found a relative risk for bladder cancer among smokers of 1.53. Most Chinese studies have shown no association between cancers of the oesophagus and stomach and smoking. A hospital-based case-control study of 137 pairs of cases of oesophageal cancer and controls carried out by Xan et al. (1988) in urban Xian, however, found an OR for oesophageal cancer due to smoking of 3.7 in univariate analysis; this was reduced to 1.8 ($p < 0.01$) after adjusting for other factors in a logistic model. In a hospital-based case–control study carried out in Harbin on 241 pairs of cases of stomach cancer matched by sex, age and residence, Hu et al. (1988b) reported an OR of 1.79 due to smoking (95% CI, 1.31-2.46), after accounting for consumption of green vegetables and alcohol drinking; the OR for smoking and drinking and lower consumption of green vegetables increased to 5.47 ($p < 0.01$).

In Shanghai, Liang et al. (1988) found an OR 1.6 (95% CI, 1.1–2.5) due to smoking in a population-based case-control study of pancreatic cancer involving 237 cases (132 males and 105 females) and the same number of controls. A significant interaction between smoking and drinking was observed.

Tu et al. (1986) followed 12 222 men aged 40 years and over in Chongming county, Shanghai, in a cohort study aimed to investigate risks factors for primary hepatic carcinoma. During a five-year follow-up of hepatitis B virus carriers, the age-adjusted incidence rate for liver cancer was 1.6 times higher among smokers ($p < 0.01$) than for nonsmokers. The incidence of primary hepatic carcinoma increased with the number of cigarettes smoked, but no effect of smoking was found among persons who were not hepatitis B virus carriers.

More studies on associations between these cancers and smoking are needed to draw convincing conclusions. However, positive relationships between smoking and cancers of the bladder, oesophagus and pancreas have been found repeatedly in other countries (IARC, 1986), and it is reasonable to assume that smoking will make some contribution to the burden of these cancers in China.

Coronary heart disease

Studies in various regions of China have shown a significant risk for coronary heart disease (CHD) associated with smoking. The Shanghai Collaborative Research Group on CHD (1981), in a matched case-control study involving 200 pairs of cases and controls, showed that the risk for myocardial infarction was 3.6 times higher for smokers than for nonsmokers. Using standards for diagnosing ischaemic heart disease established by the International Society of Heart Diseases in 1979, a matched case-control study carried out by Yao et al. (1984) in Nanjing found a relative risk of 1.6 ($p < 0.05$), for CHD among smokers in comparison with nonsmokers. A study in Ninxia (Liu et al., 1988) showed an OR of 1.91 (95% CI, 1.12-3.26) for CHD among smokers who had smoked 20 cigarettes or more per day for more than ten years. The study included 171 cases of CHD (115 males and 56 females), with the same number of controls matched by sex and age. He et al. (1988) carried out a matched case-control study in Xian involving 103 pairs of cases of CHD and controls and found a clear dose-response relationship between CHD and smoking. Among persons who smoked 21-29 cigarettes per day, the OR reached 5.4 ($p < 0.001$); the overall OR for CHD due to smoking was estimated to be 3.3. For persons who had stopped smoking for five to ten years, the OR was reduced to 1.1.

Other diseases

Smoking may also cause chronic respiratory diseases, such as chronic bronchitis, and part of its outcome — emphysema and respiratory heart disease. As shown in Table 1, these diseases accounted for 12.63% of all deaths in urban Shanghai in 1984.

This brief description is far too incomplete to evaluate completely the effect of smoking on health in China. More comprehensive investigations are needed, including cost-benefit analyses for decision-makers. The large proportion of the national income that is derived from the tobacco industry is offset by the millions of persons who will be killed by smoking in future decades in this country.

References

China Tobacco Society (1985) *Proceedings of Inaugural and First Member Congress of China Tobacco Society* (in Chinese), Beijing, Publishing House of Light Industry, p. 102

Collaborative Group for Investigating Prevalence of Smoking (1987) Results of national sampling survey on prevalence of smoking (in Chinese). *Chin. Med. J.*, **67**, 229-232

Deng, J. & Gao, Y.T. (1985) The prevalence of smoking habit among 110 000 adult residents in the Shanghai urban area (in Chinese). *Chin. J. Prev. Med.*, **5**, 271-274

Fan, D.M. (1988) International market and export of Shanghai cigarettes. *Shanghai Tobacco Industry*, **3**, 17-20

Gao, Y.T. (1986) Smoking and lung cancer in Shanghai. In: Zaridze, D. & Peto, R., eds, *Tobacco: A Major International Health Hazard* (IARC Scientific Publications No. 74), Lyon, IARC, pp. 115-121

Gao, Y.T., Blot, W.J., Zheng, W., Fraumeni, J.F. & Hsu, C.W. (1988) Lung cancer and smoking in Shanghai. *Int. J. Epidemiol.*, **17**, 277-280

He, Y., Li, L.S., Wang, Z.H., Li, L.S., Zheng, X.L., Jia, G.L. & Zheng, C.H. (1988) Relationship between smoking, hypertension and coronary heart disease (in Chinese). *Chin. J. Prev. Med.*, **22**, 131-134

Hu, J.F., Go, J.X., Liu, Y.Y. & Wu, Y.P. (1988a) Risk factors of lung cancer in Harbin (in Chinese). *Chin. J. Epidemiol.*, **9**, 88-91

Hu, J.F., Wang, G.Q., Zhang, S.F. & Liu, S.D. (1988b) Alcohol and smoking: risk factors of gastric cancer (in Chinese). *Shanghai Tumor*, **8**, 200-201

Huang, G.J., Wang, L.D., Lin, H., Jiang, L.J., Zheng, D.H., Lu, J.S. & Li, J.Y. (1981) Association of lung cancer with smoking (in Chinese). *Chin. Med. J.*, **61**, 636-637

IARC (1986) *IARC Monographs on the Evaluation of the Carcinogenic Risk of Chemicals to Humans*, Vol. 38, *Tobacco Smoking*, Lyon

Liang, J.D., Gao, Y.T., Zheng, W. & Wang, Y.S. (1988) A case-control study of pancreatic cancer in Shanghai urban area (in Chinese). *Shanghai Tumor*, **8**, 59-62

Liu, Q. & Hu, M.X. (1987) Smoking, dwelling house ventilation and lung cancer in Guangzhou (in Chinese). *Shanghai Tumor*, **7**, 256-257

Liu, T.X., Zhang, B., Zhou, H., Li, X.D. & Liu, X.Y. (1988) A matched case-control study on risk factors of coronary disease (in Chinese). *Chin. J. Epidemiol.*, **9**, 282-284

Ren, T.S. (1983) A matched case-control study on lung cancer and smoking (in Chinese). *Tianjin Med. J.*, **10**, 203-206

Shanghai Cancer Registry (1987) Cancer incidence in urban Shanghai for 1984 (in Chinese). *Shanghai Tumor*, **7**, 43-44

Shanghai Collaborative Research Group on Coronary Heart Disease (1981) A matched case-control study on coronary heart disease (in Chinese). *Chin. J. Prev. Med.*, **15**, 75-80

Shanghai Municipal Bureau of Public Health (1985) *Vital Statistics in Urban Shanghai for 1984* (in Chinese), Shanghai, Office of Shanghai Municipal Bureau of Public Health

Shanghai Tobacco Society (1986) *Proceedings of Inaugural Meeting of Shanghai Tobacco Society* (in Chinese), Shanghai

Tu, J.T., Gao, R.N. & Zhang, D.H. (1987) The risk factors of primary hepatic carcinoma in the endemic area, Chongming county – Results from a five-year cohort study. In: Oda, T., Okudu, K. & Nishioka, K., eds, *Proceedings of Japan-China Symposium on Hepatocellular Carcinoma and Viral Hepatitis*, Tokyo, Japan-China Medical Association, pp. 10-16

Wang, Q.Z., Zhou, J.J., Jin, P., Guang, W.X. & Yuan, G.L. (1988) Cancer mortality in Beijing during 1977-1983 (in Chinese). *Shanghai Tumor*, **8**, 49-52

Xan, C.R., Li, L.S. & Wang, Z.H. (1988) A case–control study of oesophageal cancer among urban residents in Xian (in Chinese). *Chin. J. Prev. Med.*, **22**, 121-125

Yao, C.L., Du, F.C., Xu, Y.C. & Hang, L.J. (1984) Smoking and coronary heart disease (in Chinese). *Chin. J. Epidemiol.*, **5**, 88-90

You, X.Y., Deng, J., Mou, D.J. & Yao, Y.F. (1981) Searching for the etiologic factors of bladder cancer (in Chinese). *Shanghai Tumor*, **4**, 12-14

Zhang, Y.B. (1988) Safe cigarette and flavour of cigarette smoke (in Chinese). *Shanghai Tobacco Industry*, **2**, 93-98

DOSIMETRY METHODS FOR DETERMINING HUMAN EXPOSURE TO *N*-NITROSO COMPOUNDS, MYCOTOXINS AND TOBACCO

BIOMONITORING OF HUMAN EXPOSURE TO ALKYLATING AGENTS BY MEASUREMENT OF ADDUCTS TO HAEMOGLOBIN OR DNA

P.B. Farmer, E. Bailey, J.A. Green, C.S. Leung & M.M. Manson

MRC Toxicology Unit, Carshalton, Surrey, UK

Recent analytical developments in the determination of adducts of DNA and protein with alkylating carcinogens are described which have considerably extended the number of carcinogens that can be examined. While sensitivity of detection equal to or better than one modified DNA base per 10^8 normal bases is now achievable for many specific alkylating carcinogens, further developments in the analytical methods are still needed for the identification and quantification of adducts derived from unknown and/or mixed exposures to carcinogens.

The monitoring of human exposure to alkylating agents by measuring the covalently linked adducts that these form with macromolecules is an established procedure (see review by Farmer *et al.*, 1987; Figure 1). The range of chemical carcinogens that can be detected has now been expanded considerably through the development of more advanced physicochemical, immunological and postlabelling analytical techniques. Thus, for example, the use of gas chromatography/mass spectrometry (GC-MS) is no longer limited to measuring exposure to very low-molecular-weight alkylating agents; exposure to polycyclic aromatic hydrocarbons may now also be determined (e.g., Manchester *et al.*, 1988). Similarly, the range of compounds to which postlabelling may be applied now includes some previously inaccessible low-molecular-weight adducts (e.g., Wilson *et al.*, 1988). The purpose of this review is to summarize these recent analytical developments.

Alkylating agent-DNA adducts

Physicochemical techniques

There are two general methods for detecting DNA-carcinogen adducts by GC-MS or other physicochemical techniques: (i) determination of alkylated purines after depurination of DNA and (ii) determination of carcinogen residues after mild hydrolysis of DNA. Method (i) was the first to be developed and has been used mostly to detect alkylpurines excreted in the urine. Early examples of its use include the determination of *N*-7-methylguanine (Shuker *et al.*, 1984) and *N*-3-methyladenine (Shuker *et al.*, 1987) by GC-MS, and of the *N*-7-guanine adduct of aflatoxin B_1 by high-performance liquid chromatography (HPLC) coupled with synchronous fluorescence spectrometry (SFS; Autrup *et al.*, 1985). More recently, a technique for detecting exposure to hydroxyethylating agents by GC-MS of N-7-(2-hydroxyethyl)guanine in lymphocyte DNA has been proposed (U. Föst, personal communication; Bolt et al., 1988). 2-Oxoethylating agents, such as vinyl chloride, yield 7-(2-oxoethyl)guanine, which can also be determined by GC-MS after appropriate derivatization (Kasemann et al., 1988; U. Föst, personal communication). The use of MS-MS to characterize and quantify N-7-alkylguanines is described by Cushnir et al. (this volume).

Figure 1. Biological monitoring of exposure to carcinogenic alkylating agents

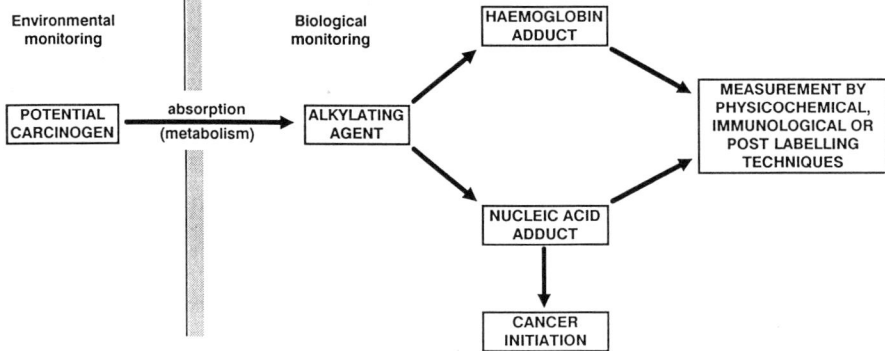

Method (ii) has been applied only recently and holds great potential for the detection of acid-labile DNA adducts, e.g., of polycyclic aromatic hydrocarbons. The reaction scheme for benzo[a]pyrene is shown in Figure 2. Conversion of the hydrolytically released benzo[a]pyrene tetrol to the trimethylsilyl derivative allows its sensitive detection by GC-MS. Alternatively, the tetrol may be detected by HPLC with SFS detection. These procedures have been applied to placental DNA samples (Manchester et al., 1988) and to peripheral lymphocyte DNA samples from coke-oven workers (Harris et al., 1985; Weston et al., 1988).

Immunological techniques

There has been continued development of immunological techniques for detecting alkylated DNA adducts. Examples of compounds studies in this way are: aflatoxin B_1, benzo[a]pyrene, 4-aminobiphenyl, 2-acetylaminofluorene, cis-diamminedichloroplatinum(II), melphalan, methylating, ethylating and hydroxyethylating agents, 8-methoxypsoralen and cyclobutadithymidine. The immunogen is usually either a derivative of a carcinogen-modified nucleoside or carcinogen-modified DNA conjugated to a carrier protein, and both polyclonal and monoclonal antibodies have been prepared (see review by Shuker, 1989). These antibodies can be used not only to quantify adducts but have been used for the preparation of immunoaffinity columns, e.g., for aflatoxin B_1 (Groopman et al., 1985), benzo[a]pyrene (Manchester et al., 1988) and, more recently, methylating agents (measured by urinary excretion of N-3-methyladenine; D. Shuker, personal communication). Use of these columns for isolating and purifying samples dramatically improves the detection limits of subsequently used physicochemical techniques, such as GC-MS and HPLC.

32P-Postlabelling techniques

In the 32P-postlabelling technique (Randerath et al., 1985), carcinogen-modified DNA is enzymatically digested to deoxynucleoside 3'-monophosphates, which are then phosphorylated at the 5'-position with 32P-ATP and T4 polynucleotide kinase. The modified deoxynucleoside 3',5'-bisphosphates are resolved from the normal ones by thin-layer chromatography (TLC) and then subjected to autoradiographic quantification.

Figure 2. Metabolism, and interaction with DNA, of benzo[a]pyrene[a]

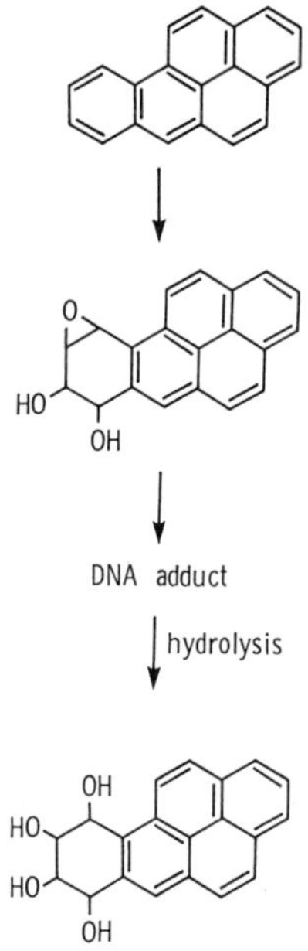

[a]The adduct may be chemically hydrolysed to yield a tetrol derivative, which may be determined as a biomonitor of exposure to benzo[a]pyrene.

The classes of compounds to which this technique has been applied include polycyclic aromatic hydrocarbons, alkenyl benzenes, heterocyclic compounds, azo and nitro compounds, arylamines and derivatives and mycotoxins. The detection of low–molecular-weight alkylated deoxynucleotides is more challenging in view of their close chromatographic similarity to the normal deoxynucleotides. However, the inclusion of an HPLC purification of the deoxynucleoside 3'-monophosphates has allowed the detection of $O6$-methyl- and $O6$-ethyl-2'-deoxyguanosine in DNA at a sensitivity of one modified base/10^7 bases using 100 μg DNA (Wilson et al., 1988). A further modification included in this analysis was treatment of the 3,5'-bisphosphates with nuclease P1, which cleaves 3'-phosphates of normal nucleotides. This has been shown to enhance the sensitivity of 32P-postlabelling for a variety of adducts (Reddy & Randerath, 1986).

The value of the data provided by postlabelling experiments would be considerably enhanced if the structural identity of the radiolabelled components on the TLC plate could be identified, and attempts to achieve this aim are now being made. For example, Schmeiser et al. (1988), using [3H]-7,12-dimethylbenz[a]anthracene, have been able to elute adducts from TLC plates after their separation, dephosphorylate them and subject the resulting nucleoside adducts to HPLC. They may then be correlated with the adducts produced from standard DNA-carcinogen interaction experiments in vitro, the chemical structures of which could be determined independently. Attempts are also being made to use fluorescence line narrowing spectrometry to assist in the identification of polycyclic aromatic hydrocarbon adducts on TLC plates (Cooper et al., 1988; Jankowiak et al., 1988). A limit of detection of three adducts per 10^8 bases appears to be feasible with this technique.

Alkylating agent-protein adducts

Physicochemical techniques

The only physicochemical technique for detecting protein adducts that is used routinely in human studies is GC-MS, and haemoglobin remains the most popular protein for study by virtue of its ready accessibility and long life (see review by Farmer and Bailey, 1989). The methods for detecting haemoglobin-alkylating agent adducts may be divided into three categories:

(i) total hydrolysis to alkylated amino acids

(ii) modified Edman degradation to thiohydantoin of alkylated N-terminal amino acid (valine); and

(iii) mild hydrolysis to release carcinogen residues from protein.

In method (i), the carcinogen-modified amino acid must be separated from (a large excess of) normal amino acids, which presents considerable practical difficulties. This method has therefore now been superseded by methods (ii) and (iii), which are considerably more sensitive. Method (ii) reaches sensitivities of about 10 pmol adduct per gram haemoglobin (Törnqvist et al., 1986) and has been used to detect adducts formed by ethylene oxide (and other hydroxyethylating agents), propylene oxide, styrene oxide and simple methylating, ethylating and other alkylating agents (see references cited by Farmer & Bailey, 1989). Exposure to aldehydes can also be estimated from measurements of their reaction products with N-terminal valines in haemoglobin. Method (iii), which has to date largely been used to estimate exposure to aromatic amines, is yet more sensitive (detection limit, <0.1 pmol adduct per gram haemoglobin; Bryant et al., 1987). It should also be applicable for the determination of carboxylic acid esters formed by reactive electrophiles (e.g., benzo[a]pyrene diol epoxide; Shugart, 1986; Wallin et al., 1987). In addition, we have recently demonstrated by GC-MS the hydrolytic release of 1-phenyl-1,2-ethanediol from haemoglobin that had been treated in vitro with styrene oxide (Figure 3). Methods are currently in development for the determination of this glycol.

Immunological techniques

Immunological techniques have not yet been used extensively to monitor alkylated proteins. Monoclonal antibodies to benzo[a]pyrene diol epoxide-modified haemocyanin (from keyhole limpet) and DNA were prepared by Santella et al. (1986) and used to quantify benzo[a]pyrene adduct levels in protein. The method used involved hydrolytic release of the benzo[a]pyrene from the protein as a tetrol, for which the antibodies showed cross-reactivity. A later development of the assay involved an insoluble protease digestion of the protein prior to enzyme-linked immunosorbent assay (ELISA) of the resulting peptides and amino acids (Lee & Santella, 1988).

Figure 3. Hydrolytic release of 1-phenyl-1,2-ethanediol from haemoglobin treated with styrene oxide

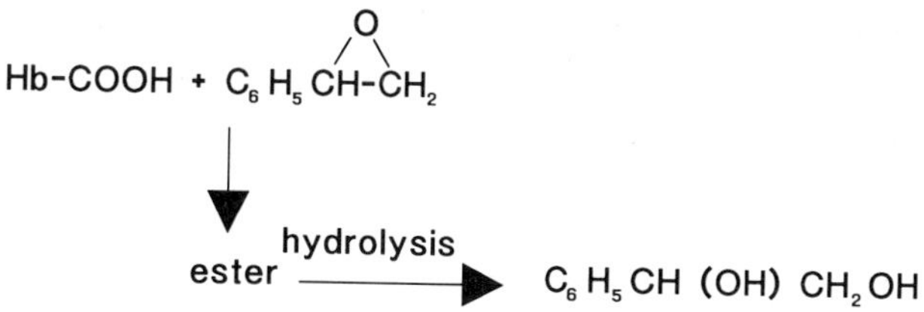

$$Hb-COOH + C_6H_5CH\underset{\underset{CH_2}{\diagdown\diagup}}{\overset{O}{}}$$

$$\downarrow$$

$$ester \xrightarrow{hydrolysis} C_6H_5CH(OH)CH_2OH$$

A polyclonal antibody to the reaction product of ethylene oxide with the N-terminal valine of haemoglobin, N-(2-hydroxyethyl)valine, has also been reported (Wraith et al., 1988). In this case, the protein was cleaved by trypsin prior to radioimmunoassay, and the results of a human study showed good correlation with those obtained by GC-MS.

The lack of widespread use of alkylated protein immunoassay techniques has prompted us to synthesize a range of N-alkylated terminal peptides of haemoglobin for generation of polyclonal and/or monoclonal antibodies. For example, N-methylation of valine was achieved using the reaction of N-trifluoroacetylvaline benzyl ester with methyl iodide in the presence of caesium carbonate. Following deprotection, the product was coupled with leucine using dicyclohexylcarbodiimide to yield N-methylvaline-leucine. N-(2-Hydroxyethyl)valine-leucine and N-(2-hydroxypropyl)valine-leucine were prepared by reacting valine-leucine methyl ester with ethylene oxide and propylene oxide, respectively. The monoalkylated products were purified as their *tert*-butoxydiphenylsilyl derivatives. These peptides were then coupled to bovine serum albumin using a carbodiimide reaction, and polyclonal antibodies to the methylated and 2-hydroxyethylated products were generated in rabbits. These antibodies have been assayed against conjugates of the alkylated dipeptides with ovalbumin using a competitive ELISA technique.

Experiments are in progress to investigate the potential of these antibodies for monitoring haemoglobin alkylations.

Future developments

The sensitivities that are now being achieved with physicochemical, immunological and postlabelling techniques are commonly in the order of one adduct per 10^7-10^8 normal bases, which is sufficient for studies of many occupational and environmental exposures to alkylating agents. However, the amount of biological sample required varies from <1 μg DNA for immunochemical or postlabelling techniques (see review by Lohman, 1988) to about 1 g haemoglobin (from 10 ml blood) for determination of aromatic amines by

GC-MS (Bryant *et al.*, 1987); clearly, further developments in sensitivity would be valuable for the latter techniques. Further improvements in MS techniques would also allow determination of more complex alkylating agent adducts; and, ultimately, it is to be hoped that MS or another physicochemical technique could be used to characterize the structures of unknown radioactive components resolved on postlabelling TLC plates. A comparison of physicochemical and immunological results should also be undertaken, as this would clarify the chemical specificity (i.e., lack of cross-reaction) of the latter technique in biomonitoring practice.

It is impossible in this short review on recent developments in biomonitoring to be comprehensive, and no mention has been made of developments in measuring other biomarkers of damage by alkylating agents to DNA, such as *hprt* mutation, haemoglobin mutation and cytogenetic analyses. In the future, an accumulation of data from all of these monitoring techniques for adducts and biomarkers on the same exposed human population should allow an assessment of their relative sensitivities and practical value.

References

Autrup, H., Wakhisi, J., Vähäkangas, K., Wasunna, A. & Harris, C.C. (1985) Detection of 8,9-dihydro-(7'-guanyl)-9-hydroxyaflatoxin in human urine. *Environ. Health Perspect.*, **62**, 105-108

Bolt, H.M., Peter, H. & Föst, U. (1988) Analysis of macromolecular ethylene oxide adducts. *Int. Arch. Occup. Environ. Health*, **60**, 141-144

Bryant, M.S., Skipper, P.L., Tannenbaum, S.R. & Maclure, M. (1987) Hemoglobin adducts of 4-aminobiphenyl in smokers and non-smokers. *Cancer Res.*, **47**, 602-608

Cooper, R.S., Jankowiak, R., Hayes, J.M., Pei-qi, L. & Small, G.J. (1988) Fluorescence line narrowing spectrometry of nucleoside-polycyclic aromatic hydrocarbon adducts on thin-layer chromatographic plates. *Anal. Chem.*, **60**, 2692-2694

Farmer, P.B. & Bailey, E. (1989) Protein-carcinogen adducts in human dosimetry. *Arch. Toxicol.*, **13**, 83-90

Farmer, P.B., Neumann, H.-G. & Henschler, D. (1987) Estimation of exposure of man to substances reacting covalently with macromolecules. *Arch. Toxicol.*, **60**, 187-191

Groopman, J.D., Donahue, P.R., Zhu, J., Chen, J. & Wogan, G.N. (1985) Aflatoxin metabolism in humans: detection of metabolites and nucleic acid adducts in urine by affinity chromatography. *Proc. Natl Acad. Sci. USA*, **82**, 6492-6496

Harris, C.C., Vähäkangas, K., Newman, M.J., Trivers, G.E., Shamsuddin, A., Sinopoli, N., Mann, D.L. & Wright, W.E. (1985) Detection of benzo[*a*]pyrene diol epoxide-DNA adducts in peripheral blood lymphocytes and antibodies to the adducts in serum from coke-oven workers. *Proc. Natl Acad. Sci. USA*, **82**, 6672-6676

Jankowiak, R., Cooper, R.S., Zamzow, D., Small, G.J., Doskocil, G. & Jeffrey, A.M. (1988) Fluorescence line narrowing-nonphotochemical hole burning spectrophotometry; femtomole detection and high selectivity for intact DNA-PAH adducts. *Chem. Res. Toxicol.*, **1**, 60-68

Kasemann, R., Föst, U. & Peter, H. (1988) Selective GC/MS analysis of 7-(2'-oxoethyl)guanine in the presence of 7-(2'-hydroxyethyl)guanine by oximation. *Arch. Toxicol.*, **61**, 245-246

Lee, B.M. & Santella, R.M. (1988) Quantitation of protein adducts as a marker of genotoxic exposure: immunologic detection of benzo[*a*]pyrene-globin adducts in mice. *Carcinogenesis*, **9**, 1773-1777

Lohman, P.H.M. (1988) Summary: adducts. In: Bartsch, H., Hemminki, K. & O'Neill, I.K., eds, *Methods for Detecting DNA Damaging Agents in Humans: Applications in Cancer Epidemiology and Prevention* (IARC Scientific Publications No. 89), Lyon, IARC, pp. 13-20

Manchester, D.K., Weston, A., Choi, J.-S., Trivers, G.E., Fennessey, P.V., Quintana, E., Farmer, P.B., Mann, D.L. & Harris, C.C. (1988) Detection of benzo[*a*]pyrene diolepoxide-DNA adducts in human placenta. *Proc. Natl Acad. Sci. USA*, **85**, 9243-9247

Randerath, K., Randerath, E., Agrawal, H.P., Gupta, R.C., Schurdak, M.E. & Reddy, M.V. (1985) Post-labeling methods for carcinogen-DNA adduct analysis. *Environ. Health Perspect.*, **62**, 57-65

Reddy, M.V. & Randerath, K. (1986) Nuclease P1-mediated enhancement of sensitivity of ^{32}P-postlabeling test for structurally diverse DNA adducts. *Carcinogenesis*, 7, 1543-1551

Santella, R.M., Lin, C.D. & Dharmaraja, N. (1986) Monoclonal antibodies to a benzo[a]pyrene modified protein. *Carcinogenesis*, 7, 441-444

Schmeiser, H., Dipple, A., Schurdak, M.E., Randerath, E. & Randerath, K. (1988) Comparison of ^{32}P-postlabeling and high pressure liquid chromatographic analyses for 7,12-dimethylbenz[a]-anthracene-DNA adducts. *Carcinogenesis*, 9, 633-638

Shugart, L. (1986) Quantifying adductive modification of hemoglobin from mice exposed to benzo[a]pyrene. *Anal. Biochem.*, 152, 365-369

Shuker, D.E.G. (1989) Nucleic acid-carcinogen adducts in human dosimetry. *Arch. Toxicol.*, 13, 55-65

Shuker, D.E.G., Bailey, E., Gorf, S.M., Lamb, J. & Farmer, P.B. (1984) Determination of N-7-[^{2}H$_3$]methyl guanine in rat urine by gas chromatography-mass spectrometry following administration of trideuteromethylating agents or precursors. *Anal. Biochem.*, 140, 270-275

Shuker, D.E..G., Bailey, E., Parry, A., Lamb, J. & Farmer, P.B. (1987) The determination of urinary 3-methyladenine in humans as a potential monitor of exposure to methylating agents. *Carcinogenesis*, 8, 959-962

Törnqvist, M., Mowrer, J., Jensen, S. & Ehrenberg, L. (1986) Monitoring of environmental cancer initiators through hemoglobin adducts by a modified Edman degradation method. *Anal. Biochem.*, 154, 255-266

Wallin, H., Jeffrey, A.M. & Santella, R.M. (1987) Investigation of benzo[a]pyrene-globin adducts. *Cancer Lett.*, 35, 139-146

Weston, A., Rowe, M.L., Manchester, D.K., Farmer, P.B., Mann, D.L. & Harris, C.C. (1988) Fluorescence and mass spectral evidence for the formation of benzo[a]pyrene anti-diolepoxide-DNA and -hemoglobin adducts in humans. *Carcinogenesis*, 10, 251-257

Wilson, V.L., Basu, A.K., Essigmann, J.M.. Smith, R.A. & Harris, C.C. (1988) O^6-Alkyldeoxyguanosine detection by ^{32}P-postlabeling and nucleotide chromatographic analysis. *Cancer Res.*, 48, 2156-2161

Wraith, M.J., Watson, W.P., Eadsforth, C.V., Van Sittert, N.J., Törnqvist, M. & Wright, A.S. (1988) An immunoassay for monitoring human exposure to ethylene oxide. In: Bartsch, H., Hemminki, K. & O'Neill, I.K., eds, *Methods for Detecting DNA Damaging Agents in Humans: Applications in Cancer Epidemiology and Prevention* (IARC Scientific Publications No. 89), Lyon, IARC, pp. 271-274

Relevance to Human Cancer of *N*-Nitroso Compounds,
Tobacco Smoke and Mycotoxins.
Ed. I.K. O'Neill, J. Chen and H. Bartsch
Lyon, International Agency for Research on Cancer
© IARC, 1991

NOVEL, SENSITIVE ASSAYS FOR O^6-ALKYLGUANINE AND ITS REPAIR AND THEIR APPLICATION TO STUDIES OF THE MOLECULAR EPIDEMIOLOGY OF THIS LESION IN HUMAN POPULATIONS

S.A. Kyrtopoulos[1,5], V.L. Souliotis[1], P. Ambatzi[1],
G. Pangalis[2], V. Bousiotou[2], G. Haritopoulos[3] & G. Davaris[4]

[1]*Programme of Chemical Carcinogenesis, National Hellenic Research Foundation, Institute of Biological Research, Athens;*
[2]*Lymphoma Unit, Haematology Clinic,*
[3]*Endoscopy Unit, First Prepedeutic Surgical Clinic, and*
[4]*Laboratory of Pathology and Anatomy,*
University of Athens Medical School, Athens, Greece

Two assays suitable for monitoring human populations are presented — one for O^6-alkylguanine-type adducts and one for the corresponding repair activity. The assay for adducts is based on competition for repair by O^6-alkylguanine-DNA-alkyltransferase of lesions in the unknown DNA and in an oligonucleotide containing a single residue of O^6-methylguanine (O^6-meGua). The assay can reliably detect as little as 0.5 fmol O^6-meGua or 3 fmol O^6-ethylguanine in 10-15 μg DNA, and it has been used to measure O^6-meGua (up to ~5 ×10^{-7} mol/mol guanine) in lymphocyte DNA from individuals receiving procarbazine or dacarbazine. The assay for repair activity utilizes the same oligonucleotide as substrate for the enzyme, in combination with immunoprecipitation for convenient separation of repaired from unrepaired substrate. Examination of repair activity in biopsy specimens of gastric mucosa has revealed a correlation with the corresponding activity in lymphocytes ($r = 0.7$; $p < 0.01$), indicating that lymphocytes could be used as surrogate markers for repair activity in gastric mucosa.

The measurement of O^6-alkylguanine accumulation and repair in human tissue may help to elucidate the role of *N*-nitroso compounds in the etiology of human cancer. The assay of this adduct is currently based mainly on immunochemical techniques, which have permitted its detection in human tissues (Umbenhauer *et al.*, 1985; Foiles *et al.*, 1988), requiring for this purpose milligram amounts of DNA. We report here a novel assay capable

---

[5] To whom correspondence should be addressed

of detecting sub-femtomole amounts of O^6-methylguanine (O^6-meGua) in microgram amounts of DNA, and its application for studying the accumulation of O^6-meGua in lymphocyte DNA from individuals exposed to the alkylating drugs procarbazine and dacarbazine. We also report the development of a sensitive assay for the corresponding repair enzyme, O^6-alkylguanine-DNA alkyltransferase (AAT), suitable for use with biopsy samples, and its application to the study of the activity of AAT in normal and diseased gastric mucosa.

Competitive repair assay for O^6-alkylguanine

This assay exploits the 'suicide' activity of AAT which permits its use as a probe for the lesion. AAT from rat liver or *Escherichia coli* (7 fmol) was incubated with DNA for 10 min (rat liver AAT) or 2 h (*E. coli* AAT) (stage 1). Subsequently (stage 2), 10 fmol of a synthetic oligonucleotide (d[CGC(O^6-meGua)AGCTCGCG]), 5'-end-labelled with 35S, were added and incubation continued overnight. Finally, repaired and unrepaired oligonucleotides were separated by immunoprecipitation with anti-O^6-meGua antibodies and the radioactivity counted. Any repair of O^6-alkylguanine during stage 1 resulted in a reduction in oligonucleotide repair during stage 2, reflected in an increase in immunoprecipitable counts. A standard curve was constructed with DNA of known methylation (Figure 1A), and the amount of O^6-alkylguanine present in unknown DNA calculated by reference to it. Incubation of methylated DNA with excess AAT followed by heat inactivation of the latter, prior to conduct of the assay, completely removed inhibition of oligonucleotide repair, confirming the specificity of the observed effect. While the competitive repair assay has been tested with methylated and ethylated DNA, in principle any type of lesion that can be repaired by a suicide mechanism by AAT can be detected.

Figure 1. Standard curves for the competitive repair assay[a]

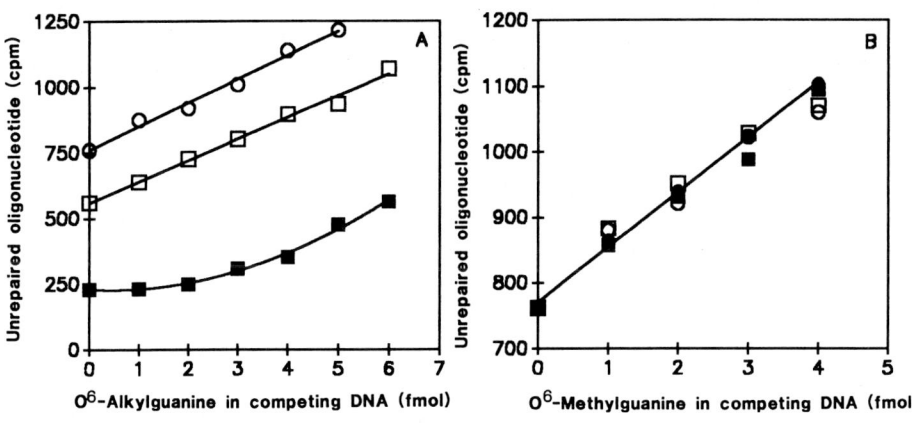

[a] A, Undigested DNA containing 76.8 μmol O^6-methylguanine (O^6-meGua)/mol guanine (open symbols) or 13.5 μmol O^6-ethylguanine/mol guanine (filled symbols); O^6-alkylguanine-DNA alkyltransferase from *Escherichia coli* (circles) and from rat liver (squares). The differences in the vertical intercepts are due to the use of oligonucleotides of different specific activities. B, DNA containing 76.8 (circles) or 0.90 (squares) μmol O^6-meGua/mol guanine, undigested (filled symbols) or exhaustively digested with MboI (open symbols).

The standard curve is not significantly affected by the extent of DNA methylation (in the range 0.9–76.8 μmol O^6-meGua/mol guanine) or the size of the DNA (in the range of 40 Kbp – 256 bp, average size; Figure 1B). The sensitivity of the assay is not significantly improved if ^{32}P-labelling is used in place of ^{32}S, and for this reason the latter is routinely used. The limits of sensitivity of the assay (Table 1) are, in practice, determined by the stability of AAT, the rate of DNA repair and the total amount of DNA that can be tolerated. Analysis of a series of DNA samples by the competitive repair assay and by radioimmunoassay indicated close agreement between the two methods.

Table 1. Sensitivity of the competitive repair assay

O^6-Alkylguanine[a]	Max DNA (μg)	Detection limit (fmol)	Sensitivity	
			fmol/μg DNA	mol/mol guanine
O^6-Methylguanine				
Rat liver AAT	15	0.8	50	8×10^{-8}
E. coli AAT	10	0.5	50	8×10^{-8}
O^6-Ethylguanine				
Rat liver AAT	15	3.0	200	3×10^{-7}

[a]AAT, O^6-alkylguanine-DNA alkyltransferase

Search for O^6-meGua in biopsy samples of human gastric mucosa using the competitive repair assay: Biopsy samples of gastric mucosa are currently being analysed using this assay for the presence of O^6-meGua. So far, samples histologically characterized as exhibiting chronic atrophic gastritis, obtained from 14 individuals, have been examined. In each case, 10 μg DNA (obtained by pooling three to five 1-mm3 samples) were analysed, and no evidence of DNA methylation was found.

O^6-MeGua in individuals exposed to procarbazine or dacarbazine: The accumulation of O^6-meGua in lymphocyte DNA has been examined in individuals taking the cytostatic drugs procarbazine or dacarbazine: O^6-meGua was detected in most of the individuals examined (Figure 2).

AAT assay

This assay is based on the use of the same oligonucleotide described above as substrate for AAT, in combination with immunoprecipitation to separate repaired from unrepaired substrate. Excess (4–5 fmol) radiolabelled oligonucleotide was incubated with tissue extract (up to 30 μg protein) until completion of the repair reaction (2 h), followed by immunoprecipitation and radiocounting (Figure 3A). The amount of AAT is calculated by reference to a standard curve constructed in parallel using known quantities of AAT. The absence of nonspecific oligonucleotide degradation (Dolan et al., 1988) is tested separately by titration of a known excess of oligonucleotide with an extract of the same type of tissue containing known amounts of AAT. The assay, which has been validated by parallel analysis of tissue extracts by a classical method, is accurate, convenient and sensitive and is currently being employed to examine AAT activity in biopsy samples of human gastric mucosa.

Results obtained so far indicate (i) no significant difference in AAT activity in gastric mucosa obtained from different regions of the stomach, in histologically normal gastric mucosa (AAT = 6.4 ± 3.0 (SD) fmol/µg DNA; n = 12) or in mucosa exhibiting chronic atrophic gastritis with or without intestinal metaplasia (AAT = 7.5 ± 4.2 fmol/µg DNA; n = 12); and (ii) a statistically significant correlation between AAT levels in gastric mucosa and in peripheral lymphocytes obtained from the same individuals (Figure 3B).

Figure 2. O^6-Methylguanine (O^6-meGua) in lymphocyte DNA from individuals exposed to procarbazine or dacarbazine

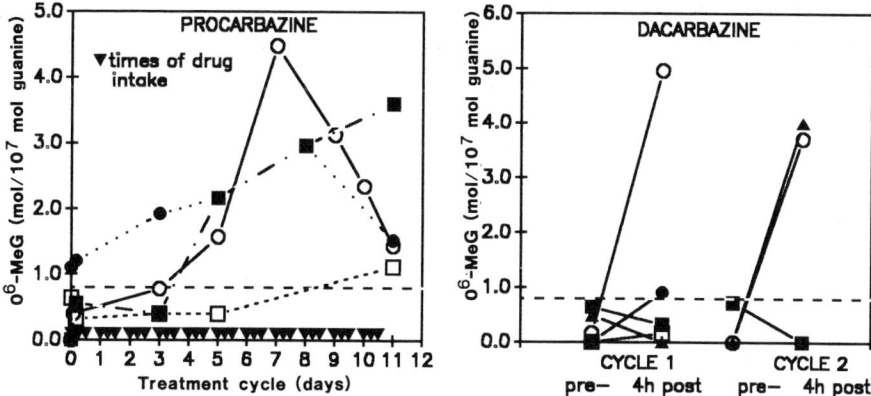

Figure 3. O^6-Alkylguanine-DNA alkyltransferase (AAT) assay[a]

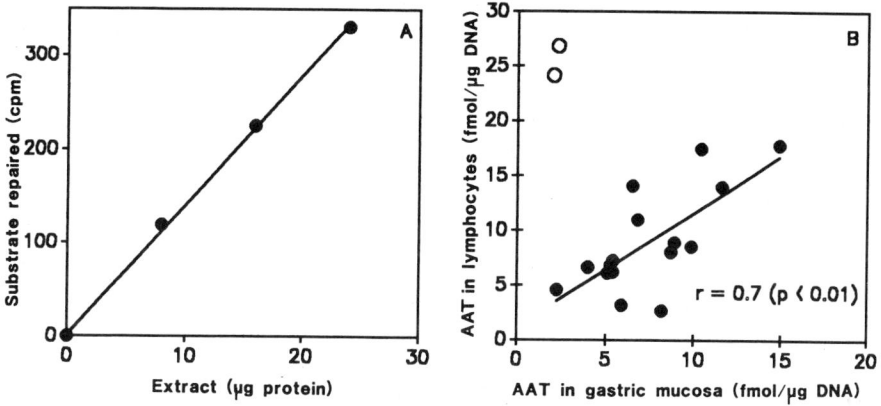

[a] A, Extent of oligonucleotide repair by AAT in different amounts of tissue extract. B, Correlation between AAT in extracts of lymphocytes and gastric mucosa. The open symbols indicate two cases of significant deviation from the correlation.

Conclusions and prospects

The sensitivity of the assays reported here is such as to make them useful for studies of the epidemiology of O^6-alkylguanine and its repair. The finding that AAT activity in lymphocytes correlates with that of gastric mucosa indicates the usefulness of lymphocytes as surrogate markers for DNA repair. Furthermore, the ability to detect O^6-meGua in individuals taking procarbazine and dacarbazine offers a possibility for examining the dosimetry relationships governing the formation of this important lesion as well as its biological consequences.

Acknowledgements

This work was supported in part by a grant from the European Economic Community (Genetic Effects of Environmental Chemicals grant no. EV-0062) to S.A.K. We thank Dr B.F. Li and Dr P.F. Swann (University College, London, UK) for the oligonucleotides, Dr A.-M. Camus (IARC, Lyon, France) for the gift of methylated DNA and Professor A.A. van Zeeland (University of Leiden, Leiden, Netherlands) for the gift of ethylated DNA.

References

Dolan, M.E., Oplinger, M. & Pegg, A.E. (1988) Use of a dodecadeoxynucleotide to study repair of the O^4-methylamine lesion. *Mutat. Res.*, **193**, 131-137

Foiles, P.G., Miglietta, L.M., Akerkar, S.A., Everson, R.B. & Hecht, S.S. (1988) Detection of O^6-methyldeoxyguanosine in human placental DNA. *Cancer Res.*, **48**, 4184-4188

Umbenhauer, D., Wild, C.P., Montesano, R., Saffhill, R., Boyle, J.M., Huh, N., Kirstein, U., Thomale, J., Rajewsky, M.F. & Lu, S.H. (1985) O^6-Methyldeoxyguanosine in oesophageal DNA in populations at high risk of oesophageal cancer. *Int. J. Cancer*, **36**, 661-665

COMPARISON OF URINARY NITRATE, N-NITROSOPROLINE, 7-METHYLGUANINE AND 3-METHYLADENINE LEVELS IN A HUMAN POPULATION AT RISK FOR GASTRIC CANCER

W.G. Stillwell[1], H.X. Xu[1], J. Glogowski[1],
J.S. Wishnok[1,4], S.R. Tannenbaum[1,2,4] & P. Correa[3]

[1]Division of Toxicology and [2]Department of Chemistry,
Massachusetts Institute of Technology, Cambridge, MA; and
[3]Department of Pathology, Louisiana State University
Medical Center, New Orleans, LA, USA

We have measured the urinary excretion of nitrate, N-nitrosoproline, 3-methyladenine and 7-methylguanine in a human population at high risk for gastric cancer. A strong correlation was observed between nitrate and N-nitrosoproline excretion ($p < 0.00000$); statistically significant correlations ($p < 0.05$) were also observed for N-nitrosoproline with 3-methyladenine ($p = 0.003$) and with 7-methylguanine ($p = 0.03$) and for 7-methylguanine with nitrate ($p = 0.02$), although these correlations were sensitive to outliers. Smokers appeared to excrete slightly higher levels of 3-methyladenine than nonsmokers; no other difference was observed between smokers and nonsmokers or between drinkers and nondrinkers. The correlation between N-nitrosoproline excretion and nitrate excretion supports the hypothesis that elevated endogenous nitrate levels may contribute to elevated endogenous formation of N-nitroso compounds.

As part of our investigations of the etiology of gastric cancer in progress in Colombia (Correa *et al.*, 1976; Cuello *et al.*, 1976; Haenszel *et al.*, 1976) and development and use of adducted protein and nucleic acids as molecular dosimeters (Bryant *et al.*, 1987, 1990; Gan *et al.*, 1989; Stillwell *et al.*, 1989), we examined the urine of over 150 people from a high-risk area for urinary excretion levels of nitrate, N-nitrosoproline (NPRO), 3-methyladenine and 7-methylguanine in order to determine if correlations exist among these various indices of endogenous nitrosation and whether any of them reflect the relative risk for gastric cancer in this group of people.

Urine samples were collected over 24 h and kept frozen until analysed. History of smoking and drinking was obtained for most of the subjects by a questionnaire; this information included type of smoking material, total number of cigarettes smoked for cigarette smokers, type of alcohol consumed and total alcohol consumption. No attempt was made to standardize the diet.

---

[4]To whom correspondence should be addressed

Quantitative analyses

Nitrate was analysed on an automated nitrate/nitrite analyser based on the Griess reagent, as described by Green et al. (1982). Urinary NPRO levels were quantified by gas chromatography-thermal energy analysis of the methyl esters following reaction with borontrifluoride-methanol (Licht et al., 1988a,b). Methods for the analysis of 3-methyladenine and 7-methylguanine were described in earlier reports (Shuker et al., 1984; Stillwell et al., 1989). In brief, 10-ml aliquots of urine were spiked with deuterated 3-methyladenine and 7-methylguanine as internal standards and applied to XAD-2 columns, which were washed with HCl. The columns were eluted with 0.001N HCl and the fractions containing the alkylated purines evaporated to near dryness and purified further by bonded-phase extraction chromatography. The methanol eluates were dried under nitrogen at 60°C and treated with acetonitrile, dry pyridine and N-(tert-butyl-dimethylsilyl)-N-methyltrifluoracetamide. The resulting tert-butyldimethyl- silyl derivatives were then quantified by gas chromatography/mass spectrometry with electron ionization using the co-eluting deuterated analogues as internal standards. The electron-ionization mass spectra of these compounds showed high-abundance ions at M-57, corresponding to loss of the tert-butyl group from the molecular ion. Single-ion chromatographs at M-57 were integrated at the retention times of the respective compounds and the concentrations determined by comparison of the areas of the analytes with the areas of the internal standards. The data were analysed by standard regression techniques using STATGRAPHICS Version 2.6. Linear regression analysis was carried out on all combinations of data sets using the raw data, on data with outliers rejected, and on some subsets of data, e.g., smokers and nonsmokers.

The results of the linear regression analyses, comprising correlation coefficients, number of subjects and p values, are summarized in Table 1. Figure 1 shows the plot of nitrate versus NPRO for the entire data set except for the highest individual excretion of NPRO (13 280 ng/day as compared to an average of approximately 1000 ng/day). Figure 2 summarizes the differences in average excretion of NPRO, nitrate, 3-methyladenine and 7-methylguanine and total alcohol consumption for smokers and nonsmokers. The median excretion of 3-methyladenine by nonsmokers was 5.5 µg/day (n = 126), while that by smokers was 8.9 µg/day (n = 41). These values are comparable to the mean values reported for a human population at normal risk for gastric cancer. There was no difference between drinkers and nondrinkers in excretion of any of the studied compounds (data not shown).

The population under investigation is known to be at high risk for gastric cancer. The progression from normal gastric pathology through gastritis and metaplasia to neoplasia is well documented and is accompanied by elevated gastric pH, increased gastric bacterial population, and increased gastric nitrate and nitrite levels. These observations, as well as the presence in the diet of 4-chloro-6-methoxyindole, which forms a potent mutagen when nitrosated (Yang et al., 1984; Buchi et al., 1986), have led to the etiological hypothesis that there is elevated endogenous formation of N-nitroso compounds in this group of people (Correa et al., 1976; Cuello et al., 1976; Haenszel et al., 1976; Correa & Tannenbaum, 1981). Urinary NPRO levels are widely considered to be a general index of intragastric nitrosation

Table 1. Linear regression analyses of urinary excretion levels of N-nitrosoproline (NPRO), nitrate, 3-methyladenine (3-meAde) and 7-methylguanine (7-meGua)

Variable set	Correlation coefficient	Probability level	No. of subjects
NPRO versus nitrate	0.43	< 0.0000	180
NPRO versus 3-meAde	0.11	0.18	166
NPRO versus 3-meAde	0.23	0.003	165[a]
NPRO versus 7-meGua	0.07	0.40	166
NPRO versus 7-meGua	0.16	0.03	165[a]
Nitrate versus 3-meAde	0.05	0.54	166
Nitrate versus 7-meGua	0.18	0.022	166
3-meAde versus 7-meGua	0.005	0.94	166

[a] Highest NPRO value (13 300 ng/day) omitted

Figure 1. Plot of urinary exretion of N-nitrosoproline (NPRO) versus that of nitrate by Colombian individuals in an area of high risk for gastric cancer[a]

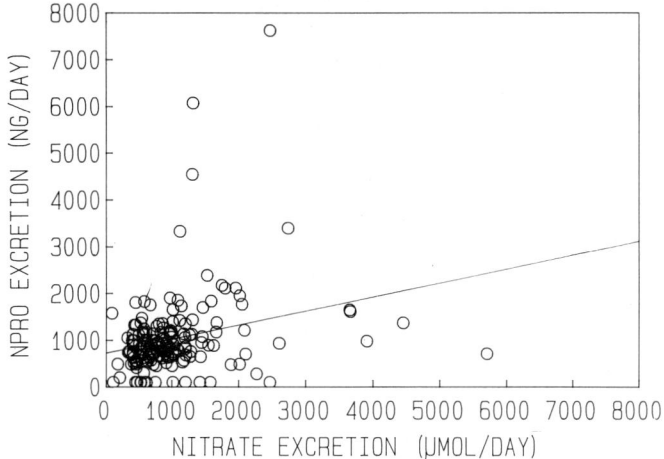

[a] One point, with a NPRO value of approximately 13 000 (10 standard deviations from the mean), was omitted from the plot. The solid line represents the best fit from the linear regression analysis.

(Ohshima & Bartsch, 1981; Lu et al., 1986), although diet contributes significantly to the background level. In this study, the subjects consumed their normal diet.

Both 7-methylguanine and 3-methyladenine have been proposed as indices of exposure to methylating agents, including endogenously formed compounds such as N-nitrosodimethylamine (Farmer et al., 1980; Gombar et al., 1983; Shuker et al., 1984; Farmer et al., 1986; Bailey et al., 1987; Stillwell et al., 1989). Our results suggest that this may be possible using 3-methyladenine in well-characterized populations in which confounding factors such as smoking and dietary contributions (D. Shuker, personal communication) can be taken into account.

In summary, excretion of NPRO was correlated with nitrate excretion and, less strongly, with 3-methyladenine excretion in a population at high risk for gastric cancer. These findings are consistent with etiological hypotheses based on the role of exposure to N-nitroso compounds by endogenous nitrosation reactions.

Figure 2. Total alcohol consumption and urinary excretion of N-nitrosoproline (NPRO), nitrate, 3-methyladenine (3-meAde), and 7-methylguanine (7-meGua) in smokers and nonsmokers[a]

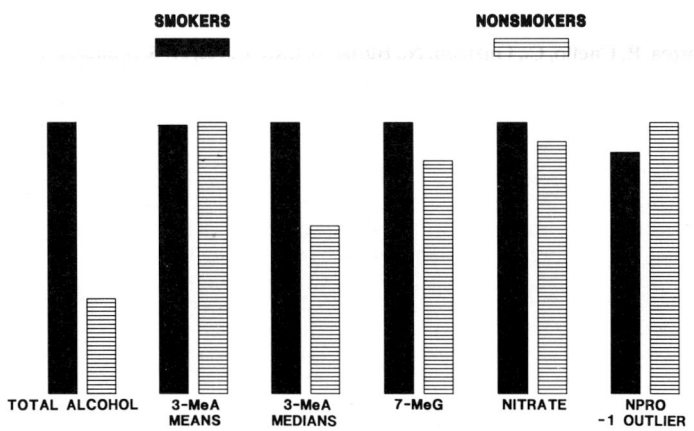

[a] The value of the higher of each of the two data sets was set at 100 in order to generate a common scale. Total alcohol in smokers, mean = 270 ml; 3-meAde in nonsmokers, mean = 9.8 μg/day; 3-meAde in smokers, mean = 8.9 μg/day; 7-meGua in smokers, mean = 6.8 mg/day; NPRO in nonsmokers, mean = 1130 ng/day; nitrate in smokers, mean = 1000 μmol/day

Acknowledgements

This work was supported by Grants CA26731 and CA28842 from the National Institutes of Health and Grant S10-RR01901 from the DHHS Shared Instrument Program.

References

Bailey, E., Farmer, P.B. & Shuker, D.E.G. (1987) Estimation of exposure to alkylating carcinogens by the GC-MS determination of adducts to hemoglobin and nucleic acid bases in urine. *Arch. Toxicol.*, **60**, 187-191

Bryant, M.S., Skipper, P.L., Tannenbaum, S.R. & Maclure, M. (1987) Hemoglobin adducts of 4-aminobiphenyl in smokers and nonsmokers. *Cancer Res.*, **47**, 602-608

Bryant, M.S., Skipper. P.L., Wishnok, J.S., Stillwell, W.G., Glogowski, J.A. & Tannenbaum, S.R. (1990) Determination of haemoglobin adducts of aromatic amines by gas-chromatography-mass spectrometry. In: *Environmental Carcinogens - Methods of Analysis and Exposure Measurement*, Vol. 12, *Indoor Air* (IARC Scientific Publications No. 109), Lyon, IARC (in press)

Buchi, G., Lee, G.C.M., Yang, D. & Tannenbaum, S.R. (1986) Direct acting, highly mutagenic alpha-hydroxy N-nitrosamines from 4-chloroindoles. *J. Am. Chem. Soc.*, **108**, 4115-4119

Correa, P. & Tannenbaum, S.R. (1981) The microecology of gastric cancer. In: Tannenbaum. S.R. & Scanlan, R.A., eds, *N-Nitroso Compounds* (ACS Symposium Series No. 174), Washington DC, American Chemical Society, pp. 319-329

Correa, P., Cuello, C., Duque, E., Burbano, L.C., Garcia, F.T., Bolanos, O., Brown, C. & Haenszel, W. (1976). Gastric cancer in Colombia. III. Natural history of precursor lesions. *J. Natl Cancer Inst.*, **57**, 1027-1035

Cuello, C., Correa, P., Haenszel, W., Gordillo, G., Brown, C., Archer, M. & Tannenbaum, S. (1976) Gastric cancer in Colombia. I. Cancer risk and suspect environmental agents. *J. Natl Cancer Inst.*, **57**, 1015-1020

Farmer, P.B., Bailey, E., Lamb, J.H. & Conners, T.A. (1980) Approach to the quantitation of alkylated amino acids in hemoglobin by gas chromatography mass spectrometry. *Biomed. Mass Spectrom.*, **7**, 41-46

Farmer, P.B., Shuker, D.E. & Bird, I. (1986) DNA and protein adducts as indicators of in vitro methylation by nitrosatable drugs. *Carcinogenesis*, **7**, 49-52

Gan, L.-S., Wishnok, J.S., Tannenbaum, S.R. & Fox, J.G. (1989) Quantitation of methylated hemoglobin via hydrolysis of methyl esters to yield methanol. *Anal. Biochem.*, **179**, 326-331

Gombar, C.T., Zubroff, J., Strahan, G.D. & Magee, P.N. (1983) Measurement of 7-methylguanine as an estimate of the amount of dimethylnitrosamine formed following administration of aminopyrine and nitrite to rats. *Cancer Res.*, **43**, 5077-5080

Green, L.C., Wagner, D.A., Glogowski, J., Skipper, P.L., Wishnok, J.S. & Tannenbaum, S.R. (1982) Analysis of nitrate and nitrite and (15N)nitrate in biological fluids. *Anal. Biochem.*, **126**, 131-138

Haenszel, W., Correa, P., Cuello, C., Guzman, N., Burbano, L.C., Lores, H. & Muñoz, J. (1976) Gastric cancer in Colombia. II. Case-control epidemiologic study of precursor lesions. *J. Natl Cancer Inst.*, **57**, 1021-1035

Licht, W.R., Fox, J.G. & Deen, W.M. (1988a) Effects of ascorbic acid and thiocyanate on nitrosation of proline in the dog stomach. *Carcinogenesis*, **9**, 373-377

Licht, W.R., Tannenbaum, S.R. & Deen, W.M. (1988b) Use of ascorbic acid to inhibit nitrosation: kinetic and mass transfer considerations for an in vitro system. *Carcinogenesis*, **9**, 365-372

Lu, S.-H., Ohshima, H., Fu, H.M., Tian, Y., Li, F.M., Blettner, M., Wahrendorf, J. & Bartsch, H. (1986) Urinary excretion of N-nitrosamino acids and nitrate by inhabitants of high- and low-risk areas for esophageal cancer in northern China: endogenous formation of nitrosoproline and its inhibition by vitamin C. *Cancer Res.*, **46**, 1485-1491

Ohshima, H. & Bartsch, H. (1981) Quantitative estimation of endogenous nitrosation in humans by monitoring N-nitrosoproline excretion in the urine. *Cancer Res.*, **41**, 3658-3666

Shuker, D.E.G., Bailey, E., Gorf, S.M., Lamb, J. & Farmer, P.B. (1984) Determination of N-7[2H3]methylguanine in rat urine by gas chromatography-mass spectrometry following administration of trideuteromethylating agents or precursors. *Anal. Biochem.*, **140**, 407-410

Stillwell, W.G., Xu, H.-X., Adkins, J.A., Wishnok, J.S. & Tannenbaum, S.R. (1989) Analysis of methylated and oxidized purines in urine by capillary gas chromatography-mass spectrometry. *Chem. Res. Toxicol.*, **2**, 94-99

Yang, D., Tannenbaum, S.R., Buchi, G. & Lee, G.C.M. (1984) 4-Chloro-6-methoxyindole is the precursor of a potent mutagen (4-chloro-6-methoxy-2-hydroxy-1-nitroso-indolin-3-one oxime) that forms during nitrosation of the fava bean (Vicia faba). *Carcinogenesis*, **10**, 1219-1224

ENDOGENOUS NITROSAMINES AND LIVER FLUKE AS RISK FACTORS FOR CHOLANGIOCARCINOMA IN THAILAND

P. Srivatanakul[1], H. Ohshima[2], M. Khlat[2], M. Parkin[2],
S. Sukarayodhin[1], I. Brouet[2] & H. Bartsch[2]

[1]*National Cancer Institute of Thailand, Bangkok, Thailand; and*
[2]*International Agency for Research on Cancer, Lyon, France*

Cholangiocarcinoma (CCA) is one of the most prevalent cancers in north-east Thailand and has been associated with infestation by the liver fluke *Opisthorchis viverrini* (OV). Two samples of 12-h overnight urine (after dosing with 500 mg proline and 200 mg ascorbic acid or 500 mg proline alone) were collected from about 100 inhabitants in five contrasting incidence areas for CCA and hepatocellular carcinoma. The incidences of CCA and hepatocellular carcinoma were not correlated with either the amount of NPRO or other nitrosamino acids, endogenous nitrosation potential (difference in NPRO levels between proline dose and proline and ascorbic acid dose), or nitrate level. However, when urinary levels of nitrosamino acids were compared in subjects living in high-risk areas, subjects who were positive for OV antibody excreted significantly more ($p < 0.01$) NPRO (12.3 ± 18.7 µg/12 h) after proline ingestion than those who were negative (3.5 ± 3.2 µg/12 h). After ingestion of ascorbic acid, the NPRO levels in the positive subjects were significantly reduced ($p < 0.01$) to 2.4 ± 2.0 µg/12 h, suggesting that endogenous nitrosation of proline was inhibited. Thus, endogenous nitrosation potential estimated from the difference of NPRO and sum of nitrosamino acids excreted in the two urine samples was significantly higher in subjects positive for the OV antibody. In addition, of the representative food samples and beverages consumed frequently in high-risk areas for CCA, fermented fish and pork contained N-nitrosodimethylamine (0–26 µg/kg), N-nitrosopyrrolidine (0–117 µg/kg) and N-nitrosopiperidine (0–23 µg/kg). These results indicate that the interaction between chemical carcinogens, especially nitrosamines, and OV infestation may play a role in the development of CCA in Thailand.

Liver cancer is the leading cause of death from malignant neoplasms in Thailand: in 1980–82, it was the most common cancer recorded by the cancer registry for males (16%) and the third most common in females (5.5%), after cervical and breast cancers (Srivatanakul *et al.*, 1988). These proportions remained more or less constant over the 12-year period 1971–82. Primary hepatocellular carcinoma (HCC) and cholangiocarcinoma (CCA) are the two major sub-types of liver cancer. There is relatively little variation in the frequency of HCC by geographical area, but the variation in the incidence of CCA is striking — more than 12-fold — and it accounts for most of the variation observed for liver cancer overall.

The occurrence of CCA in north-eastern Thailand has been found to coincide with the frequency of opisthorchiasis: north-east Thailand is endemic for *Opisthorchis viverrini* (OV) infection (Harinasuta & Vajrasthira, 1960). Although there has been no representative population survey of the prevalence of infection by province, Sadun (1955) studied the prevalence of eggs in faeces of schoolchildren in different areas and found it to be 29% in the north-east, 3.8% in the north, and practically zero in central and southern Thailand. A recent survey of the population of a village in the endemic area showed that 90% harboured the parasites (Upatham et al., 1984). Thus, the frequency of CCA appears to be correlated geographically with OV infection (Srivatanakul et al., 1988). The results of laboratory studies have also suggested that OV infection plays an important role in CCA, in association with N-nitrosamines (Thamavit et al., 1978). Patients infested with liver fluke have increased urinary excretion of nitrate and N-nitrosoproline (NPRO) (Srianujata et al., 1987), although it is not known whether the NPRO is formed endogenously or ingested in foods. It is possible, therefore, that some of the geographical variation in the occurrence of CCA might be related to differences in prevalence of OV infection and exposure in the diet to chemical carcinogens, such as N-nitrosamines or their precursors from which N-nitrosamines could be formed endogenously. We therefore applied the NPRO test (Ohshima & Bartsch, 1988) to subjects living in contrasting incidence areas for CCA and hepatocellular carcinoma in Thailand, in order to assess their exposure to endogenous and exogenous nitrosamines.

Methods

Subjects were normal healthy volunteers from among hospital and paramedical personnel, who were long-term residents of each area and who ate a diet typical of the region. The characteristics of study subjects, together with their smoking and drinking habits, are summarized in Table 1.

Antibodies to OV in serum were analysed by an indirect enzyme-linked immunosorbent assay, as described previously (Srivatanakul et al., 1985). Stools were examined for OV eggs by both the formalin–ether concentration technique (Ritchie, 1948) and Stoll's egg counting technique (Stoll, 1923). Urine samples were collected as described in footnote b to Table 1 and analysed for nitrate, N-nitrosamino acids and creatinine (Ohshima et al., 1987). The effect of ascorbic acid ingestion on endogenous nitrosation was examined using the Wilcoxon signed-ranks test for comparison of matched samples, while the effect of OV infection on endogenous nitrosation was examined using the Mann–Whitney U test for comparison of independent samples (Armitage & Berry, 1987).

Correlation between OV infection, urinary level of N-nitrosamino acids and incidence of CCA and hepatocellular carcinoma

As shown in Table 1, the proportion of subjects with an OV antibody titre $\geq$ 1:40 was significantly higher ($p < 0.001$) in high-incidence areas (23/41 in Ubon and Korat) than in the intermediate-incidence area (6/20 in Chiang Mai) and in the low-incidence areas (0/38 in Bangkok and Songkhla). Faeces from high- and intermediate-incidence areas contained OV eggs (5/41 and 2/20 subjects, respectively), but those from low-incidence areas did not. These findings confirm previous observations of a positive correlation

Table 1. Characteristics of study subjects

Study area	PIR[a]		No. of subjects		Age (years) (mean ± SD)	On day of urine collection[b]			No. of positive subjects[c]	
	CCA	HCC	Total	Male		Urine sample	No. of smokers	No. of alcohol drinkers	Anti-OV antibody titre	OV eggs in faeces
High incidence										
Ubonratchathani (Ubon)	3.08	1.24	20	10	38.8 ± 7.1	Pro	9	5	17	4
						Pro + AA	8	9		
Nakornratchasima (Korat)	1.89	0.76	21	11	32.9 ± 2.8	Pro	8	2	6	1
						Pro + AA	9	4		
Intermediate incidence										
Chiang Mai	0.90	0.84	20	10	33.4 ± 3.0	Pro	3	4	6	4
						Pro + AA	3	6		
Low incidence										
Bangkok	0.36	0.66	20	10	33.7 ± 3.3	Pro	5	3	0	0
						Pro + AA	5	4		
Songkhla	0.32	0.92	18	10	34.0 ± 3.6	Pro	1	0	0	0
						Pro + AA	2	0		

[a] Proportionate incidence ratio (relative to the whole country = 1.0); CCA, cholangiocarcinoma; HCC, hepatocellular carcinoma
[b] Pro, 12-h overnight urine collected after ingestion of 500 mg proline 1 h after evening meal; Pro + AA, 12-h overnight urine collected after ingestion of 500 mg proline and 200 mg ascorbic acid 1 h after evening meal
[c] OV, *Opisthorchis viverrini*; positive antibody titre ≥ 1:40

between OV infection and CCA incidence in Thailand (Schwartz, 1980; Srivatanakul *et al.*, 1988).

The mean amount of NPRO excreted in 12-h urine after a proline dose was in a range of 2.9–14.2 µg, and that after proline and ascorbic acid was 1.3–11.5 µg (Table 2). The incidences of CCA and hepatocellular carcinoma were not correlated with either the amount of NPRO or other *N*-nitrosamino acids, endogenous nitrosation potential (difference in NPRO level between proline dose and proline and ascorbic acid dose), or nitrate level. For example, the mean amount of NPRO excreted after the proline dose was higher in Songkhla (a low-incidence area for CCA) than in Korat (a high-risk area).

Comparison of urinary levels of N-nitrosamino acids in healthy subjects and in those with OV infection

Subjects living in the high-risk areas (Ubon and Korat) who were seropositive for OV antibody excreted significantly higher levels of NPRO after proline ingestion than those who were seronegative (Table 3). After ingestion of ascorbic acid, the NPRO levels in the seropositive subjects were significantly ($p < 0.001$) reduced from those with proline alone, suggesting that endogenous nitrosation of proline had been inhibited; a similar difference was not seen in subjects who were seronegative for OV antibody. It should also be noted that subjects who were seropositive for OV antibody excreted higher levels of creatinine in both urine samples than seronegative subjects. Therefore, when the levels of NPRO or the sum of *N*-nitrosamino acids expressed per millimole of creatinine were compared, there was no significant difference between subjects who were seropositive and seronegative for OV antibody or for OV eggs in faeces (data not shown).

Similarly, after intake of proline, the subjects who had OV eggs in their faeces excreted significantly ($p < 0.05$) more NPRO (23.2 ± 35.2 µg/12 h; n = 5) than the subjects who did not (6.4 ± 8.3 µg/12 h, n = 36). The NPRO level after intake of ascorbic acid decreased significantly, to 1.6 ± 0.9 ($p < 0.05$) and 2.5 ± 1.9 ($p < 0.001$) µg/12 h in subjects with and without OV eggs in faeces, respectively (data not shown). Although endogenous nitrosating potential was therefore about ten times higher in infected than in uninfected subjects, no such difference was observed between subjects who were seropositive and those who were seronegative for hepatitis B surface antigen (data not shown).

Volatile N-nitrosamines in Thai foods

Various fermented foods consumed frequently in high risk areas for CCA were collected and analysed for the volatile nitrosamines *N*-nitrosodimethylamine, *N*-nitrosopiperidine and *N*-nitrosopyrrolidine according to the method of Fine (1978). The results are shown in Table 4. No detectable level of volatile *N*-nitrosamines was found in five samples of Thai whiskey.

Conclusion

Our results indicate that the interaction between chemical carcinogens, especially *N*-nitrosamines, and OV infection may play an important role in the development of CCA in Thailand. Further studies are needed to investigate the mechanism(s) of the increased endogenous nitrosation and its resulting effects in subjects with liver fluke. A study is under

Table 2. Urine volumes and amounts of N-nitrosamino acids, creatinine and nitrate detected (mean ± SD)

Study area	Urine sample[a]	No. of subjects	Volume of 12-h urine (ml)	Creatinine (mmol/12 h)	N-Nitrosamino acid (µg/12 h)[b]					Nitrate (mmol/12h)
					NSAR	NPRO	NMTCA	NTCA	Sum	
Ubonratchathani (Ubon)		20								
	Pro		393 ± 183	4.1 ± 2.5	0.3 ± 0.3	14.2 ± 19.4	5.7 ± 8.0	7.5 ± 8.1	27.7 ± 31.2	0.85 ± 0.88
	Pro + AA		349 ± 173	4.5 ± 1.7	0.3 ± 0.2	3.0 ± 2.1	1.6 ± 2.2	3.5 ± 1.6	8.5 ± 4.0	1.01 ± 0.69
	p value[c]			< 0.01		< 0.001	< 0.05	< 0.05	< 0.01	
Nakornratchasima (Korat)		21								
	Pro		609 ± 266	1.1 ± 0.6	0.2 ± 0.2	2.9 ± 2.5	3.2 ± 5.8	5.5 ± 4.9	11.8 ± 10.9	1.14 ± 0.57
	Pro + AA		537 ± 234	2.1 ± 1.3	0.2 ± 0.3	1.8 ± 1.4	0.9 ± 1.9	3.4 ± 2.7	6.4 ± 4.1	1.59 ± 1.16
	p value			< 0.001			< 0.05		< 0.05	
Chiang Mai		20								
	Pro		705 ± 330	5.0 ± 1.7	0.9 ± 1.6	4.2 ± 3.5	5.4 ± 5.2	24.8 ± 28.6	35.3 ± 31.5	0.86 ± 0.79
	Pro + AA		612 ± 391	5.1 ± 2.0	0.5 ± 1.1	11.5 ± 20.2	2.5 ± 3.6	18.9 ± 21.6	33.4 ± 37.9	0.88 ± 0.58
	p value		< 0.05				< 0.05			
Bangkok		20								
	Pro		654 ± 318	2.8 ± 3.1	0.5 ± 0.8	6.3 ± 8.3	6.0 ± 14.9	9.5 ± 10.3	22.3 ± 31.5	1.03 ± 0.77
	Pro + AA		519 ± 256	3.0 ± 3.4	0.9 ± 2.0	5.3 ± 10.3	4.3 ± 6.6	10.0 ± 11.3	20.5 ± 20.8	0.94 ± 0.60
	p value		< 0.05							
Songkhla		18								
	Pro		996 ± 396	4.8 ± 2.3	2.1 ± 3.2	9.0 ± 14.8	23.7 ± 51.1	79.8 ± 95.8	109.4 ± 143.3	0.93 ± 0.90
	Pro + AA		561 ± 267	4.3 ± 2.2	0.9 ± 1.7	1.3 ± 2.0	3.4 ± 8.2	11.7 ± 19.3	17.4 ± 27.7	0.57 ± 0.45
	p value		< 0.001			< 0.01	< 0.01	< 0.01	< 0.001	

[a] Pro, 12-h overnight urine after ingestion of 500 mg proline 1 h after evening meal; Pro + AA, 12-h overnight urine after ingestion of 500 mg proline + 200 mg ascorbic acid 1 h after evening meal
[b] NSAR, N-nitrososarcosine; NPRO, N-nitrosoproline; NMTCA, N-nitroso-2-methylthiazolidine 4-carboxylic acid; NTCA, N-nitrosothiazolidine 4-carboxylic acid
[c] Statistical comparison of urine samples after ingestion of proline and after ingestion of proline plus ascorbic acid using Wilcoxon signed-ranks test; p value indicated only when the difference is significant.

Table 3. Urinary levels of N-nitrosamino acid and nitrate in subjects in high-risk areas for liver cancer who were seronegative or seropositive for antibodies to *Opisthorchis viverrini* (OV) (mean ± SD)

Subjects	Urine sample[a]	No. of subjects[b]	Volume of 12-h urine (ml)	Creatinine (mmol/12 h)	N-Nitrosamino acid (μg/12 h)[c]					Nitrate (mmol/12 h)
					NSAR	NPRO	NMTCA	NTCA	Sum	
Seronegative for OV antibody (titre < 1:40)	Pro	18	603 ± 285	1.2 ± 1.0	0.3 ± 0.4	3.5 ± 3.2	2.4 ± 4.4	4.5 ± 3.9	10.6 ± 8.6	0.98 ± 0.48
	Pro + AA		518 ± 233	2.5 ± 1.4	0.2 ± 0.3	2.5 ± 1.6	1.5 ± 2.7	3.7 ± 2.7	7.8 ± 4.4	1.57 ± 1.15
	p value[d]			< 0.001						
Positive for OV antibody (titre ≥ 1:40)	Pro	23	426 ± 195	3.6 ± 2.5	0.2 ± 0.2	12.3 ± 18.7	6.1 ± 8.2	8.0 ± 8.0	26.5 ± 29.9	1.01 ± 0.91
	Pro + AA		388 ± 205	3.8 ± 2.1	0.3 ± 0.3	2.4 ± 2.0	1.1 ± 1.4	3.3 ± 1.7	7.1 ± 4.0	1.10 ± 0.81
	p value[d]			< 0.05		< 0.001	< 0.01	< 0.01	< 0.001	
p value[e]										
Pro			< 0.05	< 0.001	NS	< 0.05	< 0.05	NS	< 0.05	NS
Pro + AA			< 0.05	< 0.05	< 0.05	NS	NS	NS	NS	NS

[a] Pro, 12-h overnight urine after ingestion of 500 mg proline 1 h after evening meal.; Pro ± AA, 12-h overnight urine after ingestion of 500 mg proline plus 200 mg ascorbic acid 1 h after evening meal
[b] Seven of 18 seronegative and 10/23 seropositive subjects smoked cigarettes on the day of urine collection.
[c] NSAR, N-nitrososarcosine; NPRO, N-nitrosoproline; NMTCA, N-nitroso-2-methylthiazolidine 4-carboxylic acid; NTCA, N-nitrosothiazolidine 4-carboxylic acid
[d] Statistical comparison of urine samples after ingestion of proline and after ingestion of proline plus ascorbic acid, using Wilcoxon signed-ranks test
[e] Statistical comparison of urine samples of individuals positive for OV antibody versus those of individuals negative for OV antibody, after ingestion of proline and after ingestion of proline plus ascorbic acid, using the Mann-Whitney test

Table 4. Volatile N-nitrosamines[a] in fermented Thai foods

Fermented food	No. of samples	NDMA			NPIP			NPYR		
		Mean ± SD (µg/kg)	Range (µg/kg)	No. of +ve samples	Mean ± SD (µg/kg)	Range (µg/kg)	No. of +ve samples	Mean ± SD (µg/kg)	Range (µg/kg)	No. of +ve samples
Fish[b]	15	3.8 ± 7.3	0–25.5	8	2.3 ± 6.4	0–23.0	3	21.1 ± 46.6	0–177	8
Pork[c]	9	1.2 ± 2.0	0–6.5	6	5.7		1	2.9 ± 7.0	0–21.4	4
Vegetables[d]	4		0–0.5	2	–				0–62	2

[a] NDMA, N-nitrosomethylamine; NPIP, N-nitrosopiperidine; NPYR, N-nitrosopyrrolidine
[b] Pla-ra, pla-chom, pla-som
[c] Nam, Thai sausage
[d] Puk-dong

way in a Syrian golden hamster model (Thamavit et al., 1988) with experimentally induced OV infection.

Acknowledgements

This study was undertaken during the tenure of an International Cancer Research Technology Transfer Programme (ICRETT) fellowship awarded to P.S. by the International Union Against Cancer. We thank Mrs E. Bayle for secretarial work.

References

Armitage P. & Berry, G. (1987) *Statistical Methods in Medical Research*, Oxford, Blackwell

Fine, D.H. (1978) Method 5 – Determination of volatile N-nitrosamines in food by chemiluminescence using the thermal energy analyser. In: Preussmann, R., Castegnaro, M., Walker, E.A. & Wassermann, A.E., eds. *Environmental Carcinogens: Selected Methods of Analysis*, Vol. 1, *Analysis of Volatile Nitrosamines in Food* (IARC Scientific Publications No. 18), Lyon, IARC, pp. 133-140

Harinasuta, C. & Vajrasthira, S. (1960) *Opisthorchiasis* in Thailand. *Ann. Trop. Med. Parasitol.*, 54, 100-105

Ohshima, H. & Bartsch, H. (1988) Urinary N-nitrosamino acids as an index of exposure to N-nitroso compounds. In: Bartsch, H., Hemminki, K. & O'Neill, I.K., eds., *Methods for Detecting DNA Damaging Agents in Humans: Applications in Cancer Epidemiology and Prevention* (IARC Scientific Publications No. 89), Lyon, IARC, pp. 83-91

Ohshima, H., Calmels, H., Pignatelli, B., Vincent, P. & Bartsch, H. (1987) N-Nitrosamine formation in urinary-tract infections. In: Bartsch, H., O'Neill, I.K. & Schulte-Hermann, R., eds, *The Relevance of N-Nitroso Compounds to Human Cancer: Exposures and Mechanisms* (IARC Scientific Publications No. 84), Lyon, IAReC, pp. 384-390

Ritchie, L.S. (1948) An ether sedimentation technique for routine stool examination. *Bull. US Army Med. Dep.*, 8, 326-331

Sadun, E.H., (1955) Studies in *Opisthorchis viverrini* in Thailand. *Am. J. Hyg.*, 62, 81-115

Schwartz D.A. (1980) Helminths in the induction of cancer: *Opisthorchis viverrini*, *Clonorchis sinensis* and cholangiocarcinoma. *Trop. Geogr. Med.*, 32, 95-100

Srianujata, S., Tonbuth, S. & Bunyaratvej, S. (1987) High urinary excretion of nitrate and N-nitrosoproline in *Opisthorchiasis* subjects. In: Bartsch, H., O'Neill, I.K. & Schulte-Hermann, R., eds, *The Relevance of N-Nitroso Compounds to Human Cancer: Exposures and Mechanisms* (IARC Scientific Publications No. 84), Lyon, IARC, pp. 544-546

Srivatanakul, P., Viyanant, V., Kurathong, S. & Tiwawech, D. (1985) Enzyme-linked immunosorbent assay for detection of *Opisthorchis viverrini* infection. *Southeast Asian J. Trop. Med. Public. Health*, 16, 234-239

Srivatanakul, P., Sontipong, S., Chotiwan, P. & Parkin D.M. (1988) Liver cancer in Thailand: temporal and geographic variations. *J. Gastroenterol. Hepatol.*, 3, 413-420

Stoll, N.R. (1923) An effective method of counting hookworm eggs in feces. *Am. J. Hyg.*, 3, 59-70

Thamavit, W., Bhamarapravati, N., Sahaphong, S., Vajrasthira, S. & Angsubhakorn, S. (1978) Effect of dimethylnitrosamine on induction of cholangiocarcinoma in *Opisthorchis viverrini*-infected Syrian golden hamsters. *Cancer Res.*, 38, 4634-4639

Thamavit, W., Moore, M.A., Hiasa, Y. & Ito, N. (1988) Generation of high yields of Syrian hamster cholangiocellular carcinomas and hepatocellular nodules by combined nitrite and aminopyrine administration and *Opisthorchis viverrini* infection. *Jpn. J. Cancer Res.*, 79, 906-916

Upatham, E.S., Viyanant, V., Kurathong, S., Rojborwonwitaya, J., Brockelman, W.Y., Ardsungnoen, S., Lee, P. & Vajrasthira, S. (1984) Relationship between prevalence and intensity of *Opisthorchis viverrini* infection, and clinical symptoms and signs in a rural community in northeast Thailand. *Bull. World Health Organ.*, 62, 451-461

POSSIBLE EFFECT OF INFECTION WITH LIVER FLUKE (*OPISTHORCHIS VIVERRINI*) ON THE MONITORING OF URINE BY ENZYME-LINKED IMMUNOSORBENT ASSAY FOR HUMAN EXPOSURE TO AFLATOXINS

K. Makarananda[1,3,] C.P. Wild[2], Y.Z. Jiang[2] & G.E. Neal[1,4]

[1]*Medical Research Council Laboratories, Toxicology Unit, Carshalton, Surrey, UK; and*
[2]*International Agency for Research on Cancer, Lyon, France*

Several laboratories have initiated studies to assess human exposure to aflatoxin at an individual level by measuring aflatoxin metabolites in the urine by immunoassay. The fact that the antibodies recognize a variety of metabolites, albeit with differing affinities, means that any environmental factor that modifies the pattern of urinary metabolites associated with a given exposure could affect quantification in immunoassay. We have examined two such possible effects: (i) the pattern of metabolites after a dose of ^{14}C-aflatoxin B_1 in rats and (ii) the pattern of metabolites in hamsters and humans with and without exposure to liver flukes. We found no dose-related effect on the pattern of urinary metabolites over a 250-fold range, but there was a significant increase in the proportion of water-soluble aflatoxin metabolites in hamsters infected with liver fluke over that in uninfected animals. In human urine samples, there also appeared to be a difference in the metabolites in individuals infected with liver fluke from those in uninfected persons. These observations are relevant to both mechanistic and monitoring aspects of research into aflatoxins.

Recent attention has been focused on measuring individual human exposure to hepatocarcinogenic aflatoxins (AF) by direct analysis of body fluids. The enzyme-linked immunosorbent assay (ELISA) appears to offer a suitable method for use in epidemiological studies for monitoring short-term exposure (one to two days) to AF, as it has the appropriate sensitivity and specificity. However, the presence of substances that are presumably not AF and which are inhibitiory in the ELISA system (e.g., Dragsted *et al.*, 1986) has necessitated the development of purification techniques, usually based on absorption onto Sep-Pak C18 cartridges and antibody affinity columns (Groopman *et al.*, 1986; 1988). In addition, because the assays are based on the abilities of the antibodies

---

[3]Present address: Department of Pharmacology, Faculty of Science, Mahidol University, Bangkok, Thailand
[4]To whom correspondence should be addressed

to recognize a range of AF metabolites, any factor that affects the spectrum of metabolites present in the urine could influence the results of affinity purification and ELISA (Groopman et al., 1986; Wild et al., 1986). Two such possible confounding factors have been considered in the work presented here: (i) the dose of AF and (ii) infection with liver fluke (a possible factor in the development of cholangiocarcinoma; Srivatanakul et al., 1988).

Urinary analysis of aflatoxins

At the IARC, urine samples (25 ml) were eluted from Sep-Pak C18 cartridges with 80% methanol and air dried. A 5-ml equivalent of urine was diluted to 50 ml in phosphate-buffered saline (PBS) and loaded onto a monoclonal antibody affinity column (Easi-extract, Oxoid). Bound AF were eluted with 2 ml of 80% methanol and, after drying, were reconstituted in 0.5 ml PBS and diluted 1:25 in PBS for ELISA. The ELISA was performed using polyclonal antibodies as described previously (Wild et al., 1987), except that a preincubation step of 1 h with 100 µl of 0.1% ovalbumin per ELISA well was employed.

At the Medical Research Council (MRC) laboratories, a 1-ml equivalent of urine eluted from Sep-pak C18 cartridges was diluted to 5 ml in PBS prior to loading onto affinity columns (antiserum-AH-Sepharose 4B). Columns were washed with 15 ml PBS, and the AF were eluted with 15 ml of 85% methanol in PBS. The first 10 ml of eluate were collected, evaporated to 0.5 ml and adjusted to 1 ml with PBS for ELISA (Makarananda & Neal, 1990). The immunogen used to raise the antibodies employed in these studies was quail hepatic microsomal protein, to which AFB_1 had been adducted by incubation of AFB_1, quail microsomes and an NADPH-generating system *in vitro*. In both affinity purifications, 50–60% of total urinary metabolites were extracted in experimental studies in rats (see Table 1 and C. Wild, unpublished data).

Animal studies

Liver fluke metacercariae were obtained from naturally infected fish in Thailand by peptic digestion. They were transported in chilled saline to the MRC, where they were administered intragastrically (30 metacercariae per animal) to female golden Syrian hamsters (eight weeks old). Successful infection was checked by counting eggs in the stools six weeks after the hamsters had been infected. Groups were injected intraperitoneally with ^{14}C-AFB_1 (5 mg/5 µCi per kg body weight in dimethyl sulfoxide) or with the vehicle alone. Urine samples were collected over a 24-h period both before and after dosing with AFB_1. Samples were stored at –40°C before analysis.

Male Wistar rats (220–240 g) were treated with ^{14}C-AFB_1 (0.13 µCi per rat) by gavage at doses ranging from 3 to 800 µg/kg body weight in olive oil. Animals were housed in metabolic cages, and 24-h urine samples were collected.

Possible influence of level of exposure on excretion rate and metabolite pattern in urine

In the rats, a constant percentage of the AF dose was excreted in the urine within 24 h, and the distribution of radioactive metabolites, on HPLC (Moss et al., 1985), did not alter

Table 1. Urinary excretion of aflatoxin metabolites by rats during the 24 h following dosing with a range of levels of ^{14}C-aflatoxin B_1 (AFB$_1$)[a]

Dose of AFB$_1$ (μg/kg body weight)	Radioactivity		ELISA of urinary aflatoxins (% of value determined by radioactivity)
	Urinary excretion (% of dose)	HPLC (% polar/ nonpolar)	
3	6.4		52
3	7.2	70.3/29.7	44
10	6.1		42
30	5.6		29
30	8.0	65.9/34.1	38
30	5.7		34
100	6.7		58
100	7.4		44
300	7.6	70.1/29.9	66
300	6.6		97
300	6.6		71
800	8.9		63
	Mean, 6.9 ± 1.0		

[a] Abbreviations: HPLC, high-performance liquid chromatography; ELISA, enzyme-linked immunosorbent assay

significantly over this range of dosing (Table 1). The shorter retention time (< 6 min) fraction on HPLC is referred to as the 'polar fraction', and that with a longer retention time (> 7 min) is referred to as the 'nonpolar fraction'. The nonpolar fraction is composed mainly of unconjugated AFM$_1$, whereas the constituents of the polar metabolite fraction are largely uncharacterized. The ELISA was sensitive enough to detect both the polar and nonpolar metabolites, which were also quantified by counting radioactivity.

Possible influence of liver fluke on the pattern of aflatoxin metabolites in hamster and human urine

Control and *Opisthorchis viverrini*-infected hamsters (Flavell & Lucas, 1983) were injected intraperitoneally with ^{14}C-AFB$_1$. Infection with liver fluke induced a change in the balance between polar and nonpolar metabolites; the results were consistent for each of the two groups (Figure 1).

AF present in 18 individual human urine samples collected from two regions of Thailand (Wild *et al.*, 1989) — one a region of high incidence of liver fluke infection (13 samples) and the other an area without infection (five samples) — were analysed by ELISA at both the IARC and the MRC (Figure 2). Although the number of individuals examined is small, it can be seen clearly that the IARC assays indicated higher levels in some individual samples from the area with low exposure to liver fluke, whereas the MRC assays generally indicated higher levels in the high-exposure area. When AF fractions purified on affinity columns in one laboratory were sent to the other for ELISA, the lower histograms were obtained: essentially, the original distributions obtained in the two laboratories were reversed. This finding indicates that, while the two antibodies used to quantify AF in ELISA had similar specificities and gave a good correlation (see below), the antibodies used to purify AF from the urine samples differed. These results are of interest in relation to both the mechanistic aspects of AF carcinogenesis and the use of urine to monitor human exposure.

Discussion

In the studies in Wistar rats, no significant change in the proportion of AFM_1 (a product of hydroxylation) or of AF polar metabolites (products of epoxidation) excreted in the urine was observed over a 250-fold dose range of AFB_1, suggesting that no dose-dependent shift in metabolism had occurred. Thus, large variations in individual human exposure are unlikely to lead to erroneous comparisons when exposure is measured by urinary analysis.

Figure 1. High-performance liquid chromatography (HPLC) fractionation of urines obtained from (A) two liver fluke-infected and (B) two uninfected hamsters during 24 h following administration of ^{14}C-aflatoxin B_1[a]

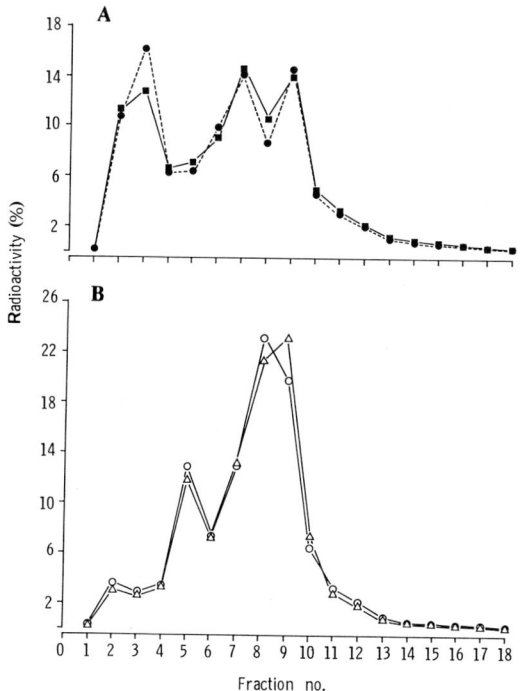

[a] Aliquots (10 μl) of urine samples were analysed by HPLC using a reverse-phase 5 ODS column (100 x 4.6 mm) and a linear 15–40% methanol/water gradient containing 0.01 phosphoric acid and 8% acetonitrile at a flow rate of 1.2 ml/min. Fractions (1 min) were collected and radioactivity determined by liquid scintillation counting. Fractions obtained from HPLC separations of urine (20 μl) from animals treated with ^{14}C-aflatoxin B_1 were analysed by both enzyme-linked immunosorbent assay (ELISA) and radioactive counting after evaporation to dryness and reconstitution in phosphate-buffered saline (see Table 1).

Figure 2. Quantification in two laboratories of aflatoxins present in human urine samples from areas of Thailand with high and low incidences of liver fluke infection[a]

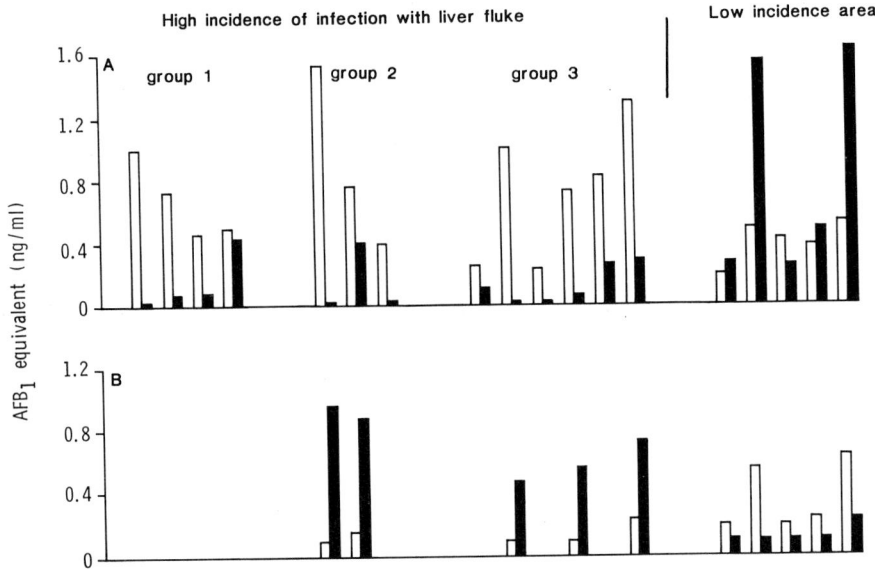

[a]A, Affinity column and enzyme-linked immunosorbent assay (ELISA) performed at IARC, filled bars; affinity column and ELISA performed at MRC, open bars. B, Affinity column at IARC and ELISA at MRC, open bars; affinity column at MRC and ELISA at IARC, filled bars. AFB_1, aflatoxin B_1

Each pair of bars represents samples from one individual analysed at the IARC and at the Medical Research Council (MRC) laboratories. The individuals in panel B are the same as those directly above in panel A. Individual exposure to liver fluke was assessed by counting eggs in the faeces and/or by the presence of antibodies to the fluke: group 1, exposure positive by both methods; group 2, exposure positive only by presence of antibodies; group 3, exposure negative by both methods.

The possible interaction between AF and other risk factors for liver cancer, such as liver fluke infestation and hepatitis B virus, is important, but little experimental information is available as yet (see Harris & Sun, 1984). We found an effect on the pattern of metabolites excreted in hamsters with and without liver fluke infestation (Figure 1). The increased proportion of polar metabolites may indicate increased activation of AFB_1 to the reactive epoxide, as opposed to hydroxylation to less carcinogenic nonpolar metabolites such as AFM_1. Clearly, detailed characterization of the various metabolites is required before any conclusion can be drawn about the possible significance of these observations in terms of AF carcinogenesis. However, if such changes in excretion pattern occur in man, and antibodies with different specificities are used to quantify these urinary metabolites, different estimates of exposure can result, as shown in our preliminary studies on human urine samples from Thailand. We believe that this is a real difference due to antibody specificity and not lack of reliability of the assay, since when affinity column-purified fractions prepared at the MRC were analysed at the IARC (and *vice versa*), quantification was similar: in comparing the IARC data in Figure 2A with the MRC data in Figure 2B and the MRC data in Figure 2A with the IARC data in Figure 2B, a highly significant correlation was seen between the two laboratories. Thus, although the antibodies used in the ELISA in the two laboratories gave similar results, the affinity column used at the MRC probably extracts a metabolite that is not extracted by the commercial column used by the IARC but which was nevertheless quantified by the polyclonal antibody used in the ELISA.

Overall, these results suggest that in order to make quantitative comparisons between studies, the metabolites being examined must be defined carefully on the basis of antibody specificity. Even within a study, using the same antibody, it is important to be aware that individual variation in total urinary AF levels, as quantified by immunoassay, may be influenced by metabolite pattern.

Acknowledgements

The authors appreciate the excellent technical assistance of B. Chapot, the assistance of D. Judah in carrying out some of the HPLC analyses and that of R. Legg for the studies on hamsters. The urine samples were collected by Dr P. Srivatanakul, National Cancer Institute, Bangkok, Thailand. YZJ acknowledges support of a fellowship from l'Association pour la Recherche sur le Cancer, Paris.

References

Dragsted, L.O., Bull, I. & Autrup, H. (1986) Substances with affinity to a monoclonal aflatoxin B_1 antibody in Danish urine samples. *Food Chem. Toxicol.*, **26**, 233-242

Flavell, D.J. & Lucas, S.B. (1983) Promotion of N-nitrosodimethylamine-initiated bile duct carcinogenesis in the hamster by the human liver fluke, Opisthorchis viverrini. *Carcinogenesis*, **4**, 927-930

Groopman, J.D., Donahue, P.R., Zhu, J.Q., Chen, J.S. & Wogan, G.N. (1986) Aflatoxin metabolism in human: detection of metabolites and nucleic acid adduct in urine by affinity chromatography. *Proc. Natl Acad. Sci. USA*, **82**, 6492-6496

Harris, C.C. & Sun, T.T. (1984) Multifactorial etiology of human liver cancer. *Carcinogenesis*, **5**, 691-701

Makarananda, K. & Neal, G.E. (1990) ELISA of aflatoxins. In *Methods in Molecular Biology*, Clifton, NJ, Humana Press (in press)

Moss, E.J., Neal, G.E. & Judah, D.J. (1985) The mercapturic acid pathway metabolites of a glutathione conjugate of aflatoxin B_1. *Chem.-Biol. Interact.*, **55**, 138-155

Srivatanakul, P., Sontipong, S., Chotiwan, P. & Parkin, D.M. (1988) Liver cancer in Thailand. Temporal and geographic variations. *J. Gastroenterol. Hepatol.*, **3**, 413-420

Wild, C.P., Umbenhauer, D., Chapot, B. & Montesano, R. (1986) Monitoring of individual human exposure to aflatoxins (AF) and N-nitrosamines (NNO) by immunoassays. *J. Cell Biochem.*, **24**, 171-179

Wild, C.P., Pionneau, F., Montesano, R., Mutiro, C.F. & Chetsanga, C.J. (1987) Aflatoxin detected in human breast milk by immunoassay. *Int. J. Cancer*, **40**, 328-333

Wild, C.P., Chapot, B., Scherer, E., Den Engelse, L. & Montesano, R. (1988) The application of antibody methodologies to the detection of aflatoxin in human body fluids. In: Bartsch, H., Hemminki, K. & O'Neill, I.K., eds, *Methods for Detection of DNA Damaging Agents in Humans: Applications in Cancer Epidemiology and Prevention* (IARC Scientific Publications No. 89), Lyon, IARC, pp. 67-74

Wild, C.P., Jiang, Y.Z., Montesano, R., Parkin, M., Khlat, M. & Srivatanakul, P. (1989) Correlation study of aflatoxin exposure and liver cancer incidence in five geographical regions of Thailand. *Proc. Am. Assoc. Cancer Res.*, **30**, 317

Relevance to Human Cancer of *N*-Nitroso Compounds,
Tobacco Smoke and Mycotoxins.
Ed. I.K. O'Neill, J. Chen and H. Bartsch
Lyon, International Agency for Research on Cancer
© IARC, 1991

A RAPID GAS CHROMATOGRAPHY-MASS SPECTROMETRY METHOD FOR THE DETERMINATION OF URINARY 3-METHYLADENINE: APPLICATION IN HUMAN SUBJECTS

D.E.G. Shuker, M.D. Friesen, L. Garren & V. Prévost

International Agency for Research on Cancer, Lyon, France

A rapid gas chromatography–mass spectrometry method, employing immunoaffinity clean-up, has been developed for the measurement of 3-methyladenine in human urine samples. A wide variation in levels of urinary 3-methyladenine was observed, indicating that at least some may be derived from the diet and not be related to endogenous nitrosation and subsequent methylation.

The determination of urinary 3-methyladenine (3-meAde) has been proposed as a noninvasive method for biomonitoring DNA methylation, including that derived from *N*-nitroso compounds (Shuker *et al.*, 1987a). The previously available method suffered from the disadvantage that it required lengthy sample clean-up. The recent development of a polyclonal antiserum to 3-meAde offered the possibility of quantification by enzyme-linked immunoabsorbent assay (ELISA) without the need for gas chromatography-mass spectrometry (GC-MS). However, the presence of cross-reactive purines in human urine rendered this approach difficult, despite extensive clean-up (Shuker & Farmer, 1988). An approach that proved successful is the combination of easily used immunoaffinity columns with low-resolution GC-MS; recent results are summarized in this paper.

Immunoaffinity columns for 3-meAde

IgG-Protein A Sepharose CL4B gel was prepared essentially according to the procedure described by Goding (1986), using an ammonium sulfate-precipitated IgG fraction from 3-meAde rabbit antiserum prepared as described previously (Shuker & Farmer, 1988). Full experimental details will be described elsewhere (Friesen *et al.*, 1991).

Aliquots (1 ml) of the antibody-modified gel were used to prepare minicolumns. The capacity of the columns was determined by elution of increasing amounts of 3-meAde (0–250 ng) containing 3H-3-meAde (200 pg, 1300 dpm). The minicolumns were found to retain quantitatively up to 100 ng 3-meAde. With human urine samples, containing 50–100 ng there was lower recovery, but this could be raised by the simple expedient of running the columns at 4°C. This phenomenon is similar to that seen with ELISA protocols (Shuker & Farmer, 1988) and is due to the increased binding of hapten to antibodies at low temperature.

The antibody-modified gel was prepared on a batch basis, and 5–20 columns were prepared at a time. This ensured that performance was consistent from column to column; quality control checks on different batches of columns confirmed that the affinity properties were reproducible. In addition, columns were found to be reusable, and several columns have been recycled up to 30 times with no apparent loss of efficiency. No detectable carry-over was found from one sample to another eluted through the same column.

GC-MS analysis of 3-meAde

An existing method for the analysis of 3-meAde by GC-MS (Shuker *et al.*, 1987a) was modified for use with a low-resolution, bench-top GC-MS system. Urine samples, which were cleaned-up by elution through the minicolumns, derivatized and injected onto the GC-MS, gave single-ion chromatograms which were free of interfering peaks (Figure 1). Quantification was carried out by use of d_3-3-meAde (50 ng/2 ml urine) as internal standard. It has been established that d_3-3-meAde is recognized by the antibody just as well as unlabelled 3-meAde (data not shown).

Figure 1. Selected ion monitoring trace of an extract of human urine purified by immunoaffinity chromatography[a]

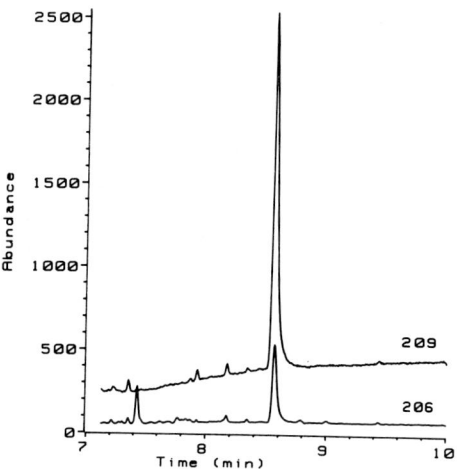

[a]3-Methyladenine, *tert*-butyldimethylsilyl derivative, $m/z = 206$ ($M^+ -57$); d_3-3-methyladenine, *tert*-butyldimethyl derivative, $m/z = 209$ ($M^+ - 57$) (internal standard)

The protocol that has been developed for the analysis of 3-meAde in humans (and also for rat and pig urine) is shown in Figure 2.

Analysis of 3-meAde in human urine samples

Good reproducibility of the immunoaffinity GC-MS analysis was observed when analysis of several aliquots of a human urine sample was undertaken (3-meAde = 80 ng/2 ml; SD = 3 ng/2 ml; n = 4).

Figure 2. Schematic summary of the immunoaffinity–gas chromatography–mass spectrometry (GC-MS) analysis of urinary 3-methyladenine

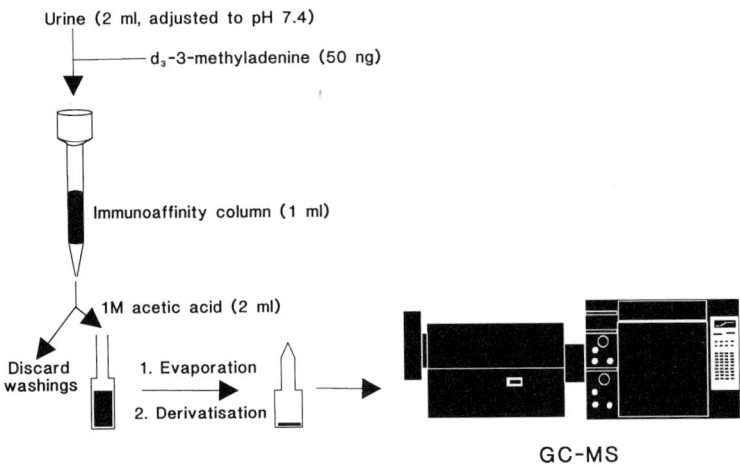

As part of larger collaborative projects on the etiology of oesophageal cancer, analyses have been undertaken of 3-meAde in human urine samples. The details of the projects are given elsewhere in this volume (Lu *et al*; Chang-Claude *et al.*); therefore, only a summary of the results will be given.

3-MeAde in subjects living in two counties in China (in collaboration with Dr S.H. Lu, Chinese Academy of Medical Sciences, Beijing, China)

Overnight urine samples (n = 43) were collected from subjects living in two counties in China, namely Linxian (n = 21) and Yuxian (n = 22), with differing incidences of oesophageal cancer. There was a slight difference in the mean values of 3-meAde when expressed as concentrations (Linxian, 3.53 ng/2 ml; SD = 3.83, n = 21; Yuxian, 6.93 ng/2 ml; SD = 4.15; n = 22), with a large variation in individual levels. When the results were expressed as a function of urinary creatinine, the results were 12.33 ($\pm$ 14.10) ng 3-meAde/mg creatinine (Linxian) and 12.99 ($\pm$ 14.00) ng 3-meAde/mg creatinine (Yuxian). Thus, the apparent difference in 3-meAde levels was due to differences in concentration.

3-MeAde in adolescent subjects from families with cases of oesophageal cancer (in collaboration with Dr J. Wahrendorf and Dr J. Chang-Claude, Deutsches Krebsforschungzentrum, Heidelberg, Federal Republic of Germany)

Overnight urine samples (n = 30) were collected from adolescents living in households where another member of the household had been diagnosed as having oesophageal cancer. Control subjects (n = 53) were matched for several factors. The results of the 3-meAde analyses are summarized in Table 1. Only slight differences in the mean levels of 3-meAde were observed when different groups (cases and controls; smokers and nonsmokers) were compared. However, as we noted for the previous study, large variations in 3-meAde levels were observed between individuals within the various groups.

Table 1. Mean urinary levels of 3-methyladenine in adolescent Chinese males selected from households having, or not, a previous case of oesophageal cancer. The same group (n = 83) was also divided into smokers and nonsmokers

Group	No. of subjects	3-Methyladenine in urine (μg/24 h; mean ± SD)
Smokers	54	5.77 ± 5.48
Nonsmokers	29	6.83 ± 7.04
Case household	30	7.25 ± 5.90
Control household	53	5.50 ± 6.10

Discussion

The immunoaffinity–GC-MS method for the analysis of urinary 3-meAde has many of the attributes required for a biomonitoring method suitable for use in epidemiological studies. It is capable of producing analytically reliable results with large sample through-put and requires small volumes of urine. In the two studies described in this paper, 126 individual urine samples were analysed; however, no striking difference in urinary levels of 3-meAde was seen with respect to parameters such as smoking status.

In order to validate thoroughly the usefulness of 3-meAde as a marker of DNA methylation, more work is now required — in particular, to determine whether this marker is derived in part from the diet. In animals, it is known that 3-meAde is excreted unchanged following oral administration (Shuker *et al.*, 1987b), and we have recently found that laboratory animal diets and human foods contain low levels of 3-meAde (D.E.G. Shuker & V. Prevost, unpublished results). Furthermore, recently published work demonstrates that 3-meAde is formed in DNA of edible plants following treatment with the widely used fumigant, methyl bromide (Starratt & Bond, 1988). If the diet proves to be a major source of 3-meAde, this may obscure the variations in DNA-derived urinary 3-meAde.

In conclusion, while the practicality of 3-meAde measurements in epidemiological studies has been demonstrated, the interpretation of the results requires further refinement.

Acknowledgements

It is a pleasure to acknowledge the continuing interest and encouragement of Dr H. Bartsch (Chief, Unit of Environmental Carcinogens and Host Factors, IARC) in this project. D.E.G.S. also acknowledges the support from an IARC Visiting Scientist Award and more recently from the US National Cancer Institute (Grant No. CA48473). The assistance of Mr K.-U. Henss is greatly appreciated.

References

Friesen, M.D., Garren, L., Prévost, V. & Shuker, D.E.G. (1991) Isolation of urinary 3-methyladenine using immunoaffinity columns prior to determination by low-resolution gas chromatography-mass spectrometry. *Chem. Res. Toxicol.* (in press)

Goding, J.W. (1986) *Monoclonal Antibodies: Principles and Practice*, London, Academic Press, pp. 226-227

Shuker, D.E.G. & Farmer, P.B. (1988) Urinary secretion of 3-methyladenine in humans as a marker of nucleic acid methylation. In: Bartsch, H., Hemminki, K. & O'Neill, I.K., eds, *Methods for Detecting DNA Damaging Agents in Humans: Applications in Cancer Epidemiology and Prevention* (IARC Scientific Publications No. 89), Lyon, IARC, pp. 92-96

Shuker, D.E.G., Bailey, E., Parry, A., Lamb, J. & Farmer, P.B. (1987a) The determination of urinary 3-methyladenine in humans as a potential monitor of exposure to methylating agents. *Carcinogenesis*, 8, 959-962

Shuker, D.E.G., Bailey, E. & Farmer, P.B. (1987b) Excretion of methylated nucleic acid bases as an indicator of exposure to nitrosatable drugs. In: Bartsch, H., O'Neill, I.K. & Schulte-Hermann, R., eds, *The Relevance of N-Nitroso Compounds to Human Cancer: Exposures and Mechanisms* (IARC Scientific Publications No. 84), Lyon, IARC, pp. 407-410

Starratt, A.N. & Bond, E.J. (1988) Methylation of DNA of maize and wheat grains during fumigation with methyl bromide. *J. Agric. Food Chem.*, 36, 1035-1039

TANDEM MASS SPECTROMETRIC APPROACHES FOR DETERMINING EXPOSURE TO ALKYLATING AGENTS

J.R. Cushnir, J.H. Lamb, A. Parry & P.B. Farmer

MRC Toxicology Unit, Carshalton, Surrey, UK

Tandem mass spectrometric techniques have been developed for analysing alkylated purines in human urine. These techniques are designed with the objectives firstly of identifying unknown adducts derived from mixed exposures to alkylating agents, and secondly of quantifying these adducts with minimal sample purification.

Exposure to alkylating agents *in vivo* results in the formation of their covalently bound adducts with nucleophilic sites in nucleic acids and proteins. A wide variety of analytical methods has now been developed for identifying these adducts in order to use them as monitors of human exposure to alkylating agents (see review by Farmer *et al.*, this volume). These methods have usually been designed to be specific for a particular adduct, or group of chemically closely related adducts, and have proved of considerable value for monitoring exposure in cases where the nature of the alkylating agent to which exposure has occurred is known (Ehrenberg & Osterman-Golkar, 1980; Farmer *et al.*, 1987). Methods are still needed, however, to identify (and ideally to quantify) alkylating agent adducts in situations where the structure is *a priori* unknown (e.g., as a result of mixed exposures due to urban pollution). We describe in this paper our current attempts to use tandem mass spectrometry (MS-MS) for this purpose.

MS-MS can be used to analyse components of complex mixtures through modes of operation which yield either 'daughter ion' or 'parent ion' spectra. To obtain a daughter ion spectrum, an ion from the component of interest (parent ion) is resolved from other ions by the first mass spectrometer and is then allowed to collide with a gas (collision gas), which causes fragmentation of the ion. The resulting daughter ions are analysed by the second mass spectrometer. This technique allows structural information to be obtained on minor components, even in cases where the sample is impure and extremely complex (e.g., urine). Quantification should also be possible using isotopically labelled internal standards and selected ion monitoring techniques.

To obtain a parent ion spectrum, the second mass spectrometer is set to transmit a particular daughter ion. Scanning the first mass spectrometer then gives information on all of the precursors of this ion. With suitable choice of daughter ion, this technique may be used to identify all the compounds that bear a common functional group which are present in a mixture. We are exploring the use of parent and daughter ion spectroscopy to analyse alkylated nucleic acid bases and protein amino acids. The examples given here are of alkylated purines, which can be detected in urine after their hydrolytic release from alkylating agent-modified nucleic acids.

Instrumentation

Spectra were obtained using a VG 70-SEQ tandem instrument, linked with a Hewlett Packard Model 5890A gas chromatograph. Chromatography was on a fused silica capillary column (25 m × 0.32 mm) coated with OV-1, using a temperature programme of 80°C for 1 min followed by a 30°C/min rise to 285°C. Samples were introduced by split injection. Compounds were ionized by electron impact (70 eV) with accelerating voltage 8 kV. Ions were separated by the electric and magnetic resolving sectors at mass resolution 1000 and allowed to collide with air prior to quadrupole mass analysis (unit mass resolution) of their daughter ions. Spectra were processed using a VG 11/250 Data System. Selective ion recording was achieved using an 80-msec dwell time for each ion.

Human excretion of 3-methyladenine

The measurement of 3-methyladenine excretion in human urine has been proposed for monitoring exposure to methylating agents, and a gas-chromatography-MS method was originally developed for this purpose (Shuker et al., 1987). Routine use of the method has proved rather laborious, and immunological approaches have more recently been preferred (Shuker & Farmer, 1988). However, the MS-MS technique has now been shown to produce acceptable analyses with only minimal purification of samples. Urine samples (2-3 ml) were first treated with d_3-3-methyladenine (200-250 ng) (Shuker et al., 1987) and then passed through a C_{18} Sep-Pak column. 3-Methyladenine was eluted with 50% aqueous methanol, and after removal of solvent under nitrogen was derivatized with N-(tert-butyldimethylsilyl)-N-methyltrifluoroacetamide. The increased selectivity of the MS analysis achieved by using daughter ions is illustrated in Figure 1.

A normal selected ion recording (using only the first mass spectrometer) of ions characteristic of d_0 and d_3-3-methyladenine (m/z 206, 209) produces an extremely complex trace which does not allow quantitative determination of 3-methyladenine to be made (Figure 1a). The trace in Figure 1b was obtained by monitoring the collision-induced decomposition of these ions (m/z 206 → 179 and 209 → 182); the 3-methyladenine signals are now clearly distinguishable and may be measured. The urine sample used for this analysis was from a patient treated with para-(3,3-dimethyl-1-triazeno)benzoic acid (CB 10-277), which is an anticancer drug believed to act via its metabolism to a methylating agent.

Although these results may be achieved very rapidly and with very little sample work-up, we have noted some lack of reproducibility caused by the complexity of the mixture injected onto the gas chromatographic column, and further purification of the extract may be necessary to improve the accuracy of the method.

Human excretion of alkylguanines

Like 3-alkyladenines, 7-alkylguanines are depurinated from nucleic acids after their formation and are excreted. We have previously demonstrated the usefulness of 7-methylguanine measurements for determining exposure to methylating agents (Farmer

Figure 1. (a) Selective ion recording trace for the *tert*-butyldimethylsilyl derivatives of d_0-3-methyladenine (m/z 206) and d_3-3-methyladenine (m/z 209) isolated from human urine by Sep-Pak chromatography (see text for details). Urine (3 ml) was spiked with 250 ng d_0-3-methyladenine and 250 ng d_3-3-methyladenine. The retention time of 3-methyladenine is 7 min 20 sec. (b) Collision-induced (daughter) fragmentation trace for the *tert*-butyldimethylsilyl derivatives of d_0-3-methyladenine (m/z 206 → 179) and d_3-3-methyladenine (m/z 209 → 182), isolated from human urine. Urine (2 ml) was spiked with 200 ng d_3-3-methyladenine (retention time, 7 min 20 s). See text for further details.

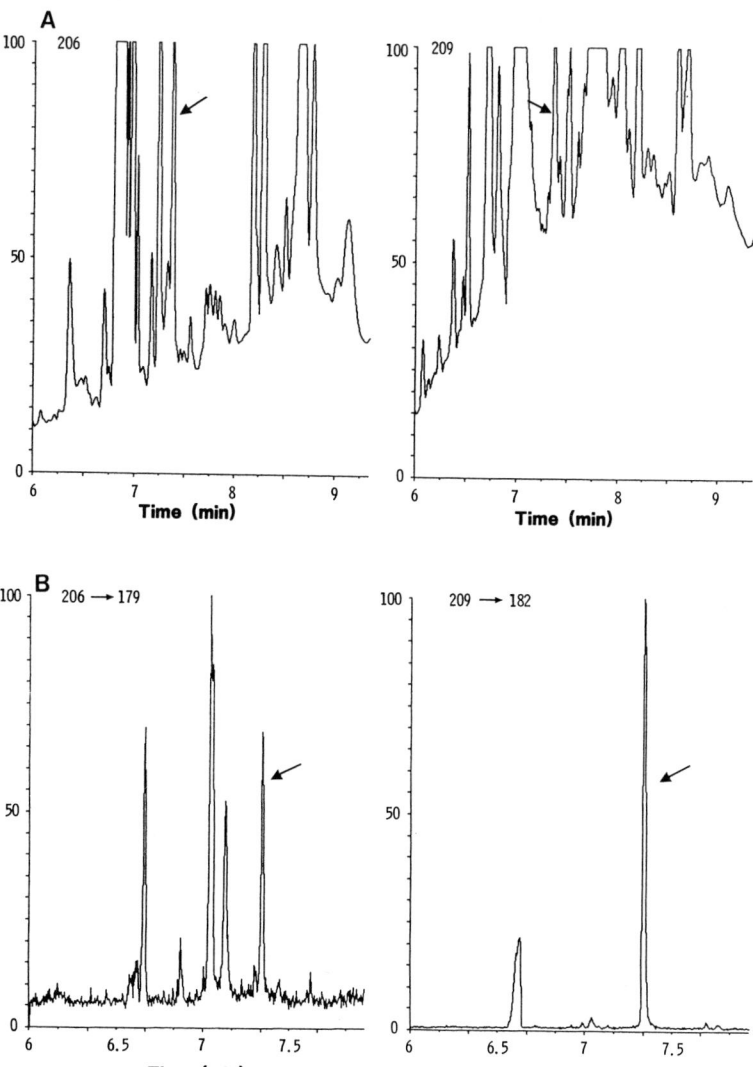

et al., 1986, 1988a), although the high background levels of this purine limit the practical use of the method. To investigate the possible presence of other alkylated guanines in urine, parent ion scans of a predicted common fragment ion (m/z 151, guanine$^+$) were initially carried out on crude urine samples. These revealed the presence of a range of components that yielded this ion and that had masses consistent with low–molecular–weight alkylated guanines (Farmer *et al.*, 1988b).

A range of substituted guanines and their stable isotope-substituted analogues has now been synthesized, so that they can be used to identify the unknown components in urine. These synthetic compounds included *N*-7, *N*2- and *O*6-substituted isomers of methyl-, ethyl- and 2-hydroxyethylguanines. *N*-7-Alkylguanines were prepared by reaction of guanosine with the appropriate alkylating agent (dimethyl sulfate, diethyl sulfate, ethylene oxide), followed by hydrolysis to the purine. *N*2-Alkylguanines were prepared by reaction of 2-bromo-6-hydroxypurine with the appropriate alkylamine (methylamine, ethylamine, ethanolamine). *O*6-Alkylguanines were synthesized by reaction of 2-amino-6-chloropurine with the alcohols methanol, ethanol and ethylene glycol.

The daughter ion spectra of the molecular ions of each of these compounds were determined and compared with the corresponding spectra for crude urine or for a silver-precipitated purine extract from urine. Evidence was obtained that methylated guanines were present (Farmer, 1988), and further characterization following high-performance liquid chromatographic analysis of the urine has confirmed these results.

The sensitivity of MS-MS detection of alkylated guanines would be greatly enhanced by the use of a gas chromatographic separation coupled with selected ion recording, and we are currently using this approach for the analysis of these compounds after formation of *tert*-butyldimethylsilyl (TBDMS) derivatives. The daughter ion spectra of the molecular ions of these derivatives generally show strong fragmentation to (M-57)$^+$ and (M-56)$^+$ ions (Table 1), which may be used to detect these compounds in relatively crude, derivatized urine extracts. Thus, for example, the daughter fragmentation m/z 279 → 222 is characteristic for the mono-TBDMS derivative of *N*-7-methylguanine, whereas m/z 279 → 223 is more abundant for *O*6-methylguanine. *N*2-Methylguanine produces a bis-TBDMS derivative which shows a collision fragmentation of m/z 393 → 336. Figure 2 shows a gas chromatography-MS-MS analysis of these fragmentations for a control sample of urine (2 ml) which had been passed through a C_{18} Sep-Pak column (with elution with 30% methanol) and converted to the TBDMS derivative. The analysis shows the presence of *N*-7-methyl- and *N*2-methylguanine, although no evidence could be seen for *O*6-methylguanine. Analogous analyses of the other alkylated guanines using stable isotope-labelled internal standards are in progress.

Of particular interest in the MS-MS studies is the investigation of the presence of hydroxyethylated guanine in control human urine. The *N*-terminal position of control haemoglobin is known to be hydroxyethylated (Törnqvist *et al.*, 1986; Bailey *et al.*, 1988), and the source of the alkyl group is unknown. These observations of 'background' hydroxy-

Table 1. Daughter fragmentations of alkylguanines after their conversion to *tert*-butyldimethylsilyl (TBDMS) derivatives[a]

Compound	Retention time (min:s)	No. of TBDMS groups added	Fragmentation monitored
N^2-Methylguanine	8:12	2	393 → 336
$N7$-Methylguanine	8:00	1	279 → 222
O^6-Methylguanine	7:21	1	279 → 223
N^2-Dimethylguanine	8:17	2	407 → 350
N^2-Ethylguanine	8:13	2	407 → 350
$N7$-Ethylguanine	8:02	1	293 → 236
O^6-Ethylguanine	7:35	1	293 → 237
N^2-Hydroxyethylguanine	10:21	3	537 → 480
$N7$-Hydroxyethylguanine	9:50	2	423 → 366
O^6-Hydroxyethylguanine	9:18	2	423 → 366

[a] Derivatization was carried out with *N*-(*tert*-butyldimethylsilyl)-*N*-methyltrifluoroacetamide (100 µl) and acetonitrile (100 µl) at 60 °C for 1 h. Gas-chromatography mass-spectrometry conditions are described in the text.

Figure 2. Collision-induced (daughter) fragmentation traces for the *tert*-butyldimethylsilyl derivative of a human urine sample (2 ml) after Sep-Pak chromatography (see text for details).

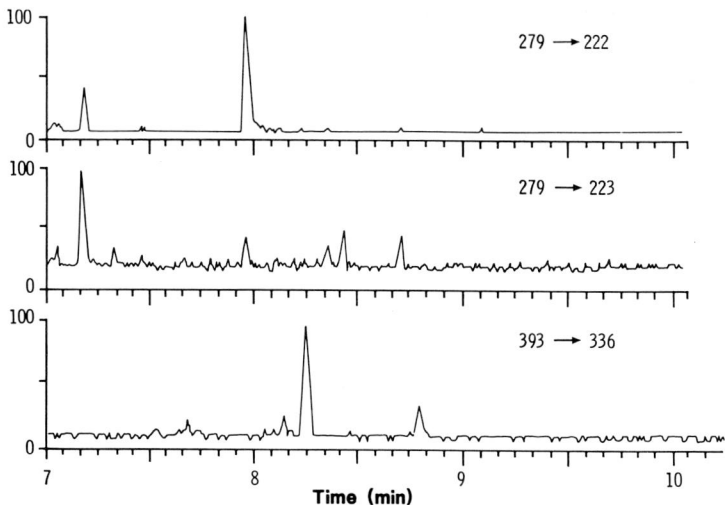

279 → 222, $N7$-methylguanine; retention time, 7 min 58 sec; 279 → 223, O^6-methylguanine; retention time, 7 min 21 sec; 393 → 336, N^2-methylguanine; retention time, 8 min 15 sec

ethylations prompted us to develop an immunoassay for these adducts so that we could do a large-scale investigation of 'control' populations in an attempt to elucidate the origin of the hydroxyethyl group (see Farmer et al., this volume). Evidence was obtained by similar MS-MS studies for the presence in urine of N^2-dimethylguanine and for an alkylated guanine of molecular weight 225, tentatively identified as a dihydroxy- propylguanine. Although the former compound was reported to be a constituent of RNA (Smith & Dunn, 1959), the significance of the latter compound is unknown.

Conclusion

As illustrated above, MS-MS techniques can potentially be used for identifying carcinogen adducts in complex mixtures without the need for extensive purification procedures. Quantification also appears to be possible, using stable isotope-labelled internal standards. However, these advantages are not gained without losing considerable sensitivity (greater than ten fold) over that of conventional mass spectral analysis. Problems may also arise with the gas chromatography, due to the use of very crude biological samples, and these might be a limiting factor in routine gas chromatography-MS-MS analysis (although they might be partly overcome by the use of high-performance liquid chromatography-MS-MS). Despite these problems, we feel that use of MS-MS may be a valuable screening procedure in situations where there are complex exposures to carcinogens.

References

Bailey, E., Brooks, A.G.F., Dollery, C.T., Farmer, P.B., Passingham, B.J., Sleightholm, M.A. & Yates, D.W. (1988) Hydroxyethylvaline adduct formation in hemoglobin as a biological monitor of cigarette smoke intake. *Arch. Toxicol.*, **62**, 247-253

Ehrenberg, L. & Osterman-Golkar, S. (1980) Alkylation of macromolecules for detecting mutagenic agents. *Teratog. Carcinog. Mutag.*, **1**, 105-127

Farmer, P.B. (1988) Tandem mass spectrometry study of urinary alkylated purines. *Biomed. Environ. Mass Spectrom.*, **17**, 143-145

Farmer, P.B., Shuker, D.E.G. & Bird, I. (1986) DNA and protein adducts as indicators of *in vivo* methylation by nitrosatable drugs. *Carcinogenesis*, **7**, 49-52

Farmer, P.B., Neumann, H.-G. & Henschler, D. (1987) Estimation of exposure of man to substances reacting covalently with macromolecules. *Arch. Toxicol.*, **60**, 187-191

Farmer, P.B., Parry, A., Franke, H. & Schmid, J. (1988a) Lack of detectable DNA alkylation for bromhexine in man. *Arzneimittel-Forsch.*, **38**, 1351-1354

Farmer, P.B., Lamb, J. & Lawley, P.D. (1988b) Novel uses of mass spectrometry in studies of adducts of alkylating agents with nucleic acids and proteins. In: Bartsch, H., Hemminki, K. & O'Neill, I.K., eds, *Methods for Detecting DNA Damaging Agents in Humans: Applications in Cancer Epidemiology and Prevention* (IARC Scientific Publications No. 89), Lyon, IARC, pp. 347-355

Shuker, D.E.G. & Farmer, P.B. (1988) Urinary excretion of 3-methyladenine in humans as a marker of nucleic acid methylation. In: Bartsch, H., Hemminki, K. & O'Neill, I.K., eds, *Methods for Detecting DNA Damaging Agents in Humans: Applications in Cancer Epidemiology and Prevention* (IARC Scientific Publications No. 89), Lyon, IARC, pp. 92-96

Shuker, D.E.G., Bailey, E., Parry, A., Lamb, J. & Farmer, P.B. (1987) The determination of urinary 3-methyladenine in humans as a potential monitor of exposure to methylating agents. *Carcinogenesis*, **8**, 959-962

Smith, J.D. & Dunn, D.B. (1959) The occurrence of methylated guanines in ribonucleic acids from several sources. *Biochem. J.*, **72**, 294-301

Törnqvist, M., Mowrer, J., Jansen, S. & Ehrenberg, L. (1986) Monitoring of environmental cancer initiators through hemoglobin adducts by a modified Edman degradation method. *Anal. Biochem.*, **154**, 255-266

QUANTIFICATION OF 4-HYDROXY-1-(3-PYRIDYL)-1-BUTANONE RELEASED FROM HUMAN HAEMOGLOBIN AS A DOSIMETER FOR EXPOSURE TO TOBACCO-SPECIFIC NITROSAMINES

S.S. Hecht, S.S. Kagan, M. Kagan & S.G. Carmella

American Health Foundation, Valhalla, NY, USA

A method was developed to quantify globin adducts of the tobacco-specific nitrosamines 4-(N-nitrosomethylamino)-1-(3-pyridyl)-1-butanone (NNK) and N'-nitrosonornicotine (NNN). Globin adducts of NNK and NNN release 4-hydroxy-1-(3-pyridyl)-1-butanone (HPB) after mild treatment with a base. HPB was analysed as its pentafluorobenzoate by capillary column gas chromatography with detection by negative-ion chemical ionization-mass spectrometry and selected-ion monitoring. The detection limit for HPB-pentafluorobenzoate was approximately 1 fmol/injection. The method was applied to haemoglobin from snuff dippers, smokers and nonsmokers. Adduct levels were highest in snuff dippers (517 ± 538 fmol HPB per gram haemoglobin), followed by smokers (79 ± 189 fmol HPB per gram haemoglobin) and nonsmokers (29.3 ± 25.9 fmol HPB per gram haemoglobin). The method will be useful in assessing the role of tobacco-specific nitrosamines as causes of cancer in smokers and snuff dippers.

We describe the results of our initial studies on the quantification of adducts between globin and 4-(N-nitrosomethylamino)-1-(3-pyridyl)-1-butanone (NNK) and N'-nitrosonornicotine (NNN) in humans. We have focused on these compounds because of their ability to induce relevant tumours in laboratory animals and because of their relatively high concentrations in tobacco and tobacco smoke. Our hypothesis is that NNK and NNN are involved in causing lung cancer, pancreatic cancer and oesophageal cancer in smokers as well as oral cavity cancer in snuff dippers (Hoffmann & Hecht, this volume).

Measurements of plasma and urinary cotinine are reliable indicators of exposure to tobacco products and of uptake of nicotine; however, they do not provide information on chronic uptake, endogenous formation or metabolic activation of NNK and NNN. Quantification of NNK and NNN globin adducts would overcome these unknowns and allow estimation of the dose of metabolically activated carcinogen which reaches the potential target macromolecules.

In previous studies, we showed that NNK and NNN form globin adducts in rats (Carmella & Hecht, 1987). Treatment of the isolated globin with aqueous NaOH at room temperature liberated the keto alcohol, 4-hydroxy-1-(3-pyridyl)-1-butanone (HPB), as illustrated in Figure 1. Although the structure of the globin adduct that liberates HPB is not known, its formation is believed to occur *via* reaction with globin of the diazohydroxide (Carmella &

Hecht, 1987). Similar reactions produce DNA adducts of NNK and NNN (Hecht *et al.*, 1988). Studies with rats have shown that liberation of HPB from globin occurs linearly with dose over four orders of magnitude. In addition, this adduct is not produced by treatment of rats with HPB itself or with nicotine. Therefore, HPB released from globin should be an effective dosimeter for uptake of NNK and NNN and for their metabolic activation in smokers, snuff dippers and other people exposed to tobacco products.

Figure 1. Metabolic activation of 4-(N-nitrosomethylamino)-1-(3-pyridyl)-1-butanone (NNK) and N'-nitrosonornicotine (NNN) to intermediates which bind to globin and release 4-hydroxy-1-(3-pyridyl)-1-butanone (HPB) upon base hydrolysis

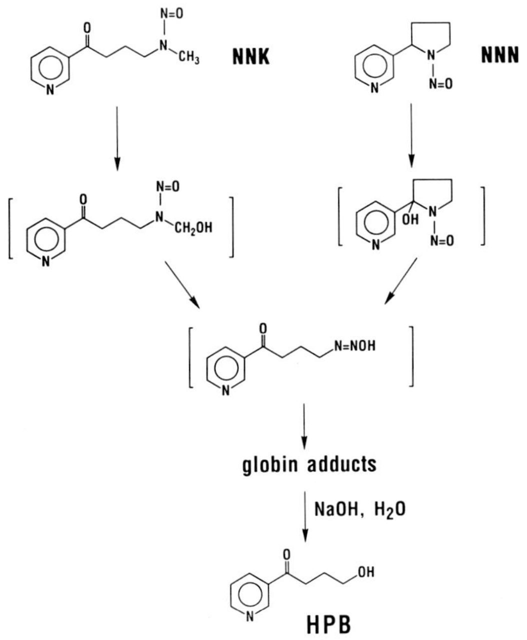

Analysis of human haemoglobin for HPB by gas chromatography-mass spectrometry

The analytical scheme is summarized in Figure 2; full details of the method will be published separately. Dialysed haemoglobin solutions are prepared from 5–10 ml of blood and treated with aqueous NaOH to release HPB. 4,4-DideuteroHPB is added as internal standard. A series of partitions is carried out to enrich the basic fraction, and this material is derivatized with pentafluorobenzoyl chloride and the derivative purified by high-performance liquid chromatography. It is then analysed by capillary column gas chromatography with detection by negative-ion chemical ionization-mass spectrometry with specific-ion monitoring at m/e 358.9, the molecular ion of HPB-pentafluorobenzoate, and m/e 360.9, the molecular ion of 4,4-dideuteroHPB-pentafluorobenzoate. HPB-tetrafluorobenzoate is used as an external standard in these analyses. The detection limit for HPB-pentafluorobenzoate was approximately 1 fmol/injection.

Figure 2. Scheme for analysis of 4-hydroxy-1-(3-pyridyl)-1-butanone (HPB) in hydrolysates of human haemoglobin

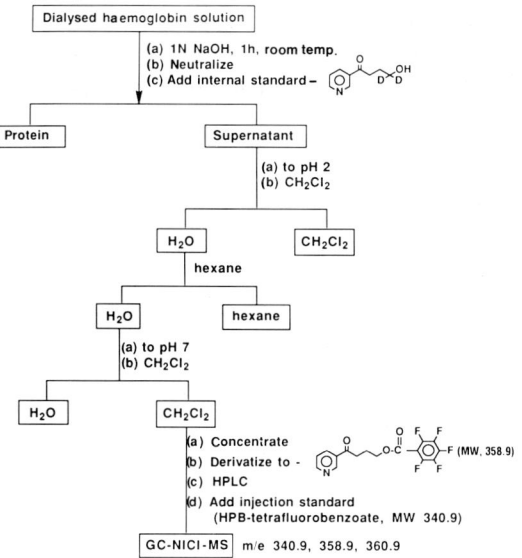

Figure 3 gives representative traces obtained from analyses of haemoglobin from smokers and snuff dippers. HPB-pentafluorobenzoate was readily detected in these samples, and its presence was confirmed by analysis under different chromatographic conditions. The method was validated by replicate analyses of haemoglobin from snuff dippers. In one sample, a value of 1480 ± 63 fmol/g Hb (n = 3) was obtained, while in another sample, which was a composite of samples with a mean value of 1100 fmol/mg, replicate analysis gave a value of 1020 ± 18 fmol/g Hb (n = 5). Analysis of 1 ml haemoglobin solution spiked with 1570 fmol HPB gave 1490 ± 420 pmol (n = 10). Recovery of internal standard was 26 ± 14% (n = 104). Analysis of blanks indicated a background HPB level of 60.7 ± 37.4 fmol (n = 11), which was subtracted from the value for each sample.

Application of the method to smokers and snuff dippers

The results are summarized in Figure 4. The largest amounts of HPB released from haemoglobin were detected in snuff dippers; these levels were significantly greater than those found in smokers ($p < 0.001$). HPB levels were not correlated with amount of snuff used per day, with salivary cotinine or with plasma cotinine. In smokers, HPB levels were not correlated with number of cigarettes smoked per day or with plasma cotinine.

Figure 3. Representative gas chromatography-mass spectrometry (MS)-negative-ion chemical ionization traces of the 4-(hydroxy)-1-(3-pyridyl)-1-butanone (HPB)-pentafluorobenzoate fraction isolated from haemoglobin of (left panels) a smoker and (right panels) a snuff dipper

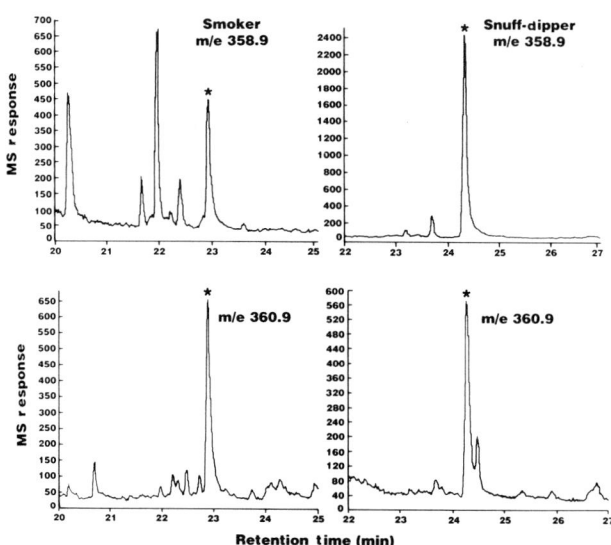

The traces were obtained by selected ion monitoring at m/e 358.9 and 360.9, the molecular ions of HPB-pentafluorobenzoate and 4,4-dideuteroHPB-pentafluorobenzoate. The peaks marked with asterisks eluted at the appropriate retention times. (The retention times in the two traces are different because slightly different conditions were used.)

A striking feature of the data in Figure 4 is the relatively high level of adducts in snuff dippers compared to smokers. The snuff used by most of the participants in this study contained 0.7 μg/g NNK and 9 μg/g NNN; average usage was 4.7 times per day. Assuming that each dip comprised 1 g dry weight of snuff, users would be exposed to approximately 3.3 μg/day NNK and 42.3 μg/day NNN. The amount of HPB released from globin of rats treated with NNN is only about 16% that from globin of rats treated with NNK (Carmella & Hecht, 1987). Therefore, the effective dose that releases HPB from NNK and NNN can be estimated as 3.3 + 0.16 × 42.3, or 10 μg/day. The smokers in our study used an average of 21.8 cigarettes/day. The typical amount of both NNK and NNN in mainstream cigarette smoke is 200 ng/cigarette. Exposure of a smoker could thus be estimated at 5 μg/day, on the basis of the same assumptions used above. Therefore, one might have expected approximately twice the levels of HPB in snuff dippers as in smokers, but HPB levels were approximately seven times higher in snuff dippers than in smokers.

The higher levels of adducts in snuff dippers than in smokers may suggest that endogenous formation of NNK or NNN occurs more readily in snuff dippers, who swallow nicotine or nornicotine, making subsequent nitrosation in the stomach likely. Binding to haemoglobin may occur more readily following oral administration of NNK, as in snuff dipping, than after inhalation. Induction of cytochrome P450 enzymes and other enzymes responsible for the metabolism of NNK is probably different in snuff dippers and smokers, which would also contribute to the observed differences in adduct levels.

Figure 4. Levels of 4-hydroxy-1-(3-pyridyl)-1-butanone (HPB) released from haemoglobin of snuff dippers, smokers and nonsmokers

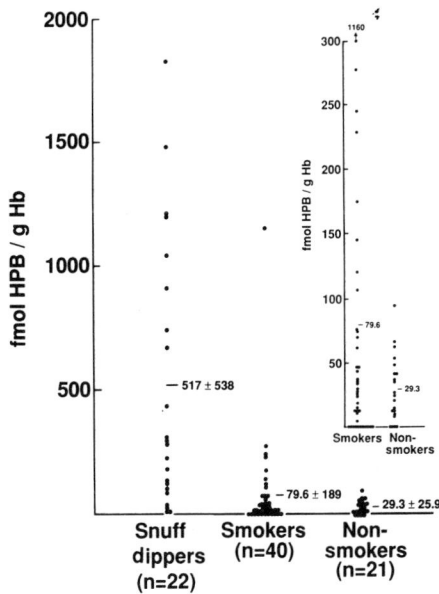

Whatever the mechanism, the high adduct levels in snuff dippers provide further evidence for the role of tobacco-specific nitrosamines in cancer induction by snuff. Previous studies showed that NNK and NNN are present in snuff dippers' saliva and are metabolically activated by cultured buccal mucosa (Hoffmann & Adams, 1981; Castonguay et al., 1983). Our results demonstrate that snuff dippers can metabolically activate NNK and NNN to intermediates which bind to globin. Since experiments in rats have shown that these same intermediates bind to DNA (Hecht et al., 1988), it can be assumed that analogous DNA damage occurs in exposed tissues of snuff dippers.

Table 1 gives the adduct levels determined in this study and those of aromatic amines, ethylene oxide and methylating agents determined in previously published comparisons of smokers and nonsmokers (Törnqvist et al., 1986; Bryant et al., 1987; Stillwell et al., 1988; Törnqvist et al., 1988). The finding that in nonsmokers HPB levels were lower than those of other adducts is consistent with the fact that HPB is derived from tobacco-specific compounds. Some of the HPB detected in samples from nonsmokers may be due to the background levels. In smokers, also, the mean HPB level was lower than those of other adducts. The measured amounts of 4-aminobiphenyl in cigarette smoke are only about 5 ng/cigarette; however, 5% of the dose is bound to haemoglobin in rats, leading to its release (Green et al., 1984) compared with 0.02% released as HPB in rats treated with NNK

(Carmella & Hecht, 1987). The high levels of ethylene oxide adducts in smokers are related to the relatively high concentration of ethylene — approximately 250 µg per cigarette — in mainstream smoke (Törnqvist et al., 1986). Haemoglobin methylation occurs endogenously and after exposure to numerous exogenous agents and may not be a useful dosimeter of methylating agents in tobacco smoke (Törnqvist et al., 1988).

Table 1. Haemoglobin adduct levels in smokers and nonsmokers

Compound analysed	Adduct levels (fmol/g Hb)		Reference
	Smokers	Nonsmokers	
4-Hydroxy-1-(3-pyridyl)-1-butanone	79.6 ± 189	29.3 ± 25.9	This study
4-Aminobiphenyl	911 ± 278	166 ± 77	Bryant et al. (1987)
2-Aminonaphthalene	100 ± 50	40 ± 20	Stillwell et al. (1988)
o-Toluidine	930 ± 280	320 ± 90	Stillwell et al. (1988)
m-Toluidine	4600 ± 1600	6400 ± 1900	Stillwell et al. (1988)
p-Toluidine	1200 ± 470	650 ± 370	Stillwell et al. (1988)
Aniline	47 000 ± 25 000	41 000 ± 22 000	Stillwell et al. (1988)
N-(2-Hydroxyethyl)valine	389 000 ± 138 000	58 000 ± 25 000	Törnqvist et al. (1986)
Methylvaline	540 000 ± 90 000	500 000 ± 10 000	Törnqvist et al. (1988)
N^τ-Methylhistidine	9 200 000 ± 6 300 000	25 000 000 ± 14 000 000	Törnqvist et al. (1988)

The data presently available show that approximately 15% of smokers have elevated HPB levels compared to nonsmokers. It will be important to determine which factors contribute to these higher adduct levels and whether such factors are related to risk.

Acknowledgements

This study was supported by Grant No. CA-29580 from the US National Cancer Institute.

References

Bryant, M.S., Skipper, P.L., Tannenbaum, S.R. & Maclure, M. (1987) Hemoglobin adducts of 4-aminobiphenyl in smokers and nonsmokers. *Cancer Res.*, **47**, 602-608

Carmella, S.G. & Hecht, S.S. (1987) Formation of hemoglobin adducts upon treatment of F344 rats with the tobacco-specific nitrosamines 4-(methylnitrosamino)-1-(3-pyridyl)-1-butanone and N'-nitrosonornicotine. *Cancer Res.*, **47**, 2626-2630

Castonguay, A., Stoner, G.D., Schut, H.A.J. & Hecht, S.S. (1983) Metabolism of tobacco-specific N-nitrosamines by cultured human tissues. *Proc. Natl Acad. Sci. USA*, **80**, 6694-6697

Green, L., Skipper, P.L., Turesky, R.J., Bryant, M. & Tannenbaum, S.R. (1984) In vivo dosimetry of 4-aminobiphenyl in rats via a cysteine adduct in hemoglobin. *Cancer Res.*, **44**, 4254-4259

Hecht, S.S., Spratt, T.E. & Trushin, N. (1988) Evidence for 4-(3-pyridyl)-4-oxobutylation of DNA in F344 rats treated with the tobacco specific nitrosamines 4-(methylnitrosamino)-1-(3-pyridyl)-1-butanone and N'-nitrosonornicotine. *Carcinogenesis*, **9**, 161-165

Hoffmann, D. & Adams, J.D. (1981) Carcinogenic tobacco-specific N-nitrosamines in snuff and saliva of snuff-dippers. *Cancer Res.*, **41**, 4305-4308

Stillwell, W.G., Bryant, M.S. & Wishnok, J.S. (1988) GC/MS analysis of biologically important aromatic amines. Application to human dosimetry. *Biomed. Environ. Mass Spectr.*, **14**, 221-227

Törnqvist, N., Osterman-Golkar, S., Kautiainen, A., Jensen, S., Farmer, P.B. & Ehrenberg, L. (1986) Tissue doses of ethylene oxide in cigarette smokers determined from adduct levels in hemoglobin. *Carcinogenesis*, **7**, 1519-1521

Relevance to Human Cancer of N-Nitroso Compounds,
Tobacco Smoke and Mycotoxins.
Ed. I.K. O'Neill, J. Chen and H. Bartsch
Lyon, International Agency for Research on Cancer
© IARC, 1991

PROMUTAGENIC LESIONS PERSIST IN THE DNA OF TARGET CELLS FOR NITROSAMINE-INDUCED CARCINOGENESIS

C.Y. Fan[1], W.H. Butler[2] & P.J. O'Connor[1,3]

[1]*CRC Department of Carcinogenesis, Paterson Institute for Cancer Research, Christie Hospital and Holt Radium Institute, Manchester, and* [2]*The British Industrial Biological Research Association, Carshalton, Surrey, UK*

Immunohistochemical procedures for the location of O^6-methylguanine (O^6meGua) permit detection of cells proficient for the metabolism of N-nitrosodimethylamine (NDMA) and deficient for the repair of this DNA lesion. Such cells are potentially at high risk for cancer induction and are present in various tissues. In animals maintained on a protein-deficient diet, the distribution and intensity of alkylation of individual cells is altered, particularly in liver where fewer cells apparently retain the capacity to metabolize the nitrosamine, thereby permitting increased levels of alkylation in other tissues. In the renal cortex, specific, O^6-meGua-positive target cells for renal cancer induced by a single dose of NDMA in weanling rats persist at least up to the appearance of early lesions. Persistence of alkylated cells in several tissues indicates prospects for the detection of environmental exposure.

The modification of mammalian tissues by environmental alkylating results in the dose-dependent formation of promutagenic lesions in DNA (Saffhill *et al.*, 1985), and the potential importance of these observations for man is now indicated by the detection of O^6-methylguanine (O^6-meGua) in the DNA of human tissues from widely separated populations (Saffhill *et al.*, 1988). Until recently, data on the alkylation of tissue DNA has consisted mainly of estimates of averages based on analysis of DNA extracted from tissue homogenates. Use of immunohistochemical staining for these DNA adducts now permits the identification of potential target cells for toxicity and carcinogenicity and should therefore provide a clearer understanding of dose-response relationships.

Treatment of animals and staining of tissues

Weanling, male outbred Wistar rats (40-50 g bodyweight at four weeks of age) from the Paterson Institute colony were maintained for three days either on a normal diet or on a protein-depleted diet comprised of starch (70%) and sucrose (30%), prepared as a solidified cake. N-Nitrosodimethylamine (NDMA; 30 mg/kg bw) was administered intraperitoneally

---

[3] To whom correspondence should be addressed

on the morning of the fourth day. Tissues were sampled at various times later, fixed in ethanol (70%), sectioned in paraffin wax and stained for the presence of O^6-meGua in nuclear DNA using a polyclonal anti-O^6-methyldeoxyguanosine, as previously described (O'Connor et al., 1988). Primary antibody binding was visualized using a peroxidase-anti-peroxidase complex with 3,3′-diaminobenzidine.

Alkylation of liver

After treatment with NDMA, nuclei containing O^6-meGua were prominent in the hepatic lobules of animals fed a normal diet. Only relatively few unreactive or near-unreactive cells were present in the vicinity of the portal vessels, and three-quarters of the remaining hepatocytes were stained quite intensely (Figure 1a). Epithelial cells lining the central veins and the radical vessels were also strongly stained. However, in rats maintained on a protein-deficient diet, positively staining nuclei were restricted to the vascular epithelium and a relatively limited zone of hepatocytes (one-quarter to one-half) of the lobule close to the central vein (Figure 1b).

Figure 1. O^6-Methylguanine in nuclei of rat tissues[a]

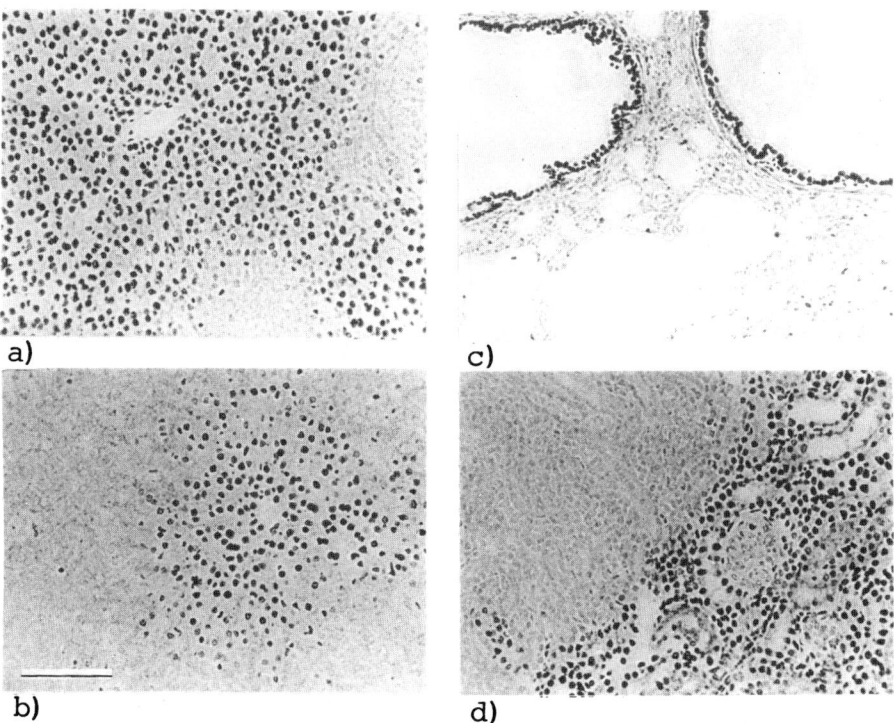

[a] Immunohistochemical staining of 3-μm sections of various rat tissues for the presence of O^6-methylguanine in nuclear DNA. Rats were treated with N-nitrosodimethylamine (30 mg/kg bw), and tissue samples were taken 3 h later; a, liver of a normal rat given a normal diet; and b, liver, c, lung and d, kidney of rats maintained on a protein-deficient diet (by phase contrast; bar equals 100 μm)

Alkylation of extrahepatic tissues

In the lungs and kidneys of animals maintained on either the normal or the protein-deficient diet, the pattern of alkylation was generally similar but was remarkable in that only the nuclei of specific cell types were stained for O^6-meGua. In lung, by far the most heavily alkylated cells were the epithelia of some, but not all, of the bronchioles, while relatively few cells of the stroma or alveoli contained O^6-meGua-positive nuclei (Figure 1c). The bronchiolar epithelium appeared to contain many repair-deficient cells, as many of these were still strongly stained 12 days later. Loss of O^6-meGua staining capacity does not seem to be due to cell replacement, as animals maintained for 18 h on drinking-water containing bromodeoxyuridine (BUdR) did not show any BUdR-labelled nuclei, while such cells were seen in the liver and especially the renal cortex (detected using a monoclonal antibody supplied by Dako Ltd and an immunohistochemical procedure similar to that described for O^6-meGua; see above).

In the kidneys of these animals, many cells in the cortex were positively stained for O^6-meGua, but the cells of the renal medulla were unstained. The intensity and numbers of positively stained nuclei were detectably higher in the protein-deficient animals than in animals maintained on a normal diet. The nuclei of the proximal tubules were prominently stained, but the glomeruli were not (Figure 1d). The small mesenchymal cells lying between the proximal tubules were also strongly stained, although initially they were not very obvious in the presence of the heavily stained nuclei of the proximal tubules. Over the course of the next few days, the number of positively staining nuclei in proximal tubules decreased due to a combination of DNA repair and cell turnover, as many of these cells were BUdR-positive when BUdR was included in the drinking-water (see above).

Persistence of O^6-meGua in target cells

After about one week, the staining capacity of the nuclei of the proximal tubules was either very low or absent, and by this time the nuclei of the mesenchymal elements could be clearly seen as O^6-meGua-positive cells. These are the target cells for the dose-dependent induction of mesenchymal tumours of the kidney when protein-deprived weanling rats are given single doses of NDMA (Driver et al., 1987). Their nuclei remain strongly stained at least up to the time of appearance of the early lesions, when the animals are virtually fully grown, ten weeks after treatment (Figure 2). Although many of these cells continued to stain strongly for O^6-meGua, it is evident that this is a slowly cycling population of cells, as BUdR incorporation (see above) can always be detected in a few of these cells at various times during this period.

Relevance of observations

These studies reveal the presence in tissues of cells that are both proficient for the metabolism of an environmentally relevant nitrosamine and deficient for the repair of promutagenic lesions formed in DNA by these agents. Such cells, therefore, exhibit a double risk factor which could predispose to cancer. Importantly, as such cells clearly have a high capacity for nitrosamine metabolism, they offer better prospects for the detection of low-level, environmentally induced DNA alkylation than do analyses of DNA from whole tissue homogenates, provided that appropriate, simple cell separation procedures can be developed.

Finally, if the DNA lesion O^6-meGua is indeed a critical factor in carcinogenesis induced by alkylating agents, then it should persist in target cells, at least until cell division occurs, to permit fixation of an initiating event. The metabolically proficient-repair deficient mesenchymal target cells for NDMA-induced renal tumours appear to satisfy these criteria.

Figure 2. O^6-Methylguanine persists in target-cell DNA[a]

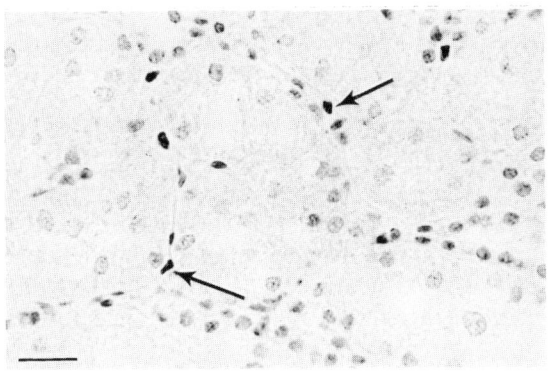

[a] Immunohistochemical staining for the presence of O^6-methylguanine in nuclei of mesenchymal cells in the renal cortex of a rat ten weeks after treatment with N-nitrosodimethylamine (30 mg/kg bw). The animal was maintained on the protein-deficient diet prior to nitrosamine treatment. Counterstained with haematoxylin (by phase contrast; bar equals 25 μm)

Acknowledgements

This work was supported by the Cancer Research Campaign (CRC); C.Y. Fan is the recipient of a British Council Studentship.

References

Driver, H.E., White, I.N.H. & Butler, W.H. (1987) Dose response relationships in chemical carcinogenesis: renal mesenchymal tumours induced in the rat by single dose nitrosamine. *Br. J. Exp. Pathol.*, **68**, 133-143

O'Connor, P.J., Fida, S., Fan, C.Y., Bromley, M. & Saffhill, R. (1988) Phenobarbital: a non-genotoxic agent which induces the repair of O^6-methylguanine from hepatic DNA. *Carcinogenesis*, **9**, 2033-2038

Saffhill, R., Margison, G.P. & O'Connor, P.J. (1985) Mechanisms of carcinogenesis induced by alkylating agents. *Biochim. Biophys. Acta*, **823**, 111-145

Saffhill, R., Badawi, A.G. & Hall, C.N. (1988) Detection of O^6-methylguanine in human DNA. In: Bartsch, H., Hemminki, K. & O'Neill, I.K., eds, *Methods for Detecting DNA Damaging Agents in Humans: Applications for Cancer Epidemiology and Prevention* (IARC Scientific Publications No. 89), Lyon, IARC, pp. 301-305

Relevance to Human Cancer of N-Nitroso Compounds,
Tobacco Smoke and Mycotoxins.
Ed. I.K. O'Neill, J. Chen and H. Bartsch
Lyon, International Agency for Research on Cancer
© IARC, 1991

NITRITE-TRAPPING CAPACITY OF THIOPROLINE IN THE HUMAN BODY

M. Tsuda[1] & Y. Kurashima

Biochemistry Division, National Cancer Center Research Institute, Tokyo, Japan

Human nitrosating capacity has been monitored using proline; however, *N*-nitrosothiazolidine 4-carboxylic acid (NTCA; *N*-nitrosothioproline), one of the predominant *N*-nitroso compounds in human urine, is also nonmutagenic and, presumably, noncarcinogenic. Thioproline is nitrosated about 1000 times faster than proline *in vitro*, and NTCA is excreted into the urine without being metabolized. We have therefore proposed thioproline as an effective nitrite-trapping agent in the human body. Recently, we found thioproline in various cooked foods, including cod and dried *shiitake* mushrooms. In the study reported here, we evaluate the nitrite trapping capacity of thioproline in a male nonsmoking volunteer ingesting NO_3^- and eating a controlled diet. The highest level of NTCA excreted, 5.89 μmol, was measured after the subject ingested 6 mmol NO_3^- followed by 0.45 mmol (60 mg) thioproline. We estimated the effective amount of nitrite, defined as the actual amount of nitrite participating in nitrosation in the stomach, to be 0.3% of the NO_3^- ingested. Thus, the effective amount of NO_2^- for 6 mmol NO_3^- ingested was calculated to be 18 μmol, and 33% of this nitrite was trapped by ingestion of 0.45 mmol thioproline. We conclude that thioproline is a most sensitive probe for evaluating human nitrosating capacity and an effective nitrite-trapping agent.

Endogenous formation of *N*-nitroso compounds is thought to be an important factor in human carcinogenesis (National Academy of Sciences, 1981). Human nitrosating capacity has been monitored by using the proline test (Ohshima & Bartsch, 1981). *N*-Nitrosothiazolidine 4-carboxylic acid (NTCA; *N*-nitrosothioproline), a predominant *N*-nitroso compound in human urine (Ohshima *et al.*, 1983; Tsuda *et al.*, 1983, 1987a), is, like *N*-nitrosoproline (NPRO), nonmutagenic and presumably noncarcinogenic (Mirvish *et al.*, 1973). Nitrosation of thioproline occurs about 1000 times faster than that of proline *in vitro* (Tahira *et al.*, 1984), and it is excreted into the urine without undergoing metabolism (Ohshima *et al.*, 1984). Thus, we have proposed thioproline as an effective nitrite-trapping agent in the human body. Recently, we found this compound in various cooked foods; large amounts were found in cod (Tsuda *et al.*, 1988) and in dried *shiitake* mushrooms (*Lentinus edodes*) (Tsuda *et al.*, 1987b). The purpose of the study reported here was to evaluate the trapping capacity of thioproline in the human body, and to establish a sensitive method for monitoring the capacity for endogenous nitrosation.

[1]To whom correspondence should be addressed

Volunteer experiments

A male nonsmoker ingested nitrate (max, 6 mmol) with or without thiocyanate in 100 ml tap-water at 07:00 h. After 30 min, he ingested thioproline with or without ascorbic acid in 100 ml tap-water; during this interval, he ate a low-nitrate, standard breakfast. For 3 h after intake of thioproline, the subject did not eat or drink; he then ate lunch consisting mainly of cooked rice containing low nitrate and milk. Urine was collected for 12 h starting at 07:00 h, and an aliquot of 12-h urine was stored with NaOH (2 g/l) at $-20°C$ until analysis. Saliva samples (~1 ml) were collected in a polypropylene tube containing 20 µl 10N NaOH and stored as above.

Salivary nitrite and urinary NTCA levels after ingestion of nitrate and thioproline

Salivary nitrite levels usually increased with the amount of nitrate intake (Figure 1). Since NTCA is excreted into urine without metabolic change, the amount of urinary NTCA corresponds directly to the amount formed endogenously. The amounts of NTCA in 12-h urine after ingestion of 0.45 mmol thioproline are shown in Figure 2. Endogenous formation of NTCA and of salivary nitrite was less effective when less than 2 mmol nitrate were ingested. The amounts of urinary NTCA and salivary nitrite increased markedly with intake of more than 4 mmol nitrate, although the variation also increased.

Figure 1. Changes in salivary nitrite level in a male volunteer after ingestion of various amounts of nitrate with and without thiocyanate

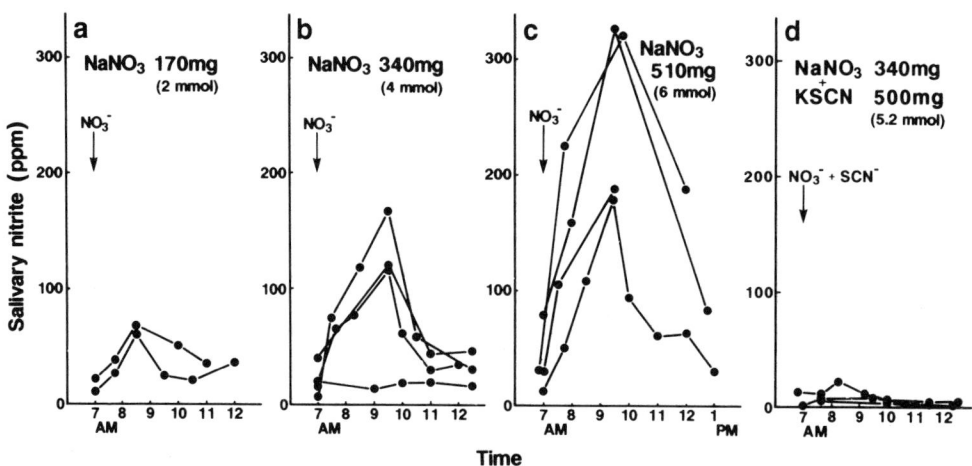

Salivary nitrite was determined by the Griess method.

Effective amount of nitrite participating in nitrosation in the human stomach

We calculated the effective amount of nitrite (NO_2^- eff), defined as the actual amount of nitrite participating in nitrosation in the stomach, on the basis of the following assumptions:

(1) The effective nitrosation reaction period in stomach after ingestion of nitrate (or a nitrate-rich meal) is about 2 h.

(2) The conversion of nitrate to nitrite in the oral cavity is estimated to be 6.3% per 24 h (Stephany & Schuller, 1980); and 3% of ingested nitrate is converted to nitrite in the oral cavity during the initial 2 h after ingestion of nitrate. Thus, 6 mmol NO_3^- × 0.03 = 0.18 mmol (NO_2^-).

Figure 2. *N*-Nitrosothiazolidine 4-carboxylic acid (NTCA) in the urine of a male volunteer after ingestion of various amounts of nitrate with 0.45 mmol (60 mg) thioproline.

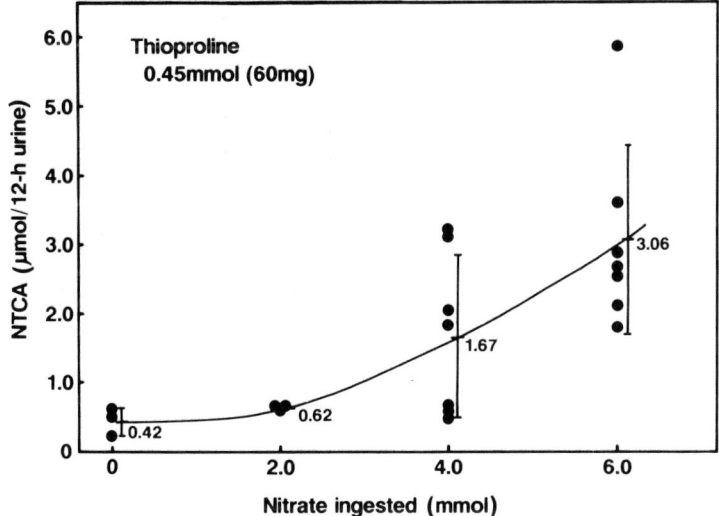

Urinary NTCA was analysed by gas chromatography-thermal energy analysis as described previously (Tsuda *et al.*, 1986).

(3) Absorption of nitrite by the stomach mucosa is rapid: the half-life of nitrite in an empty stomach is < 10 min (Friedman *et al.*, 1972; Mirvish *et al.*, 1975; Licht *et al.*, 1986; Yamamoto *et al.*, 1987). Thus, about 90% of nitrite from saliva disappears within 1 h and only about 10% participates in nitrosation.

(4) Nitrite also reacts with gastric juice and food components, such as ascorbic acid, phenols and amino acids; it undergoes autodegradation at acidic pH; and it is transferred to the duodenum (Licht *et al.*, 1986). The NO_2^- *eff* in the stomach can therefore be calculated approximately as follows:

NO_2^- *eff* (mmol) = NO_3^- mmol (ingested) × (2) × (3)
            = NO_3^- mmol (ingested) × 0.03 × 0.1.

For example,

NO_2^- *eff* (mmol) = 6 mmol NO_3^- × 0.003 = 0.018 mmol
            = 18 μmol.

In Table 1, the amounts of N-nitroso compounds formed and the percentage of NO_2^- eff utilized for endogenous nitrosation are summarized, with reported values for the proline test (Ohshima & Bartsch, 1981; Ladd et al., 1984) for comparison. Thioproline can apparently trap the NO_2^- eff in stomach roughly several hundred times more effectively than proline. In this study, the largest amount of NTCA formed endogenously was 5.89 μmol (954 μg), after ingestion of 6 mmol nitrate and 0.45 mmol thioproline; thus, about 35% of the NO_2^- eff from 6 mmol NO_3^- ingested was trapped by thioproline.

Our method of calculation was also used on the results of experiments in which subjects ingested 4 mmol nitrate and 1.2 mmol aminopyrine with ethanol (Spiegelhalder & Preussmann, 1985); this showed that almost 50% of the NO_2^- eff (12 μmol) was utilized for maximal formation of N-nitrosodimethylamine by nitrosation. Clearly, the level of carcinogenic N-nitroso compound formation in the body could be significant in human carcinogenesis.

Table 1. Nitrite-trapping capacities of thioproline and proline in humans

Precursors ingested by volunteers			Products formed endogenously			NO_2^- eff trapped by amino acid (%)	
NO_3^-		Amino acid	NO_2^- eff[a] (μmol)	N-Nitrosamino acid[b] (μmol)			
mmol mg		mmol mg		Mean	Maximun	Mean	Maximum
		Thioproline		**NTCA**			
4.0 250		0.075 10	12	1.02	1.98	8.5	16.5
4.0 250		0.45 60	12	1.67	3.18	13.9	26.5
6.0 370		0.45 60	18	3.06	5.89	17.0	32.7
		Proline		**NPRO**			
5.2 325		4.35 500	16	0.16	0.2[b]	1.0	1.3
5.2 325		4.35 500	16	0.12	0.33[c]	0.8	2.1

[a] Estimated effective amount of nitrite in stomach, as determined in this study, calculated from NO_2^- eff (mmol) = 0.003 × NO_3^- ingested (mmol)
[b] Data from Ohshima & Bartsch (1981)
[c] Data (nonsmokers) from Ladd et al. (1984)

Effects of ascorbic acid and thiocyanate on amount of urinary NTCA after ingestion of nitrate and thioproline

Urinary NTCA levels after intake of 4 mmol nitrate and 0.075 mmol thioproline taken together with ascorbic acid or thiocyanate are shown in Table 2. As expected (Oshima & Bartsch, 1981; Ohshima et al., 1984), 1.7 mmol ascorbic acid inhibited the formation of NTCA, but not completely. An additional 1.7 mmol dose after a 2-h interval did not significantly change the level of urinary NTCA excretion, suggesting that no further formation of NTCA occurs 2 h after ingestion of thioproline.

Co-ingestion of nitrate (4 mmol) with thiocyanate (5.2 mmol) resulted in almost complete inhibition of salivary nitrite formation, as shown in Figure 1d. A consequent decrease in NTCA formation was also observed (see row 4, Table 2). Salivary nitrate levels

2 h after nitrate intake were reduced remarkably, from 494.5 ± 18.5 ppm to 21.9 ± 7.9 ppm, by co-administration of thiocyanate. It is well known that thiocyanate competitively inhibits the secretion of nitrate into the oral cavity from the salivary glands (Edwards et al., 1954). Our results clearly support that finding in the human body.

The salivary thiocyanate level in smokers is two to three times higher than that in nonsmokers (Butts et al., 1974). Ladd et al. (1984) reported that when volunteers ingested nitrate (5.2 mmol) and proline (4.35 mmol) the urinary excretion level of NPRO was significantly higher in smokers than in nonsmokers, but the salivary nitrate levels in smokers were lower than those in nonsmokers. These results were attributed to a catalytic effect of thiocyanate on nitrosation in vivo (Boyland & Walker, 1974), even though the salivary nitrite level was lower in smokers. Our finding (row 4, Table 2) is an extreme case due to intake of large amounts of thiocyanate with nitrate and thioproline. Studies of variations with dose are in progress.

Table 2. Effects of ascorbic acid and thiocyanate on urinary excretion of *N*-nitrosothiazolidine 4-carboxylic acid (NTCA) in a male volunteer after ingestion of nitrate and thioproline

Compounds ingested (mmol)	Urinary NTCA (nmol/12-h urine; mean ± SD; n = 3)
$NaNO_3$ (4.0) + thioproline (0.075)	475 ± 204
$NaNO_3$ (4.0) + thioproline (0.075) + ascorbic acid (1.7)	138 ± 24.7
$NaNO_3$ (4.0) + thioproline (0.075) + ascorbic acid (1.7) + ascorbic acid (1.7) 2 h after first dose	133 ± 78.4
$NaNO_3$ (4.0) + thioproline (0.075) + KSCN (5.2)	64.8 ± 26.5
$NaNO_3$ (4.0) alone	54.9 ± 40.1
Thioproline (0.075) alone	40.1 ± 28.4
Control diet alone	6.2 ± 1.9

Acknowledgements

We thank Dr H. Esumi and Dr T. Sugimura for encouraging discussions and Miss T. Kobayashi for typing the manuscript. This work was partly supported by Grants-in-Aid for Cancer Research from the Ministry of Health and Welfare and the Ministry of Education, Science and Culture of Japan, and by a grant from the Smoking Research Foundation, Japan.

References

Boyland, E. & Walker, S.A. (1974) Effect of thiocyanate on nitrosation of amines. *Nature*, **248**, 601-602

Butts, W.C., Kuehneman, M. & Widdowson, G.M. (1974) Automated method for determining serum thiocyanate, to distinguish smokers from nonsmokers. *Clin. Chem.*, **20**, 1344-1348

Edwards, D.A.W., Fletcher, K. & Rowlands, E.N. (1954) Antagonism between perchlorate, iodine, thiocyanate, and nitrate for secretion in human saliva. *Lancet*, **i**, 498-499

Friedman, M.A., Greene, E.J. & Epstein, S.S. (1972) Rapid gastric absorption of sodium nitrite in mice. *J. Pharm. Sci.*, **61**, 1492-1494

Ladd, K.F., Newmark, H.L. & Archer, M.C. (1984) N-Nitrosation of proline in smokers and nonsmokers. *J. Natl Cancer Inst.*, **73**, 83-87

Licht, W.R., Schultz, D.S., Fox, J.G., Tannenbaum, S.R. & Deen, W.M. (1986) Mechanisms for nitrite loss from the stomach. *Carcinogenesis*, **7**, 1681-1687

Mirvish, S.S., Bulay, O., Runge, R.G. & Patil, K. (1973) Study of the carcinogenicity of large doses of dimethylnitramine, N-nitroso-L-proline, and sodium nitrite administered in drinking water to rats. *J. Natl Cancer Inst.*, **64**, 1435-1442

Mirvish, S.S., Patil, K., Ghadirian, P. & Kommineni, V.R.C. (1975) Disappearance of nitrite from the rat stomach: contribution of emptying and other factors. *J. Natl Cancer Inst.*, **54**, 869-875

National Academy of Sciences (1981) *The Health Effects of Nitrate, Nitrite and N-Nitroso Compounds*, Committee on Nitrite and Alternative Curing Agents in Foods, Washington DC

Ohshima, H. & Bartsch, H. (1981) Quantitative estimation of endogenous nitrosation in human by monitoring N-nitrosoproline excretion in the urine. *Cancer Res.*, **41**, 3658-3662

Ohshima, H., Friesen, M., O'Neill, I.K. & Bartsch, H. (1983) Presence in human urine of a new N-nitroso compound, *N*-nitrosothiazolidine 4-carboxylic acid. *Cancer Lett.*, **20**, 183-190

Ohshima, H., O'Neill, I.K., Friesen, M., Béréziat, J.-C. & Bartsch, H. (1984) Occurrence in human urine of new sulphur-containing N-nitrosamino acids, N-nitrosothiazolidine 4-carboxylic acid and its 2-methyl derivative, and the formation. *J. Cancer Res. Clin. Oncol.*, **108**, 121-128

Spiegelhalder, B. & Preussmann, R. (1985) In vivo nitrosation of aminopyrine in humans: use of 'ethanol effect' for biological monitoring of N-nitrosodimethylamine in urine. *Carcinogenesis*, **6**, 545-548

Stephany, R.W. & Schuller, P.L. (1980) Daily dietary intakes of nitrate, nitrite and volatile N-nitrosamines in the Netherlands using the duplicate portion sampling technique. *Oncology*, **37**, 203-210

Tahira, T., Tsuda, M., Wakabayashi, K., Nagao, M. & Sugimura, T. (1984) Kinetics of nitrosation of thioproline, the precursor of a major nitroso compound in human urine, and its role as a nitrite scavenger. *Gann*, **75**, 889-894

Tsuda, M., Hirayama, T. & Sugimura, T. (1983) Presence of N-nitroso-L-thioproline and N-nitroso-L-methylthioprolines in human urine as major N-nitroso compounds. *Gann*, **74**, 331-333

Tsuda, M., Niitsuma, J., Sato, S., Hirayama, T., Kakizoe, T. & Sugimura, T. (1986) Increase in the levels of N-nitrosothioproline and N-nitroso-2-methylthioproline in human urine by cigarette smoking. *Cancer Lett.*, **30**, 117-124

Tsuda, M., Nagai, A., Suzuki, H., Hayashi, T., Ikeda, M., Kuratsune, M., Sato, S. & Sugimura, T. (1987a) Effect of cigarette smoking and dietary factors on the amounts of *N*-nitrosothiazolidine 4-carboxylic acid and *N*-nitroso-2-methylthiazolidine 4-carboxylic acid in human urine. In: Bartsch, H., O'Neill, I.K. & Schulte-Hermann, R., eds, *The Relevance of N-Nitroso Compounds to Human Cancer: Exposures and Mechanisms* (IARC Scientific Publications No. 84), Lyon, IARC, pp. 446-450

Tsuda, M., Kurashima, Y., Sato, S. & Sugimura, T. (1987b) Thioproline as a nitrite trapping agent in the human body and its ubiquitous occurrence in foods [in Japanese]. *Sulfur Amino Acid*, **10**, 271-277

Tsuda, M., Frank, N., Sato, S. & Sugimura, T. (1988) Marked increase in the urinary level of N-nitrosothioproline after ingestion of cod with vegetables. *Cancer Res.*, **48**, 4049-4052

Yamamoto, M., Ishiwata, H., Yamada, T., Yoshihira, K. & Tanimura, A. (1987) Studies in the guinea-pig stomach on the formation of N-nitrosourea, from methylurea and sodium nitrite, and its disappearance. *Food Chem. Toxicol.*, **25**, 663-668

Relevance to Human Cancer of *N*-Nitroso Compounds,
Tobacco Smoke and Mycotoxins.
Ed. I.K. O'Neill, J. Chen and H. Bartsch
Lyon, International Agency for Research on Cancer
© IARC, 1991

PRACTICAL ASPECTS OF APPLICATION OF DOSIMETRY METHODS: SUMMARY OF A WORKSHOP DISCUSSION

Prepared by

D.E.G. Shuker (IARC, Lyon, France) & D. Forman (ICRF, Oxford, UK)

Introduction

The objective of this workshop was to discuss available dosimetry methods related to some aspects of the three main themes of the meeting — *N*-nitroso compounds, mycotoxins and tobacco smoke. The appropriate application of these methods to human studies, epidemiological requirements in study design, the strengths and limitations of existing techniques, and probable future developments and needs were considered. Short presentations by invited speakers were followed by a discussion open to all of the participants of the meeting.

This report summarizes the presentations and subsequent discussions and is not meant to be a comprehensive review of methods in dosimetry.

The speakers were W. Blot (USA), P.B. Farmer (UK), S. Fischer (FRG), D. Forman (UK), Y.T. Gao (China), S.S. Hecht (USA), J. Hotchkiss (USA), H. Ohshima (France), D.E.G. Shuker (France), A.R. Tricker (FRG), Y. Ueno (Japan) and C.P. Wild (France).

Perspective What do epidemiologists need from biochemical techniques of measurement? (W. Blot)

There is no doubt that epidemiological studies will rely more and more on laboratory-based measurement techniques in order to improve assessment of exposure (preferably at the level of organs or tissues). Such techniques can indicate genetic susceptibility and indicate markers of early changes, i.e., precancerous conditions.

With regard to exposure, two fundamental study designs can be distinguished: those that are retrospective and require information about past exposures (ideally over a lifetime), and those that are prospective which require information about current exposures. Most available biochemical measurement techniques, with some notable exceptions, such as viral antibody determinants, are in the latter category, although what is often the most efficient epidemiological study (the case–control comparison) requires retrospective information. This discrepancy is undoubtedly a major problem, since it implies that the epidemiological studies in which biochemical markers can be most readily employed are either relatively weak in terms of analytical power (ecological studies) or very large, time-consuming and costly (prospective cohort studies). A further problem is that many exposures that are now of concern, such as to endogenously synthesized N-nitroso compounds, are likely to be experienced by everyone. Thus, qualitative estimates of exposure may not be sufficient to discriminate between people who do and those who do not develop cancer; the actual dose to which they are exposed must be assessed.

Despite these difficulties, progress is being made. Biochemical methods are now being used extensively to validate questionnaires and to measure exposures that cannot easily be assessed by questionnaire, e.g., to micronutrients and environmental tobacco smoke. They are also being used in experimental studies, as in studies to assess whether intended interventions are having any detectable biochemical effects, and in occupational biomonitoring to provide an 'early warning' of potential hazards, such as N-nitroso compounds in biological samples from workers exposed to certain cutting fluids.

N-*Nitroso compounds*

– Do we need to measure in-vivo nitrosation in humans? (D. Forman)
– How can in-vivo nitrosation be measured? (H. Ohshima)
– What we have and have not learned from N-nitrosoproline dosimetry (J. Hotchkiss)
– What are the requirements for new tests of in-vivo nitrosation? (A.R. Tricker)
– What are the requirements for new tests to assess exposure to N-nitroso compounds? (D. Shuker)

Human beings are potentially exposed to N-nitroso compounds from a variety of exogenous and endogenous sources. Exposure to preformed N-nitroso compounds has been clearly demonstrated in many studies; however, due to the limited time available, the presentations and discussion focused on endogenous nitrosation.

The two major reasons for understanding and hence measuring in-vivo nitrosation in humans are, firstly, to establish whether nitrosation is critically related to human cancer, and, secondly, since nitrosation can be considered an endpoint in itself, to assess the relative importance of factors affecting the level of nitrosation. A further consideration is the appropriateness of the application of monitoring methods to epidemiological studies, since, for example, many tests of endogenous nitrosation have been used in ecological studies, which are the weakest from an epidemiological standpoint.

The most widely used test of endogenous nitrosation is the N-nitrosoproline (NPRO) test, developed by Ohshima and Bartsch. It was developed to meet the requirement for a simple measure of endogenous nitrosation that circumvented the need for biopsy or blood samples, which are often difficult to obtain from subjects in the most interesting populations. Samples of overnight or 24-h urine are analysed for NPRO by a simple extraction followed by gas chromatography–thermal energy analysis. Depending on the nature of the test, differences in levels of NPRO are a measure of endogenous nitrosation potential. This test has permitted the unequivocal demonstration of endogenous nitrosation in humans and experimental animals. In a number of studies, increased levels of endogenous nitrosation have been related to specific risk factors, such as intake of nitrate and bacterial and parasitical infections, as well as to protective effects of ascorbic acid and vegetable intake. The NPRO test has been used in studies of groups at high cancer risk, and subsequent studies strengthened the evidence for the role of N-nitroso compounds (e.g., the detection of DNA adducts in tissue obtained from cases of oesophageal cancer in Linxian in China).

Nonetheless, the presence of variable amounts of preformed NPRO in human diets often leads to substantial confounding of the results of this test. In some studies, a modest degree of dietary control resulted in relatively constant basal levels of NPRO excretion. A number of alternatives are available, including the use of, for example, stable isotopes (ethically acceptable in humans) or different nitrosatable amino acids as probes (provided that they are not mutagenic or carcinogenic and that the nitrosamino acid is excreted as completely as NPRO). The choice of suitable new substrates should be governed by factors such as their optimal pH for nitrosation, occurrence in the diet and sensitivity to different nitrosation pathways.

Despite some shortcomings of the NPRO test, it constitutes a measure of the result of a complex series of interactions and often provides the basis for further studies. Many of the studies now under way incorporate additional and complementary markers of endogenous nitrosation.

The consequences of endogenous nitrosation or exposure to exogenous N-nitroso compounds also provide opportunities for biomonitoring. Most N-nitroso compounds or their metabolites and related (diazo- and diazonium) compounds are potent alkylating agents which modify DNA and/or proteins. Current methods have demonstrated the presence of methylated bases (O^6-methyldeoxyguanosine and 7-methyldeoxyguanosine) in human DNA following presumed or actual exposure to N-nitroso compounds. Methods that are capable of identifying new adducts arising from hitherto unknown exposures would be desirable, preferably with the subsequent possibility of using more routine methods in epidemiological studies. At present, methods involving mass spectrometry are suitable for identifying alkylated DNA bases, their subsequent quantification being carried out by immunochemical methods. A novel approach, which may eventually be applicable to dosimetry, is the use of semipermeable microcapsules containing surrogate target molecules, which are recovered in faeces following transit through the gastrointestinal tract.

A further consideration is the use of biological endpoints such as mutations, cytogenetic markers and oncogene activation. The advantage of this methodology over biochemical approaches is that the biological damage is detectable long after the initial lesion has disappeared. In consequence, however, information about the nature of the DNA damaging agent is lost. The use of, for example, mutational spectra may redress the balance and make the use of biological endpoints more attractive in dosimetry.

Mycotoxins

– What are the strengths and limitations of current measures of aflatoxin exposure? (C.P. Wild)
– What other mycotoxins are of interest and how can they be measured? (Y. Ueno)

A number of possibilities exist for the measurement of aflatoxin exposure. In various biological samples, one can measure either (i) the parent aflatoxins and/or their metabolites, or (ii) aflatoxin adducts (with DNA or protein). The latter approach has the potential advantage of providing information about genotoxic damage, directly or indirectly, because of the common pathway of metabolic activation of aflatoxins in the formation of DNA and protein adducts. The same should be true for the metabolites formed from the breakdown of glutathione conjugates, which may be excreted in urine; but, in contrast to DNA and protein adducts, these are as yet largely uncharacterized in man.

In urine samples, aflatoxins and their metabolites have been measured using chromatographic and immunochemical methods. Current research has focused on a measure of either 'total' urinary aflatoxins or specific metabolites such as aflatoxin M- and aflatoxin B_1-guanine. Whatever the target molecule for urinary analysis, a major limitation is that urinary excretion reflects exposure only over the preceding 24 h.

In blood, metabolites of aflatoxins bind almost exclusively to albumin, with little or no modification of either haemoglobin or lymphocyte DNA. Current immunoassay methods permit the specific and sensitive detection of aflatoxin–albumin adducts in 50 µl of blood, i.e., a finger-prick blood sample. The great advantage of this method is that, for example, 1–3% of a single dose of aflatoxin B_1 binds to albumin, and adducts can accumulate owing to the relatively long half-life of the protein. In addition, aflatoxin–albumin adducts may reflect DNA adduct levels in hepatocytes. This method appears to be the most promising for incorporation into field studies.

Analysis of DNA adducts in tissue samples is limited not so much by the available technology, which is similar to that for blood and urine, but by the availability of samples and their relevance to epidemiological studies.

Many other mycotoxins are produced and are likely to be present in human foods as a consequence of fungal contamination. For the majority of these, no method has been developed to measure human exposure using biological samples. Two compounds that have attracted attention are nivalenol (from *Fusarium* strains) and ochratoxin A (from *Penicillium* and *Aspergillus* genera). Sensitive gas chromatography–mass spectrometry methods have been developed for analysing nivalenol at levels as low as 1–2 ng/kg in foodstuffs. Many cereal samples from several countries have been analysed, and about half contained more than 250 ng/kg of the mycotoxin. Samples of scabby wheat from China, which are heavily contaminated with mould and normally used only for animal food, contained up to 4 µg/kg nivalenol.

Many studies, involving high-performance liquid chromatography with fluorimetric detection, have established that ochratoxin A contamination of human foodstuffs is widespread. In addition, ochratoxin A has been detected in human sera from blood banks and in human breast milk. In order to enable large-scale screening of foodstuffs, reliable, rapid assays of ochratoxin A are required. The recent preparation of highly specific monoclonal antibodies against this mycotoxin has permitted the development of such an assay, which may also be applied to biological samples; the limit of detection of ochratoxin A in serum is about 0.1 ng/ml. Thus, an international survey of ochratoxin A contamination in foodstuffs and its occurrence in human blood is now technically feasible and might be considered desirable.

Tobacco smoke

– What is the best way to measure exposure to tobacco smoke in epidemiological studies? (Y.T. Gao)
– Do we need dosimetry for risk evaluation of tobacco smoke? (S. Fischer)

- Are there markers of exposure to tobacco smoke that can be used to predict cancer risk? (S.S. Hecht)
- What are the most appropriate uses of new mass spectral techniques to detect smoking-related DNA damage? (P.B. Farmer)

In contrast to the situation with N-nitroso compounds and mycotoxins, exposure to tobacco smoke has been measured reliably for many years, by means of simple questionnaires. By asking appropriate questions about duration of smoking and average amount smoked each day, it has been possible to make estimates of total individual cumulative exposure, and this has permitted a detailed understanding of the epidemiology of smoking-related cancers. Although there may be a tendency for some individuals to underreport their own smoking patterns, the consistency between studies and the reproducibility of results would indicate that this is not a major problem.

Numerous reasons exist, however, for developing and refining tools for the biochemical estimation of exposure to tobacco smoke. At one level, such tools can be used to validate and extend findings obtained from questionnaires. A reliable biochemical assay could, for example, be used to quantify the extent to which underreporting was taking place; it could also provide a basis for assessing passive exposure to smoke, a measure that is made only extremely crudely by questionnaires. On a different level, it is known that different types of tobacco and cigarette construction can result in smoke with different degrees of carcinogenic potential. Also, individuals obviously vary in their susceptibility to tobacco smoke carcinogenesis. By developing methods to assess the interaction between smoke products and host cells or cell products, it may be possible to increase our understanding of the mechanisms involved.

A diversity of techniques is available for assessing exposure to tobacco smoke. These range from assays for carbon monoxide in exhaled air, thiocyanate in blood and cotinine in urine to assays for smoke-related adducts in globin and DNA. The former group tends to represent very short-term exposures, and their measurement does not provide specific information on individual cancer risk. Because the products measured are not quantitatively related to the carcinogenic agents in tobacco smoke, these tests can discriminate between smokers and nonsmokers but can provide only relatively insensitive measures of smokers' risk. This is particularly true for nicotine products, which tend to be held at a constant level within individuals depending on the strength of their addiction.

Markers with the greatest potential value for predicting risk would be those which indicate the binding to DNA or protein of specific compounds with well-established carcinogenic properties. These carcinogenic properties should also relate to the cancers that are induced in humans by tobacco products. Methods for measuring specific, chemically characterized adducts for which the analytical parameters can be well-defined would be the most useful. In this respect, presently available mass spectrometric methods for measuring haemoglobin adducts of aminobiphenyl, ethylene oxide and tobacco-specific N-nitrosamines are of special interest. The extension of such methods to measurements of DNA would be important, as the relationship of haemoglobin adducts to DNA adducts in target tissues is likely to be complex. Postlabelling methods for measuring DNA adducts are presently the most sensitive, but the quantitative aspects of these methods require further development, as does their specificity in terms of identifying adducts of known structure. Immunoassays may have wide applicability when fully developed, but in many cases they are still limited in terms of sensitivity and specificity. The most appropriate method for measuring DNA adducts thus depends on a number of factors and must be chosen for each adduct individually.

Much remains to be done to improve our understanding of the importance of different smoking-related adducts, i.e., how they are related to each other and to biological endpoints, how specific they are for exposure to smoke and how predictive they are of overall cancer risk. Because of the enormous number of biologically active agents in tobacco smoke and the very high cancer risk associated with exposure to smoke, there are numerous biochemical markers and a well-defined epidemiological data base. This is, therefore, an area in which it should be possible to clarify the relationship between biochemical parameters and epidemiologically relevant exposure.

EXPOSURES AND ETIOLOGY

ENDOGENOUS N-NITROSATION AND CANCER OF THE BILIARY TRACT

C.P.J. Caygill[1], S.A. Leach[2], F. Fernandez[2], S. Duncan[2], M.J. Hill[2], R. Massey[3] & R. Hall[4]

[1]Public Health Laboratory Service, Communicable Disease Surveillance Centre, London; [2]Pathology Division, Public Health Laboratory Service, Centre for Applied Microbiological Research, Porton Down, Wilts; [3]Ministry of Agriculture, Fisheries and Food, Norwich; and [4]York District Hospital, York, UK

Evidence is presented that N-nitroso compounds occur in bile from patients who have undergone surgery for gallstones or had gastrectomy and from unoperated persons. It is unlikely, therefore, that local formation of nitrosamines can account for the excess risk for gallbladder cancer in the first two groups. Gastric formation remains the likeliest hypothesis.

Biliary tract cancer is commoner, after a 20-year latency, in persons who have undergone gastric operations (Caygill et al., 1984, 1988), and the risk for cancer of the gallbladder was increased by 15.8 fold among people treated for gastric ulcer. The number of sites at excess risk (Caygill et al., 1987) and the geographical correlation between gastric and biliary-tract cancers (Caygill et al., 1983) suggested a common etiology involving local endogenous formation of N-nitroso compounds in the stomach.

Table 1. Analytical findings in bile samples taken from three groups of patients

Patient group	No.	pH	Range	No. of samples with bacteria			Apparent total N-nitroso compounds		Nitrate (μM)	Nitrite (μM)
				Sterile	Anaerobes	Mixed	μg/kg	No. of samples		
Gastric surgery	5	6.7	6.3–7.0	0	3	3	35	1	12.5	15.0
Gallstones	20	6.6	6.3–7.0	7	6	5	25	2	28.6	8.0
No gallstones	18	6.4	6.0–6.6	1	9	9	73	3	27.7	16.6

Bile was collected at surgery for gallstones from five patients who had previously undergone gastrectomy and from 20 patients who had not, two groups known to be at excess risk for biliary-tract cancer. Bile was also collected during post-mortem examination of persons who had not had gallstones. The samples were analysed for bacteria, nitrate, nitrite, apparent total N-nitroso compounds and pH (Table 1). N-Nitroso compounds were analysed by the method of Massey *et al.* (1984); other analyses were carried out as described by Caygill *et al.* (1984).

N-Nitroso compounds were detected in all bile samples assayed. There is thus no evidence from this study that N-nitroso compounds are formed more readily in association with gallstones or after partial gastrectomy than in the normal gallbladder. Of the gastric surgery patients, 60% had a mixed bacterial flora, compared with 20% of patients with gallstones only. This was reflected in the nitrate reducing activity ($NO_2/NO_2 + NO_3$), which was 55% in the gastric surgery patients, 23% in those with gallstones only and 37% in the controls. We had expected the controls to be free of bacteria, but this was not the case, presumably due to bacterial proliferation before sampling.

As all samples contained N-nitroso compounds, and it is unlikely that they were formed locally, gastric formation remains the likeliest hypothesis. This needs confirmation by a larger study and further investigation of absorption from the stomach and biliary secretion of N-nitroso compounds.

Acknowledgements

We would like to thank Dr J.M. Hopkinson and Mrs K. Stirk for collecting the bile samples and Mrs S. Dickson for the organization.

References

Caygill, C., Hall, R. & Hill, M.J. (1983) Association between biliary tract cancer and gastric cancer. *Lancet*, ii, 1204

Caygill, C.P.J., Hill, M., Craven, J., Hall, R. & Miller, C. (1984) Relevance of gastric achlorhydria to human carcinogenesis. In: O'Neill, I.K., von Borstel, R.C., Miller, C.T., Long, J. & Bartsch, H., eds, N-*Nitroso Compounds: Occurrence, Biological Effects and Relevance to Human Cancer* (IARC Scientific Publications No. 57), Lyon, IARC, pp. 895-900

Caygill, C.P.C., Hill, M.J., Hall, C.N., Kirkham, J.S. & Northfield, T.C. (1987) Increased risk of cancer at multiple sites after gastric surgery for peptic ulcer. *Gut*, **28**, 924-928

Caygill, C., Hill, M., Kirkham, J. & Northfield, T.C. (1988) Increased risk of biliary tract cancer following gastric surgery. *Br. J. Cancer*, **57**, 434-436

Massey, R.C., Key, P.E., McWeeny, D.J. & Knowles, M.E. (1984) The application of a chemical denitrosation and chemiluminescence detection procedure for estimation of the apparent concentration of total N-nitroso compounds in foods and beverages. *Food Addit. Contamin.*, **1**, 11-16

EFFECT OF ASCORBIC ACID ON THE INTRAGASTRIC ENVIRONMENT IN PATIENTS AT INCREASED RISK OF DEVELOPING GASTRIC CANCER

P.I. Reed[1,4], B.J. Johnston[1], C.L. Walters[2] & M.J. Hill[3]

[1]*Lady Sobell Gastrointestinal Unit, Wexham Park Hospital, Slough, Berks;*
[2]*Department of Biochemistry, University of Surrey, Guildford; and*
[3]*Public Health Laboratory Service, Centre for Applied Microbiological Research, Porton Down, Salisbury, Wilts, UK*

Ascorbic acid has been shown to decrease nitrosation *in vivo*, and epidemiological data suggest that the consumption of foods rich in this vitamin is associated with a reduced risk for gastric cancer. In order to study this suggestion further, fasting gastric juice samples were obtained from 62 high-risk patients (seven with atrophic gastritis, ten with pernicious anaemia, ten with partial gastrectomy, 21 with vagotomy and drainage and 14 with highly selective vagotomy), before, during four weeks' treatment with 1 g ascorbic acid four times daily, and four weeks after treatment. Samples were analysed for pH, total and nitrate-reducing bacterial counts, nitrite and *N*-nitroso compounds. Treatment with ascorbic acid lowered the median pH only in the vagotomized patients ($p < 0.001$) but resulted in a reduction in median nitrate-reducing bacterial counts and in nitrite and *N*-nitroso compound concentrations in all groups, except for an increase in the nitrate-reducing bacterial count in atrophic gastritis patients and in nitrite in those with pernicious anaemia. These data suggest that treatment with a high dose of ascorbic acid reduces the intragastric formation of nitrite and *N*-nitroso compounds.

Alterations in the intragastric environment are probably important in the stages leading to the development of gastric cancer, and *N*-nitroso compounds may be implicated. As ascorbic acid, a powerful antioxidant, may reverse some of these environmental changes, thereby making progression to gastric cancer less likely, we have investigated the effect of the administration of high doses of ascorbic acid on the intragastric environment in patients at increased risk of developing gastric cancer.

Sixty-two patients (Table 1) with conditions leading to an increased risk for gastric cancer, taking a normal diet but avoiding vegetable and fruit juices, were treated for four weeks with ascorbic acid at 4 g daily. Gastric juice samples were obtained, either at endoscopy or through a nasogastric tube, one week apart, just prior to the start of treatment, at the end of four weeks' treatment and four weeks after the vitamin had been discontinued. The

---
[4]To whom correspondence should be addressed

pH, nitrite and total extractable N-nitroso compounds (Walters et al., 1978) were estimated and the juice cultured for total and nitrate-reducing bacteria. Blood samples were taken and serum ascorbic acid levels estimated.

Table 1. Characteristics of patients in the study

Clinical condition	Number	Male	Female	Age (years)		Patients with serum ascorbic acid < 2.0 mg/l (%)
				Mean	Range	
Atrophic gastritis	7	4	3	66.6	38–77	28
Pernicious anaemia	10	7	3	65.7	48–84	30
Partial gastrectomy	10	10	0	57.5	40–72	80
Vagotomy and drainage	21	16	5	56.4	41–73	57
Highly selective vagotomy	14	11	3	50.6	28–65	57

Nonparametric analysis of variance was used to examine the effects of treatment and disease group. Friedman's two-way analysis of variance was used to compare treatment periods, and Kruskal-Wallis' one-way analysis of variance was used to compare baseline values between patient groups.

Effects of treatment with ascorbic acid

Many patients had low baseline serum levels of ascorbic acid, and compliance with treatment was confirmed by a marked rise in serum levels during treatment (Figure 1). The levels were still high four weeks after the patients had been instructed to stop taking ascorbic acid, implying that treatment had often not been discontinued.

Figure 1. Median serum levels of ascorbic acid[a]

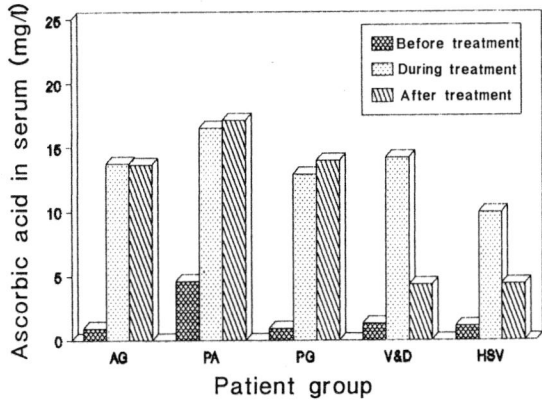

[a] AG, atrophic gastritis; PA, pernicious anaemia; PG, partial gastrectomy; V & D, vagotomy and drainage; HSV, highly selective vagotomy

There was no significant difference in the baseline pH between the groups. Pretreatment median pH was raised in all groups: atrophic gastritis, 5.5; pernicious anaemia, 6.0; partial gastrectomy, 5.8. The only alteration was seen in vagotomized patients: vagotomy and drainage, 4.1 pre- and 2.5 during; highly selective vagotomy, 4.6 pre- and 3.4 during treatment ($p < 0.05$).

Baseline levels of *N*-nitroso compounds were high in all groups compared with the levels found in normal controls (n = 50; mean, 0.14 μmol/l; Reed *et al.*, 1981), but a highly significant fall ($p < 0.001$) was observed during treatment (Figure 2). Baseline nitrite levels were also higher than expected in all patient groups (Reed *et al.*, 1981). Ascorbic acid treatment resulted in decreased levels in all groups ($p < 0.05$) except those with pernicious anaemia (Figure 3).

Figure 2. Median levels of *N*-nitroso compounds in gastric juice[a]

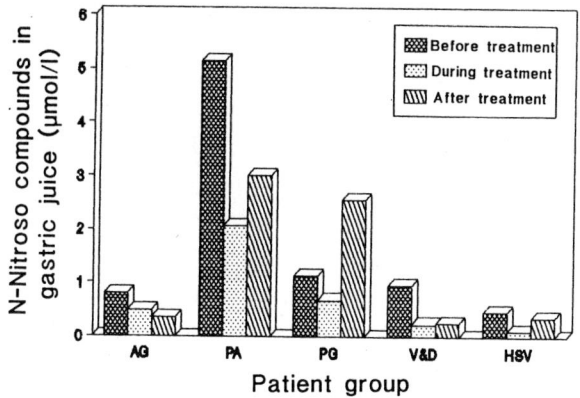

[a] AG, atrophic gastritis; PA, pernicious anaemia; PG, partial gastrectomy; V & D, vagotomy and drainage; HSV, highly selective vagotomy

Figure 3. Median levels of nitrite in gastric juice[a]

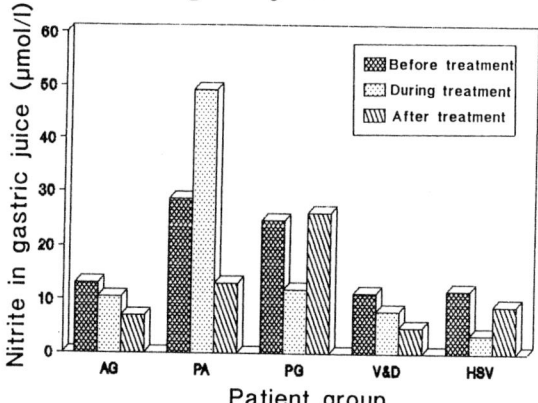

[a] AG, atrophic gastritis; PA, pernicious anaemia; PG, partial gastrectomy; V & D, vagotomy and drainage; HSV, highly selective vagotomy

Ascorbic acid treatment led to a significant reduction in nitrate-reducing bacterial counts ($p < 0.05$), which paralleled the fall in pH in vagotomized patients (Figure 4). There was no significant change in total bacterial counts during the study.

Figure 4. Median counts of nitrate-reducing bacteria in gastric juice[a]

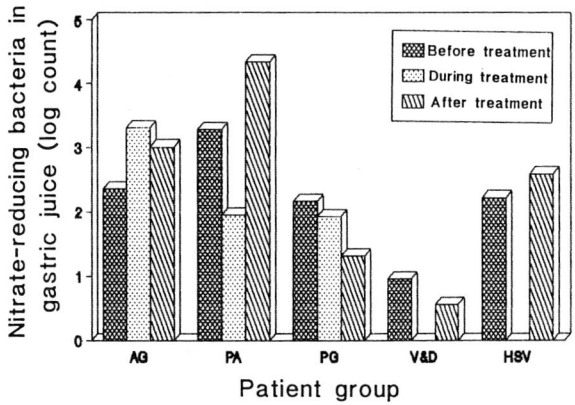

[a] AG, atrophic gastritis; PA, pernicious anaemia; PG, partial gastrectomy; V & D, vagotomy and drainage; HSV, highly selective vagotomy

All the patient groups studied thus had elevated baseline levels of both N-nitroso compounds and nitrite in gastric juice. This may explain the increased risk for gastric cancer in these patients, since it has been hypothesized that increased levels of these compounds are risk factors for gastric cancer.

Proposed future studies

Ascorbic acid treatment altered the intragastric environment, returning it closer to that seen in normal individuals. Thus, long-term prophylactic treatment with high doses of ascorbic acid may reduce the risk for gastric cancer in patients at increased risk. This concept requires testing in large-scale, long-term intervention studies in high-risk geographical areas.

Acknowledgements

BJJ was supported by the Wexham Gastrointestinal Trust.

References

Reed, P.I., Smith, P.L.R., Haines, K., House, F.R. & Walters, C.L. (1981) Gastric juice N-nitrosamines in health and gastroduodenal disease. *Lancet*, **ii**, 550-552

Walters, C.L., Downes, M.J., Edwards, M.W.K. & Smith, P.L.R. (1978) Determination of a non-volatile N-nitrosamine on a food matrix. *Analyst*, **103**, 1127-1133

AN APPROACH TO ESTABLISHING N-NITROSO COMPOUNDS AS THE CAUSE OF GASTRIC CANCER

K.X. Shi, D.J. Mao, W.F. Cheng, Y.S. Ji & L.Z. Xu

Department of Public Health, Shanghai Second Medical University, Shanghai 200025, China

Although gastric cancer is the most common cause of mortality from cancer, its etiology is not yet clear. To elucidate the role of N-nitroso compounds, we investigated 30 cases of gastric cancer by determination of the contents of nitrate and ascorbic acid and by detection of mutagens in urine. Cases were paired 1:1 with patients with dysplasia and normal controls of the same sex and age group. In comparison with normal controls, the gastric cancer group had higher nitrate and lower ascorbic acid levels in urine. Mutagenicity was observed in the urines of 83.3% of the gastric cancer cases and in 16.6% of the dysplasia group, but in none of those from normal controls. When the N-nitroso compound content of gastric juice was determined, the levels in control subjects were significantly lower than those in persons with gastric cancer. These results support the hypothesis that the cause of gastric cancer may be N-nitroso compounds.

Reports of epidemiological surveys show that dietary factors are closely related to gastric cancer (Shi *et al.*, 1988), and it has been suggested that N-nitroso compounds may be the cause of gastric cancer (Zhang, 1988). We have compared a group of gastric cancer patients with dysplastic and normal controls in an approach to elucidating the cause of gastric cancer.

Methods

We selected 30 cases of gastric cancer and paired them 1:1 with dysplastic patients and normal controls of the same sex and age group. All cases were diagnosed by endoscopy. The content of reduced ascorbic acid in urine was determined by titration with 2,6 dichlorophenol indophenol. For detection of mutagens in urine, the Ames test was employed. Determination of the content of nitrate in urine was done by using the KDK-2 type ion-selected electrode method. N-Nitroso compounds were determined in gastric juice by gas chromatography–thermal energy analysis.

Results

The reduced ascorbic acid content of urine was lower in the gastric cancer group than in normal controls, while the content of nitrate and the nitrate:ascorbic acid ratio in urine was significantly higher in the cancer patients than in controls (Table 1).

In the Ames test in *Salmonella typhimurium* TA98, in the presence of a 9000 g fraction from rat liver, 83.3% of urine samples from gastric cancer patients and 16.6% of those from dysplastic patients were mutagenic, while no mutagenicity was seen in urine from normal controls (Table 2).

Table 1. Contents of ascorbic acid and nitrate and nitrate:ascorbic acid in urine of gastric cancer patients and normal controls (mean ± SD)

Group	No.	Ascorbic acid (mg/g creatinine)	Nitrate (g/g creatinine)	Nitrate:ascorbic acid
Gastric cancer	30	21.8 ± 16.4	44.8 ± 76.2	1912 ± 1685
Normal controls	30	28.7 ± 27.4	14.7 ± 14.6	706 ± 596
p		>0.05	<0.05	<0.01

Table 2. Results of the Ames test with urine from gastric cancer and from dysplastic patients and normal controls

Group	No.	Positive rate (%)			
		TA98		TA100	
		-S9	+S9	-S9	+S9
Gastric cancer	30	23.3	83.3	30	30
Dysplasia	30	3.3	16.6	10	6.6
Normal controls	30	0	0	0	0

Using gas chromatography–thermal energy analysis, we determined the content of N-nitroso compounds in gastric juice of 12 cases of gastric cancer, six cases of dysplasia, 17 cases of chronic atrophic gastritis and 12 normal controls (Table 3). The content in normal controls was significantly lower than that in the gastric cancer patients ($p < 0.05$).

Table 3. Content of N-nitroso compounds[a] in gastric juice (ppb; mean ± SD)

Group	No.	NDMA	NDEA	NPYR	NPIP	Others	Total
Normal controls	12	2.0±5.9	0.21±0.72	0.57±0.86	0.01±0.05	0	2.8±5.8
Chronic atrophic gastritis	17	13.8±35.1	3.3 ±10.8	3.5 ±5.7	0.59±1.8	0.06±0.20	20.9±52.4
Dysplasia	6	17.9±40.2	0.46±1.1	0.82±1.7	0	0.44±1.1	24.6±51.7
Gastric cancer	12	11.0±8.8	5.2 ±13.2	6.0 ±12.9	0.76±1.5	0	25.4±39.0

[a] NDMA, N-nitrosodimethylamine; NDEA, N-nitrosodiethylamine; NPYR, N-nitrosopyrrolidine; NPIP, N-nitrosopiperidine

Discussion

The lower excretion of ascorbic acid in the urine of gastric cancer patients than in normal controls reflects a lower intake of this vitamin. Decreased intake of ascorbic acid could increase the synthesis of N-nitroso compounds in the stomach. The higher excretion of nitrate in urine from gastric cancer patients reflected a higher nitrate intake; and an increased level of nitrate in the stomach could be reduced by bacteria to nitrite, which would in turn increase the synthesis of N-nitroso compounds. We infer that the increased mutagenicity of the urine of gastric cancer patients might be related to increased synthesis of N-nitroso compounds in the stomach, since the gastric juice of the cancer patients contained significantly more N-nitroso compounds than that from normal controls. These results support the hypothesis that N-nitroso compounds may be the cause of gastric cancer.

References

Shi, K.X., Tao, Z., Qin, X.Y. & Jiao, G.S. (1988) Epidemiology of gastric cancer in China. In: *First Shanghai International Symposium on Gastrointestinal Cancers*, pp. 34-44

Zhang, R.F. (1988) The possible high risk factors for stomach cancer in China. In: *First Shanghai International Symposium on Gastrointestinal Cancers*, pp. 45-51

Relevance to Human Cancer of N-Nitroso Compounds,
Tobacco Smoke and Mycotoxins.
Ed. I.K. O'Neill, J. Chen and H. Bartsch
Lyon, International Agency for Research on Cancer
© IARC, 1991

THE N-NITROSOPROLINE TEST AS A MEASURE OF CANCER RISK IN GEOGRAPHICAL COMPARISON STUDIES: RESULTS FROM ITALY AND AN OVERALL COMPARISON

T. Knight[1], D. Forman[1], S.A. Leach[2], P. Packer[2], G. Cocco[3], D. Palli[4] & R. Pirastu[5]

[1]Imperial Cancer Research Fund, Epidemiology Unit, Radcliffe Infirmary, Oxford; [2]Bacterial Metabolism Group, Pathology Division, Public Health Laboratory Service, Centre for Applied Microbiological Research, Porton Down, Salisbury, Wilts, UK; [3]Institute of Occupational Medicine, Cagliari, Sardinia; [4]Centre for the Study and Prevention of Cancer, Florence; and [5]University of Rome, Rome, Italy

The N-nitrosoproline (NPRO) test has been used in studies in which populations at high risk of cancer have been compared with equivalent populations at lower risk, to examine whether the geographical variation in cancer risk correlates with propensity for endogenous nitrosation, as assessed by the NPRO test. The usual method employed has been to determine NPRO in 12- or 24-h urine samples, after ingestion of L-proline, in a representative sample of the general population. We present results from one such geographical study conducted in two regions of Italy (Florence and Cagliari) with an approximately three-fold variation in gastric cancer mortality. The nonsignificant difference in mean NPRO excretion between the two populations was insufficient to explain the difference in cancer risk. The fact that there are appreciable international differences in formation of NPRO suggests, firstly, that nitrosation may be of relevance to cancer risk in some countries but not in others and, secondly, that variations within one country may not be large enough for significant geographical differences to be evident. Multivariate analysis of individual, rather than grouped, results from our Italian study made it possible to quantify the relevance of different factors to NPRO formation: a major factor is exposure to nitrate. Important relationships may be missed by analysing only grouped data.

Nitrosation ability, as measured by the N-nitrosoproline (NPRO) test, has been compared in a number of studies (Bartsch *et al.*, 1989) in populations at different risks for cancer. Most of these studies have been carried out in Asian countries in populations at ages at which they are likely to have a high prevalence of gastric hypochlorhydria; under these conditions, the NPRO test performs suboptimally. We have studied the nitrosating ability of young, healthy males from two regions of Italy with contrasting rates of gastric cancer: the mortality rate from this cancer in males of all ages in 1975–77 was 63.8/100 000 population in Florence and 22.0/100 000 in Cagliari, adjusted to the European standard population (De Carli *et al.*, 1986).

In Florence and in Cagliari, 40 local male residents aged 25–40 years were invited to participate in the study by their family doctors. None had ever been diagnosed with gastric disease. On two consecutive days, 12-h urine samples were collected from each subject starting 1 h after the evening meal. The second sample was taken immediately after ingestion of 500 mg L-proline dissolved in water. To preserve the urine, 5 g NaOH were placed inside the containers; within 12 h of collection, sample volumes were recorded and two 100-ml aliquots were extracted and stored. One aliquot was stored at $-20\,°C$ for analysis of NPRO (Ohshima & Bartsch, 1981) within four months of collection; the other was stored at $4\,°C$ for analysis of nitrate (Phizackerley & Al-Dabbagh, 1983) within three months of collection.

Urinary NPRO and nitrate excretion

The levels of excreted NPRO and nitrate are given in Table 1. Background NPRO excretion (i.e., with no L-proline loading dose) was significantly higher in Florence than in Cagliari ($p = 0.038$). The levels of NPRO excretion rose significantly after ingestion of L-proline (test NPRO excretion) in both Florence and Cagliari ($p = 0.002$ and 0.0001, respectively), but there was no significant difference in mean urinary test NPRO excretion between the two areas. Mean urinary nitrate excretion did not vary significantly between the provinces, and there was no significant change in nitrate excretion after ingestion of L-proline in either province.

The interrelationships between background and test excretion of NPRO and nitrate (combining data from the two provinces) are examined in Table 2. At both low and high background exposure to NPRO, test NPRO excretion increased significantly with increasing exposure to nitrate. To some extent, test NPRO excretion was also dependent on background NPRO excretion, especially when urinary nitrate levels were low.

Table 1. Urinary excretion of N-nitrosoproline (NPRO) ($\mu g/12h$)^a and nitrate (mg/12h)^a in groups of 40 samples taken in two cities in Italy

Compound excreted	Florence		Cagliari	
	Background	With L-proline	Background	With L-proline
NPRO	0.94 (0.72–1.24)	1.59 (1.21–2.08)*	0.58 (0.41–0.83)	1.44 (1.10–1.87)*
Nitrate	37.1 (29.7–46.3)	36.0 (29.7–43.6)	33.7 (27.0–42.0)	34.4 (28.8–41.0)

^{a} Geometric mean (with 95% confidence interval)
* Different from background level at $p < 0.05$ (matched t-test)

Geographical comparisons

Our results contrast with those from a similar study in Japan (Kamiyama et al., 1987) in which NPRO excretion after L-proline dosing was significantly higher in a high-risk population for gastric cancer than in a low-risk population. In that study, nitrate excretion was significantly higher in the low-risk area, and it may be that the critical difference between the two populations was in dietary intake of modifying agents that inhibit nitrosation. If nitrosating ability, as indicated by the NPRO test, is partly responsible for

Table 2. Mean urinary N-nitrosoproline (NPRO) excretion[a], after L-proline loading dose, at different levels of excretion of background NPRO and of excretion of nitrate after the loading dose

Background NPRO (per 12 h)	No. of samples	Excretion of nitrate after L-proline loading dose (per 12 h)							p values[b]
		Low (<26.2 mg)	No. of samples	Medium (26.3–44.4 mg)	No. of samples	High (>44.5 mg)	No. of samples	All	
Low (<1 μg)	16	0.74 (0.54–1.02)	14	1.05 (0.65–1.68)	12	2.42 (1.40–4.18)	44	1.16 (0.88–1.53)	<0.001
High (≥1 μg)	10	1.37 (0.95–1.98)*	10	1.96 (1.39–2.75)	14	3.25 (2.53–4.19)	35	2.14 (1.73–2.64)*	<0.001
All	26	0.94 (0.72–1.22)	25	1.34 (0.97–1.83)	26	2.84 (2.13–3.77)	80	1.51 (1.25–1.83)	<0.001

[a] Geometric mean (with 95% confidence interval)
[b] Significance of univariate regression analysis of test NPRO versus test nitrate
* Significantly different at $p < 0.05$ from low background NPRO excretion (t-test)

the geographical difference in gastric cancer frequency within Japan, the same factor would not appear to discriminate between populations within Italy. Several explanations are possible.

Firstly, in correlation studies such as this, no account is taken of latency effects. Thus, current cancer rates relate to exposures that may have occurred several decades previously. In Japan, geographical differences in diet may not be large, so that current nitrosating ability is much the same as that in the past. In Italy, dietary patterns may have changed, especially among young people, so that geographical differences in gastric cancer within Italy may diminish, consistent with the patterns of nitrosation.

Secondly, the striking feature of similar studies (Table 3) is the substantial absolute difference in NPRO excretion between populations in Japan and China and those in Europe. The gross differences in diet (and possibly bacterial composition) between Asiatic and European populations are clear, but perhaps only in Asian countries is there sufficient variability in nitrosating ability to discriminate between populations.

Table 3. Studies in which urinary N-nitrosoproline (NPRO) excretion has been measured in population samples following a loading dose of L-proline

Country	L-Proline dose (mg)	No. of samples	Median NPRO (μg/24 h)	Median nitrate (mg/24 h)	Reference
Japan – Akita	3 × 100	52	12.6	116	Kamiyama et al. (1987)
– Iwate		52	7.1	177	
China – Linxian	3 × 100	52	8.3	85	Lu et al. (1986)
– Fanxian		52	4.4	57	
China[a] – 26 counties	500	Pooled from ~1000 males	24.8	360	Chen et al. (1987)
Poland – rural	3 × 100	43	2.8	118	Zatonski et al. (1989)
– urban		46	2.4	76	
Denmark[a] – low-nitrate water	500	Total, 281	0.0	63	Møller et al. (1989)
– medium-nitrate water			1.4	101	
– high-nitrate water			1.7	127	
Italy[a] – Florence	500	40	3.2	72	This study
– Cagliari		40	2.9	69	

[a] In these studies, urine was collected for 12 h; therefore, for comparison, the values presented are double those in the original paper.

Finally, the prevalence of atrophic gastritis in different populations is likely to increase in relation to the age of the population and give rise to hypochlorhydria. Decreased gastric acidity has the dual effect of enhancing reduction of nitrate to nitrite, an active nitrosating agent, while also reducing the efficiency of the chemically catalysed formation of NPRO. It has been established that NPRO formation is diminished in populations with atrophic gastritis (Crespi et al., 1987; Knight et al., 1988), and NPRO is not a good marker of nitrosation under such conditions. It has, however, been argued (Bartsch et al., 1989) that nitrosation may also play a role in carcinogenesis in the healthy normoacidic stomach, and in this case NPRO would be a valid indicator. Our evidence weighs against such an association.

Determinants of NPRO formation

Two factors appear to influence the rate of NPRO excretion after a proline dose: the levels of background NPRO and of nitrate excreted. Background NPRO probably derives from ingestion of preformed NPRO in the diet, especially from cured meats (Stich et al., 1984); this may be the explanation for the significantly higher background NPRO levels in residents of Florence, and it is necessary to adjust for preformed dietary NPRO when assessing NPRO as an indicator of endogenous nitrosation. Background NPRO may also derive from nitrosation of dietary L-proline. The occurrence of other, carcinogenic N-nitroso compounds in cured meat may also be of etiological significance for gastric cancer.

The pronounced and highly significant association between excretion of NPRO and of nitrate in individuals indicates that nitrate intake is an important determinant of NPRO formation. This finding contrasts with those of the population comparisons carried out in Japan, mentioned above; and the international differences in NPRO excretion (Table 3) cannot readily be explained by differences in nitrate intake. A dependency on nitrate level is, however, consistent with results obtained in nonsmokers in an individual-based study in Denmark (Møller et al., 1989). In our data, adjustment for smoking had no effect on the overall results. It would be interesting to test whether the relationship between nitrate and NPRO excretion is comparable in Chinese and Japanese individuals in whom the mean levels of NPRO excretion are considerably higher than those in Europe. It is possible that comparison of groups obscures effects that would be seen at an individual level.

References

Bartsch, H., Ohshima, H., Pignatelli, B. & Calmels, S. (1989) Human exposure to endogenous N-nitroso compounds: quantitative estimates in subjects at high risk for cancer of the oral cavity, esophagus, stomach and urinary bladder. *Cancer Surv.*, 8, 335-362

Chen, J., Ohshima, H., Yang, H., Li, J., Campbell, T.C. & Bartsch, H. (1987) A correlation of urinary excretion of N-nitroso compounds and cancer mortality in the People's Republic of China: interim results. In: Bartsch, H., O'Neill, I.K. & Schulte-Hermann, R., eds, *The Relevance of N-Nitroso Compounds to Human Cancer: Exposures and Mechanisms* (IARC Scientific Publications No. 84), Lyon, IARC, pp. 503-506

Crespi, M., Ohshima, H., Ramazzotti, V., Muñoz, N., Grassi, A., Casale, V., Leclerc, H., Calmels, S., Cattoen, C., Kaldor, J. & Bartsch, H. (1987) Intra-gastric nitrosation and precancerous lesions in the gastrointestinal tract: testing of an etiological hypothesis. In: Bartsch, H., O'Neill, I.K. & Schulte-Hermann, R., eds, *The Relevance of N-Nitroso Compounds to Human Cancer: Exposures and Mechanisms* (IARC Scientific Publications No. 84), Lyon, IARC, pp. 511-517

DeCarli, A., La Vecchia, C., Cislaghi, C., Mezzanotte, G. & Marubina, E. (1986) Descriptive epidemiology of gastric cancer in Italy. *Cancer*, 58, 2560-2569

Kamiyama, S., Ohshima, H., Shimada, A., Saito, N., Bourgade, M.C., Ziegler, P. & Bartsch, H. (1987) Urinary excretion of N-nitrosoamino acids and nitrate by inhabitants in high and low risk areas for stomach cancer in northern Japan. In: Bartsch, H., O'Neill, I.K. & Schulte-Hermann, R., eds, *The Relevance of N-Nitroso Compounds to Human Cancer: Exposures and Mechanisms* (IARC Scientific Publications No. 84), Lyon, IARC, pp. 497-502

Knight, T.M., Forman, D., Leach, S., Packer, P., Vindigni, C., Minacci, C., Lorenzini, L., Tosi, P., Frosini, G., Marini, M. & Carnicelli, N. (1988) N-Nitrosoproline excretion in patients with gastric lesions and in a control population. In: Bartsch, H., Hemminki, K. & O'Neill, I.K., eds, *Methods for Detecting DNA Damaging Agents in Humans: Applications in Cancer Epidemiology and Prevention* (IARC Scientific Publications No. 89), Lyon, IARC, pp. 97–101

Lu, S., Ohshima, H., Fu, H., Tian, Y., Li, F., Blettner, M., Wahrendorf, J. & Bartsch, H. (1986) Urinary excretion of N-nitrosoamino acids and nitrate by inhabitants in high and low risk areas for oesophageal cancer in northern China: endogenous formation of nitrosoproline and its inhibition by vitamin C. *Cancer Res.*, **46**, 1485-1491

Møller, H., Landt, J., Pedersen, E., Jensen, P., Autrup, H. & Jensen, O.M. (1989) Endogenous nitrosation in relation to nitrate exposure from drinking water and diet in a Danish rural population. *Cancer Res.*, **49**, 3117-3121

Ohshima, H. & Bartsch, H. (1981) Quantitative estimation of endogenous nitrosation in humans by monitoring N-nitrosoproline excreted in the urine. *Cancer Res.*, **41**, 3658-3662

Phizackerley, P.J.R. & Al-Dabbagh, S.A. (1983) The estimation of nitrate and nitrite in saliva and urine. *Anal. Biochem.*, **131**, 242-245

Stich, H.F., Hornby, A.P. & Dunn, B.P. (1984) The effect of dietary factors on nitrosoproline levels in human urine. *Int. J. Cancer*, **33**, 625-628

Zatonski, W., Ohshima, H., Przewozniak, K., Drosik, K., Mierzwinska, J., Krygier, M., Chmielarczyk, W. & Bartsch, H. (1989) Urinary excretion of N-nitrosamino acids and nitrate by inhabitants of high- and low-risk areas for stomach cancer in Poland. *Int. J. Cancer*, **44**, 823-827

ROLE OF NITROSAMIDES IN THE HIGH RISK FOR GASTRIC CANCER IN CHINA

R.F. Zhang, D.J. Deng, Y. Chen, H.Y. Wu & C.S. Chen

*Beijing Institute for Cancer Research,
Da Hong Luo Chang Street, Beijing 100034, China*

Gastric cancer is the leading cause of death from cancer in China. Samples of fish sauce, a traditional seasoning, were collected in the high-risk area for gastric cancer in the Fuzhou area, Fujian Province. When fish sauce samples were nitrosated at pH 2.0, direct mutagenicity and high contents of N-nitrosamide were detected (30.9–78.0 μM); the N-nitrosamide content of three samples of fish sauce made in Guangdong and purchased from a market outside Fujian were low (2.1–6.0 μm). When the nitrosated fish sauce extract was given to newborn rats by gavage, dysplasia and adenocarcinoma were induced in the glandular stomach in the 4th and 16th experimental week, respectively. N-Nitrosamides were also found in fasting gastric juice from patients with chronic gastritis in the high-risk area of Putian. The mean concentration of total N-nitrosamides in the extracts correlated with the severity of gastritis in the stomach. These findings indicate that N-nitrosamides may play an important role in causing gastric cancer in China.

Gastric cancer is the leading cause of death from cancer in China, accounting for 23% of all cancer deaths. In th Fuzhou area in Fujian province, the risk is particularly high, with standardized mortality rates in males of 113 per 100 000 in Changle and 68 per 100 000 in Putian counties.

The role of directly acting mutagenic/carcinogenic N-nitrosamides in gastric cancer in the Fuzhou area of China was studied by:

(i) measuring the mutagenicity of extracts of fish sauce, a traditional seasoning in the area, before and after nitrosation;

(ii) measuring the carcinogenicity of these extracts at pH 2 in newborn rats;

(iii) determining total N-nitrosamide in these extracts; and

(iv) correlating the concentration of N-nitrosamides in gastric juice samples from patients with chronic gastritis in Putian county with the severity of pathological changes in the stomach.

Testing of samples of fish sauce

A total of 52 samples of fish sauce were collected from villages in Changle county with different rates of mortality from gastric cancer. These were pooled into six samples: Zhanggang home-made, Zhanggang factory-made, Guhuai factory-made, Wuhang factory-made, Meihua factory-made and Meihua home-made. Another three samples purchased at markets were made in Guangdong province where the mortality rate for gastric cancer is relatively low.

Samples of 10 ml were extracted with ethyl acetate before and after nitrosation with 4000 ppm $NaNO_2$ under simulated gastric conditions at pH 2.0 and incubated at 25°C for 1 h. Sulfamic acid (20 mg/ml) was added to terminate the reaction, and 2,6-dimethylmorpholine was added to check for artefactual formation of N-nitroso compounds. The extracts were dried over anhydrous sodium sulfate and concentrated in a rotary evaporator to 1.0 ml for the Ames test (*Salmonella typhimurium* TA100; Maron & Ames, 1983), or were evaporated to dryness under a stream of nitrogen and the residue dissolved in dimethyl sulfoxide and deionized water to 2.0 ml for sister chromatid exchange (in V79 cells; Sandberg, 1980) and micronucleus (Heddle et al., 1983) tests.

For the carcinogenicity test, 0.1 ml extract (equivalent to 1.0 ml fish sauce) of Zhanggang home-made fish sauce nitrosated at pH 2.0 was given to ten newborn Wistar rats of each sex by gavage on three consecutive days. N-Methyl-N'-nitro-N-nitrosoguanidine (total dose, 150 μg/rat) was used as the positive control, and non-nitrosated fish sauce as the negative control. The rats were autopsied and the stomachs examined in the 4th and 16th experimental weeks. Paraffin sections of the stomach were stained with haematoxylin and eosin and examined under the light microscope.

N-Nitrosamides were determined using photohydrolysis system–thermal energy analysis (PHS-TEA; Chen et al., 1989). The photolytic apparatus was constructed according to Shuker and Tannenbaum (1983); no column was used. Freshly prepared nitrous acid was used as standard. The conditions for analysis were: mobile phase, deionized water; flow speed, 1.0 ml/min; oven temperature, 490–500°C; flow speed to carrier gas, 20 ml/min; temperature of cold trap 1, 0°C (iced water) and cold trap 2, –78° (dry ice/acetone); flow speed of O_2 in ozonator, 10 ml/min: vacuum, 0.4 torr; sample size, 10 μl; concentration of standards, 5 ppm. Under these conditions, no response was observed with N-nitrosamide standards (N-methyl-N-nitrosourethane and N-methyl-N-nitrosourea) when the lamp was off, but they gave good responses when the lamp was on. Seven volatile N-nitrosamine standards gave good responses with gas chromatography–TEA but no response with PHS-TEA, with the lamp off or on. The sensitivity was 0.5 ng or less.

All procedures were carried out in the dark.

Testing of gastric juice

A total of 46 gastric juice samples were collected from patients with chronic gastritis in Putian, a high-risk area for gastric cancer, after an overnight fast, *via* polyethylene tubes through the side channel of a fibroendoscope into containers of sulfamic acid (20 mg/ml), thoroughly mixed and centrifuged. The supernatant (5 ml) was extracted with acetone:dichloromethane (1:5), evaporated and concentrated to dryness. The residue was dissolved in 0.5 ml deionized water and analysed for N-nitrosamides.

The effective number of gastric juice samples was 39; they comprised samples from 13 patients with normal mucosa or chronic superficial gastritis, 16 with chronic atrophic gastritis and 17 with chronic atrophic gastritis plus intestinal metaplasia or dysplasia.

Direct mutagenicity of acidic, nitrosated fish sauce extracts

In the absence of nitrosation, none of the fish sauce extracts was mutagenic, whether an exogenous metabolic activation system was present or not. After nitrosation, however, all six pooled samples had marked direct mutagenic activity in *S. typhimurium* TA100 and induced sister chromatid exchange with a dose-response relationship. In the micronucleus test, direct mutagenicity was observed only for the extracts from Zhanggang and Guhuai villages (Table 1).

Table 1. Direct mutagenicity of nitrosated fish sauce extracts

Sample[a]		No. of his^+ revertants[b] (mean ± SD)	V79 cells[c]		
			SCE/cell (mean ± SD)	MNC (%)	MN (%)
Control		167±42.8	6.50±2.61	0.7	0.7
Zhanggang	(H)	414±70.0	8.72±3.34*	1.6[d]	2.2[d]
Zhanggang	(F)	406±57.3	8.72±3.59*	1.0	1.1
Guhuai	(F)	377±47.2	8.36±3.19**	1.4[d]	1.4[d]
Wuhang	(F)	375±51.0	9.20±4.92**	1.2	1.3
Meihua	(F)	397±77.4	8.00±3.11***	1.1	1.1
Meihua	(H)	361±40.1	8.16±3.60***	0.9	1.1

[a] H, home-made; F, factory-made
[b] 100 μl extract/plate; 159±14.8 spontaneous revertants
[c] 25 μl extract/ml medium; total extract, 125 μl in 5 ml medium; SCE, sister chromatid exchange; MNC, micronucleated cell rate; MN, micronucleus rate
[d] Positive (greater than twice the value of the negative control)
* $p < 0.001$
** $p < 0.01$
*** $p < 0.05$

Carcinogenicity of acidic, nitrosated fish sauce extracts

All five rats autopsied in the 4th experimental week after receiving extracts of fish sauce from Zhanggang showed marked precancerous dysplasia in the mucosa of the glandular stomach (Figure 1). In the 16th week, an adenocarcinoma was observed in the pyloric region in one of the five rats autopsied. The dysplastic glands and cells had penetrated the mucosa and infiltrated into the submucosa and muscular layers of the gastric wall (Figure 2). A cancerous ulceration was also seen in the glandular stomach, with scattered neoplastic cells in the submucosa and muscular layer beneath the ulcer.

N-Nitrosamides in extracts of acidic, nitrosated fish sauce and gastric juice

On PHS-TEA, no response was observed with non-nitrosated fish sauce extracts, but high levels of *N*-nitrosamides were detected in all six pooled fish sauce samples after acidic nitrosation at concentrations of 30.9–78.0 μM, with a mean value of 59.9 μM. No marked difference was seen among the samples. The concentrations of *N*-nitrosamides in the three samples from Guangdong were significantly lower, with a mean value of 3.5 μM (Table 2). No artefactual formation of *N*-nitroso compounds was seen.

High levels of N-nitrosamides were detected in the gastric juice samples from patients with chronic gastritis in Putian county. The positive rates and the mean concentrations in the three patient groups were positively correlated with the severity of pathological changes in the stomach (Table 3).

Figure 1. Dysplasia in rat glandular stomach after administration of nitrosated extract of fish sauce from a high-incidence area for gastric cancer in China; H & E; × 60

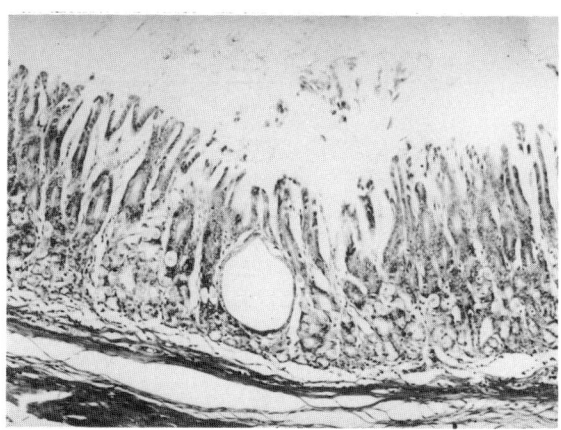

Table 2. Concentrations (μM) of total N-nitrosamides in extracts of fish sauce from low- and high-risk areas for gastric cancer after nitrosation at pH 2.0

High-risk area for gastric cancer (Changle)			Low-risk area for gastric cancer (Guangdong)	
Zhanggang	(H)	60.4	(F)	2.4
Zhanggang	(F)	54.6	(F)	2.1
Guhuai	(F)	76.1	(F)	6.0
Wuhang	(F)	30.9		
Meihua	(F)	78.0		
Meihua	(H)	59.1		
Mean ± SD		59.9±33.5		3.5±4.3
p value		<0.01		

H, home-made; F, factory-made

Discussion

Samples of the commonly used seasoning, fish sauce, collected from a high-risk area for gastric cancer in the Fuzhou area of Fujian province contained high levels of precursors of directly acting mutagens/carcinogens. After nitrosation at pH 2.0, the extracts induced direct mutagenicity and dysplasia and adenocarcinoma in rat glandular stomach. PHS-TEA analysis indicated that the compounds in nitrosated fish sauce extracts were mostly nonvolatile N-nitrosamides.

In previous studies, we found that the nitrate contents of drinking-water, grains and vegetables were higher in Changle county than in other areas of China (Zhang et al., 1984, 1985).

Figure 2. Adenocarcinoma in rat glandular stomach 16 weeks after administration of nitrosated extract of fish sauce from a high incidence area for gastric cancer in China; H & E; × 120

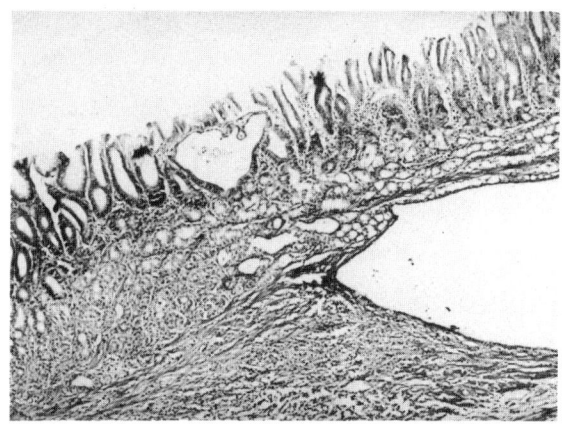

Table 3. Relationship between presence and concentrations of N-nitrosamide in gastric juice with pathological changes in gastric mucosa

Gastric mucosal status[a]	No. of cases	Conc. total N-nitrosamides		Positive rate[b] (%)
		Mean ± SD	Range	
N and CSG	12	2.8±13.8	ND-23.6	16.7
CAG	14	26.7±43.7[c]	ND-60.9	71.4[c]
CAG and IM or Dys	13	55.2±77.3[d]	ND-137.9	92.3

[a] N, normal mucosa; CSG, chronic superficial gastritis; CAG, chronic atrophic gastritis; IM, intestinal metaplasia; DYS, dysplasia
[b] Total positive rate, 61.5%
[c] Significantly different from N and CSG at $p < 0.01$
[d] Significantly different from CAG at $p < 0.05$
ND, not detected

The nitrite content of gastric juice samples from patients with chronic gastritis in this county was also the highest measured (Zhang et al., 1984). If the precursors of N-nitrosamides present in locally made fish sauce are nitrosated in the stomachs of residents, this preparation may play an important role in the development of precancerous lesions for gastric cancer and even of gastric cancer in people in that area.

Using the new PHS-TEA technique, we detected nonvolatile N-nitrosamides directly at high levels in gastric juice samples and found the total concentration to be positively correlated with the severity of chronic gastritis.

These data strongly supported the hypothesis that the etiology of human gastric cancer is closely related to intragastric formation of directly acting mutagenic/carcinogenic N-nitrosamides from exogenous and endogenous nitrosatable precursors.

Acknowledgements

The authors would like to thank Miss S.X. Zhu for her help in conducting the animal experiments and for preparing slides and Miss T. Zhou and Ms R.M. Wang for collecting and processing the gastric juice samples. This work was supported by a grant from the National Commission of Science and Technology, China, and in part by financial support from the International Agency for Research on Cancer, Lyon, France.

References

Chen, Y., Wu, H.Y., Chen, C.S. & Zhang, R.F. (1989) Establishment of a N-nitrosamide detection system and investigation of total N-nitrosamide in fasting gastric juice of chronic gastritic patients. In: *4th National Conference on Gastric Cancer, Shenyang, 19-21 October 1989, Liaoning Province, China*

Heddle, J.A., Hite, M., Kirkhart, B., Mavournin, K., MacGregor, J.T., Newell, G.W. & Salamone, M.F. (1983) The induction of micronuclei as a measure of genotoxicity. A report of the US Environmental Protection Agency Gene-Tox Program. *Mutat. Res.*, **123**, 61-118

Maron, D.M. & Ames, B.N. (1983) Revised methods for the Salmonella mutagenicity test. *Mutat. Res.*, **11**, 173-215

Sandberg, A.A. (1980) Method for sister chromatid exchange. In: Sandberg, A.A., ed., *The Chromosomes in Human Cancer and Leukemia*, New York, Elsevier, pp. 106-109

Shuker, D.E.G. & Tannenbaum, S.R. (1983) Determination of nonvolatile N-nitroso compounds in biological fluids by liquid chromatography with postcolumn photohydrolysis detection. *Anal. Chem.*, **55**, 2152-2155

Zhang, R.F., Sun, H.L., Jin, M.L. & Li, S.N. (1984) A comprehensive survey of etiologic factors of stomach cancer in China. *Chin. Med. J.*, **97**, 322-332

Zhang, R.F., Zhou, X.M., Shen, W.Y., Liu, D.X. & Song, P.J. (1985) The relationship between nitrate and nitrite contents in foodstuffs and in saliva and gastric juice of patients with chronic gastric disorders and its significance in the etiology of gastric cancer. *Chin. J. Cancer*, **3**, 143-147

Relevance to Human Cancer of N-Nitroso Compounds,
Tobacco Smoke and Mycotoxins.
Ed. I.K. O'Neill, J. Chen and H. Bartsch
Lyon, International Agency for Research on Cancer
© IARC, 1991

MICROFLORA OF THE NASOPHARYNX IN CAUCASIAN AND MAGHREBIAN SUBJECTS WITH AND WITHOUT NASOPHARYNGEAL CARCINOMA

M. Charrière[1,3], S. Poirier[1,2], S. Calmels[2], H. De Montclos[3],
C. Dubreuil[4], R. Poizat[5], M. Hamdi Cherif[6] & G. de Thé[1]

[1]*Faculté Alexis Carrel,* [2]*International Agency for Research on Cancer,*
[3]*Institut Pasteur,* [4]*Jules Courmont Hospital, and* [5]*Croix Rousse Hospital, Lyon, France; and* [6]*Sétif Hospital, Sétif, Algeria*

The possible role of bacteria in the etiology of nasopharyngeal carcinoma (NPC) was studied by bacteriological and biochemical analyses of nasopharyngeal swabs collected in the cavum and in the fossa of Rosenmüller of NPC patients and healthy controls in France and Algeria. Counts of total bacteria and of total nitrate-reducing bacteria, mainly enterobacteria, were higher in the Maghrebians than in the Caucasians. The composition of the bacterial flora was different: in Maghrebians, enterobacteria were present in five of 17 control subjects and eight of 15 NPC patients, while the prevalence was only one out of 15 control subjects in Caucasians. Twelve of 32 bacterial species isolated from Caucasians and Maghrebians with normal or tumorous nasopharyngeal microflora were able to catalyse nitrosation of morpholine *in vitro*. This result suggests that colonization of the nasopharynx by microflora that contain nitrate-reducing microorganisms which can form N-nitroso compounds might represent a risk factor for NPC in Maghrebian populations.

Nasopharyngeal carcinoma (NPC) is rare in Europe and in North America but represents one of the major causes of mortality from cancer in south-east Asia (Waterhouse *et al.*, 1976; de Thé, 1982), in Arctic areas (Lanier *et al.*, 1980) and in the Maghreb (North Africa; Muir, 1971). The site of the initial growth of NPC is either the posterior, superior walls of the cavum or the fossa of Rosenmüller. These fossae could represent an ecological 'niche' with suitable conditions for colonization by a specific microflora, which may differ according to diet and therefore between ethnic groups. Some bacterial flora could result in (i) production of fatty acids, such as butyric acid, which is known to increase the reactivation of Epstein-Barr virus in Raji cells (Ito *et al.*, 1981), and/or (ii) the formation of N-nitroso compounds *via* reduction of nitrate into nitrite and subsequent catalysis of the nitrosation of amino precursors, possibly also formed by bacterial metabolism (Calmels *et al.*, 1987; Leach *et al.*, 1987).

In order to examine the potential role of bacteria in the etiology of NPC, bacteriological analyses were carried out on swabs collected in the cavum and in the fossa of Rosenmüller of both NPC patients and control subjects in France and Algeria and analysed for their ability to produce butyric acid, to reduce nitrate into nitrite and to catalyse nitrosation.

Caucasian controls (n = 15) and NPC cases (n = 3) were selected in hospitals in Lyon, France, and Maghrebian control subjects (n = 17) and NPC patients (n = 15) were selected in Sétif, Algeria. Samples were collected using a cotton bud from the cavum and from the right and left fossa of Rosenmüller and were stored on a Difco swab medium (which does not allow growth of bacteria) during transport of the samples at ambient temperature. Samples were cultured (i) under 10% CO_2 on blood agar for 24 h to isolate aerobic flora; (ii) on 5% horse blood agar under anaerobiosis for four days to isolate anaerobic flora; and (iii) on Sabouraud agar in the presence of chloramphenicol at 30°C for four days to detect the presence of yeasts.

Bacterial count and composition of microflora

The total mean bacterial count was ten times greater in Maghrebian controls (10^7 bacteria/sample) than in the Caucasian controls (10^6 bacteria/sample) in both cavum and fossa (Table 1); however, the bacterial counts varied widely among the three Caucasian NPC cases, so it was difficult to compare this group with the others.

Table 1. Microflora of the cavum and of the fossa of Rosenmüller in Caucasian and Maghrebian populations (mean ± SD)

Subjects	Bacteria/sample $\times 10^6$		No. of samples with		Nitrate-reducing bacteria/sample $\times 10^6$	
	Cavum	Fossa	Entero-bacteria	Yeasts	Cavum	Fossa
Caucasians						
Controls	1.3±6.6	3.5±6.5	4/15	1/15	1.0±1.3	2.7±6.4
NPC cases	2.1±3.0	12.3±17.4	0/3	1/3	1.5±2.1	11.3±16.0
Maghrebians						
Controls	39.6±85.5^a	42.6±74.1^b	7/17	1/17	16.2±24.3	24.0±24.0
NPC cases	23.5±27.1	36.7±28.9	11/15	6/15	9.4±9.3	21.8±16.2

^{a} $p < 0.001$ in comparison with Caucasian controls
^{b} $p < 0.005$ in comparison with Caucasian controls

The anaerobic bacterial count was low in both groups. Two genera, *Propionibacterium* and *Bacteroides* predominated, which are normally present in the nasopharynx. None produced butyric acid.

The aerobic flora tended to be more numerous and diversified (see Table 2), particularly in the Maghrebian populations; eight genera were identified in the Caucasian groups and 12 in the Maghrebian groups. *Streptococcus* and *Staphylococcus* were found in both populations; the other genera belonged mainly to the Enterobacteriaceae family (Table 1). In the Maghrebian groups, the enterobacteria *Klebsiella pneumoniae*, *K. oxytoca* and *Morganella morganii* generally predominated in both the cavum and the fossa of Rosenmüller. *Proteus* species were detected in both populations, but in the Maghrebian groups development of this species masked the growth of other microorganisms, so that five samples had to be discarded. Gram-negative bacteria (*Moraxella* and *Pasteurella*) were identified only in Maghrebian populations.

Yeasts (10-10² cells/sample) were observed mainly in Maghrebian NPC patients (Table 1); they were also present in one Caucasian NPC patient treated with radiotherapy. The yeasts belonged to the genera of *Candida* and *Torulopsis*, which are normal hosts of the mucosae.

Nitrate reductase and nitrosating activities of nasopharyngeal microflora

Most of the bacterial species, except *Streptococcus* and *Neisseria*, reduced nitrate to nitrite (Table 2). Nitrate-reducing bacteria constituted 75% of the total microflora in Caucasian and Maghrebian control groups and 86% of that in NPC patients (Table 1), however, the percentage in the Maghrebian study subjects is greater because the total microflora count was larger.

Table 2. Bacterial genera found in the study populations, with nitrate reductase and nitrosating capacities

Bacteria	No. of strains tested	Nitrate reductase activity (present (+)/absent (-))	Nitrosating activity[a]
Staphylococcus aureus	4	+	ND
Staphylococcus epidermis	4	+	ND
Streptococcus	3	-	ND
Proteus	3	+	2–127
Klebsiella pneumoniae	3	+	ND – 4
Klebsiella oxytoca	2	+	ND – 32
Morganella	6	+	ND – 54
Enterobacter	1	+	ND
Escherichia	1	+	15
Serratia		+	NA
Citrobacter		+	NA
Haemophilus	3	+	ND
Neisseria		-	NA
Moraxella		+	NA
Pasteurella	1	+	ND
Pseudomonas stutzeri	2	+	35–89

[a] nmol N-nitrosomorpholine formed/mg protein per hour; ND, not detected; NA, not assayed. The assay for nitrosation was performed as reported previously (Calmels *et al.*, 1987) in the presence of a number of bacteria equivalent to 1-5 mg protein per assay containing 0.1 ml/l Tris buffer pH 7.2, 25 mmol/l $NaNO_2$, 25 mmol/l morpholine (final volume, 5 ml). After 1 h incubation at 37°C, the reaction was stopped by adding 1 ml 5N NaOH to 1 ml of the reaction mixture, 500 ng/l N-nitrosomethylpentylamine were added as internal standard, and volatile nitrosamines were extracted in dichloromethane and analysed by gas chromatogrpahy-thermal energy analysis. The presence or absence of nitrate reductase activity was determined by measuring nitrite in the nitrate growth media in which the bacterial strain had been cultured overnight.

Most of the bacterial species were studied for their ability to catalyse nitrosation of morpholine by nitrite at neutral pH *in vitro* at 37°C: 12 out of 32 bacterial strains belonging to the Enterobacteriaceae (*Klebsiella, Proteus, Escherichia coli, Morganella*) and Pseudomonadaceae families had both nitrate reductase and nitrosating activities (Table 2).

Bacterial strains of *Streptococcus*, with no nitrate reductase, were unable to catalyse nitrosamine formation *in vitro*; however, *Staphylococcus* and *Haemophilus* strains, which have strong nitrate reductase activity, were also devoid of nitrosating capacity.

The question is raised of whether bacterial colonization of the nasopharynx represents a cause or a consequence of NPC development. Our results show that in the Maghrebian subjects the typical nasopharyngeal microflora was already abundant before clinical manifestation of NPC. This suggests that colonization of the nasopharynx by a microflora that contains nitrate-reducing organisms, which can result in the formation of *N*-nitroso compounds *via* nitrite, could represent a risk factor for NPC in Maghrebian populations. The general validity of this hypothesis should be verified by similar studies in Chinese and Greenland Eskimo populations, which are also at high risk for NPC.

Conclusion

The results of this preliminary study suggest that nasopharyngeal microflora may be involved in the etiology of NPC. Although we studied only a limited number of Caucasian NPC patients, the total count of nasopharyngeal microflora was higher in the control Maghrebian group than in the Caucasian control groups. The microflora in the Maghrebian groups was of more diverse composition, and many of the bacterial species (approximately 80%) could reduce nitrate to nitrite. Most importantly, some of the nitrate-reducing genera, mainly enterobacteria, were able to catalyse the formation of nitrosamines *in vitro*.

References

Calmels, S., Ohshima, H., Rosenkranz, H., McCoy, E. & Bartsch, H. (1987) Biochemical studies on the catalysis of nitrosation by bacteria. *Carcinogenesis*, **8**, 1085-1088

Ito, Y., Kawanishi, M., Hirayama, T. & Takabayashi, S. (1981) Combined effect of the extracts from *Croton tiglium*, *Euphorbia lathyris*, or *Euphorbia tirucalli* and *n*-butyrate on Epstein-Barr virus expression in human lymphoblastoid P3 HR-1 and Raji cells. *Cancer Lett.*, **12**, 175-180

Lanier, A., Bender, T., Talbot, M., Wilmeth, S., Tschopp, C., Henle, W., Henle, G., Ritter, D. & Terasaki, P. (1980) Nasopharyngeal carcinoma in Alaskan Eskimos, Indians and Aleuts: a review of cases and study of Epstein-Barr virus HLA and environmental risk factors. *Cancer*, **46**, 2100-2106

Leach, S., Cook, A., Challis, B., Hill, M. & Thompson, M. (1987) Bacterially mediated *N*-nitrosation reactions and endogenous formation of *N*-nitroso compounds. In: Bartsch, H., O'Neill, I.K. & Schulte-Hermann, R., eds, *The Relevance of N-Nitroso Compounds to Human Cancer: Exposures and Mechanisms* (IARC Scientific Publications No. 84), Lyon, IARC, pp. 396-398

Muir, C.S. (1971) Nasopharyngeal carcinoma in non-Chinese populations with special reference to south-east Asia and Africa. *Int. J. Cancer*, **8**, 351-363

de Thé, G. (1982) Epidemiology of Epstein-Barr virus and associated disease in man. In: Ruizman, B., ed., *The Herpes Viruses*, Vol. I, New York, Plenum Press, pp. 25-103

Waterhouse, J.A.H., Muir, C., Correa, P. & Powell, J., eds (1976) *Cancer Incidence in Five Continents, Vol. III* (IARC Scientific Publications No. 15), Lyon, IARC, pp. 21-25

Relevance to Human Cancer of *N*-Nitroso Compounds,
Tobacco Smoke and Mycotoxins.
Ed. I.K. O'Neill, J. Chen and H. Bartsch
Lyon, International Agency for Research on Cancer
© IARC, 1991

EXPOSURE TO *N*-NITROSAMINES AND OTHER RISK FACTORS FOR GASTRIC CANCER IN COSTA RICAN CHILDREN

R. Sierra[1], H. Ohshima[2], N. Muñoz[2], S. Teuchmann[2], A.S. Peña[3],
C. Malaveille[2], B. Pignatelli[2], A. Chinnock[1], F. El Ghissassi[2],
C. Chen[2], A. Hautefeuille[2], C. Gamboa[1] & H. Bartsch[2]

*[1]University of Costa Rica, San José, Costa Rica;
[2]International Agency for Research on Cancer, Lyon,
France; and [3]Department of Gastroenterology, University
Hospital, Leiden, The Netherlands*

The hypothesis that endogenous chemical nitrosation in the normal stomach in early life could play a crucial role in inducing chronic atrophic gastritis/intestinal metaplasia in later life was tested by applying the *N*-nitrosoproline (NPRO) test to 12-h urine samples from about 50 children (aged 8–14 years) living in high- and low-risk areas for stomach cancer. The median values of NPRO and the sum of four nitrosamino acids analysed were 0.28–0.84 µg/12 h and 0.75–1.75 µg/12 h, respectively. The NPRO level after proline intake was significantly higher in children from a high-risk area than in those from a low-risk area ($p < 0.04$), and markedly reduced after ingestion of ascorbic acid and proline ($p < 0.05$). Urinary nitrate level was lower than that of adults. NPRO levels on the day of proline intake, however, correlated well with nitrate levels ($p < 0.001$), indicating that children in a high-risk area in Costa Rica have high endogenous nitrosation potential. Blood samples were also collected from about 300 children (aged 7–20 years) and analysed for antibodies against *Campylobacter pylori*, a suspected gastritis-causing bacteria. About 71% of children in both high- and low-risk areas for stomach cancer had antibodies. In addition, raw and cooked beans, which are consumed very frequently in Costa Rica, were collected from families in both areas and analysed for levels of nitrite/nitrate, total *N*-nitroso compounds and genotoxicity in the SOS chromotest. Mean levels of total *N*-nitroso compounds in an aqueous extract (pH 2) of cooked bean samples from high- and low-incidence areas were similar (0.4–0.6 nmol/g of cooked beans). Acid-catalysed nitrosation of the same aqueous extracts produced levels up to 2.4 µmol/g of cooked beans. There was no difference in mean levels of nitrosation-dependent total *N*-nitroso compounds between samples from the two areas. Only two out of 11 extracts from the low-incidence area and two out of 14 from the high-incidence area showed weak direct genotoxicity. After acid-catalysed nitrosation, all samples were genotoxic at similar levels.

Stomach cancer is the most common cause of death from cancer in Costa Rica: in 1984, it caused 33% of deaths from cancer in males and 22% in females. For the period of 1980–83, the age-adjusted incidence rate was 58.8 per 100 000 for men and 25.2 per

100 000 for women. Although these incidence rates are declining, they are the highest in Latin America, and internationally they are second only to those observed in Japan. There is, however, great variation in rates in this small country (51 000 km2) which has a relatively homogeneous population: for men, the rate varies from 84.2 per 100 000 in the highlands of the centre of the country to 25.4 in the coastal areas (Sierra et al., 1989). These data suggest that environmental factors and habits in the different regions could be important risk factors.

It has been postulated that N-nitroso compounds, in particular N-nitrosamides, are formed intragastrically by bacteria colonized in the stomach, on the basis of studies of patients with chronic atrophic gastritis (a precancerous condition of the stomach; Correa et al., 1975; Correa, 1988). However, it is a matter of controversy whether the level of N-nitroso compounds formed in the stomach of such patients is higher than that of normal subjects (Bartsch et al., 1989); patients with chronic atrophic gastritis did not excrete more N-nitrosoproline than patients with a normal stomach after ingestion of proline and nitrate (Crespi et al., 1987). On the basis of these results and epidemiological data, in particular that from studies of migrants, it has been hypothesized that N-nitroso compounds formed by acid-catalysed chemical nitrosation in the normal stomach in early life could play a crucial role in gastric carcinogenesis (Crespi et al., 1987; Ohshima et al., 1988).

We have postulated that bacterial colonization in the stomach, especially by the bacterium *Campylobacter pylori*, recently identified in human stomach mucosa as a possible etiological agent of gastritis, may contribute to gastric carcinogenesis by inducing chronic irritation or inflammation. We tested these hypotheses in Costa Rican children living in high- and low-risk areas for stomach cancer, using the N-nitrosoproline test and serological examination for *C. pylori* infection to assess exposures.

One community in a high-risk area (Turrubares; age-adjusted incidence rate of gastric cancer, 61.3/100 000) and one in a low-risk area (Hojancha; 18.7/100 000) for stomach cancer were selected on the basis of the following criteria: rural community, a large enough population of school children, similar ethnic characteristics and accessibility. A list was produced of all children aged 8–14 years in each community, and 25 subjects were selected at random from each list. Two samples of 12-h overnight urine were collected from children after they had ingested 500 mg proline and 200 mg ascorbate or 500 mg proline alone 1 h after the evening meal. The urine samples were analysed for N-nitrosamino acids and nitrate as exposure markers, as described previously (Ohshima et al., 1987).

Blood samples were also collected from about 200 children aged 7–13 years and from about 80 volunteers aged 14–20 in the same communities and analysed for antibodies to *C. pylori*, using a method described by Peña et al. (1990).

Levels of N-nitrosamino acids and nitrate

The results of analyses for N-nitrosoamino acids and nitrate are given in Table 1. The levels of N-nitrosamino acids were much lower (1/5–1/10) than those detected in adult urines collected in Japan, China, Poland and other countries (Bartsch et al., 1989). It is not known whether the low levels in Costa Rican children are due to their low body weight or to other factors related to age, or to geographical, environmental and dietary factors, such

Table 1. Median (95% confidence intervals) amounts of N-nitrosamino acids (μg/12 h) and nitrate (mmol/12 h) excreted in 12-h urine[a]

Urinary component[b]	Low-risk area		High-risk area	
	Day 1 (proline plus vitamin C)	Day 2 (proline)	Day 1 (proline plus vitamin C)	Day 2 (proline)
Volume (ml)	415.00 (395–542)	445.00 (396–559)	327.00 (293–473)	377.00 (286–460)
N-Nitrosamino acids				
NPRO	0.28 (0.21–0.51)	0.54 (0.43–1.09)	0.66 (0.44–1.29)	0.84 (0.89–2.44)
NTCA	0.48 (0.38–1.27)	0.52 (0.37–1.48)	0.40 (0.33–1.64)	0.77 (0.64–1.80)
NMTCA	0.05*(0.05–0.10)	0.05 (0.04–0.51)	0.05 (0.04–0.40)	0.05 (0.04–0.30)
Sum	0.75 (0.73–1.80)	1.25 (1.08–2.82)	1.62 (1.00–3.13)	1.75 (1.67–4.44)
Nitrate	0.23 (0.22–0.38)	0.20 (0.21–0.39)	0.29 (0.26–0.66)	0.23 (0.17–0.87)

[a] 26 children (19 boys and 7 girls) and 25 children (15 boys and 10 girls) from high- and low-risk areas for stomach cancer, respectively, participated in the study; their mean ages were 11.3 and 10.5 years, and mean body weights were 36 and 34 kg, respectively
[b] NPRO, N-nitrosoproline; NTCA, N-nitrosothiazolidine 4-carboxylic acid; NMTCA, N-nitroso-2-methythiazolidine 4-carboxylic acid

as a high intake of ascorbic acid (dietary data obtained in these areas are currently being analysed; Sierra et al., in preparation). All comparisons of levels of N-nitrosoproline were statistically significant. Thus, the level was reduced by ingestion of ascorbic acid with proline compared with that seen with proline alone in both areas (low-risk area, $p < 0.01$; high-risk area, $p < 0.05$; Wilcoxon test); and the level was significantly higher in children in the high-risk area than in those in the low-risk area after ingestion of ascorbic acid plus proline ($p < 0.02$) and after ingestion of proline alone ($p < 0.04$; both Mann-Whitney test). Thus, the level of N-nitrosoproline after proline intake was higher in children from the high-risk area than in those from the low-risk area and markedly reduced after ingestion of ascorbic acid and proline. The nitrate level was also lower (1/2–1/3) than that in adult urines; however, the N-nitrosoproline levels on the day of proline intake correlated well with the nitrate levels ($p < 0.001$). These results indicate that children living in this high-risk area in Costa Rica have a higher potential for endogenous nitrosation than those living in the low-incidence area, although the absolute values are lower than those of adult subjects in other countries.

Infection with Campylobacter pylori

As shown in Table 2, seropositivity for *C. pylori* in young subjects was 62.6-71.7% for IgG antibody and 38.7–73.0% for IgA antibody. There was no difference between subjects living in high- and low-risk areas; however, seropositivity and average antibody values were greater in some groups of children in the age range 14–20 years than in that of 7–13 years. These results indicate that the majority of the children studied were infected with *C. pylori*

Table 2. Prevalences (%) of seropositive subjects and average (±SD) optical densities (OD) of serum antibodies against *Campylobacter pylori* in young people from two areas of different risks of gastric cancer[a]

Area	Age range (years)	No. of subjects	IgG		IgA	
			% Positive	OD	% Positive	OD
High-risk	7–13	107	62.6	0.48 (±0.26)	73.0	0.38 (±0.39)
	14–20	46	71.7	0.53 (±0.25)	43.5	0.41 (±0.36)
Low-risk	7–13	93	69.9	0.51 (±0.26)	38.7	0.37 (±0.30)
	14–20	30	70.0	0.64 (±0.26)*	56.7	0.46 (±0.32)
Total	7–13	200	66.0	0.49 (±0.26)	41.0	0.38 (±0.35)
	14–20	76	71.1	0.58 (±0.26)*	48.7*	0.43 (±0.35)

[a] IgG- and IgA- specific antibodies agianst *C. pylori* were measured by a modified enzyme–linked immunosorbent assay (Peña et al., 1990).
*, Significant difference at $p < 0.05$ between children aged 7–13 years and those aged 14–20 years

at less than 20 years, in contrast to findings in young people in developed countries, where the infection has been found to be relatively uncommon in healthy children and the prevalence rates in teenagers are much lower than in Costa Rica. For example, Perez-Perez *et al.* (1988) studied a healthy population in Denver, CO, USA, and found no case of seropositivity among 40 children under ten years of age; the prevalence in a group of subjects aged 10–19 years was less than 20%. Only after 60 years of age was the prevalence of seropositivity above 50%.

Genotoxicity and levels of total N-nitroso compounds, nitrate and nitrite in samples of beans consumed frequently in Costa Rica

Raw and cooked beans consumed very frequently in Costa Rica were collected from families in the two areas and analysed for nitrite, nitrate, total *N*-nitroso compounds and genotoxicity in the SOS chromotest (Table 3). The mean levels of total *N*-nitroso compounds in aqueous extracts (pH 2) of cooked bean samples (stored at room temperature up to nine days) from the high- and low-incidence areas were similar. Acid-catalysed nitrosation of the extracts produced a maximal level of total *N*-nitroso compounds of 1.8 μmol/g wet weight of cooked beans, but there was no difference in the mean level of nitrosation-dependent total *N*-nitroso compounds between the two areas. Only two out of 11 extracts prepared from cooked beans from the low-incidence area and two out of 14 from the high-incidence area showed weak, direct genotoxicity; however, after acid-catalysed nitrosation, all bean samples were genotoxic. The mean genotoxicity was also similar in samples from the two areas. Although the mean nitrate concentration of uncooked beans from the high-risk area (0.76 μmol/g dry weight, n = 9) was about twice as high as that of beans from the low-risk area (0.36 μmol/g, n = 6), no significant difference was seen in the nitrate levels in cooked beans from the two areas (0.97 μmol/g, n = 32, versus 0.89 μmol/g, n = 32, respectively). Furthermore, there was no clear increase or decrease in nitrite or nitrate concentrations in cooked beans stored in a refrigerator or at room temperature for up to nine days.

Table 3. Concentrations of total *N*–nitroso compounds (NOC) and genotoxicity in the SOS chromotest (SOS induction potency; SOSIP) of acidic aqueous extracts of cooked beans collected in high- and low-risk areas for stomach cancer in Costa Rica

Area	No. of samples analysed[a]	Mean (range) of total NOC (nmol/g cooked beans)		Mean (range) of SOSIP (per g cooked beans)		SOSIP/µmol total NOC (After nitrosation)
		Before nitrosation	After nitrosation	Before nitrosation	After nitrosation	
High-risk	14/13	0.57 (0–1.7)	1231 (560–1600)	4 (0–28)	96 (38–162)	58 (26–92)
Low-risk	11/11	0.42 (0–1.8)	1322 (798–1810)	8 (0–44)	80 (22–189)	45 (12–80)

[a] Before (first figure) and after nitrosation (second figure). Nitrosation was carried out *in vitro* at pH 2 at 37°C for 60 min with 100 mmol/l $NaNO_2$. The reaction was stopped by adding an excess of sulfamic acid. Total NOC concentration was determined by the method of Pignatelli *et al.* (1987) and genotoxicity by the SOS chromotest as described by Malaveille *et al.* (1989).

Conclusion

We have demonstrated that Costa Rican children living in a high-risk area for stomach cancer have a greater potential for endogenous nitrosation than children living in a low-risk area; the levels of *N*-nitrosamino acids detected in their urine were, however, much lower than those in adults living in other countries. *C. pylori* infection, assessed by serological examination, was found to be very prevalent in children in both areas, indicating that gastritis occurs very early in Costa Rican children and persists for a long time. The genotoxicity and total concentrations of *N*-nitroso compounds in beans consumed very frequently in Costa Rica, before and after nitrosation *in vitro*, were similar for the two areas. Further studies are required to identify other factors that determine the risk for developing gastric cancer in the inhabitants of the high-risk area.

Acknowledgements

This study was undertaken during the tenure of an International Cancer Research Technology Transfer Programme (ICRETT) fellowship awarded to R.S. by the International Union Against Cancer. We thank Mrs E. Bayle for secretarial work.

References

Bartsch, H., Ohshima, H., Pignatelli, B. & Calmels, S. (1989) Human exposure to endogenous N-nitroso compounds: quantitative estimates in subjects at high risk for cancer of the oral cavity, oesophagus, stomach and urinary bladder. *Cancer Surv.*, 8, 335-362

Correa, P. (1988) A human model of gastric carcinogenesis. *Cancer Res.*, 48, 3354-3560

Correa, P., Haenszel, W., Cuello, C., Tannenbaum, S. & Archer, M. (1975) A model for gastric cancer epidemiology. *Lancet*, ii, 58-60

Crespi, M., Ohshima, H., Ramazzotti, V., Muñoz, N. Grassi, A., Casale, V., Leclerc, H., Calmels, S., Carroen, S., Kaldor, J. & Bartsch, H. (1987) Intragastric nitrosation and precancerous lesions of the gastrointestinal tract: testing of an etiological hypothesis In: Bartsch, H., O'Neill, I. & Schulte-Hermann, R., eds, *The Relevance of N-Nitroso Compounds to Human Cancer. Exposures and Mechanisms* (IARC Scientific Publications No. 84), Lyon, IARC, pp. 511-517

Malaveille, C., Vineis, P., Estève, J., Ohshima, H., Brun, G., Hautefeuille, A., Gallet, P., Ronco, G., Terracini, B. & Bartsch, H. (1989) Levels of mutagens in the urine of smokers of black and blond tobacco correlate with their risk of bladder cancer. *Carcinogenesis*, **10**, 577-586

Ohshima, H., Calmels, S., Pignatelli, B., Vincent, P. & Bartsch, H. (1987) *N*-Nitrosamine formation in urinary-tract infections. In: Bartsch, H., O'Neill, I. & Schulte-Hermann, R., eds, *The Relevance of N-Nitroso Compounds to Human Cancer. Exposures and Mechanisms* (IARC Scientific Publications No. 84), Lyon, IARC, pp. 384-390

Ohshima, H., Pignatelli, B., Malaveille, C., Friesen, M., Calmels, S., Shuker, D., Muñoz, N. & Bartsch, H. (1988) Markers for intragastric nitrosamine formation and resulting DNA damage. In: Reed, P.I. & Hill, M.J., eds, *Gastric Carcinogenesis*, Amsterdam, Elsevier, pp. 175-185

Peña, A.S., Endtz, H.P., Offerhaus, G.J.A., Hoogenboom-Veregaal, A., van Duin, W., de Vargas, N., den Hartog, G., Kreuning, J., Reijden, J., Mouton, R.P. & Lamers, C.B.H.W. (1990) The value of serology (ELISA) and immunoblotting for the diagnosis of Campylobacter pylori infection. *Digestion* (in press)

Perez-Perez, G., Dworkin, B., Chodas, J. & Blases, M. (1988) Campylobacter pylori: antibodies in humans. *Ann. Intern. Med.*, **109**, 11

Pignatelli, B., Richard, I., Bourgade, M.-C. & Bartsch, H. (1987) Improved group determination of total *N*-nitroso compounds (NOC) in human gastric juice by chemical denitrosation and thermal energy analysis. *Analyst*, **112**, 945-949

Sierra, P., Parkin, D.M., Muñoz, N. & Leiva, G. (1989) Cancer in Costa Rica, *Cancer Res.*, **49**, 717-724

Relevance to Human Cancer of N-Nitroso Compounds,
Tobacco Smoke and Mycotoxins.
Ed. I.K. O'Neill, J. Chen and H. Bartsch
Lyon, International Agency for Research on Cancer
© IARC, 1991

URINARY EXCRETION OF N-NITROSOPROLINE IN RELATION TO CONSUMPTION OF RAW AND COOKED VEGETABLES IN A DANISH RURAL POPULATION

H. Møller[1,5], J. Landt[2], E. Pedersen[3], P. Jensen[4], H. Autrup[2] & O.M. Jensen[1]

[1]Danish Cancer Registry, Institute of Cancer Epidemiology, Danish Cancer Society, Copenhagen; [2] Fibinger Institute, Danish Cancer Society, Copenhagen; [3]National Food Agency, Søborg; and [4]Institution of Medical Health Officers, Nordjyllands Amt, Denmark

Several recent case–control studies of gastric cancer have demonstrated the protective effect of consumption of vegetables. According to Correa's model of gastric carcinogenesis, the initiating agent is N-nitroso compounds either ingested or formed *in vivo*. In our study of endogenous nitrosation, we measured intragastric formation of N-nitroso compounds in 285 individuals by the nitrosation of proline; in this presentation we analysed the effect of consumption of vegetables on urinary excretion of N-nitrosoproline (NPRO). When adjustment was made for the dominating determinants of NPRO excretion (total nitrate intake and tobacco smoking), a marked difference in the effects of consumption of raw and cooled vegetables was seen: consumption of cooked vegetables increased endogenous nitrosation of proline, while consumption of raw vegetables had only a marginal effect. We suggest that the difference between raw and cooked vegetables is due to destruction of ascorbate in the cooking of the vegetables. The lack of a protective effect of consumption of raw vegetables on the rate on endogenous nitrosation of proline indicates, however, that the determinants of nitrosation of proline and the determinants of gastric cancer risk may be different.

It is well established that consumption of vegetables has a protective effect on the risk for gastric cancer (Graham *et al.*, 1972; Risch *et al.*, 1985; Trichopoulos *et al.*, 1985: Jedrychowski *et al.*, 1986; La Vecchia *et al.*, 1987; Hu *et al.*, 1988). According to the model of gastric carcinogenesis advanced by Correa (1988), endogenous formation of carcinogenic N-nitroso compounds is an intermediate step in the development of gastric carcinoma. Within the framework of this model, the protective effect of vegetables may be attributable to the presence of chemical inhibitors of nitrosation, such as ascorbate (Bartsch *et al.*, 1988). Analyses of the effect of consumption of vegetables on the rate of endogenous nitrosation, measured by urinary excretion of N-nitrosoproline (NPRO) after a 500-mg oral

---

[5] To whom all correspondence should be addressed

dose of L-proline, indicated an increase in NPRO excretion in Danish individuals who consumed vegetables (Møller et al., 1989a). We suggested that the enhancing effect of vegetables on NPRO formation was due to boiling of most vegetables in the Danish diet, since cooking of vegetables could destroy ascorbate but does not greatly affect the nitrate content. The present paper describes further analysis in which the preparation of vegetables is taken into account specifically.

Data were available for 285 individuals on 12-h urinary N-nitrosoproline (NPRO) excretion, nitrate in 24-h duplicate diets (liquids and solids collected separately), smoking of tobacco, the amount and type of dietary items taken during the 24-h period of collection of the duplicate samples, and other background variables as described in detail elsewhere (Møller et al., 1989a,b). For the analysis reported here, all vegetables in the categories 'vegetables on sandwiches', 'potatoes', 'cabbage', 'green vegetables' and 'other vegetables' were reclassified as to whether they were eaten raw or cooked.

Statistical analysis was performed by unconditional logistic regression, which produces estimates of the odds ratio (OR) for high NPRO excretion (≥ 1 µg/12 h) being associated with a given level of consumption, while adjusting for other covariables.

The strongest determinant of urinary NPRO excretion in nonsmokers was total nitrate intake (OR, 1.9 for an increase of 50 mg NO_3^-/24 h; 95% confidence interval, 1.4–2.6). In smokers, urinary NPRO was generally elevated but quite independent of nitrate intake. Because of this interaction between nitrate intake and smoking, the effects of consumption of raw and cooked vegetables were analysed in nonsmokers and in smokers separately.

Within the group of nonsmokers, consumption of cooked vegetables increased the rate of NPRO formation, and a clear dose–response relationship was apparent when adjustment was made for nitrate derived from drinking-water, the dominant source of variation in total nitrate intake in this material (Table 1; Møller et al., 1989b). This finding indicates the role of cooked vegetables as a source of nitrate and the fact that the ascorbate in the vegetables, which in theory should reduce intragastric formation of NPRO, was destroyed in the preparation of the vegetables. In smokers, a similar effect was seen (Table 2), but the trend is not statistically significant.

Table 1. Consumption of boiled vegetables and excretion of N-nitrosoproline (NPRO) by nonsmokers

Boiled vegetables (servings/ 24 h)	12-h NPRO excretion		Crude odds ratio (95% confidence interval)	Adjusted odds ratio[a] (95% confidence interval)	Adjusted odds ratio[b] (95% confidence interval)
	≥ 1 µg	< 1 µg			
0–1	13	33	1.0	1.0	1.0
2	26	28	2.4 (1.0–5.4)	2.6 (1.0–6.7)	2.1 (0.8–5.6)
≥ 3	27	25	2.7 (1.2–6.4)	2.9 (1.1–7.5)	1.8 (0.7–5.1)
	Trend:		$p = 0.02$	$p = 0.03$	$p = 0.27$

[a] Adjusted for consumption of fresh vegetables, nitrate intake from drinking-water, sex, age, history of gastric disease, batch of NPRO analysis, urinary creatinine concentration and consumption of cured meat
[b] Adjusted for consumption of fresh vegetables, total nitrate intake, sex, age, history of gastric disease, batch of NPRO analysis, urinary creatinine concentration and consumption of cured meat

A protective effect for consumption of raw vegetables was expected when adjusting for total nitrate intake, but no significant effect was observed in nonsmokers (Table 3) or smokers (Table 4). A tendency to reduced NPRO excretion was seen in nonsmokers who

Table 2. Consumption of boiled vegetables and excretion of *N*-nitrosoproline (NPRO) by smokers

Boiled vegetables (servings/24 h)	12-h NPRO excretion		Crude odds ratio (95% confidence interval)	Adjusted odds ratio[a] (95% confidence interval)	Adjusted odds ratio[b] (95% confidence interval)
	$\geq 1\ \mu g$	$< 1\ \mu g$			
0-1	23	35	1.0	1.0	1.0
2	17	25	1.0 (0.5-2.3)	1.3 (0.5-3.1)	1.3 (0.5-3.2)
≥ 3	16	17	1.4 (0.6-3.4)	2.0 (0.8-5.4)	2.1 (0.8-5.7)
	Trend:		$p = 0.44$	$p = 0.17$	$p = 0.16$

[a] Adjusted for consumption of fresh vegetables, nitrate intake from drinking-water, sex, age, history of gastric disease, batch of NPRO analysis, urinary creatinine concentration and consumption of cured meat
[b] Adjusted for consumption of fresh vegetables, total nitrate intake, sex, age, history of gastric disease, batch of NPRO analysis, urinary creatinine concentration and consumption of cured meat

Table 3. Consumption of fresh vegetables and excretion of *N*-nitrosoproline (NPRO) by nonsmokers

Fresh vegetables (servings/24 h)	12-h NPRO excretion		Crude odds ratio (95% confidence interval)	Adjusted odds ratio[a] (95% confidence interval)	Adjusted odds ratio[b] (95% confidence interval)
	$> 1\ \mu g$	$< 1\ \mu g$			
0	36	51	1.0	1.0	1.0
1	16	14	1.6 (0.7-3.7)	1.8 (0.7-4.7)	1.7 (0.6-4.8)
≥ 2	14	21	0.9 (0.4-2.1)	0.9 (0.4-2.3)	0.6 (0.2-1.7)
	Trend:		$p = 0.91$	$p = 0.98$	$p = 0.48$

[a] Adjusted for consumption of boiled vegetables, nitrate intake from drinking-water, sex, age, history of gastric disease, batch of NPRO analysis, urinary creatinine concentration and consumption of cured meat
[b] Adjusted for consumption of boiled vegetables, total nitrate intake, sex, age, history of gastric disease, batch of NPRO analysis, urinary creatinine concentration and consumption of cured meat

Table 4. Consumption of fresh vegetables and excretion of *N*-nitrosoproline (NPRO) by smokers

Fresh vegetables (servings/24 h)	12-h NPRO excretion		Crude odds ratio (95% confidence interval)	Adjusted odds ratio[a] (95% confidence interval)	Adjusted odds ratio[b] (95% confidence interval)
	$\geq 1\ \mu g$	$< 1\ \mu g$			
0	44	51	1.0	1.0	1.0
1	6	13	0.5 (0.2-1.5)	0.6 (0.2-1.8)	0.6 (0.2-1.9)
≥ 2	6	13	0.5 (0.2-1.5)	0.5 (0.2-1.8)	0.5 (0.2-1.8)
	Trend:		$p = 0.15$	$p = 0.21$	$p = 0.24$

[a] Adjusted for consumption of boiled vegetables, nitrate intake from drinking-water, sex, age, history of gastric disease, batch of NPRO analysis, urinary creatinine concentration and consumption of cured meat
[b] Adjusted for consumption of boiled vegetables, total nitrate intake, sex, age, history of gastric disease, batch of NPRO analysis, urinary creatinine concentration and consumption of cured meat

consumed two or more servings of raw vegetables per day (OR = 0.6) and in smokers who consumed raw vegetables (OR = 0.5-0.6), but none of these estimates reached statistical significance.

Our findings point to a fundamental difference between raw and cooked vegetables as determinants of endogenous nitrosation of proline. Given the well-established facts about gastric cancer risks, however, we would have expected consumption of vegetables, and especially raw vegetables, to convey some protection against endogenous nitrosation. We do not know whether our findings indicate a fundamental discrepancy between the determinants of nitrosation of proline and the true determinants of gastric cancer risk. It is noteworthy that few of the participants in our study consumed raw vegetables: of a total of 825 servings of vegetables recorded, only 221 (27%) were classified as raw vegetables, and only 103 of 285 participants (36%) consumed them. Other studies have shown that relatively high doses of ascorbate are necessary in order to control intragastric formation of NPRO (Leaf et al., 1987). Unfortunately, we do not know how well our discrimination between raw and cooked vegetables translates into ascorbate intake. It is interesting, however, that one case-control study of gastric cancer risk and dietary factors performed in a north-western European country (the UK) failed to show any effect of consumption of vegetables on gastric cancer risk (Acheson & Doll, 1964). A detailed investigation of the epidemiology and etiology of gastric cancer in Denmark is desirable in order to resolve this issue.

References

Acheson, E.D. & Doll, R. (1964) Dietary factors in carcinoma of the stomach: a study of 100 cases and 200 controls. *Gut*, **51**, 126-131

Bartsch, H., Ohshima, H. & Pignatelli, B. (1988) Inhibitors of endogenous nitrosation: mechanisms and implications in human cancer prevention. *Mutat. Res.*, **202**, 307-324

Correa, P. (1988) A human model of gastric carcinogenesis. *Cancer Res.*, **48**, 3554-3560

Graham, S., Schotz, W. & Martino, P. (1972) Alimentary factors in the epidemiology of gastric cancer. *Cancer*, **30**, 927-938

Hu, J., Zhang, S., Jia, E., Wang, Q., Liu, Y., Wu, Y. & Cheng, Y. (1988) Diet and cancer of the stomach: a case-control study in China. *Int. J. Cancer*, **41**, 331-335

Jedrychowski, W., Wahrendorf, J., Popiela, T. & Rachtan, J. (1986) A case-control study of dietary factors and stomach cancer risk in Poland. *Int. J. Cancer*, **37**, 837-842

La Vecchia, C., Negri, E., Decarli, A., D'Avanzo, B. & Franceschi, S. (1987) A case-control study of diet and gastric cancer in Northern Italy. *Int. J. Cancer*, **40**, 484-489

Leaf, C.D., Vecchio, A.J., Roe, D.A. & Hotchkiss, J.H. (1987) Relationship between ascorbic acid dose and N-nitrosoproline excretion in humans on controlled diets. In: Bartsch, H., O'Neill, I.K. & Schulte-Hermann, R., eds, *The Relevance of N-Nitroso Compounds to Human Cancer: Exposure and Mechanisms* (IARC Scientific Publications No. 84), Lyon, IARC, pp. 299-303

Møller, H., Landt, J., Pedersen, E., Jensen, P., Autrup, H. & Jensen, O.M. (1989a) Endogenous nitrosation in relation to nitrate exposure from drinking water and diet in a Danish rural population. *Cancer Res.*, **49**, 3117-3121

Møller, H., Landt, J., Jensen, P., Pedersen, E., Autrup, H. & Jensen, O.M. (1989b) Nitrate exposure from drinking water and diet in a Danish rural population. *Int. J. Epidemiol.*, **18**, 206-212

Risch, H.A., Jain, M., Choi, N.W., Fodor, J.G., Pfeiffer, C.J., Howe, G.R., Harrison, L.W., Craib, K.J.P. & Miller, A.B. (1985) Dietary factors and the incidence of cancer of the stomach. *Am. J. Epidemiol.*, **122**, 947-959

Trichopoulos, D., Ouranos, G., Day, N.E., Tzonou, A., Manousos, O., Papadimitriou, C. & Trichopoulos, A. (1985) Diet and cancer of stomach: a case-control study in Greece. *Int. J. Cancer*, **36**, 291-297

Relevance to Human Cancer of N-Nitroso Compounds,
Tobacco Smoke and Mycotoxins.
Ed. I.K. O'Neill, J. Chen and H. Bartsch
Lyon, International Agency for Research on Cancer
© IARC, 1991

N-NITROSO COMPOUNDS, GENOTOXINS AND THEIR PRECURSORS IN GASTRIC JUICE FROM HUMANS WITH AND WITHOUT PRECANCEROUS LESIONS OF THE STOMACH

B. Pignatelli[1], C. Malaveille[1], C. Chen[1], A. Hautefeuille[1],
P. Thuillier[1], N. Muñoz[1], B. Moulinier[2], F. Berger[2],
H. De Montclos[3], H. Ohshima[1], R. Lambert[2] & H. Bartsch[1]

[1]*International Agency for Research on Cancer,*
[2]*E. Herriot Hospital, and* [3]*Pasteur Institute, Lyon, France*

We are investigating the interrelationships between levels of total N-nitroso compounds (NOC), genotoxic activity (both before and after nitrosation), degree of bacterial colonization in gastric juice and degree of severity or absence of precancerous lesions of the stomach. The mean level of constitutive total NOC in gastric juice was similar in the different groups of patients, but it was higher in acidic gastric juice (n = 30) than in gastric juice at pH > 4.5 (n = 12). Acid-catalysed nitrosation of gastric juice *in vitro* increased the concentration of total NOC by up to several thousand fold, to a maximum of 1330 μmol/l. Genotoxicity, expressed as SOS-inducing potency per 100 μl of gastric juice was measurable in only 20% of gastric juice samples tested. After acid-catalysed nitrosation, however, all samples showed genotoxic activity, the mean SOS-inducing potency being four to seven times greater than the corresponding constitutive value. There was no association between the mean SOS-inducing potency of gastric juice and the severity of precancerous lesions. The mean SOS-inducing potency of neutral or basic gastric juice was slightly greater than that of acidic samples. In a kinetic study on N-nitrosation of gastric juice *in vitro*, a mixture of amino and amido substrates was nitrosated; both qualitative and quantitative individual differences in nitrosatable substrates in gastric juice were seen. Fractionation of acidic, neutral and basic nitrosated gastric juice samples revealed a preponderance of nonvolatile, unknown NOC with varying polarities. The results of our study suggest that only pH determines the nature and level of precursors of NOC and of nitrosation-dependent genotoxins in gastric juice.

Patients with chronic atrophic gastritis (CAG) or pernicious anaemia and those who have undergone gastric surgery are at high risk for stomach cancer (reviewed by Hill, 1986). Endogenous formation of N-nitroso compounds (NOC), possibly increased by the bacteria that colonize the achlorhydric stomach, has been implicated in its etiology (Correa *et al.*, 1975; Mirvish, 1983; Hill, 1986; Miwa *et al.*, 1987; Correa, 1988). In a study in progress, we are investigating the interrelationships between level of total NOC, genotoxic activity (both before and after nitrosation), degree of bacterial colonization in gastric juice and the degree of severity or absence of precancerous lesions of the stomach.

Apart from some volatile nitrosamines, the NOC that are present or formed after nitrosation in human gastric juice have not been identified. In order to obtain information about the nature of the nitrosatable substrates that occur in gastric juice and about the factors that influence formation of their N-nitroso derivatives, we studied nitrosation reactions in acidic, neutral and basic gastric juice *in vitro*.

Subjects

Two groups of patients were studied. In the first (group I), patients 30–70 years of age were selected from people undergoing gastroscopy without premedication as part of a routine diagnostic procedure at E. Herriot Hospital. They were grouped according to histologically confirmed diagnosis as follows: dysplasia (n=4), CAG (n=14), diffuse interstitial gastritis (n=15) and normal gastric mucosa (n=9). After an anamestic questionnaire had been completed, routine gastroscopy was performed and fasting gastric juice (1.7–50 ml) was collected for pH measurement, bacteriological analysis and determination of nitrate, nitrite, total NOC and genotoxic activity. Biopsies were collected for histopathological evaluation.

The second group (group II) consisted of patients undergoing an operation for duodenal ulcers. Gastric juice samples were obtained before and after the operation and were pooled according to their pH.

pH and bacterial flora in gastric juice of group I patients.

Only 12 out of the 42 patients in the first group had gastric juice with a pH in the range 4.5–8.5: two patients out of nine with normal gastric mucosa, three out of 15 with diffuse interstitial gastritis, five out of 14 with CAG and two out of four with dysplasia. Three samples contained $>10^5$/ml bacteria with nitrate-reducing activity (indicating bacterial colonization); the remainder had a mean bacterial density of 1.8×10^4/ml.

Levels of nitrate, nitrite and total NOC (before and after nitrosation) in gastric juice of group I patients

Nitrate and nitrite were analysed according to a method described by Green et al. (1982). Total NOC were determined by a group-selective method using chemical denitrosation and a thermal energy analyser (Pignatelli et al., 1987). The mean levels of nitrate (283 µmol/l) and nitrite (48 µmol/l) in gastric juice at pH 4.5–8.5 were about 70% and 200% of those in acidic gastric juice, respectively. No difference was found between the patients, possibly because of the small number of subjects investigated so far and the low prevalence of patients with basic gastric juice containing abundant nitrate-reducing flora. The mean level of total NOC was similar in the different groups of patients (Table 1), but it was lower in gastric juice at pH > 4.5 than in acidic gastric juice (Table 2), owing perhaps to the lack of abundant gastric flora in the samples tested.

Acid-catalysed nitrosation ([NaNO$_2$], 80 mmol/l; pH 1.5, 60 min at 37°C; residual nitrite destroyed by adding 20 mg/ml sulfamic acid) of gastric juice *in vitro* increased the concentration of total NOC by up to several thousand fold, with a maximum of 1330 µmol/l (Tables 1 and 2). The mean concentration of total NOC formed was higher in gastric juice from CAG and dysplasia patients than in that from patients with diffuse interstitial gastritis or normal gastric mucosa (Table 1). High concentrations of NOC were thus more frequently found in gastric juice from patients with precancerous lesions of the stomach. Since (i) the mean NOC concentration in neutral or basic gastric juice was more than two fold that of acidic samples (Table 2) and (ii) the frequency of neutral or basic gastric juice in patients with normal mucosa, diffuse interstitial gastritis, CAG and dysplasia was 14, 18, 40 and 50%, respectively, the difference in the level of nitrosatable precursors might be attributable to pH.

Table 1. Concentration of total N-nitroso compounds (NOC) in human gastric juice (before and after nitrosation) in relation to degree of severity or absence of precancerous lesions of the stomach

Diagnosis[a]	Total NOC (µmol/l)						
	Constitutive			After nitrosation[b]			
	No. of patients	Range	Mean	No. of patients	Range	Mean	Patients with [NOC] >250 µmol/l (%)
NGM	9	ND-2.3	0.54	7	142-529	252	28
DIG	15	0.06-1.2	0.30	11	73-556	278	45
CAG	14	ND-1.0	0.29	10	111-1332	406	70
Dys	4	0.2-0.4	0.27	4	432-560	490	100

[a] NGM, normal gastric mucosa; DIG, diffuse interstitial gastritis; CAG, chronic atrophic gastritis; Dys, dysplasia
[b] Nitrosation at pH 1.5 for 1 h at 37°C in the presence of 80 mmol/l nitrite

Table 2. Concentration of total N-nitroso compounds (NOC) in human gastric juice in relation to pH range before and after nitrosation

pH range	Total NOC (µmol/l)					
	Constitutive			After nitrosation[a]		
	No. of patients	Range	Mean	No. of patients	Range	Mean
1.1-3.5	30	0.01-2.29	0.41	23	73-560	249
4.5-8.5	12	ND-0.50	0.18	9	303-1332	569

[a] Nitrosation at pH 1.5 for 1 h at 37° in the presence of 80 mmol/l nitrite

Factors affecting the levels of genotoxins in gastric juice of group I patients

The genotoxicity of gastric juice was measured using the SOS chromotest (Quillardet & Hofnung, 1985), with modifications as described previously (Malaveille et al., 1989). SOS-inducing potency, expressed per 100 µl of gastric juice, was measurable in only 20% of samples tested, and genotoxicity was not affected by pH (Table 3). The number of positive samples was too low to allow comparison of the different groups on the basis of SOS-inducing potency. After acid-catalysed nitrosation (with 10 mg/ml sulfamic acid), all of the samples showed genotoxic activity, the mean inducing potency increasing by four to seven fold as compared to the corresponding constitutive value. There was no association between the mean SOS-inducing potency of gastric juice and the severity of precancerous lesions. The mean potency of neutral or basic gastric juice was slightly higher than that of acidic samples. A three-fold difference was seen between the two groups, however, in the prevalence of patients who had a SOS-inducing potency > 10 (Table 3).

Table 3. SOS-inducing potency of human gastric juice (before and after nitrosation) in relation to pH

pH range	SOS-inducing potency per 100 µl gastric juice							
	Constitutive				After nitrosation[a]			
	No. of patients	No activity	Range	Mean	No. of patients	Range	Mean	Patients with potency > 10 (%)
1.1-3.5	26	73	0.57-2.46	1.59	27	2.43-74.50	10.48	26
4.5-8.5	10	70	1.42-7.00	3.97	10	5.68-22.00	14.36	80

[a] Nitrosation at pH 1.5 for 1 h at 37°C in the presence of 80 mmol/l nitrite

Assuming that the genotoxic activity measured is attributable to NOC, we can express SOS-inducing potency per nanomole of NOC and compare it with that of well known carcinogenic nitrosamides. For all samples, the SOS-inducing potency was higher than that of N-methyl-N-nitrosourea (potency, 0.015), and for 32% of samples it was similar to or higher than that of N-methyl-N'-nitro-N-nitrosoguanidine (potency, 0.9). All of the latter samples were acidic (pH 1.5-3.5), although a basic pH was shown earlier to favour the presence of higher concentrations of both NOC and genotoxins. These findings suggest that pH affects the precursors of genotoxic NOC quantitatively and qualitatively, which may explain the lack of correlation between level of NOC and genotoxic activity in gastric juice.

Kinetics of N-nitrosation of gastric juice of group II patients in vitro

The concentration of NOC increased up to 60 min and then reached a plateau during nitrosation of acidic, neutral and basic gastric juice (pH 1.5; [NaNO$_2$], 5-10 mmol/l). The relationship between NOC formation in the presence of 80 mmol/l nitrite (60 min) and pH, with the presence or the absence of a marked optimal pH, suggests that both amino- and amido-type substances were nitrosated. The rate of nitrosation at pH 1.5 increased linearly with concentrations of nitrite from 5 to 80 mmol/l, indicating the involvement of amido-type compounds (Fig. 1). This finding is in accordance with the observed direct genotoxicity of nitrosated gastric juice under identical nitrosation conditions. It is interesting to note that for the two pooled samples of acidic gastric juice, the slopes of the lines were larger than those for the pooled neutral and basic samples. These findings indicate both qualitative and quantitative individual differences in the nitrosatable substrates present in gastric juice.

Diversity of NOC and contribution of volatile N-nitrosamines and N-nitrosamino acids in nitrosated gastric juice of group II patients

After 1 h nitrosation ([NaNO$_2$], 80 mmol/l, pH 1.5-7.5) at 37°C, sulfamic acid (20 mg/ml) was added to destroy unreacted nitrite. The mixture was then adjusted to pH 7, and volatile N-nitrosamines were extracted with dichloromethane and analysed in a gas chromatograph coupled to a thermal energy analyser. The aqueous phase was acidified to pH 2, and N-nitrosamino acids were extracted with 10% methanol in dichloromethane and analysed by gas chromatography-thermal energy analyser after derivatization with diazomethane. Other NOC were extracted from the acidic aqueous phase with ethyl acetate. The levels of total NOC present in the samples were measured in the reaction mixture, in all organic solvent extracts and in the remaining aqueous phase after organic solvent extractions (Pignatelli et al., 1987,1989).

Figure 1. Rate of formation of total N-nitroso compounds (NOC) as a function of nitrite concentration

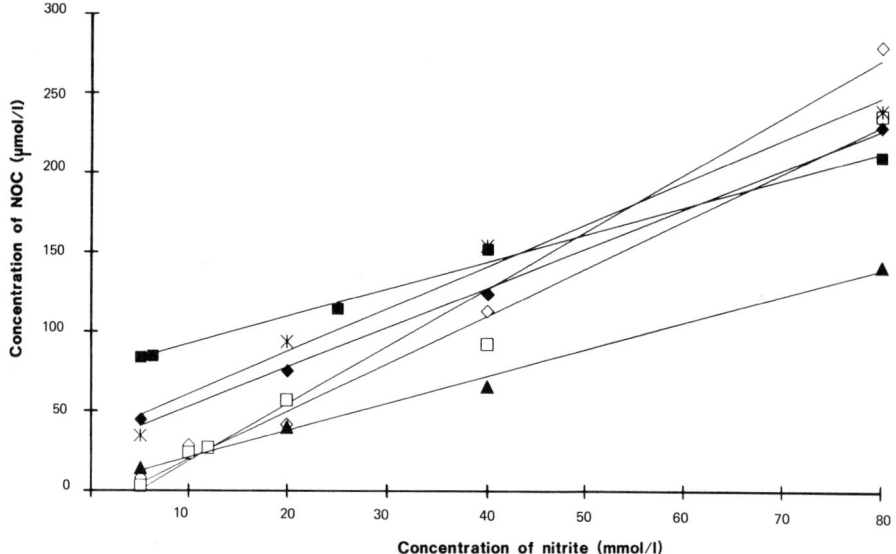

◇, pooled acidic gastric juice (1); □, pooled acidic gastric juice (2); ◆, pooled neutral gastric juice; ■, pooled basic gastric juice (1); ▲, pooled basic gastrice juice (2); *, pooled basic gastric juice (3); nitrosation conditions: pH 1.5, 60 min at 37°C

The levels of total NOC, volatile N-nitrosamines and N-nitrosamino acids found in pooled samples of acidic, neutral and basic gastric juice are shown in Table 4. About 40–50% of total NOC was not extractable in organic solvents. NOC extractable with dichloromethane from the basic aqueous solution accounted for only 1–2% of total NOC. The volatile N-nitrosamines identified, N-nitrosodimethylamine and N-nitrosopyrrolidine, accounted for less than 10% of the dichloromethane-extractable NOC. NOC extractable with methanol:dichloromethane represented 10–17% of the total NOC, of which 1–10% were N-nitrosamino acids; the most abundant, N-nitrosoproline, accounted for 2–6% of total NOC. A further 15–16% of the total NOC was extractable in ethyl acetate. These results revealed that a preponderance of nonvolatile, unknown NOC with varying polarities is formed in nitrosated gastric juice.

Table 4. Volatile nitrosamines (VNA), N-nitrosamino acids (NNA) and total N-nitroso compounds (NOC) found in nitrosated pooled samples of human gastric juice and their organic extractable/nonextractable NOC

pH range (No. of pooled samples)	pH and range of pH of nitrosation[a] (No. of assays)	Total NOC ± SD (%)[b]						Total NOC in extracts and water (%)
		DCM extract		DCM/MeOH extract		EtAc extract	Water	
		Total	NDMA	Total	NPRO	(total)	(total)	
1.5-2.0 (4)	1.5 (1)	1.8±0.9	0.1±0.1	13.6±2.6	2.8±0.8	16.4±4.4	51.6±3.8	72.4±7.9
1.5-2.0 (2)	3.3-7.5 (4)	1.9±1.4	0.2±0.1	9.4±4.2	0.8±1.0	10.3±8.0	52.5±7.1	69.5±7.2
4.20 (1)	1.5 (1)	1.2	0.1	*	*	18.2*	43.4	62.8
7.7-7.8 (3)	1.5 (1)	2.0±1.6	0.01±0.02	16.8±2.8	5.7±8.1	12.6±6.0	42.8±13.6	74.2±10.1
7.7 (1)	3.2-5.0 (2)	2.5±0.6	0.3±0.2	9.6±8.8	3.1±3.6	4.7±2.3	36.6±0.6	53.4±6.5

[a] Conditions of nitrosation: [NaNO$_2$], 80 mmol/l; 60 min at 37°C; unreacted nitrite destroyed by adding 20 mg/ml sulfamic acid
[b] DCM, dichloromethane; DCM/MeOH, dichloromethane:methanol (9:1); EtAc, ethyl acetate; NDMA, N-nitrosodimethylamine; NPRO, N-nitrosoproline
* DCM:MeOH extraction not performed before that by EtAC

The results of our study, which is in progress, suggest that only pH determines the nature and level of precursors of NOC and of nitrosation-dependent genotoxins in gastric juice. Work is in progress to elucidate the structure of these various substances.

Acknowledgement

The work reported in this paper was supported by a grant from the US National Cancer Institute, No. 1 RO1-CA 47591-01A1.

References

Correa, P. (1988) A human model of gastric carcinogenesis. *Cancer Res.*, **48**, 3354-3560
Correa, P., Haenszel, W., Cuello, C., Tannenbaum, S. & Archer, M. (1975) A model for gastric cancer epidemiology. *Lancet*, **ii**, 58-60
Green, L.C., Wagner, D.A., Glogowski, J., Skipper, P.L., Wishnok, J.S. & Tannenbaum, S.R. (1982) Analysis of nitrate, nitrite and ^{15}N nitrate in biological fluids. *Anal. Biochem.*, **126**, 131-138
Hill, M.J. (1986) *Microbes and Human Carcinogenesis*, London, Edward Arnold
Malaveille, C., Vineis, P., Estève, J., Ohshima, H., Brun, G., Hautefeuille, A., Gallet, P., Ronco, G., Terracini, B. & Bartsch, H. (1989) Levels of mutagens in the urine of smokers of black and blond tobacco correlate with their risk of bladder cancer. *Carcinogenesis*, **10**, 577-586
Mirvish, S.S. (1983) The etiology of gastric cancer. Intragastric nitrosamide formation and other theories. *J. Natl Cancer Inst.*, **71**, 629-647
Miwa, M., Stuehr, D.J., Marletta, M.A., Wishnok, J.S. & Tannenbaum, S.R. (1987) Nitrosation of amines by stimulated macrophages. *Carcinogenesis*, **8**, 955-958
Pignatelli, B., Richard, I., Bourgade, M.-C. & Bartsch, H. (1987) Improved group determination of total N-nitroso compounds (NOC) in human gastric juice by chemical denitrosation and thermal energy analysis. *Analyst*, **112**, 945-949
Pignatelli, B., Chen, C.-S., Thuillier, P. & Bartsch, H. (1990) An improved method for the analysis of total N-nitroso compounds (NOC) in biological matrices including human gastric juice. In: Proceedings of the Titisee Symposium on the Significance of N-Nitrosation of Drugs, 2-4 October 1988 (in press)
Quillardet, P. & Hofnung, M. (1985) The SOS-chromotest, a colorimetric bacterial assay for genotoxins: procedures. *Mutat. Res.*, **147**, 65-78

Relevance to Human Cancer of *N*-Nitroso Compounds,
Tobacco Smoke and Mycotoxins.
Ed. I.K. O'Neill, J. Chen and H. Bartsch
Lyon, International Agency for Research on Cancer
© IARC, 1991

URINARY NITRATE, NITRITE AND *N*-NITROSO COMPOUNDS IN BLADDER CANCER PATIENTS WITH SCHISTOSOMIASIS (BILHARZIA)

A.R. Tricker[1,3], M.H. Mostafa[2], B. Spiegelhalder[1] & R. Preussmann[1]

[1]*Institute of Toxicology and Chemotherapy, German Cancer Research Centre, Heidelberg, Germany; and* [2]*Institute of Graduate Studies and Research, University of Alexandria, Alexandria, Egypt*

Urinary excretion of nitrite and of volatile and nonvolatile *N*-nitroso compounds is increased in schistosomiasis (bilharzia) patients with *Schistosoma haematobium* infection. This observation suggests that the formation of nitrite and *N*-nitroso compounds *in vivo* in the urinary bladder of bilharzial patients may be an important etiological factor in the induction of bladder cancer associated with *S. haematobium* infection.

Schistosomiasis (bilharzia) ranks as one of the most important health problems in subtropical and tropical countries, with an estimated 200 million infected people in 72 countries and a further 500–600 million people exposed to infection. In Egypt, bilharzial infection by *Schistosoma haematobium* and *S. mansoni* is the most frequent endemic parasitic disease, with an estimated 33 million people at risk from infection. Considerable evidence (World Health Organization, 1985) suggests that *S. haematobium* infection may be a cause of bladder cancer; this cancer accounts for 28.8% of all cancers in Egyptian males and 11.7% in females (Aboul Nasr *et al.*, 1986). It has also been suggested that endogenous nitrosamine formation in the urinary bladder may be involved in the etiology of biharzial bladder cancer (Hicks *et al.*, 1976).

Sample collection and analysis

Samples of 24-h urine and saliva (mid-morning and afternoon) were collected from volunteer Egyptian male controls (n = 27; mean age, 35; range, 25–57 years) and from clinically examined *S. haematobium*-infected bilharzial patients (n = 27; mean age, 43; range, 27–60 years) and bilharzial bladder cancer patients (n = 23; mean age, 53; range, 33–65 years). All samples were stabilized with alkali and flown to the Federal Republic of Germany for analysis of nitrate, nitrite and volatile and nonvolatile *N*-nitrosamines, as described previously (Tricker *et al.*, 1989). Statistical analysis was performed using the Wilcoxon rank sum test (Lehmann, 1975).

---

[3]To whom correspondence should be addressed

Exposure to nitrate and nitrite

Table 1 shows the mean levels and nitrate and nitrite in saliva and urine. The concentrations in saliva showed wide intra-group variation but no significant difference between groups. Nitrite was detected in two control urine samples, and significant increases were seen in bilharzial patients and in bilharzial bladder cancer patients. Urinary nitrate concentrations were increased in both bilharzia groups but not significantly.

Table 1. Nitrate and nitrite in saliva and urine of controls, bilharzial patients and bilharzial bladder cancer patients

Patient group	No.	Saliva (mg/l)				Urine (mg/day)			
		NO_3^-		NO_2^-		NO_3^-		NO_2^-	
		Mean±SD	Range	Mean±SD	Range	Mean±SD	Range	Mean±SD[a]	Range
Controls	27	41.7±52.3	5.9-258.0	8.8±9.0	0.9-34.7	139.3±82.2	35-402	1.7,0.3 (2)	
Bilharzia	27	48.9±68.7	8.3-274.0	5.95±5.99	1.3-25.0	143.6±136.3	16-506	5.2±9.1 (16)**	ND-45.0
Bilharzial bladder cancer	23	48.9±38.6	3.9-132.0	6.8±4.8	0.4-19.0	175±190	35-855	1.8±2.8 (16)*	ND-9.8

[a] In parentheses, number of positive samples; *, $p<0.05$, **, $p<0.02$ in comparison with controls

Urinary N-nitrosamine excretion

The levels of volatile and nonvolatile nitrosamines found in urine are presented in Table 2. N-Nitrosodimethylamine (NDMA) was detected in 11 of 27 control urines, and one sample also contained N-nitrosopiperidine and N-nitrosopyrrolidine. Bilharzial patients had higher urinary concentrations of NDMA, N-nitrosopiperidine ($p<0.05$), N-nitrosopyrrolidine ($p<0.02$) and total volatile N-nitroso compounds ($p<0.02$) than controls; and significant increases in the levels of NDMA ($p<0.05$) and total volatile N-nitroso compounds ($p<0.02$) were found in bilharzial bladder cancer patients. N-Nitrosodiethylamine was found in one control urine and in that of six bilharzial patients and four bilharzial bladder cancer patients. One bilharzial patient also excreted 0.40 µg/day N-nitrosodipropylamine and 0.50 µg/day N-nitrosomorpholine.

No significant difference was found with regard to the concentrations of N-nitrososarcosine in the controls, bilharzial patients and bilharzial bladder cancer patients; and the level of excretion of N-nitroso-2-methylthiazolidine 4-carboxylic acid in the control group and in bilharzial bladder cancer patients was almost identical, although a nonsignificant increase was seen in bilharzial patients. The level of N-nitrosothiazolidine 4-carboxylic acid was increased over that in controls in both bilharzial patients ($p<0.05$) and bilharzial bladder cancer patients. N-Nitrosoproline excretion was also increased in bilharzial patients ($p<0.02$) and in bilharzial bladder cancer patients.

Total urinary excretion of nonvolatile N-nitrosamino acids and of N-nitrosamines was increased in both bilharzial and bilharzial bladder cancer patients.

Table 2. N-Nitrosamines in urine of controls, bilharzial patients and bilharzial bladder cancer patients[a]

N-Nitrosamine (μg/day)	Controls (n=27)	Bilharzial patients (n=27)	Bilharzial bladder cancer patients (n=23)
Volatile			
NDMA	0.27±0.47 (11) [ND-1.7]	2.7±6.1 (24) [ND-32.1]	1.3±1.7 (20)* [ND-6.5]
NDEA	0.4 (1) [ND-0.4]	0.26±0.71 (16) [ND-3.1]	0.17±0.47 (4) [ND-1.9]
NPIP	0.6 (1) [ND-0.6]	0.21±0.43 (10)* [ND-1.7]	0.09±0.19 (4) [ND-0.6]
NPYR	0.7 (1) [ND-0.7]	0.26±0.35 (12)** [ND-1.3]	0.14±0.36 (5) [ND-1.5]
Total	0.32±0.64 (11) [ND-2.7]	3.5±6.4 (25)** [ND-32.5]	1.7±2.0 (20)** [ND-6.8]
Nonvolatile			
NSAR	3.4±5.9 (24) [ND-29.6]	6.0±3.5 (24) [ND-13.4]	6.2±9.9 (23) [2.1-9.9]
NPRO	7.2±5.1 (27) [1.2-20.1]	17.0±7.2 (27)** [2.9-33.6]	12.6±3.4 (23) [2.7-17.2]
NTCA	11.7±7.6 (27) [3.5-33.4]	24.4±9.3 (27)* [3.7-41.4]	15.9±3.8 (23) [9.4-23.7]
NMTCA	9.1±5.6 (27) [3.1-23.2]	15.1±6.8 (27) [1.8-29.2]	9.5±4.6 (23) [4.9-24.1]
Total	31.2±22.1 (27) [9.7-94.8]	62.9±22.0 (27)* [13.2-108.7]	44.9±7.3 (23) [24.6-65.5]
Total N-nitrosamines	31.5±22.5 (27) [9.7-96.0]	65.9±25.8 (27)* [13.4-118.0]	46.0±11.2 (23) [24.6-65-5]

[a] Mean ± SD or individual values (no. of positive samples); [range]; *$p<0.05$; **; $p<0.02$ in comparison with controls
[b] NDMA, N-nitrosodimethylamine; NDEA, N-nitrosodiethylamine; NPIP, N-nitrosopiperidine; NPYR, N-nitrosopyrrolidine; NSAR, N-nitrososarcosine; NPRO, N-nitrosoproline; NTCA, N-nitrosothiazolidine 4-carboxylic acid; NMTCA, N-nitroso-2-methylthiazolidine 4-carboxylic acid

Discussion

Bacterial infection of the urinary tract increases the risk for developing bladder cancer (Kantor et al., 1984) in patients with chronic or repeated cystitis (Radomski et al., 1978), paraplegia (Melzak, 1966) and S. haematobium infection (Hicks et al., 1976), as well as increasing the frequency of all cancers (Nordenstram et al., 1986). Urine from S. haematobium-infected patients contained several nitrate-reducing strains of bacteria, including Staphylococcus aureus, haemolytic Staphylococcus albus, Proteus mirabilis, Klebsiella and Escherichia coli, free nitrite, NDMA and other N-nitroso compounds (Hicks et al.). The bacterial strains identified can mediate nitrosation reactions between secondary amines and nitrate under the physiological pH conditions normally encountered in the urinary bladder (Calmels et al., 1988).

Our results clearly demonstrate that bilharzial patients have significantly greater exposure of the urinary bladder and greater body burdens of nitrite, volatile and nonvolatile N-nitrosamines and N-nitrosamino acids than controls. The exceptionally high concentrations of nitrate (up to 855 mg/day) found in the urine of bilharzial patients may be due to macrophage synthesis (Stuehr & Marletta, 1985) as a result of chronic bladder inflammation.

In populations not exposed to bilharzial infection, the peak incidence of bladder cancer occurs during the sixth decade of life, and only 12% of cases occur in people under the age of 50. In Egypt, the mean age for diagnosis of bilharzial bladder cancer is 46 years (El-Bolkainy et al., 1972), and 73% of all cases occur in persons under the age of 50 (Aboul Nasr et al., 1962). We speculate that this reduction in the peak age of diagnosis of bladder cancer results from a complex interaction of repeated bacterial infections of the urinary tract starting at an early age, macrophage involvement in the synthesis of nitrate, nitrite and N-nitroso compounds (chemical initiators), and rapidly proliferating cells in areas of urothelial hyperplasia caused by physical irritation induced by eruption of $S.$ $haematobium$ ova through the bladder mucosa. These factors might promote a synergistic initiation of tumour foci and their progression to bladder tumours.

References

Aboul Nasr, A.L., Gazayerli, M.F., Fawazi, R.M. & El-Sebai, I. (1962) Epidemiology and pathology of cancer of the bladder in Egypt. *Acta Unio Int. Cancrum*, **18**, 528-537

Aboul Nasr, A.L., Boutros, S.G. & Husein, M.H. (1986) Egypt. In: Parkin, D.M., ed., *Cancer Occurrence in Developing Countries* (IARC Scientific Publications No. 75), Lyon, IARC, pp. 37-41

Calmels, S., Ohshima, H. & Bartsch, H. (1988) Nitrosamine formation by denitrifying and non-denitrifying bacteria: implication of nitrite reductase and nitrate reductase in nitrosamine analysis. *J. Gen. Microbiol.*, **134**, 221-226

El-Bolkainy, M.N., Ghoneim, M.A. & Mansour, M.A. (1972) Carcinoma of the bilharzial bladder in Egypt: clinical and pathological features. *Br. J. Urol.*, **44**, 461-570

Hicks, R.M., Walters, C.L., El-Sebai, I., El-Aaser, A., El-Merzabani, M.M. & Gough, T.A. (1976) Determination of nitrosamines in human urine: preliminary observations as the possible etiology for bladder cancer in association with chronic urinary tract infections. *Proc. R. Soc. Med.*, **70**, 413-416

Hicks, R.M., Ismail, M.M., Walters, C.L., Beecham, P.T., Rabie, M.T. & El-Alamy, M.A. (1982) Association of bacteriuria and urinary nitrosamine formation with *Schistosomiasis haematobium* infection in the Qalyub area of Egypt. *Trans. R. Soc. Trop. Med. Hyg.*, **76**, 519-527

Kantor, A.F., Hartge, P., Hoover, R.N., Naragana, A.S., Sullivan, J.W. & Fraumeni, J.F. (1984) Urinary tract infection and risk of bladder cancer. *Am. J. Epidemiol.*, **119**, 510-515

Lehman, E.L. (1975) *Neoparametrics: Statistical Methods Based on Ranks*, Oakland, CA, Holden-Day

Melzak, J. (1966) The incidence of bladder cancer in paraplegia. *Puruplegia*, **4**, 85-96

Nordenstam, G.R., Ake Brandberg, C., Odén, A.S., Svanborg Edén, C.M. & Svanborg, A. (1986) Bacteriuria and mortality in an elderly population. *New Engl. J. Med.*, **314**, 1152-1156

Radomski, J.L., Greenwald, D., Hearn, W.L., Block, N.L. & Woods, F.M. (1978) Nitrosamine formation in bladder infections and its role in the etiology of bladder cancer. *J. Urol.*, **120**, 48-50

Stuehr, D.J. & Marletta, M.A. (1985) Mammalian nitrate biosynthesis: mouse macrophages produce nitrite and nitrate in response to *Escherichia coli* lipopolysaccharides. *Proc. Natl Acad. Sci. USA*, **82**, 7738-7742

Tricker, A.R., Mostafa, M.A., Spiegelhalder, B. & Preussmann, R. (1989) Urinary excretion of nitrate, nitrite and N-nitroso compounds in *Schistosomiasis* and bilharzial bladder cancer patients. *Carcinogenesis*, **10**, 547-552

World Health Organization (1985) *The Control of Schistosomiasis* (*Tech. Rep. Ser.*, **728**), Geneva

FORMATION OF N-TRIMETHYL-N-NITROSOUREA IN THE GASTRIC LUMEN OF THE FISTULATED PIG

C.M. Maragos[1], J.H. Hotchkiss & S.L. Fubini

Institute of Food Science and College of Veterinary Medicine, Cornell University, Ithaca, NY, USA

The formation of N-trimethyl-N-nitrosourea (TMNU) in the stomach of full-sized gastric-fistulated pigs was determined. Nitrite was added to the stomach in amounts that have been reported to occur in the human stomach (25-375 μmol). The vehicle was an 'artificial meal' made up of synthetic gastric juice. The total amount of TMNU formed (not concentration) was estimated by determining the concentration and gastric volume using multiple additions of a nonabsorbable marker (polyethylene glycol). At an initial pH of 3, 23-1000 μg of TMNU were found after 10-25 min. The rate of loss of polyethylene glycol and TMNU from the stomach were the same, indicating that TMNU was not directly absorbed. These data suggest that formation of TMNU under conditions similar to those found in the human stomach is possible.

It has been suggested that human gastric cancer may be related to the formation of N-nitroso compounds, particularly N-nitrosamides, within the stomach (Mirvish, 1983). Several groups have assessed the formation of N-nitrosamines in the stomachs of a variety of species, including dogs (Mysliwy *et al.*, 1974; Lintas *et al.*, 1982; Licht *et al.*, 1988) and guinea-pigs (Yamamoto *et al.*, 1987). However, the gastric formation of N-nitrosamines appears less likely than that of N-nitrosamides (Licht & Dean, 1988). We have developed a full-sized fistulated pig model and an analytical method for determining the rate and amount of formation of N-trimethyl-N-nitrosourea (TMNU) within the stomach.

Methods

Pigs were surgically fitted with a solid, expandable Teflon stomach cannula. After 24 h, normal feeding and care were resumed. The cannula remained in the animal for up to eight months. Polyethylene glycol (PEG) was used as a marker of gastric emptying (Buxton *et al.*, 1979). TMNU was extracted from gastric fluid using C_{18} solid-phase extraction columns and was quantified by high-performance liquid chromatography/thermal energy analysis using a photolysis interface and reverse-phase chromatography (acetonitrile:water; Conboy & Hotchkiss, 1989).

---

[1]Present addess: National Cancer Institute, Frederick Cancer Research and Development Center, Building 538, Frederick, MD 21702, USA

TMNU or its precursor and PEG were administered by fistula after a 24-h fast in 200 ml of an artificial gastric fluid 'meal', which was formulated to match porcine gastric fluid closely. The pH was adjusted to a desired value (1-7) with $NaHCO_3$ or HCl. Samples of 7 ml were collected every 15 min after administration of precursors and 2-ml samples (for additional PEG analyses) collected every 5 min. The 7-ml samples were collected into vials containing 25 mg NaN_3 to inhibit nitrosation. All samples were kept on ice until the end of the experiment, at which time the 7-ml samples were subdivided into 5-ml samples for TMNU analysis and 2-ml samples for PEG analysis. The TMNU samples were extracted within 2 h of the last experimental time point. PEG samples were kept frozen until analysis, generally within three days of the experiment.

TMNU absorption was determined by comparing gastric TMNU and PEG concentrations after a single dose of TMNU and PEG in artificial gastric juice pH 4.0. Initial concentrations, obtained by analysis of a reserved portion, were 9.1 mg PEG/ml and 560 ng TMNU/ml, respectively.

To measure the gastric formation of TMNU, fasted animals were dosed with artificial gastric fluid containing PEG. Trimethylurea (TMU) was prepared in water within 45 min of use. $NaNO_2$ was prepared in water 1 min before administration. Immediately preceding an experiment, the stomach contents were drained through the fistula; the volume collected was 50-200 ml. Figure 1 indicates the time course for administration of test solutions. Typically, 200 ml of 5 mg/ml PEG in artificial gastric juice, followed by TMU, followed by nitrite, were administered at time 0. The first sample was collected at 5 min. Volume at 5 min was estimated by the extent to which the initial PEG dose had been diluted. Subsequently, gastric volume was determined by addition of either 750 or 500 mg PEG in 10 ml water. The increase in gastric PEG concentration following an addition was used to determine volume (Hurwitz, 1981), using the equation:

$$\text{Volume} = \frac{A - (C_p X V_a)}{(C_p - C_{bp})},$$

where A is the amount of PEG added, V_a is the volume of PEG solution added, C_{bp} is the PEG concentration before the addition, and C_p is the PEG concentration after the addition. Final volume at 65 min was estimated from the total volume that could be collected. The total amount of TMNU present was obtained from the TMNU concentration and the gastric volume.

Absorption of TMNU

TMNU absorption was determined following a single dose of 560 ng TMNU/ml and 9.1 mg PEG/ml in artificial gastric juice pH 4.0. When expressed as percentage of the initial concentration, the curves for TMNU and PEG are very similar (Figure 2), suggesting that gastric absorption or decomposition of TMNU is minimal under these conditions. There was, however, a nonsignificant difference between the PEG and TMNU curves at times longer than 30 min. Loss of both PEG and TMNU due to gastric emptying was complete at 60 min.

Figure 1. Protocol for the addition of test solutions to pig stomach[a]

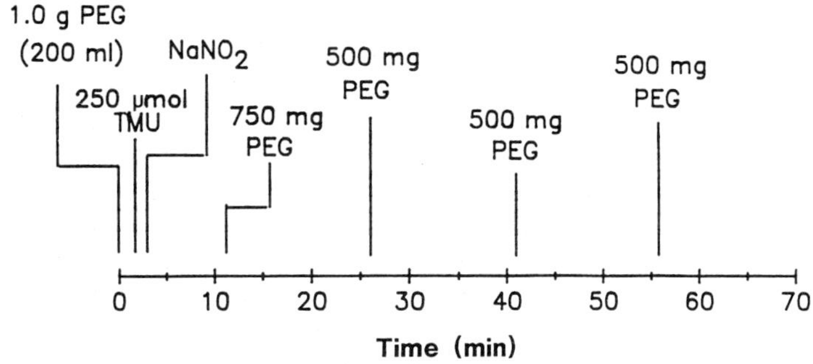

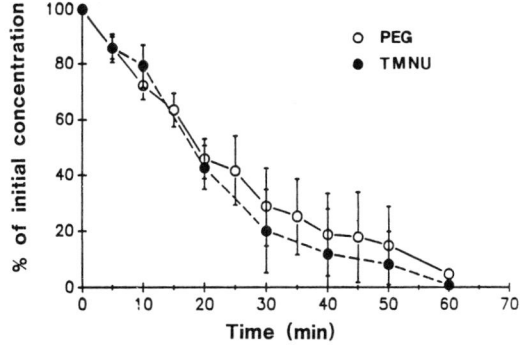

[a] A bolus dose (200 ml) of 5 mg/ml polyethylene glycol (PEG) in artificial gastric juice was given at time zero. Trimethylurea (TMU) and $NaNO_2$ in 10 ml water were added shortly after. Further additions of PEG were 10 ml of the appropriate concentration in 10 ml artificial gastric juice.

Figure 2. N-Trimethyl-N-nitrosourea (TMNU) content of pig stomach after administration of 200 ml artificial gastric fluid containing 9.1 mg/ml polyethylene glycol (PEG) and 560 ng/ml TMNU

Formation of TMNU

The formation of TMNU was followed after addition of 250 μmol TMU to the stomach with nitrite at 25-375 μmol (100-1500 μM). Estimates of nitrite concentration in gastric fluid range from 1-5 μM (fasted) up to 340 μM after eating (Gatehouse & Tweats, 1982; Walters et al., 1976). In order to determine the total amount of TMNU in the stomach, the gastric volume was estimated through multiple additions of PEG. At the highest dose of nitrite (375 μmol), 1000 μg TMNU were found in the stomach and at the lowest nitrite dose (25 μmol), 23 μg TMNU (Figure 3). The largest amounts of TMNU occurred 10-25 min after dosing. The highest concentration of TMNU found was 4.4 μg/ml, which occurred after 375 μmol of nitrite were added to the stomach. No TMNU was detected over a 65-min span when 10 ml distilled water were added instead of nitrite.

Mysliwy et al. (1974) described the formation of N-nitrosopyrrolidine in chronically fistulated dogs following the addition of 724 μmol $NaNO_2$ and 141 μmol pyrrolidine. Up to 1000 ng/ml N-nitrosopyrrolidine were formed within 3 min. Similar effects were observed by Lintas *et al.* (1982) with N-nitrosodimethylamine in dogs. In the latter report, the concentration of nitrosamine peaked at 450 ng/ml, 3 min after addition of 870 μmol $NaNO_2$ and 400 μmol dimethylamine through the fistula. In both studies, nitrite and nitrosamine disappeared rapidly from the stomach. However, the recent model developed by Licht and Deen (1988) suggests that the concentrations of nitrite and dialkylamine in gastric fluid and the kinetics of nitrosation are all too low to result in the formation of significant amounts of N-nitrosamines under normal physiological conditions.

Similarly, the formation of TMNU *in vivo* in pig stomach is relatively rapid. The kinetics of formation in gastric fluid *in vitro* is currently being examined in our laboratory. The unexpectedly high amounts of TMNU formed at physiologically relevant nitrite concentrations is probably due to the ease of nitrosation of substituted ureas (Mirvish, 1971). The prolonged presence of TMNU in the stomach in this study was a result of its stability and the absence of significant direct gastric absorption.

Figure 3. Gastric formation of N-trimethyl-N-nitrosourea (TMNU) following the protocol for administration of test solutions depicted in Figure 1; 250 μmol trimethylurea in 10 ml water were used in each experiment. $NaNO_2$, also in water, was prepared immediately before addition.

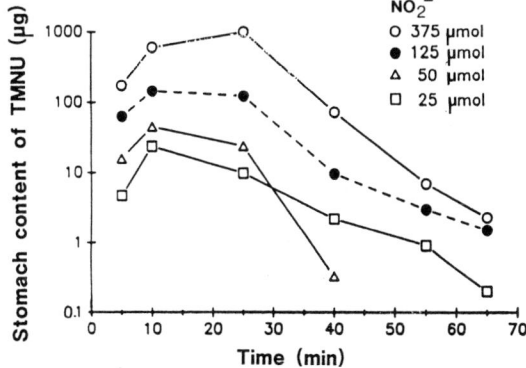

TMNU is carcinogenic to rats when present in drinking-water at 1 mM for 47 weeks (Lijinsky *et al.*, 1980). The extent of dietary exposure to TMU is unknown but is probably very low or nonexistent. TMU was chosen for these experiments because of the stability of TMNU.

In summary, the endogenous formation of a model nitrosourea, TNMU, was demonstrated in pigs. Gastric absorption appears to be minor, but TMNU was produced when as little as 25 μmol nitrite were administered immediately after a 200-ml 'meal' of artificial gastric fluid at pH 3. These data suggest that N-nitrosamides can form in the stomach under conditions that might be normally found in humans.

References

Buxton, T.B., Crockett, J.K., Moore, W.L., III, Moore, W.L., Jr & Rissing, J.P. (1979) Protein precipitation by acetone for the analysis of polyethylene glycol in intestinal perfusion fluid. *Gastroenterology*, **76**, 820-824

Conboy, J.J. & Hotchkiss, J.H. (1989) Photolytic interface for high performance liquid chromatography-chemiluminescence detection of non-volatile N-nitroso compounds. *Analyst*, **114**, 155-159

Licht, W.R. & Deen, W.M. (1988) Theoretical model for predicting rates of nitrosamine and nitrosamide formation in the human stomach. *Carcinogenesis*, **9**, 2227-2237

Licht, W.R., Fox, J.G. & Deen, W.M. (1988) Effects of ascorbic acid and thiocyanate on nitrosation of proline in the dog stomach. *Carcinogenesis*, **9**, 373-377

Lijinsky, W., Reuber, M.D. & Blackwell, B.N. (1980) Carcinogenicity of nitrosotrialkylureas in Fischer 344 rats. *J. Natl Cancer Inst.*, **65**, 451-453

Lintas, C., Clark, A., Fox, J., Tannenbaum, S.R. & Newberne, P.M. (1982) *In vivo* stability of nitrite and nitrosamine formation in the dog stomach: effect of nitrite and amine concentration and of ascorbic acid. *Carcinogenesis*, **3**, 161-165

Mirvish, S.S. (1971) Kinetics of nitrosamide formation from alkylureas, *N*-alkylurethans, and alkylguanidines: possible implications for the etiology of human gastric cancer. *J. Natl Cancer Inst.*, **46**, 1183-1193

Mirvish, S.S. (1983) The etiology of gastric cancer: intragastric nitrosamide formation and other theories. *J. Natl Cancer Inst.*, **71**, 631-644

Mysliwy, T.S, Wick, E.L., Archer, M.C., Shank, R.C. & Newberne, P.M. (1974) Formation of *N*-nitrosopyrrolidine in a dog's stomach. *Br. J. Cancer*, **30**, 279-283

Yamamoto, M., Ishiwata, H., Yamada, T., Yoshihira, K. & Tanimura, A. (1987) Studies in the guinea-pig stomach on the formation of *N*-nitrosomethylurea, from methylurea and sodium nitrite, and its disappearance. *Food Chem. Toxicol.*, **25**, 663-668

Relevance to Human Cancer of N-Nitroso Compounds,
Tobacco Smoke and Mycotoxins.
Ed. I.K. O'Neill, J. Chen and H. Bartsch
Lyon, International Agency for Research on Cancer
© IARC, 1991

BACTERIAL FORMATION OF N-NITROSO COMPOUNDS IN THE RAT STOMACH AFTER OMEPRAZOLE-INDUCED ACHLORHYDRIA

S. Calmels, J.-C. Béréziat, H. Ohshima & H. Bartsch

International Agency for Research on Cancer, Lyon, France

N-Nitrosamine formation by bacteria in the achlorhydric stomach has been proposed as an important factor in the development of gastric cancer. Thus, the effect of the presence of bacteria in the stomach on endogenous nitrosation was investigated in rats given omeprazole (an inhibitor of gastric H^+,K^+-ATPase) which reduces gastric secretion sufficiently to allow survival of a bacterial suspension of *Escherichia coli* or *Pseudomonas*. When rats were given both thiazolidine 4-carboxylic acid and nitrate, greater endogenous nitrosamine formation was observed in rats receiving omeprazole and an *E. coli* suspension than in control or omeprazole-treated rats. A similar result was obtained when rats were given morpholine and nitrate. Since the endogenous formation of N-nitrosomorpholine (NMOR) can be evaluated more precisely from the levels of its urinary metabolites, N-nitrosohydroxyethylglycine (NHEG), the metabolism of NMOR was studied in omeprazole-treated rats. In this preliminary study, we showed that 60% of an oral dose of NMOR was excreted as NHEG, while in rats with a higher gastric pH 20% was excreted as NHEG. The amount of endogenously formed NMOR was increased in omeprazole-treated rats given morpholine and nitrite together with bacteria, and greater excretion of unchanged urinary NMOR was observed. Thus, as shown in this in-vivo model, bacteria efficiently reduce nitrate to nitrite and catalyse nitrosation, resulting in increased endogenous formation of N-nitroso compounds in the achlorhydric stomach.

A number of disorders of the gastrointestinal tract are accompanied by gastric achlorhydria, allowing bacterial colonization of the stomach, which in healthy persons is virtually sterile (Simon & Gorbach, 1981). Some of the microorganisms isolated from the achlorhydric stomach possess nitrate-reductase and nitrosating activity (Calmels *et al.*, 1987; Leach *et al.*, 1987; O'Donnell *et al.*, 1987). The enhanced gastric formation of nitrite and the bacterially catalysed formation of N-nitroso compounds (NOC) have been studied intensively, in view of the increased risk for gastric cancer in some subjects suffering from achlorhydria (Jones *et al.*, 1978). However, controversy still exists with regard to whether or not the catalysis of endogenous nitrosation by bacteria is increased in subjects with achlorhydria (Reed *et al.*, 1981; Milton-Thompson *et al.*, 1982; Keighley *et al.*, 1984; Hall *et al.*, 1986). To study further the effect of bacterial colonization of the stomach on endogenous nitrosation, experiments were conducted in rats treated with omeprazole (2-pyridinyl-2-sulfinylbenzimidazole), a drug belonging to a new class of antisecretory agents which act by inhibiting the proton pump of the gastric mucosa (H^+,K^+-ATPase) selectively (Fellenius *et al.*, 1981).

Effect of omeprazole on gastric pH

Six-week-old male Sprague-Dawley rats weighing 200 g were gavaged with 3 mg omeprazole per rat after an overnight fast. They were sacrificed at different time intervals, and the pH of the gastric content was measured. An increase in gastric pH from 2 to 6.7 was observed during the first 3 h after treatment, and the pH remained high for the following 5 h, then decreased slowly. Therefore, rats were gavaged with precursors of NOC and/or a bacterial suspension 3 h after treatment with omeprazole, i.e., when gastric achlorhydria had been established.

Endogenous formation of N-nitrosothiazolidine 4-carboxylic acid and N-nitrosomorpholine

Thiazolidine 4-carboxylic acid (20 µmol) or morpholine (20 µmol) was used as the nitrosatable model substrate, and both were given with or without nitrite (20 µmol) or nitrate (20 µmol – 1.6 mmol) in the presence or absence of the nitrosation-proficient strains *Escherichia coli* A10 and *Pseudomonas aeruginosa* D3375 (approximately 10^{13} bacteria). These strains have been isolated from human intestinal contents (Suzuki & Mitsuoka, 1984) and infections (Calmels *et al.*, 1985). The same rats were used repeatedly over three weeks. Endogenous nitrosation was quantified by measuring the levels of *N*-nitrosothiazolidine 4-carboxylic acid (NTCA), *N*-nitrosomorpholine (NMOR) and its metabolite, *N*-nitrosohydroxyethylglycine (NHEG; Hecht *et al.*, 1984) excreted in 24-h urine.

When rats were given both thiazolidine 4-carboxylic acid and nitrate, more NTCA was formed endogenously in rats receiving *E. coli* A10 suspension than in control rats (Table 1). Although there were large interindividual variations, this difference was statistically significant. When nitrite was administered together with thiazolidine 4-carboxylic acid, endogenous nitrosation was slightly, but not significantly, higher in control rats with an acidic stomach than in those treated with omeprazole with or without *E. coli* A10 (Table 1). Because thiazolidine 4-carboxylic acid is nitrosated rapidly at acidic pH and relatively

Table 1. Urinary excretion of *N*-nitrosothiazolidine 4-carboxylic acid (nmol, mean ± SEM) in groups of ten rats receiving thiazolidine 4-carboxylic acid (TCA) and 20 µmol nitrate or nitrite[a]

Treatment	TCA + NO_2^-	TCA + NO_3^-
Controls	7100 ± 1080	4.82 ± 1.19
Omeprazole	4960 ± 2170	2.75 ± 0.68
Omeprazole plus *E. coli*	3200 ± 1390	19.21 ± 3.48[b]

[a] Rats were given omeprazole (3 mg/rat) orally after an overnight fast and 3 h later received nitrosoamino precursors by gavage with or without bacterial suspension in 0.9% NaCl ($\sim 10^{13}$ bacteria/rat). Bacteria were cultured overnight in a 10 mM $NaNO_3$ minimum milieu (Calmels *et al.*, 1988), harvested and resuspended in 0.9% NaCl. Rats were placed in metabolic cages and kept without food for 24 h; drinking-water was given *ad libitum*. 24-h urine was collected in the presence of 1 ml NaOH and made up to 50 ml with distilled water. *N*-Nitrosothiazolidine 4-carboxylic acid was analysed as described previously (Ohshima *et al.*, 1984).
[b] Significant in comparison with controls at $p < 0.005$

high concentrations of precursors were used in this experiment, it was considered that small variations in gastric pH might result in large differences in nitrosation rate. Thus, in our experiment, the chemical nitrosation totally masked the contribution of bacterial nitrosation in the achlorhydric stomach.

Similar experiments were performed with morpholine (Table 2). In order to quantify the endogenous formation of NMOR, NHEG, a major urinary metabolite of NMOR, was measured as a marker. Three times as much of an oral dose of NMOR was excreted as NHEG in the urine of control rats as in that of rats treated with omeprazole with or without *E. coli* or *P. aeruginosa*, indicating that the endogenous formation of NMOR can be evaluated more precisely by measuring the rate of metabolism of NMOR to NHEG. As shown in Table 3, more NMOR was formed endogenously (estimated from levels of NHEG) and more unchanged NMOR was excreted in omeprazole-treated rats given morpholine and nitrite together with bacteria than in control rats or in rats receiving omeprazole alone.

Table 2. Urinary excretion of *N*-nitrosohydroxyethylglycine (NHEG) and unchanged *N*-nitrosomorpholine (NMOR) (mean ± SEM) in 24-h urine of groups of five rats after administration of NMOR (5 μmol)[a]

Treatment	NHEG (μmol) (%)	Unchanged NMOR (nmol) (%)
Controls	3.01 ± 0.07 (60.2%)	5.1 ± 1.63 (0.1%)
Omeprazole	1.00 ± 0.14 (20.6%)	5.65 ± 0.59 (0.11%)
Omeprazole plus bacteria	1.03 ± 0.08 (20.6%)	7.68 ± 1.38 (0.15%)
E. coli	0.97 ± 0.04	4.8 ± 0.9
P. aeruginosa	1.11 ± 0.10	10.6 ± 1.87

[a]NMOR (5 μmol/rat in 1 ml distilled water) was administered orally to rats after an overnight fast; 24-h urine was collected as described in footnote *a* to Table 1. NHEG was analysed as follows: *N*-nitrosohydroxyethyl-β-alanine was added as an internal standard to 10 ml urine that had been treated with excess sulfamic acid. NHEG was extracted three times with 25 ml ethylacetate, and the extracts, dried over anhydrous sodium sulfate, were concentrated to dryness. The residue was then derivatized in the presence of *N*-methyl-*N*-(*t*-butyldimethylsilyl)trifluoroacetamide (90 μl) and dimethylformamide (10 μl) at 70°C for 20 min. The derivatized sample (10 μl) was analysed on a gas chromatograph equipped with a 2 m × 3 mm i.d. glass column containing 3% OV-17 on Chromosorb w (60-80 mesh) at a column oven temperature of 240°C. A thermal energy analyser (model 543) was used as detector.

When rats were given 20 μmol morpholine and 1.6 mmol nitrate, higher levels of unchanged NMOR ($p < 0.01$) were excreted in 24-h urine of rats gavaged with omeprazole and bacteria (1 ± 0.18 nmol NMOR per day with *E. coli* A10 and 0.92 ± 0.23 nmol NMOR per day with *P. aeruginosa* D3375), as compared to control rats (0.37 ± 0.12 nmol NMOR per day) or rats treated with omeprazole alone (0.35 ± 0.07 nmol NMOR per day). The analysis of NHEG in urine in these experiments was made difficult by the high concentrations of nitrate.

Conclusions

Our data imply that bacteria in the achlorhydric stomach are responsible for increased endogenous formation of NTCA and NMOR from nitrate and the respective amino precursor *via* reduction of nitrate into nitrite followed by bacterial or chemical nitrosation. Furthermore, bacterial catalysis of intragastric nitrosation was observed when rats were given morpholine and nitrite together with a bacterial suspension.

We have established an animal model for studying the role of bacteria in intragastric nitrosation using omeprazole to block gastric acid secretion. Bacteria were shown to reduce nitrate to nitrite efficiently and to catalyse nitrosation of morpholine by nitrate or nitrite, resulting in significantly increased endogenous formation of NOC in the achlorhydric stomach. Unexpectedly, it was found that the excretion/metabolism of NMOR was different in rats treated with omeprazole and in controls. Whether this is true for other nitrosamines needs to be investigated. We plan to study the consequences of bacterial colonization of the stomach using DNA alkylation in the gastric mucosa as a marker in this animal model.

Table 3. Urinary excretion of *N*-nitrosohydroxyethylglycine (NHEG) and *N*-nitrosomorpholine (NMOR) (mean $\pm$ SEM) in groups of five rats receiving morpholine and nitrite[a]

Treatment	NHEG in urine (nmol)	NMOR in urine (nmol)	Endogenously formed NMOR (nmol)
Controls	233 $\pm$ 35	0.58 $\pm$ 0.17	388 $\pm$ 60
Omeprazole	139 $\pm$ 74	0.63 $\pm$ 0.17	690 $\pm$ 370
Omeprazole plus bacteria	204 $\pm$ 88	0.95 $\pm$ 0.19[b]	1025 $\pm$ 330[b]
E. coli	193 $\pm$ 78	1.03 $\pm$ 0.33	972 $\pm$ 46
P. aeruginosa	215 $\pm$ 98	0.87 $\pm$ 0.20	1110 $\pm$ 118

[a] Morpholine-HCl (20 μmol in 0.5 ml distilled water) and sodium nitrite (20 μmol in 0.5 ml distilled water) were administered sequentially to rats treated with omeprazole, as described in footnote *a* to Table 1. 24-h urine samples were collected and analysed for NHEG as described in footnote *a* to Table 2 and NMOR as reported previously (Calmels *et al.*, 1985). The amount of endogenously formed NMOR was estimated from that of NHEG excreted in 24-h urine and corrected for excretion rates given in Table 2.
[b] Significant in comparison with levels detected in controls at $p < 0.05$

Acknowledgement

The work reported in this paper was supported by a grant from the US National Cancer Institute, No. 1 RO1-CA 47591-01A1.

References

Calmels, S., Ohshima, H., Vincent, P., Gounot, A.-M. & Bartsch, H. (1985) Screening of microorganisms for nitrosation catalysis at pH 7 and kinetic studies on nitrosamine formation from secondary amines by *E. coli* strains. *Carcinogenesis*, **6**, 911-915

Calmels, S., Ohshima, H., Crespi, M., Leclerc, H. Cattoen, C. & Bartsch, H. (1987) *N*-Nitrosamine formation by microorganisms isolated from human gastric juice and urine: biochemical studies on bacteria-catalysed nitrosation. In: Bartsch, H., O'Neill, I.K. & Schulte-Hermann, R., eds, *The Relevance of* N-*Nitroso Compounds to Human Cancer: Exposures and Mechanisms* (IARC Scientific Publications No. 84), Lyon, IARC, pp. 396-398

Calmels, S., Ohshima, H. & Bartsch, H. (1988) Nitrosamine formation by denitrifying and non-denitrifying bacteria: implication of nitrite reductase and nitrate reductase in nitrosation catalysis. *J. Gen. Microbiol.*, **134**, 221-226

Fellenius, E., Berglindh, T., Sachs, G., Olbe, L., Elander, B., Sjöstrand, S.-E. & Wallmark, B. (1981) Substituted benzimidazoles inhibit gastric acid secretion by blocking H+, K+ -ATPase. *Nature*, **290**, 159-161

Hall, C., Darkin, D., Brimblecombe, R., Cook, A., Kirkham, J. & Northfield, T. (1986) Evaluation of the nitrosamine hypothesis of gastric carcinogenesis in precancerous conditions. *Gut*, **27**, 491-498

Hecht, S., Morrison, B. & Young, R. (1984) N-Nitroso(2-hydroxyethyl)glycine, a urinary metabolite of N,N'-dinitropiperazine with potential utility as a monitor for its formation *in vivo* from piperazine. *Carcinogenesis*, **5**, 979-981

Jones, S., Davies, P. & Savage, A. (1978) Gastric juice nitrite and gastric cancer. *Lancet*, **i**, 1355

Keighley, M., Younger, D., Poxon, V., Morris, D., Muscroft, T., Burdon, D., Barnard, J., Bavin, P., Brimblecombe, R., Darkin, D., Moore, P. & Viney, N. (1984) Intragastric N-nitrosation is unlikely to be responsible for gastric carcinoma developing after operations for duodenal ulcer. *Gut*, **25**, 238-245

Leach, S., Cook, A., Challis, B., Hill, M. & Thompson, M. (1987) Bacterially mediated N-nitrosation reaction and endogenous formation of N-nitroso compounds. In: Bartsch, H., O'Neill, I.K. & Schulte-Hermann, R., eds, *The Relevance of N-Nitroso Compounds to Human Cancer: Exposures and Mechanisms* (IARC Scientific Publications No. 84), Lyon, IARC, pp. 396-399

Milton-Thompson, G., Lightfoot, N., Ahmet, Z., Hunt, R., Barnard, J., Bavin, P., Brimblecombe, R., Darkin, D., Moore, P. & Viney, N. (1982) Intragastric acidity, bacteria, nitrite and N-nitroso compounds, before, during and after cimetidine treatment. *Lancet*, **i**, 1091-1095

O'Donnell, C., Edwards, C., Corcoran, G., Ware, J. & Edwards, P. (1987) In vitro production of nitrosamines by bacteria isolated from the operated stomach. In: Bartsch, H., O'Neill, I.K. & Schulte-Hermann, R., eds, *The Relevance of N-Nitroso Compounds to Human Cancer: Exposures and Mechanisms* (IARC Scientific Publications No. 84), Lyon, IARC, pp. 400-403

Ohshima, H., O'Neill, I.K., Friesen, M., Pignatelli, B. & Bartsch, H. (1984) Presence in human urine of new sulfur-containing N-nitrosamino acids: N-nitrosothiazolidine 4-carboxylic acid and N-nitroso 2-methylthiazolidine 4-carboxylic acid. In: Bartsch, H., O'Neill, I.K. & Schulte-Hermann, R., eds, *The Relevance of N-Nitroso Compounds to Human Cancer: Exposures and Mechanisms* (IARC Scientific Publications No. 84), Lyon, IARC, pp. 77-85

Reed, P., Smith, P., Haines, K., House, F. & Walters, C. (1981) Gastric juice N-nitrosamines in health and gastroduodenal disease. *Lancet*, **ii**, 550-552

Simon, G.L. & Gorbach, S. (1981) Intestinal flora in health and disease In: Johnson, L., ed., *Physiology of Gastrointestinal Bacteria*, Vol. 2, New York, Raven Press, pp. 1361-1380

Suzuki, K. & Mitsuoka, T. (1984) N-Nitrosamine formation by intestinal bacteria. In: O'Neill, I.K., von Borstel, R.C., Miller, C.T., Long, J. & Bartsch, H., eds, *N-Nitroso Compounds: Occurrence, Biological Effects and Relevance to Human Cancer* (IARC Scientific Publications No. 57), Lyon, IARC, pp. 275-281

Relevance to Human Cancer of *N*-Nitroso Compounds,
Tobacco Smoke and Mycotoxins.
Ed. I.K. O'Neill, J. Chen and H. Bartsch
Lyon, International Agency for Research on Cancer
© IARC, 1991

EPIDEMIOLOGICAL STUDY OF PRECURSOR LESIONS OF OESOPHAGEAL CANCER AMONG YOUNG PERSONS IN HUIXIAN, CHINA

J. Chang-Claude[1], J. Wahrendorf[1], S.L. Qiu[2], G.R. Yang[2]
N. Muñoz[3], M. Crespi[4], R. Raedsch[5], D.I. Thurnham[6] & P. Correa[7]

[1]*German Cancer Research Center, Institute of Epidemiology and Biometry, Heidelberg, Germany;*
[2]*Henan Institute of Medical Sciences, Zhengzhou, Henan Province, China;*
[3]*International Agency for Research on Cancer, Lyon, France;*
[4]*National Cancer Institute (Regina Elena), Rome, Italy;*
[5]*Department of Medicine, University of Heidelberg, Heidelberg, Germany*
[6]*University of Cambridge and Medical Research Council, MRC Dunn Nutrition Unit, Cambridge, UK; and*
[7]*Louisiana State University Medical Center, New Orleans, LA, USA*

A total of 538 subjects (354 males, 184 females) were investigated in a high-risk area for oesophageal cancer in China to provide large-scale data on the prevalence of chronic oesophagitis among persons below 25 years of age. The survey included an oesophagoscopy with guided biopsies and cytology, a physical examination, an interview and collection of a 10-ml blood sample and overnight urine. Histologically confirmed oesophagitis was found in 43.5% of male and 35.6% of female subjects. Mild and moderate oesophagitis was associated positively with consumption of burning hot beverages, family history of oesophageal cancer, frequent consumption of cottonseed oil, cigarette smoking, a clinical diagnosis of oral leukoplakia and seborrhoeic dermatitis; negative associations were seen with frequent consumption of fresh fruit, meat and eggs. A significantly higher level of urinary nitrate was found in the diseased group, but no difference was seen with regard to any of the N-nitrosamino acids measured.

Chronic oesophagitis is considered to precede epithelial dysplasia, a major precancerous lesion of oesophageal cancer. Thus, prevention of these precursor lesions may be an efficient method for preventing oesophageal cancer. The aim of this study was to determine the prevalence of precursor lesions for oesophageal cancer at early ages (15-26) in a high-risk area for oesophageal cancer in China and to identify risk factors associated with the development of these lesions in a case-control study.

Subjects and methods

Household of oesophageal cancer cases diagnosed after 1981 and twice the number of control persons, frequency-matched by age and sex to the cases, were identified in seven production brigades of Meng Zhuang Commune, Huixian, Henan Province. Eligible subjects were young adults aged between 15 and 25 years who had lived in a case or control household during the previous five years. Of 887 subjects, 545 (62%) participated in the field study, conducted in May 1988.

Participants were interviewed using a questionnaire designed to obtain information on exposures in early childhood and recently, with regard to known and suspected risk factors, such as diet, nutritional deficiencies, consumption of mouldy foods, pickled vegetables, tobacco, alcohol and hot food and beverages, environmental exposures and family history of oesophageal cancer. A physical examination was conducted, and a 10-ml blood sample was taken. Overnight urine was collected from a random sample of 85 subjects for the analysis of certain N-nitrosamino acids as an indirect indicator of nitrosation. Endoscopic examination of the oesophagus was performed, and biopsy specimens were taken from the middle and lower third of the oesophagus or from macroscopic lesions.

Slides of biopsy specimens were read by three pathologists (Q.S.L., N.M., P.C.) who had no knowledge of the clinical data, and lesions were graded as very mild, mild, moderate or severe (Crespi et al., 1979). Several N-nitrosamino acids, nitrate and creatinine in urine were analysed at the IARC.

Prevalence odds ratios were calculated for all variables of interest, and logistic regression models were fitted in order to evaluate different risk variables simultaneously using the LOGIST procedure in SAS. The Wilcoxon rank sum test was used for comparing the urinary excretion levels of nitrate and N-nitrosamino acids in two groups.

Prevalence of chronic oesophagitis

In 538 study subjects for whom a histological diagnosis was available, the prevalence of very mild oesophagitis showed negligible differences by sex and by two age groups, whereas mild oesophagitis was more common among males (Table 1). No case of severe oesophagitis or dysplasia was found. Significantly more oesophagitis was seen among young persons from case households than from control households, predominantly attributable to the occurrence in males.

Risk factors for chronic oesophagitis (case-control analysis)

Cases were defined as those with mild or moderate oesophagitis and controls those with very mild oesophagitis and normal oesophagus. The consumption of beverages at burning hot temperatures gave the highest odds ratio for people of each sex (Table 2), although it was not statistically significant for females. The possibility that the temperature of foods and beverages plays a role in the etiology of oesophageal cancer has been suggested by the results of population correlation and case-control studies in several countries; experimental work has also shown an effect of thermal injury (Yioris et al., 1984; Ghadirian, 1987). Muñoz et al. (1987) observed a two-fold increase in the prevalence of precancerous lesions of the oesophagus in people who drink hot *mate*.

Table 1. Histological findings among young persons in Huixian, 1988

Subjects	No. of subjects	Oesophagitis (%)			
		Normal	Very mild	Mild	Moderate
Male	354	56.5	31.6	10.7	1.1
15-20 years[a]	167	57.5	32.3	8.4	1.8
21-26 years	187	55.6	31.0	12.8	0.5
Case household	99	41.4	40.4	16.2	2.0
Control household	255	62.4	28.2	8.6	0.8
Female	184	64.1	30.4	4.3	1.1
15-20 years[a]	140	66.4	28.6	3.6	1.4
21-26 years	44	56.8	36.4	6.8	0
Case household	67	64.2	29.9	4.5	1.5
Control household	117	64.1	30.8	4.3	0.9

[a] Included one 14-year-old

Table 2. Prevalence odds ratios[a] of single risk variables for oesophagitis

Factor	Category	Males		Females	
		Ca/Co[b]	Odds ratio	Ca/Co[b]	Odds ratio
Risk					
Temperature of beverages	Warm or moderate	35/302	1.0	8/163	1.0
	Burning hot	7/10	5.8*	2/11	3.7
Family history	Negative	19/213	1.0	5/101	1.0
	Positive	23/99	3.0*	5/73	1.9
Cigarettes/day	0	10/122	1.0	10/173	-
	1-15	21/138	1.8	0/1	-
	>15	11/52	2.4*	0/0	-
Cottonseed oil used most often	No	19/208	1.0	5/108	1.0
	Yes	23/104	2.3*	5/66	1.6
Pickled vegetable juice	≤ twice/week	40/306	1.0	10/168	-
	> twice/week	2/6	3.1	0/6	-
Protective					
Fresh fruit in summer	< once/week	28/143	1.0	10/84	-
	≥ once/week	14/169	0.4*	0/90	-
Green vegetables or fresh fruit	< once/week	6/25	1.0	3/11	1.0
	≥ once/week	36/287	0.6	7/163	0.1*
Raw vegetables	< once/month	11/96	1.0	6/42	1.0
	≥ once/month	31/216	1.3	4/132	0.2*
Wheat flour products	≤ twice/day	20/98	1.0	8/57	1.0
	> twice/day	22/214	0.5*	2/117	0.1*
Meat or eggs	< twice/week	27/149	1.0	8/98	1.0
	≥ twice/week	15/163	0.5	2/76	0.3

[a] Adjusted for household
[b] Ca/Co, number of cases/number of controls
* $p < 0.05$

Familial aggregation has been observed with regard to oesophageal cancer (Ghadirian, 1985); thus, a positive association between family history of oesophageal cancer and chronic oesophagitis suggests the involvement of chronic oesophagitis in the natural history of oesophageal cancer. However, although this variable persisted among males, even after some risk variables had been accounted for in logistic regression models, inadequate adjustment for other environmental risk factors cannot be ruled out.

A poor diet with low intake of animal protein, fruit and green vegetables has been observed repeatedly among persons with oesophageal cancer in both high- and low-risk areas (Cook-Mozaffari et al., 1979; Li & Cheng, 1984; Tuyns et al., 1987; Yu et al., 1988). In our study, a relatively high intake of vegetables and animal protein was protective against the occurrence of oesophagitis; and, conversely, low consumption was associated with disease.

Although cigarette smoking has not been identified as a strong risk factor for oesophageal cancer in China (Li & Cheng, 1984), it was positively associated with the occurrence of chronic oesophagitis among males and showed a statistically significant dose-response relationship. Alcohol, on the other hand, was not found to be associated with disease.

The association with cottonseed oil cannot be explained adequately before the oil used locally has been characterized chemically. However, cottonseed oil may contain gossypol, a phenolic toxicant, and malvalic acid, a promotor of the hepatocarcinogenic activity of aflatoxin B (National Research Council, 1973).

Both oral leukoplakia and seborrhoeic dermatis were diagnosed three times more frequently among male oesophagitis cases than in controls. Since 80% of those with oral leukoplakia were current smokers, compared with 57% of those without, cigarette smoking may be a common risk factor even if oral leukoplakia is not related to oesophagitis. Similarly, seborrhoeic dermatitis, which may be a sign of vitamin A deficiency, would be more common among cases of oesophagitis if oesophagitis were associated with lower vitamin A levels.

Urinary excretion of N-nitrosamino acids

The mean levels of N-nitroso-2-methylthiazolidine 4-carboxylic acid, N-nitrosothiazolidine 4-carboxylic acid, N-nitrososarcosine and N-nitrosoproline, corrected for creatinine (Ohshima et al., 1987), were not significantly different between persons with chronic oesophagitis and those with a normal oesophagus (Table 3). Only nitrate levels were significantly different between the two groups. No difference was observed between smokers and nonsmokers; however, urinary excretion levels of total N-nitrosamino acids were significantly higher among drinkers than nondrinkers.

Conclusions

The results of this study support the idea that poor cellular nutrition, thermal injury and chronic irritation by tobacco constituents and other chemicals may be responsible for inflammation and atrophy (Correa, 1982). Additional studies, such as measurement of vitamins and trace elements in plasma, studies of cell proliferation in oesophageal biopsy material, micronuclei tests with oesophageal cells and chemical analysis of cottonseed oil, may provide further clues to the significance of the present findings and suggestions for additional risk factors. In order to evaluate the relevance of oesophagitis as a precursor condition and to allow identification of the factors involved in the progression of oesophageal disease to carcinoma, a follow-up of the present study population is warranted.

Acknowledgement

Supported in part by the Stiftung Volkswagenwerk, Hanover, Germany.

Table 3. Levels of nitrate and N-nitrosamino acids[a] in overnight urine (mean ± SD in mg/mmol creatinine)

Group	No. of subjects	Nitrate	NMTCA	NTCA	NSAR	NPRO	Sum
Normal[b]	71	0.33±0.27	1.71±3.30	16.70±23.83	0.23±0.60	6.95±10.66	25.60±31.20
Oesophagitis[c]	12	0.56±0.45*	0.88±0.85	11.11±12.25	0.16±0.22	5.22±6.62	17.37±17.27
Nonsmoker	33	0.31±0.25	1.25±2.07	14.67±22.40	0.13±0.27	4.95±5.85	20.99±25.92
Current smoker	50	0.40±0.34	1.82±3.59	16.70±22.85	0.28±0.69	7.86±12.13	26.66±31.93
Nondrinker	22	0.39±0.32	0.52±0.84	10.09±13.38	0.18±0.40	4.13±4.93	14.92±16.50
Ever drinker	61	0.35±0.31	1.98±3.48*	17.99±24.81	0.23±0.62	7.63±11.37	27.83±32.5*

[a] NMTCA, N-nitroso-2-methylthiazolidine 4-carboxylic acid; NTCA, N-nitrosothiazolidine 4-carboxylic acid; NSAR, N-nitrososarcosine; NPRO, N-nitrosoproline
[b] Includes very mild oesophagitis
[c] Mild and moderate oesophagitis
* Wilcoxon rank sum test, $p < 0.05$

References

Cook-Mozaffari, P.J., Azordegan, F., Day, N.E., Ressicaud, A., Sabai, C. & Aramesh, B. (1979) Oesophageal cancer studies in the Caspian littoral of Iran: results of a case-control study. *Br. J. Cancer*, **39**, 293-309

Correa, P. (1982) Precursors of gastric and esophageal cancer. *Cancer*, **50**, 2554-2565

Crespi, M., Muñoz, N., Grassi, A., Aramesh, B. & Mojtabai, A. (1979) Oesophageal lesions in northern Iran; a premalignant condition? *Lancet*, **ii**, 217-221

Ghadirian, P. (1985) Familial history of esophageal cancer. *Cancer*, **56**, 2112-2116

Ghadirian, P. (1987) Thermal irritation and oesophageal cancer in northern Iran. *Cancer*, **60**, 1909-1914

Li, M.X. & Cheng, S.J. (1984) Etiology of carcinoma of the esophagus. In: Huang, G.J. & Wu, Y.K., eds, *Carcinoma of the Esophagus and Gastric Cardia*, Berlin, Springer Verlag, pp. 26-47

Muñoz, N., Victora, C.G., Crespi, M., Saul, C., Braga, N.M. & Correa, P. (1987) Hot mate drinking and precancerous lesions of the oesophagus: an endoscopic survey in southern Brazil. *Int. J. Cancer*, **39**, 708-709

National Research Council (1973) *Toxicants Occurring Naturally in Foods*, Washington DC, National Academy of Sciences

Ohshima, H., Calmels, S., Pignatelli, B., Vincent, P. & Bartsch, H. (1987) N-Nitrosamine formation in urinary-tract infections. In: Bartsch, H., O'Neill, I.K. & Schulte-Hermann, R., eds, *The Relevance of N-Nitroso Compounds to Human Cancer: Exposures and Mechanisms* (IARC Scientific Publications No. 84), Lyon, IARC, pp. 384-390

Tuyns, A.J., Riboli, E., Doornbos, G. & Péquignot, G. (1987) Diet and esophageal cancer in Calvados (France). *Nutr. Cancer*, **9**, 81-92

Yioris, N., Ivankovic, S. & Lehnert, T. (1984) Effect of thermal injury and oral administration of N-methyl-N-nitro-N-nitrosoguanidine on the development of esophageal tumors in Wistar rats. *Oncology*, **41**, 36-38

Yu, M.C., Garabrant, D.M., Peters, J.M. & Mack, T.M. (1988) Tobacco, alcohol, diet, occupation and carcinoma of the esophagus. *Cancer Res.*, **48**, 3843-3848

GLIOMAS AND MENINGIOMAS IN MEN IN LOS ANGELES COUNTY: INVESTIGATION OF EXPOSURES TO N-NITROSO COMPOUNDS

S. Preston-Martin & W. Mack

Department of Preventive Medicine, University of Southern California, School of Medicine, Los Angeles, CA, USA

We conducted a case-control study of primary tumours of the brain and cranial meninges in Los Angeles County to investigate the hypothesis that these tumours are related to occupational exposures. We also collected limited data on diet and personal habits that are likely to involve exposure to N-nitroso compounds (NOC), NOC precursors and modulators of NOC metabolism. Interviews were conducted with 272 men with a brain tumour diagnosed in 1980–84 and with 272 individually matched neighbourhood controls. The study was of sufficient size to allow for separate analyses of the 202 pairs of glioma patients and of the 70 pairs of meningioma. Six glioma cases and one control had worked in the rubber industry, in which excesses of brain tumour have been shown in previous studies and where there are high levels of volatile NOC at various work sites. Ten meningioma patients and five controls had used cooling, cutting or lubricating oils, and most had used these daily (eight cases; four controls). Cases and controls were not different, however, with respect to other occupations known to involve exposures to NOC. Cases and controls also did not differ in their consumption of alcoholic beverages or cigarettes or in their passive exposure to cigarette smoke. The most striking dietary finding was a significant protective effect among glioma pairs of use of vitamin supplements, which increased with increasing frequency of use (p for trend = 0.04; odds ratio for use at least twice a day = 0.4 (95% confidence interval = 0.24–0.77)). Among meningioma pairs, a protective effect was seen for consumption of citrus fruit (odds ratio for eating citrus five or more times/week = 0.4; p = 0.07). The risk for meningioma (but not for glioma) appeared to increase with increasing consumption of all types of cured meats, but this finding was not statistically significant. In summary, the brain tumour–NOC hypothesis gains limited support from this study. The many difficulties involved in investigating this hypothesis epidemiologically are discussed.

It is well established that brain tumours can be caused experimentally by N-nitroso compounds (NOC). This effect has been studied extensively in a variety of species using various NOC, and the N-nitrosoureas appear to be the most effective (Bogovski & Bogovski, 1981). The hypothesis that NOC cause brain tumours in humans has received only limited attention in epidemiological studies, however. The strongest support for this hypothesis was the result of a case–control study of young brain tumour patients (Preston-Martin *et al.*, 1982), in which increased risk was associated with maternal contact during pregnancy with NOC-containing substances, such as burning incense, sidestream cigarette smoke and face make-up. Increased risk was also associated with maternal use of diuretics and antihistamines and with the level of maternal consumption of cured meats. Diuretics and antihistamines contain nitrosatable amines and amides, and cured meats contain nitrites – chemicals which are precursors of NOC.

Studies in adults have provided limited support for the hypothesis and have been somewhat inconsistent. Level of consumption of cured meats was found to be related to the occurrence of intracranial meningiomas in women (Preston-Martin et al., 1980) but not in men (Preston-Martin et al., 1983) or with malignant brain tumours (mostly gliomas) in men and women (Burch et al., 1987). This most recent study did show a significant increase in risk associated with consumption of spring water, wine and pickled fish, and it showed a significant protective effect of regular consumption of fruit.

We report here on an additional case–control study of primary brain tumours in men in Los Angeles County. This study was large enough to conduct separate analyses for the two major histological types of brain tumours — gliomas, which are predominantly malignant, and meningiomas, which are usually benign.

Study subjects

The patients were black and white men with a glioma or meningioma first diagnosed during 1980–84. Any man who was resident of Los Angeles County and 25–69 years of age at the time his brain tumour was diagnosed was eligible for inclusion if he was alive and could be interviewed. The cases were identified through the Los Angeles County Cancer Surveillance Program (Hisserich et al., 1975). All diagnoses had been confirmed microscopically. We interviewed 277 (74%) of the 374 patients contacted about the study.

We sought a neighbourhood control for each of these 277 patients using a procedure that defines a sequence of houses on specified neighbourhood blocks. Our goal was to interview the first male resident in the sequence who corresponded to the patient with regard to race and age (birth year within five years of birth year of the patient). In all, we identified and interviewed 272 controls, and these controls and the corresponding 272 cases were included in the analysis. Further details on the selection of subjects and on other methods used have been published elsewhere (Preston-Martin et al., 1989).

The interview

A questionnaire sought information on various life experiences that had occurred two years or more prior to the year of diagnosis of the case. The first and longest section elicited a detailed job history, including information on specific tasks and materials used. A list of chemicals and other exposures (radiation, radioactive materials) was used in relation to all jobs. Limited information was collected on possible sources of NOC exposure, such as diet and exposure to tobacco smoke. In the diet section, we asked only about cured meats, citrus fruit and use of vitamin supplements; the questionnaire did not ask about other possible sources of NOC precursors such as drugs and drinking-water.

Statistical analysis

In the analysis of dichotomous questionnaire data, we used matched odds ratios (OR; the ratio of discordant pairs) to estimate relative risks and used the exact binominal test to calculate associated p values and confidence intervals (CIs). Conditional logistic regression models were used for the dose-response analysis of a single variable considered at more than two levels, for trend tests and for multivariate analyses (Breslow & Day, 1980). If, for any variable, the information was not known for either the patient or the control, we excluded the pair from the relevant analysis. All statistical significance levels (p values) cited are based on a likelihood ratio test and are one-sided because we had one-directional a-priori hypotheses for all factors related to the brain tumour–NOC hypothesis. However, the 95% CIs shown are two-sided by definition.

Results

The level of consumption of various groups of cured meats in cases and controls is shown in Table 1. Among glioma pairs, there was little difference between cases and controls. Compared to their controls, meningioma patients appeared to eat more ham and bacon and more of all types of cured meats combined, but these differences are not statistically significant.

A significant protective effect was seen among glioma pairs for use of vitamin supplements (Table 2). Protection increased with increasing frequency of use (p for trend = 0.004, and the OR for use at least twice a day was 0.4 (95% CI = 0.24-0.77). Vitamin use appeared to be unrelated to socioeconomic status when either level of education or occupational category was used to estimate that status. Among meningioma pairs, a protective effect was seen for consumption of citrus fruit but not for use of vitamin supplements.

More patients than controls answered 'yes' when asked if they had ever worked in rubber processing (discordant pairs: glioma, 6:1; meningioma, 3:2). In addition, when giving their job histories, more patients had volunteered previously (in the occupational history section of the questionnaire) that they had been exposed to rubber or rubber dust when asked about substances they came in contact with (discordant pairs: glioma, 7:2; meningioma, 4:0). Four subjects had worked in the manufacture of rubber tyres: two glioma patients had worked for 24 and nine years, and one meningioma patient and one meningioma control had each worked for two years.

Another finding relevant to occupational factors previously suggested to involve exposure to NOC was related to use of cooling, cutting or lubricating oils. Ten meningioma patients and five controls had used such oils, and most had used them daily (eight cases; four controls).

Patients and controls were not different in their consumption of alcoholic beverages or cigarettes or in their passive exposure to cigarette smoke. The findings related to head trauma and head X-rays are reported elsewhere (Preston-Martin et al., 1989); in summary, head trauma was clearly a risk factor for meningiomas, but not for gliomas, and frequent dental X-rays appeared to be a risk factor for both histological types.

Discussion

The brain tumour–NOC hypothesis receives only limited support from this study. We failed to find an association between consumption of cured meats and gliomas, which are the most common histological type of intracranial tumour caused experimentally and are the most common type of primary brain tumour in humans. We did not ask about consumption of pickled fish, which was related to risk for brain tumour in an earlier study (Burch et al., 1987). We failed to confirm associations found previously with wine consumption and with smoking of non-filter cigarettes (Burch et al., 1987). We did not ask, however, about the type of cigarette smoked.

Table 1. Comparison of glioma and meningioma cases and controls by frequency of consumption of various cured meats, Los Angeles County, 1980–84[a]

Consumption of cured meats	Glioma						Meningioma					
	No. of cases	No. of controls	RR	p	95% CI	p for trend	No. of cases	No. of controls	RR	p	95% CI	p for trend
Fried ham or bacon												
< once a month	25	24	1.0			0.50	6	13	1.0			0.19
Once a month to once a week	100	95	1.0	0.50	0.5–2.0		39	32	2.8	0.04	0.9–9.3	
> once a week	38	51	0.7	0.18	0.3–1.5		15	17	2.2	0.14	0.5–9.0	
Daily or almost daily	39	32	1.2	0.36	0.5–2.5		10	8	2.7	0.08	0.7–10.8	
Salami, pastrami or corned beef												
< once a month	65	75	1.0			0.17	24	22	1.0			0.36
Once a month to once a week	116	109	1.2	0.17	0.8–1.9		40	41	0.9	0.61	0.4–1.8	
> once a week	18	15	1.4	0.21	0.7–2.8		4	5	0.8	0.65	0.2–3.0	
Daily or almost daily	3	3	1.2	0.43	0.2–5.9		2	2	0.9	0.53	0.1–7.1	
Bologna or other lunch meats												
< once a month	47	46	1.8			0.13	19	17	1.0			
Once a month to once a week	99	83	1.2	0.29	0.7–1.9		31	32	0.9	0.63	0.4–2.0	
> once a week	31	46	0.6	0.07	0.3–1.2		13	14	0.8	0.64	0.3–2.3	
Daily or almost daily	24	26	0.8	0.30	0.4–1.7		7	7	0.9	0.56	0.2–4.0	
Hot dogs or Polish sausage												
< once a month	52	47	1.0			0.40	22	21	1.0			0.27
Once a month to once a week	134	141	0.9	0.28	0.6–1.4		41	39	1.0	0.50	0.5–2.1	
> once a week	14	12	1.1	0.45	0.4–2.6		6	8	0.7	0.70	0.2–2.5	
Daily or almost daily	2	2	1.0	0.48	0.1–7.0		1	2	0.5	0.71	0.0–5.8	
Any of the above												
< once a month	5	3	1.0			0.26	2	3	1.0			0.26
Once a month to once a week	85	75	1.2	0.42	0.3–4.7		28	31	1.3	0.38	0.2–8.3	
> once a week	42	51	0.8	0.40	0.2–3.4		19	17	1.7	0.30	0.3–11.1	
Daily or almost daily	69	70	1.0	0.50	0.2–4.2		21	19	1.6	0.31	0.3–10.2	

[a] RR, relative risk; CI, confidence interval; p values are one-sided

Table 2. Comparison of glioma and meningioma cases and controls by frequency of consumption of vitamin supplements and citrus fruit, Los Angeles County, 1980-84[a]

Consumption of vitamins or citrus fruit	Glioma						Meningioma					
	No. of cases	No. of controls	RR	p	95% CI	p for trend	No. of cases	No. of controls	RR	p	95% CI	p for trend
Vitamin supplements												
Never	119	101	1.0			<0.01	30	35	1.0			0.33
< 5 times/week	20	18	1.0	0.46	0.5–1.9		8	3	2.9	0.93	0.7–11.2	
> 5 times/week	63	83	0.6	0.02	0.4–1.0		32	32	1.2	0.66	0.6–2.3	
Citrus fruit												
Never	19	15	1.0			0.29	8	3	1.0			0.12
< 5 times/week	97	98	0.8	0.27	0.4–1.6		29	30	0.4	0.09	0.1–1.5	
> 5 times/week	86	89	0.8	0.25	0.4–1.6		33	37	0.4	0.07	0.1–1.4	
Vitamins and/or citrus fruit												
Never	13	11	1.0			0.05	5	2	1.0			0.28
< 5 times/week	73	57	1.1	0.45	0.5–2.5		15	17	0.4	0.13	1.1–2.1	
> 5 times/week	116	134	0.7	0.20	0.3–1.6		50	51	0.4	0.14	0.1–2.1	

[a] RR, relative risk; CI, confidence interval; p values are one-sided

We found only a weak suggestion that occupational exposures to NOC are related to risk for brain tumour. Six glioma patients (*versus* one control) had worked in the manufacture of rubber, and two had worked in tyre plants for several years. Tyre plants have been found to contain high levels of various NOC, including N-nitrosodimethylamine, N-nitrosodiethylamine, N-nitrosomorpholine and N-nitrosodiphenylamine, in the air, floor scrapings, steam condensate and compounding chemicals (Rounbehler & Fajen, 1983). Hot processes such as tyre curing emitted the highest levels of volatile NOC, in particular N-nitrosomorpholine. More meningioma patients than controls had used metal working fluids, which have been shown to contain N-nitrosodiethanolamine (Rounbehler & Fajen, 1983).

Our most striking finding was the protective effect seen among glioma pairs for use of vitamin supplements. We see this as consistent with the protective effect of fruit consumption reported in a recent study of brain cancer (Burch *et al.*, 1987). Vitamins C and E have both been shown to be effective inhibitors of nitrosation and, when present in the stomach at the same time as precursors of NOC, they block the endogenous formation of these compounds (Tannenbaum & Mergens, 1980; Mirvish, 1981). Among our glioma pairs, the strongest protection was related to use of vitamin C. The vitamin supplements used most commonly by the men in our study were ascorbic acid (i.e., vitamin C) and multivitamin tablets (which contain both C and E). We did not ask questions about the timing of ingestion of various foods or of vitamin pills; however, the strongest effect was seen among those who took vitamins two or more times a day, and it is likely that these men took vitamins around meal times. Vitamins would be most effective in blocking the formation of NOC from precursors contained in foods and beverages when taken with meals. The protective effect of citrus fruit intake among the meningioma pairs in our study and of fruit intake in an earlier study of brain tumours (Burch *et al.*, 1987) can also be interpreted as consistent with the brain tumour–NOC hypothesis. Citrus fruit contains high levels of vitamin C, and most other fruits also contain vitamin C and/or E. Although we interpret the protective effect of vitamin (or fruit) intake as relevant to the brain tumour–NOC hypothesis, it is entirely possible that some other mechanisms may explain this protective effect.

Our study was, in fact, quite limited in its ability to test the hypothesis. Given the extent of current knowledge about exposures to NOC in the human environment, good epidemiological studies to test this hypothesis would be difficult do do. Nonetheless, some limitations of the present study could be overcome, as discussed below.

Because our study focused on occupational exposure, we did not attempt also to make a complete dietary survey but asked instead about only a few food items. A questionnaire designed to collect information on total intake of food, beverages and drugs would allow calculation of total nitrate, nitrite, preformed NOC, vitamin C, vitamin E, polyphenols, etc. In order to estimate levels of endogenously formed NOC, questions would also have to be asked about timing of intake of various items. Such questions could focus on likely combinations (e.g., 'Do you take your vitamin pills around meal times?'; 'When you eat fish, how often do you eat it with lemon?').

Our study is also limited because there are many other sources of NOC, NOC precursors and modulators of the nitrosation reaction (including catalysts as well as inhibitors) which we did not ask about. We did not ask about drugs, for example, which, in the USA, can be an important source of NOC precursors; however, for many drugs and for many foods it has not yet been fully determined how all the constituents relate to the NOC hypothesis. For example, foods that are known to contain NOC precursors may later be shown to contain nitrosation inhibitors. For drugs, it must be determined both how readily the NOC precursor, such as secondary amine, is nitrosated in the human stomach and how stable the resulting NOC is. Many other consumer products, such as personal hygiene products and rubber products, may also be a source of exposure to NOC; only some of these have been tested. NOC have also been found in various industrial environments, but most work sites have not been tested for contamination with NOC.

In summary, we are far from being able to quantify an individuals's total exposure to NOC in epidemiological interview studies. If appropriate biochemical markers of the relevant exposure to NOC were available, epidemiological studies of the brain tumour–NOC hypothesis would be strengthened considerably.

References

Bogovski, P. & Bogovski, S. (1981) Animal species in which *N*-nitroso compounds induce cancer. *Int. J. Cancer*, **27**, 471-474

Breslow, N.E. & Day, N.E. (1980) *Statistical Methods in Cancer Research*, Vol. 1, *The Analysis of Case-control Studies* (IARC Scientific Publications No. 32), Lyon, IARC, p. 211

Burch, J.D., Craib, K.J.P., Choi, B.C.K., Miller, A.B., Risch, H.A. & Howe, G.R. (1987) An exploratory case-control study of brain tumours in adults. *J. Natl Cancer Inst.*, **78**, 601-609

Hisserich, J.C., Preston-Martin, S. & Henderson, B.E. (1975) An areawide reporting network. *Public Health Rep.*, **90**, 15-17

Mirvish, S.S. (1981) Inhibition of the formation of carcinogenic *N*-nitroso compounds by ascorbic acid and other compounds. In: Burchenal, J.H. & Oettgen, H.F., eds, *Cancer: Achievements, Challenges and Prospects for the 1980's*, New York, Grune & Stratton

Preston-Martin, S., Paganini-Hill, A., Henderson, B.E., Pike, M.C. & Wood, C. (1980) Case-control study of intracranial meningiomas in women in Los Angeles county. *J. Natl Cancer Inst.*, **75**, 67-73

Preston-Martin, S., Yu, M.C., Benton, B. & Henderson, B.E. (1982) *N*-Nitroso compounds and childhood brain tumors: a case-control study. *Cancer Res.*, **42**, 5240-5245

Preston-Martin, S., Yu, M.C., Henderson, B.E. & Roberts, C. (1983) Risk factors for meningiomas in men in Los Angeles County. *J. Natl Cancer Inst.*, **70**, 863-866

Preston-Martin, S., Mack, W. & Henderson, B.E. (1989) Risk factors for gliomas and meningiomas in men in Los Angeles County. *Cancer Res.*, **49**, 6137-6143

Rounbehler, D.P. & Fajen, J.M. (1983) N-*Nitroso Compounds in the Factory Environment* [Contract No. 210-77-0100], Cincinnati, OH, National Institute for Occupational Safety and Health

Tannenbaum, S.R. & Mergens, W. (1980) Reaction of nitrite with vitamins C and E. *Ann. N.Y. Acad. Sci.*, **355**, 267-277

EPSTEIN-BARR VIRUS ACTIVATORS, MUTAGENS AND VOLATILE NITROSAMINES IN PRESERVED FOOD SAMPLES FROM HIGH-RISK AREAS FOR NASOPHARYNGEAL CARCINOMA

G. Bouvier[1,2], S. Poirier[1,2], Y.M. Shao[1,2,4] C. Malaveille[1],
H. Ohshima[1], A. Polack[3], G.W. Bornkamm[3],
Y. Zeng[4], G. de-Thé[2] & H. Bartsch[1]

[1]*International Agency for Research on Cancer;*
[2]*Laboratoire d'Epidémiologie et d'Immunovirologie des Tumeurs,
Faculté de Médecine A. Carrel, Lyon; and*
[3]*Institut für Sozial Klinische Molekular Biologie und Tumor Genetik,
Hämatologikum der Gesellschaftsforschung, Munich,
Germany*

Representative samples of preserved foods collected from high-risk areas for nasopharyngeal carcinoma were assayed for capacity to induce Epstein-Barr virus (EBV) by measuring induction of the DR promoter gene or of genes for early antigens in Raji cells. The two assays gave concordant results, but the DR induction assay detected EBV-inducing substances at lower concentrations and more reproducibly. Three of 17 preserved food items were active in both assays; they were also weakly mutagenic in the SOS chromotest and contained low or moderate levels of volatile N-nitrosamines. After in-vitro nitrosation, the levels of mutagens and nitrosamines increased, whereas EBV-inducing activity was unchanged or decreased. Thus, EBV inducers appear to be a different class of substances from mutagens and volatile N-nitrosamines but could act with them in the etiology of nasopharyngeal carcinoma.

Nasopharyngeal carcinoma (NPC) is closely associated with the ubiquitous Epstein-Barr virus (EBV) (de-Thé, 1982). The particular geographical distribution of NPC, which is common among populations of southern China, North Africa and the Arctic, has been shown to be related to a combination of genetic predisposition (HLA haplotypes), environmental factors, especially food habits (Yu et al., 1988), and the Epstein-Barr virus. We showed previously that traditional preserved food items from high-risk areas for NPC (southern China, Tunisia and Greenland) contain volatile N-nitrosamines (Poirier et al., 1987) and that some of these food samples exhibit EBV-inducing activity *in vitro* (Shao et al., 1988).

[4]Present address: Institute of Virology, Chinese Academy of Preventive Medicine, Beijing, China

We now report the results of a study in which some food items were screened for EBV-inducing activity using an assay to measure induction of the EBV early promoter DR (DR induction assay), in parallel with an earlier assay in which the induction of EBV early antigens (EA) is measured (EA induction assay). We also screened food extracts for the presence of mutagens and volatile N-nitrosamines before and after in-vitro nitrosation.

Methods

Aqueous food extracts were screened for EBV-inducing activity in Raji cells carrying an autoreplicative plasmid in which the EBV-DR promoter controls the gene for bacterial chloramphenicol acetyltransferase (Gorman *et al.*, 1982). DR induction in Raji cells was then compared with EA induction assayed in Raji cells by an immunological technique (Ito *et al.*, 1981; Shao *et al.*, 1988).

The mutagenic activity of *n*-hexane and ethyl acetate extracts of each of 15 food items (see Table 2) collected from high-risk areas for NPC was measured in *Salmonella typhimurium* strains TA98 and TA100 in the absence and in the presence of a rat liver metabolic activation system (data not shown). Concentrations tested ranged from 75 to 300 mg wet weight equivalent of food per ml of assay; higher concentrations could not be tested due to the limit of solubility in dimethyl sulfoxide. Aqueous food extracts were not tested in *S. typhimurium* strains, because the presence of histidine can lead to artefacts, and the SOS chromotest was used.

Three volatile N-nitrosamines were measured: N-nitrosodimethylamine (NDMA), N-nitrosopyrrolidine (NPYP) and N-nitrosopiperidine (NPIP), as described previously (Poirier *et al.*, 1987, 1989). In-vitro nitrosation is described in the legend to Table 3.

EBV-inducing activity

Table 1 presents the results obtained in these two assays with aqueous extracts of foods previously found to contain EBV inducers (Shao *et al.*, 1988); 12-*O*-tetradecanoylphorbol 13-acetate (TPA), a tumour promoter, was used as the positive control. The DR induction assay detected EBV-inducing substances at lower concentrations and more reproducibly; furthermore, the background levels were lower and showed less fluctuation than those observed with the immunoenzymatic test. Food extracts that were active in the DR induction assay were also active in the EA induction assay, except for the second sample of squid, which was weakly active in the EA induction assay but not active in the DR induction assay. Only *harissa* and Japanese mackerel, of the food items listed in Table 2, showed EBV-inducing activity.

Mutagens in food extracts before and after nitrosation

Only the ethyl acetate extract of berries preserved in seal oil from Greenland (at 300 mg wet weight equivalent of food per ml assay medium and in the presence of metabolic activation) doubled the number of revertants above the spontaneous mutation level in TA98 strain. The induction factor (an index of the induction of SOS DNA repair functions in *Escherichia coli*) was greater than 1.5 in 13 of 15 food samples; the highest values were observed in salted anchovies and in sap from the mastic tree in samples from Tunisia. Although most of the tested samples were cytotoxic, no association was observed between the induction factor of the samples and their toxicity. The addition of a rat liver metabolic activation system (up to 9.5%) decreased the induction factor (data not shown). Thus, the results revealed the presence of weakly active, directly acting genotoxic substances in many of the food items tested, mostly in organic extracts (Table 2).

Table 1. Comparison of DR induction assay and early antigen (EA) induction assay in Raji cells for the capacity of preserved food from high-risk areas for nasopharyngeal carcinoma to induce Epstein-Barr virus (EBV)

Food sample	DR induction assay (%)[a]	EA induction assay (%)[b]
Japanese mackerel	14.6	17.8
Squid	15.2	13.6
Squid	6.4	16.6
Harissa	39.9	20.7
TPA	98	43
Control	7.4 (6.7–8.1)	4.0 (1.3-9)

[a] Raji cells carrying chloramphenicol acetyltransferase (CAT) gene under the control of EBV-DR promoter and extracts were incubated with ^{14}C-chloramphenicol and acetyl-coenzyme A as described by Gorman *et al.* (1982). After separation of the acetylated and unacetylated forms of chloramphenicol by thin-layer chromatography, chloramphenicol was detected by autoradiography and quantified by counting radioactivity. CAT activity was expressed as the percentage of radioactivity of total acetylated chloramphenicol over total radioactivity used in the assay.

[b] Induction of EBV EA was measured by an enzyme-linked immunoassay after incubation with the test sample for two days (Shao *et al.*, 1988). The immunological assay was performed with serum from a patient with nasopharyngeal carcinoma (IgG/EA titre, 1:20), stained with peroxidase linked to protein A (1:100), followed by a diaminobenzidine reaction with hydrogen peroxide. The results are expressed as percentage of positive cells calculated from at least 500 cells counted under the microscope.

In order to determine the presence of nitrosamine precursors in the preserved food extracts, the genotoxicity of the aqueous extracts was measured after acid-catalysed nitrosation *in vitro* (Table 3). Nine of the 15 food samples were found to contain precursors that upon nitrosation yielded directly acting genotoxic substances. Nitrosation led to formation of genotoxic compounds not only in fish and meat – known to contain high concentrations of secondary and tertiary amines – but also in vegetables fermented in brine and in a spice mixture from Tunisia. It is therefore conceivable that these DNA damaging compounds are formed intragastrically from dietary precursors and, because of their relative stability, also reach distant targets such as the nasopharynx *via* the blood stream.

Volatile nitrosamines

Table 3 lists the levels of three volatile nitrosamines detected before and after in-vitro nitrosation of aqueous extracts of food samples. Before nitrite treatment, volatile nitrosamines were detected in 12 of 15 samples; no detectable level was found in soft salted dried fish or fermented soya bean paste from China or dried Atlantic cod from Greenland. The highest level was observed in hard salted dried fish from China; concentrations over 100 µg/kg were seen in hard salted dried fish and fermented shrimp/fish past from China, *harissa* and salted anchovies from Tunisia and dried seal meat from Greenland.

Table 2. Genotoxicity of aqueous, hexane and ethyl acetate extracts of food as determined in the SOS chromotest

Origin and food sample	Aqueous extract	Hexane extract	Ethyl acetate extract	Overall evaluation of genotoxicity
China				
Hard salted dried fish	−	+	−	+
Soft salted dried fish	−	+	+	+
Fermented shrimp/fish paste	−	−	+	+
Fermented soya bean paste	−	−	+	+
Cabbage fermented in brine	−	−	−	−
Tunisia				
Harissa spice mixture	−	−	+	+
Dried mutton preserved in oil	−	+	−	+
Stewing base	−	+	+	+
Turnip fermented in brine	−	−	−	−
Salted anchovies	−	+	+	+
Sap from mastic tree	−	+	+	+
Greenland				
Dried Atlantic cod	−	+	−	+
Dried Polar cod	+	−	−	+
Dried capelin	+	+	+	+
Dried fiord seal meat	+	−	+	+

Concentrations tested ranged rom 1.8 to 54 (mg wet weight equivalent of food per ml assay medium) for the aqueous phase and from 6.7 to 202 for the hexane phase and the ethyl acetate phase. Four different concentrations were assayed for each solvent. −, all three extracts were negative; +, at least one was positive. The SOS chromotest was performed as described (Quillardet & Hofnung, 1985; Poirier *et al.*, 1989). Results were judged positive when the induction factor was over 1.5.

After nitrosation, volatile nitrosamines were detected in all samples; the increase above the preformed level observed in 11 out of 15 samples ranged from two fold (hard salted dried fish from China) to more than 100 fold (soft salted dried fish and fermented soya bean paste from China, turnips fermented in brine from Tunisia and dried Atlantic cod from Greenland). These results demonstrate that high concentrations of nitrosatable precursors for volatile nitrosamines are present in preserved foods, especially in soft salted dried fish and cabbage fermented in brine from China and dried Polar cod and dried capelin from Greenland. Carcinogenic volatile nitrosamines and other genotoxic substances could be formed endogenously in the human stomach after ingestion of such preserved foods.

A comparison of EBV-inducing activity, total volatile nitrosamine levels and genotoxicity in the aqueous food extracts before and after in-vitro nitrosation revealed that this activity was not correlated with the presence of volatile nitrosamines or genotoxic substances; however, the formation of volatile nitrosamines paralleled the increase in genotoxic activity after nitrosation of food samples. Thus, EBV-inducing substances seem to be a different class of substances from genotoxins and nitrosamines.

Table 3. Genotoxicity (in the SOS chromotest), volatile nitrosamine (VNA) levels and capacity to induce Epstein-Barr virus (EBV) of food samples before and after nitrosation[a]

Origin and food sample	Genotoxicity in aqueous phase		Sum of VNA[b] (µg/kg)		EBV inducing activity	
	Before	After	Before	After	Before	After
China						
Hard salted dried fish	-	↑	523	↑	-	(o)
Soft salted dried fish	-	↑	0	↑↑↑	+	↓
Fermented shrimp/fish paste	-	(o)	105	↓	-	(o)
Fermented soya bean paste	-	(o)	0	↑↑↑	-	(o)
Cabbage fermented in brine	-	↑	85	↑	-	(o)
Tunisia						
Harissa spice mixture	-	↑	196	↑↑↑	+	↓
Dried mutton preserved in oil	-	↑	13	↑↑↑	-	(o)
Stewing base	-	↑	5	↑	-	(o)
Turnips fermented in brine	-	↑	62	↑↑↑↑	-	(o)
Salted anchovies	-	(o)	300	↓↓	-	(o)
Sap from mastic tree	-	(o)	12	(o)	-	(o)
Greenland						
Dried Atlantic cod	-	(o)	0	↑↑↑↑	-	(o)
Dried Polar cod	+	(o)	17	↑↑↑	-	(o)
Dried capelin	+	↓	24	↑↑↑	-	(o)
Dried fiord seal meat	+	(o)	190	↓	-	(o)

[a] For nitrosation, 3-ml aliquot of food extract (equivalent to 5 g food sample) was adjusted to pH 1.5 with HCl and centrifuged at 2000 g for 20 min at 4°C. The volume of the supernatant was made up to 4 ml with HCl/KCl buffer 0.2M pH 1.5. NaNO$_2$ solution was added to 3 ml of this solution to a final concentration of 26 mmol/l or 6.5 mmol/l, respectively. After 20 min incubation in a shaking water-bath at 37°C in the dark, ammonium sulfamate was added to a final concentration of 94 mmol/l or 23.5 mmol/l. Then, pH was adjusted to 7. Control assays without NaNO$_2$ were carried out. Aliquots at time 0 and 20 min were tested for genotoxicity. VNA were analysed as described previously (Poirier et al., 1987). EBV early antigen induction was assayed as described previously (Shao et al., 1988). −, negative and +, positive response in assay before nitrosation of food samples (see legend to Table 2) for genotoxicity in the SOS chromotest or EA induction assay; o, unchanged; ↑, increase; ↓, decrease; ↑, <2 fold; ↑↑, <10 fold; ↑↑↑, <30 fold; ↑↑↑↑, >100 fold
[b] N-Nitrosodimethylamine, N-nitrosopyrrolidine and N-nitrosopiperidine

In conclusion, we have identified carcinogenic nitrosamines, mutagens and EBV-inducing substances that could be etiological factors in the pathogenesis of NPC. Some volatile nitrosamines are known to induce nasal cavity tumours in experimental animals and could be formed endogenously in man after ingestion of certain food items. Work is in progress to isolate and characterize EBV-inducing substances.

References

Gorman, G.M., Moffat, L.F. & Howard, B.M. (1982) Recombinant genomes which express chloramphenicol acetyltransferase in mammalian cells. *Mol. Cell. Biol.*, 2, 1044-1051

Ito, Y., Yanase, S., Harayama, M., Khashima, M. & Imanaka, H. (1981) A short-term *in vitro* assay for promoter substances using human lymphoblastoid cells latently infected with Epstein-Barr virus. *Cancer Lett.*, 19, 113-117

Poirier, S., Ohshima, H., de-Thé, G., Hubert, A., Bourgade, M.C. & Bartsch, H. (1987) Volatile nitrosamine levels in common foods from Tunisia, South China and Greenland, high risk areas for nasopharyngeal carcinoma (NPC). *Int. J. Cancer*, **39**, 293-296

Poirier, S., Bouvier, G., Malaveille, C., Ohshima, H., Shao, Y.M., Hubert, A., Zeng, Y., de-Thé, G. & Bartsch, H. (1990) Volatile nitrosamine levels and genotoxicity of food samples from high-risk areas for nasopharyngeal carcinoma before and after nitrosation. *Int. J. Cancer*, **44**, 1088–1094

Quillardet, P. & Hofnung, M. (1985) The SOS chromotest, a colorimetric bacterial assay for genotoxins: procedures. *Mutat. Res.*, **147**, 65-78

Shao, Y.M., Poirier, S., Ohshima, H., Malaveille, C., Zeng, Y., de-Thé, G. & Bartsch, H. (1988) Epstein-Barr virus activation in Raji cells by extracts of preserved foods from high risk areas for nasopharyngeal carcinoma. *Carcinogenesis*, **9**, 1455-1457

de-Thé, G. (1982) Epidemiology of the Epstein-Barr virus and associated diseases. In: Roizman, B., ed., *The Herpesviruses*, Vol. 1A, New York, Plenum Press, pp. 25-103

Yu, M.C., Mo, C.S., Chong, W.-X., Yeh, F.S. & Henderson, B.E. (1988) Preserved foods and nasopharyngeal carcinoma: a case-control study in Guangxi, China. *Cancer Res.*, **48**, 1945-1959

DIETARY SOURCES OF N-NITROSAMINES IN A HIGH-RISK AREA FOR OESOPHAGEAL CANCER — KASHMIR, INDIA

M.A. Siddiqi[1,3], A.R. Tricker[2], R. Kumar[1], Z. Fazili[1] & R. Preussmann[2]

[1]*Department of Biochemistry, University of Kashmir, Srinagar, Jammu and Kashmir, India; and* [2]*Institute of Toxicology and Chemotherapy, German Cancer Research Centre, Heidelberg, Germany*

Samples of foods consumed frequently in Kashmir, a high-risk area for oesophageal cancer, were analysed for the presence of volatile N-nitrosamines. Relatively high levels of N-nitrosodimethylamine (NDMA), N-nitrosopiperidine (NPIP) and N-nitrosopyrrolidine (NPYR) were detected in smoked fish (20 μg/kg NDMA), sun-dried spinach (5.8 μg/kg NDMA; 23.8 μg/kg NPIP), sun-dried pumpkin (24.6 μg/kg NPYR) and dried, mixed vegetables (10 μg/kg NDMA). The possible role of N-nitrosamines in the etiology of oesophageal cancer in Kashmir is discussed.

A high occurrence of oesophageal cancer has been reported from the Kashmir region in India (Maqbool & Ahad, 1976). Owing mainly to improved diagnostic facilities in recent years, it has been estimated that 20–30% of all cancer cases in this region are of the oesophagus (Table 1). A high prevalence of chronic oesophagitis in the general population (Goswami *et al.*, 1987) indicates a considerable predisposition to oesophageal cancer. While some of the well-known risk factors for this disease, such as alcohol consumption (Tuyns *et al.*, 1979), use of tobacco (Martinez, 1969) and betel-nut, are absent in the region, distinct, stable dietary habits are notable (Siddiqi & Preussmann, 1989). Use of sun-dried and pickled vegetables, high consumption of red chillies and spices in food and of hot-salted tea are some of the local features that are suspected to have a strong bearing on the occurrence of oesophageal cancer.

Since exposure to N-nitroso compounds is suspected to be involved in the etiology of human oesophageal cancer, we have investigated the presence of preformed volatile N-nitrosamines in raw foods commonly consumed in Kashmir.

Food samples and analysis of volatile N-nitrosamines

The most common types of food (dried, pickled and smoked) unique to the area were collected from families or from local markets in Srinagar, Kashmir. Samples were analysed by mineral-oil vacuum distillation and gas chromatography-thermal energy analysis, as reported previously (Siddiqi *et al.*, 1988).

---
[3]To whom correspondence should be addressed

Table 1. Oesophageal cancer incidence in India

Area	Period	All sites (no. of cases)	Oesophagus	
			No. of cases	Frequency (%)
Ahmedabad[a]	1982	7740	582	7.5
Bangalore[a]	1982	2137	131	6.1
Chandigarh[a]	1982	2455	139	5.7
Dibrugarh[a]	1982	1226	188	15.3
Madras[a]	1982	2297	115	5.0
Trivandrum[a]	1982	3493	134	3.8
Bombay[b]	1978–82	18 690	1660	8.9
Kashmir[c]	1962–71	1413	304	21.5
Kashmir[d]	1984–85	2808	826	29.4

[a] From Parkin (1986)
[b] From Waterhouse et al. (1982)
[c] From Mattoo & Kaul (1974)
[d] From B. Sanyal (personal communication)

Our preliminary data (Table 2) suggest relatively high exposure to preformed *N*-nitrosamines from consumption of common foods in Kashmir: several volatile *N*-nitrosamines were detected in foods that constitute a major part of the Kashmiri diet. It appears that consumption of preserved vegetables and fish may provide considerable exposure to volatile *N*-nitrosamines. The levels of *N*-nitrosamines present in Kashmiri foods are relatively higher than those reported for most western foods (reviewed by Tricker & Preussmann, 1988).

Table 2. Occurrence of volatile *N*-nitrosamines[a] in Kashmiri foods

Food item	No. of samples analysed	NDMA (µg/kg; range)	NPIP (µg/kg; range)	NPYR (µg/kg; range)
Sun-dried carp	8	1.3 (0.7–1.9)	ND	ND
Dried *Schiziothorax* (fish)	8	1.4 (0.7–1.8)	ND	ND
Smoked *Schiziothorax*	8	6.7 (2.0–20)	ND	ND
Sun-dried spinach	3	3.0 (0.8–5.8)	13.6 (4.8–23.8)	ND
Sun-dried pumpkin	3	1.0 (ND–2.4)	ND	9.8 (ND–24.6)
Sun-dried turnip	3	1.6 (ND–4.0)	ND	2.9 (ND–6.9)
Dried mixed vegetables	8	5.1 (1.9–10.0)	ND	ND
Mixed pickled vegetables	8	1.3 (ND–6.1)	ND	ND
'Wur' (mixed spice cake)	2	2.8 (1.7, 3.9)	ND	ND
'Mawal' (food colourant)	1	2.6	ND	ND
Saffron	1	ND	ND	ND
Lotus stem	1	ND	ND	1.2
Red chillies	1	ND	ND	0.8

[a] NDMA, *N*-nitrosodimethylamine; NPIP, *N*-nitrosopiperidine; NPYR, *N*-nitrosopyrrolidine; ND, not detected

Possible role of volatile N-nitrosamines and other risk factors in the etiology of oesophageal cancer in Kashmir

Dietary habits, such as the consumption of hot, salted tea (which is specific to Kashmir), may result in physical irritation of the oesophageal epithelium, rendering it more susceptible to chemical initiation by N-nitroso compounds and other classes of carcinogens in foods. Consumption of copious amounts of hot tea is a local feature common to several high-risk areas for oesophageal cancer (Ghadirian, 1987). Other commonly encountered dietary risk factors for oesophageal cancer appear to be absent in Kashmir: low fruit intake and lack of animal proteins are not prevalent in this population, where fruit, fish and lamb are widely eaten. Kashmir is one of the greatest fruit producing areas in India. The major N-nitrosamine-related risk factor evident in Kashmir is therefore the dietary dependence on pickled and sun-dried vegetables and sun-dried and smoked fish. This practice is due to the climatic conditions in Kashmir, where a long, cold winter and short, hot summer result in uncertain annual harvests of vegetables, and to its distance from the sea. Foods may be stored for years prior to consumption, which may account for the relatively high levels of volatile N-nitrosamines found.

The only other potential risk factor may be the use of copper utensils for cooking. High blood levels of copper and ceruloplasmin have been reported in Kashmir (Narang et al., 1987). Although copper has not been implicated in oesophageal cancer, preferential binding of copper to metallothioneins may result in zinc deficiency, a known risk factor for N-nitrosamine-induced oesophageal cancer.

On the basis of our preliminary results, it is not possible to establish whether or not volatile N-nitrosamines are involved in the etiology of oesophageal cancer in Kashmir; however, relatively high exposure to these compounds occurs. We are currently involved in establishing a field study in this area in which a more extensive investigation of exposure to N-nitroso compounds and other risk factors will be made.

References

Ghadirian, P. (1987) Thermal irritation and oesophageal cancer in Iran. *Cancer*, **60**, 1909-1914

Goswami, K.C., Khuroo, M.S., Zargar, S.A. & Pathania, A.G.S. (1987) Chronic oesophagitis in a population (Kashmir) with high prevalence of oesophageal carcinoma. *Indian J. Cancer*, **24**, 232-241

Maqbool, M. & Ahad, A. (1976) Carcinoma of esophagus in Kashmir. *Indian J. Otolaryngol.*, **28**, 118-122

Martinez, I. (1969) Factors associated with cancer of esophagus, mouth and pharynx in Puerto Rico. *J. Natl Cancer Inst.*, **42**, 1069-1094

Mattoo, A.R. & Kaul, H.K. (1974) Incidence of malignant neoplasm in Kashmir. *J. Indian Med. Assoc.*, **62**, 309-311

Narang, A.P.S., Verma, A., Sanyal, B., Qadri, A. & Mattoo, R.L. (1987) Evaluation of serum copper and ceruloplasmin in prognosis of gastrointestinal tract cancers. *Trace Elem. Med.*, **4**, 25-27

Parkin, D.M., ed. (1986) *Cancer Occurrence in Developing Countries* (IARC Scientific Publications No. 78), Lyon, IARC, pp. 203-222

Siddiqi, M. & Preussmann, R. (1989) Esophageal cancer in Kashmir — an assessment. *J. Cancer Res. Clin. Oncol.*, **115**, 111-117

Siddiqi, M., Tricker, A.R. & Preussmann, R. (1988) The occurrence of preformed N-nitroso compounds in food samples from a high risk area of esophageal cancer in Kashmir, India. *Cancer Lett.*, **39**, 37-43

Tricker, A.R. & Preussmann, R. (1988) N-Nitroso compounds and their precursors in the human environment. In: Hill, M.J., ed., *Nitrosamines — Toxicology and Microbiology*, Chichester, Ellis Harwood Ltd, pp. 88-116

Tuyns, A.J., Péquignot, G. & Abbatucci, J.C. (1979) Oesophageal cancer and alcohol consumption: importance of types of beverage. *Int. J. Cancer*, **23**, 443-447

Waterhouse, J., Muir, C., Shanmugaratnam, K. & Powell, J., eds (1982) *Cancer Incidence in Five Continents, Vol. IV* (IARC Scientific Publications No. 42), Lyon, IARC, pp. 390-397

Relevance to Human Cancer of N-Nitroso Compounds,
Tobacco Smoke and Mycotoxins.
Ed. I.K. O'Neill, J. Chen and H. Bartsch
Lyon, International Agency for Research on Cancer
© IARC, 1991

VOLATILE N-NITROSAMINES IN FISH MEAL, WITH SPECIAL REFERENCE TO THE MECHANISM OF FORMATION OF N-NITROSOTHIAZOLIDINE

H. Tozawa[1] & T. Kawabata[2]

[1]*Technical Society of Fish Jelly Product, Tokyo; and* [2]*Department of Biomedical Research on Foods, National Institute of Health, Tokyo, Japan*

Volatile N-nitrosamines in 32 commercial Japanese fish meal samples were analysed, and three compounds, N-nitrosodimethylamine, N-nitrosopyrrolidine and N-nitrosothiazolidine (NT), were detected. We also examined the mechanism of formation of NT during fish meal production. Both cysteamine and thiazolidine were found to be precursors of NT. The cysteamine content decreased during boiling and drying of the fish; but that of thiazolidine increased during boiling. N-Nitrosothiazolidine 4-carboxylic acid (NTCA) was formed in sardine meal after treatment with nitrogen dioxide gas, and the rate of formation of NT from NTCA added to sardine meal was as high as 10% when the meal was heated at 160°C for 60 min. We propose two pathways for NT formation during fish meal manufacture: (1) cysteamine → thiazolidine → NT; and (2) cysteine → (thiazolidine 4-carboxylic acid) → NTCA → NT.

Levels of volatile N-nitrosamines (VNA) in commercial fish meals have been reported to be much higher than those in human foods (Sen *et al.*, 1972; Hurst, 1976; Juszkiewicz & Kowalski, 1978; Helgason *et al.*, 1984). In addition, N-nitrosodimethylamine (NDMA) in fish meal that was fed to hens was found to migrate into eggs (Juszkiewicz & Kowalski, 1978). More than one million tonnes of fish meal are consumed annually by poultry, hogs and fish in Japan, and much attention has been paid to the presence of VNA in commercial fish meals. We have made a series of studies on the occurrence and formation of VNA in commercial fish meals (Tozawa & Kawabata, 1986, 1987a,b,c). We report here on the mechanism of formation of N-nitrosothiazolidine (NT) during fish meal preparation.

Occurrence of VNA in commercial fish meals

A survey was conducted on the occurrence of VNA and nitrite in 32 commercial Japanese fish meal samples. VNA were analysed by the modified method of Kawabata *et al.* (1982). NDMA, N-nitrosopyrrolidine (NPYR) and NT were detected in meal samples dried by hot air from an oil burner (direct firing or direct drying), while no appreciable amount of NT was detected in a sample dried by a steam drier (indirect drying). Furthermore, the levels of VNA in directly dried meals were higher than those in indirectly dried meals (Table 1). Only 1-3 mg/kg nitrite were detected in the directly dried meals, indicating that the VNA in the meal samples were formed by nitrosation with nitric oxides during the drying process.

Formation of VNA by treatment of fish meal with nitrogen dioxide gas

The formation of VNA in sardine meal samples was examined in our laboratory by treating them with about 1000 μg/kg gaseous nitrogen dioxide, followed by heating at 120°C for 60 min (Tozawa & Kawabata, 1987a). A typical gas chromatography-thermal energy analysis chromatogram is shown in Figure 1. The three peaks were identified as NDMA, NPYR and NT, and their pattern coincided exactly with the VNA observed in commercial fish meal samples. The mean levels in meal samples prepared from small sardines were 3000 μg/kg NDMA, 20 μg/kg NPYR and 30 μg/kg NT. Much higher values were recorded in samples made from large sardines: 16 000 μg/kg NDMA, 60 μg/kg NPYR and 220 μg/kg NT.

Table 1. Volatile nitrosamine levels (μg/kg) in commercial fish meal samples dried by different methods

Nitrosamine[a]	Direct drying[b]		Indirect drying[c]	
	Maximum	Average	Maximum	Average
NDMA	1400	370	7.0	3.9
NPYR	20.8	9.7	1.3	0.3
NT	14.9	4.9	Not detected	

[a] NDMA, N-nitrosodimethylamine; NPYR, N-nitrosopyrrolidine; NT, N-nitrosothiazolidine
[b] Sixteen sardine meal samples dried by hot air from an oil burner
[c] Ten sardine meal samples dried by a steam drier

Changes in the amounts of cysteamine and thiazolidine in fresh sardine meat during fish meal preparation

We have developed a new method for the determination of cysteamine and thiazolidine, in which cysteamine is transformed into thiazolidine by reaction with formaldehyde, the thiazolidine is nitrosated with $NaNO_2$, and the resulting NT is determined by gas chromatography-thermal energy analysis (Tozawa & Kawabata, 1987b). Using this method, we analysed raw fish and meal samples for these two compounds during fish meal preparation (Figure 2). The amount of cysteamine was reduced from 200 μg/kg in fresh sardine to 28% after boiling and to 5% in the final dried product. In contrast, only 5-6 μg/kg thiazolidine were detected in fresh sardine, and the content increased remarkably after boiling, to reach about 130 μg/kg; this level did not change even after drying.

The amount of formaldehyde was found to decrease after boiling and to decrease even further after freeze-drying. These results indicate that cysteamine in fresh sardines reacts with formaldehyde during the boiling step to yield thiazolidine. Formaldehyde in fish meat is derived from enzymic or thermal degradation of trimethylaminoxide, a characteristic component of marine fish tissues. The type of thiazolidine formation shown here would thus be specific to marine fish and invertebrates.

Figure 1. Typical gas chromatography-thermal energy analysis chromatogram obtained with a sardine meal sample after treatment with gaseous nitrogen dioxide[a]

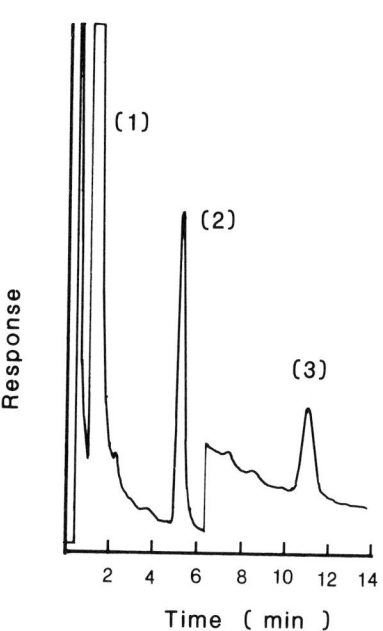

[a](1) N-nitrosodimethylamine; (2) N-nitrosopyrrolidine; (3) N-nitrosothiazolidine

Cysteamine and thiazolidine contents of meal samples and formation of NT after treatment with gaseous nitrogen dioxide

About 700 µg/kg thiazolidine were detected in meal samples made from fresh sardine, while about 100 µg/kg were detected in samples made from decomposed material; cysteamine levels were lower in the former than in the latter. After treatment with gaseous nitrogen dioxide, about 220 µg/kg NT were detected in meal prepared from fresh material, in contrast to about 50 µg/kg in meal made from decomposed fish.

Authentic thiazolidine and cysteamine and coenzyme A were added to sardine meal samples and reacted with gaseous nitrogen dioxide. The amount of NT formed from thiazolidine was higher than that formed from the other two compounds (Table 2). Thiazolidine thus appears to be one of the most important precursors of NT in sardine meal.

Figure 2. Changes in cysteamine (CYA) and thiazolidine (THZ) contents at different stages of sardine meal preparation[a]

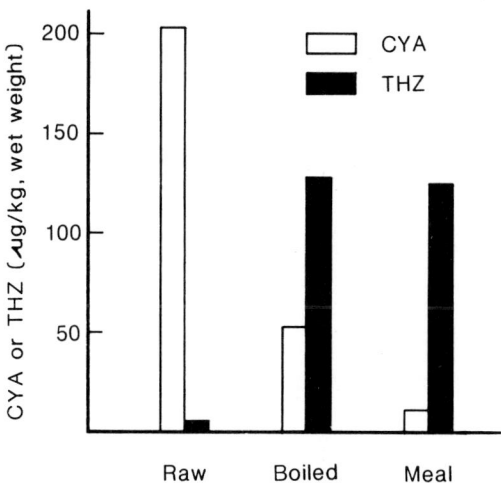

[a] CYA and THZ contents expressed in terms of wet weight. Sampling stages: raw = ~600 g very fresh sardine minced twice and mixed thoroughly; boiled = boiling for 15 min of minced raw fish packed in plastic pouches; meal = boiled fish lyophilized, then crushed to fine powder

Table 2. Formation of N-nitrosothiazolidine (NT) from authentic thiazolidine (THZ) and cysteamine (CYA) and coenzyme A (CoA) added to fish meal samples and reacted with gaseous nitrogen dioxide[a]

Meal sample	Reagent added			NT formed	
	THZ (μg)	CYA (μg)	CoA (μg)[b]	Net weight[c] (μg)	Yield (%)
Meat meal[d]	10.6			4.99	35.5
Meal-L[e]	10.6			3.09	22.0
Meal-L		9.88		0.34	2.2
Meal-L			8.59	1.07	8.1

[a] Meal samples (5 g) in sealed vessels containing 100 μl aqueous solution of THZ, CYA or CoA were flushed with gaseous nitrogen dioxide (2.5 ml at 30°C) and heated at 120°C for 60 min; control samples containing 100 μl distilled water were treated with nitrogen dioxide in the same way
[b] Expressed in terms of μg CYA equivalent
[c] NT content of respective control sample subtracted
[d] Prepared as described in footnote to Figure 2
[e] Made from whole, large sardines

Formation of NT from N-nitrosothiazolidine 4-carboxylic acid (NTCA) and mechanism of NT formation

NTCA formed in meal samples after treatment with gaseous nitrogen dioxide was determined by the method of Tozawa and Kawabata (1987b). The mean level found in meal samples prepared from fresh sardines was about 800 μg/kg, which was three to four times higher than that of NT formed in the same samples. The amount of NT formed from authentic NTCA added to sardine meal samples largely depended on the heating conditions: up to 10% was observed with heating at 160°C for 60 min (Table 3).

Table 3. Formation of *N*-nitrosothiazolidine (NT) from authentic *N*-nitrosothiazolidine 4-carboxylic acid (NTCA) in meal samples heated at different temperatures with and without gaseous nitrogen dioxide[a]

Meal sample	NT formation (%)[b]			
	At 120°C		At 160°C	
	With NO_2	Without NO_2	With NO_2	Without NO_2
Small sardines	4.1	4.2	2.0	0.2
Large sardines	2.8	2.6	10.0	2.8
Large sardines	2.2	-	8.7	-

[a] NTCA at 455 μg in 100 μl distilled water added to 5 g fish meal; control samples received distilled water only. Samples were flushed with gaseous nitrogen dioxide (2.5 ml at 30°C) and heated at 120°C or 160°C for 60 min.
[b] Percentage of NT derived from NTCA; control value subtracted

These results strongly suggest that NTCA is also an important precursor of NT. We propose the following two possible pathways for the formation of NT during fish meal preparation: (1) cysteamine → thiazolidine → NT, and (2) cysteine → (thiazolidine 4-carboxylic acid) → NTCA → NT.

References

Helgason, T., Ewen, S.W.B., Jaffray, B., Stowers, J.M., Outram, J.R. & Pollock, J.R.A (1984) *N*-Nitrosamines in smoked meats and their relation to diabetes. In: O'Neill, I.K., von Borstel, R.C., Miller, C.T., Long, J. & Bartsch, H., eds, N-*Nitroso Compounds: Occurrence, Biological Effects and Relevance to Human Cancer* (IARC Scientific Publications No. 57), Lyon, IARC, pp. 911-920

Hurst, R.E. (1976) Dimethylnitrosamine levels in untreated herring meals. *J. Sci. Food Agric.*, **27**, 600-602

Juszkiewicz, T. & Kowalski, B. (1978) Absorption, tissue deposition and passage into eggs of *N*-nitrosodimethylamine in hens. In: Walker, E.A., Castegnaro, M., Griciute, L. & Lyle, R.E., eds, *Environmental Aspects of* N-*Nitroso Compounds* (IARC Scientific Publications No. 19), Lyon, IARC, pp. 433-439

Kawabata, T., Matsui, M., Ishibashi, T., Hamano, M. & Ino, M. (1982) Formation of *N*-nitroso compounds during cooking of Japanese food. In: Bartsch, H., O'Neill, I.K., Castegnaro, M. & Okada, M., eds, N-*Nitroso Compounds: Occurrence and Biological Effects* (IARC Scientific Publications No. 41), Lyon, IARC, pp. 287-297

Sen, N.P., Schwinghamer, L.A., Donaldson, B.A. & Miles, W.F. (1972) N-Nitrosodimethylamine in fish meal. *J. Agric. Food Chem.*, **20**, 1280-1281

Tozawa, H. & Kawabata, T. (1986) Volatile *N*-nitrosamines detected in commercial fish meals [in Japanese]. *Bull. Jpn. Soc. Sci. Fish.*, **52**, 1969-1974

Tozawa, H. & Kawabata, T. (1987a) Factors affecting the formation of volatile *N*-nitrosamines in the preparation process of fish meals [in Japanese]. *Bull. Jpn. Soc. Sci. Fish.*, **53**, 1449-1456

Tozawa, H. & Kawabata, T. (1987b) Mechanism of *N*-nitrosothiazolidine formation at the stage of fish meal preparation [in Japanese]. *Bull. Jpn. Soc. Sci. Fish.*, **53**, 2209-2216

Tozawa, H. & Kawabata, T. (1987c) Formation of *N*-nitroso-2-methylthiazolidine in fish meals [in Japanese]. *Bull. Jpn. Soc. Sci. Fish.*, **53**, 2259-2262

REGIONAL DIFFERENCES IN N-NITROSAMINE CONTENT OF TRADITIONAL CHINESE FOODS

J. Gao[1], J.H. Hotchkiss[1] & J. Chen[2]

[1]*Institute of Food Science, Cornell University, Ithaca, NY, USA; and*
[2]*Institute of Nutrition and Food Hygiene, Chinese Academy of Preventive Medicine, Beijing, China*

Traditional uncooked and cooked Chinese foods from six provinces with different cancer patterns were analysed for volatile N-nitrosamines (VNA) by gas chromatography-thermal energy analysis. Selected samples were also analysed for N-nitrosamino acids (NAA). N-Nitrosodiethylamine was the VNA found most often (in 91/108 samples; < 0.2-17.5 µg/kg), followed by N-nitrosodimethylamine (in 75/108 samples; < 0.2-17.3 µg/kg). N-Nitrosodipropylamine was detected in 11/108 samples (0.9-6.2 µg/g). Samples selected for analysis of NAA on the basis of their VNA content contained one or more NAA, including N-nitrosoproline and N-nitrosothiazolidine 4-carboxylic acid (range, 1.1-115 µg/kg; average, 35 µg/kg). Both the incidence and pattern of occurrence of VNA suggest that Chinese foods differ from western foods, which would not be expected to contain NAA as frequently nor to contain N-nitrosodipropylamine. The highest percentages of positive samples were from Zhejiang and Fujian provinces (89 and 91%, respectively), and they had the highest average VNA content. Beijing city and Sichuan province gave the lowest percentages of positive samples (75 and 56%, respectively) and the lowest average VNA content.

Volatile N-nitrosamines (VNA) have been reported in western foods in several publications, but their presence in traditional Chinese foods has been analysed rarely (Song & Hu, 1988). In a preliminary study of traditional Chinese foods from six areas with different cancer risks, we analysed food samples for their content of VNA and N-nitrosamino acids (NAA).

A total of 108 food samples were collected from three high-risk areas, Chanangle county in Fujian province, Futuo county in Zhejiang province and Hainan Island in Guangdong province, and from three areas where the cancer risk is lower. All samples were of traditional Chinese foods obtained from local retail markets and manufacturers. Samples were stored at -30°C before analysis.

The methods of Hotchkiss *et al.* (1980), Hotchkiss and Vecchio (1985) and Sen *et al.* (1983, 1985) were used with some modifications, to determine VNA and NAA. The identity of the VNA found in foods was confirmed by oxidation to nitramines, with reanalysis by gas chromatography-thermal energy analysis against nitramine standards, by ultraviolet irradiation, and by gas chromatography-mass spectrometry. Potential artefact formation was investigated by adding morpholine to several samples and analysing for N-nitrosomorpholine.

The NAA found in the 15 samples analysed were N-nitrosoproline (at 1.1-115 µg/kg) and N-nitrosothiazolidine 4-carboxylic acid (at 3.9-62 µg/kg; Table 1).

Table 1. N-Nitrosoamino acids[a] in traditional Chinese foods (µg/kg)

Food	NPRO	NTCA
Ham	1.1	ND
Dried shrimps	57	ND
Roasted fish	3.0	3.9
Wet salted spatchcock	ND	14
Wet salted spatchcock	21	62
Spanish mackerel	21	10.1
Salted river dolphin	3.3	20
Salted anchovy hairtail	93	ND
Salted fish	29	ND
Ham	114	ND
Smoked cured pork	12	ND
Soya cheese (red)	115	ND
Smoked Chinese sausage	6.6	23
Bean paste	39	ND
Roast pig	11	ND

ND, not detected (< 0.2 µg/kg)
[a] NPRO, N-nitrosoproline; NTCA, N-nitrosothiazolidine 4-carboxylic acid

VNA were detected in 91 of the 108 samples (Table 2). The concentrations of N-nitrosodimethylamine ranged from < 0.2 to 17.3 µg/kg, those of N-nitrosodiethylamine from < 0.2 to 17.5 µg/kg, and those of N-nitrosodipropylamine from < 0.2 to 40.3 µg/kg. N-Nitrosothiazolidine was detected in one smoked seafood sample; N-nitrosopyrrolidine was found in one sample of heated seafood and one of preserved seafood. The contents of N-nitrosodimethylamine of some foods (such as dried shrimps and salted fish) were lower than those reported previously. We found that N-nitrosodimethylamine was formed artefactually in these foods, especially in dried shrimps, during vaccum distillation with mineral oil; these foods formed N-nitrosomorpholine from added morpholine during distillation. In order to overcome this problem, we added a considerable excess of acidic ammonium sulfamate. The N-nitrosodiethylamine contents of some foods were higher than expected, averaging 3.1 µg/kg. The finding of N-nitrosodipropylamine at this same average level in several seafoods, particularly in wet samples, cannot be explained, as this nitrosamine is seldom if ever found in western foods.

When the VNA content was compared in foods that had undergone different types of processing, considerable variation was found, heated and preserved foods having higher contents than fermented foods (Table 2).

Samples from the three areas of higher cancer risk more often contained VNA and at higher concentrations than samples from lower-risk areas (Table 3). This was especially true for the provinces of Zhejiang and Fujian, implying that exposure to VNA is higher there than in other areas. The VNA were derived mainly from seafood.

Table 2. Volatile N-nitrosamines[a] in traditional Chinese foods processed by different methods (µg/kg)

Food	No. of samples	NDMA	NDEA	NDPA
Heated foods				
Seafood	8	1.6 (7)	4.7 (7)	6.2. (2)
Pork	7	1.1 (6)	3.3 (7)	1.2 (1)
Smoked bean curd	5	0.4 (4)	4.8 (5)	ND
Preserved foods				
Dried seafood	26	2.1 (22)	4.0 (26)	0.9 (6)
Wet seafood	19	1.3 (13)	3.5 (18)	4.9 (5)
Meat	10	0.7 (9)	2.6 (9)	ND
Vegetables	6	0.5 (4)	1.6 (6)	ND
Fermented foods				
Fish sauce	11	ND (1)	0.2 (3)	ND
Soya bean sauce	4	ND (1)	0.3 (1)	ND
Bean paste	3	0.8 (1)	1.8 (2)	ND
Soya cheese	4	0.2 (2)	0.4 (3)	ND
Vinegar	2	ND (1)	ND	ND
Sauerkraut	3	0.3 (3)	2.6 (3)	ND

Numbers in parentheses are numbers of positive samples
ND, not detected (< 0.02 µg/kg)
[a] NDMA, N-nitrosodimethylamine; NDEA, N-nitrosodiethylamine; NDPA, N-nitrosodipropylamine

Table 3. Volatile N-nitrosamine[a] content (µg/kg) in Chinese foods from various areas

Area	No. of samples	NDMA	NDEA	NDPA	NPYR	NT
Zhejiang	28	1.9 (24)	4.7 (25)	5.5 (10)	ND (2)	ND
Fujian	39	1.0 (22)	2.6 (32)	0.3 (3)	ND	ND
Guangdong	8	1.2 (7)	2.4 (7)	ND	ND	ND
Sichuan	0	0.7 (7)	2.1 (8)	ND	ND	0.3 (1)
Shanxi	9	0.4 (13)	3.3 (16)	ND (1)	ND	ND
Beijing	4	0.1 (2)	0.3 (3)	ND	ND	ND
Mean	108	1.1 (75)	3.1 (91)	1.5 (14)	ND (2)	ND (1)

ND, not detected (< 0.02 µg/kg)
[a] NDMA, N-nitrosodimethylamine; NDEA, N-nitrosodiethylamine; NDPA, N-nitrosodipropylamine; NPYR, N-nitrosopyrrolidine; NT, nitrosothiazolidine

The production and processing of traditional Chinese foods thus results in the formation of VNA and NAA, particularly in dried shrimps, salted fish and heated foods. The fact that many of these foods are made in the home, dried in air and fermented may result in higher VNA contents than in western foods. Most Chinese prefer preserved, fermented and heated foods, which are those in which most of the *N*-nitroso compounds occur. In order to reduce their exposure, the Chinese should be encouraged to eat fewer preserved foods and more fresh foods. Producing and processing facilities should also improve their methods of preservation so as to reduce the formation of *N*-nitroso compounds.

References

Hotchkiss, J.H., Libbey, L.M. & Scanlan, R.A. (1980) Confirmation of low µg/kg amounts of volatile *N*-nitrosamines in foods by low resolution mass spectrometry. *J. Assoc. Off. Anal. Chem.*, **63**, 74-79

Hotchkiss, J.H. & Vecchio, A.J. (1985) An update on nitrosamines in fried bacon: the use of fried-out fat as an edible oil. *Food Technol.*, **39**, 67-73

Sen, N.P., Tessier, L. & Seaman, S.W. (1983) Determination of N-nitrosoproline and N-nitrososarcosine in malt and beer. *J. Agric. Food Chem.*, **32**, 1033-1036

Sen, N.P., Seaman, S.W. & Baddoo, P.L. (1985) N-Nitrosothiazolidine and nonvolatile N-nitroso compounds in foods. *Food Technol.*, **39**, 84-88

Song, P.J. & Hu, J.F. (1988) N-Nitrosamines in Chinese foods. *Food Chem. Toxicol.*, **26**, 205-208

Relevance to Human Cancer of *N*-Nitroso Compounds,
Tobacco Smoke and Mycotoxins.
Ed. I.K. O'Neill, J. Chen and H. Bartsch
Lyon, International Agency for Research on Cancer
© IARC, 1991

TRACE ANALYSIS OF *N*-NITROSOUREAS BY THEIR ALKYLATING ACTIVITY

P. Mende, B. Spiegelhalder & R. Preussmann

German Cancer Research Centre, Institute of Toxicology and Chemotherapy, Heidelberg, Germany

N-Nitrosoureas can be analysed indirectly by gas chromatography with chemoluminescence detection after reaction of the corresponding diazoalkanes with the scavenger reagent *N*-nitroso-*N*-*tert*-butylglycine. A lower determination limit of 2 ng *N*-methyl-*N*-nitrosourea per sample is achieved. The method is specific for *N*-nitroso compounds, which release nonpolar diazoalkanes upon alkaline treatment.

Few methods are available for the trace analysis of carcinogenic *N*-alkyl-*N*-nitrosoureas. Unlike the well established analysis of *N*-nitrosamines, direct gas chromatographic determination of *N*-nitrosoureas requires a modified chemoluminescence (thermal energy analyser) detector (Fine *et al.*, 1987).

N-*Nitrosourea analysis by diazoalkanes*

We have used a method in which *N*-nitrosoureas are analysed indirectly after conversion to diazoalkanes, followed by reaction with a scavenger reagent. *N*-Nitroso-*N*-*tert*-butylglycine (NTBG) is used for this purpose (Mende *et al.*, 1989), in which the carboxyl group is the reactive site towards diazoalkanes. Owing to the presence of an *N*-nitroso group in the molecule, the NTBG esters formed by this reaction can be determined with the sensitive thermal energy analysis detector. In addition, a tertiary branching adjacent to the *N*-nitroso group is introduced into the molecule in order to make the reagent safer to handle: nitrosamines containing this structural element are known to be devoid of carcinogenic activity (Druckrey *et al.*, 1967; Gold *et al.*, 1981).

Both the alkaline decomposition of *N*-nitrosoureas to diazoalkanes and the reaction with NTBG are performed on a small column (Figure 1). The scavenger reagent is bound to a polar silica gel phase, and the upper layer is finely grouped potassium carbonate. The sample containing the *N*-nitrosourea, in any volume of dichloromethane up to 10 ml, is applied to the column. The diazoalkane released in the potassium carbonate layer reacts with NTBG within less than 5 min to form the corresponding NTBG esters, which are subsequently eluted with dichloromethane:ether (9:1), whereas excess scavenger reagent is retained on the silica gel. The eluate is then analysed directly by gas chromatography-thermal energy analysis using NTBG esters as reference substances.

Figure 1. Structure of column used for derivatization of *N*-nitrosoureas; NTBG, *N*-nitroso-*N*-*tert*-butylglycine; R, ester

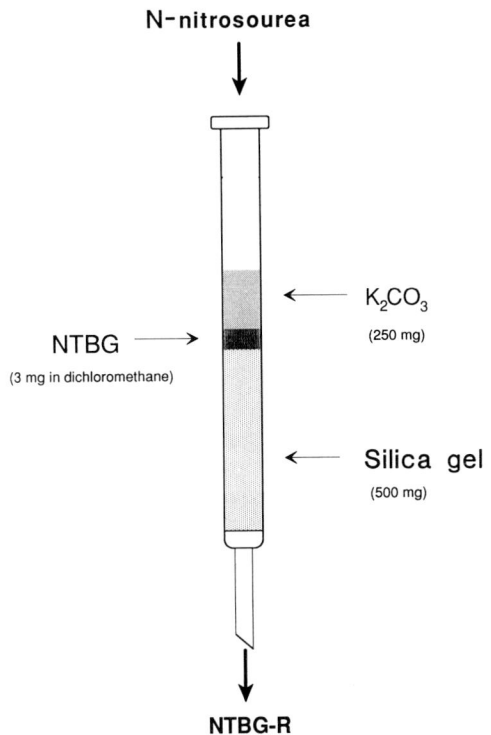

Sensitivity and specificity of the test system

A linear correlation between the concentration of *N*-methyl-*N*-nitrosourea and formation of NTBG methyl ester is seen up to 100 µg *N*-methyl-*N*-nitrosourea. A minimum of 2 ng *N*-methyl-*N*-nitrosourea can be determined per sample, corresponding to a detection limit of 25 pg NTBG methyl ester per gas chromatographic injection. Analysis of *N*-ethyl, *N*-propyl- and *N*-butyl-*N*-nitrosourea as NTBG esters is also possible (Figure 2), with lower determination limits of 5 ng, 5 ng and 10 ng per sample, respectively. The yields were 70% for *N*-methyl-*N*-nitrosourea and about 40-50% for the other *N*-nitrosoureas. The lower yields are due to a reduced reactivity of the corresponding diazoalkanes with NTBG or/and increasing stability of these nitrosoureas to the potassium carbonate treatment, as confirmed by the finding that the yield of NTBG methyl ester from the more stable *N,N*-dimethyl-*N*-nitrosourea was only 20%. Other direct alkylating compounds like methyl- and ethylmethanesulfonate and iodomethane, did not react with NTBG under the conditions of the test system, even when applied at amounts up to 10 mmol per sample. The method is therefore specific for compounds which release nonpolar diazoalkanes after treatment with alkali.

Figure 2. Gas chromatogram of esters of N-nitroso-N-*tert*-butylglycine (NTGB) obtained after simultaneous application of N-methyl-, N-ethyl-, N-propyl- and N-butyl-N-nitrosoureas (10 nmol each) to the test system[a]

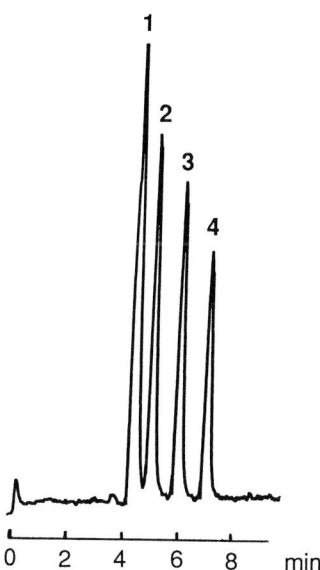

[a]1, NTBG methyl ester; 2, NTBG ethyl ester; 3, NTBG propyl ester; 4, NTBG butyl ester; OV 17 column (10% on Chromosorb WHP), 1.2 m, 130-220 °C at 10°/min; injection volume, 5 µl

Thermal decomposition of N-nitrosoureas also yields diazoalkanes. When NTBG (3 mg) was heated with an N-nitrosourea in an inert solvent like ethyleneglycol diethylether (90°C, 1 h), the corresponding NTBG esters were formed. Excess scavenger reagent was removed before analysis by chromatography on silica gel. Yields of 40% N-methyl- and N-ethyl-N-nitrosourea and 30% N-propyl- and N-butyl-N-nitrosourea were determined. N,N-Dimethyl-N-nitrosourea did not form the NTBG derivative under these conditions.

These results demonstrate that such methods may be useful for the trace analysis of certain N-nitrosoureas and other compounds that release nonpolar diazoalkanes after alkaline or thermal treatment.

References

Druckrey, H., Preussmann, R., Ivankovic, S. & Schmähl, D. (1967) Organotrope carcinogene Wirkungen bei 65 verschiedenen N-Nitroso-Verbindungen an BD-Ratten. *Z. Krebsforsch.*, 69, 103-201

Fine, D.H., Goff, E.U., Johnson, R.A. & Rounbehler, D.P. (1987) Thermal energy analysis of N-nitroso compounds, including ureas, urethanes and guanidines. In: Bartsch, H., O'Neill, I.K. & Schulte-Hermann, R., eds, *The Relevance of N-Nitroso Compounds to Human Cancer: Exposures and Mechanisms* (IARC Scientific Publications No. 84), Lyon, IARC, pp. 216-217

Gold, B., Salmasi, S., Linder, W. & Althoff, J. (1981) Biological and chemical studies involving methyl-t-butylnitrosamine, a non-carcinogenic nitrosamine. *Carcinogenesis*, 2, 529-532

Mende, P., Spiegelhalder, B. & Preussmann, R. (1989) A sensitive analytical procedure for detection of N-nitrosamides *via* their alkylating activity. *Food Chem. Toxicol.*, 27, 475-478

Relevance to Human Cancer of *N*-Nitroso Compounds,
Tobacco Smoke and Mycotoxins.
Ed. I.K. O'Neill, J. Chen and H. Bartsch
Lyon, International Agency for Research on Cancer
© IARC, 1991

POTENTIAL OCCUPATIONAL EXPOSURE ASSOCIATED WITH PARENTERAL ADMINISTRATION OF *N*-NITROSODIETHYLAMINE TO *MACACA MULATTA*: EVIDENCE OF POST-INJECTION RELEASE TO THE ATMOSPHERE

E.B. Sansone & S.D. Keimig

Environmental Control and Research Program, NCI-Frederick Cancer Research Facility, Frederick, MD, USA

In order to assess the exposure of workers administering *N*-nitrosodiethylamine (NDEA) parenterally to *Macaca mulatta*, air samples were drawn through Thermosorb/N cartridges. Samples were analysed by gas chromatography–thermal energy analysis; the limit of detection was 0.02 µg/m³. Significant amounts of NDEA were found in those samples taken in the animal holding room. The NDEA recovered may be accounted for by its expiration by the animals (the contribution from excreta and leakage from the injection site is probably minor). On the basis of the total amount of NDEA administered (840 mg during the first experiment and 250 mg in the second) and the rate at which the animal holding room was ventilated, and assuming that the samples were representative, we estimate that 0.9% of the NDEA administered was released to the atmosphere in 5 h in the first experiment and that 2.7% and 0.8% were released in the first and second 24-h periods, respectively, in the second experiment. It should be noted that this potential source of exposure may be significant not only for workers but also for control or other experimental animals housed in the same room.

The potential hazards of work with *N*-nitroso compounds are well known: many of them are toxic and carcinogenic (Magee *et al.*, 1976; IARC, 1978). The precautions to be taken in working safely with these materials have also been identified (e.g., Montesano *et al.*, 1979; Castegnaro & Sansone, 1986). In this study, we assessed the potential exposure of workers administering *N*-nitrosodiethylamine (NDEA) to *Macaca mulatta* by parenteral injection.

Animals and treatment

In the first experiment, 14 animals weighing 2.2–4.6 kg were anaesthetized with Ketamine (10–15 mg/kg body weight) and brought, two at a time, from an animal holding room to a laboratory, where they were put in a fume hood and injected parenterally with a solution of NDEA in sterile saline (20.4 mg/kg body weight). Each pair of animals was returned to the holding room following administration of the NDEA. The procedure required about 65 min in all; 842.5 mg NDEA were administered.

In the second experiment six months later, the same procedure was followed for the four surviving animals. A total of 253 mg NDEA were administered.

Sample collection

Samples of air (~100 l) were drawn through Thermosorb/N cartridges using Dupont Alpha-1, MSA model G, or Sipin pumps. Pumps were located in the fume hood, in the fume hood exhaust system downstream of a charcoal filter, in the room in which the animals were dosed with NDEA, in the holding room containing dosed animals, in a holding room containing undosed animals, in the corridors outside the animal holding rooms, and in the smoking and nonsmoking employee sitting rooms (stationary samples). Samples of air in the immediate vicinity of the workers were obtained before, during and after the preparation and dosing and dosing of the animals (breathing zone samples). Samples collected over a period longer than 8 h were obtained using a programmable air sampling pump which operated for 20 min every hour. After sampling, the cartridges were capped, wrapped in aluminium foil and stored at $-20°C$ until analysis.

Sample elution and analysis

The cartridges were eluted with 1.5 ml methanol:dichloromethane (25:75, analytical grade) at a flow rate of 0.5 ml/min. Aliquots were analysed on a Hewlett Packard model 5710A gas chromatograph equipped with a thermal energy analyser model 502A detector and an automatic peak area integrator. The silanized glass column was 1.8 m × 2 mm i.d. and packed with 10% Carbowax 20 M + 2% KOH on 80/100 Supelcoport. The injection temperature was 250°C, and the column temperature was 120°C; the helium carrier gas flow rate was 30 ml/min. The quantity of NDEA present was determined by comparing its peak area with that of a four-point calibration curve prepared over the range of 0.2–2 ng NDEA and after correcting for desorption efficiency. The limit of detection with this procedure was 0.02 µg/m3.

The sample with the largest peak area was also analysed using a VG Analytical Model 70–250 magnetic mass spectrometer and data system. The molecular ion and base peak at $m/e = 102$, together with the fragmentation pattern, confirmed that the compound was NDEA.

Results and discussion

In the first experiment, four breathing zone and five stationary air samples were taken. No NDEA was detected in the two breathing zone air samples obtained during animal preparation and dosing; the two obtained while the workers were performing other duties, including checking the dosed animals in the holding room for recovery from anaesthesia, and while they were at lunch in a room with smokers yielded 0.088 and 0.092 µg/m3 NDEA. No NDEA was detected in the stationary air samples obtained in the control animal holding room or in the laboratory in which the animals were prepared and dosed. The air sample taken downstream of the charcoal filter in the exhaust system of the fume hood in which the animals were dosed yielded no detectable NDEA; 0.056 µg/m3 was found in the air sample taken upstream of the charcoal filter. The 5-h sample taken in the holding room where the dosed animals were housed yielded 4.24 µg/m3 NDEA.

In the second experiment, four breathing zone and nine stationary air samples were taken. The breathing zone samples, taken before, during and after administration of NDEA to animals, contained no NDEA. Seven of the stationary air samples – in the laboratory in which the animals were dosed, in the control animal holding room, in the corridors outside the control and dosed animal holding rooms (two samples), and in smoking (one sample) and nonsmoking sitting rooms (two samples) – yielded no detectable NDEA. The samples taken from the dosed animal holding room in the first and second 24-h periods following the dosing of the animals yielded air concentrations of 0.78 and 0.23 $\mu g/m^3$ NDEA, respectively.

We believe that the source of the NDEA detected is expired air from the experimental animals (except for the sample taken in the fume hood, where the NDEA probably came from the stock solution and from materials contaminated with NDEA). Leakage from the injection site cannot have contributed much to the air concentrations of NDEA because the animals remained in the laboratory for only a few minutes before they were returned to the holding room. Neither can NDEA in the excreta have made much of a contribution. The total amount excreted in the urine was less than 1% in 24 h following administration of an intraperitoneal dose of 20.4 mg/kg (Palmer, unpublished data); the vapour pressure of pure NDEA is only about 2 mm Hg at 37°C, and the air:water distribution ratio of a saturated solution of NDEA in water at 37°C is 30×10^{-5} (Mirvish *et al.*, 1976). From the total amounts of NDEA administered, the rate at which the animal holding room was ventilated (5.9 m3/min), and the amounts of NDEA recovered, we calculate that 0.9% of the NDEA administered was released into the atmosphere over 5 h in the first experiment, and 2.7% and 0.8% were released in the first and second 24-h periods, respectively, in the second experiment.

Issenberg and Swanson (1980) reported that eight rats and four hamsters that had received *N*-nitrosodiallylamine subcutaneously at 80 mg/kg bw excreted 12.9 and 4.4% of the dose in 24 h in the expired air, respectively. Heath (1962) injected groups of five rats intraperitoneally with *N*-nitrosodimethylamine, NDEA, *N*-nitrosobutylmethylamine, and *N*-nitroso-*tert*-butylmethylamine at 50, 200, 100 and 300 mg/kg bw, respectively, and recovered the nitrosamines in the expired air, although it is not clear what the percentage recovery was.

We recognize that the anaesthetic used in our experiments may have affected the metabolism of NDEA by the animals. Nevertheless, it is clear that our work and that of Issenberg and Swanson and of Heath shows that several nitrosamines administered to different species of animals over a range of dosages may be excreted unchanged into the expired air. This potential source of exposure may be significant not only for laboratory workers but also for control and other experimental animals housed in the same room.

Acknowledgements

Research supported by the National Cancer Institute under Contract No. NO1-CO-74102 with Program Resources, Inc. We thank Dr A. Palmer, Dr J. Phillips and Dr H. Issaq for their assistance.

References

Castegnaro, M. & Sansone, E.B. (1986) *Chemical Carcinogens: Some Guidelines for Handling and Disposing in the Laboratory*, Berlin, Springer

Heath, D.F. (1962) The decomposition and toxicity of dialkylnitrosamines in rats. *Biochem. J.*, 85, 72-91

IARC (1978) *IARC Monographs on the Evaluation of the Carcinogenic Risk of Chemicals to Humans*, Vol. 17, *Some N-Nitroso Compounds*, Lyon

Issenberg, P. & Swanson, S.E. (1980) Monitoring exposure of personnel to volatile nitrosamines in the laboratory environment. In: Walker, E.A., Griciute, L., Castegnaro, M. & Börzsönyi, M., eds, *N-Nitroso Compounds: Analysis, Formation and Occurrence* (IARC Scientific Publications No. 31), Lyon, IARC, pp. 531-540

Magee, P.N., Montesano, R. & Preussmann, R. (1976) N-Nitroso compounds and related carcinogens. In: Searle, C.E., ed., *Chemical Carcinogens* (ACS Symposium Series 173), Washington DC, American Chemical Society, pp. 491-625

Mirvish, S.S., Issenberg, P. & Sornson, H.C. (1976) Air-water and ether-water distribution of *N*-nitroso compounds: implications for laboratory safety, analytic methodology, and carcinogenicity for the rat esophagus, nose, and liver. *J. Natl Cancer Inst.*, 56, 1125-1129

Montesano, R., Bartsch, H., Boyland, E., Della Porta, G., Fishbein, L., Griesemer, R.A., Swan, A.B. & Tomatis, L., eds (1979) *Handling Chemical Carcinogens in the Laboratory: Problems of Safety* (IARC Scientific Publications No. 33), Lyon, IARC

Relevance to Human Cancer of *N*-Nitroso Compounds,
Tobacco Smoke and Mycotoxins.
Ed. I.K. O'Neill, J. Chen and H. Bartsch
Lyon, International Agency for Research on Cancer
© IARC, 1991

ANALYSIS OF VOLATILE *N*-NITROSAMINES IN BEER AND FOOD IN CHINA

H.X. Xu, X.H. Shen, Z.L. Jin & X.B. Xu

*Research Center for Eco-environmental Sciences
(previously Institute of Environmental Chemistry), Academia Sinica, Beijing, China*

Capillary gas chromatography-thermal energy analysis was used for the study of volatile *N*-nitrosamines in comestibles. The finding of a comparatively high content of *N*-nitrosodimethylamine in some Chinese beers confirms the results of other authors. We describe studies of Chinese foods and environmental samples carried out in this laboratory in collaboration with epidemiological groups. The concentrations of *N*-nitrosamines could be correlated with mortality from digestive cancer.

Capillary gas chromatography coupled with thermal energy analysis was used to measure volatile *N*-nitrosamines in about 40 kinds of beer. *N*-Nitrosodimethylamine (NDMA) was found in 27 samples at 0.05-0.55 µg/kg. The levels were similar in domestic and foreign beers; thus, < 0.05 µg/kg was found in a light Beijing beer, < 0.09 in a beer from Qingdao; 0.51 in another Beijing brand and 0.55 in a beer from Guangzhou, while a beer from Brazil contained 0.23 µg/kg and one from the USA, 0.17 µg/kg. In view of the increasing consumption of various brands of beer in China, any potential carcinogenic hazard should be avoided.

We also investigated more than 1100 food and environmental samples from the Hetao area (Xu & Shen, 1989). The concentrations of *N*-nitrosamines were correlated with death rates from digestive cancer in a ten-year survey of this area, and it was postulated that these compounds in both food and environmental samples might be the main etiological factor for this cancer. The levels of total nitrosamines in various kinds of food are given in Table 1.

Very high concentrations of NDMA (not detected – 203 µg/kg) and *N*-nitrosodiethylamine (9.4-493 µg/kg) were found in pickled vegetables (63 samples) from southern China (Cai *et al.*, 1986), where a high death rate from nasopharyngeal cancer has been recorded.

Acknowledgements

Supported by the Academia Sinica.

Table 1. Total *N*-nitrosamines in some Chinese foods

Food	No. of samples	% positive	*N*-Nitrosamines (μg/kg)	
			Mean	Range
Water	152	82	0.77	0.02–4.4
Pickled cabbage	146	82	7.3	0.05–90
Sour rice soup	52	52	22.4	0.43–1106
Bacon	155	72	3.8	0.45–32.2
Potato	120	25	1.4	0.15–14.4
Flour	119	35	0.49	0.02–14.4
Maize	118	38	1.0	0.02–8.3
Sour milk	35	94	34.6	0.8–119.5
Cheese	35	91	22.8	2.9–97.5
Dried meat	35	88	17.7	0.6–60.2
Tomato	10	20	0.14	1.87–2.7
Fresh cabbage	10	70	0.52	0.14–1.9
Liquor	10	70	0.74	0.14–3.6
Fresh meat	10	70	1.3	0.14–6.6
Eggplant	10	0	ND	
Total	1017	61		

ND, not detected

References

Cai, H.-Y. & Shen, X.-H. (1986) Preliminary report on the determination in the home-made fish sauce (FS) and in the urine of persons after taking FS. *Chin. J. Cancer*, 5, 137-

Xu, H.-X. & Shen, X.-H. (1989) Correlation study of nitrite, *N*-nitrosamines with digestive cancer (to be published)

Relevance to Human Cancer of *N*-Nitroso Compounds,
Tobacco Smoke and Mycotoxins.
Ed. I.K. O'Neill, J. Chen and H. Bartsch
Lyon, International Agency for Research on Cancer
© IARC, 1991

RECENT STUDIES IN CANADA ON THE OCCURRENCE AND FORMATION OF *N*-NITROSO COMPOUNDS IN FOODS AND FOOD CONTACT MATERIALS

N.P. Sen

Food Research Division, Bureau of Chemical Safety, Food Directorate, Health Protection Branch, Health and Welfare Canada, Ottawa, Canada, K1A OL2

We present data on the levels of both volatile and nonvolatile *N*-nitroso compounds in various smoked meats, including bacon, and in food contact materials (e.g., baby bottle rubber nipples and pacifiers). Evidence presented suggests that the formation of *N*-nitrosothiazolidine and *N*-nitrosothiazolidine 4-carboxylic acid in smoked meats and bacon and that of *N*-nitroso-*N*-methylaniline in Icelandic smoked mutton, can be minimized by changing or modifying the smoking methods. The presence of two other nonvolatile *N*-nitroso compounds in these products is also reported.

We present data (Table 1) on the levels of various *N*-nitroso compounds in foods and food contact materials and report the identification of three previously undetected *N*-nitroso compounds, namely *N*-nitroso(2-hydroxymethyl)thiazolidine, *N*-nitroso-*N*-methylaniline and *N*-nitrosodibenzylamine, in such products. Considerable knowledge of the mechanism of formation of some of these compounds has been gained that will be useful in minimizing their formation in foods.

Conclusions

Bacon and meats processed by direct smoking methods contain more *N*-nitroso thiazolidine 4-carboxylic acid (NTCA) than those processed with liquid smoke. Since NTCA can undergo heat-induced decarboxylation, fried bacon from directly smoked raw bacon may contain extremely high levels of both NTCA (up to 13 700 µg/kg) and *N*-nitrosothiazolidine (up to 241 µg/kg; Table 2). *N*-Nitroso-*N*-methylaniline appears to be formed in Icelandic smoked mutton by nitrosation of smoke generated by burning sheep dung – the fuel used for smoking such meats. We suggest that the formation of NTCA, *N*-nitrosothiazolidine and *N*-nitroso-*N*-methylaniline in smoked meats and bacon can be minimized by changing or modifying the method of smoking.

Due to the introduction of newer types of elastic rubber netting, the levels of *N*-nitrosodi-*n*-butylamine in cured pork products have decreased markedly during the past two years; instead, the new nettings have promoted the formation of *N*-nitrosodibenzylamine. Varying levels of this compound have also been detected in recent samples

N-Nitroso compounds in foods

Table 1. Levels of volatile and nonvolatile N-nitroso compounds detected in various foods and food–contact materials[a]

Sample analysed	\multicolumn{18}{c}{No. of samples (n) analysed and levels detected (μg/kg)}

Sample analysed	NT			NTCA			NHMT			NMA			NDBA			NDBzA		
	n	Mean	Range	n	Mean	Range	n	Mean	Range	n	Mean	Range	n	Mean	Range	n	Mean	Range
Raw bacon (direct and indirect smoking)	18	4.5	N-14.4	16	1382	N-9000	7	N	–	–	–	–	–	–	–	–	–	–
Fried bacon (from directly and indirectly smoked varieties)	72	21.3	N-241	19	2265	20-13 700	16	5.3	N-13	–	–	–	–	–	–	–	–	–
Miscellaneous cured meats (e.g., frankfurters, sausages, salami, bologna)	124	1.8	N-27	23	423	(N-3900)	9	mostly N	–	–	–	–	–	–	–	–	–	–
Fish and seafoods (smoked and unsmoked)	32	<1	N-1.8	20	147	N-1600	–	–	–	–	–	–	–	–	–	–	–	–
Icelandic smoked mutton[b] (cooked)	9	0.6	N-2.4	9	180	N-390	–	–	–	9	8.5	N-34.7	–	–	–	–	–	–
Cured pork products (e.g., hams, smoked cottage rolls, pork picnic shoulders) packaged in elastic rubber nettings																		
Analysed 1985-86	–			–			–			–			23	14.7	N-56	–		
Analysed 1987-88	–			–			–			–			66	1.5	N-9.8	39	28.8	N-112
Baby bottle rubber nipples[b]	–			–			–			–			–			24	41	N-146

[a] Summary of some new data plus those reported previously; the identity of the N-nitroso compounds in selected samples was confirmed by gas chromatography–mass spectrometry. Abbreviations: NT, N-nitrosothiazolidine; NTCA, N-nitrosothiazolidine 4-carboxylic acid; NHMT, N-nitroso-2-(hydroxymethyl)thiazolidine; NMA, N-nitroso-N-methylaniline; NDBA, N-nitrosodi-n-butylamine; NDBzA, N-nitrosodibenzylamine; N, negative; –, not analysed for
[b] In collaboration with T. Helgason

of baby bottle rubber nipples and pacifiers. These findings warrant a more detailed evaluation of the toxicity of *N*-nitrosodibenzylamine.

Table 2. Levels of *N*-nitrosothiazolidine (NT) and *N*-nitrosothiazolidine 4-carboxylic acid (NTCA) (μg/kg) in raw and fried bacon processed by different smoking methods

Bacon	Direct smoking				Liquid smoking			
	NT		NTCA		NT		NTCA	
	No. of samples	Mean	No. of samples	Mean	No. of Samples	Mean	No. of samples	Mean
Raw	8	5.5	8	3760	11	3.0	10	105
Fried	8	67.0	8	5390	13	7.8	11	158

Acknowledgement

The author wishes to thank the Field Operations Directorate, Quebec Region Laboratories, for analysing some of the samples, and Dr T. Helgason for collaboration in the Icelandic smoked mutton project.

OCCURRENCE OF AND EXPOSURE TO N-NITROSAMINES IN SWEDEN: A REVIEW

B.-G. Österdahl

Swedish National Food Administration, Uppsala, Sweden

During the last ten years, over 900 samples of foods, snuff and other products on the Swedish market were analysed for N-nitrosamines. The average daily intake of volatile N-nitrosamines from foods was estimated to be 0.29 µg/person, and the daily intake of tobacco-specific N-nitrosamines from snuff was calculated to be up to 110 µg/snuff user.

Research into the occurrence of and exposure to N-nitrosamines has been carried out in Sweden for about ten years. The present short review summarizes these results.

Foods

Volatile N-nitrosamines (VNA) have been detected in a wide variety of foods (Josefsson & Nygren, 1981; Österdahl, 1988a; Table 1). N-Nitrosodimethylamine was the VNA occurring most frequently. The highest levels of VNA were found in fried bacon. The formation of N-nitrosopyrrolidine in fried bacon depends on a number of factors; we found that soaking of bacon in water before frying (Österdahl, 1988b) and microwave-cooking as compared to pan-frying (Österdahl & Alrikson, 1989) significantly lowered the levels. The average per-caput daily intake of VNA has been calculated to be 0.29 µg (Österdahl, 1988a). It is not possible at present to estimate reliably the carcinogenic risk for man of exposure to such very low levels of VNA.

Snuff

All samples of snuff tested have been found to contain detectable levels of both VNA and tobacco-specific N-nitrosamines (TSNA) (Österdahl & Slorach, 1983, 1984; Table 2). The levels of VNA in Swedish snuff have decreased considerably since 1979 (mean, 110 µg/kg) and are now normally less than 10 µg/kg; however, snuff contains much higher concentrations of TSNA. Total TSNA levels of 3.8–11 mg/kg were found. Considerable amounts of TSNA (up to 240 µg/g) have also been detected in the saliva of habitual snuff users (Österdahl & Slorach, 1988). The daily intakes of VNA and TSNA were calculated to be 0.22 and up to 110 µg per snuff user, respectively. The association between oral cancer and snuff dipping could well be due to the presence of TSNA in the saliva. Snuff users should therefore be warned about the possible carcinogenic risk involved in their habit.

Table 1. Volatile N-nitrosamines in foods on the Swedish market, 1979-88

Food	No. of samples	No. positive[a]	Mean level (μg/kg)[b]			
			NDMA	NDBA	NPIP	NPYR
Bacon and pork, fried	152	146	1.2	ND	0.1	6.1
Other meat products	68	28	0.4	Trace	0.3	Trace
Smoked fish	76	63	1.2	ND	ND	ND
Other fish and seafood	33	14	0.6	ND	ND	Trace
Beer	291	165	0.3	ND	Trace	ND
Whiskey	15	11	1.2	ND	ND	ND
Other spirits and wine	73	2	Trace	ND	ND	ND
Dried milk	27	3	Trace	ND	ND	ND
Other dried products	38	27	0.9	ND	0.1	0.1
Cocoa	12	11	0.5	ND	ND	1.0
Chocolate products	17	12	0.1	ND	ND	0.1
Tea and coffee	22	18	0.2	ND	ND	0.1
Cheese	26	8	ND	ND	0.1	ND
Miscellaneous products	25	4	Trace	ND	Trace	ND

[a] >0.1-0.3 μg/kg
[b] NDMA, N-nitrosodimethylamine; NDBA, N-nitrosodibutylamine; NPIP, N-nitrosopiperidine; NPYR, N-nitrosopyrrolidine; ND, not detected

Table 2. Volatile and tobacco-specific N-nitrosamines in snuff and chewing tobacco on the Swedish market, 1983-86

Product	Country of origin	No. of samples[a]	Mean level (μg/kg wet weight)[b]							
			NNN	NAT	NAB	NNK	NDMA	NPIP	NPYR	NMOR
Snuff	Sweden	67 (32)	4000	2700	170	770	0.7	Trace	5.1	Trace
	USA	11 (5)	15 000	18 000	150	7400	27	ND	31	ND
	Norway	2 (2)	21 000	12 000	1700	3300	130	8.9	180	32
Dried snuff	UK	2 (0)	1800	850[c]		260				
	Germany	4 (0)	680	310[c]		100				
Chewing tobacco	Sweden	6 (4)	1200	1200	ND	160	0.2	ND	0.8	0.4
	USA	9 (6)	820	670	20	100	64	ND	0.8	0.6
	Denmark	16 (8)	430	690	30	20	5.5	ND	16	ND

[a] Numbers in parentheses, numbers of samples analysed
[b] NNN, N'-nitrosonornicotine; NAT, N-nitrosoanatabine; NAB, N-nitrosoanabasine; NNK, 4-(N-nitrosomethylamino)-1-(3-pyridyl)-1-butanone; NDMA, N-nitrosodimethylamine; NPIP, N-nitrosopiperidine; NPYR, N-nitrosopyrrolidine; NMOR, N-nitrosomorpholine
[c] NAT and NAB
ND, not detected

References

Josefsson, E. & Nygren, S. (1981) Volatile N-nitroso compounds in foods in Sweden. *Vår Föda*, **33** (Suppl. 2), 147-165

Österdahl, B.-G. (1988a) Volatile nitrosamines in foods on the Swedish market and estimation of their daily intake. *Food Addit. Contam.*, **5**, 587-595

Österdahl, B.-G. (1988b) Effect of water on nitrosamine formation in fried bacon. *Food Addit. Contam.*, **5**, 33-37

Österdahl, B.-G. & Alriksson, E. (1989) Volatile nitrosamines in microwave-cooked bacon. *Food Addit. Contam.* **7**, 51-54

Österdahl, B.-G. & Slorach, S.A. (1983) Volatile N-nitrosamines in snuff and chewing tobacco on the Swedish market. *Food Chem. Toxicol.*, **21**, 759-762

Österdahl, B.-G. & Slorach, S.A. (1984) N-Nitrosamines in snuff and chewing tobacco on the Swedish market in 1983. *Food Addit. Contam.*, **1**, 299-305

Österdahl, B.-G. & Slorach, S.A. (1988) Tobacco-specific N-nitrosamines in the saliva of habitual male snuff dippers. *Food Addit. Contam.*, **5**, 581-586

Relevance to Human Cancer of N-Nitroso Compounds,
Tobacco Smoke and Mycotoxins.
Ed. I.K. O'Neill, J. Chen and H. Bartsch
Lyon, International Agency for Research on Cancer
© IARC, 1991

N-NITROSOALKANOLAMINES IN COSMETICS

G. Eisenbrand, M. Blankart, H. Sommer & B. Weber

*Department of Food Chemistry and Environmental Toxicology,
University of Kaiserslautern,
Kaiserslautern, Germany*

A method has been developed for the determination of the N-nitrosoalkanolamines, N-nitrosodiethanolamine (NDELA) and N-nitrosobis(2-hydroxypropyl)amine (NDHPA) in cosmetics. In model systems, we studied nitrosation of the most relevant precursors by $NaNO_2$, by the preservatives Bronopol and Bronidox and by nitric oxides. Secondary amines were most rapidly nitrosated, and Bronopol, Bronidox and atmospheric nitric oxides appeared to be the most relevant nitrosating agents. In order to remove the most important sources of contamination, the Federal Health Office issued an official recommendation to producers to avoid use of secondary amines in cosmetics (March, 1987). Analysis of cosmetics taken from the German market 6-18 months later showed that only 19/126 samples were contaminated with NDELA (12-235 µg/kg or NDHPA (40-215 µg/kg). The results reflect a strong downward trend in contamination.

Alkanolamines such as bis- and tris(hydroxyethyl)amine (di- and triethanolamine), bis(2-hydroxypropyl)amine and cocoa fatty acid diethanolamides are widely used as cosmetic ingredients. The high reactivity of these ingredients towards nitrosating agents represents a potential health hazard, because the resulting nitrosamines, N-nitrosobis-(2-hydroxyethyl)amine (N-nitrosodiethanolamine, NDELA) and N-nitrosobis(2-hydroxypropyl)amine (NDHPA), are potent carcinogens.

A method has been developed for the determination of N-nitrosoalkanolamines in cosmetics (Sommer & Eisenbrand, 1988). It consists of on-column extraction (kieselguhr) of the sample with *n*-butanol and purification on a silica gel column. Trimethylsilyl derivatives of N-nitrosodialkanolamines are determined by gas chromatography-thermal energy analysis. Artefact formation during clean-up of cosmetics containing free amines is inhibited by using kieselguhr containing 50% sodium ascorbate. Collaborative studies showed that NDELA can be analysed with high accuracy at the 50 µg/kg level using this method. With few exceptions, a detection limit of 5 µg/kg NDELA is achieved (Sommer *et al.*, 1989).

Samples of cosmetics taken from the German market during 1986 were found frequently to be contaminated (40% positive; Table 1). Secondary and tertiary alkanolamines, cocoa fatty acid alkanolamides and laurylethersulfates or cetylphosphoric acid esters neutralized with alkanolamines appeared to be the most relevant precursors. Moreover, in several products, the nitrosating preservatives Bronopol (2-bromo-2-nitropropane-1,3-diol) and Bronidox (5-bromo-5-nitro-1,3-dioxan) (Figure 1) had been used with alkanolamines.

Nitrosation studies were carried out with di- and triethanolamine in aqueous solution using $NaNO_2$, the preservatives Bronopol and Bronidox and atmospheric nitrogen oxides as nitrosating agents. Under all conditions, diethanolamine was far more rapidly nitrosated than triethanolamine (Table 2). The high reactivity of diethanolamine with nitrosating agents in cosmetics is also shown in Table 3.

Table 1. Contamination of commercial cosmetics with N-nitrosoalkanolamines[a] (summer 1986)

Cosmetic	No. of samples	No. positive	NDELA (µg/kg)	NDHPA (µg/kg)
Hair shampoos	10	3	70; 110; 150	-
Foam bath and shower gels	8	4	7; 35; 60; 90	-
Body lotions	6	3	7; 10; 130	-
Creams	7	3	85; 275	30
Suntan lotions	9	2	7; 2000	-
Skin oils	1	-	-	-
Washing emulsion	1	1	-	20
Synthetic soap	2	2	25; 175	-
Total	44	18		

[a] NDELA, N-nitrosodiethanolamine; NDHPA, N-nitrosobis(2-hydroxypropyl)amine; -, not detected

Figure 1. Formulae of Bronopol and Bronidox

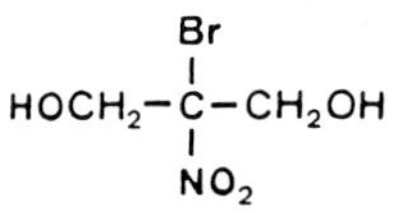

Bronopol
(2-bromo-2-nitropropane-1,3-diol)

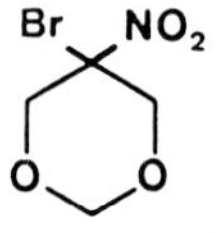

Bronidox
(5-bromo-5-nitro-1,3-dioxan)

Table 2. Nitrosation of di- and triethanolamines in aqueous solution by various nitrosating agents at pH optima

Nitrosating agent[a]	pH	Diethanolamine		Triethanolamine	
		4 h	1 year	4 h	1 year
$NaNO_2$	4	66.0%	100%	1.6%	1.6%
Bronopol	12	-	50%	-	10.0%
Bronidox	10	-	9%	-	3.5%
Nitrogen oxides	11	0.7%	ND	3.5×10^{-5}%	ND

[a] $NaNO_2$, Bronopol and Bronidox at 80 nmol were reacted with 20 mmol di- or triethanolamine; 100 ppm nitrogen oxides were reacted with 100 mg di- or triethanolamine in 10 ml water in 7.2-cm petri dishes
ND, no data; -, not detected

The German Federal Health Office (Bundesgesundheitsamt, 1987) issued an official recommendation to stop using secondary amines in cosmetics production, and this recommendation was adopted by the German manufacturers' association (Industrieverband Körperpflege und Waschmittel, 1988). Further specifications were that fatty acid diethanolamides contain as low as achievable contamination by residual diethanolamine; triethanolamine with a purity of more than 99% and containing less than 1.0% diethanolamine, less than 0.5% monoethanolamine and less than 50 µg/kg NDELA should be used.

Table 3. Nitrosation of 1% diethanolamine in emulsion (pH 6.8) by various nitrosating agents

Nitrosating agent	Time	N-Nitrosodiethanolamine (µg/kg)
50 ppm NaNO$_2$	50 days	7 550
1% Bronopol	50 days	11 000
1% Bronidox	50 days	1 200
10 ppm nitrogen oxide[a]	4 h	2 500

[a] 0.7 g sample applied onto glass plates (7.5 × 15 cm)

Table 4. Contamination of commercial cosmetics with N-nitrosodialkanolamines[a] (summer 1987-88)

Cosmetic	No. of samples	No. positive	NDELA (µg/kg)	NDHPA (µg/kg)
Hair shampoos	16	4	12; 15; 40; 220	–
Foam baths and shower gels	24	6	20; 20; 70; 85; 95; 235	–
Body lotions	22	3	30; 90[b]	70; 215[b]
Creams	32	1	–	50
Suntan lotions	15	2	10; 65	–
Washing emulsions	8	3	60; 105	40
Synthetic soap	4	–	–	–
Deodorant sticks	2	–	–	–
Hair and shaving creams	4	–	–	–
Total	127	19		

[a] NDELA, N-nitrosodiethanolamine; NDHPA, N-nitrosobis(2-hydroxypropyl)amine; -, not detected
[b] Sample contained both NDELA and NDHPA

Commercially available products from the German market analysed six to 18 months after the recommendation had been issued showed that only 15% were contaminated with NDELA or NDHPA (Table 4). Shampoos, foam baths, shower gels and washing emulsions contributed a relatively high percentage (25-30%) to the overall contamination. Information on the formulations of the contaminated samples obtained from the producers indicated that bis(2-hydroxypropyl)amine was still used in two cases, and triethanolamine and cocoa fatty acid diethanolamide of unknown purity were identified as possible sources of the contamination. The overall results of this study demonstrate, however, a strong downward trend in both levels and frequency of contamination. They prove that nitrosamine contamination of cosmetics can be minimized by simple preventive measures.

References

Bundesgesundheitsamt (1987) Empfehlung zur Vermeidung von Nitrosaminen in kosmetischen Mittel. *Bundesgesundheitsblatt*, **30**, 114

Industrieverband Körperpflege und Waschmittel (1988) *Empfehlung zur Vermeidung von Nitrosaminen in kosmetischen Mitteln*, Frankfurt

Sommer, H. & Eisenbrand, G. (1988) A method for determination of N-nitrosoalkanolamines in cosmetics. *Z. Lebensmittelunters. Forsch.*, **186**, 235-238

Sommer, H., Blankart, M. & Eisenbrand, G. (1989) Determination of N-nitrosodiethanolamines in cosmetics and in alkanolamines: results of collaborative studies. *Z. Lebensmittelunters. Forsch.*, **189**, 144-146

Relevance to Human Cancer of *N*-Nitroso Compounds,
Tobacco Smoke and Mycotoxins.
Ed. I.K. O'Neill, J. Chen and H. Bartsch
Lyon, International Agency for Research on Cancer
© IARC, 1991

N-NITROSODIMETHYLAMINE CONTENT OF US AND CANADIAN BEERS

R.A. Scanlan & J.F. Barbour

*Department of Food Science and Technology,
Oregon State University, Corvallis, OR, USA*

A total of 194 beers (148 US and 46 Canadian) were analysed for volatile *N*-nitrosamines. The sampling was designed to include different types of beers brewed from malts at all commercial-scale malt houses by the major beer manufacturers in both countries. *N*-Nitrosodimethylamine (NDMA), the only volatile *N*-nitrosamine detected, was found in 56% of the beers (detection level, 0.05 µg/kg), at a mean level of 0.074 µg/kg. The results indicate that NDMA levels in present-day US and Canadian beers are approximately 1-5% of what they were a decade ago.

Immediately following the initial reports of the occurrence of *N*-nitrosodimethylamine (NDMA) in beer ten years ago, investigators in many countries reported that NDMA levels of approximately 1-5 µg/kg were common (Scanlan, 1983). In the USA and Canada, as well as many other countries, maltsters moved quickly to reduce NDMA formation in malt, initially by use of sulfur dioxide and subsequently by conversion to indirect-fired drying processes. This report provides current information on the levels of NDMA in US and Canadian beers.

Commercial beers (148 US, 46 Canadian) were analysed for volatile *N*-nitrosamines according to the Celite procedure described by Marinelli (1981), except that the sample size was increased from 25.0 to 50.0 g and gas chromatography-thermal energy analysis was used for detection. The beers included the four types indicated in Table 1 and were brewed with malts from all commercial-scale malting houses in both countries. Bottled and canned beers were obtained on the open market throughout both countries and represent the major domestic brands manufactured in each country in 1988.

The results of analysis are shown in Table 1. NDMA, the only volatile *N*-nitrosamine detected, was found in 56% of the beers at or above the lowest detection level of 0.05 µg/kg. The data indicate that NDMA levels in beers currently produced in the USA and Canada are approximately 1-5% of the levels found in beers a decade ago. These results indicate that efforts by the malting and brewing industries to reduce NDMA formation have been successful.

In 1981, a committee of the National Academy of Sciences estimated a daily exposure of 0.97 µg NDMA per person in the USA on the basis of an NDMA level in beer of 2.8 µg/l. Using the same beer consumption figures as used by the National Academy of Sciences in 1981 and the mean NDMA level of 0.074 µg/kg from Table 1, the daily exposure of

NDMA per person now can be estimated to be 0.026 µg, which is 2-3% of the exposure to NDMA from beer estimated in 1981. The results of this study indicate that exposure to NDMA in present-day beer is very significantly less than a decade ago.

Table 1. *N*-Nitrosodimethylamine in US and Canadian beers (µg/kg)

Type of beer	No. of samples	No. of samples with levels of					Mean level[a]	Median level
		<0.05	0.05-0.09	0.10-0.19	0.20-0.39	0.40-0.59		
Lager	98	41	25	26	5	1	0.07	0.06
Light	62	36	18	4	2	2	0.06	<0.05
Malt liquor	17	3	3	8	3	0	0.12	0.12
Ale	17	6	4	3	3	1	0.11	0.06
Total	194	86	50	41	13	4	0.074	0.050

[a] <0.05 was considered to be 0.

Acknowledgement

Supported in part by the Alcoholic Beverages Medical Research Foundation.

References

Marinelli, L. (1981) N-Nitrosamines in malt and beer. *J. Am. Soc. Brew. Chem.*, 39, 99-106

National Academy of Sciences (1981) *The Health Effects of Nitrate, Nitrite, and N-Nitroso Compounds*, Washington DC, National Academy Press, pp. 8-30

Scanlan, R.A. (1983) Formation and occurrence of nitrosamines in foods. *Cancer Res.*, 43, 2435-2440

NITROSATION OF TERTIARY AROMATIC AMINES RELATED TO SUNSCREEN INGREDIENTS

R.N. Loeppky[1], R. Hastings[1], J. Sandbothe[1],
D. Heller[1], Y. Bao[1] & D. Nagel[2]

[1]Department of Chemistry, University of Missouri, Columbia, MO;
and [2]The Eppley Institute for Cancer Research, Omaha, NE, USA

Possible routes to the formation of the sunscreen contaminant, 2-ethylhexyl 4-N-methyl-N-nitrosoaminobenzoate, have been investigated in a study of the nitrosation chemistry of 2-ethylhexyl 4-N,N-dimethylaminobenzoate (Padimate-O) and related tertiary and secondary amines. Padimate-O and the corresponding ethyl ester nitrosate rapidly at 25°C in either N_2O_3:ether or HNO_2:HOAc to produce a mixture of alkyl 4-N-methyl-N-nitrosoaminobenzoate and alkyl 4-N,N-dimethylamino-3-nitrobenzoate, the former of which is the major product. The nitrosative dealkylation of these amines at this low temperature is unusual. Asymmetrical amines exhibit a preference for nitrosative demethylation (methyl *versus* ethyl or benzyl), but the cleavage ratios in N_2O_3:ether are time-dependent, suggesting competing mechanisms with different reactant kinetic orders. A radical cation route would explain the unusual reactivity, which may compete with the established nitrosative dealkylation mechanism. 2-Ethylhexyl 4-N-methyl-N-nitrosoaminobenzoate was mutagenic in two strains of *Salmonella typhimurium* in the Ames assay.

A knowledge of what types of compounds can easily produce carcinogenic nitrosamines is essential to successful cancer prevention. Unfortunately, our knowledge of the chemistry of nitrosamine formation is far from complete. Although much less attention has been given to the occurrence of N-nitroso compounds in the US environment over the past eight years, a recent report produced evidence for the occurrence of a nitrosamine in commercial products containing the sunscreen agent 2-ethylhexyl 4-N,N-dimethylaminobenzoate (Padimate-O, PABAO). The nitrosamine, 2-ethylhexyl 4-methylnitrosaminobenzoate (NPABAO), is not expected to be produced readily from PABAO (Fig. 1).

We report here that PABAO and related compounds nitrosate unusually rapidly. The data suggest that another mechanism of tertiary amine nitrosation, which involves radical cations and has implications that apply to the nitrosatability of aromatic amines, may be operative. We have also found that NPABAO exhibits significant mutagenicity in the Ames assay. The data not only argue for remedial steps to eliminate nitrosamine contamination of products containing PABAO and a further investigation into the possible biological effects of NPABAO, but reveal again the important connection between nitrosamine occurrence and mechanistic chemical investigations which reveal other possible problems.

Fig. 1. Constituents of sunscreen and tanning lotions

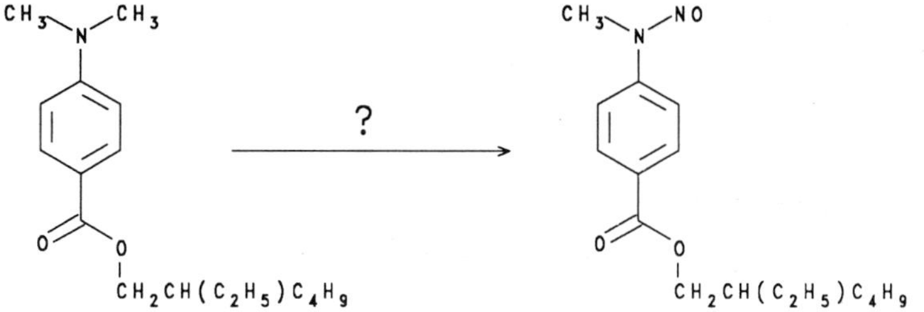

PABAO →[?] NPABAO

Nitrosation rates

Previous work on the nitrosation of tertiary amines led us to anticipate a very slow rate of nitrosation for PABAO. Smith and Loeppky (1967) found that temperatures of 70-90°C were required to effect the nitrosative dealkylation of a variety of benzylic amines. Gowenlock et al. (1979) found rate constants in the range of 6.4-40 × 10^{-3} l/mol per min for the nitrosation of a series of trialkylamines at pH 3.8 at 75°C. An E_a of 21.2 kCal/mol was determined for the triethylamine nitrosation. PABAO is not very basic (estimated pKa, 2.8). While it is well known that the observed rate of secondary amine nitrosation varies inversely with amine basicity, Gowenlock et al. concluded that tertiary amine rates are altered very little with pKa changes. If their observations on triaklyamines can be applied to an aromatic dialkylamine like PABAO, then its lower basicity should not greatly increase its nitrosation rate.

In order to test our hypothesis that PABAO would nitrosate very slowly at ambient temperatures, we prepared its ethyl ester analogue, ethyl 4-N,N-dimethylaminobenzoate (PABAE), in a highly pure state (free of its analogous secondary amine at a level of 0.001%), and examined its nitrosation. The reaction proceeds as given in Eq. 1 with formation of ethyl 4-N-methylnitrosaminobenzoate and ethyl 4-N,N-dimethylamino-3-nitrobenzoate. The time course of the transformation is illustrated in Figure 2, and details are given in Table 1. As can be seen, N_2O_3:ether nitrosation occurs readily at 23°C. There is little published information on this nitrosation medium. We estimate that the nitrosating agent could be present in amounts 15 times greater than the amine under these conditions. As a result, we have no basis for deciding whether this nitrosation is unusual.

Equation 1

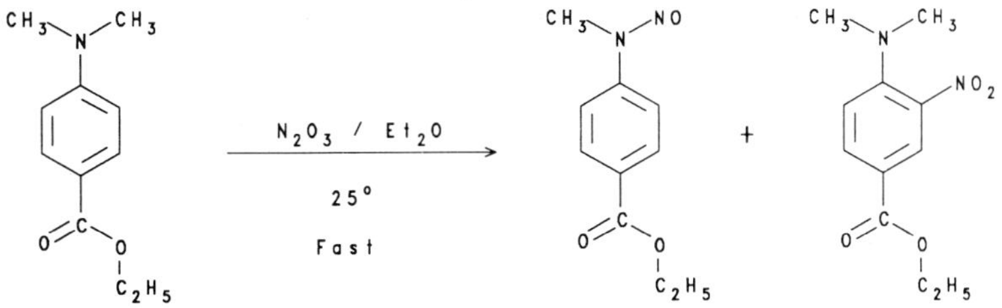

PABAE —[N_2O_3 / Et_2O, 25°, Fast]→ product + product

Table 1. Compounds, products and conditions of nitrosation reactions

$$\underset{1}{\underset{R_1}{\overset{R_2\diagdown N\diagup CH_3}{\text{[aryl-COOR}_1\text{]}}}} \xrightarrow[\text{or } NO_2^-/HOAc]{N_2O_3/Et_2O} \underset{2}{\overset{CH_3\diagdown N\diagup NO}{\text{[aryl-COOR}_1\text{]}}} + \underset{3}{\overset{R_2\diagdown N\diagup NO}{\text{[aryl-COOR}_1\text{]}}} + \underset{4}{\overset{CH_3\diagdown N\diagup R_2}{\text{[aryl(NO}_2\text{)-COOR}_1\text{]}}}$$

Compound no.	Reactant			Conditions					Products		Rate
		R_1	R_2	Nitrosation mixture	[Amine]	['NO+']	T	pH	2/3	2+3/4	(k_{obs})
1	1a	C_2H_5	CH_3	N_2O_3/Et_2O	0.005	–	23	–	–	1.8	0.00470
2	1a	C_2H_5	CH_3	$NO_2^-/HOAc$	0.052	0.40	30	–	–	0.7	–
3	1b	2-EtHx	CH_3	$NO_2^-/HOAc/$ NaOAc/Dioxane	0.014	0.16	30	3.7	–	1.9	0.00051
4	1c	C_2H_5	CH_2Ph	N_2O_3/Et_2O	0.110	–	23	–	Var.	–	0.00220
5	1c	C_2H_5	CH_2Ph	$NO_2^-/HOAc$	0.110	0.98	30	2.3	–	–	0.00160
6	1d	C_2H_5	C_2H_5	$NO_2^-/HOAc$	0.120	1.10	30	–	1.5	5.0	–

To probe this point, we have examined the nitrosation behaviour of PABAE and related amines under both these and more familiar acid nitrosation conditions and investigated the nitrosation of tribenzylamine, a representative trialkylamine studied by both Smith and Loeppky (1967) and Gowenlock et al. (1979), in N_2O_3:ether. Tribenzylamine does not undergo nitrosative dealkylation in N_2O_3 ether at 23°C. PABAE, however, is nitrosated smoothly in glacial acetic acid to give the same products as found in ether. The rate of nitrosation of PABAO in acetic acid buffer (containing dioxane to permit solubility) was determined carefully. Rate constants and product ratios are given in Table 1. It is seen that this transformation at 30°C has a rate constant similar to those observed for trialkylamines at 80°C.

Amines analogous to PABAO and PABAE but with different N-bound substituents also nitrosate easily under all conditions (see Table 1). The nitro compound is formed in all nitrosations, except for the substrate with the N-benzyl group. Nevertheless, the smooth nitrosation of this compound in N_2O_3:ether demonstrates that nitrosative debenzylation can occur under these conditions. Benzaldehyde is the product of the nitrosative debenzylation under all conditions examined.

Fig. 2. Time course of the nitrosation of ethyl 4-N,N-dimethylaminobenzoate with N_2O_3: ether at 23°C. The products are those depicted in Eq. 1.

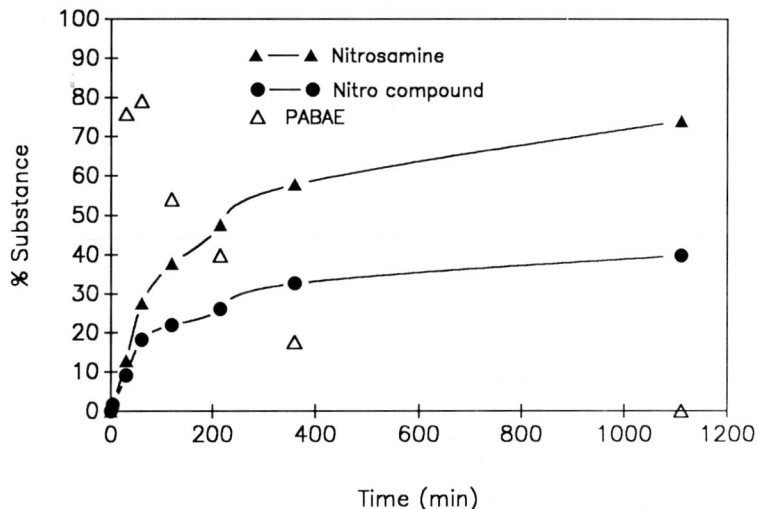

The data presented strongly suggest that aromatic dialkylamines nitrosate unusually rapidly. This conclusion is supported by unpublished observations from our laboratory as well as by anecdotal citings in the older literature. The question, 'Is the nitrosation of PABAO and analogues unusually fast for an amine of such low basicity?' cannot be answered definitively at this time. More kinetic experiments and mechanistic details are required. It is clear, however, that the nitrosation of PABAO is 10^{-5} to 10^{-7} times slower than that of the secondary amine under comparable conditions. We have used Mirvish's (1975) equation to estimate the rate of secondary amine nitrosation at 10^4 l/mol per min at pH 3.7. Although it is difficult to know whether the details of acidic nitrosation will hold under the conditions of formulation of sunscreens and related products, it is clear that the major source of the nitrosamine contamination is likely to be the secondary amine; the tertiary amine may play a role due to its high concentration and apparent susceptibility to nitrosation.

Mechanistic details

One of the goals of our research is to make predictions regarding the nitrosatability of tertiary nitrogen compounds on the basis of their molecular structure. Nitrosation rates are determined by reaction mechanisms, which are related to structure and nitrosation conditions. The studies cited above have led to the expectation that tertiary nitrogen compounds should nitrosate slowly and be of minimal importance in environmental and in-vivo nitrosamine formation. Notable important exceptions include aminopyrine (Lijinsky & Greenblatt, 1972) and gramine in malted beverages (Mangino *et al.*, 1981). Our goal is to identify new nitrosation mechanisms and establish their structure-reactivity profiles. The knowledge deduced from this type of study is useful in the prevention of nitrosamine carcinogenesis.

Our investigations related to the gramine nitrosation problem (Loeppky et al., 1983) have led to the finding (to be published shortly) that a class of tertiary amines are capable of producing nitrosamines very rapidly (much more rapidly than PABAO) even *in vivo*. A requirement of this mechanism is that the amine have an electron-rich aromatic ring attached through a saturated carbon to nitrogen (as in gramine). The more electron-rich the ring, the faster the transformation. PABAO does not possess these structural requirements. Nevertheless, the apparently relatively rapid rate of its nitrosation, along with the formation of an aromatic nitro compound, suggests a different mechanism of nitrosation for PABAO and other aromatic dialkylamines (Scheme 1).

Scheme 1

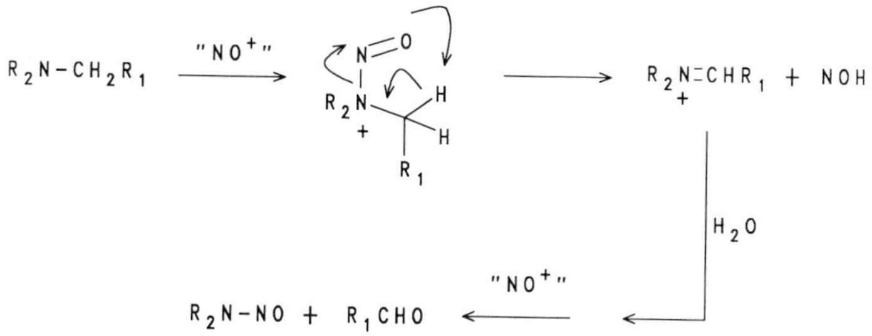

$$R_2N-CH_2R_1 \xrightarrow{"NO^+"} \left[\begin{array}{c} N=O \\ | \\ R_2N^+ \\ | \\ R_1 \end{array} \right] \longrightarrow R_2\overset{+}{N}=CHR_1 + NOH$$

$$R_2N-NO + R_1CHO \xleftarrow{"NO^+"} \xleftarrow{H_2O}$$

The regiochemistry of alkyl group cleavage from nitrogen is a useful tool for testing new mechanisms of amine nitrosation. For example, our class of highly reactive tertiary amines gives only a single nitrosamine instead of the two that could be produced. Benzyldimethylamine, however, behaves 'classically'. Methyl is cleaved 1.5 times more readily than benzyl, due to steric factors that influence the rate of elimination of NOH from R_3NNO^+ (Smith & Loeppky, 1967). We have attempted to use this regiochemical tool to test for a new mechanism of nitrosamine formation in compounds related to PABAO by determining the relative extent of benzyl/methyl cleavage from ethyl 4-*N*-benzyl-*N*-methylaminobenzoate under several nitrosation conditions. The reaction proceeds as shown in Eq. 2. In glacial acetic acid, the ratio of methyl to benzyl cleavage is 2.3 and is invariant with time, as shown in Figure 3(A). The regiochemistry of the cleavage in N_2O_3:ether surprisingly varies with time at a ratio of from 1.1 to 0.6 (Fig. 3B). While the acidic nitrosation is 'well behaved' and appears to follow the classical mechanism (Scheme 1), the results with N_2O_3 nitrosation suggest a switching of mechanism during the transformation. It is notable that no nitro compound is formed during either of these transformations. The absence of the nitro compound when benzylic amines are nitrosated has been observed and explained previously (Loeppky & Tomasik, 1983).

Because we believe that the formation of nitro compounds in the nitrosation of PABAO and its analogues is strongly suggestive of a radical cation mechanism, we sought another regiochemical probe that would produce a nitro compound as a coproduct. Ethyl 4-*N*-ethyl-*N*-methylaminobenzoate served us well in this respect. Nitrosation in glacial acetic acid resulted in both nitration and nitrosative dealkylation of the ethyl and methyl substituents. The ratio of methyl to ethyl cleavage, as deduced from the nitrosamine yields, is 1.5. This is not significantly different from what is expected of the classical mechanism.

Equation 2

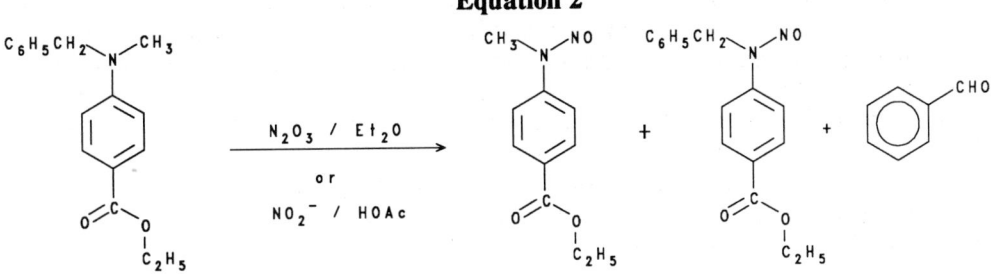

Fig. 3. Comparison of the extent of methyl and benzyl cleavage under different nitrosating conditions: A, acidic nitrosation; B, N_2O_3 nitrosation. With N_2O_3:ether, the cleavage preference changes as the reaction progresses, indicating a change in mechanism.

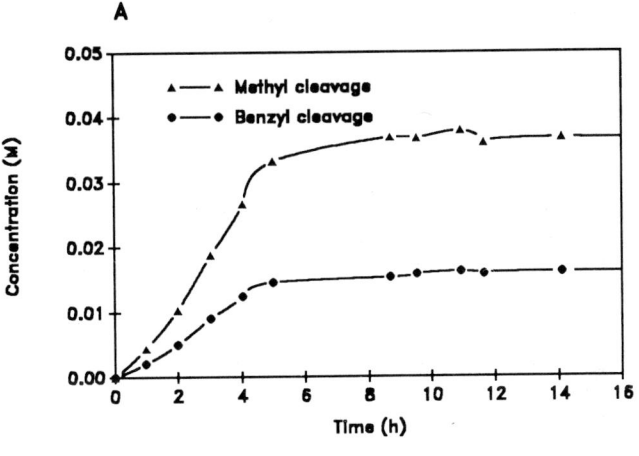

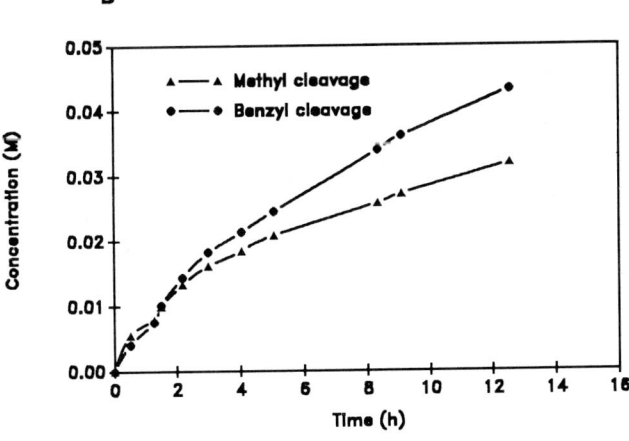

Does nitrosation of PABAO and analogues occur by the mechanism shown in Scheme 2? The only evidence for this pathway is formation of the nitro compound and the suggestion of the competing mechanisms in the N_2O_3 nitrosation. A comparison of the reduction potentials of aromatic amines with that of NO^+ suggests that almost all aromatic amines should be easily oxidized to radical cations by this species. The ease of this oxidation could influence the behaviour of a nitrosammonium ion, R_3NNO^+. Homolytic dissociation of this species would generate two relatively stable radicals, $NO^{\cdot}$ and $R_3N^{+\cdot}$. This process could compete with loss of NOH (Scheme 1). In the case of the aromatic amines, the radical cation would be stabilized by delocalization, and this could provide a less energetic, more rapid pathway to nitrosative dealkylation. Reaction at the aromatic ring with $NO_2^{\cdot}$ would competitively produce the nitro compound. Thus, the pathway shown in Scheme 2 is a reasonable one for PABAO and its analogues, but no strong evidence to differentiate the mechanistic routes in this case has been produced so far. This problem is under active investigation.

Scheme 2

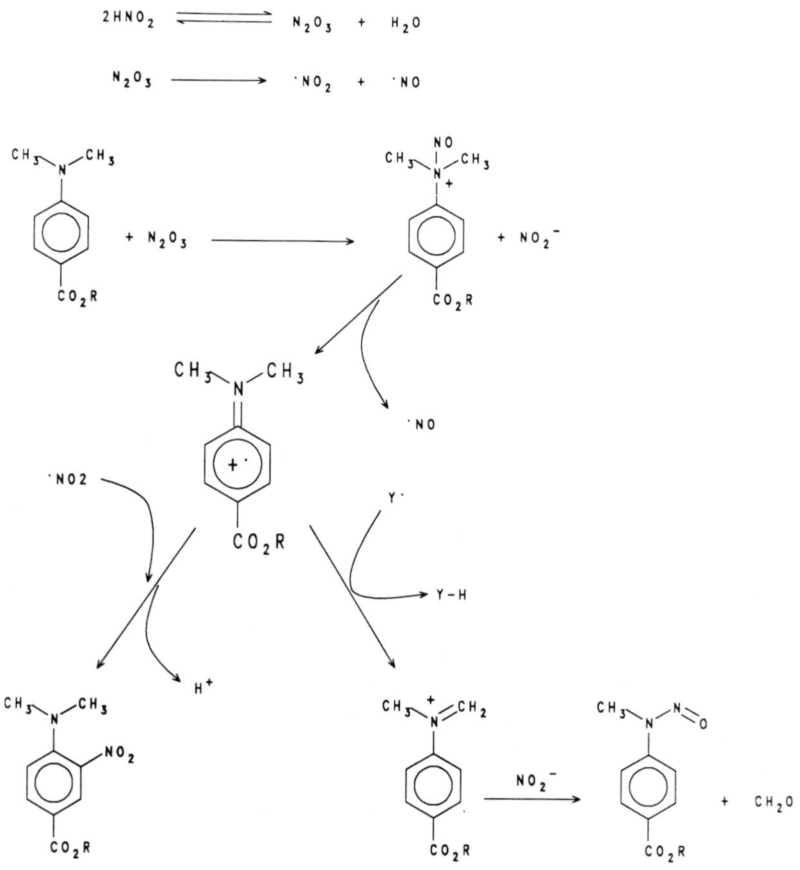

Mutagenicity

A highly purified sample of NPABAO was submitted to mutagenicity assay in *Salmonella typhimurium* strains TA100 and TA1535. NPABAO exhibited dose-dependent mutagenicity in both strains, in the presence of exogenous metabolic activation; it was inactive without activation. The results are presented in Figure 4. Although nitrosamines do not exhibit the same high degree of mutagenicity in the Ames system as other indirect carcinogens, they often display a mutagenicity analogous to that which has been found for NPABAO. These data suggest that steps should be taken to block the formation of NPABAO in products containing PABAO and that its biological activity should be investigated further.

Fig. 4. Mutagenicity of 2-ethylhexyl 4-methylnitrosaminobenzoate in the presence of an exogenous metabolic activation system; NDMA, *N*-nitrosodimethylamine, positive control

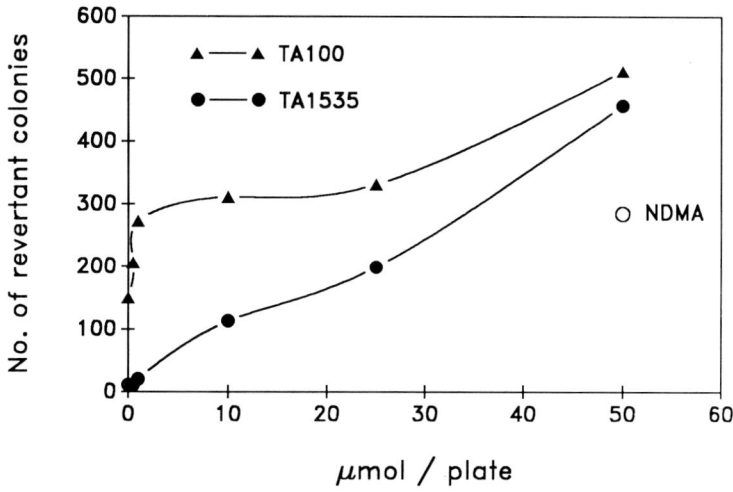

Conclusions

The active principal ingredient of sunscreens, PABAO, is nitrosated readily in buffered acetic acid or N_2O_3:ether to give the nitrosamine NPABAO and 3-nitroPABAO. The nitrosation rate suggests that aromatic dialkylamines nitrosate more rapidly than anticipated, which may result from intervention of a radical cation mechanism. The variable cleavage rates of an unsymmetrical amine in N_2O_3:ether and formation of the nitro compound provide tentative support for this supposition. A better understanding of aromatic amine nitrosation is required better to assess the potential problems resulting from nitrosamine formation from these compounds. The rate of PABAO nitrosation is significantly less than that of its secondary amine. Since commercial samples of PABAO contain as much as 0.5% secondary amine, it is likely that nitrosamine formation results chiefly from its presence. The fact that NPABAO exhibits dose-dependent mutagenicity in the presence of metabolic activation strongly suggests that it could be carcinogenic and should be evaluated further, while steps are taken to block its formation in products containing PABAO.

References

Gowenlock, B.G., Hutchenson, R.J., Little, J. & Pfab, J. (1979) Nitrosative dealkylation of some symmetrical tertiary amines. *J. Chem. Soc. Perkin Trans. II*, 1110-1114

Lijinsky, W. & Greenblatt, M. (1972) Carcinogen dimethylnitrosamine produced in vivo from nitrite and aminopyrine. *Nature-New Biol.*, **236**, 177-178

Loeppky, R.N. & Tomasik, W. (1983) Stereoelectronic effects in tertiary amine nitrosation: nitrosative cleavage vs. aryl ring nitration. *J. Org. Chem.*, **48**, 2751-2757

Loeppky, R.N., Outram, J.R., Tomasik, W. & Faulconer, A. (1983) Rapid nitrosamine formation from a tertiary amine: the nitrosation of 2-(N,N-dimethylaminomethyl)pyrrole. *Tetrahedron Lett.*, **24**, 4271-4274

Mangino, M.M., Scanlan, R.A. & O'Brien, T.J. (1981) N-Nitrosamines in beer, In: Scanlan, R.A. & Tannenbaum, S.R., eds, N-*Nitroso Compounds*, Washington DC, American Chemical Society, pp. 229-246

Mirvish, S.S. (1975) Formation of N-nitroso compounds: chemistry, kinetics, and in vivo occurrence. *Toxicol. Appl. Pharmacol.*, **31**, 325-351

Smith, P.A.S. & Loeppky, R.N. (1967) Nitrosative cleavage of tertiary amines. *J. Am. Chem. Soc.*, **89**, 1147-1157

MUTAGENICITY OF *ALTERNARIA ALTERNATA* AND *PENICILLIUM CYCLOPIUM* ISOLATED FROM GRAINS IN AN AREA OF HIGH INCIDENCE OF OESOPHAGEAL CANCER — LINXIAN, CHINA

Y.Z. Zhen, Y.M. Xu, G.T. Liu, J. Miao, Y.D. Xing,
Q.L. Zheng, Y.F. Ma, T. Su, X.L. Wang, L.R. Ruan,
J.F. Tian, G. Zhou & S.L. Yang

*Esophageal Cancer Research Laboratory, Henan Medical University,
Zhengzhu, Henan Province, China*

In order to study the relationship between daily consumption of mouldy food and the incidence of oesophageal cancer, we examined the mutagenicity of *Alternaria alternata* and *Penicillium cyclopium*, which seriously contaminate grain in Linxian county, China. We first examined extracts of cultured strains of *A. alternata*. In the reverse mutation test, positive results were obtained in 85% of strains; positive results were seen in 84% of 19/20 strains in the *rec* assay. Eight of ten strains induced sister chromatid exchange, and two of eight strains induced chromatid breaks. Six of seven strains induced unscheduled DNA synthesis and DNA synthesis inhibition. One extract induced a higher frequency of sister chromatid exchange in lymphocytes from normal persons than in those from patients with oesophageal cancer, and the spontaneous break-points in patients were related to fragile sites and neoplasia-associated break-points. The toxins alternariol and its monomethyl ether, produced by *A. alternata*, were examined in the reverse mutation assay and for unscheduled DNA synthesis. The results were similar to those obtained with extracts of the different strains. Alternariol had a four to eight times greater effect than its monomethyl ether. Of 24 strains of *P. cyclopium* isolated from cereals in Linxian, four were cultured with rice and 19 in Raulin-Thom medium. Cultures in Raulin-Thorn medium, solution and hyphae were then extracted. The strains cultured with rice induced sister chromatid exchange, unscheduled DNA synthesis and DNA synthesis inhibition. The solution extracts of 14 strains were positive in the *rec* assay, and five strains were positive in the reverse mutation test. The hyphae extracts of five and 15 of the 19 strains were positive in the *rec* assay and the reverse mutation test, respectively.

In areas of high incidence of oesophageal cancer, people often eat mouldy food. Furthermore, we have induced oesophageal cancer in rats by feeding them mouldy food (Henan Medical College, 1983). Between 1974 and 1980, more than 1000 samples of cereal were obtained from areas with high and low incidences of oesophageal cancer. Six species of fungi, including *Alternaria alternata* and *Penicillium cyclopium*, were more prevalent in cereals from the high-incidence areas ($p < 0.001$); *A. alternata* was found particularly frequently in wheat (Zhen et al., 1984). We report here the results of tests for the mutagenicity of extracts of *A. alternata* and preliminary results of mutagenicity tests on *P. cyclopium*.

Mutagenicity of A. alternata

Nineteen strains of *A. alternata* were isolated from cereals in Linxian county in 1980, cultured in corn powder and extracted with alcohol or ether. The extracts were examined for mutagenicity in the *Bacillus subtilis rec* assay and in the *Escherichia coli* ND160 reversion test. A close correlation was found between the two methods, with mutation rates of 84.2% and 85%, respectively, in the absence of an exogenous metabolic activation system, indicating that direct mutagenicity and DNA damage could be induced by most of the strains tested.

Extracts of ten of the strains in ether were tested for their ability to induce sister chromatid exchange and chromosomal aberrations in normal human peripheral blood lymphocytes *in vitro* (Chen et al., 1985). Eight strains induced a high frequency of sister chromatid exchange ($p < 0.01$), and two induced chromosomal aberrations. When two of the strains were tested in mice *in vivo*, however, sister chromatid exchange but not chromosomal aberration was induced (Liu et al., 1988).

Ether extracts of seven of the strains were tested for induction of unscheduled DNA synthesis (UDS) and inhibition of DNA synthesis in human amniotic cells (Yu et al., 1982). Six of the strains gave positive results in both tests, five in the absence of metabolic activation and one only in the presence of metabolic activation (Figure 1). A pattern was observable in relation to dose (Zhen et al., 1988).

Figure 1. Induction of unscheduled DNA synthesis in human amniotic fluid cells by ether extracts of six strains of *Alternaria alternata*[a]

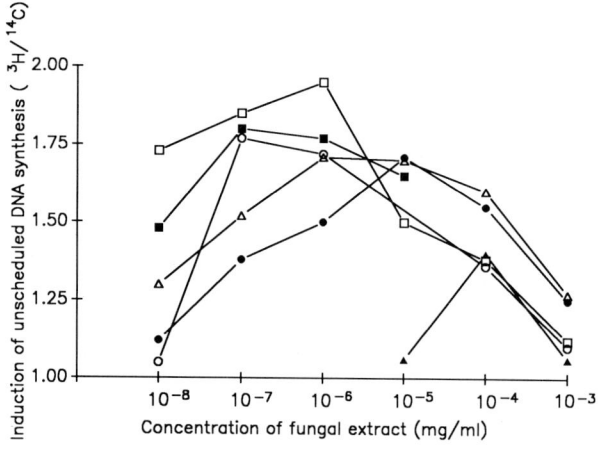

[a] x, strain requiring exogenous metabolic activation

In the test for inhibition of DNA synthesis, the six strains induced 20-90% inhibition in the absence of metabolic activation (Figure 2). In order to confirm whether DNA damage was being induced, N-methyl-N'-nitro-N-nitrosoguanidine was introduced as a control. DNA damage was seen with all compounds during the test, but when the extracts were separated from the cells 30 min later, DNA damage persisted in the cells treated with the nitrosamide. DNA synthesis in those treated with extracts gradually recovered, indicating that the DNA damage induced by the fungal extracts was not severe.

Figure 2. Inhibition of DNA synthesis in human amniotic fluid cells by ether extracts of five strains of *Alternaria alternata*a

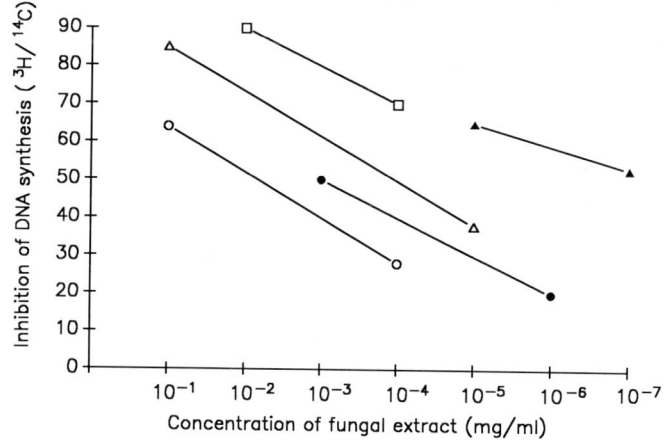

^{a} x, strain requiring exogenous metabolic activation

The results of these two tests indicate that the mutagenic products of the various strains might be different, some producing indirect and some producing direct mutagens.

Mutagenicity of alternariol monomethyl ether and alternariol from A. alternata

Alternariol and its monomethyl ether, which are mycotoxins produced by *Alternaria* species, were isolated and purified from two strains of *A. alternata* (An et al., 1985) and tested for activity in the bacterial reversion assay, the *rec* assay and for induction and inhibition of DNA synthesis.

In the absence of metabolic activation, alternariol monomethyl ether induced mutation in *E. coli* ND160, was active in the *rec* assay, induced UDS at doses of 10^{-3}–10^{-5} mg/ml and inhibited DNA synthesis by 20–80% at doses of 5×10^{-7} mg/ml. No effect on DNA synthesis was seen in the presence of a metabolic activation system.

Alternariol gave similar results but was four to eight times more potent than its ether in inducing reversion and in the *rec* assay.

Mutagenicity of P. cyclopium

During 1986-87, 24 strains of *P. cyclopium* were isolated from 700 samples of cereal in Linxian county and extracted using different media and methods. Four strains isolated from

200 samples of cereal in 1986 were incubated in rice medium and extracted with chloroform:methanol; and 19 strains isolated from 500 samples of cereal in 1987 were cultured in Raulin-Thom medium and the culture medium and fungal membrane were extracted with chloroform and with chloroform:acetidine:alcohol respectively.

The extracts cultured in rice medium did not induce mutagenicity in the reversion assay or in the *rec* assay, although they induced UDS and inhibited DNA synthesis. The extracts of solutions of the 19 strains cultured in Raulin-Thom medium were also inactive in the Ames test in the absence of metabolic activation. With a metabolic activation system, one strain induced mutation in *Salmonella typhimurium* TA98 and TA100 (Table 1). Five strains induced twice as many reversions as the control group in *E. coli* ND160, but no dose-response relationship was seen (Zhou & Zhen, 1989); in the reversion test three strains inhibited *E. coli* ND160. In the *rec* assay, 13 strains had positive, dose-related effects.

The extracts of the fungal membranes of 15 strains induced mutation in the *E. coli* reversion assay, and three of the 19 strains induced UDS, inhibited DNA synthesis and were active in the *rec* assay.

Table 1. Mutagenicity of two strains of *Penicillium cyclopium* in *Salmonella typhimurium* in the presence of metabolic activation

Sample	Dose/plate	Revertants	
		TA100	TA98
Strain 1	0.25 mg	94	50
	0.5 mg	155	69.7
	1.0 mg	304	128
	2.0 mg	99	48.3
Strain 2	0.25 mg	146.5	50
	0.5 mg	223	47
	1.0 mg	122	70
Dimethyl sulfoxide	0.1 mg	83	44.4
CP	200.0 μg	660	
AO	39.0 μg		1085

P. cyclopium cultured in Raulin-Thom medium thus induced bacterial mutation and DNA damage in prokaryotic and human epithelial cells. The mutagens in the fungal membrane and in the culture medium, however, appeared to be different.

The extracts of the four strains cultured in rice medium induced a high frequency of sister chromatid exchange in human lymphocytes ($p < 0.01$) and blocked cell division at the M_1 phase. This finding indicates that the mycotoxins in this fungus may affect immune function in human beings.

Conclusion

Out studies of the mutagenicity of DNA damaging properties of *A. alternata* and *P. cyclopium* in prokaryotic, animal and human cells indicate that the mycotoxins produced by the fungi are mutagenic and toxic. The relationship between exposure to these fungi and tumours of the oesophagus requires further study.

References

An, Y.H., Zhao, T.D., Miao, J., Shi, W.H., Hu, D.P., Zhen, Y.Z., Xu, Y.M. & Liu, G.T. (1986) The studies of isolation, identification and mutagenicity of alternariol monomethyl ether. *J. Henan Med. Univ.*, 21, 204-207

Chen, Y.F., Wang, J.G., Deng, Z.Z., Zheng, Y.Z., Liu, G.T., Ding, L.P., Miao, J., Guan, L. & Guo, P. (1985) *J. Henan Med. Univ.*, 20, 162-165

Henan Medical College (1983) *Esophageal Cancer*, Henan, Press of People's Health No. 1, pp. 50-83

Liu, G.T., Miao, J., Zheng, Y.Z. & Xu, Y. (1988) The progress of research on the carcinogenicity of fungi in the esophagus in Henan Province, China. *J. Henan Med. Univ.*, 2, 4-11

Yu, Y.N., Ding, C., Wang, H.-X. & Chen, X.-R. (1982) *J. Zhejiang Med. Univ.*, 11, 101-104

Zhen, Y.Z., Yang, S.-L., Han, F.M., Ding, L.P., Yang, W.X. & Liu, J.S. (1984) Isolation and culture of fungi from the cereals in counties of Henan Province – 5 with high and 3 with low incidences of esophageal cancer [in Chinese]. *Chin. J. Oncol.*, 6, 27-29

Zhen, Y.Z., Han, S.-Y., Wang, X.L., Xu, M. & Ding, L.P. (1988) *J. Mycol.*, 245-251

Zhou, G. & Zhen, Y.Z. (1989) Postgraduate Academic Paper (in press)

RELATIONSHIPS BETWEEN *ALTERNARIA ALTERNATA* AND OESOPHAGEAL CANCER

G.T. Liu, Y.Z. Qian, P. Zhang, Z.M. Dong, Z.Y. Shi,
Y.Z. Zhen, J. Miao & Y.M. Xu

Henan Medical University, Zhengzhou, China

Although it is uncertain whether fungi can produce the mycotoxins that induce oesophageal cancer, the tumorigenicity of fungi isolated from grain in an area of high incidence of oesophageal cancer has been confirmed. *Alternaria alternata* is a fungus of importance in this respect. The contamination rate with *A. alternata* in corn from areas of high morbidity from oesophageal cancer is higher than that in low-morbidity areas. Extracts of *A. alternata* induced reverse mutation in *Escherichia coli*, unscheduled DNA synthesis in cultured human amnion FL cells, chromosomal aberrations and sister chromatid exchange in human peripheral blood lymphocytes, mutation in V79 cells and transformation of NIH3T3 cells. Alternariol methyl ether, which is an active compound produced by this fungus, also induced transformation of NIH3T3 cells, and the transformed cells grew in soft agar and were tumorigenic in nude mice. Food mildewed by this fungus induced forestomach tumours in rats. Thus, (i) *A. alternata* is tumorigenic, (ii) it has been isolated from corn in an area of high incidence of oesophageal cancer, and (iii) contamination with *A. alternata* in the area of high incidence is higher than that in the area of low incidence. We therefore believe that *A. alternata* is one of the causes of human oesophageal cancer.

The relation between fungi and cancers received much attention as early as the 1960s, and now about ten kinds of mycotoxin are recognized as carcinogenic. However, studies on the carcinogenicity of fungi in the oesophagus are rare. Zhang *et al.* found that grain infected by toxigenic *Fusarium* could induce forestomach papillomas in rats fed it for a long time. In the mid-1960s, Yang *et al.* studied *Geotrichum candidum* in pickled vegetables in Linxian, an area of high incidence of oesophageal cancer, and Zhang *et al.* induced epithelial hyperplasia and papilloma of the oesophagus and forestomach in mice with mouldy, dried sweet potatoes and pickled vegetable juice. Liu *et al.* reported that naturally mouldy food could induce oesophageal papillomas and early squamous-cell carcinomas in Wistar rats, which promoted studies of the carcinogenicity of fungi in the oesophagus (Liu *et al.*, 1988).

The Microbiology Department of Henan Medical College investigated fungal contamination of grain from Linxian and also from Fanxian, an area of low incidence of oesophageal cancer. It was found that the contamination rate with *Fusarium* in maize from Linxian was two to 15 times higher than that from Fanxian; the contamination rate with

Alternaria alternata was 12.9% in Linxian and 5% in Fanxian, and the difference was statistically significant. Zhen isolated fungi from grain samples collected from five counties of high incidence of oesophageal cancer and found contamination rates as follows: *Penicillium cyclopium*, 0.39%; *A. alternata*, 6.5%; *Fusarium moniliforme*, 6.8%; *Aspergillus nidulans*, 0.23%; and *Aspergillus fumigatus*, 0.18%. The rates were all higher than those in areas of low incidence of oesophageal cancer (Liu et al., 1988).

Fungi were isolated from wheat, maize and millet from areas of high and low incidence of oesophageal cancer and identified twice. Rates of contamination with *A. alternata* in grain from the area of high incidence were higher than those from the area of low incidence. Thus, the mutagenicity, carcinogenicity and effective constituents of *A. alternata* were studied systematically. The present paper summarizes the results of our research.

Mutagenicity of Alternaria alternata

Xu et al. (1985a,b) observed that these fungi were mutagenic to *Escherichia coli* ND160 in 73.3-85% of tests. Of 19 strains of *A. alternata*, 16 damaged DNA of *Bacillus subtilis* to various extents at a rate of 84.2%.

Chen et al. (1985) determined the effects of extracts of strains of *A. alternata* with ethanol and ethyl ether on the induction of sister chromatid exchange (SCE) and chromosomal aberrations in human peripheral blood cells. Ethanol extracts of eight strains out of ten increased the rate of SCE to 12.7-20.7/cell versus 6.6-7.6/cell in controls. Two of ten ethyl ether extracts increased the rate of chromosomal aberrations to 0.22/cell, from zero in the control group. The ethanol extracts were less mutagenic than those with ethyl ether. Lei and Liu et al. obtained similar results with regard to the effects of *A. alternata* on SCE rate. Dong, W. et al. (1987) examined the effects of the extracts of strains 261 and C_{12}-B_3 on SCE rate in human peripheral blood lymphocytes in vitro and in vivo. Similar results to those reported above were obtained in vitro. For testing in vivo, the extracts were injected into the peritoneal cavity of mice and the sera separated 40-50 min after injection. The rate of SCE was increased significantly by the extracts after metabolism in the mouse body, but no effect on chromosomal aberration rate was observed.

Using ethyl ether extracts of strain 261 from maize, Dong Ziming et al. (1988) induced 8-azaguanine-resistant mutation in V79 cells only after addition of rat liver microsomal fraction (S9), indicating that the extract contained a premutagen that required metabolic activation. An et al. (1986) extracted 261-B_2-3 strain with ethanol, methanol, acetone and ethyl ether, and Dong Zigang et al. (1987) confirmed that the extract could induce 6-thioguanine-resistant mutation in V79 cells in the absence of metabolic activation, showing that this extract contains a direct mutagen.

Zhang and Zhen studied the effect of strains 261-B_2-3, $C_{12}B_3$ and 219 on unscheduled DNA synthesis (UDS) in human amnion epithelial cells (FL) using the 3H-thymidine incorporation method. UDS was induced by 261-B_2-3 and $C_{12}B_3$. With 261-B_2-3 at 10^{-3}-10^{-7} mg/ml, the effect increased with concentration in the absence of S9 but decreased at concentrations greater than 10^{-5} mg/ml. At all concentrations, the effect was significantly different from that in the control group. These results indicate that extract 261-B_2-3 contains substances that damage DNA directly.

Dong Zigang et al. (1987) treated NIH3T3 cells in vitro with the 261-B$_2$-3 extract at doses of 32, 64 and 128 µg/ml. The rates of transformation were 35.5%, 37.6% and 43.1%, respectively; the rate in the control group was 8.2%. The difference was significant. Dong, W. et al. (1987) also determined the effects of extracts of two strains of A. alternata on the nucleolar organization and the proliferation cycle of human peripheral lymphocytes. Both strains suppressed nucleolar organization, increased the rate of G$_1$ and reduced that of S and G$_2$ plus M.

An et al. (1986) carried out thin-layer chromatography and column chromatography with 14 solvent systems after extraction of strain 261 from maize. Two constituents, toxin A and toxin B, were obtained. Toxin A was isolated, purified and crystallized, and the chemical structure was confirmed as alternariol monomethyl ether (AME). Its melting-point is 276-277°C and its molecular weight 272 kD.

Mutagenicity and carcinogenicity of alternariol monomethyl ether

AME induced reverse mutations in E. coli ND160 and induced 6-thioguanine-resistant mutants in V79 cells in the presence or absence of S9. The mutagenic effect in the absence of S9 was stronger than that in its presence, indicating that AME is a direct mutagen. AME caused morphological transformation of NIH3T3 cells at rates of 9.5%, 16.4% and 30.4%, at concentrations of 12.5, 25 and 50 µg/ml, respectively, which was significantly different from that in the control, giving a clear dose-response relationship. The transformed NIH3T3 cells could grow into colonies in soft agar and into tumours in the subcutaneous tissues of nude mice two weeks after implantation (Zhang et al., 1988). These results indicate that AME produced by A. alternata isolated from grain in an area of high incidence of oesophageal cancer is directly mutagenic, can transform cells and is carcinogenic. N-(4-Carboxyphenyl)retinamide, haemin and garlicin inhibited the mutagenicity of AME.

Effects of alteriol monomethyl ether and alternariol on lipid peroxidation in the epithelium of fetal oesophagus and stomach

We studied the effects of AME and alternariol, another mycotoxin produced by A. alternata, on lipid peroxidation in the epithelium of human fetal oesophagus and stomach in vitro. Lipid peroxidation could be initiated with both compounds, and their levels in oesophageal tissues increased with length of treatment. The levels of malondialdehyde in oesophageal epithelia were higher than those in stomach epithelia treated with the same dosage of AME or alternariol (Dong Ziming et al., 1988).

Distribution of alternariol monomethyl ether in rats and mice

The distribution of AME labelled with tritium was studied in rats and mice. Animals were given 3H-AME at 800 µCi/kg bw by peritoneal injection and killed by bleeding at various intervals after dosing. The radioactivity in the thymus, lung, lower part of the oesophagus, forestomach, glandular stomach, liver, spleen, kidney and colon was determined by liquid scintillation counting. Two hours after treatment, radioactivity was highest in the liver; activity in the oesophagus was only third highest in mice and fourth highest in rats. Radioactivity was highest in this tissue after 24 h in mice and after 72 h in rats. The percentage decrease in radioactivity, from the maximum to the final value counted, was 71% in oesophagus, 78% in forestomach, 77% in liver and 80% in kidney in mice; and 38%

in oesophagus, 53% in forestomach, 61% in liver and 52% in kidney in rats. The rate of decline of radioactivity in oesophagus was thus the slowest in the four organs with higher radioactivity, suggesting that elimination of 3H-AME and its metabolites was slowest in oesophagus, i.e., they have the highest affinity for oesophagus. Since similar results were obtained in the two species, it may be assumed that AME and its metabolites also have high affinity for human oesophagus.

Binding between alternariol monomethyl ether and human DNA

In order to examine the role of the fungus in the causation of oesophageal cancer, binding between AME and human fetal oesophageal DNA was studied by means of a spectrophotometric method. The λ_{max} (210 nm) of AME shifted towards a longer wavelength with decreased absorbance after incubation of AME with DNA, and the shift in absorbance was recovered in the presence of NaCl or phosphate but was not influenced by urea. The λ_{max} of A, T, G and C shifted towards a longer wavelength after addition of AME. The changes in the ultraviolet absorption spectra of nucleosides induced by AME were similar to those of DNA. These results indicate that AME can bind to human oesophageal DNA without base specificity; the maximal molar radio in the binding of AME to DNA nucleotide is 0.005, and the balance constant is equivalent to 2.2×10^3. This suggests that the binding between AME and DNA might be through an ionic bond. AME may therefore be one of the genotoxic etiological factors in oesophageal cancer in Linxian.

Effect of alternariol monomethyl ether on cultured human fetal oesophageal epithelium

Cultured human fetal oesophageal epithelial tissue was exposed to AME, further cultured for one week and examined. The basal cells of the oesophageal epithelium were found to have undergone hyperplasia, disorderly arrangement and papillary growth. The nuclei of the cells were larger and heavily stained, and a nuclear karyokinetic phase was found. These changes were clearly different from those in the control groups. The same changes were observed in groups treated with N-methyl-N'-nitro-N-nitrosoguanidine or N-nitrophenylsomethylamine, suggesting that AME is one of the causes of human oesophageal cancer.

References

An, Y.H., Zhao, T.Z., Miao, J., Shi, W.H., Hu, D.P., Zhen, Y.Z., Xu, Y.M. & Liu, G.T. (1986) The studies of isolation, identification and mutagenicity of alternariol monomethyl ether. *J. Henan Med. Univ.*, **21**, 204-207

Chen, Y.F., Wang, J.G., Deng, Z.Z., Zhen, Y.Z., Liu, G.T., Ding, L.P., Miao, J., Guan, L. & Guo, P.Z. (1985) Study on sister-chromatid exchanges and chromosome aberration induced by *Alternaria alternata*. *Chin. J. Oncol.*, Suppl., 40-41

Dong, W.H., Liu, G.T., Chen, Y.F., Zheng, Z.M., Zhen, Y.Z., An, Y.H. & Miao, J. (1987) The effect of extracts of *Alternaria alternata* on nucleolus organizer region and cell proliferating cycle of lymphocytes. *Chin. J. Pathophysiol.*, **3**, 18-21

Dong, Z.G., Liu, G.T., Dong, Z.M., Qian, Y.Z., An, Y.H., Miao, J. & Zhen, Y.Z. (1987) Induction of mutagenesis and transformation by the extract of *Alternaria alternata* isolated from grains in Linxian, China. *Carcinogenesis*, **8**, 989-991

Dong, Z.G., Liu, G.T., Qian, Y.Z., Dong, Z.M., Yang, H.Y., An, Y.H., Miao, J. & Zhen, Y.Z. (1988) Study on mutation in V79 cells and transformation in NIH/3T3 cells induced by the extract 261-B_2-3 of *Alternaria alternata*. *Chin. J. Pathophysiol.*, **4**, 204-207

Dong, Z.M., Zhang, P., Liu, G.T., Yang, H.Y., Hao, H.L., Wang, N.C., Miao, J. & Zhen, Y.Z. (1988) The effect of alternariol on the lipid peroxidation in the epithelium of the fetal oesophagus and stomach. *J. Henan Med. Univ.*, **23**, 314-318

Liu, G.T., Miao, J., Zhen, Y.Z. & Xu, Y.M. (1988) The progress of research on the carcinogenicity of fungi in the esophagus in Henan Province, China. *J. Henan Med. Univ.*, **2**, 4-11

Xu, Y.M., Zheng, Q.L. & Tian, J.F. (1985a) Study on mutagenicity of the cultures of 90 *Alternaria alternata* strains. *Acta Acad. Med. Henan*, **20**, 1-4

Xu, Y.M., Su, D., Ruan, L.R., Tian, J.F., Yang, S.L., Zhen, Y.Z., Han, S.Y., Guan, L., Miao, J. & Guo, P.Z. (1985b) Study on the extract of *Alternaria alternata* induced reversed mutation and rec-effect of the bacteria. *Chin. J. Oncol.*, Suppl., 35-37

Zhang, X.-Y., Liu, G.T., Yang, B., Gong, Y. & Qian, Y.Z. (1988) The carcinogenicity of alternariol monomethyl ether transformed cells in nude mice. *Beijing Lab. Anim. Sci.*, **5**, 25-26

MECHANISMS AND BIOLOGICAL EFFECTS OF
N-NITROSO COMPOUNDS AND MYCOTOXINS

ENZYME MECHANISMS IN THE METABOLISM OF NITROSAMINES

C.S. Yang, T. Smith, H. Ishizaki & J.Y. Hong

*Department of Chemical Biology & Pharmacognosy,
College of Pharmacy, Rutgers University, Piscataway, NJ, USA*

Many nitrosamines are metabolized by cytochromes P450, one of which (P450IIE1) has received much attention because of its role in the metabolic activation of N-nitrosodimethylamine. This enzyme exists in man, rat, mouse, hamster and other animal species. It is inducible by fasting, diabetes and exposure to ethanol, acetone, isoniazid, benzene and other chemicals. P450IIE1 is responsible for the low K_m form of N-nitrosodimethylamine demethylase and is the major enzyme catalysing the metabolic activation of this carcinogen. In addition, P450IIE1 is the most active P450 species known in the metabolism of N-nitrosoethylmethylamine and N-nitrosopyrrolidine. In the metabolism of N-nitrosobutylmethylamine, P450IIE1 preferentially oxidizes the methyl group over the butyl group, whereas P450IIB1 efficiently oxidizes both the methyl and butyl groups. P450IIB1 also catalyses the α-oxygenation of both the pentyl and methyl groups of N-nitrosopentylmethylamine, forming pentaldehyde and formaldehyde at a rate ratio of 2:1, as well as oxygenation at other carbons of the pentyl group. Many nitrosamines are effectively activated in nonhepatic target tissues. The metabolism of 4-(N-nitrosomethylamino)-1-(3-pyridyl)-1-butanone in lung and nasal microsomes is discussed.

Since the discovery of the carcinogenicity of N-nitrosodimethylamine (NDMA), the metabolism of nitrosamines has been studied extensively (Magee & Barnes, 1967; Lai & Arcos, 1980). Although α-hydroxylation was shown many years ago to be a key step in the activation of nitrosamines, the enzymatic mechanisms of the metabolic activation of some of these compounds are not clearly understood. Lack of this information has hindered our progress in understanding the tissue- and species-specific carcinogenicity of nitrosamines. In this communication, the enzymes and mechanisms involved in the activation of NDMA and other nitrosamines are discussed.

Roles of cytochrome P450IIE1 in NDMA metabolism

The involvement of P450 in NDMA metabolism was demonstrated by Czygan *et al.* (1973) and others (Lotlikar *et al.*, 1975; Guengerich *et al.*, 1982); however, the role of P450 in the metabolic activation of NDMA *in vivo* and *in vitro* has been questioned (reviewed by Lai & Arcos, 1980; Yoo & Yang, 1985) because of the following observations: (i) NDMA demethylase is not induced by classical inducers such as phenobarbital and 3-methylcholanthrene; (ii) the activity is not inhibited by well-recognized P450 inhibitors such as metyrapone and SKF-525A; and (iii) there are multiple K_m values for NDMA demethylase, and unrealistically high K_m values (> 100 mM) were observed with phenobarbital-induced microsomes.

We initiated our work in 1980 with the hypothesis that the multiplicity of the K_m of NDMA demethylation is due to the catalytic activity of the multiple forms of P450, and the P450 species showing the lowest K_m and highest V_{max} is likely to be the enzyme responsible for activation of this carcinogen. In an effort to identify this key P450 for the metabolism of NDMA, we reexamined the kinetic parameters of NDMA demethylase and studied the induction of the low K_m form of NDMA. With uninduced rat liver microsomes, we observed a much lower K_m value of 50-70 μM (Yang, 1982; Peng et al., 1982; Tu et al., 1983), in addition to K_m values of 0.3-0.5 mM and 30-50 mM, which corresponded to the NDMA demethylases I and II, respectively (Lai & Arcos, 1980; Lake et al., 1976). This low K_m form of activity was induced by pretreatment of rats with ethanol, acetone, isopropanol, pyrazole and other chemicals as well as by fasting and diabetes (Tu et al., 1981; Peng et al., 1982, 1983; Tu & Yang, 1983; Tu et al., 1983); however, it was not inducible by classic P450 inducers such as phenobarbital and 3-methylcholanthrene (Hong & Yang, 1985). This enzyme, P450ac, was purified from acetone-induced rat liver microsomes (Patten et al., 1986). It is probably identical to P450j purified from isoniazid-induced rat liver microsomes (Ryan et al., 1985) and is orthologous to rabbit P450LM3a (Koop et al., 1982, 1985). Similar orthologues probably exist in man (Wrighton et al., 1986; Yoo et al., 1988), mice, hamsters, guinea-pigs and other animal species (Yang et al., 1985a; Hong et al., 1989). In this communication, we use the systematic name P450IIE1 for this type of P450 in all species.

The alcohol-inducible P450LM3a (P450IIE1) was the most efficient of six purified rabbit liver P450 forms in catalysing the demethylation and denitrosation of NDMA (Yang et al., 1985b). Some of the results are summarized in Table 1. Studies with purified rat liver P450 isozymes in our and other laboratories also demonstrated that P450IIE1 was more active than other forms in catalysing the metabolism of NDMA (Tu & Yang, 1985; Levin et al., 1986; Patten et al., 1986). Other P450 forms showed substantial activity only at high substrate concentrations, reflecting the high K_m values involved (Yang et al., 1985b; Tu & Yang, 1985). The results are consistent with the concept that the multiple K_m values for microsomal NDMA demethylase are due to the catalytic activity of multiple forms of P450 in microsomes. The P450IIE1-dependent NDMA demethylase was inhibited by nonclassical inhibitors such as 2-phenylethylamine, 3-amino-1,2,4-triazole and pyrazole, but not effectively by the well-recognized P450 inhibitor SKF-525A (Tu & Yang, 1985; Yang et al., 1985b).

Although in a reconstituted NDMA demethylase system, P450IIE1 displayed K_m values of 0.35 mM or higher (Patten et al., 1986), P450IIE1 is believed to be the enzyme responsible for the low K_m form of NDMA demethylase in liver microsomes. The higher K_m values in the reconstituted systems are probably due to the assay conditions used (Yoo et al., this volume). Antibodies against P450IIE1 inhibited NDMA demethylase activity in both control and acetone-induced microsomes, almost to completion, and they inhibited the denitrosation reaction to about the same extent (Figure 1). This result is different from that of Amelizad et al. (1988) and is consistent with our previous proposal that the demethylation and denitrosation pathways share an initial common intermediate (Lorr et al., 1982; Yang et al., 1985b; Patten et al., 1986; Wade et al., 1987).

Table 1. Metabolism of nitrosamines by rabbit liver P450 isozymes[a]

N-Nitrosamine	Dose (mM)	Cytochrome P450 isozymes (turnover number/min)					
		2	3a	3b	3c	4	6
NDMA	4	0.10	5.89	0.03	0.08	<0.03	0.31
NDMA	100	2.21	6.71	1.45	0.79	2.01	4.35
NEMA	4	0.07	0.94	-	-	0.07	-
NBMA	4	1.73	0.23	-	-	0.10	-
NMBzA	4	0.50	0.28	-	-	0.08	-
NMA	4	5.57	2.51	-	-	1.57	-
NPYR	5	1.11	6.13	-	-	0.93	2.97
NPYR	20	2.50	7.05	2.22	1.48	3.14	7.31
NDMMOR	1	1.63	0.25	0.04	0.04	0.04	0.04

[a] NDMA, N-nitrosodimethylamine; NEMA, N-nitrosoethylmethylamine; NBMA, N-nitrosobutylmethylamine; NMBzA, N-nitrosomethylbenzylamine; NMA, N-nitrosomethylaniline; NPYR, N-nitrosopyrrolidine; NDMMOR, N-nitroso-2,6-dimethylmorpholine. Results for NDMA, NEMA, NBMA, NMA and NMBzA from Yang et al. (1985b), for NPYR from McCoy & Koop (1988), and for NDMMOR from Kokkinakis et al. (1985). -, studies not performed

The key role of NDMA demethylase in the activation of NDMA was demonstrated in several systems: (i) P450IIE1 was more efficient than other P450 species in activating NDMA to a mutagen in Chinese hamster V79 cells (Yoo & Yang, 1985); (ii) some of the previously reported species and age differences in rats and hamsters concerning the ability to activate NDMA could be interpreted on the basis of the quantity of P450IIE1 present in the microsomes (Yoo et al., 1987); (iii) pretreatment of rats with ethanol or acetone increased NDMA-induced methylation of DNA in vivo and hepatotoxicity; however, the enhancement was observed only when high doses of NDMA (> 25 mg/kg body weight) were given to the rats (Lorr et al., 1984; Hong & Yang, 1985), suggesting that there is a sufficient amount of P450IIE1 in untreated rats to metabolize low doses of NDMA.

Regulation and functions of cytochrome P450IIE1

The cDNAs and the complete gene sequence for rat and human P450IIE1 have been determined (Song et al., 1986; Umeno et al., 1988a,b). The human enzyme shares 78% amino acid similarity with the rat enzyme and immunochemically cross-reacts with antibodies prepared against the rat enzyme (Yoo et al., 1988). P450IIE1 is inducible by a variety of environmental chemicals and by metabolic conditions such as fasting and diabetes (Figure 2). Induction of NDMA demethylase activity under different conditions was due to increases in the quantity of this enzyme rather than to post-translational modifications. The induction of P450IIE1 by acetone as well as by isopropanol, pyrazole and 4-methylpyrazole was not accompanied by an elevation of the P450IIE1 mRNA level (Song et al., 1986; Hong et al., 1987a,b). The induction of P450IIE1 by fasting and diabetes was accompanied by an elevation of P450IIE1 mRNA (Hong et al., 1974a,b; Dong et al., 1988). P450IIE1 is regulated developmentally; transcription activation after birth has been demonstrated (Song et al., 1986). In several mouse strains, kidney P450IIE1 is regulated by testosterone (Hong et al., 1989). The higher P450IIE1 and NDMA demethylase activity found in male mice than in female mice may be related to the higher NDMA toxicity observed in males (Mohla et al., 1981).

Figure 1. Inhibition of demethylation and denitrosation of N-nitrosodimethylamine (NDMA) by antibodies against P450IIE1

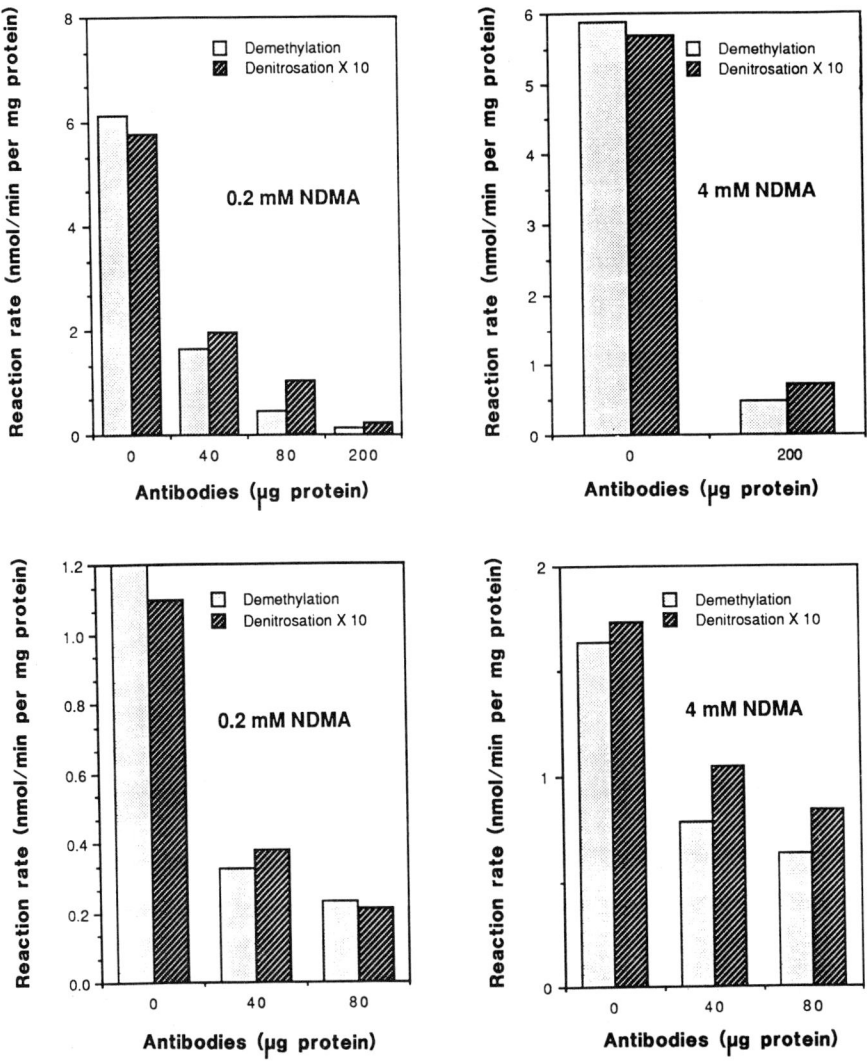

Two substrate concentrations (0.2 and 4.0 mM NDMA) were used. The reactions were catalysed by acetone-induced rat liver microsomes (upper panels) and control microsomes (lower panels). The reaction rates in the presence and absence of antibodies (in micrograms protein) are shown in nanomoles per minute per milligram protein, except that the denitrosation rate was at one-tenth of the scale.

Figure 2. Inducers and substrates for cytochrome P450IIE1

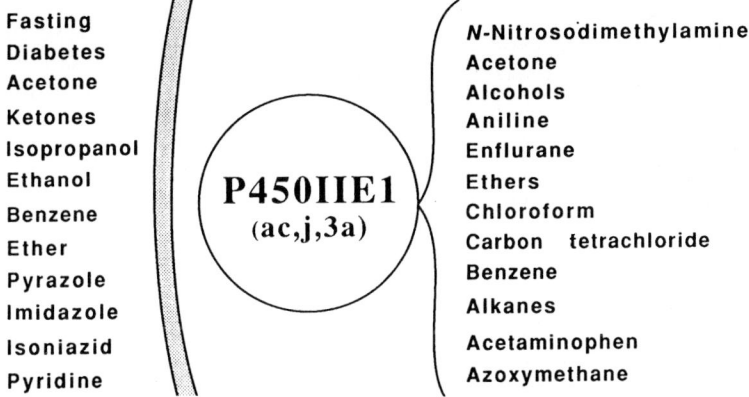

P450IIE1 also catalyses the metabolism of other important environmental chemicals (Brady *et al.*, 1988; and references cited), such as diethyl ether, enflurane, acetaminophen, carbon tetrachloride, benzene, alcohols, aniline, *para*-nitrophenol and other chemicals (Figure 2). It is worth noting that many compounds are both substrates and inducers of P450IIE1. In addition, one substrate can serve as a competitive inhibitor of another substrate. Understanding of the induction and function of P450IIE1 helps to explain, at least in part, many of the previously observed interactions between chemicals – for example, the inhibition of NDMA metabolism and carcinogenesis by ethanol.

Metabolism of other nitrosamines

Fasting or treatment of rats with P450IIE1 inducers enhanced microsomal demethylase activities with several nitrosamines (Peng *et al.*, 1982; Tu & Yang, 1983; Tu *et al.*, 1983); however, this result does not mean that P450IIE1 is the enzyme responsible for the metabolic activation of all nitrosamines. The enzyme and substrate specificities in the metabolism of several nitrosamines are summarized in Table 1. Because of structural similarities, *N*-nitrosodimethylamine and perhaps *N*-nitrosodiethylamine are preferentially metabolized by P450IIE1. P450IIE1 is also more active than other forms in catalysing the α-oxygenation of *N*-nitrosopyrrolidine (McCoy & Koop, 1988). For nitrosamines with larger alkyl chains, the situation may be very different. For example, *N*-nitroso-2,6-dimethylmorpholine is metabolized more efficiently by the phenobarbital-inducible LM_2 (IIB4) than by LM_{3a} (IIE1) and other P450 isozymes from rabbit liver microsomes (Kokkinakis *et al.*, 1985). Demethylation of *N*-nitrosobutylmethylamine, *N*-nitrosomethylbenzylamine and *N*-nitrosomethylaniline was more effectively catalysed by LM_2 than by LM_{3a}. Furthermore, the phenobarbital-inducible rat P450IIB1 is more active in oxygenating the butyl and pentyl groups of *N*-nitrosobutyl- and *N*-nitrosopentylmethylamine, respectively (Lee *et al.*, 1989; Ji *et al.*, 1989). Debutylation and depentylation of these compounds leads to the formation of methyldiazonium, which is probably the reactive species for carcinogenesis.

In the metabolism of asymmetrical *N*-nitrosodialkylamines, a key question is selectivity between the two different alkyl groups. This is determined by the composition of the enzymes in a specific tissue and the substrate specificities of the enzymes. For example, if P450IIE1 is the major enzyme form, then demethylation is favoured over debutylation. The butyl group was proposed to bind to a hydrophobic pocket in the active site of P450IIE1, leaving the methyl group at the oxygenation site. The K_m for the debutylation was higher than that for the demethylation, and only the latter was markedly decreased by the presence of cytochrome b_5 (Lee et al., 1989).

In collaboration with Dr C. Ji and Dr S.S. Mirvish, the position selectivity of P450IIB1 in the metabolism of N-nitrosopentylmethylamine was studied. P450IIB1 catalysed the α-oxygenation of the pentyl and methyl groups at a rate ratio of 2:1. It also oxygenated other carbons; the rates of the formation of 2-HO, 3-HO, 4-HO and 5-HO metabolites, however, were only 0.6, 1.5, 26.5 and 2.4% that of the pentaldehyde formation, respectively. The rate of denitrosation was 25% that of the depentylation reaction. Similarly, N-nitrosodibutylamine was activated by a presumably P450-catalysed α-oxygenation pathway leading to the formation of a butylating species. In addition, 2-HO, 3-HO and 4-HO derivatives were formed, and glucuronide conjugates were isolated from urine (Okada, 1983).

Whereas the P450 concentration in the liver is much higher than those in nonhepatic tissues and the liver usually plays an important role in the metabolism of nitrosamines, many nitrosamines can be effectively activated in nonhepatic target tissues. For example, 4-(N-nitrosomethylamino)-1-(3-pyridyl)-1-butanone (NNK), a potent tobacco-specific lung carcinogen, is metabolized efficiently by lung microsomes. In rats (Sprague-Dawley), liver microsomes had activity similar to that of lung microsomes in the metabolism of NNK to a keto alcohol derivative (a product of methyl oxygenation) but had 200 times more activity in the formation of 4-(N-nitrosomethylamino)-1-(3-pyridyl)-1-butanol (NNal), a reductive product (Table 2). Nasal microsomes were about 100 times more active than liver and lung microsomes in converting NNK to keto alcohol. Nasal microsomes were also active in the formation of many other metabolites, but not of NNal. In A/J mice, liver and lung microsomes showed similar activities in the formation of keto alcohol. The lung had lower activity in catalysing NNK to NNal but had much higher activity in the formation of NNK N-oxide (Table 2). With mouse lung microsomes, a K_m of 5 μM and V_{max} of 56 pmol/min per mg protein were obtained for the formation of keto alcohol (as well as NNK N-oxide), but a much higher K_m was observed for the formation of NNal.

The involvement of P450IIB1 in the metabolism of NNK was suggested by Devereux et al. (1988) and confirmed in our laboratory (in collaboration with Dr P. Thomas). Antibodies against P450IIB1 inhibited keto alcohol formation (from NNK) by 40% in rat lung microsomes. With mouse lung microsomes, anti-P450IIB1 and anti-P450IA1 inhibited keto alcohol formation by 25 and 15%, respectively. Anti-P450IA2 showed no inhibitory effect. The results suggest that more than half of the activity was due to other enzymes; their identities remain to be established. Cyclooxygenase inhibitors, indomethacin and aspirin, inhibited the metabolism of NNK when present at 100 and 300 μM. The N-oxidation pathway appeared to be inhibited to a greater extent (up to 50%) than other pathways. More work is needed to study the identities of the nonhepatic enzymes responsible for the metabolism of NNK and nitrosamines.

Table 2. Metabolism of 4-(N-nitrosomethylamino)-1-(3-pyridyl)-1-butanone (NNK) in mice and rats[a]

Microsomes	Metabolites (pmol/min per mg)						
	NNAl	Keto alcohol	NNK N-oxide	Diol	NNAl N-oxide	Keto acid[b]	Hydroxy acid
Mouse							
Liver	65.5	50.0	-	-	-	-	-
Lung	14.7	37.1	33.3	-	-	2.0	-
Rat							
Liver	950.1	5.9	-	-	-	-	-
Lung	5.5	3.9	2.9	-	-	-	-
Nasal	-	451.0	7.7	15.5	7.5	281.7	6.6
Rat, acetone treated							
Nasal	-	150.3	2.8	0.9	-	11.6	0.7

[a] Microsomes (0.1–0.4 mg protein) were incubated with 10 or 50 μM [5-^{3}H]NNK in the presence of a NADPH generating system for 20 or 30 min. The metabolites were separated by reverse-phase high performance liquid chromatography. The abbreviations used for the metabolites are the same as those used by Hecht et al. (1980): NNAl, 4-(N-nitrosomethylamino)-1-(3-pyridyl)-1-butanol; keto alcohol, 4-oxo-4-(3-pyridyl-N-oxide)-1-butanol; NNK N-oxide, 4-(N-nitrosomethylamino)-1-(3-pyridyl-N-oxide)-1-butanol; keto acid, 4-oxo-4-(3-pyridyl)butyric acid; hydroxy acid, 4-hydroxy-4-(3-pyridyl)butyric acid; diol, 4-hydroxy-4-(3-pyridyl)-1-butanol
[b] Identity remains to be confirmed

Concluding remarks

Although cytochrome P450 enzymes are responsible for the metabolism of a major portion of the known nitrosamines, they are not responsible for the metabolism of all nitrosamines. For example, prostagladin H synthetase has been suggested to be involved in the metabolism of NDEA by pulmonary endocrine cells (Schuller et al., 1987). P450-independent oxidative pathways may also be important in the metabolism of other nitrosamines. A well established P450-independent non-oxidative pathway has been illustrated in the activation of N-nitrosodiethanolamine and other β-hydroxyalkylnitrosamines (Michejda et al., 1987). These compounds are refractory to microsomal oxidation but activated by a conjugation reaction with sulfate.

The aforementioned examples show that nitrosamines can be metabolically activated by reactions catalysed by different species of P450 and other enzymes. The difference in the distribution of these enzymes may be a key factor in determining the tissue specificity in carcinogenesis.

Acknowledgements

This study was supported by NIH Grants CA-37037, CA-46535, and ES-03938. T.S. was supported by a postdoctoral fellowship from the New Jersey Commission for Cancer Research.

References

Amelizad, A., Appel, K.E., Oesch, F. & Hildebrandt, A.G. (1988) Effect of antibodies against cytochrome P-450 on demethylation and denitrosation of *N*-nitrosodimethylamine and *N*-nitrosomethylaniline. *Cancer Res.*, **114**, 380-384

Brady, J.F., Lee, M.J., Li, M., Ishizaki, H. & Yang, C.S. (1988) Diethyl ether as a substrate for acetone/ethanol-inducible cytochrome P-450 and as an inducer for cytochrome P-450. *Mol. Pharmacol.*, **33**, 148-154

Czygan, P., Greim, H., Garro, A.J., Hutterer, F., Schaffner, F., Popper, H., Rosenthal, O. & Cooper, D.Y. (1973) Microsomal metabolism of dimethylnitrosamine and the cytochrome P-450 dependency of its activation to a mutagen. *Cancer Res.*, **33**, 2983-2986

Devereux, T.R., Anderson, M.W. & Belinsky, S.A. (1988) Factors regulating activation and DNA alkylation by 4-(*N*-methyl-*N*-nitrosamino)-1-(3-pyridyl)-1-butanone and nitrosodimethylamine in rat lung and isolated lung cells, and the relationship to carcinogenicity. *Cancer Res.*, **48**, 4215-4221

Dong, Z., Hong, J.Y., Ma, Q., Li, D., Bullock, J., Gonzalez, F.J., Park, S.S., Gelboin, H.V. & Yang, C.S. (1988) Mechanisms of induction of cytochrome P-450ac (P-450j) in chemically induced and spontaneously diabetic rats. *Arch. Biochem. Biophys.*, **263**, 29-35

Guengerich, F.P., Dannan, G.A., Wright, S.T., Martin, M.V. & Kaminsky, L.S. (1982) Purification and characterization of liver microsomal cytochrome P-450: electrophoretic, spectral, catalytic, and immunochemical properties and inducibility of eight isozymes isolated from rat treated with phenobarbital of β-naphthoflavone. *Biochemistry*, **21**, 6019-6030

Hecht, S.S., Young, R. & Chen, C.B. (1980) Metabolism in the F344 rat of 4-(*N*-methyl-*N*-nitrosamino)-1-(3-pyridyl)-1-butanone, a tobacco-specific carcinogen. *Cancer Res.*, **40**, 4144-4150

Hong, J. & Yang, C.S. (1985) The nature of microsomal *N*-nitrosodimethylamine demethylase and its role in carcinogen activation. *Carcinogenesis*, **6**, 1805-1809

Hong, J., Pan, J., Dong, Z. & Yang, C.S. (1987a) Regulation of *N*-nitrosodimethylamine demethylase in rat liver and kidney. *Cancer Res.*, **47**, 5948-5953

Hong, J., Pan, J., Gonzalez, F.J., Gelboin, H.V. & Yang, C.S. (1987b) The induction of a specific form of cytochrome P-450 (P-450j) by fasting. *Biochem. Biophys. Res. Commun.*, **142**, 1077-1083

Hong, J.-Y., Pan, J., Ning, S. & Yang, C.S. (1989) Molecular basis for the sex-related difference in renal *N*-nitrosodimethylamine demethylase in C3H/HeJ mice. *Cancer Res.*, **49**, 2973-2979

Ji, C., Mirvish, S.S., Nickols, J., Ishizaki, H., Lee, M.J. & Yang, C.S. (1989) Formation of hydroxy derivatives, aldehydes and nitrite from *N*-nitrosomethyl-n-amylamine by rat liver microsomes and by purified cytochrome P-450 IIB1. *Cancer Res.*, **49**, 5299-5304

Kokkinakis, D.M., Koop, D.R., Scarpelli, D.G., Coon, M.J. & Hollenberg, P.F. (1985) Metabolism of *N*-nitroso-2,6-dimethylmorpholine by isozymes of rabbit liver microsomal cytochrome P450. *Cancer Res.*, **45**, 619-624

Koop, D.R., Morgan, E.T., Tarr, G.E. & Coon, M.J. (1982) Purification and characterization of a unique isozyme of cytochrome P-450 from liver microsomes of ethanol-treated rabbits. *J. Biol. Chem.*, **257**, 8472-8480

Koop, D.R., Crump, B.L., Nordblom, G.D. & Coon, M.J. (1985) Immunochemical evidence for induction of the alcohol-oxidizing cytochrome P-450 of rabbit liver microsomes by diverse agents: ethanol, imidazole, trichloroethylene, acetone, pyrazole, and isoniazid. *Biochemistry*, **82**, 4065-4069

Lai, D.Y. & Arcos, J.D. (1980) Dialkylnitrosamine bioactivation and carcinogenesis. *Life Sci.*, **27**, 2149-2165

Lake, B.G., Phillips, J.C., Heading, C.E. & Gangolli, S.D. (1976) Studies on the *in vitro* metabolism of dimethylnitrosamine by rat liver. *Toxicology*, **5**, 297-309

Lee, M.J., Ishizaki, H., Brady, J.F. & Yang, C.S. (1989) Substrate specificity and alkyl group selectivity in the metabolism of *N*-nitrosodialkylamines. *Cancer Res.*, **49**, 1470-1474

Levin, W., Thomas, P.E., Oldfield, N. & Ryan, D.E. (1986) *N*-Demethylation of *N*-nitrosodimethylamine by purified rat hepatic microsomal cytochrome P-450: isozyme specificity and role of cytochrome b_5. *Arch. Biochem. Biophys.*, **248**, 158-165

Lorr, N.A., Tu, Y.Y. & Yang, C.S. (1982) The nature of nitrosamine denitrosation by rat liver microsomes. *Carcinogenesis*, **3**, 1039-1043

Lorr, N.A., Miller, K.W., Chung, H.R. & Yang, C.S. (1984) Potentiation of the hepatotoxicity of N-nitrosodimethylamine by fasting, diabetes, acetone and isopropanol. *Toxicol. Appl. Pharmacol.*, 73, 423-431

Lotlikar, P.D., Baldy, W.J., Jr & Dwyer, E.N. (1975) Dimethylnitrosamine demethylation by reconstituted liver microsomal cytochrome P-450 enzyme system. *Biochem. J.*, 152, 705-708

Magee, P.N. & Barnes, J.M. (1967) Carcinogenic nitroso compounds. *Adv. Cancer Res.*, 10, 163-246

McCoy, G.D. & Koop, D.R. (1988) Reconstitution of rabbit liver microsomal N-nitrosopyrrolidine α-hydroxylase activity. *Cancer Res.*, 48, 3987-3992

Michejda, C.J., Koepke, S.R., Kroeger-Koepke, M.B. & Bosan, W. (1987) Recent findings on the metabolism of β-hydroxyalkylnitrosamines. In: Bartsch, H., O'Neill, I.K. & Schulte-Hermann, R., eds, *The Relevance of* N-*Nitroso Compounds to Human Cancer: Exposures and Mechanisms* (IARC Scientific Publications No. 84), Lyon, IARC, pp. 77-82

Mohla, S., Ampy, F.R., Sanders, K.J. & Criss, W.E. (1981) Hormonal regulation of the metabolism of carcinogens in renal tissue of BALB/c mice. *Cancer Res.*, 41, 3821-3823

Okada, M. (1983) Comparative metabolism of N-nitrosamines in relation to their organ and species specificity. In: O'Neill, I.K., von Borstel, R.C., Miller, C.T., Long, J. & Bartsch, H., eds, N-*Nitroso Compounds: Occurrence, Biological Effects and Relevance to Human Cancer* (IARC Scientific Publications No. 57), Lyon, IARC, pp. 401-409

Patten, C., Ning, S.M., Lu, A.Y.H. & Yang, C.S. (1986) Acetone-inducible cytochrome P-450: purification, catalytic activity and interaction with cytochrome b5. *Arch. Biochem. Biophys.*, 251, 629-638

Peng, R., Tu, Y.Y. & Yang, C.S. (1982) The induction and competitive inhibition of a high affinity microsomal nitrosodimethylamine demethylase by ethanol. *Carcinogenesis*, 3, 1457-1461

Peng, R., Tennant, P., Lorr, N.A. & Yang, C.S. (1983) Alterations of microsomal monooxygenase system and carcinogen metabolism by streptozotocin-induced diabetes in rats. *Carcinogenesis*, 4, 703-708

Ryan, D.E., Ramanthan, L., Iida, S., Thomas, P.E., Haniu, M., Shively, J.E., Lieber, C.S. & Levin, W. (1985) Characterization of a major form of rat hepatic microsomal cytochrome P-450 induced by isoniazid. *J. Biol. Chem.*, 260, 6385-6393

Schuller, H.M., Falzon, M., Gazdar, A.F. & Hegedus, T. (1987) Cell type-specific differences in metabolic activation of N-nitrosodimethylamine by human lung cancer cell lines. In: Bartsch, H., O'Neill, I.K. & Schulte-Hermann, R., eds, *The Relevance of* N-*Nitroso Compounds to Human Cancer: Exposures and Mechanisms* (IARC Scientific Publications No. 84), Lyon, IARC, pp. 138-140

Song, B.J., Gelboin, H.V., Park, S.S., Yang, C.S. & Gonzalez, F. (1986) Complementary DNA and protein sequence of ethanol-inducible rat and human P-450: transcriptional and post-transcriptional regulation of the rat enzyme. *J. Biol. Chem.*, 261, 16689-16697

Tu, Y.Y. & Yang, C.S. (1983) A high affinity nitrosamine dealkylase system in rat liver microsomes and its induction by fasting. *Cancer Res.*, 43, 623-629

Tu, T.Y. & Yang, C.S. (1985) Demethylation and denitrosation of nitrosamines by cytochrome P-450 isozymes. *Arch. Biochem. Biophys.*, 242, 32-40

Tu, Y.Y., Sonnenberg, J., Lewis, K.F. & Yang, C.S. (1981) Pyrazole-induced cytochrome P-450 in rat liver microsomes: an isozyme with high affinity for dimethylnitrosamine. *Biochem. Biophys. Res. Commun.*, 103, 905-912

Tu, Y.Y., Peng, R.X., Chang, Z.F. & Yang, C.S. (1983) Induction of a high affinity nitrosamine demethylase in rat liver microsomes by acetone and isopropanol. *Chem.-biol. Interactions*, 44, 247-260

Umeno, M., McBride, O.W., Yang, C.S., Gelboin, H.V. & Gonzalez, F.J. (1988a) Human ethanol-inducible P450IIE1: complete gene sequence, promoter characterization, chromosome mapping, and cDNA-directed expression. *Biochemistry*, 27, 9006-9013

Umeno, M., Song, B.J., Kozak, C., Gelboin, H.V. & Gonzalez, F.J. (1988b) The rat P450IIE1 gene: complete intron and exon sequence, chromosome mapping, and correlation of developmental expression with specific 5' cytosine demethylation. *J. Biol. Chem.*, 263, 4956-4962

Wade, D., Yang, C.S., Metral, C.J., Roman, J.M., Hrabie, J.A., Riggs, C.W., Anjo, T., Keefer, L.K. & Mico, B.A. (1987) Deuterium isotope effect on denitrosation and demethylation of N-nitrosodimethylamine by rat liver microsomes. *Cancer Res.*, 47, 3373-3377

Wrighton, S.A., Thomas, P.E., Molowa, D.T., Haniu, M., Shively, J.E., Maines, S.L. & Watkins, P.S. (1986) Characterization of ethanol-inducible human liver N-nitrosodimethylamine demethylase. *Biochemistry*, **25**, 6731-6735

Yang, C.S. (1982) Nitrosamines and other etiological factors in the esophageal cancer in northern China. In: Magee, P.N., ed., *Nitrosamines and Human Cancer*, Cold Spring Harbor, NY, CSH Press, pp. 487-501

Yang, C.S., Koop, D.R., Wang, T. & Coon, M.J. (1985a) Immunochemical studies on the metabolism of nitrosamines by ethanol-inducible cytochrome P-450. *Biochem. Biophys. Res. Commun.*, **128**, 1007-1013

Yang, C.S., Tu, Y.Y., Koop, D.R. & Coon, M.J. (1985b) Metabolism of nitrosamines by purified rabbit liver microsomal P-450 isozymes. *Cancer Res.*, **45**, 1140-1145

Yoo, J.S.H. & Yang, C.S. (1985) Enzyme specificity in the metabolic activation of N-nitrosodimethylamine to a mutagen for Chinese hamster V79 cells. *Cancer Res.*, **45**, 5569-5574

Yoo, J.S.H., Ning, S.M., Patten, C. & Yang, C.S. (1987) Metabolism and activation of N-nitrosodimethylamine by hamster and rat microsomes: a comparative study with weanling and adult animals. *Cancer Res.*, **47**, 992-998

Yoo, J.S.H., Guengerich, F.P. & Yang, C.S. (1988) Metabolism of N-nitrosodialkylamines by human liver microsomes. *Cancer Res.*, **48**, 1499-1504

MODULATION OF MYCOTOXIN AND NITROSAMINE CARCINOGENESIS BY INDOLE-3-CARBINOL: QUANTITATIVE ANALYSIS OF INHIBITION *VERSUS* PROMOTION

G.S. Bailey, R.H. Dashwood, A.T. Fong, D.E. Williams,
R.A. Scanlan & J.D. Hendricks

*Department of Food Science and Technology, Oregon State
University, Corvallis, OR, USA*

The value of chemopreventive agents for reducing human response to mycotoxins and N-nitrosamines remains uncertain, especially since many such agents also can act as tumour promoters. Indole-3-carbinol (I3C) from cruciferous vegetables can inhibit DNA adduction and hepatocarcinogenesis induced by aflatoxin B_1 (AFB_1) or N-nitrosodiethylamine in trout if given before and with the carcinogen but promotes carcinogenesis when given after initiation. Similar results have been obtained with I3C and AFB_1 in rats. In detailed studies using 10 000 trout, inhibition of AFB_1 carcinogenesis was found to be saturable at high doses of I3C, approximately proportional to dose of I3C through the range of human intake and, within this range, quantitatively predicted by I3C-mediated reduction of AFB_1-DNA adduction in liver. In a second study, post-initiation promotion of AFB_1 carcinogenesis was approximately proportional to I3C dose, increased with duration of exposure, decreased with delayed onset of exposure, and reduced but still significant when I3C was given on alternate months or weeks or twice per week only. Hence, promotion by this common component of cruciferous vegetables required prolonged exposure but not necessarily on a daily basis.

Several commonly consumed dietary factors are now known to act as anticarcinogens or promoters of experimental carcinogenesis, depending on the model system and exposure protocol chosen (Williams *et al.*, 1989). In order to evaluate the probable impact of such agents on human cancer rates, a clear understanding of these opposing mechanisms is needed, with quantitative assessment of their relative potencies for inhibitory *versus* promotional behaviour. We have begun to address these issues by examining modulation of hepatocarcinogenesis induced by aflatoxin B_1 (AFB_1) and N-nitrosodiethylamine (NDEA) in rainbow trout and rats by dietary indole-3-carbinol (I3C).

Inhibition of DNA adduction and carcinogenesis

Prior exposure of rainbow trout (Table 1) or Fischer 344 rats (not shown) to dietary I3C was found to reduce AFB_1-DNA adduction in liver and subsequent hepatocarcinogenesis. Similar results were seen in trout when NDEA was used as the carcinogen (Table 1) and when β-naphthoflavone or Aroclor 1254 was used as the anticarcinogen (Shelton et al., 1986; Goeger et al., 1988). These studies show that I3C is typical of many inhibitors, termed 'blocking agents' (Wattenberg, 1985), which reduce carcinogen-DNA adduction in target organs.

Table 1. Influence of dietary indole-3-carbinol (I3C) on carcinogen-induced DNA adduction and hepatocarcinogenic response in rainbow trout

Dietary pretreatment	Carcinogen treatment[a]	DNA adduction (mean ± SD)		Tumour incidence	
		O^6-Ethylguanine (μmol/mol guanine)	AFB_1-$N7$-guanine (μmol/mol guanine)	No.	%
Experiment 1[b]					
Control	–	–		0/138	0
Control	NDEA	184 ± 12		105/131	80.2
I3C	NDEA	108 ± 28[d]		38/128	27.5[f]
Experiment 2[c]					
Control	–	–		0/118	0
Control	AFB_1		25.9 ± 3.2[e]	45/118	38
I3C	AFB_1		14.7 ± 1.7	5/118	4[g]

[a] NDEA, N-nitrosodiethylamine; AFB_1, aflatoxin B_1
[b] Data on NDEA taken from Fong et al. (1988); aqueous gill uptake from 250 ppm NDEA for 24 h; I3C fed at 2000 ppm in control diet for six weeks prior to NDEA
[c] Data on AFB_1 taken from Nixon et al. (1984). For carcinogenicity studies, AFB_1 was given at 20 ppb in the diet for two weeks; I3C was given at 1000 ppm for eight weeks before, two weeks during and six weeks after AFB_1 treatment. For DNA binding studies, AFB_1 was given as a single intraperitoneal injection at 3.67 μg/kg bw, I3C was given at 1000 ppm for 12 weeks prior to AFB_1, and DNA adducts were determined 24 h after AFB_1 treatment
[d] Three pools of ten fish each; significantly different from control at $p < 0.02$
[e] Four pools of three fish each; significantly different from control at $p < 0.001$
[f] Significantly different from positive NDEA control at $p < 0.000001$
[g] Significantly different from positive AFB_1 control at $p < 0.01$

A detailed study was designed to test if inhibition of AFB_1-DNA adduction by I3C could quantitatively predict reduced carcinogenic response. Tumour dose-response curves were generated for several levels of I3C, each curve generated by experiments with four doses of AFB_1 with three replicates of 150 trout at each dose. I3C was given four weeks prior to and two weeks with AFB_1, and fish were taken randomly to determine AFB_1-DNA adduct levels. The remaining fish in each group were reared for nine months to determine tumour incidence. Details of the study are reported elsewhere (Dashwood et al., 1989). With each increasing dose of I3C, the curves for logit incidence *versus* log AFB_1 dose were shifted toward higher AFB_1 dose. However, when plotted as logit incidence *versus* AFB_1-DNA adducts (i.e., dose received), all curves up to 2000 ppm I3C described the same line (Figure 1), indicating that I3C inhibition involves solely reduced AFB_1-DNA adduction and no additional mechanism need be invoked. An important consequence of this finding is that reduced DNA adduction can be used in place of full tumour studies to predict anticarcinogenic effects in this system, using doses of I3C up to that at which toxicity is evident.

Figure 1. Logit tumour incidence *versus* log aflatoxin B_1 (AFB_1)-DNA adducts formed in trout liver[a]

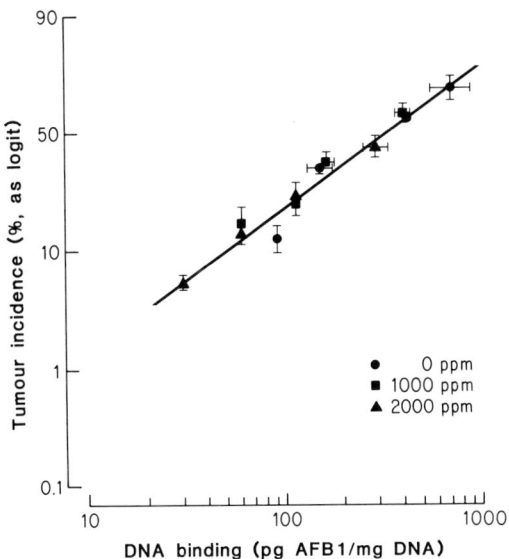

[a]Data points and bars represent means and standard errors of triplicate determinations of tumour incidence and DNA adduction for a given combination of treatment with indole-3-carbinol (I3C). See text for details of exposure. Adapted from Dashwood *et al.* (1989)

Potency of I3C inhibition versus *promotion*

A quantitative measure of the inhibitory potency of I3C can be obtained from the above study on the basis of the dose of AFB_1 required for 50% tumour response (i.e., TD_{50} value) for each I3C curve, compared to that in the absence of I3C. As shown in Figure 2, 50% inhibition of AFB_1 tumorigenicity (TD_{50} control/TD_{50} I3C = 0.5) occurred with approximately 1400 ppm I3C. An analogous study was conducted to assess the potency of I3C to promote tumours, in which trout were fed I3C continuously at 0, 750 or 1500 ppm for 24 weeks, beginning five days *after* AFB_1 treatment. This (unpublished) study is also summarized in Figure 2, where 50% 'tumour promotion' (TD_{50} I3C/$_{50}$ control = 0.5) was obtained with 1000 ppm I3C.

The capacity of I3C to promote prior initiation thus equals or exceeds its potency to inhibit concurrent treatment with AFB_1, depending on dose. However, the duration of I3C treatment was not the same in the two studies. Additional studies were therefore conducted to investigate the effects of duration, delay and intermittent treatment with I3C on promotion (Figure 3). The results can be summarized as follows: (i) promotion was less effective when I3C treatment was reduced from nine to six or three months; (ii) promotion was still significant when the onset of I3C treatment was delayed for up to six months; indeed, one to three months' delay actually enhanced the extent of promotion; (iii) promotion was reduced, but still significant, when I3C was given on alternate months; and

(iv) I3C treatments on alternate weeks or twice per week only were equally effective and resulted in significant promotion. These studies were all conducted using a high (2000 ppm) level of I3C; Figure 1 indicates that proportionately less promotion should be expected at lower doses. The effects of intermittent exposure to I3C on inhibition of carcinogenesis have not been examined.

Figure 2. Percent inhibition and percent promotion of hepatocarcinogenesis due to aflatoxin B_1 (AFB_1) in trout at various dietary concentrations of indole-3-carbinol (I3C)[a]

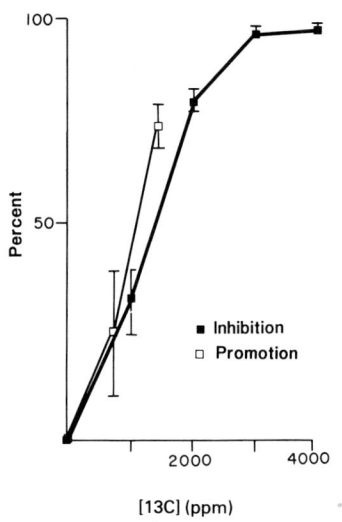

[a]Data points and bars represent means and standard errors of triplicate determinations. Percent inhibition was defined as:
$$100 (1-TD_{50control}/TD_{50x}),$$
where x = each I3C concentration. The analogous definition of percent promotion is
$$100 (1-TD_{50x}/TD_{50control}).$$
See text for details. Adapted in part from Dashwood et al. (1989)

Conclusions and future studies

Inhibition and promotion of hepatocarcinogenesis were similarly dependent on I3C dose within the range of possible human consumption; however, the dependence of these contrasting behaviours on duration of exposure to I3C may differ. Studies of inhibition mechanisms (unpublished results) predict that protection against nitrosamines or aflatoxins present in a meal can occur within hours of occasional ingestion of I3C. By contrast, promotion requires extended periods of regular (but not necessarily daily) exposure. Future studies will be carried out to understand the mechanism of promotion by I3C and its modulating effects, by experiments involving blind randomized exposure in order to mimic the irregular human patterns of carcinogen and anticarcinogen intake.

Figure 3. Influence of frequency and total duration of post-initiation treatment with indole-3-carbinol (I3C) on promotion of hepatocarcinogenesis by aflatoxin B_1 (AFB$_1$) in trout[a]

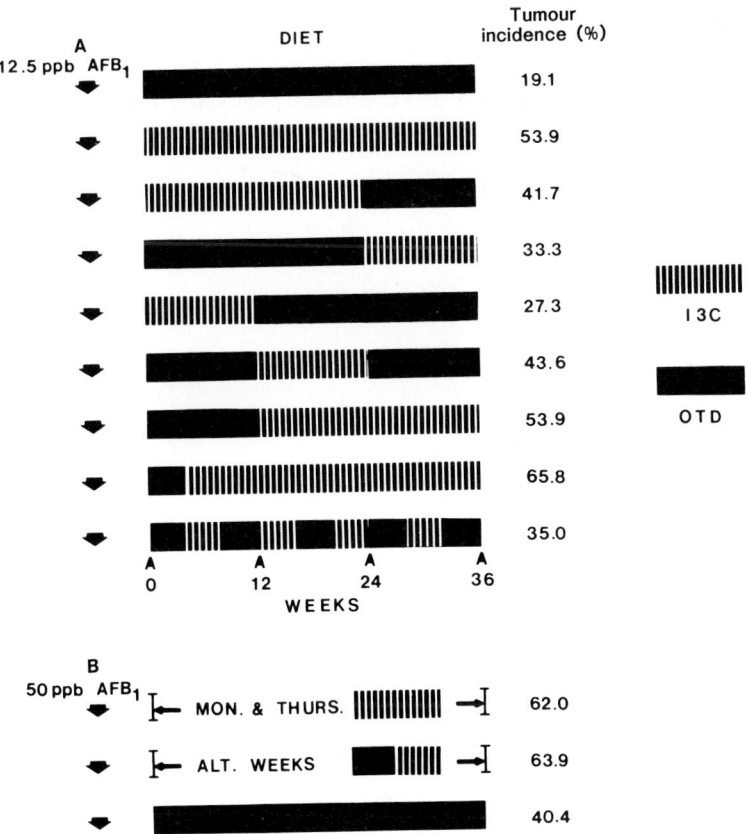

[a] Stippled bars represent periods of exposure to I3C at 2000 ppm in test diet; solid bars represent feeding of control diet only. Tumours were initiated by treatment at 12.5 ppb (A) or 50 ppb (B) AFB$_1$ for 30 min, control diet was fed for five days, and I3C treatments were then begun. Tumour incidences were determined after 36 more weeks.

Acknowledgements

Supported in part by DHHS grants ES03850, ES00210, CA34732, and CA44317

References

Dashwood, R., Arbogast, D.N., Fong, A.T., Pereira, C., Hendricks, J.D. & Bailey, G.S. (1989) Quantitative inter-relationships between aflatoxin B1 carcinogen dose, indole-3-carbinol anti-carcinogen dose, target organ DNA adduction and final tumor response. *Carcinogenesis*, **10**, 175-181

Fong, A.T., Hendricks, J.D., Dashwood, R.H., Van Winkle, S., Lee, B.C. & Bailey, G.S. (1988) Modulation of diethylnitrosamine-induced hepatocarcinogenesis and O⁶-ethylguanine formation in rainbow trout by indole-3-carbinol, β-naphthoflavone, and Aroclor 1254. *Toxicol. Appl. Pharmacol.*, **96**, 93-100

Goeger, D.E., Shelton, D.W., Hendricks, J.D., Pereira, C. & Bailey, G.S. (1988) Comparative effect of dietary butylated hydroxyanisole and β-naphthoflavone in aflatoxin B1 metabolism, DNA adduct formation, and carcinogenesis in rainbow trout. *Carcinogenesis*, **9**, 1793-1800

Nixon, J.E., Hendricks, J.D., Pawlawski, N.E., Pereira, C.B., Sinnhuber, R.O. & Bailey, G.S. (1984) Inhibition of aflatoxin B1 carcinogenesis in rainbow trout by flavone and indole compounds. *Carcinogenesis*, **5**, 615-619

Shelton, D.W., Goeger, D.E., Hendricks, J.D. & Bailey, G.S. (1986) Mechanisms of anticarcinogenesis: the distribution and metabolism of aflatoxin B1 in rainbow trout fed Aroclor 1254. *Carcinogenesis*, **7**, 1065-1072

Wattenberg, L.W. (1985) Chemoprevention of cancer. *Cancer Res.*, **45**, 1-8

Williams, D.E., Dashwood, R.D., Hendricks, J.D. & Bailey, G.S. (1989) Anti-carcinogens and tumour promoters in foods. In: Taylor, S. & Scanlan, R., eds, *Food Toxicology: A Perspective on the Relative Risks* (IFT Basic Symposium Series), New York, Marcel Dekker, pp. 101-150

/ Relevance to Human Cancer of N-Nitroso Compounds,
Tobacco Smoke and Mycotoxins.
Ed. I.K. O'Neill, J. Chen and H. Bartsch
Lyon, International Agency for Research on Cancer
© IARC, 1991

ARECA-NUT TOXICITY IN CULTURED HUMAN BUCCAL EPITHELIAL CELLS

K. Sundqvist[1], Y. Liu[1], P. Erhardt[1], J. Nair[2], H. Bartsch[2] & R.C. Grafström[1]

[1]Department of Toxicology, Karolinska Institute, Stockholm, Sweden; and [2]International Agency for Research on Cancer, Lyon, France

In cultured human buccal epithelial cells, at doses of 3-540 µg/ml, areca-nut extract significantly decreased viability, as determined by colony-forming efficiency, clonal growth rate, ability to take up neutral red and ability to exclude trypan blue, and also caused significant formation of DNA single-strand breaks and DNA protein cross-links. Comparisons of the areca nut-related compounds, 3-(N-nitrosomethylamino)propionaldehyde (NMPA), 3-(N-nitrosomethylamino)propionitrile (NMPN), N-nitrosoguvacoline, N-nitrosoguvacine, arecoline, arecaidine, guvacoline and guvacine, in terms of the above endpoints, indicate that NMPA is ten times more cytopathic to buccal cells than the other agents on a molar basis. Because metabolism of NMPA can potentially yield several reactive breakdown products, including aldehydes, this study indicates that both the parent compound and its metabolites may contribute to the observed pathobiological effects. Taken together, the observed pathobiological effects of areca-nut extract and certain related compounds in cultured human buccal epithelial cells indicate that these agents may contribute to the oral carcinogenicity associated with chewing betel quid.

Areca nut is a major ingredient of betel quid. Although several studies using laboratory animals and isolated animal cells have investigated the effects of compounds and extracts of areca nut, the possible contribution of areca nut to the carcinogenicity of betel quid has not been clarified (IARC, 1985). Several N-nitroso compounds have been found in the saliva of betel-quid chewers (Nair et al., 1985; Prokopczyk et al., 1987), and it has been shown that they can arise through nitrosation of areca-nut alkaloids (Wenke & Hoffmann, 1983; Nair et al., 1985). One of the most frequent sites of oral cancers in betel-quid chewers is the buccal epithelium (IARC, 1985). Using recently developed techniques for obtaining serum-free, replicative cultures of human buccal epithelial cells from normal non-tumorous tissue (Sundqvist et al., 1990), several pathobiological effects that may be associated with the carcinogenicity of four areca nut-specific N-nitroso compounds, four of their parent alkaloids and an aqueous areca-nut extract have now been studied directly in this target cell.

Results and discussion

The cytotoxicity of an aqueous areca-nut extract in cultured human buccal epithelial cells was determined using three assays. Survival and growth were assessed by measuring colony-forming efficiency and clonal growth rate, respectively, as described by Grafström et al. (1988). Two dye toxicity assays, neutral red uptake and trypan blue exclusion, provided information about energy status and membrane integrity, respectively, as described by Sundqvist et al. (1989).

The extract, dissolved in thiol-free growth medium without pituitary extract immediately before exposure of the cells for 3 h (Sundqvist et al., 1989), caused a dose-dependent decrease in both colony-forming efficiency and clonal growth rate at 0.3-100 µg/ml. The concentration that resulted in a 50% reduction of colony-forming efficiency was 3 µg/ml, and the concentration that reduced clonal growth rate to 50% of control was 7 µg/ml.

Higher doses were required to decrease neutral red uptake and trypan blue exclusion than for the growth-related assays. The concentration that reduced neutral red uptake to 50% of control was 160 µg/ml. The concentration that decreased trypan blue exclusion by 50% as compared to control was 540 µg/ml. Marked adherence of the extract to the cells, even after extensive rinsing, points to the possibility of continuous exposure to remaining extract during the eight days of the clonal growth assay and may explain why substantially lower concentrations of extract decreased clonal growth.

The effect of the extract on cellular thiol content was of interest, as thiols are involved in a multitude of cellular functions, e.g., detoxification of reactive agents, including active oxygen species, which can be generated by aqueous areca-nut extracts (Nair et al., 1987). The effect on the content of total low-molecular-weight thiols, a major fraction of which in cultured buccal epithelial cells is glutathione (K. Sundqvist, unpublished observation), was found to be significantly decreased at 540 µg/ml, the concentration that reduced intracellular free thiols to 75% of the control level. Because this concentration is relatively toxic, the content of sulfhydryl-reactive compounds in the extract is probably of minor importance to its cytotoxicity.

Extracts of areca nut have been found to be genotoxic in various systems *in vitro* (IARC, 1985). In human buccal epithelial cells, the extract caused both DNA single-strand breaks and DNA protein cross-links, as determined using the sensitive alkaline elution assay, as described by Grafström et al. (1988). After exposure to 300 µg/ml extract, 1.2 DNA single-strand breaks and 2.2 DNA protein cross-links per 10^{10} Da of DNA were formed in cultured cells. Dose-dependent formation of both types of lesion (Sundqvist et al., 1989) clearly indicates the presence of agents that interact with DNA, and further studies will be needed to provide information on the chemical nature of these genetic lesions and the possible involvement of DNA repair.

The effects of the areca-nut N-nitroso compounds N-nitrosoguvacoline, N-nitrosoguvacine, 3-(N-nitrosomethylamino)propionitrile (NMPN) and 3-(N-nitrosomethylamino)propionaldehyde (NMPA) on survival, thiol content and frequency of DNA single-strand breaks in buccal epithelial cells were investigated in parallel with the effects of their precursor alkaloids, arecoline, arecaidine, guvacoline and guvacine (Table 1). Direct comparisons of 50% inhibitory doses in terms of survival and of the doses that decreased the thiol content to 75% clearly indicate that NMPA is the most potent compound on a molar basis. This agent decreased colony-forming efficiency, clonal growth rate and cellular thiols in a parallel, dose-response fashion, concomitantly with the formation of DNA single-strand breaks (Figure 1); moreover, it caused DNA protein cross-links (K. Sundqvist, unpublished observation). N-Nitrosoguvacoline, arecoline and guvacoline also decreased survival, clonal growth rate and thiols, but at concentrations about ten times higher than those required for NMPA, and these compounds had only minor genotoxic effects, measured as formation of DNA single-strand breaks. N-Nitrosoguvacine and arecaidine, neither of which has been found to be carcinogenic to animals (IARC, 1985), had no significant effect on cellular growth, thiol content or DNA integrity. In contrast, NMPN, a known carcinogen in rats (Prokopczyk et al., 1987), produced none of these genotoxic effects. This finding could be due to a low level of the specific cytochrome P450-associated enzymes that are required for metabolic activation of NMPN in buccal epithelial cells.

Table 1. Cytotoxic and genotoxic effects of areca-nut N-nitroso compounds and their precursor alkaloids in human buccal epithelial cells

Compound[a]	CFE_{50}[b]	CGR_{50}[c]	$Thiol_{75}$[d]	DNA SSB[e]
N-Nitroso compounds				
N-Nitrosoguvacoline	1.7	> 5.0	2.9	0.5
N-Nitrosoguvacine	> 5.0	> 5.0	3.8	0.4
NMPA	0.15	0.36	0.08	2.6[e]
NMPN	> 5.0	> 5.0	> 5.0	0.2
Alkaloids				
Arecoline	1.6	4.7	0.57	0.5
Arecaidine	> 5.0	> 5.0	> 5.0	0.2
Guvacoline	2.1	5.0	0.88	0.5
Guvacine	> 5.0	> 5.0	> 5.0	0.3

[a] NMPA, 3-(N-nitrosomethylamino)propionaldehyde; NMPN, 3-(N-nitrosomethylamino)propionitrile
[b] CFE_{50}, concentration that resulted in a 50% reduction in colony-forming efficiency
[c] CGR_{50}, concentration that decreased clonal growth rate by 50%
[d] $Thiol_{75}$, concentration that decreased cellular free thiol content to 75% of untreated control cells
[e] Number of DNA single-strand breaks/10^{10} Da of DNA formed after exposure to 5.0 mM of the compound indicated, except for NMPA where 0.3 mM was used due to high toxicity

This comparison of the cytopathic effects of N-nitroso compounds with those of their non-nitrosated analogues indicates that the methylester group is of greater importance than the nitroso group in causing cytotoxicity and thiol depletion in buccal epithelial cells. Of the defined areca-nut compounds investigated, NMPA was the most potent on a molar basis in decreasing survival and thiol content, and was the only one to cause significant DNA damage. The aldehyde moiety of NMPA, or possible aldehyde breakdown products (Figure 2), may be causative in this regard, as aldehydes such as formaldehyde and acrolein are cytotoxic, thiol reactive and also cause several types of DNA damage in cultured human cells (Grafström et al., 1987; Grafström, 1990). However, further studies are required to elucidate to what extent areca nut-related *N*-nitroso compounds are metabolized by cultured human buccal epithelial cells. Our results indicate the usefulness of serum-free cultures of normal human buccal epithelial cells for investigating the relationships of areca nut-related pathobiological effects, betel-quid chewing and carcinogenesis.

Figure 1. Effect of 3-(*N*-nitrosomethylamino)propionaldehyde on colony survival, clonal growth, thiol content and formation of DNA damage. (▲) Total free cellular thiols); (●) colony-forming efficiency; (■) clonal growth rate; (◆) DNA single-strand breaks.

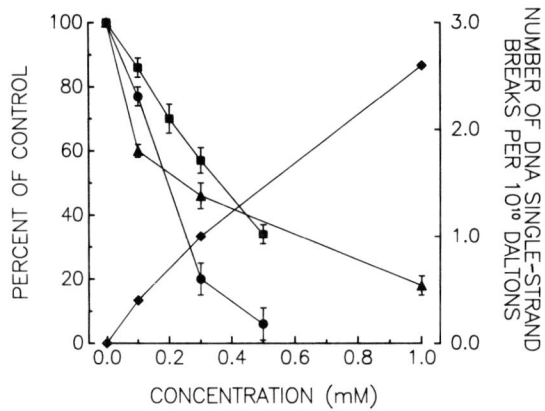

Figure 2. Potential metabolic breakdown products of 3-(*N*-nitrosomethylamino)-propionaldehyde (NMPA)

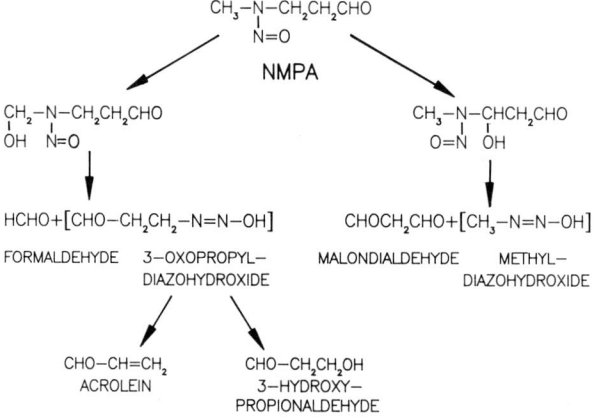

Acknowledgements

This work was supported in part by grants from the Swedish National Board of Laboratory Animals, the Swedish Tobacco Company, the Swedish Cancer Society, the Swedish Natural Science Resource Council, the Swedish Fund for Scientific Research without Animal Experiments, and by NIH grant No. 1001 CA 43176-01 from the US Public Health Service.

References

Grafström, R.C. (1990) In vitro studies of aldehyde effects related to human respiratory carcinogenesis. *Mutat. Res.*, **238**, 175-184

Grafström, R.C., Sundqvist, K., Dypbukt, J.M. & Harris, C.C. (1987) Pathobiological effects of aldehydes in cultured human bronchial cells. In: Bartsch, H., O'Neill, I.K. & Schulte-Hermann, R., eds, *The Relevance of N-Nitroso Compounds to Human Cancer: Exposures and Mechanisms* (IARC Scientific Publications No. 84), Lyon, IARC, pp. 443-445

Grafström, R.C., Dypbukt, J.M., Willey, J.C., Sundqvist, K., Edman, C.C., Atzori, L. & Harris, C.C. (1988) Pathobiological effects of acrolein in cultured human bronchial epithelial cells. *Cancer Res.*, **48**, 1717-1721

IARC (1985) *IARC Monographs on the Evaluation of the Carcinogenic Risk of Chemicals to Humans*, Vol. 37, *Tobacco Habits Other than Smoking; Betel-quid and Areca-nut Chewing; and Some Related Nitrosamines*, Lyon

Nair, J., Ohshima, H., Friesen, M., Croisy, A., Bhide, S.V. & Bartsch, H. (1985) Tobacco-specific and betel nut-specific N-nitroso compounds in saliva and urine of betel quid chewers and formation in vitro by nitrosation of betel quid. *Carcinogenesis*, **6**, 295-303

Nair, U.J., Floyd, R.A., Nair, J., Bussachini, V., Friesen, M. & Bartsch, H. (1987) Formation of reactive oxygen species and 8-hydroxydeoxyguanosine in DNA in vitro with betel quid ingredients. *Chem.-biol. Interactions*, **63**, 157-169

Prokopczyk, B., Rivenson, S., Bertinato, P., Brunnemann, K.D. & Hoffmann, D. (1987) 3-(Methylnitrosamino)propionitrile: occurrence in saliva of betel quid chewers, carcinogenicity and DNA methylation in F344 rats. *Cancer Res.*, **47**, 467-471

Sundqvist, K., Liu, Y., Nair, J., Bartsch, H., Arvidson, K. & Grafström, R.C. (1989) Cytotoxic and genotoxic effects of areca nut-related compounds in cultured human buccal epithelial cells. *Cancer Res.*, **49**, 5294-5298

Sundqvist, K., Liu, Y., Arvidson, K., Ormstad, K., Nilsson, L. & Grafström, R.C. (1990) Growth regulation of serum-free cultures of epithelial cells from normal human buccal mucosa (submitted)

Wenke, G. & Hoffmann, D. (1983) A study of betel quid carcinogenesis. 1. On the in vitro nitrosation of arecoline. *Carcinogenesis*, **4**, 169-172

BIOACTIVATION OF ASYMMETRIC N-DIALKYL-NITROSAMINES IN RAT TISSUES DERIVED FROM THE VENTRAL ENTODERM

B. Ludeke, T. Meier & P. Kleihues

Institute of Pathology, University of Zurich, CH-8091 Zurich, Switzerland

Aliphatic methylalkylnitrosamines with a chain length of three to six carbon atoms are powerful oesophageal carcinogens in rats and have been shown to methylate target organ DNA preferentially. This class of carcinogens is efficiently metabolized not only in the oesophageal mucosa but also in the mucosa of the nasal and oral cavity, trachea and bronchioli, i.e., tissues derived from the rat ventral entoderm. In order to determine whether more than one cytochrome P450 isozyme is involved in the bioactivation of asymmetrical aliphatic dialkylnitrosamines in these tissues, we have studied the effects of various modulators of nitrosamine metabolism, including dietary zinc deficiency, ethanol and disulfiram, on DNA alkylation by N-nitrosomethyl-n-butylamine (NMBA) and its ethyl analogue N-nitrosoethyl-n-butylamine (NEBA). Formation of O^6-methyl- and O^6-ethyldeoxyguanosine by a single dose of NMBA and NEBA, respectively, was quantified after a survival time of 6 h by immuno-slot-blot assay. In control rats, methylation of DNA by NMBA was highest in oesophagus, followed by nasal mucosa, liver and lung. Formation of O^6-ethyldeoxyguanosine from NEBA, however, was twice as high in liver as in nasal mucosa and lung and four times as high in liver as in oesophagus. In oesophagus, trachea and bronchioli, both nitrosamines were selectively metabolized in mucosal cells. Bioactivation of NMBA and NEBA was almost completely inhibited in nasal mucosa by ethanol. In contrast, a striking interorgan shift in DNA methylation by NMBA from liver (−50%) to lung (+100%), oesophagus (+300%) and nasal mucosa (+400%) was obtained with dietary disulfiram, whereas only 20–50% increases in extrahepatic DNA ethylation were determined for NEBA. Similarly elevated levels of extrahepatic DNA alkylation by NMBA and NEBA were observed in zinc-deficient rats. The variations in the extent to which bioactivation of NMBA and NEBA was modulated suggest the presence of a family of closely related P450 isozymes with similar but not identical substrate specificities and kinetic characteristics.

The extensive, systematic studies of Druckrey and co-workers (Druckrey *et al.*, 1967) showed that asymmetrical N-nitrosomethylalkylamines are the most powerful oesophageal carcinogens in rats. Decomposition of these compounds to the ultimate carcinogen, an alkyldiazonium ion, is initiated by α-C hydroxylation of one of the alkyl groups in a reaction catalysed by cytochrome P450 (P450) isozymes. Although enzymatic attack can occur at either of the substituents, studies on DNA alkylation by N-nitrosomethylalkylamines have shown that by far the most significant modification is methylation (Schweinsberg & Kouros, 1979; von Hofe & Kleihues, 1986; von Hofe et al., 1986a). Extensive hydroxylation can also occur at any of the other carbon atoms of aliphatic nitrosamines prior to conversion to a methylating agent, but the more polar intermediates so formed may be secreted more readily in the urine than the parent compound (Mirvish et al., 1985; von Hofe et al., 1986b). In rat oesophagus, DNA methylation by aliphatic N-nitrosomethyl-n-alkylamines closely parallels carcinogenicity, at least with respect to the propyl, butyl and pentyl derivatives but not to N-nitrosomethyl-n-hexylamine (von Hofe et al., 1987), suggesting that the oesophageal mucosa contains a distinct P450 isozyme capable of hydroxylating asymmetrical nitrosamines at the α-methylene of alkyl groups bearing three to five carbon atoms.

We have now extended these investigations to compare the organ-specific bioactivation of N-nitrosomethyl-n-butylamine (NMBA) and its ethyl homologue N-nitrosoethyl-n-butylamine (NEBA). These compounds induce primarily oesophageal and nasal neoplasms and, at a lower incidence, hepatic and pulmonary (with NEBA) tumours in rats (Druckrey et al., 1967, 1968; Druckrey & Landschütz, 1971), presumably by DNA methylation or ethylation (Lijinsky et al., 1980). In order to examine whether more than one P450 isozyme with activity for asymmetrical alkylnitrosamines is present in these tissues, we compared the effects of dietary zinc deficiency, ethanol and disulfiram (tetraethylthiuram disulfide), known modulators of nitrosamine metabolism, on DNA methylation by NMBA and ethylation by NEBA. In addition, we have used immunohistochemical techniques to identify specific cell populations within the target tissues that are involved in bioactivation of these carcinogens.

Materials and methods

Cremophor EL solubilizer, disulfiram and NEBA were obtained from commercial sources. N-Nitrosomethylamylamine (NMAA) and NMBA were synthesized by Dr W. Lijinsky, Frederick Cancer Research Facility, Frederick, MD (USA). N-Nitroso[methyl-14C]amylamine ([methyl-14C]NMAA) was synthesized by Dr M. Wiessler, German Cancer Center, Heidelberg (FRG). Standard and zinc-deficient rat chows were supplied by Klingentalmühle AG, Basel (Switzerland). All commercial chemicals were of analytical grade.

Male Fischer 344 rats weighing 110–150 g were maintained on commercial diets with distilled water ad libitum. The experiments with [methyl-14C]NMAA are detailed in a previous report (Koenigsmann et al., 1988). The effect of dietary ethanol, zinc deficiency and disulfiram on DNA methylation and ethylation by NMBA and NEBA, respectively, was examined in four groups of ten rats for each compound. The control group was maintained on a standard diet containing 40 ppm zinc; a second group of animals was also fed a standard diet but received 1 ml of a 5% (v/v) solution of ethanol/100 g body weight (bw) by gavage 20 min prior to administration of the nitrosamine. The third group was maintained on a diet containing only 5 ppm zinc for ten weeks, while the fourth group was fed 0.2 g disulfiram/kg of standard chow for four weeks.

Nitrosamines were dissolved in distilled water containing 0.25% (v/v) Cremophor EL solubilizer and administered by intraperitoneal injection at a dose of 0.1 mmol NMBA/kg bw or 1 mmol NEBA/kg bw in a volume of 1 ml/100 g bw. After 6 h, the animals were killed by exsanguination under ether anaesthesia. Tissues were removed rapidly, frozen in liquid nitrogen and stored at −70°C until analysis.

DNA was isolated by phenolic extraction and adsorption onto hydroxylapatite as described previously (von Hofe et al., 1986a). The amounts of O^6-methyldeoxyguanosine (MedGuo) and O^6-ethyldeoxyguanosine (EtdGuo) produced by NMBA and NEBA, respectively, were determined by immuno-slot-blot assay using rabbit antisera raised against keyhole limpet haemocyanin conjugates of O^6-methylguanosine (NPZ 193-1; 1:8'000) and O^6-ethylguanosine (NPZ 105-1; 1:6'000), as described elsewhere (Ludeke & Kleihues, 1988), except that an alkaline phosphatase-conjugated goat-anti-rabbit IgG antibody, 5-bromo-4-chloro-3-indolylphosphate and nitroblue tetrazolium were used to visualize bound anti-O^6-alkyldeoxyguanosine antibodies (Ey & Ashman, 1986). The rabbit anti-O^6-alkyldeoxyguanosine antisera were used at dilutions of 1:6000 and 1:8000.

Immunohistochemistry for O^6-methyl- and O^6-ethyldeoxyguanosine

The organs were removed rapidly and quickly frozen onto small aluminium plates placed directly on slabs of dry ice. Staining for MedGuo and EtdGuo was carried out according to the procedure of Heyting et al. (1983) and Menkveld et al. (1985) with the modifications outlined by Koenigsmann et al. (1988). The antisera were diluted 1:20'000 and used without prior absorption.

Autoradiography

As reported previously (Koenigsmann et al., 1988), animals received an intraperitoneal injection of hydroxyurea (500 mg/kg) to inhibit metabolic incorporation of ^{14}C into the C-1 pool, and a single dose of [methyl-^{14}C]NMAA (9.14 mCi/mmol; 0.1 mmol/kg) was injected 30 min later. After an additional interval of 2 h, animals were sacrificed and tissues fixed in buffered 4% formaldehyde. Paraffin-embedded sections were used to produce contact autoradiographs (LKB 3H-Ultrofilm).

DNA methylation by N-nitrosomethylamylamine (NMAA)

Following a single intraperitoneal administration of [methyl-^{14}C]NMAA at a dose of 13 mg/kg, DNA methylation was detectable in all tissues investigated (Koenigsmann et al., 1988). Concentrations of 7-methylguanine were highest in oesophagus (798 μmol/mol guanine), nasal cavity (672 μmol) and liver (624 μmol), followed by trachea (214 μmol) and lung (101 μmol/mol). Levels in forestomach and kidney were approximately 30-40 times lower than those in oesophagus. In the glandular stomach, spleen and duodenum, the amount of 7-methylguanine was close to the limit of detection. Quantifiable amounts of O^6-methylguanine were present only in oesophagus, nasal cavity, liver, trachea and lung. In these tissues, the O^6-:7-methylguanine ratio was close to 0.11, except in liver (0.08), indicating that in extrahepatic tissues no significant repair of O^6-methylguanine occurred during the interval of 2 h.

In order to study the site of NMAA bioactivation at the cellular level, histoautoradiographic examinations were carried out in rats that had received a similar dose of [methyl-14C]NMAA. In the upper thoracic cavity, including the mediastinum, incorporation of ^{14}C into macromolecules was most accentuated in the oesophageal mucosa. Other target sites easily identifiable were the trachea and bronchial tree; the remaining lung parenchyma showed diffuse and considerably less marked incorporation of radioactivity. In the nasal cavity, intense radiolabelling was present both in the olfactory and respiratory epithelia and in the lateral nasal glands (Steno's glands). Unexpectedly, we also observed a considerable amount of radioactivity in the oral mucosa (Figure 1). These cell types are known to be susceptible to malignant transformation by NMAA (Bulay & Mirvish, 1979).

Modulation of N-nitrosomethyl-n-butylamine (NMBA) and N-nitrosoethyl-n-butylamine (NEBA) metabolism in rat tissues

In control animals, concentrations of MedGuo formed by a single intraperitoneal dose of NMBA (0.1 mmol/kg) were highest in the oesophagus (107 μmol/mol deoxyguanosine), followed by nasal mucosa, liver and lung (Figure 2, top). Animals received a higher dose of NEBA (1 mmol/kg bw) because the yield of alkylated bases was much lower with this nitrosamine. In contrast to NMBA, the extent of base alkylation by NEBA did not correlate with the preferred sites of tumour induction: the levels of EtdGuo in liver (65 μmol/mol deoxyguanosine) were twice as high as those in lung and nasal mucosa and four times as high as those in the oesophagus (Figure 2, bottom). Both compounds produced only trace amounts of O^6-alkylguanines in renal DNA (data not shown).

Figure 1. Histoautoradiograph of a frontal section through the nose and oral cavity of a male F344 rat following a single intraperitoneal dose of N-nitroso[methyl-14C]amylamine (0.1 mmol/kg bw; survival time, 2 h)[a]

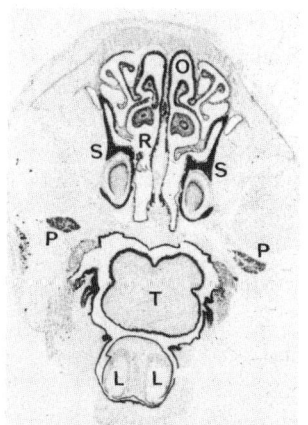

[a]In order to suppress DNA synthesis, the animal was given a single intraperitoneal injection of hydroxyurea 30 min before nitrosamine application. × 3.8

L, lower incisors; O, olfactory epithelium; P, parotid gland; R, respiratory epithelium; S, Steno's (lateral nasal) glands; T, tongue

Dietary zinc deficiency, ethanol and disulfiram have been reported to cause significant increases in the incidence of tumour induction by oesophageal carcinogens (Gibel, 1967; Fong et al., 1978; Schweinsberg & Bürkle, 1981). Pretreatment with these modulators of nitrosamine metabolism, except ethanol, led to significantly elevated levels of DNA alkylation by NMBA and NEBA in rat oesophagus and nasal cavity (Figure 2). Chronic dietary zinc deficiency led to increases of 25-50% in the formation of MedGuo from NMBA in all tissues. Similar increases were observed in the extent of ethylation by NEBA in oesophagus and nasal mucosa, but there was no significant effect on DNA alkylation in lung or liver. The elevated levels of oesophageal DNA alkylation agree well with the nearly ten-fold increase in the rate of N-nitrosomethylbenzylamine metabolism by oesophageal microsomes from zinc-deficient rats (Barch et al., 1984). Those authors also observed a 1.5-fold increase in the rate of hepatic nitrosamine metabolism, which correlates well with the increase in liver DNA alkylation seen with dietary zinc deficiency that we observed with NMBA but not with NEBA. It has been suggested that zinc can interact with P450s to decrease substrate binding (Jeffrey, 1983).

Figure 2. Modulation of the metabolism of *N*-nitrosomethyl-*n*-butylamine (NMBA) and *N*-nitrosoethyl-*n*-butylamine (NEBA) in various rat tissues[a]

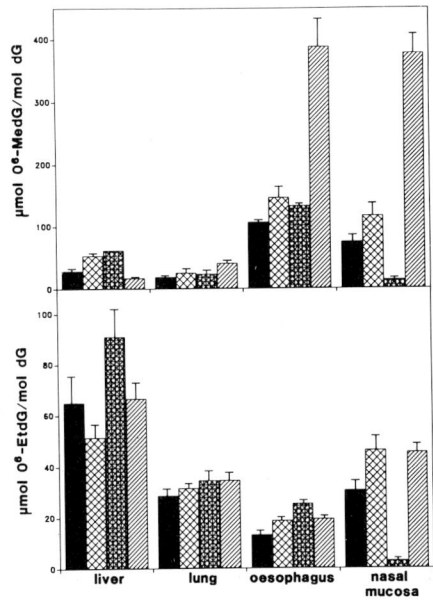

[a]Rats were maintained on a standard diet containing 40 ppm zinc (■), on a diet containing 5 ppm zinc (▩), on a standard diet but received 1 ml/100 g bw of a 5% (v/v) solution of ethanol 20 min before administration of the respective nitrosamine (▨), or on a standard diet containing disulfiram (0.2 g/kg; ▥). The nitrosamines were administered at doses of 0.1 mmol/kg (NMBA) and 1 mmol/kg (NEBA) by intraperitoneal injection, and animals were killed 6 h later. Concentrations of O^6-methyldeoxyguanosine (MedGuo) and O^6-ethyldeoxyguanosine (EtdGuo) are expressed as μmol/mol deoxyguanosine (dGuo)

Administration of 1 ml/100 g bw of a 5% (v/v) solution of ethanol 20 min before the nitrosamine resulted in a decrease of more than 80% in the extent of both methylation by NMBA and ethylation by NEBA in DNA of nasal mucosa, with concomitant increases in DNA alkylation of 25-100% in the oesophagus and liver. This result was unexpected, as ethanol has been shown to be a competitive inhibitor of hepatic nitrosamine metabolism (Peng et al., 1982), which shifts the balance of metabolism to extrahepatic P450 isozymes that are not inhibited by low concentrations of ethanol (Swann et al., 1984). Thus, the increase in DNA alkylation in oesophagus by N-nitrosomethylethlamine and N-nitrosomethylbenzylamine that is induced by ethanol is paralleled by a significant decrease in alkylation in liver (Wiestler et al., 1987).

In animals maintained on a diet containing disulfiram, we observed a striking interorgan shift in the metabolism of NMBA toward extrahepatic tissues, with one- to five-fold increases in methylation in lung, oesophagus and nasal mucosa that were offset by a 50% decrease in liver. In contrast, the effects of disulfiram on NEBA metabolism were of the same magnitude as those of dietary zinc deficiency. Clearly elevated levels of EtdGuo were observed in oesophagus and nasal mucosa in treated animals, but levels in pulmonary and hepatic DNA were not significantly different from those in controls. Dietary disulfiram has been shown to inhibit bioactivation of N-nitrosomethylbenzylamine in rat liver but leads to the induction of pulmonary enzyme activity, to result in a similar shift in DNA methylation from liver to lung (Schweinsberg et al., 1984).

Cell-specific bioactivation of NMBA and NEBA in extrahepatic tissues

In contrast to biochemical determinations, which usually yield data on alkylation in total DNA isolated from homogenized tissues, immunocytochemical localization of O^6-alkyldeoxyguanosines allows the identification of specific cell populations involved in the enzymatic activation of N-nitroso compounds. In rat oesophagus, DNA methylation and ethylation by NMBA and NEBA, respectively, was observed in the mucosa proper. In the lower respiratory tract, staining was confined to tracheal, bronchial and bronchiolar epithelia. Staining for EtdGuo after treatment with NEBA was similar in tracheal mucosa and bronchioli and was much more intense than in the oesophageal mucosa (Figure 3), whereas methylation by NMBA was more uniform in these epithelia (data not shown).

The organ-specific distribution, the relative extent of DNA alkylation by NMBA and NEBA and the different responses to the modulatory effect of disulfiram on their activation indicate that the P450s expressed in tissues derived from the ventral entoderm belong to a family of closely related but distinct isozymes with individual substrate specificities and kinetic characteristics. Immunohistochemical localization of O^6-alkyldeoxyguanosines strongly suggests that several isozymes capable of activating asymmetrical N-nitrosodialkylamines are expressed in extrahepatic target cell populations.

Figure 3. Immunocytochemical localization of O^6-ethyldeoxyguanosine in the oesophageal mucosa (left) and bronchiolus (right) of a male zinc-deficient F344 rat 6 h after a single intraperitoneal injection of N-nitrosoethyl-n-butylamine (1 mmol/kg)[a]

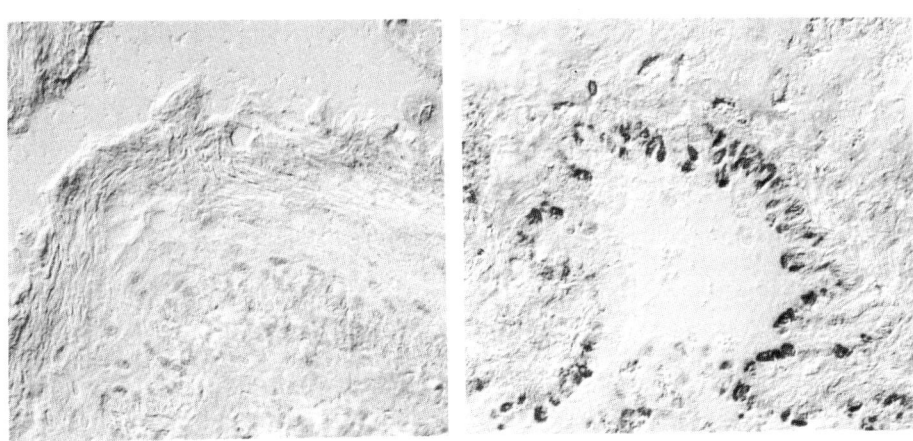

[a]Nomarsky interference contrast microscopy, × 210

References

Barch, D.H., Kuemmerle, S.C., Hollenberg, P.F. & Iannaccone, P.M. (1984) Esophageal microsomal metabolism of N-nitrosomethylbenzylamine in the zinc-deficient rat. *Cancer Res.*, **44**, 5629-5633

Bulay, O. & Mirvish, S. (1979) Carcinogenesis in rat oesophagus by intraperitoneal injection of different doses of methyl-n-amylnitrosamine. *Cancer Res.*, **39**, 3644-3646

Druckrey, H. & Landschütz, C. (1971) Carcinome der Nase bei Ratten nach chronischer Inhalation von 0.05 ppm Methyl-butylnitrosamin. *Z. Krebsforsch.*, **75**, 221-224

Druckrey, H., Preussmann, R., Ivankovic, S. & Schmähl, D. (1967) Organotrope carcinogene Wirkungen bei 65 verschiedenen N-Nitroso-Verbindungen an BD-Ratten. *Z. Krebsforsch.*, **69**, 103-201

Druckrey, H., Landschütz, C. & Preussmann, R. (1968) Oesophagus-Carcinome nach Inhalation von Methyl-butyl-nitrosamin (MBNA) an Ratten. *Z. Krebsforsch.*, **71**, 135-139

Ey, P.L. & Ashman, L.K. (1986) The use of alkaline phosphatase-conjugated anti-immunoglobulin with immunoblots for determining the specificity of monoclonal antibodies to protein mixtures. In: Langone, J.J. & Van Vunakis, H., eds, *Methods in Enzymology*, Vol. 121, New York, Academic Press, pp. 497-509

Fong, L.Y.Y., Sivak, A. & Newberne, P.M. (1978) Zinc deficiency and methylbenzylnitrosamine-induced esophageal cancer in rats. *J. Natl Cancer Inst.*, **61**, 145-150

Gibel, W. (1967) Experimentelle Untersuchungen zur Synkarzinogese beim Ösophaguskarzinom. *Arch. Geschwulstforsch.*, **30**, 181-189

Heyting, C., Van der Laken, C.J., Van Raamsdonk, W. & Pool, C.W. (1983) Immunohistochemical detection of O^6-ethyldeoxyguanosine in the rat brain after in vivo applications of N-ethyl-N-nitrosourea. *Cancer Res.*, **43**, 2935-2941

von Hofe, E. & Kleihues, P. (1986) Comparative studies on hepatic DNA alkylation in rats by N-nitrosomethylethylamine and N-nitrosodimethylamine plus N-nitrosodiethylamine. *J. Cancer Res. Clin. Oncol.*, **112**, 205-209 von Hofe, E., Grahmann, F., Keefer, L.K., Lijinsky, W., Nelson, V. & Kleihues, P. (1986a) Methylation *versus* ethylation of DNA in target and nontarget tissues of Fischer 344 rats treated with N-nitrosomethylethylamine. *Cancer Res.*, **46**, 1038-1042

von Hofe, E., Kleihues, P. & Keefer, L.K. (1986b) Extent of DNA 2-hydroxyethylation by N-nitrosomethylethylamine and N-nitrosodiethylamine *in vivo*. *Carcinogenesis*, **7**, 1335-1337

von Hofe, E., Schmerold, I., Lijinsky, W., Jeltsch, W. & Kleihues, P. (1987) DNA methylation in rat tissues by a series of homologous aliphatic nitrosamines ranging from N-nitrosodimethylamine to N-nitrosomethyldodecylamine. *Carcinogenesis*, **8**, 1337-1341

Jeffrey, E.H. (1983) The effect of zinc on NADPH oxidation and monooxygenase activity in rat hepatic microsomes. *Mol. Pharmacol.*, **23**, 467-473

Koenigsmann, M., Schmerold, I., Jeltsch, W., Ludeke, B., Kleihues, P. & Wiessler, M. (1988) Organ and cell specificity of DNA methylation by N-nitrosomethylamylamine in rats. *Cancer Res.*, **48**, 5482-5486

Lijinsky, W., Reuber, M.D., Saavedra, J.E. & Blackwell, B.-N. (1980) The effect of deuterium on the carcinogenicity of nitroso-methyl-*n*-butylamine. *Carcinogenesis*, **1**, 157-160

Ludeke, B.I. & Kleihues, P. (1988) Formation and persistence of O^6-(2-hydroxyethyl)-2'-deoxyguanosine in DNA of various rat tissues following a single dose of N-nitroso-N-(2-hydroxyethyl)urea. An immuno-slot-blot study. *Carcinogenesis*, **9**, 147-151

Menkveld, G.J., Van Der Laken, C.J., Hermsen, T., Kriek, E., Scherer, E. & Den Engelse, L. (1985) Immunohistochemical localization of O^6-ethyldeoxyguanosine and deoxyguanosin-8-yl(acetyl)-aminofluorene in liver sections of rats treated with diethylnitrosamine, ethylnitrosourea or N-acetylaminofluorene. *Carcinogenesis*, **6**, 263-270

Mirvish, S.S., Wang, M.-Y., Smith, J.W., Deshpande, A.D., Makary, M.H. & Issenberg, P. (1985) β- to ω-Hydroxylation of the esophageal carcinogen methyl-*n*-amylnitrosamine by the rat esophagus and related tissues. *Cancer Res.*, **45**, 577-583

Peng, R., Yong, T.Y. & Yang, C.S. (1982) The induction and competitive inhibition of a high affinity microsomal nitrosodimethylamine demethylase by ethanol. *Carcinogenesis*, **3**, 1457-1461

Schweinsberg, F. & Bürkle, V. (1981) Wirkung von Disulfiram auf die Toxizität und Carcinogenität von N-Methyl-N-nitrosobenzylamin bei Ratten. *J. Cancer Res. Clin. Oncol.*, **102**, 43-47

Schweinsberg, F. & Kouros, M. (1979) Reactions of N-methyl-N-nitrosobenzylamine and related substrates with enzyme-containing cell fractions isolated from various organs of rats and mice. *Cancer Lett.*, **7**, 115-120

Schweinsberg, F., Weissenberger, I., Brückner, B., Schweinsberg, E., Bürkle, V., Wittenberg, H. & Reinecke, H.-J. (1984) Effect of disulfiram on N-nitroso-N-methylbenzylamine metabolism. Biochemical aspects. In: O'Neill, I.K., von Borstel, R.C., Miller, C.T., Long, J. & Bartsch, H., eds, N-*Nitroso Compounds: Occurrence, Biological Effects and Relevance to Human Cancer* (IARC Scientific Publications No. 57), Lyon, IARC, pp. 525-532

Swann, P.F., Coe, A.M. & Mace, R. (1984) Ethanol and dimethylnitrosamine and diethylnitrosamine metabolism in the rat. Possible relevance to the influence of ethanol on human cancer incidence. *Carcinogenesis*, **5**, 1337-1343

Wiestler, O.D., von Deimling, A., von Hofe, E., Schmerold, I., Wiestler, E. & Kleihues, P. (1987) Interorgan shift of nitrosamine metabolism by dietary ethanol. *Arch. Toxicol.*, Suppl. 11, 53-65

Relevance to Human Cancer of *N*-Nitroso Compounds,
Tobacco Smoke and Mycotoxins.
Ed. I.K. O'Neill, J. Chen and H. Bartsch
Lyon, International Agency for Research on Cancer
© IARC, 1991

ROLE OF ONCOGENES AND TUMOUR SUPPRESSOR GENES IN HUMAN LUNG CARCINOGENESIS

C.C. Harris[1,3], R. Reddel[1,4], A. Pfeifer[1,5], D. Iman[1], M. McMenamin[1], B.F. Trump[3] & A. Weston[1]

[1]*Laboratory of Human Carcinogenesis, Division of Cancer Etiology, National Cancer Institute, National Institutes of Health, Bethesda, MD; and* [2]*Department of Pathology, University of Maryland School of Medicine, Baltimore, MD, USA*

Six families of activated protooncogenes, *ras, raf, fur, neu, jun* and *myc* have so far been associated with human lung cancer. Human bronchial epithelial cells *in vitro* are being used to investigate the functional role of these specific oncogenes and growth regulatory genes in carcinogenesis and tumour progression. When transferred into normal human bronchial epithelial cells by the highly efficient protoplast fusion method, the v-Ha-*ras* oncogene initiates a cascade of events leading to decreased responsiveness of these cells to inducers of squamous differentiation, aneuploidy and, less frequently, 'immortality' and tumorigenicity with metastasis in athymic nude mice. Transfection of the SV40 T antigen gene results in nontumorigenic cell lines that have a nearly normal pathway of terminal squamous differentiation and can be transformed into malignant cells by transfected Ha-*ras*, N-*ras* or Ki-*ras*. The combination of transfected c-*myc* and c-*raf*-1 also transforms human bronchial epithelial cells into neoplastic cells that exhibit some phenotypic traits found in small-cell carcinomas. These and other results indicate that proto-oncogenes dysregulate the pathways of growth and differentiation of human bronchial epithelial cells and play an important role in human carcinogenesis.

Analyses of allelic deletion and somatic cell hybrids are being used to identify the chromosomal localization of tumour suppressor genes. We have examined 54 non-small-cell bronchogenic carcinomas with 13 polymorphic markers. Loss of heterozygosity was more frequent than among 23 squamous-cell carcinomas than among 23 adenocarcinomas or eight large-cell carcinomas. Loss of heterozygocity for chromosome 17p was found in 89% of cases of squamous-cell carcinoma and 18% of adenocarcinomas. Analysis of chromosome 11 for allelic deletions revealed two commonly deleted regions (11p13 and 11p15.5). Somatic cell hybrids between normal human bronchial epithelial cells

---
[3]To whom correspondence should be addressed
[4]Present address: Children's Medical Research Foundation, PO Box 61, Camperdown, NSW 2050, Australia
[5]Present address: Genetic Toxicology Section, Nestle Research Centre, Vers-Chez-Leblanc, 1026 Lausanne, CH-1800, Vevey, Switzerland

and Hut292DM, a lung carcinoma cell line, had a finite lifespan *in vitro* and were nontumorigenic in athymic nude mice. Tumour suppressive effects of individual or combinations of specific human chromosomes on Hut292DM are being examined by formation of microcell-cell hybrids. Chromosome 11 has tumour suppressor activity in these hybrids. Both of these studies suggest that tumour suppressor genes play a dominant role in lung carcinogenesis and provide in-vitro model systems for isolating these genes by subtraction library and insertional mutagenesis techniques.

Neoplastic transformation of human bronchial epithelial cells

The strategy that we have formulated for investigating the neoplastic transformation of normal human bronchial epithelial cells is as follows:

(i) Select activated proto-oncogenes associated with human lung cancer.

(ii) Transfer activated proto-oncogenes into the progenitor epithelial cells of bronchogenic carcinoma.

(iii) Select preneoplastic and neoplastic cells from putative suppressive normal cells.

(iv) Investigate dysregulation of molecular controls of growth and terminal differentiation.

(v) Determine tumorigenic potential in athymic nude mice.

Six families of activated proto-oncogenes, *ras*, *raf*, *jun*, *fur*, *neu* and *myc*, have so far been associated with human lung cancer. Since association does not necessarily indicate causation, human bronchial epithelial cells *in vitro* are being used to investigate the functional role of these genes in carcinogenesis. When transferred into normal human bronchial epithelial cells by protoplast fusion, the v-Ha-*ras* oncogene initiates a cascade of events that results in their decreased responsiveness to inducers of squamous differentiation, aneuploidy, increased lifespan and, less frequently, 'immortality' and tumorigenicity with metastasis in athymic nude mice (Yoakum *et al.*, 1985). Genomic instability and apparent trans-activation of certain genes, e.g., procollagenase IV, may be important mechanisms leading to the phenotypic changes observed in v-Ha-*ras* transformed cells.

The experiments described above and others (Rhim *et al.*, 1985; Namba *et al.*, 1986; Pater & Pater, 1986; Amstad *et al.*, 1988; Byrd *et al.*, 1988; Clark *et al.*, 1988; Reddel *et al.*, 1988a; Reznikoff *et al.*, 1988; Seremetis *et al.*, 1989) indicate that 'immortalization' is a rate-limiting step in the multistage process of carcinogenesis in human cells *in vitro*. Therefore, we have developed human bronchial epithelial cell lines by transferring the SV40 T antigen gene (Reddel *et al.*, 1988b). Aneuploidy is a common feature of these and other 'immortalized' human epithelial cells, and alterations in growth and in squamous differentiation pathways can also be observed frequently.

Activated *ras* oncogenes have been identified in human bronchial carcinomas, especially adenocarcinomas (Rodenhuis *et al.*, 1987). In most cases, the activated *ras* genes have been Ki-*ras*, but activated N-*ras* and Ha-*ras* oncogenes have been described in human lung cancer cell lines (Yuasa *et al.*, 1983, 1984). In an experimental model system, a mutated *ras* (activated human Ha-*ras* from the EJ bladder carcinoma cell line) gene linked to an immunoglobulin gene enhancer/SV40 T promoter induced multicentric adenomatous tumours (comparable to well-differentiated adenocarcinomas of the lung in man) in transgenic mice (Suda *et al.*, 1987).

Immortalized human bronchial epithelial cell lines have been developed in order to define conditions under which *ras* and other oncogenes reproducibly cause neoplastic transformation of human bronchial epithelial cells (Reddel *et al.*, 1988b). This investigation and others have shown that human cells immortalized by a variety of means may be transformed by *ras* oncogenes (Rhim *et al.*, 1985; Namba *et al.*, 1986; O'Brien *et al.*, 1986; Pater & Pater, 1986; Amstad *et al.*, 1988; Reddel *et al.*, 1988a). Induction of tumorigenicity in the BEAS-2B immortalized human bronchial epithelial cell line following infection with a v-Ha-*ras* recombinant containing the retrovirus has been observed (Amstad *et al.*, 1988). The anaplastic carcinomas produced in athymic nude mice by these cells have been shown to be of human epithelial origin and to have the isoenzyme phenotype and marker chromosomes characteristic of the BEAS-2B cell line. In addition, cell lines from these tumours also express an abundant p21 that is protein immunoreactive with antibodies specific for the codon 12 mutation found in v-Ha-*ras*, and which can be autophosphorylated, in contrast to cellular *ras* oncogenes. This indicates expression of the v-Ha-*ras* gene rather than activation of endogenous *ras* genes.

Activated *ras* oncogenes may act in both early (Guerrero *et al.*, 1984; Zarbl *et al.*, 1985) and late (Bondy *et al.*, 1985; Kasid *et al.*, 1985) stages of the process of multistage carcinogenesis. Although exact parallels cannot be drawn between the multistep carcinogenesis *in vitro* described here and multistep carcinogenesis occurring *in vivo*, the role of *ras* in inducing malignancy in the already immortalized BEAS-2B cell line is most probably analogous to a late event.

A number of studies have demonstrated the presence of activated c-Ki-*ras* oncogenes in human lung tumours (Pulciani *et al.*, 1982; Winter *et al.*, 1985; Rodenhuis *et al.*, 1987) and lung tumour cell lines (Der *et al.*, 1982; Pulciani *et al.*, 1982; Santos *et al.*, 1984; Winter *et al.*, 1985; Valenzuela & Groffen, 1986). The oncogene used in our recently published study (Reddel *et al.*, 1988a), v-Ki-*ras*, codes for serine at codon 12, and this mutation has also been found in a human bronchial adenocarcinoma cell line, A549 (Valenzuela & Groffen, 1986); the viral oncogene also has a significant mutation at codon 59. Transfer of this oncogene into an immortalized human bronchial epithelial cell line, BEAS-2B, either by infection of the cells with Ki-Moloney simian virus or by transfection of a plasmid, PHaKi, containing the v-Ki-*ras* coding region, resulted in neoplastic transformation.

A common feature of Ha-*ras*- and Ki-*ras*-transformed BEAS-2B cells was altered responsiveness to factors that induce terminal squamous differentiation in normal bronchial epithelial cells (Amstad *et al.*, 1988; Reddel *et al.*, 1988a). As previously reported (Ke *et al.*, 1988), BEAS-2B cells exhibited decreased colony–forming efficiency and clonal growth rate in response to either transforming growth factor-β_1 or fetal bovine serum. In contrast, BEAS-2B cells containing either v-Ki-*ras* or v-Ha-*ras* were unaffected by transforming growth factor-β_1, and their clonal growth rate was increased by fetal bovine serum. These fundamental changes in behaviour appear to be related to *ras*-induced neoplastic transformation rather than to the experimental protocols, for the following reasons. The altered responsiveness was not due to the process of selection in nude mice since a mitogenic response to fetal bovine serum was exhibited by BEAS-2B/v-Ha-*ras* cells before injection into athymic nude mice. The G418 selection used in some of the experimental protocols was not a major factor, since tumour cell lines established from Ki-Moloney simian virus-infected BEAS-2B cells (that were not G418-selected) also showed enhanced clonal growth in the presence of serum. The processes of gene transfer used, i.e., strontium phosphate coprecipitation or retroviral infection, were not responsible for this change, since BEAS-2B cells transfected with control plasmid DNA, including pRSVneo, or infected with ZipNeoSV(X) retrovirus and selected for G418 resistance retained an inhibitory response to serum.

Overexpression of c-*raf*-1 and the *myc* family of proto-oncogenes is primarily associated with small-cell carcinoma (Graziano *et al.*, 1987; Gu *et al.*, 1988; Krystal *et al.*, 1988; Rapp *et al.*, 1988), which accounts for about 25% of human lung cancer. To determine the functional significance of c-*raf*-1 and/or c-*myc* gene expression in lung carcinogenesis and to delineate the relationship between proto-oncogene expression and tumour phenotype, we introduced both proto-oncogenes, alone or in combination, into human bronchial epithelial cells (Pfeifer *et al.*, 1989). Two retroviral recombinants, pZip-*raf* and pZip-*myc*, containing the complete coding sequences of the human c-*raf*-1 and murine c-*myc* genes, respectively, were constructed and transfected into SV40 T antigen-immortalized BEAS-2B cells, followed by the selection for G418 resistance. BEAS-2B cells expressing both the transfected c-*raf*-1 and c-*myc* sequences formed large-cell carcinomas in athymic nude mice with a latency of 4-21 weeks, whereas both pZip-*raf*- and pZip-*myc*-transfected cells were nontumorigenic after 12 months. Cell lines established from tumours contained the co-transfected c-*raf*-1 and c-*myc* sequences and expressed morphological, chromosomal and isoenzyme markers which showed the BEAS-2B cells to be the progenitor cells of the tumours. Increased levels of neuron-specific enolase were detected in BEAS-2B cells containing both the c-*raf*-1 and c-*myc* genes and in the derived tumour cell lines. These findings demonstrate that the concomitant overexpression of the c-*raf* and c-*myc* proto-oncogenes causes neoplastic transformation of human bronchial epithelial cells, resulting in large-cell carcinomas with particular neuroendocrine markers.

Tumour suppression in neoplastic human lung cells

Despite the occurrence of a large number of progenitor cells, clinically evident cancer is a pathobiological event of exceedingly low probability. Although systemic host factors, e.g., the immune system, may largely account for its rarity, the lack of convincing reports of 'spontaneous' transformation of human cells *in vitro* and the difficulty in inducing their neoplastic transformation *in vitro* with chemical, physical and viral oncogenic agents attest to the presence of inherent suppressing factors at the biological level of the progenitor cells. Evidence for these presumably dominant 'tumour suppressor' genes has arisen primarily from epidemiological studies (Knudson, 1985); molecular analysis of genetic loci exhibiting DNA-restriction fragment length polymorphism show reduction to homozygosity of chromosome 13 found in retinoblastoma and osteosarcoma (Hansen *et al.*, 1985) and of chromosome 11 in Wilms' tumour (Fearon *et al.*, 1984; Koufos *et al.*, 1984; Orkin *et al.*, 1984; Reeve *et al.*, 1984) and bladder cancer (Fearon *et al.*, 1985). The latter findings are corroborated by studies of somatic cells using human cell hybrids (Stanbridge *et al.*, 1982; Sager, 1985; Kaelbling & Klinger, 1986). The tumour suppressor gene, RB-1, involved in retinoblastoma and a variety of other tumours, including small-cell carcinoma of the lung (Harbour *et al.*, 1988) and breast carcinoma (T'Ang *et al.*, 1988), has recently been cloned (Friend *et al.*, 1986; Fung *et al.*, 1987; Lee *et al.*, 1987) and shown to suppress the tumorigenicity of a cell line with a defective RB-1 (Huang *et al.*, 1988).

Our strategy for identifying and studying additional tumour suppressor genes that are involved in human lung carcinogenesis is as follows:

(1) Identify chromosomal location of putative tumour suppressor genes by:
 (i) allelic deletion analysis of tumour DNA *versus* germline DNA; and
 (ii) genetic analysis of lifespan and tumorigenicity of somatic cell-cell hybrids and monochromosome-cell hybrids.
(2) Isolate genes by subtraction library approach:
 (i) in hybrids; and
 (ii) by terminal differentiation.
(3) Isolate genes by insertion mutagenesis approach.
(4) Determine structure, function and tumour suppressive potential of isolated genes.

Tumour suppressive genes may have multiple functions. They can:

(i) induce terminal differentiation;
(ii) trigger senescence;
(iii) regulate growth;
(iv) inhibit proteases;
(v) modulate histocompatibility antigens;
(vi) regulate angiogenesis;
(vii) facilitate cell-cell communication; and
(viii) maintain chromosomal stability.

They may also be located on several different chromosomes. Therefore, a comprehensive approach is required. In our initial study, we have used allelic sequence deletion analysis to identify the location of chromosomal regions that may harbour putative tumour suppressor genes.

Carcinomas of the lung appear to be the result of the gross damage to DNA and chromosomes that follows exposure to mutagenic agents found primarily in cigarette smoke and the urban environment (Doll, 1985). Few karyotypic data are available for adenocarcinomas, large-cell carcinomas and squamous-cell carcinomas of the lung. The available data show a considerable degree of complexity with regard to the presence of deletions, translocations, marker chromosomes and aneuploidy (Zech *et al.*, 1985). These changes are consistent with the extensive molecular changes observed in allelic sequence deletion analysis of six chromosomes, and particularly with respect to the molecular changes observed in more than 50% of informative tests for squamous-cell carcinomas (Weston *et al.*, 1989). Moreover, for each chromosome studied, with the exception of 13q, loss of heterozygocity was more extensive for squamous-cell carcinomas than for adenocarcinomas. Interestingly, the risk for squamous-cell carcinoma is greater than the risk for adenocarcinoma in heavy smokers (more than one pack per day) (Fraumeni & Blot, 1982; IARC, 1986). Additional studies are required to determine the relationship between the clastogenic effects of tobacco smoke and the widespread genetic deletions found in non-small-cell carcinomas and the time of the occurrence of the deletions during the multistage process of carcinogenesis.

Previous investigations (Brauch et al., 1987; Kok et al., 1987; Naylor et al., 1987; Yokota et al., 1987) of lung cancer have focused on small-cell carcinoma, and only a limited number of non-small-cell carcinomas have been studied by DNA sequence deletion analysis (Brauch et al., 1987; Naylor et al., 1987; Shiraishi et al., 1987; Yokota et al., 1987). We have recently analysed a sufficient number of non-small-cell carcinomas, i.e., squamous-cell carcinoma, large-cell carcinoma and adenocarcinoma, to allow comparisons of DNA sequence deletions in different histological types (Weston et al., 1989). For example, the data consistently showed a loss of heterozygosity at 17p13 in squamous-cell carcinomas but not in adenocarcinomas of the lung. Frequent loss of heterozygosity at this region of chromosome 17 has also been observed in cases of colorectal cancer (Fearon et al., 1987) and small-cell lung cancer (Yokota et al., 1987). The putative tumour suppressor gene, p53, may be inactivated in several different types of human cancer (Fearon et al., 1987).

The finding of loss of heterozygosity for markers on chromosome 3 (3p25-p21) was in agreement with other reports in which DNA-restriction fragment length polymorphisms were used to examine genetic loci on chromosome 3 in non-small-cell lung cancer (Brauch et al., 1987; Naylor et al., 1987; Yokota et al., 1987). However, several cases remained heterozygous, and therefore, loss of heterozygosity was found to be substantially less than 100%, which is not in agreement with one previous report (Kok et al., 1987).

For the three different histological types that were studied, it was possible to demonstrate two commonly deleted regions on chromosome 11: 11pter-p15.5 and 11p13-q13 (Weston et al., 1989). In no case was there a conflict, in that if both of these regions were absent then all of the polymorphic markers between them were also lost. Theoretically, it would be possible to isolate a genetic locus in which loss of heterozygosity could be shown to occur both distally and proximally to that locus; for example, the calcitonin locus would be a candidate. This could occur as the result of a translocational event or through formation of a marker chromosome; however, this was not observed. These findings are consistent with observations of paediatric tumours in which two separate gene regions have been described on chromosome 11 that may harbour distinctly different putative tumour suppressor genes – 11p13 in Wilms' tumour (Turleau et al., 1981; van Heyningen et al., 1985) and 11pter-15.5 in rhabdomyosarcoma (Scrable et al., 1987).

The results with regard to non-small-cell lung cancer are very complex and suggest that there may be some differences in the spectrum of genetic deletions occurring in the various histological types of lung cancer (Weston et al., 1989). These genetic changes might also be involved in the pathogenesis of lung cancer within a 'multi-hit' framework, as has been described for other cancers (Gusella, 1986; Hansen & Cavenee, 1987), which may include loss of genes, loss of elements that affect gene expression or loss of DNA sequences that affect chromatin structure. Of the 54 tumours studied, only 20 did not show loss of heterozygosity for chromosome 11, and, of those, ten showed deletions at other chromosomal loci (Weston et al., 1989). In addition, in squamous-cell carcinomas, loss of

DNA sequences from chromosome 17 was associated with loss of DNA sequences from chromosome 11 in seven of eight cases in which information was available for both chromosomes. Similarly, loss of genes on chromosome 17 was associated with loss of genes on chromosome 3 in five of seven cases. Therefore, it is argued that consistent loss of specific genes (putative tumour suppressor genes) that have been recognized in other diseases may act independently or in combination in non-small-cell lung cancer.

Somatic cell genetic analyses of hybrids between normal and neoplastic human cells have generally shown that the normal cell has a dominant effect in suppressing the tumorigenicity of the malignant cell (Stanbridge et al., 1982; Kaelbling & Klinger, 1986; Harris, 1988). Although a rodent-human cell hybrid has been studied (Carney et al., 1979), in no previous study has an examination been made of the phenotypic properties of bronchogenic carcinomas fused with either normal or SV40 immortalized (BEAS-2B) human bronchial epithelial cells. The results of our studies are summarized in Figure 1 (Kaighn et al., 1989). The hybrids between the normal human bronchial epithelial and lung carcinoma cells (HuT292DM) had a finite lifespan, i.e., about 40 population doublings, and were nontumorigenic (5×10^6 cells, subcutaneously, per athymic nude mouse, >6-month observation period). Although the somatic cell hybrids between BEAS-2B and HuT292DM appear to have an infinite lifespan, the majority of the cloned hybrids were nontumorigenic. This result indicates that genes other than those involved in senescence can have tumour suppressive properties.

Figure 1. Somatic cell genetic analyses of lifespan and tumorigenicity of human bronchial epithelial cell hybrids

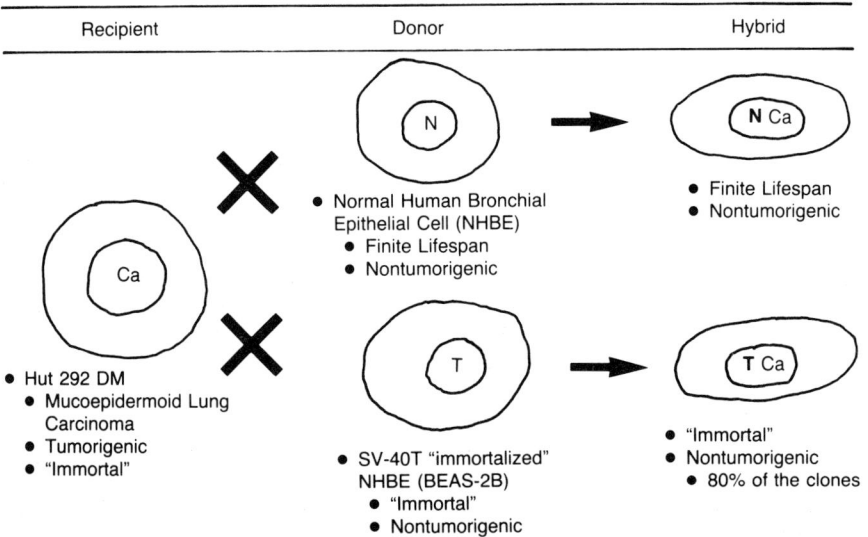

Chromosome 11 from normal cells can be introduced by microcell transfer into carcinomas derived from the kidney or cervix, and the monochromosome-cell hybrids are less tumorigenic (Saxon et al., 1986; Weissmann et al., 1987; Koi et al., 1989). Guided by the results of DNA sequence deletion analysis in non-small-cell carcinomas of the lung (Weston et al., 1989), we are systematically transferring single or combinations of human chromosomes 3, 11, 13 and 17 into human bronchogenic carcinoma cell lines. Our initial findings indicate that chromosome 11-carcinoma cell hybrids have a substantially increased tumour latency (McMenamin et al., 1989).

Acknowledgement
The editorial aid of Robert Julia is appreciated.

References

Amstad, P., Reddel, R.R., Pfeifer, A., Malan-Shibley, L., Mark, G.E. & Harris, C.C. (1988) Neoplastic transformation of a human bronchial epithelial cell line by a recombinant retrovirus encoding viral Harvey ras. *Mol. Carcinog.*, **1**, 151-160

Bondy, G.P., Wilson, S. & Chambers, A.F. (1985). Experimental metastatic ability of H-ras-transformed NIH3T3 cells. *Cancer Res.*, **45**, 6005-6009

Brauch, H., Johnson, B., Hovis, J., Yano, T., Gazdar, A.F., Pettengill, O.S., Graziano, S.L., Sorenson, G.D., Poiesz, B.J., Minna, J., Linehan, M. & Zbar, B. (1987) Molecular analysis of the short arm of chromosome 3 in small-cell and non-small-cell carcinoma of the lung. *New Engl. J. Med.*, **317**, 1109-1113

Byrd, P.J., Grand, R.J. & Gallimore, P.H. (1988) Differential transformation of primary human embryo retinal cells by adenovirus E1 regions and combinations of E1A + ras. *Oncogene*, **2**, 477-484

Carney, D.N., Edgell, C.J., Gazdar, A.F. & Minna, J.D. (1979) Suppression of malignancy in human lung cancer (A549/8) times mouse fibroblast (3P3-4E) somatic cell hybrids. *J. Natl Cancer Inst.*, **69**, 411-415

Clark, R., Stampfer, M.R., Milley, R., O'Rourke, E., Walen, K.H., Kriegler, M., Kopplin, J. & McCormick, F. (1988) Transformation of human mammary epithelial cells by oncogenic retroviruses. *Cancer Res.*, **48**, 4689-4694

Der, C.J., Krontiris, T.G. & Cooper, G.M. (1982) Transforming genes of human bladder and lung carcinoma cell lines are homologous to the ras genes of Harvey and Kirsten sarcoma viruses. *Proc. Natl Acad. Sci. USA*, **79**, 3637-3640

Doll, R. (1985) Cancer: a worldwide perspective. In: Harris, C.C., ed., *Biochemical and Molecular Epidemiology of Cancer*, New York, Alan R. Liss, pp. 111-125

Fearon, E.R., Vogelstein, B. & Feinberg, A.P. (1984) Somatic deletion and duplication of genes on chromosome 11 in Wilms' tumours. *Nature*, **309**, 176-178

Fearon, E.R., Feinberg, A.P., Hamilton, S.H. & Vogelstein, B. (1985) Loss of genes on the short arm of chromosome 11 in bladder cancer. *Nature*, **318**, 377-380

Fearon, E.R., Hamilton, S.H. & Vogelstein, B. (1987) Clonal analysis of human colorectal tumors. *Science*, **238**, 193-197

Fraumeni, J.F., Jr & Blot, W.J. (1982) Lung and pleura. In: Schottenfeld, D. & Fraumeni, J.F., Jr, eds, *Cancer Epidemiology and Prevention*, Philadelphia, W.R. Saunders Co., pp. 564-582

Friend, S.H., Bernards, R., Rogelj, S. Weinberg, R.A., Rapaport, J.M., Albert, D.M. & Dryja, T.P. (1986) A human DNA segment with properties of the gene that predisposes to retinoblastoma and osteosarcoma. *Nature*, **323**, 643-646

Fung, Y.K., Murphree, A.L., T'Ang, A., Qian, J., Hinrichs, S.H. & Benedict, W.F. (1987) Structural evidence for the authenticity of the human retinoblastoma gene. *Science*, **236**, 1657-1661

Graziano, S.L., Cowan, B.Y., Carney, D.N., Bryke, C.R., Mitter, N.S., Johnson, B.E., Mark, G.E., Planas, A.T., Catino, J.J., Comis, R.L. & Poiesz, B.J. (1987) Small cell lung cancer cell line derived from a primary tumour with a characteristic deletion of 3p. *Cancer Res.*, **47**, 2148-2155

Gu, J., Linnoila, R.I., Seibel, N.L., Gadzar, A.F., Minna, J.D., Brooks, B.J., Jr, Hollis, G.F. & Kirsch, I.R. (1988) A study of myc-related gene expression in small cell lung cancer by in situ hybridization. *Am. J. Pathol.*, **132**, 13-17

Guerrero, I., Calzada, P., Mayer, A. & Pellicer, A. (1984) A molecular approach to leukemogenesis: mouse lymphomas contain an activated c-ras oncogene. *Proc. Natl Acad. Sci. USA*, **81**, 202-205

Gusella, J.F. (1986) DNA polymorphism and human disease. *Ann. Rev. Biochem.*, **55**, 831-854

Hansen, M.F. & Cavenee, W.K. (1987) Genetics of cancer predisposition. *Cancer Res.*, **47**, 5518-5527

Hansen, M.F., Koufos, A., Gallie, B.L., Phillips, R.A., Fodstad, O., Brogger, A., Gedde-Dahl, T. & Cavenee, W.K. (1985) Osteosarcoma and retinoblastoma: a shared chromosomal mechanism revealing recessive predisposition. *Proc. Natl Acad. Sci. USA*, **82**, 6216-6220

Harbour, J.W., Lai, S.L., Whang-Peng, J., Gazdar, A.F., Minna, J.D. & Kaye, F.J. (1988) Abnormalities in structure and expression of the human retinoblastoma gene in SCLC. *Science*, **241**, 353-357

Harris, H. (1988) The analysis of malignancy by cell fusion: the position in 1988. *Cancer Res.*, **48**, 3302-3306

van Heyningen, V., Boyd, P.A., Seawright, A., Fletcher, J.M., Fantes, J.A., Buckton, K.E., Spowart, G., Porteous, D.J., Hill, R.E., Newton, M.S. & Hastie, N.D. (1985) Moleculr analysis of chromosome 11 deletions in aniridia-Wilms tumor syndrome. *Proc. Natl Acad Sci. USA*, **82**, 8592-8596

Huang, H.J., Yee, J.K., Shew, J.Y., Chen, P.L., Bookstein, R., Friedmann, T., Lee, E.Y. & Lee, W.H. (1988) Suppression of the neoplastic phenotype by replacement of the RB gene in human cancer cells. *Science*, **242**, 1563-1566

IARC (1986) *IARC Monographs on the Evaluation of the Carcinogenic Risk of Chemicals to Humans*, Vol. 38, *Tobacco Smoking*, Lyon

Kaelbling, M. & Klinger, H.P. (1986) Suppression of tumorigenicity in somatic cell hybrids. III. Cosegregation of human chromosome 11 of a normal cell and suppression of tumorigenicity in intraspecies hybrids of normal diploid x malignant cells. *Cytogenet. Cell Genet.*, **41**, 65-70

Kaighn, M.E., Gabrielson, E.W., Iman, D.S., Pauls, E.A. & Harris, C.C. (1989) Suppression of tumorigenicity of a human lung carcinoma line by nontumorigenic bronchial epithelial cells in somatic cell hybrids (Abstract). *Proc. Am. Assoc. Cancer Res.*, **30**, 445

Kasid, A., Lippman, M.E., Papageorge, A.G., Lowy, D.R. & Gelmann, E.P. (1985) Transfection of v-rasH DNA into MCF-7 human breast cancer cells bypasses dependence on estrogen for tumorigenicity. *Science*, **228**, 725-728

Ke, Y., Reddel, R.R., Gerwin, B.I., Miyashita, M., McMenamin, M.G., Lechner, J.F. & Harris, C.C. (1988) Human bronchial epithelial cells with integrated SV40 virus T antigen genes retain the ability to undergo squamous differentiation. *Differentiation*, **38**, 60-66

Knudson, A.G., Jr (1985) Hereditary cancer, oncogenes, and antioncogenes. *Cancer Res.*, **45**, 1437-1443

Koi, M., Morita, H., Yamada, H., Satoh, H., Barrett, J.C. & Oshimura, M. (1989) Normal human chromosome 11 suppresses tumorigenicity of human cervical tumor cell line SiHa. *Mol. Carcinog.*, **2**, 12-21

Kok, K., Osinga, J., Carritt, B., Davis, M.B., van der Hout, A.H., van der Veen, A.Y., Landsvater, R.M., de Leij, L.F., Berendsen, H.H., Postmus, P.E., Poppema, S. & Buys, C.H. (1987) Deletion of a DNA sequence at the chromosomal region 3p21 in all major types of lung cancer. *Nature*, **330**, 578-581

Koufos, A., Hansen, M.F., Lampkin, B.C., Workman, M.L., Copeland, N.G., Jenkins, N.A. & Cavenee, W.K. (1984) Loss of alleles at loci on human chromosome 11 during genesis of Wilms' tumour. *Nature*, **309**, 170-172

Krystal, G., Birrer, M.J., Way, J., Nau, M.M., Sausville, E., Thompson, C., Minna, J.D. & Battey, J.F. (1988) Multiple mechanisms for transcriptional regulation of the myc gene family in small-cell lung cancer. *Mol. Cell Biol.*, **8**, 3373-3381

Lee, W.H., Bookstein, R., Hong, F., Young, L.J., Shew, J.Y. & Lee, E.Y. (1987) Human retinoblastoma susceptibility gene: cloning, identification, and sequence. *Science*, **235**, 1394-1399

McMenamin, M.G., Iman, D.S., Stanbridge, E.J., Shows, T.B. & Harris, C.C. (1989) Tumor suppressor activity of microcell transferred human chromosomes in lung carcinoma cells (Abstract). *Proc. Am. Assoc. Cancer Res.*, **30**, 189

Namba, M., Nishitani, K., Fukushima, F., Kimoto, T. & Nose, K. (1986) Multistep process of neoplastic transformation of normal human fibroblasts by 60Co gamma rays and Harvey sarcoma viruses. *Int. J. Cancer*, **37**, 419-423

Naylor, S.L., Johnson, B.E., Minna, J.D. & Sakaguchi, A.Y. (1987) Loss of heterozygosity of chromosome 3p markers in small-cell lung cancer. *Nature*, **329**, 451-454

O'Brien, W., Stenman, G. & Sager, R. (1986) Suppression of tumor growth by senescence in virally transformed human fibroblasts. *Proc. Natl Acad. Sci. USA*, **83**, 8659-8663

Orkin, S.H., Goldman, D.S. & Sallan, S.E. (1984) Development of homozygosity for chromosome 11p markers in Wilms' tumour. *Nature*, **309**, 172-174

Pater, A. & Pater, M.M. (1986) Transformation of primary human embryonic kidney cells to anchorage independence by a combination of BK virus DNA and the Harvey-ras oncogene. *J. Virol.*, **58**, 680-683

Pfeifer, A.M.A., Mark, G.E., Malan-Shibley, L., Graziano, S., Amstad, P. & Harris, C.C. (1989) Cooperation of c-raf-1 and c-myc protooncogenes in the neoplastic transformation of SV40 T antigen immortalized human bronchial epithelial cells. *Proc. Natl Acad. Sci. USA*, **86**, 10075-10079

Pulciani, S., Santos, E., Lauver, A.V., Long, L.K., Aaronson, S.A. & Barbacid, M. (1982) Oncogenes in solid human tumours. *Nature*, **300**, 539-542

Rapp, U.R., Huleihel, M., Pawson, T., Linnoila, I., Minna, J.D., Heidecker, G., Cleveland, J.L., Beck, T., Forchhammer, J. & Storm, S.M. (1988) Role of raf oncogenes in lung carcinogenesis. *Lung Cancer*, **4**, 162-167

Reddel, R.R., Ke, Y., Kaighn, M.E., Malan-Shibley, L., Lechner, J.F., Rhim, J.S. & Harris, C.C. (1988a) Human bronchial epithelial cells neoplastically transformed by v-Ki-ras: altered response to inducers of terminal squamous differentiation. *Oncogene Res.*, **3**, 401-408

Reddel, R.R., Ke, Y., Gerwin, B.I., McMenamin, M.G., Lechner, J.F., Su, R.T., Brash, D.E., Park, J.B., Rhim, J.S. & Harris, C.C. (1988b) Transformation of human bronchial epithelial cells by infection with SV40 or adenovirus-12 SV40 hybrid virus, or transfection via strontium phosphate coprecipitation with a plasmid containing SV40 early region genes. *Cancer Res.*, **48**, 1904-1909

Reeve, A.E., Housiaux, P.J., Gardner, R.J., Chewings, W.E., Grindley, R.M. & Millow, L.J. (1984) Loss of a Harvey ras allele in sporadic Wilms' tumour. *Nature*, **309**, 174-176

Reznikoff, C.A., Loretz, L.J., Christian, B.J., Wu, S.Q. & Meisner, L.F. (1988) Neoplastic transformation of SV40-immortalized human urinary tract epithelial cells by in vitro exposure to 3-methylcholanthrene. *Carcinogenesis*, **9**, 1427-1436

Rhim, J.S., Jay, G., Arnstein, P., Price, F.M., Sanford, K.K. & Aaronson, S.A. (1985) Neoplastic transformation of human epidermal keratinocytes by AD12-SV-40 and Kirsten sarcoma viruses. *Science*, **227**, 1250-1252

Rodenhuis, S., Van de Wetering, M.L., Mooi, W.J., Evers, S.G., van Zandwijk, N. & Bos, J.L. (1987) Mutational activation of the K-ras oncogene. A possible pathogenetic factor in adenocarcinoma of the lung. *New Engl. J. Med.*, **317**, 929-935

Sager, R. (1985) Genetic suppression of tumour formation. *Adv. Cancer Res.*, **44**, 43-68

Santos, E., Martin-Zanca, D., Reddy, E.P., Pierotti, M.A., Della Porta, G. & Barbacid, M. (1984) Malignant activation of a K-ras oncogene in lung carcinoma but not in normal tissue of the same patient. *Science*, **223**, 661-664

Saxon, P.J., Srivatsan, E.S. & Stanbridge, E.J. (1986) Introduction of human chromosome 11 via microcell transfer controls tumorigenic expression of HeLa cells. *EMBO J.*, **5**, 3461-3466

Scrable, H.J., Witte, D.P., Lampkin, B.C. & Cavenee, W.K. (1987) Chromosomal localization of the human rhabdomyosarcoma locus by mitotic recombination mapping. *Nature*, **329**, 645-647

Seremetis, S., Inghirami, G., Ferrero, D., Newcomb, E.W., Knowles, D.M., Dotto, G.P. & Dalla-Favera, R. (1989) Transformation and plasmocytoid differentiation of EBV-infected human B lymphoblasts by ras oncogenes. *Science*, **243**, 660-663

Shiraishi, M., Morinaga, S., Noguchi, M., Shimosato, Y. & Sekiya, T. (1987) Loss of genes on the short arm of chromosome 11 in human lung carcinomas. *Jpn. J. Cancer Res.*, **78**, 1302-1308

Stanbridge, E.J., Der, C.J., Doersen, C.J., Nishimi, R.Y., Peehl, D.M., Weissman, B.E. & Wilkinson, J.E. (1982) Human cell hybrids: analysis of transformation and tumorigenicity. *Science*, **215**, 252-259

Suda, Y., Aizawa, S., Hirai, S., Inoue, T., Furuta, Y., Suzuki, M., Hirohashi, S. & Ikawa, Y. (1987) Driven by the same Ig enhancer and SV40 T promoter ras induced lung adenomatous tumors, myc induced pre-B cell lymphomas and SV40 large T gene a variety of tumors in transgenic mice. *EMBO J.*, **6**, 4055-4065

T'Ang, A., Varley, J.M., Chakraborty, S., Murphree, A.L. & Fung, Y.K. (1988) Structural rearrangement of the retinoblastoma gene in human breast carcinoma. *Science*, **242**, 263-266

Turleau, C., de Grouchy, J., Dufier, J.L., Phuc, L.H., Schmelck, P.H., Rappaport, R., Nijoul-Fekete, C. & Diebold, N. (1981) Aniridia, male pseudohermaphroditism, gonadoblastoma, mental retardation, and del 11p13. *Hum. Genet.*, **57**, 300-306

Valenzuela, D.M. & Groffen, J. (1986) Four human carcinoma cell lines with novel mutations in position 12 of c-K-ras oncogene. *Nucleic Acids Res.*, **14**, 843-852

Weissman, B.E., Saxon, P.J., Pasquale, S.R., Jones, G.R., Geiser, A.G. & Stanbridge, E.J. (1987) Introduction of a normal human chromosome 11 into a Wilms' tumor cell line controls its tumorigenic expression. *Science*, **236**, 175-180

Weston, A., Willey, J.C., Modali, R., Sugimura, H., McDowell, E.M., Resau, J., Light, B., Haugen, A., Mann, D.L., Trump, B.F. & Harris, C.C. (1989) Differential DNA sequence deletions from chromosomes 3, 11, 13 and 17 in squamous cell carcinoma, large cell carcinoma and adenocarcinoma of the human lung. *Proc. Natl Acad. Sci. USA*, **86**, 5099-5103

Winter, E., Yamamoto, F., Almoguera, C. & Perucho, M. (1985) A method to detect and characterize point mutations in transcribed genes: amplification and overexpression of the mutant c-Ki-ras allele in human tumor cells. *Proc. Natl Acad. Sci. USA*, **82**, 7575-7579

Yoakum, G.H., Lechner, J.F., Gabrielson, E.W., Korba, B.E., Malan-Shibley, L., Willey, J.C., Valerio, M.G., Shamsuddin, A.K.M., Trump, B.F. & Harris, C.C. (1985) Transformation of human bronchial epithelial cells transfected by Harvey ras oncogene. *Science*, **227**, 1174-1179

Yokota, J., Wada, M., Shimosato, Y., Terada, M. & Sugimura, T. (1987) Loss of heterozygosity on chromosomes 3, 13, and 17 in small-cell carcinoma and on chromosome 3 in adenocarcinoma of the lung. *Proc. Natl Acad. Sci. USA*, **84**, 9252-9256

Yuasa, Y., Srivastava, S.K., Dunn, C.Y., Rhim, J.S., Reddy, E.P. & Aaronson, S.A. (1983) Acquisition of transforming properties by alternative point mutations within c-bas/has human proto-oncogene. *Nature*, **303**, 775-779

Yuasa, Y., Gol, R.A., Chang, A., Chiu, I.M., Reddy, E.P., Tronick, S.R. & Aaronson, S.A. (1984) Mechanism of activation of an N-ras oncogene of SW-1271 human lung carcinoma cells. *Proc. Natl Acad. Sci. USA*, **81**, 3670-3674

Zarbl, H., Sukumar, S., Arthur, A.V., Martin-Zanca, D. & Barbacid, M. (1985) Direct mutagenesis of Ha-ras-1 oncogenes by N-nitroso-N-methylurea during initiation of mammary carcinogenesis in rats. *Nature*, **315**, 382-385

Zech, L., Bergh, J. & Nilsson, K. (1985) Karyotypic characterization of established cell lines and short-term cultures of human lung cancers. *Cancer Genet. Cytogenet.*, **15**, 335-347

ALKYLATION OF DNA RELATED TO ORGAN-SPECIFIC CARCINOGENESIS BY N-NITROSO COMPOUNDS

W. Lijinsky

*National Cancer Institute - Frederick Cancer Research Facility,
Bionetics Research Inc., Basic Research Program, Frederick, MD, USA*

Alkylation of DNA by a number of methylating and ethylating carcinogens, mainly N-nitroso compounds, has been examined in target and non-target organs of rats and Syrian hamsters. Six hours after administration by gavage of small doses identical to those given twice weekly for several months to elicit tumours, animals were killed and dissected. DNA was isolated from several organs and hydrolysed, and the content of methyl- and ethylguanines was measured using high-performance liquid chromatography for separation. In most experiments, radiolabelled carcinogen was used, but in some cases measurement of alkylguanines was by fluorescence. Methylation, O^6- and $N7$-, by methylating compounds was much more extensive than ethylation by the corresponding ethyl compounds, irrespective of their relative potencies in inducing tumours. Similar patterns of alkylation were found in target organs and in non-target organs of the carcinogens. Only marginal differences in methylation were seen with N-nitrosobis(2-oxopropyl)amine between male and female rat livers, although liver tumours are induced only in females, in feminized males and in old males. Deuterium labelling of the methylene of N-nitrosoethylmethylamine had little effect on methylation or ethylation of DNA in rat liver, although the deuterated compound was a much more potent liver carcinogen. The conclusion is that reactions of the carcinogen other than alkylation of DNA are important in giving rise to tumours.

The chemical property of N-nitroso compounds most relevant to their biological activity, mutagenesis and carcinogenesis is considered to be their conversion to electrophiles that are alkylating agents (Magee & Farber, 1962). This has led us to examine and compare the pattern and extent of DNA alkylation in target and non-target organs of rats and hamsters, and to correlate these findings with the carcinogenesis of a number of compounds that give rise to the same alkylating agent. For these experiments, low doses of a carcinogenic agent similar to those used in tumour-induction experiments were used, and the arbitrary time of 6 h following treatment was chosen for examination of DNA alkylation.

The compounds studied are methylating and ethylating agents. The methylating compounds include N-nitrosodimethylamine (NDMA), N-nitrosoethylmethylamine (NEMA) and the similar compound azoxymethane; also included is N-nitrosobis(2-oxopropyl)amine, which is not obviously a methylating compound but gives rise to methylation of nucleic acids in animals *in vivo*. Methylation by the nitrosamines has been compared with that by *N*-methyl-*N*-nitrosourea and *N*-methyl-*N*-nitrosoethylurea, which have very different carcinogenic activities from the nitrosamines. Comparisons of ethylating compounds have involved those few *N*-nitroso compounds that can be radiolabelled at sufficiently high specific activity to permit the small yields of ethylated bases to be counted.

Animals were treated by gavage with the alkylating compound dissolved in corn oil:ethyl acetate (2:1). CO_2 excretion was measured during 6 h in the case of the ^{14}C-labelled compounds. At 6 h, animals were killed and their organs dissected and immediately frozen in liquid N_2. DNA, RNA and soluble protein were isolated and counted. DNA was hydrolysed in mild acid at 75°C and chromatographed by high-performance liquid chromatography on an ion-exchange column, with phosphate buffer as eluant; radioactive fractions were collected and counted, and those corresponding to standard O^6- and N^7-alkylguanines were totalled. The alkylguanines arising from non-radiolabelled compounds were measured by fluorescence, according to the method of Herron and Shank (1979). Both methods were used in some cases, and there was good agreement.

Table 1 shows the extent of O^6- and N^7-methylation of guanine in the DNA of liver and other organs of rats and hamsters 6 h after treatment with approximately 20 μmol of various methylating agents, including NDMA, NEMA, azoxymethane, *N*-methyl-*N*-nitrosourea and *N*-methyl-*N*-nitrosoethylurea. In the case of those compounds that were labelled with ^{14}C, the rate of excretion of the radiolabel as $^{14}CO_2$ was similar – 50-60% of the dose in 6 h — for all compounds. This similarity in the extent of CO_2 excretion indicates that most of the original compound, if not all, had been metabolized by this time and that formation of the alkylating agent and alkylation of DNA was at or past its maximum.

The extent of O^6-methylguanine formation in liver DNA of rats and of hamsters differs somewhat, especially in that more was found in hamsters than in rats, which can be related to the larger size of rats. However, there are some discrepancies from the carcinogenic effects of the compounds, since NEMA is a more effective inducer of liver tumours in rats after gavage than are NDMA and azoxymethane, while it is less effective than NDMA and azoxymethane in inducing liver tumours in hamsters (Lijinsky *et al.*, 1987).

N-Methyl-*N*-nitrosourea induced a similar extent of methylation of DNA in the liver and other organs of rats and hamsters as NDMA and azoxymethane, although it induced only tumours of the forestomach and of the nervous system in rats and of the spleen in hamsters (Lijinsky, 1987). Azoxymethane methylated liver DNA in rats and hamsters as effectively as NDMA and NEMA, although it did not induce liver tumours in rats after gavage. Since the pattern of methylation is similar among the methylating compounds but the pattern of tumours induced by them is so different, methylation of DNA seems to be only one of several reactions induced by the carcinogen; other reactions may be of equal or greater importance in leading to induction of tumours.

Comparison of the ethylation of DNA by some ethylating agents with their effects in inducing tumours leads to the same conclusion (Table 2). In all cases, the extent of ethylation was almost two orders of magnitude lower than by an equivalent dose of the corresponding methylating agent. *N*-Nitrosodiethylamine and NEMA induce tumours of the liver and

Table 1. Methylation of DNA by alkylating carcinogens 6 h after treatment

Compound	Dose (µmol)	Species	Organ	Methylation of DNA (pmol/mg DNA)		Tumours
				$N7$-Methyl-guanine	O^6-Methyl-guanine	
N-Nitrosodi-methylamine	18	Rat	Liver	380	23	+
			Kidney	51	2.4	+
		Hamster	Liver	420	29	+
Azoxymethane	27	Rat	Liver	740	48	-
			Kidney	21	2.2	+
			Colon	19	1	+
	27	Hamster	Liver	1180	110	+
N-Nitrosoethyl-methylamine	20	Rat	Liver	270	24	+
			Kidney	44	5.5	-
			Spleen	84	7.8	-
			Lung	22	5	+
	20	Hamster	Liver	1000	71	+
			Kidney	40	5	-
			Lung	30	5	-
N-Methyl-N-nitrosourea	20	Rat	Liver	103	11	-
			Kidney	111	16	-
			Lung	54	4.8	-
			Brain	23	2.4	+
			Spleen	180	14	-
	20	Hamster	Liver	130	16	-
			Kidney	108	11	-
			Lung	22	2.7	-
N-Methyl-N-nitrosoethylurea	20	Rat	Liver	54	3.4	-
			Lung	21	2.2	+
			Brain	17	1.8	+

oesophagus in rats and of the liver in hamsters, while N-ethyl-N-nitrosourea does not induce liver tumours in rats but induces tumours in several other organs; in hamsters, N-ethyl-N-nitrosourea induced only tumours of the forestomach and haemangiosarcomas of the spleen. O^6-Ethylation of DNA in liver by N-ethyl-N-nitrosourea at the dose used was too small to measure. Equivalent doses of NDEA and NEMA produce similar levels of O^6- and $N7$-ethylation of guanine in the livers of rats and of hamsters; NDEA was considerably more potent in inducing liver tumours in rats than NEMA, but NEMA was more potent than NDEA in inducing liver tumours in hamsters. This finding again supports the likelihood that ethylation of DNA is not the prime factor in liver tumour induction by these N-nitroso compounds. Deuterium labelling of the α position of the ethyl group in NEMA led to a large increase in the potency of NEMA inducing liver tumours in rats (Lijinsky & Reuber, 1980), but the effect on alkylation of DNA in rat liver (ethylation or methylation) is not great. The increase in carcinogenic potency of deuterium-labelled NEMA, therefore, is not paralleled by an increase in the extent of DNA alkylation, as is the case with increased doses of the nitrosamine, which lead to increased tumour incidence and increased extent of DNA alkylation.

Table 2. Alkylation of DNA by ethylating carcinogens 6 h after treatment

Compound	Dose (μmol)	Species	Organ[a]	Alkylation of liver DNA (pmol/mg DNA)			
				$N7$-Ethyl-guanine	O^6-Ethyl-guanine	$N7$-Methyl-guanine	O^6-Methyl-guanine
N-Nitrosoethyl-methylamine	17	Rat	Liver +	3.3	1.4	182	17
			Kidney –	1	0.14	38	1.7
			Lung +	1.5	0.2		
			Spleen –	1.3	0.15		
			Pancreas –	2.6	0.6		
			Colon –	2.5	0.5		
	17	Hamster	Liver +	3.9	1.2		
			Kidney –	1.4	0.3		
			Lung –	2.9	0.4		
N-Nitrosoethyl-d_2-methylamine	17	Rat	Liver +	6.7	0.7	145	8
N-Ethyl-N-nitrosourea	21	Rat	Liver –	4.5	ND		
			Kidney –	1.9	0.4		
			Lung +	3.6	0.4		
			Spleen +	2.9	0.2		
			Colon +	4.6	0.9		
	21	Hamster	Liver –	2.9	ND		
			Brain –	6.7	1.4		
N-Nitrosodi-ethylamine	20	Rat	Liver +	5.3	1.8		
	20	Hamster	Liver +	14	5		

[a] +, tumours induced; –, tumours not induced

N-Nitrosobis(2-oxopropyl)amine has been studied extensively, because it induces liver tumours in female rats but tumours of the thyroid, bladder and renal pelvis, but no liver tumours, in males; a high incidence of lung tumours occurs in animals of each sex (Lijinsky et al., 1988). Feminized male rats showed the same response as female rats – liver tumours at high incidence. Such a large gender-related difference is unusual in carcinogenesis induced by N-nitroso compounds and suggests a hormone-modulated effect on the metabolism and activation of this compound. Male rats treated with N-nitrosobis(2-oxopropyl)amine in 'middle age' (65 weeks old) behaved like female rats and developed liver and lung tumours but few of the other tumours induced in young male rats (Lijinsky & Kovatch, 1986). The possibility was investigated that the hormonal or other influences on activation in the liver, leading to induction of liver tumours, manifested themselves by increasing alkylation of liver DNA compared with that in young male rats. There was no large difference in the extent of methylation of liver DNA (Table 3), after correction for different body weights, between young and old males or females that would itself explain the great differences in tumour response.

Table 3. Methylation of DNA by ^{14}C-N-nitrosobis(oxypropyl)amine by gavage (2.5 mg/5 µCi) measured 6 h after treatment

Species	Sex	Age (weeks)	Tumours[a]	Methylation of DNA (pmol/mg DNA)		
				$N7$-Methyl-guanine	O^6-Methyl-guanine	O^6-Methyl-guanine/dose[b]
Rat	M	20	Liver −	43	5.7	0.85
	F	20	Liver +	180	14	1.0
	M (castrated)	20	Liver +	120	12	1.2
	M	65	Liver +	40	4.3	0.74
	M	65	Kidney −	19	2.1	0.37
Hamster	F	20	Liver +	840	3.5	2.3

[a] +, tumours induced; −, tumours not induced
[b] Corrected for dose of N-nitrosobis(oxypropyl)amine in mg/kg bw

It can be concluded that there are differences in the pattern and extent of alkylation by both direct and indirect alkylating mutagens. The quantitative differences between methylation and ethylation are large, and do not match the relative carcinogenic potency of ethylating and methylating agents in rats or hamsters, where the carcinogenic effects are often opposite. It has been shown by others (Silinskas et al., 1984), as well as by us, that methylation occurs according to chemical expectation, but that methylation in a particular organ, even at a presumed mutagenic locus, does not determine that tumours will ensue; other factors are more important. Cells that do not develop tumours following methylation of DNA by one alkylating agent are susceptible to tumour development when an alkylating agent of different structure produces methylation of the DNA. Different rates of DNA repair might be important in explaining the differences between alkyl groups of different structure, but this is not a logical explanation of the differences in tumour pattern produced by compounds forming the same alkylating moiety and giving the same pattern of alkylation in several organs.

Acknowledgements

I thank Dr J.E Saavedra for synthesizing the isotopically labelled NEMA and Grace Stafford for the analyses by fluorescence. Research sponsored by the National Cancer Institute, Department of Health and Human Services, under contract No. NO1-CO-74101 with Bionetics Research, Inc. The contents of this publication do not necessarily reflect the views or policies of the Department of Health and Human Services, nor does mention of trade names, commercial products or organizations imply endorsement by the US Government. By acceptance of this article, the publisher or recipient acknowledges the right of the US Government to retain an nonexclusive, royalty-free license in and to any copyright covering the article.

References

Herron, D.C. & Shank, R.C. (1979) Quantitative high pressure liquid chromatographic analysis of methylated purines in DNA of rats treated with chemical carcinogens. *Anal. Biochem.*, **100**, 58-63

Lijinsky, W. (1987) Structure-activity relations in carcinogenesis by *N*-nitroso compounds. *Cancer Metastasis Rev.*, **6**, 301-356

Lijinsky, W. & Kovatch, R.M. (1986) The effect of age on susceptibility of rats to carcinogenesis by two nitrosamines. *Gann*, **77**, 1222-1226

Lijinsky, W. & Reuber, M.D. (1980) Carcinogenicity in rats of nitrosomethylethylamines labeled with deuterium in several positions. *Cancer Res.*, **40**, 19-20

Lijinsky, W., Kovatch, R.M. & Riggs, C.W. (1987) Carcinogenesis by nitrosodialkylamines and azoxyalkanes given by gavage to rats and hamsters. *Cancer Res.*, **47**, 3968-3972

Lijinsky, W., Thomas, B.J. & Kovatch, R.M. (1988) Effects of feminization of male F344 rats on tumour induction and on nucleic acid alkylation by nitrosobis(2-oxopropyl)amine. *Chem.-Biol. Interact.*, **66**, 111-119

Magee, P.N. & Farber, E. (1962) Toxic liver injury and carcinogenesis. Methylation of rat-liver nucleic acids by dimethylnitrosamine in vivo. *Biochem. J.*, **83**, 114-124

Silinskas, K.C., Zucker, P.F., Labuc, G.E. & Archer, M.C. (1984) Formation of O^6-methylguanine in regenerating rat liver by N-nitrosomethylbenzylamine is not sufficient for initiation of neoplastic foci. *Carcinogenesis*, **5**, 541-542

RELATIONSHIP BETWEEN DOSE AND RISK REDUCTION: STATISTICAL EVALUATION OF A COMBINATION EXPERIMENT WITH THREE HEPATOCARCINOGENIC N-NITROSAMINES IN RATS

M.R. Berger[1], D. Schmähl[1] & L. Edler[2]

[1]*Institute of Toxicology and Chemotherapy; and* [2]*Institute of Epidemiology and Biometry, German Cancer Research Centre, Heidelberg, Germany*

Data from an experiment on the single and combination effects of very low doses of N-nitrosodiethylamine (NDEA), N-nitrosopyrrolidine (NPYR) and N-nitrosodiethanolamine (NDELA) in 1800 male Sprague-Dawley rats were analysed for age-specific incidence (time to death with liver tumour) by Cox's proportional hazards model. The model revealed a linear relationship between daily exposure to low levels of N-nitrosamine and time to death with liver tumour within the dose range investigated: an increase in individual dose resulted in a proportional decrease in liver tumour-free survival. The finding was established for the three individual carcinogens as well as for their combination. NDEA was 40 times more active than NDELA. Extrapolation to N-nitrosamine exposure levels lower than those used in the experiment revealed only a minor reduction in age-specific liver tumour incidence compared to that achieved by an equivalent reduction within the experimental dose range. In rats at advanced age, a further reduction in carcinogen-induced liver tumour incidence did not contribute to longer overall survival, due to competitive, probably independent causes of death. The data thus support the idea of a quasi-threshold in terms of a 'no observed effect level'.

In a study of the combination effects of very low doses of three N-nitrosamines, N-nitrosodiethylamine (NDEA), N-nitrosopyrrolidine (NPYR) and N-nitrosodiethanolamine (NDELA), tumour formation in the liver, the common target organ, was dose-dependent, even at very low levels of exposure to the single N-nitrosamines or to their combination (Table 1). The carcinogenic effects of the three N-nitroso compounds were additive when they were given in combination, in an apparently linear manner. Doses that would presumably not have been carcinogenic when given alone were carcinogenic when administered in combination (Table 2; Berger *et al.*, 1987). Since that experiment involved 1800 rats, it seemed large enough to ensure a reasonable data base for statistical evaluation of the underlying relationship between dose and carcinogenic hazard. The data were thus fit to a model that can be used to estimate the long-term toxic effects of dosages different from those of the experiment, and, in addition, to delineate some general aspects of carcinogenesis near the 'no observed effect level'.

Table 1. Experimental design for administering N-nitrosodiethylamine (NDEA), N-nitrosopyrrolidine (NPYR) and N-nitrosodiethanolamine (NDELA) to male SD rats

No. of animals	Abbreviation[a]	Single dose administered/day[b]			Median total dose[b,c]		
		NDEA	NPYR	NDELA	NDEA	NPYR	NDELA
500	-	-	-	-	-	-	-
80	HD	0.1	-	-	61.1	-	-
80	MD	0.032	-	-	20.3	-	-
80	LD	0.01	-	-	6.5	-	-
80	HD	-	0.4	-	-	272	-
80	MD	-	0.133	-	-	85.2	-
80	LD	-	0.04	-	-	26.4	-
80	HD	-	-	2.0	-	-	1327
80	MD	-	-	0.63	-	-	420.8
80	LD	-	-	0.20	-	-	130.9
100	HCD	0.032	0.13	0.63	21.0	85	409
240	MCD	0.01	0.04	0.2	6.7	27	135
240	LCD	0.0032	0.013	0.063	2.1	8.7	42.2

[a]HD, high dose; MD, medium dose; LD, low dose; HCD, high combination dose; MCD, medium combination dose; LCD, low combination dose
[b]mg/kg body weight
[c]N-nitrosamines were administered via the drinking-water 5 × per week

Table 2. Median survival time, incidence of malignant tumours and incidence of liver tumours in male SD rats exposed to N-nitrosodiethylamine (NDEA), N-nitrosopyrrolidine (NPYR) or N-nitrosodiethanolamine (NDELA) or their combination

Treatment[a]	Median survival time (days) (95% confidence interval)	Incidence of malignant tumours (%)	Incidence of liver tumours (%)	Significance from control[b]
Control	931 (910-945)	29	0.6	-
NDEA HD	854 (816-892)[c]	65	45.0	0.001
NDEA MD	879 (812-939)[d]	34	3.8	0.001
NDEA LD	914 (879-1017)	29	2.5	0.043
NPYR HD	951 (854-1001)	44	21.3	0.001
NPYR MD	879 (875-948)	35	5.0	0.001
NPYR LD	926 (907-989)	29	1.3	NS
NDELA HD	928 (897-985)	35	7.5	0.001
NDELA MD	934 (908-990)	36	1.3	NS
NDELA LD	916 (889-968)	36	2.5	0.038
HCD	914 (875-983)	48	16.0	0.001
MCD	944 (908-983)	33	4.2	0.001
LCD	937 (903-965)	31	1.7	0.059

[a] HD, high dose; MD, medium dose; LD, dow dose; HCD, high combination dose; MCD, medium combination dose; LCD, low combination dose
[b] Pooled, age-adjusted analysis for heterogeneity (two-tailed) between observed and expected liver tumour incidences according to Peto et al. (1980)
[c] $p = 0.001$ versus control according to the Wilcoxon rank sum test
[d] $p = 0.003$ versus control according to the Wilcoxon rank sum test

Age-specific incidence

Time from start of the experiment until age at death was analysed by methods of censored survival times (Kalbfleisch & Prentice, 1980), taking death with liver tumour as the event of incidence and death without liver tumour as a censoring event. For the investigation of a combination effect of the three compounds, two multivariate proportional hazards models were fitted to the data:

$$IR\ (t/d_1,d_2,d_3) = IR_0(t)\exp(\beta_1 d_1 + \beta_2 d_2 + \beta_3 d_3)\ \text{and} \tag{1}$$

$$IR\ (t/d_1,d_2,d_3) = IR_0(t)\exp(\beta_1 d_1 + \beta_2 d_2 + \beta_3 d_3 + \beta_4 d_1 d_2 d_3), \tag{2}$$

where IR is incidence rate; t is time; d is dosage; and β is a regression parameter. The presence of a third-order combination effect — the only type of combination envisaged by the design — was tested by the null hypothesis, $H_0: \beta_4 = 0$ versus $H_1: \beta_4 \neq 0$. Relative risks, obtained as $RR_i = \exp(\beta_i)$, gave the factors by which one has to multiply the instantaneous risk of dying with a tumour if the dose is increased by one unit. The results of the fit are shown in Table 3; they indicate no significant third-order combination effect, since the product term of the three dosages ($d_1 \times d_2 \times d_3$) does not improve the quality of the fit. This is confirmed by the deviance of the two models, which equals 1.34 with a corresponding nonsignificant p value of 0.25.

Table 3. Results of fitting multivariate models (1)[a] and (2)[b]

	Regression parameter	β[c]	SE[d]	RR[e]	p value[f]
Equation (1)					
NDEA	β_1	48.1	3.0	7.4×10^{20}	10^{-5}
NPYR	β_2	6.9	0.8	914.7	10^{-5}
NDELA	β_3	0.8	0.2	2.2	0.0008
Combination	β_4	–	–	–	–
Equation (2)					
NDEA	β_1	48.6	3.1	1.2×10^{21}	10^{-5}
NPYR	β_2	7.1	0.8	1.2×10^3	10^{-5}
NDELA	β_3	0.9	0.2	2.5	0.0001
Combination	β_4	-127.0	111.5	0.0	0.3

[a] (1): $IR\ (t/d_1,d_2,d_3) = IR_0(t) \exp(\beta_1 d_1 + \beta_2 d_2 + \beta_3 d_3)$
[b] (2): $IR\ (t/d_1,d_2,d_3) = IR_0(t) \exp(\beta_1 d_1 + \beta_2 d_2 + \beta_3 d_3 + \beta_4 d_1 d_2 d_3)$
[c] Statistically estimated value of the respective regression parameter
[d] Standard error
[e] Relative increase in the risk of dying with liver tumour if the dosage is increased by one unit
[f] According to Wald's test (Kalbfleisch & Prentice, 1980)
NDEA, N-nitrosodiethylamine; NPYR, N-nitrosopyrrolidine; NDELA, N-nitrosodiethanolamine

Tumour-free survival

On the basis of the final model (1), curves of survival free of liver tumour were estimated for different combinations of dosages of the three compounds (d_1, d_2, d_3) according to Kalbfleisch and Prentice (1980) and plotted for ages over 500 days and survival probabilities within the range of 100-75%. Extrapolation to doses lower than those used in the experiment was performed by calculating the survival curves for doses $d_1 < 0.01$ mg/kg, $d_2 < 0.04$ mg/kg and $d_3 < 0.2$ mg/kg of NDEA, NPYR and NDELA, respectively.

The curves of the highest and lowest experimental dosages (scaled by a factor of 10), together with an extrapolated ten-fold lower dose, are depicted in Figure 1. It is clear that extrapolation to a dosage lower than that used experimentally results in a minor reduction in age-specific liver tumour incidence as compared to that achieved by an equivalent reduction within the experimental dose range. Thus, the additional gain in liver tumour-free survival if the dose 3 were reduced to one-tenth of the lowest experimental dose would be maximally one-fifth of that gain in liver tumour-free survival observed with the reduction from the highest to the lowest experimental dose. This finding is due to the reduced life expectancy of the animals, which were already at an advanced age when liver tumours were observed. All liver tumours in this experiment occurred in rats older than two years; the lowest experimental dosages are associated with corresponding times to liver tumour of 1100-1250 days, i.e., a time when fewer than 20% of the control rats were still alive (Berger *et al.*, 1987). Therefore, if the incidence of liver tumours were reduced by lowering exposure to *N*-nitroso compounds, no measurable increase in liver tumour-free survival could be expected, due to the limitation in overall survival. In these dose ranges of *N*-nitrosamines, allowing a median survival of more than 900 days, a further reduction in carcinogen-induced liver tumour occurrence would not contribute to a longer overall survival of the animals due to competitive, probably independent causes of death. In other words, a quasi-threshold of carcinogen exposure has been reached below which further reduction may alter the cause of death or the incidence of non-fatal liver neoplasms but not life expectancy.

Carcinogenic potency

For each nitrosamine and for their combination, a potency measure was calculated, as follows. The univariate proportional hazards model was fitted to the data for the three experimental groups and the control group, and liver tumour-free survival was calculated on the basis of the fitted model. The three quantiles, $t_{90}(d)$, $t_{95}(d)$ and $t_{99}(d)$ of liver tumour-free survival time were estimated and plotted against dose (see Abel *et al.*, 1986), $t_{95}(d)$, e.g., being the time at which 5% of the animals had died with a liver tumour found at necropsy. The decrease in these quantiles with increasing dose was further analysed by linear regression by fitting the straight line $t_p(d) = a_p - bd$ to the observed curves of the 90%, 95% and 99% quantiles, as shown in Figure 2. The slope of the regression line was used as a potency measure, and the confidence interval of the slope was calculated tentatively in order to assess variability. For the observed range of tumour incidences, the potency of the three *N*-nitroso compounds was quantified: relative to NDELA (the least potent carcinogen), the time to death with liver tumour per dose unit (0.1 mg/kg) was decreased by factors of 40 and four for NDEA and NPYR, respectively, and by a factor of three for the combination (Table 4).

The empirically derived potency measure obtained from the slope of these curves (b) was used to quantify the differences in risk inherent to the three carcinogens and to assess the potency of the combination (at least for small percentages of tumours), which was derived directly from the univariate regression model applied to the dose-response data for each

Relationship between dose and risk reduction

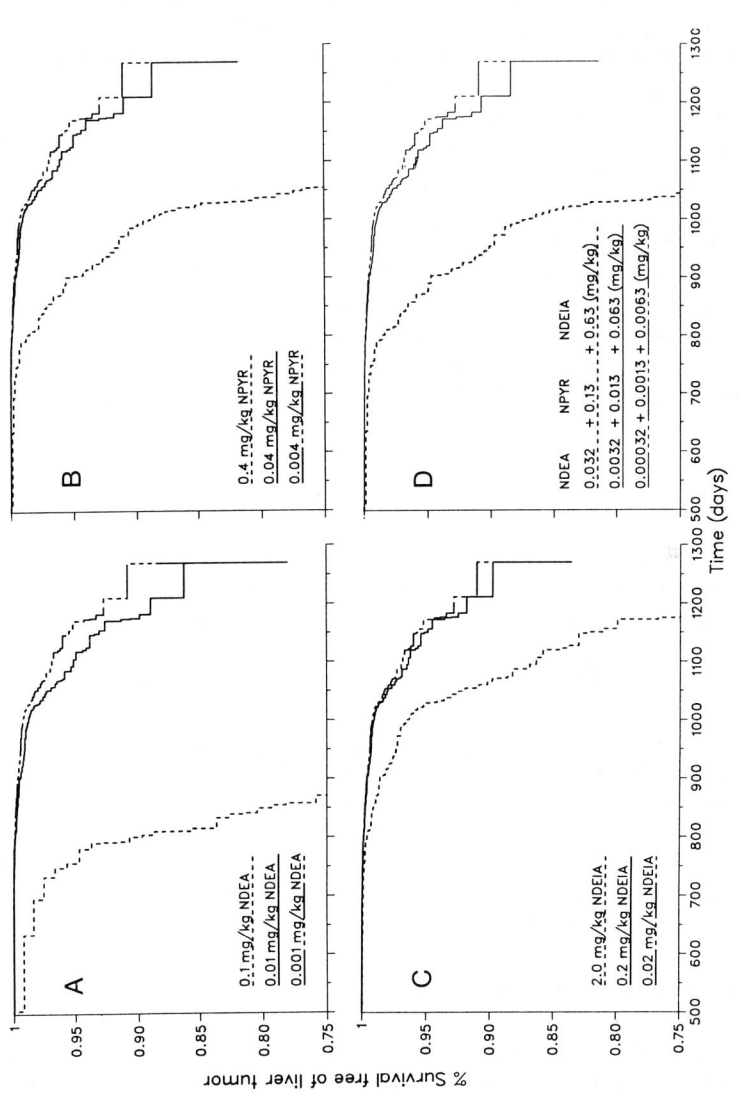

Figure 1. Survival without a liver tumour of rats older than 500 days with survival probabilities of 75–100% for A, N-nitrosodiethylamine (NDEA); B, N-nitrosopyrrolidine (NPYR); C, N-nitrosoethanolamine (NDELA); and D, their combination[a]

[a]The combination was tested at doses within the experimental range and, additionally at a ten-fold lower dose.

compound separately. It can be related mathematically to the parameters of Cox's regression model (Cox, 1972). It is tempting to extrapolate the quantile lines to doses lower than those used experimentally, and even to zero exposure, which is equal to the intercept of the regression lines. A relatively high background incidence of liver cancer is indicated by such extrapolations.

If similar mechanisms are active in man after exposure to low levels of carcinogens, several conclusions can be drawn.

Figure 2. Relationship between individual, daily dose of *N*-nitrosodiethylamine and time at which 1% (■), 5% (●) and 10% (▲) of animals had died with a liver tumour

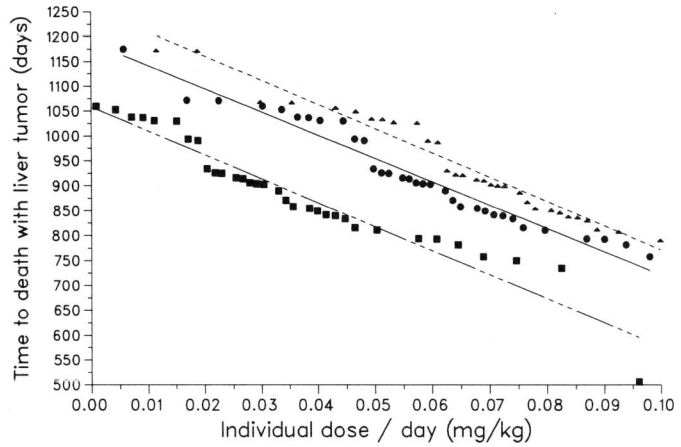

Table 4. Results of linear regression applied to quantile plots[a]

Carcinogen[b]	Quantile (%)	a (days)	b (days)	95% confidence interval for b	Correlation coefficient
NDEA	90	1258	486	450-524	-0.98
	95	1188	466	439-594	-0.98
	99	1058	480	432-528	-0.96
NPYR	90	1237	55	44-67	-0.94
	95	1163	46	35-58	-0.91
	99	1047	60	39-80	-0.93
NDELA	90	1295	14	8-20	-0.96
	95	1204	12	9-16	-0.96
	99	1042	11	6-15	-0.96
Combination	90	1219	31	29-33	-0.99
	95	1158	31	28-33	-0.99
	99	1048	36	32-41	-0.99

[a] Linear regression: $t_p = a_p - bd$, where t is time in days, d is individual dose per day, a is the estimated time to liver tumour occurrence at zero exposure and b is the decrease in manifestation time of liver tumours that corresponds to a 0.1 mg/kg increase in *N*-nitrosamine dose; the ratio of the combination is 0.05:0.2:1.
[b] NDEA, *N*-nitrosodiethylamine; NPYR, *N*-nitrosopyrrolidine; NDELA, *N*-nitrosodiethanolamine

(i) A reduction in or exclusion of genotoxic carcinogens from the human environment can contribute to reduced tumour occurrence even if no causal relationship with cancer has been established, since this approach reduces the effect of the summation of small carcinogenic effects (Schmähl et al., 1989).

(ii) Such a reduction will not necessarily translate into longer tumour-free survival if the neoplasms usually occur at the end of an individual's lifetime.

(iii) A background of cancer incidence probably exists, independent of exogenous induction, which seems to be age-related only and is therefore not amenable to primary prevention. This latter view is in accord with recent epidemiological estimates of the proportion of preventable cancers in certain organs (Wahrendorf, 1987).

References

Abel, U., Berger, J. & Edler, L. (1986) A method for analyzing the dependence of failure-time statistics on quantitative covariates. *Elektron. Datenverarb. Med. Biol.*, **17**, 90-92

Berger, M.R., Schmähl, D. & Zerban, H. (1987) Combination experiments with very low doses of three genotoxic N-nitrosamines with similar organotropic carcinogenicity in rats. *Carcinogenesis*, **8**, 1635-1643

Cox, D.R. (1972) Regression models and lifetables (with discussion). *J.R. Stat. Soc. B.*, **34**, 187-202

Kalbfleisch, J.D. & Prentice, R.L. (1980) *The Statistical Analysis of Failure Time Data*, New York, John Wiley

Peto, R., Pike, M.C., Day, N.E., Gray, R.G., Lee, P.N., Parish, S., Peto, J., Richards, S. & Wahrendorf, J. (1980) Annex. Guidelines for simple, sensitive significance tests for carcinogenic effects in long-term animal experiments. In: *IARC Monographs on the Evaluation of the Carcinogenic Risk of Chemicals to Humans*, Suppl. 2, *Long-term and Short-term Screening Assays for Carcinogens: A Critical Appraisal*, Lyon, IARC, pp. 311-426

Schmähl, D., Preussmann, R. & Berger, M.R. (1989) The causes of cancer – an alternative view to Doll and Peto (1981). *Klin. Wochenschr.*, **67**, 1169-1173

Wahrendorf, J. (1987) An estimate of the proportion of colorectal and stomach cancers which might be prevented by certain changes in dietary habits. *Int. J. Cancer*, **40**, 625-628

Relevance to Human Cancer of *N*-Nitroso Compounds,
Tobacco Smoke and Mycotoxins.
Ed. I.K. O'Neill, Chen and H. Bartsch
Lyon, International Agency for Research on Cancer
© IARC, 1990

CARCINOGENIC ACTIVITY OF ENDOGENOUSLY SYNTHESIZED *N*-NITROSOBIS(2-HYDROXYPROPYL)AMINE IN RATS

Y. Konishi[1,4], K. Yamamoto[1,2], H. Eimoto[1], M. Tsutsumi[1], M. Sugimura[2], H. Nii[3] & Y. Mori[3]

[1]*Department of Oncological Pathology, Cancer Center,*
[2]*Department of Oral and Maxillofacial Surgery,
Nara Medical College; and* [3]*Laboratory of Radiochemistry,
Gifu Pharmaceutical University, Gifu, Japan*

The carcinogenic activity of endogenously synthesized *N*-nitroso-bis(2-hydroxypropyl)amine (NDHPA) was investigated in male Wistar rats administered bis(2-hydroxypropyl)amine (DHPA), mixed into a powdered diet at a concentration of 1%, and $NaNO_2$ dissolved in distilled water at concentrations of 0.15% and 0.3%, for 94 weeks. Urinary excretion of NDHPA clearly demonstrated its endogenous synthesis in rats given 1% DHPA and 0.3% $NaNO_2$, but not in the groups receiving either of these precursors alone. Tumours of the nasal cavity, lung, oesophagus, liver and urinary bladder were found in rats treated with 1% DHPA and 0.15% or 0.3% $NaNO_2$. The incidences of nasal cavity and lung tumours reached 74% and 58%, respectively, in rats given 1% DHPA and 0.3% $NaNO_2$. The tumour distribution was almost the same as that seen in rats given NDHPA. These results indicate that endogenously synthesized NDHPA has similar carcinogenic activity to exogenously administered NDHPA in rats.

N-Nitrosobis(2-hydroxypropyl)amine (NDHPA) is known to be a potent carcinogen, affecting mainly the nasal cavity, lung, oesophagus, liver, thyroid, urinary bladder and kidney (Konishi *et al.*, 1976; Mohr *et al.*, 1977). In the present study, we investigated the carcinogenic activity of endogenously synthesized NDHPA on the basis of our previous finding (Konishi *et al.*, 1987; Mori *et al.*, 1988) that NDHPA could be synthesized *in vivo* and *in vitro* from bis(2-hydroxypropyl)amine (DHPA) and $NaNO_2$, which occur commonly.

Experimental protocol

Groups of 20–28 male Wistar rats (Shizuoka Laboratory Animal Center, Shizuoka, Japan), six weeks old, were given DHPA (Tokyo Kasei Co., Tokyo, Japan) mixed in a commercial powdered diet, Oriental M (Oriental Yeast Co., Tokyo, Japan), at a concentration of 1%. $NaNO_2$ (Wako Pure Chemical Industry, Kyoto, Japan) was dissolved in distilled water at concentrations of 0.15% and 0.3%. Rats had free access to diet and water continuously throughout the experimental period.

---

[4]To whom correspondence should be addressed

Urinary excretion of endogenously synthesized NDHPA

Three to five rats were used to investigate urinary excretion of endogenously synthesized NDHPA. Rats were placed in individual metabolic cages, and urine samples were collected for 24h. The 24-h urine and the washings after filtration were adjusted to 200 ml with water and extracted three times with 200 ml ethyl acetate. The combined extracts were evaporated to dryness at < 40°C, cleaned up by thin-layer chromatography, taken up in acetonitrile to a final volume of 1 ml and analysed by high-performance liquid chromatography. The calculated recovery of NDHPA from the urine was 68.0 ± 2.4% (mean ± SE), and the results were corrected accordingly. The results (Table 1) clearly demonstrate endogenous synthesis of NDHPA in rats given DHPA with $NaNO_2$.

Table 1. Urinary excretion of endogenously synthesized *N*-nitrosobis(2-hydroxypropyl)amine (NDHPA) in rats given $NaNO_2$ and bis(2-hydroxypropyl)amine (DHPA)

Treatment	No. of rats examined	Experimental weeks	Urinary excretion of NDHPA (μmol/rat per day)	
			Average	Range
Untreated	3	24	ND[a]	-
0.3% $NaNO_2$	3	24	ND	-
1% DHPA	3	24	ND	-
1% DHPA + 0.3% $NaNO_2$	3	24	1.51	1.04 - 2.20
	3	34	0.97	0.69 - 1.47
	5	80	1.51	1.11 - 2.20

[a]Not detectable (limit of detection, 50 nmol (200 ml))

Carcinogenicity of endogenously synthesized NDHPA

The incidences of tumours at the target organs of NDHPA are presented in Table 2. Most of the nasal cavity tumours developed in the respiratory region, a few being found in the squamous region; none was found in the olfactory region. This finding is in line with previous reports (Mohr *et al.*, 1977; Lijinsky & Reuber, 1984) concerning administration of exogenous NDHPA and metabolically related compounds to rats. The distribution and histological types of lung tumours were also similar to those seen in rats given NDHPA exogenously (Konishi *et al.*, 1976). The low incidence of tumour induction in the other NDHPA target organs is presumably a reflection of the relatively low amount of NDHPA synthesized. Tumours at sites that are not target organs of NDHPA were found only at levels similar to those previously reported for spontaneous tumours in male Wistar rats (Maekawa *et al.*, 1983). Endogenously synthesized NDHPA thus has similar carcinogenic activity to exogenously administered NDHPA in rats.

Since the precursors investigated and other nitrosatable amines are widely distributed in the human environment, the results suggest that endogenously synthesized nitrosamines may play an important role in cancer development in man.

Table 2. Incidences of tumours at the target organs of N-nitrosobis(2-hydroxypropyl)amine in rats treated with bis(2-hydroxypropyl)amine (DHPA) and/or $NaNO_2$

Treatment	Effective no. of rats	Incidence of tumours[a]								Oesophagus (PAP)	Liver (HCC)	Urinary bladder (TCC)	Thyroid adenoma	Kidney
		Nasal cavity			Lung									
		CA	PAP	Total	AC	A	SCC	PAP	Total					
Untreated	19	0	0	0	0	0	0	0	0	0	0	0	1(5)	0
0.15% $NaNO_2$	18	0	0	0	0	0	0	0	0	0	0	0	1(6)	0
0.3% $NaNO_2$	16	0	0	0	0	0	0	0	0	0	0	0	2(13)[b]	0
1% DHPA	16	0	0	0	0	0	0	0	0	0	0	0	0	0
1% DHPA + 0.15% $NaNO_2$	19	0	0	0	0	0	0	3(16)	3(16)	0	0	1(5)	1(5)	0
1% DHPA + 0.3% $NaNO_2$	19	10(53)	11(58)	14(74)[c]	1(5)	2(11)	2(11)	10(53)	11(58)[d]	2(11)	1(5)	0	0	0

[a] CA, carcinoma; PAP, papilloma; AC, adenocarcinoma; A, adenoma; SCC, squamous-cell carcinoma; HCC, hepatocellular carcinoma; TCC, transitional-cell carcinoma; in parentheses, %
[b] Including one adenocarcinoma
[c] $p < 0.01$ compared with groups 1–5
[d] $p < 0.01$ compared with groups 1–4, $p < 0.05$ compared with group 5

Acknowledgements

Supported by a Grant-in-Aid from the Ministry of Health and Welfare, and by a Grant-in-Aid from the same source for the Comprehensive 10-year Strategy for Cancer Control, Japan.

References

Konishi, Y., Denda, A., Kondo, H. & Takahashi, S. (1976) Lung carcinomas induced by oral administration of N-nitrosobis(2-hydroxypropyl)amine in rats. *Jpn. J. Cancer Res. (Gann)*, **67**, 773-780

Konishi, Y., Yokose, Y., Mori, Y., Yamazaki, H., Yamamoto, K., Nakajima, A. & Denda, A. (1987) Lung carcinogenesis by *N*-nitrosobis(2-hydroxypropyl)amine-related compounds and their formation in rats. In: Bartsch, H., O'Neill, I.K. & Schulte-Hermann, R., eds, *The Relevance of N-Nitroso Compounds to Human Cancer: Exposures and Mechanisms* (IARC Scientific Publications No. 84), Lyon, IARC pp. 250-252

Lijinsky, W. & Reuber, M.D. (1984) Dose-response study with *N*-nitrosodiethanolamine in F344 rats. *Food chem. Toxicol.*, **22**, 23-26

Maekawa, A., Onodera, H., Taniguchi, H., Furuta, K., Kodama, Y., Horiuchi, S. & Hayashi, Y. (1983) Neoplastic and non-neoplastic lesions in aging Slc: Wistar rats. *J. Toxicol. Sci.*, **8**, 279-290

Mohr, U., Reznik, G. & Pour, P. (1977) Carcinogenic effect of diisopropanolnitrosamine in Sprague-Dawley rats. *J. Natl Cancer Inst.*, **58**, 361-366

Mori, Y., Yamazaki, H., Yamamoto, K., Nakajima, A., Yokose, Y. & Konishi, Y. (1988) Mutagenesis of isopropanolamine and ethanolamine after reaction with sodium nitrite. *Exp. Oncol. (Life Sci. Adv.)*, **7**, 35-40

EFFECTS OF LONG-TERM INHALATION OF N-NITROSODIMETHYLAMINE IN RATS

R.G. Klein, I. Janowsky, B.L. Pool-Zobel, P. Schmezer, R. Hermann,
F. Amelung, B. Spiegelhalder & W.J. Zeller

*Institute of Toxicology and Chemotherapy,
German Cancer Research Center, Heidelberg,
Germany*

The toxicological evaluation and histopathological findings of a long-term inhalation study in progress with N-nitrosodimethylamine (NDMA) are presented. Exposure was to three concentrations of NDMA: 0.04, 0.2 and 1.0 ppm (corresponding to 120, 600 and 3000 µg/m3). A significant reduction in median survival time (nine months) was seen in animals treated with the highest concentration of NDMA. Tumours occurred mainly in the nasal cavity, with the highest incidences in the groups receiving 1.0 and 0.2 ppm NDMA (19/36 and 31/36 tumour-bearing animals); in the lowest exposure group, 13/36 nasal tumours have been observed. The survival time of this treatment group, however, was about two months longer than that of the controls.

Exposure by inhalation is similar to a major route of human exposure, especially for volatile compounds and particulate matter. N-Nitrosodimethylamine (NDMA) is a highly potent carcinogen to experimental animals, inducing high yields of tumours following application by many routes, including inhalation. Druckrey *et al.* (1967) observed exclusively tumours of the olfactory region (aesthesioneuroblastomas) in BD rats, and no liver tumour, following chronic inhalation (2 × 30 min/week) of 50 and 100 ppm NDMA, corresponding to 150 and 300 mg/m3, respectively. Moiseev and Benemanskij (1975) reported only tumours in organs other than the nose in white rats, namely liver, kidneys and lungs, following continuous inhalation of 0.2 mg/m3 NDMA. Benemanskij *et al.* (1981) confirmed the latter findings using a continuous concentration of 0.66 mg/m3 NDMA. Humans may be exposed in several occupational situations *via* inhalation (Rounbehler & Fajen, 1982; Spiegelhalder & Preussmann, 1982), in which amounts up to several hundred micrograms per cubic metre have been found. Furthermore, NDMA has been identified in tobacco smoke (Brunnemann *et al.*, 1977; Hoffmann *et al.*, 1984). For this reason, we initiated a long-term inhalation study using exposure schedules in which the lowest concentration was comparable to peak concentrations at certain work places. Established analytical methods for determining nitrosamines were modified and applied for current measurements and for safety control of personnel involved in the experiment, NDMA-contaminated laboratory equipment, and animals' fur and faeces after exposure.

Materials and methods

Female Sprague-Dawley rats (Hanover, FRG), aged eight weeks at the beginning of the experiment were housed in Makrolon cages and fed Altromin pellets (FRG) and tap-water *ad libitum*. Four groups of 36 animals each were randomized and received 0 ppm (NDMA concentration below analytical detection limit), 1.0 ppm, 0.2 ppm or 0.04 ppm (Table 1). Tumour-bearing animals were sacrificed by ether inhalation, and their organs were excised and fixed in formalin. Paraffin-embedded sections (5 µm) were prepared for microscopic analysis by haematoxylin and eosin and periodic acid-Schiff staining. In addition to the organs taken during routine histopathology, sagittal sections of the skull and cross-sections are prepared (modified from Young, 1981).

Table 1. Exposure of rats to *N*-nitrosodimethylamine (NDMA) by long-term inhalation

Concentration (nominal)		No. of animals	Daily uptake (effective mean) (µg/kg)	Total uptake (mg/kg)	Length of exposure (days)	Age at death (days)	
ppm	µg/m³					Median	Range
0	0	36	0	0	77-207	795	341-1128
0.04	120	36	10	1.3-2.0	95-207	860	167-1200
0.2	600	36	40	3-8	57-207	772	268-1102
1.0	3000	36	180	13-37	49-207	524	220-710

Animals were exposed to NDMA (four times per week, 4-5 h/day, 207 days) in stainless-steel inhalation boxes and cages without food, water or bedding material in order to avoid contamination with NDMA vapours. Animals were transferred to wire cages just before exposure and returned after exposure. Control animals were handled in the same way but exposed to clean air. The wire cages inside the inhalation boxes were rotated for each exposure in order to avoid local flow asymmetries that might influence NDMA uptake. The 4-h exposure was followed by 2 h of enhanced air flow to diminish NDMA contamination of the boxes, cages and animals' fur. The daily uptake of NDMA was estimated to be 10 µg/kg body weight for animals exposed to 0.04 ppm NDMA, 40 µg/kg bw for animals exposed to 0.2 ppm and 180 µg/kg bw for those exposed to 1.0 ppm, on the basis of actual nitrosamine concentrations and the mean breathing volume of all rats in each group (Table 1).

The concentrations of NDMA and exhaled CO_2 were determined continuously for each box. Even at room temperature, NDMA is very volatile (Klein, 1982) and is readily absorbed in water and by moist surfaces, as well as by plastic materials; it may also be exhaled to some extent from the animals' respiratory tract (Klein & Schmezer, 1984). Three analytical methods were used to determine the concentrations of NDMA and are listed below according to decreasing sensitivity,

(i) *ThermoSorb/N tubes (Thermo Electron Corporation, USA).* Air samples from the inhalation boxes are drawn through magnesium silicate (Rounbehler *et al.*, 1980), and volatile nitrosamines are absorbed quantitatively. The sorbent is eluted with solvent and then analysed with a thermal energy analyser (Fine *et al.*, 1975). The detection limits of this method are in the nanogram range for NDMA. This most sensitive method was used to calibrate the other procedures and was used preferentially to determine the dose of the lowest exposure group.

(ii) *Ultraviolet absorption.* Air from the boxes is drawn through impinger washing bottles with 20 ml water. NDMA is absorbed to more than 90% in the first impinger. Samples are analysed directly with an LKB spectrophotometer (Ultrospec II). The ultraviolet spectrum of an air sample taken from an exposure chamber shows that volatile substances other than NDMA (e.g., animal excretions) do not contribute much to the signal, indicating that the method is suitable for control measurements after calibration with an NDMA standard at 230 nm. Depending on the molar extinction of NDMA, the content of impurities in the air sample and the water volume in the impingers (two in series), the sensitivity of this method was determined to be in the range of 10 µg/m³ NDMA in air.

(iii) *Tecan chemoluminescence analyser (Nucletron, Munich, FRG).* For analysis of nitrosamine vapours, samples were pyrolysed at 450°C before insertion into the Tecan. NDMA is cracked thermally to produce gaseous NO and other compounds; NO is detected quantitatively by chemoluminescence. The accuracy of the Tecan device is in the range of 100 µg/m³ NDMA.

The agreement among the three methods was satisfactory and lies within methodological variation. Differences in NO_x detection during daytime (range, up to 100 ppb) may be a result of the variation in background NO_x in the atmosphere.

Safety precautions

Negative pressure must be maintained between exposure chambers, the safety box and the laboratory area. Charcoal filters (50 × 40 × 20 cm) should be used with sufficient contact with the exhaust air stream (> 10 sec). The efficiency of the filters is monitored analytically. Measurements resulted in an effectivity of 10^3-10^4 for a decrease in the nitrosamine concentration.

One-way protective clothing, gas masks with a fresh-air supply and personal monitoring were provided for technicians. The stainless-steel walls of the safety area were additionally covered with charcoal filter mats with a surface of 5 m² and an absorption capacity of 1 kg/m² organic material reported from the manufacturer (Technopor 5111, Sorbexx Gefrees, FRG).

Survival and histopathological findings

The median survival time of animals given 1 ppm NDMA was nine months less than that of untreated controls, owing to toxicity, the median survival of animals given 0.04 ppm was two months longer than that of control rats (Fig. 1, Table 1). Body weight curves additionally document the toxic effect of the highest concentration of NDMA (Fig. 2).

The histopathological findings in the nasal cavity are given in Table 2. A remarkable difference in the histological types of nasal tumours was seen between the highest dose group and the other groups: at 1 ppm NDMA, 47% of tumours were aesthesioneuroblastomas, with a median manifestation time of 320 days, whereas only 6% and 15% of this tumour type were observed following inhalation of 0.2 and 0.04 ppm NDMA, respectively. In the latter two groups, mucoepidermoid tumours represented the greatest proportion.

Figure 1. Survival of rats exposed by inhalation to *N*-nitrosodimethylamine at 0 (—), 0.04 ppm(----), 0.2 ppm (- -) and 1.0 ppm (---)

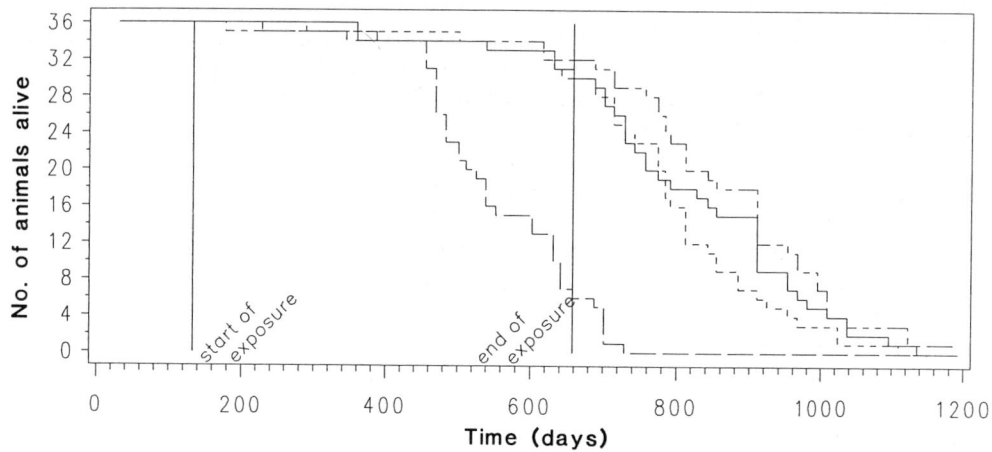

Figure 2. Body weight of rats exposed by inhalation to *N*-nitrosodimethylamine at 0(—), 0.04 ppm(----), 0.2 ppm (- -) and 1.0 ppm (---)

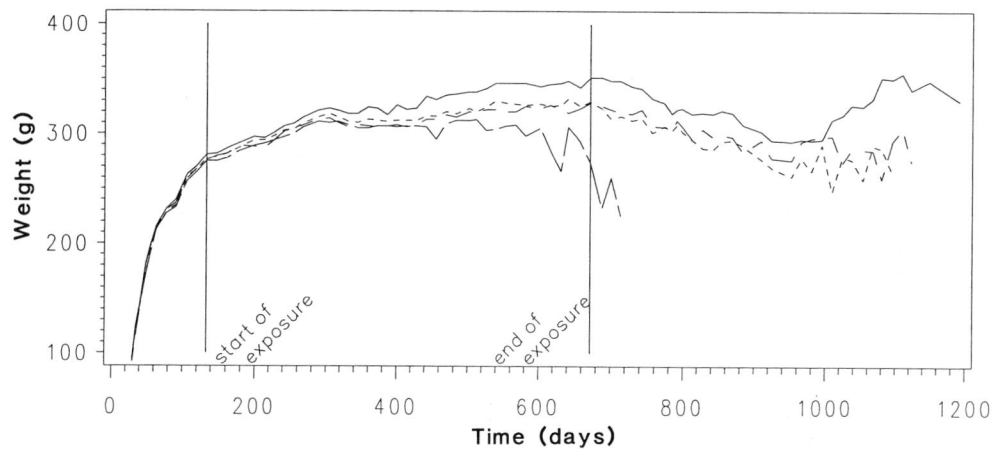

Mucoepidermoid tumours and carcinomas were located mainly in the respiratory part of the nasal mucosa, approximately half in the anterior and middle part and less than 25% in the caudal part of the respiratory epithelium. Most of mucoepidermoid carcinomas included epithelial elements and showed a destructive growth pattern. The aesthesioneuroblastomas were located exclusively in the olfactory part of the nasal mucosa. Approximately half exhibited destruction of the cribiform plate with invasion of the olfactory bulb; most developed only a small amount of neurofibrillary matrix as well as pseudo-rosettes and rosette forms. Squamous-cell carcinomas were found in the anterior nasal cavity, like the mucoepidermoid tumours; sarcomas were localized in the bones of the nasal or basal skull. No tumour was seen in respiratory or olfactory regions of the septal mucosa. The median time to manifestation of nasal tumours was shorter in the group given 1 ppm NDMA than in the other groups. These observations indicate that both tumour manifestation and tumour type depend on the concentration and/or the total amount of inhaled NDMA.

The occurrence of tumours other than nasal tumours is summarized in Table 3. Hepatic tumours (except cholangiomas) were seen exclusively in the groups given 0.04 ppm and 0.2 ppm NDMA; this result might be attributable to the longer survival in these two groups compared to the highest dose group.

Table 2. Histopathological findings in rats after long-term inhalation of *N*-nitrosodimethylamine (NDMA): nasal tumours

Finding	NDMA treatment[a]			
	0	0.04 ppm	0.2 ppm	1.0 ppm
No. of tumour-bearing animals	-	13	31	19
Aesthioneuroblastomas		2 (671)	2 (562)	9 (320)
Mucoepidermoid tumours		11 (770)	30 (662)	7 (491)
(including carcinomas)		2 (594)	8 (583)	3 (498)
Squamous-cell carcinomas			2 (498)	1 (398)
Neurogenic sarcomas				1 (454)
Osteogenic sarcomas				2 (510)
First and last tumour (days)	-	568-897	356-972	198-579
Uptake (mg/kg)	0	2.0	7.8-8.0	20-37

[a]Median time to manifestation in days, in parentheses. One animal receiving 0.2 ppm and one receiving 1.0 ppm developed both a mucoepidermoid tumour and an aesthesioneuroblastoma; one animal receiving 0.2 ppm developed both a mucoepidermoid carcinoma and a squamous-cell carcinoma, and another had both a mucoepidermoid tumour and a carcinoma.

These findings demonstrate that exposure of rats to 0.04 ppm NDMA, corresponding to 120 μg/m3, can induce a high rate of nasal tumours. This result is of particular significance with regard to observed peak concentrations of NDMA at certain work places.

Acknowledgements

This study was supported by Projekt Europäisches Forschungszentrum für Massnahmen zue Luftreinhaltung, Kernforschungszentrum Karlsruhe. We thank G. Bielefeld for translating the Russian references.

References

Benemanskij, V.V., Prusakov, V.M, & Leshchenko, M.E. (1981) Study on blastogene effect of low concentrations of dimethylnitrosamine, dimethylamine and nitrogen dioxide. *Vopr. Onkol.*, 27 (10), 56-62

Brunnemann, K.D., Yu, L. & Hoffmann, D. (1977) Assessment of carcinogenic volatile N-nitrosamines in tobacco and in mainstream and sidestream smoke from cigarettes. *Cancer Res.* 37, 3218-3222

Druckrey, H., Preussmann, R., Ivankovic, S. & Schmähl, D. (1967) Organotrope carcinogene Wirkungen bei 65 verschiedenen N-Nitroso-Verbindungen an BD-Ratten. *Z. Krebsforsch.*, 69, 103-201

Fine, D.H., Lieb, D. & Rufeh, F. (1975) Principle of operation of the Thermal Energy Analyzer for the trace analysis of volatile and non-volatile N-nitroso compounds. *J. Chromatogr.*, 107, 351-357

Table 3. Histopathological findings in rats after long-term inhalation of N-nitrosodimethylamine (NDMA): other tumours

Finding[a]	NDMA treatment			
	0	0.04 ppm	0.2 ppm	1.0 ppm
Respiratory tract				
Adenocystic lung carcinoma				1
Tracheal adenoma			1	
Digestive tract				
Hepatocellular carcinoma		1	1	
Hepatic adenoma		2	1	
Cholangiocarcinoma				1
Cholangioma	12	11	7	8
Pancreatic carcinoma	2			
Pancreatic insuloma	2	1		
Intestinal tumour	1 (adenocarcinoma)	1 (myoma)		
Endocrine glands				
No. of tumour-bearing animals	30	30	28	12
Pituitary adenoma	19	20	19	
Suprarenal gland cortical adenoma	18	19	14	10
Suprarenal gland phaeochromocytoma	9	11	7	
Thyroid adenoma	12	15	12	3
Mammary gland				
No. of tumour-bearing animals	24	22	18	6
Adenoma, fibroma, fibroadenoma	26	28	15	3
Adenocarcinoma	14	9	8	
Fibrosarcoma		2		
Squamous-cell carcinoma			1	
Other tumours	Neurogenic sarcoma in the abdominal cavity Skin (2 squamous-cell carcinomas, 1 haemangioma, 1 sebaceous adenoma) Rhabdomyoma Squamous-cell carcinoma of the oral mucosa	Neurogenic sarcoma in the abdominal cavity (2) Leukaemia Lymphoma Uterine myoma Squamous-cell carcinoma of the oral mucosa (2)	Histiocytic sarcoma of the abdominal cavity	Neurogenic sarcoma of the abdominal cavity Ependymoma of the cerebrum Astrocytoma of the cerebrum Theca-cell tumour of the ovary Adenocarcinoma of the oral mucosa

[a]Multiple tumours were seen in ten controls, nine animals receiving 0.04 ppm NDMA and five receiving 0.2 ppm

Hoffmann, D., Brunnemann, K.D., Adams, J.D. & Hecht, S.S. (1984) Formation and analysis of N-nitrosamines in tobacco products and their endogenous formation in tobacco consumers. In: O'Neill, I.K., von Borstel, R.C., Miller, C.T., Long, J. & Bartsch H., eds, N-*Nitroso Compounds: Occurrence, Biological Effects and Relevance to Human Cancer* (IARC Scientific Publications No. 57), Lyon, IARC, pp. 743-762

Klein, R.G. (1982) Calculations and measurements on the volatility of N-nitrosamines and their aqueous solutions. *Toxicology*, **23**, 135-147

Klein, R.G. & Schmezer, P. (1984) Quantitative measurements of the exhalation rate of volatile N-nitrosamines in inhalation experiments with anaesthetized Sprague-Dawley rats. In: O'Neill, I.K., von Borstel, R.C., Miller, C.T., Long J. & Bartsch, H., eds, N-*Nitroso Compounds: Occurrence, Biological Effects and Relevance to Human Cancer* (IARC Scientific Publications No. 57), Lyon, IARC, pp. 513-517

Moiseev, G.E. & Benemanskij, V.V. (1975) Carcinogenic effect of low concentrations of nitrosodimethylamine on inhalation *Vopr. Onkol.*, **21**(6), 107-109

Rounbehler, D.P. & Fajen, J.M. (1982) N-*Nitroso Compounds in the Factory Environment* (Report, NIOSH Contract No. 210-77-0100), Cincinnati, OH, National Institute for Occupational Safety and Health

Rounbehler, D.P., Reisch, J.W., Coombs, J.R & Fine, D.H. (1980) Nitrosamine air sampling sorbents compared for quantitative collection and artifact formation. *Anal. Chem.*, **52**, 273-276

Spiegelhalder, B. & Preussmann, R. (1982) Nitrosamines and rubber. In: Bartsch, H., O'Neill, I.K., Castegnaro, M. & Okada, M., eds, N-*Nitroso Compounds: Occurrence and Biological Effects (IARC Scientific Publications* No. 41), Lyon, IARC, pp. 231-243

Young, J.T. (1981) Histopathologic examination of the rat nasal cavity. *Fundam. Appl. Toxicol.*, **1**, 309-312

Relevance to Human Cancer of *N*-Nitroso Compounds,
Tobacco Smoke and Mycotoxins.
Ed. I.K. O'Neill, J. Chen and H. Bartsch
Lyon, International Agency for Research on Cancer
© IARC, 1991

SPECIFICITY IN THE METHYLATION OF DNA BY *N*-NITROSO COMPOUNDS

J.R. Milligan, S. Skotnicki, S.J. Lu & M.C. Archer

*Department of Medical Biophysics, University of Toronto,
Ontario Cancer Institute, Toronto, Ontario, Canada*

A sequencing assay was used to determine the reactivity of *N*-nitroso compounds that are simple methylating agents with individual nucleotides in a defined DNA sequence. The maximal difference in reactivity between guanines is about five fold. DNA in the Z, cruciform and H conformations was shown to be methylated by *N*-methyl-*N*-nitrosourea in a manner which was indistinguishable from the reactivity of B-DNA. Electronic factors rather than steric factors appear to dominate the methylation reaction. Transcriptionally active genes were shown to be methylated by *N*-nitroso compounds *in vivo* more extensively than untranscribed genes. The results suggest that local sequence, secondary conformation and transcriptional activity may all influence the carcinogenic potential of *N*-nitroso compounds.

Although it has been known for many years that *N*-nitroso compounds alkylate DNA, there is little information on the selectivity of this reaction with respect to base sequence, conformation or particular genes within the genome. We have determined the effect of these parameters on DNA methylation by simple *N*-nitroso compounds.

Sequence selectivity

We have shown that *N*-nitroso(acetoxymethyl)methylamine (AcO-NDMA) and *N*-nitrosomethyl(acetoxybenzyl)amine (AcO-NMBzA) in the presence of esterase, and *N*-methyl-*N*-nitrosourea (MNU), produce the same pattern of methylation, as determined by formation of $N7$-methylguanine and O^6-methylguanine when these nitroso compounds react with various DNA substrates (Milligan *et al.*, 1990). These data suggest that the three compounds methylate DNA *via* a common intermediate such as the methyl diazonium ion. With calf thymus DNA, a supercoiled plasmid and poly(dG).poly(dC), $N7$-methylguanine was formed in about ten times the yield of O^6-methylguanine, as expected. For poly(dGdC).poly(dGC), however, the yield of O^6-methylguanine was the same as that of the other DNA substrates, but the $N7$-methylguanine yield was only about one-third of what we expected. Our result confirms the observations of Briscoe and Cotter (1984, 1985) for MNU.

In order to determine whether sequence specificity plays a role in these observations, we first methylated the plasmid pBR322, then isolated the 345 bp BamHI-HindIII restriction fragment. After heating at neutral pH to remove $N7$-methylguanine and

N3-methyladenine residues, the samples were end-labelled with ^{32}P, then treated with spermidine to cleave the DNA at apurinic sites. Mixtures of small fragments were fractioned by gel electrophoresis. There was no difference between lanes in which the DNA was reacted with AcO-NDMA, AcO-NMBzA or MNU, confirming our original observations, described above. There was, however, clearly some preference for reaction at particular guanines within the restriction fragment. In order to quantify this observation, we analysed the sequencing gel autoradiograph by densitometry. Integration of the areas of the peaks showed that individual guanines can differ by about five fold in their reactivity with the nitroso compounds. A similar variation exists in the reactivity of guanines with dimethyl sulfate, but there was no correlation between the reactivity of the three N-nitroso compounds and dimethyl sulfate. These results suggest that the sequence selectivity depends on both the stereoelectronic environment of the guanines and the nature of the alkylating agent.

Effects of DNA conformation

In addition to the effects of sequence selectivity discussed above, it was possible that the 20% dimethyl sulfoxide, which we used to aid the solubility of the carcinogens, promoted the B to Z transition in poly(dGdC).poly(dGdC) (Van de Sande & Jovin, 1982; Saenger *et al.*, 1987), which in turn may have affected the reactions. We therefore examined the reaction of MNU with B- and Z-DNA, and extended the studies to include DNA in the cruciform conformation and a structure known as H-DNA, in which dA:dT Watson–Crick base pairs alternate with Hoogsteen (syn) dG:GC pairs (Wells, 1988). We used plasmids in which inserts in the non-B topological isomeric forms could be readily prepared by topoisomerase I-catalysed relaxation in the presence of ethidium (Peck *et al.*, 1982). After methylating the plasmids, a fragment containing the insert was isolated by restriction endonuclease digestion. Methylation sites were determined by sequencing as described above.

The sequencing gels showed that DNA sequences in the Z, cruciform and H conformations are methylated by MNU in a manner that is indistinguishable from the reaction of MNU with the same sequences in the B conformation. We conclude that electronic factors rather than steric factors dominate methylation of DNA by MNU. Furthermore, any conformational change induced in poly(dGdC).poly(dGdC) by the co-solvent in our previous studies had no effect on the methylation reaction.

Non-B-DNA structures may exist *in vivo* and play an important role in various genetic processes (Wells, 1988). Our results show that simple carcinogenic methylating agents, typified by MNU, interact with these structures in a manner similar to B-DNA and may have important implications for chemical carcinogenesis. For example, the inability of some repair enzymes to act on methylated guanines in non-B-DNA (Lagravere *et al.*, 1984; Boiteux *et al.*, 1985) could greatly contribute to the carcinogenic potential of the lesion.

Alkylation of individual genes

We recently described a method for quantifying alkylated bases in defined gene sequences of genomic DNA following treatment of animals with a carcinogen (Milligan & Archer, 1988). DNA isolated from a particular tissue is first digested with a restriction endonuclease to generate a fragment of the gene of interest. The DNA is then heated to

65°C at pH 8 to cause depurination, followed by reaction with spermidine to generate strand breaks at the apurinic sites. Gel electrophoreses followed by Southern transfer allow sequences of interest to be visualized using specific probes. The presence of strand breaks in methylated restriction fragments reduces their intensity in comparison to that of unmethylated fragments. Using this method, we showed that the transcriptionally active albumin gene is methylated by *N*-nitrosodimethylamine to a much greater extent than the untranscribed immunoglobulin gene IgE (Milligan & Archer, 1988). We showed more recently that the Ha-*ras*-1 proto-oncogene is extensively methylated in DNA isolated from the breast tissue of female rats treated with MNU. The Ki-*ras* gene appears to be considerably less methylated, possibly because it is not expressed to the same extent as the Ha-*ras* gene. In this model, the Ha-*ras*-1 gene has been shown to be activated by a point mutation to a transforming gene, while the Ki-*ras* gene is not mutated (Zarbl *et al.*, 1985). Thus, the extent of alkylation of a given gene may be a factor in determining whether the gene is mutated by a chemical carcinogen.

Acknowledgements

This work was supported by Grants MT-7025 and MT-10491 from the Medical Research Council of Canada. We thank Dr D. Pulleyblank (University of Toronto) for supplying the plasmids for studies on DNA conformation.

References

Boiteux, S., Costa de Olivera, R. & Laval. J. (1985) The *Escherichia coli* O^6-methylguanine-DNA methyl transferase does not repair pro-mutagenic O^6-methylguanine residues when present in Z-DNA. *J. Biol. Chem.*, **260**, 8711-8715

Briscoe, W.T. & Cotter, L.E. (1984) The effects of neighboring bases on N-methyl-N-nitrosourea aklylation of DNA. *Chem.-Biol. Interactions*, **52**, 103-110

Briscoe, W.T. & Cotter, L.E. (1985) DNA sequence has an effect on the extent and kinds of alkylation of DNA by a potent carcinogen. *Chem.-Biol. Interactions*, **56**, 321-331

Lagravere, C., Malfoy, B., Leng, M. & Laval, J. (1984) Ring-opened alkylated guanine is not repaired in Z-DNA. *Nature*, **30**, 798-800

Milligan, J.R. & Archer, M.C. (1988) Alkylation of individual genes in rat liver by the carcinogen N-nitrosodimethylamine. *Biochem. Biophys. Res. Commun.*, **155**, 14-17

Milligan, J.R., Hirani-Hojatti, S., Catz-Biro, L. & Archer, M.C. (1990) Methylation of DNA by three N-nitroso compounds: evidence for sequence specific methylation by a common intermediate. *Chem.-Biol. Interactions* (in press)

Peck, L.J., Nordheim, A., Rich, A. & Wang, J.C. (1982) Flipping of cloned d(pCpG)n.d(pCpG)n DNA sequences from right to left-handed helical structure by salt, Co(III), or negative supercoiling. *Proc. Natl Acad. Sci. USA*, **79**, 4560-4564

Saenger, W., Hunter, W.N. & Kennard, O. (1986) DNA conformation is determined by economics in the hydration of phosphate groups. *Nature*, **324**, 385-388

Van de Sande, J.H. & Jovin, T.M. (1982) Z DNA, the left-handed helical form of poly[d(G-C)] in $MgCl_2$-ethanol, is biologically active. *EMBO J.*, **1**, 115-120

Wells, R.D. (1988) Unusual DNA structures. *J. Biol. Chem.*, **263**, 1095-1098

Zarbl, H., Sukumar, S., Arthur, A.V., Martin-Zanca, D. & Barbacid, M. (1985) Direct mutagenesis of Ha-ras-1 oncogenes by N-nitroso-N-methylurea during initiation of mammary carcinogenesis in rats. *Nature*, **315**, 382-385

MECHANISM OF ACTION OF THE URINARY BLADDER CARCINOGEN N-NITROSOBUTYL-3-CARBOXYPROPYLAMINE

C. Janzowski[1], D. Jacob[1], I. Henn[2], H. Zankl[2],
B.L. Pool-Zobel[3], M. Wiessler[3] & G. Eisenbrand[1]

[1]*Department of Food Chemistry and* [2]*Department of Biology,
University of Kaiserslautern, Kaiserslautern;
and* [3]*Institute for Toxicology and Chemotherapy,
German Cancer Research Center, Heidelberg,
Germany*

The carcinogenic action of N-nitrosodibutylamine in the urinary bladder is related to ω-oxidation of a butyl chain. N-Nitrosobutyl-4-hydroxybutylamine and its proximate metabolite N-nitrosobutyl-3-carboxypropylamine (NBCPA) selectively induce urinary bladder tumours in different animal species. The mechanism by which NBCPA exerts its carcinogenic action is not known. We found a small but significant dealkylation of NBCPA with microsomes from rat liver or pig urinary bladder, which could be inhibited by SKF 525A. NBCPA was not mutagenic to *Salmonella typhimurium* (with or without external metabolizing systems from rat liver or pig urinary bladder) and did not induce DNA strand breaks in tumour cell lines (with or without external activation) or primary cells (rat hepatocytes, pig urinary bladder epithelia). Significant induction of sister chromatid exchange and micronuclei, however, was observed in human tumour cells. N-Nitrosoureas that generate the same electrophiles as NBCPA after α- or *via* β-oxidation (N-butyl-N-nitrosourea, N-3-carboxypropyl-N-nitrosourea and N-2-oxopropyl-N-nitrosourea) induced single-strand breaks in Namalva cells, the oxopropyl compound being more potent than the butyl or carboxypropyl compounds. Our data suggest that NBCPA is activated *via* α-oxidation in the urinary bladder, even though the activation rate *in vitro* is so low that a positive response is not detectable by classical short-term tests. Provided that β-oxidation to a highly genotoxic agent proceeds at an adequate rate, it might also be a relevant activation pathway.

N-Nitrosodibutylamine (NDBA) and certain N-nitrosomethylalkylamines, known to be environmental contaminants (Hecht *et al.*, 1982; Spiegelhalder & Preussmann, 1982; Billedeau *et al.*, 1986), characteristically induce tumours in the urinary bladder of rodents and other laboratory animals (Okada, 1976; Preussmann & Stewart, 1984). The organotropic effect of these nitrosamines is related to metabolic formation of an N-nitrosoalkyl-3-carboxypropylamine (Okada *et al.*, 1975; Singer *et al.*, 1981). In the case of NDBA, ω-oxidation generates N-nitrosobutyl-4-hydroxybutylamine (NBHBA), which is rapidly converted into N-nitrosobutyl-3-carboxypropylamine (NBCPA; Okada, 1976; Airoldi et al., 1987; Pastorelli et al., 1988). Both metabolites act as specific urinary bladder carcinogens in different animal species (Okajima et al., 1981; Preussmann & Stewart, 1984; Ohtani et al., 1986). The mechanism by which NBCPA, the more proximate carcinogen, induces urinary bladder tumours is not known.

We used different in-vitro tests to find out whether NBCPA is activated (see Fig. 1) to a genotoxic/mutagenic agent. In some experiments, NDBA, NBHBA and the nongenotoxic N-nitroso-*tert*-butyl-3-carboxypropylamine (N-t-BCPA) were included. Additionally, N-nitrosoureas were tested which spontaneously generate the same ultimate electrophiles as NBCPA after α-C-hydroxylation (N-butyl-N-nitrosourea, BNU; N-3-carboxypropyl-N-nitrosourea, CPNU) or *via* β-oxidation (N-2-oxopropyl-N-nitrosourea, OPNU). N-2-Carboxyethyl-N-nitrosourea (CENU) was used for comparison.

Figure 1. Proposed pathway of activation of N-nitrosobutyl-3-carboxypropylamine by α-oxidation

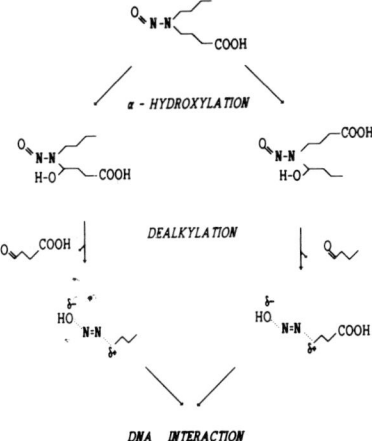

Microsomal metabolism of nitrosamines

Dealkylation of nitrosamines was monitored by high-performance liquid chromatography-336-nm analysis of the corresponding aldehyde 2,4-dinitrophenyl-hydrazones, as described previously (Janzowski et al., 1982; Pool et al., 1988; Janzowski et al., 1989). NBCPA was dealkylated by microsomal fractions from rat liver and pig urinary bladder (see Table 1) at low rates. In bladder microsomes, debutylation of NDBA and NBHBA was similarly low, in contrast to the high rates found in liver microsomal fractions. Addition of the monooxygenase inhibitor SKF 525A (1 mM) to the incubation mixture reduced the dealkylation rate of NBCPA (10 mM) to 50% in phenobarbital-induced rat liver and in pig urinary bladder microsomal fractions.

Biological activity of nitrosamines

DNA single-strand breaks were determined in human tumour cell lines (lymphoblastoid Namalva cells, urinary bladder epithelial RT4 cells; external activation with microsomes from rat liver or pig urinary bladder) and in primary cells (rat hepatocytes, urinary bladder epithelial cells from pigs and bovines) as described previously (Sterzel et al., 1985; Pool et al., 1988; Janzowski et al., 1989).

Table 1. Dealkylation of N-nitrosodibutylamine (NDBA), N-nitrosobutyl-4-hydroxybutylamine (NBHBA) and N-nitrosobutyl-3-carboxypropylamine (NBCPA) at incubation with microsomal fraction

Origin of microsomal fraction	Compound	Generated aldehydes (nmol/mg protein/60 min)			
		Butyraldehyde		4-Oxobutyric acid	
		Mean ± SD	n	Mean ± SD	n
Phenobarbital-induced rat liver	NDBA	330 ±22	4		
	NBHBA	68 ± 4	5		
	NBCPA	1.5 ± 0.3	7	2.8 ± 0.7	4
Uninduced rat liver	NDBA	60 ± 9	6		
	NBHBA	34 ± 8	5		
	NBCPA	1 ± 0.3	2	2.0 ± 0.6	6
Pig urinary bladder	NDBA	ND	2		
	NBHBA	1 ± 0.5	2		
	NBCPA	0.4 ± 0.01	2	0.8 ± 0.2	5

ND, not detectable (< 0.7); blanks, not applicable. Substrate concentration, 10 mM; incubation, 1 h at 37°C

DNA single-strand breaks were not observed with NBCPA in Namalva or RT4 cells in the presence or absence of external activation. NDBA (> 0.5 mM) and NBHBA (> 25 mM), however, showed distinct dose-dependent induction of strand breaks in Namalva cells when incubated with phenobarbital-induced liver microsomal fractions. In primary rat hepatocytes, similarly positive results were obtained with NDBA (Schmezer et al., 1989), whereas NBHBA and NBCPA gave negative results. In the presence of pig urinary bladder microsomes or in primary bladder epithelial cells, none of the nitrosamines induced detectable DNA damage.

Similarly negative findings were obtained in bacterial mutagenicity tests with *Salmonella typhimurium* (TA1535, TA100; preincubation, 30 min) (Pool et al., 1988). NBCPA was not mutagenic in the presence or absence of external metabolizing systems (rat liver postmitochondrial fraction and microsomes, bovine and pig urinary bladder postmitochondrial fraction or epithelial cells).

Cytogenetic effects were determined according to standard procedures (Latt, 1973; Perry & Wolff, 1974; Fenech & Morley, 1985). NBCPA showed significant induction of sister chromatid exchange and micronuclei in Namalva cells without external activation (Fig. 2). In the presence of SKF 525A (50 μM), the rate of micronuclei induction was reduced. Significant induction of micronuclei by NBCPA (50 μM) was also observed in the RT4 human bladder carcinoma cell line. The nongenotoxic N-t-BCPA gave negative results under identical experimental conditions. With NDBA, significant rates of sister chromatid exchange and micronuclei formation were detected at substantially lower concentrations (1 µM and 10 µM, respectively).

Biological activity of model compounds (N-nitrosoureas)

N-Nitrosoureas exhibited concentration-dependent induction of single-strand breaks when incubated (1 h) with Namalva cells; CENU and OPNU were substantially more potent than BNU and CPNU (Fig. 3). Rapid decomposition of the test compounds in RPMI medium ($t_{1/2}$, < 12 min, except for CPNU; c > 3mM) indicates almost complete liberation of the respective electrophiles within the first 60 min of incubation. With prolonged incubation (4 h), induction of single-strand breaks by CENU and CPNU tended to increase with time, whereas the number of BNU-induced single-strand breaks decreased.

Figure 2. Induction of micronuclei and sister chromatid exchanges by N-nitrosobutyl-3-carboxypropylamine (NBCPA) in Namalva cells[a]

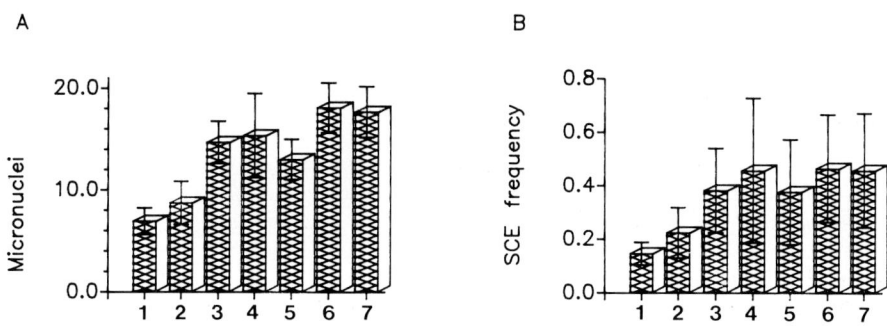

[a] A, no. of micronuclei in 500 cytokinesis-blocked cells (n = 5); means of 2-3 experiments; B, no. of sister chromatid exchanges (SCE) per chromosome (n = 50); one experiment; 1, control; 2, control plus dimethyl sulfoxide; 3, 12.5 µM NBCPA; 4, 25 µM NBCPA; 5, 50 µM NBCPA; 6, 100 µM NBCPA; 7, 150µM NBCPA. Mean + SD; incubation, 1 h at 37°C, 2 × 10^6 cells/ml serum-free medium; cells washed by centrifugation after incubation and recultivated

The mutagenicity of BNU, CENU and CPNU in *S. typhimurium* (Fig. 4) correlates with their potential to induce single-strand breaks. The potent inducer OPNU was described previously as a potent mutagen in *S. typhimurium* TA1535 (Lijinsky *et al.*, 1987). β-Oxidation with subsequent decarboxylation of NBCPA to N-nitrosobutyl-2-oxopropylamine is known to occur at low rates *in vivo* (Irving & Daniel, 1987) and might therefore also be important.

Figure 3. Induction of single-strand breaks by N-butyl-N-nitrosourea (BNU), N-3-carboxypropyl-N-nitrosourea (CPNU), N-2-carboxyethyl-N-nitrosourea (CENU) and N-2-oxopropyl-N-nitrosourea in Namalva cells[a]

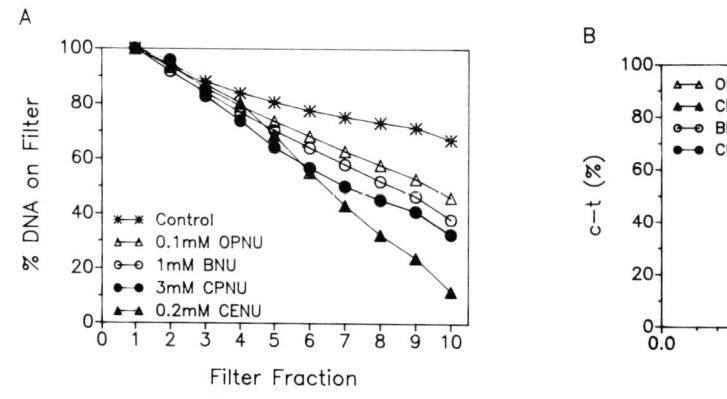
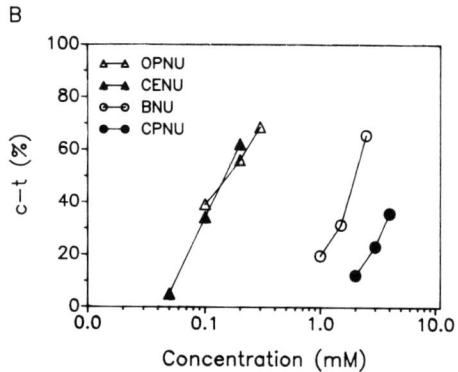

[a] A, characteristic elution profiles; B, concentration-dependent induction of single-strand breaks; c-t%, % DNA retained on filter in control minus % DNA retained on filter in treated cells at fraction 10; mean of two experiments. Incubation: 1 h at 37°C; 2-3 × 10⁶ cells/ml RPMl medium with HEPES

Figure 4. Mutagenicity of N-butyl-N-nitrosourea (BNU), N-3-carboxypropyl-N-nitrosourea (CPNU) and N-2-carboxyethyl-N-nitrosourea (CENU) in *Salmonella typhimurium*; A, TA1535; B, TA100[a]**; 30-min preincubation**

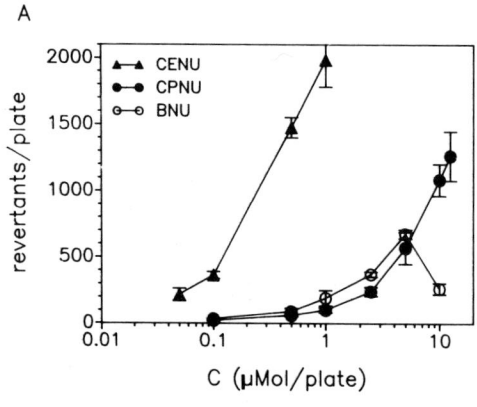
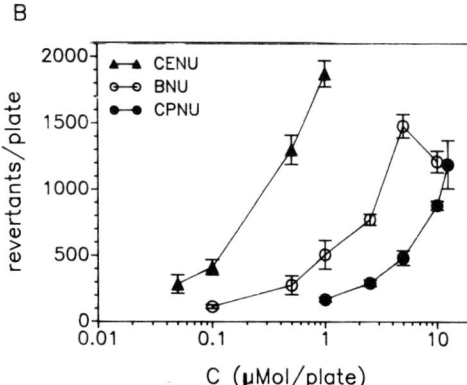

[a] Means ± SD were obtained from 2-3 experiments, each performed in triplicate.

Conclusions

NDBA and NBHBA show low dealkylation rates and fail to induce DNA single-strand breaks when incubated with urinary bladder epithelial cells or microsomes, in contrast to the distinct activation rates and genotoxic effects obtained in the presence of liver microsomal fractions. NBCPA is dealkylated at low rates by microsomes from both urinary bladder epithelium and liver. Electrophiles presumably generated from dealkylation of NBCPA are mutagenic and induce single-strand breaks. The activation rates are so low, however, that several short-term tests do not respond. In contrast, NBCPA induced micronuclei and sister chromatid exchange. These highly sensitive cytogenetic endpoints indicate that oxidative metabolism is important in the biological activity of NBCPA, even though it occurs at low rates.

Acknowledgements

We thank M. Litmianski, R. Aulenbacher, W. Köhl, D. Kohlmüller and R. Gliniorz for excellent assistance, O. Zelesny for synthesizing nitrosamines, and Dr B. Ames, Dr W. Lijinsky and Smith, Kline & French Ltd, for the gift of *S. typhimurium* strains, OPNU, and SKF 525A, respectively. This work was supported by the Bundesministerium für Forschung und Technologie, grant No 0704861 2.

References

Airoldi, L., Bonfanti, M., Magagnotti, C. & Fanelli, R. (1987) Development of an experimental model for studying bladder carcinogen metabolism using the isolated rat urinary bladder. *Cancer Res.*, 47, 3697-3700

Billedeau, S.M., Thompson, H.C., Miller, B.J. & Wind, M.L. (1986) Volatile N-nitrosamines in infant pacifiers sold in the United States as determined by gas chromatography/thermal energy analysis. *J. Assoc. off. anal. Chem.*, 69, 31-34

Fenech, M. & Morley, A.A. (1985) Measurement of micronuclei in human lymphocytes. *Mutat. Res.*, 149, 29-36

Hecht, S.S., Morrison, J.B. & Wenninger, J.A. (1982) N-Nitroso-N-methyldodecylamine and N-nitroso-N-methyltetradecylamine in hair-care products. *Food Chem. Toxicol.*, 20, 165-169

Irving, C. & Daniel, D. (1987) Influence of disulfiram on the metabolism of the urinary bladder carcinogen N-butyl-N-(4-hydroxybutyl)nitrosamine in the rat. *Carcinogenesis*, 8, 1309-1315

Janzowski, C., Gottfried, J., Eisenbrand, G. & Preussmann, R. (1982) Fluorosubstituted N-nitrosamines. 3. Microsomal metabolism of N-nitrosodibutylamine and of fluorinated analogs in liver microsomal fractions. *Carcinogenesis*, 3, 777-780

Janzowski, C., Jacob, D., Pool, B.L. & Eisenbrand, G. (1989) Investigations on organ-specific metabolism and genotoxic effects of the urinary bladder carcinogen N-nitrosobutyl-3-carboxypropylamine (BCPN) and its analogs N-nitrosodibuytylamine (NDBA) and N-nitrosobutyl-4-hydroxybutylamine (4-OH-NDBA). *Toxicology*, 59, 195-209

Latt, S.A. (1973) Microfluorometric detection of deoxyribonucleic acid replication in human chromosomes. *Proc. Natl Acad. Sci. USA*, 70, 3395-3399

Lijinsky, W., Elespuru, R.K. & Andrews, A.W. (1987) Relative mutagenic and prophage-inducing effects of mono- and di-alkyl nitrosoureas. *Mutat. Res.*, 178, 157-165

Ohtani, M., Kakizoe, T., Nishio, Y., Sato, S., Sugimara, T., Fukushima, S. & Niijima, T. (1986) Sequential changes of mouse bladder epithelium during induction of invasive carcinomas by N-butyl-N-(4-hydroxybutyl)nitrosamine. *Cancer Res.*, 46, 2001-2004

Okada, M. (1976) Metabolic aspects in organotropic carcinogenesis by dialkylnitrosamines. In: Magee, P.N. *et al.*, eds, *Fundamentals in Cancer Prevention*, Baltimore, University Park Press, pp. 251-266

Okada, M., Suzuki, E., Aoki, J., Iiyoshi, M. & Hashimoto, Y. (1975) Metabolism and carcinogenicity of N-butyl-N-(4-hydroxybutyl)nitrosamine and related compounds, with special reference to induction of urinary bladder tumours. *Gann Monogr. Cancer Res.*, 17, 161-176

Okajima, E., Hiramatsu, T., Hirao, K., Ijuin, M., Hirao, Y., Babaya, K., Ikuma, S., Ohara, S., Shiomi, T., Hijioka, T. & Ohishi, H. (1981) Urinary bladder tumours induced by N-butyl(4-hydroxybutyl)nitrosamine in dogs. *Cancer Res.*, **41**, 1958-1966

Pastorelli, R., Ancidei, A., Benfenati, E., Fanelli, R. & Airoldi, L. (1988) Effect of butylated hydroxyanisole added in vitro or administered to rats on N,N-dibutylnitrosamine and N-butyl-N-(4-hydroxybutyl)nitrosamine metabolism by post-mitochondrial supernatant of liver homogenates. *Toxicology*, **48**, 71-80

Perry, P. & Wolff, S. (1974) New Giemsa method for the differential staining of sister chromatids. *Nature*, **251**, 156-158

Pool, B.L., Gottfried-Anacker, J., Eisenbrand, G. & Janzowski, C. (1988) N-Nitrosobutyl-3-carboxypropylamine (BCPN), a urinary bladder carcinogen, non-mutagenic in S. typhimurium. *Mutat. Res.*, **209**, 79-81

Preussmann, R. & Stewart, B.W. (1984) N-Nitroso carcinogens. In: Searle, C.E., ed., *Chemical Carcinogens* (ACS Monograph 182), Vol. 2, Washington DC, American Chemical Society, pp. 742-868

Schmezer, P., Preussmann, R., Schmähl, D. & Pool, B.L. (1989) In vitro effect of hepatic and extrahepatic carcinogenic N-nitrosamines in primary hepatocytes isolated from rat, hamster and porcine liver (in press)

Singer, G.M., Lijinsky, W., Buettner L. & McClusky, G.A. (1981) Relationship of rat urinary metabolites of N-nitrosomethyl-N-alkylamine to bladder carcinogenesis. *Cancer Res.*, **41**, 4929-4946

Spiegelhalder, B. & Preussmann, R. (1982) Nitrosamines and rubber. In: Bartsch, H., O'Neill, I., Castegnaro, M. & Okada, M., eds, *N-Nitroso Compounds: Occurrence and Biological Effects* (IARC Scientific Publications No. 41), Lyon, IARC, pp. 231-242

Sterzel, W., Bedford, P. & Eisenbrand, G. (1985) Automated determination of DNA using the Fluorochrome Hoechst 33258. *Anal. Biochem.*, **147**, 462-467

MUTAGENICITY, DNA DAMAGE AND DNA ADDUCT FORMATION BY N-NITROSO-2-HYDROXYALKYLAMINE AND CORRESPONDING ALDEHYDES

G. Scherer[1], B. Ludeke[2], P. Kleihues[2], R.N. Loeppky[3] & G. Eisenbrand[1]

[1]*Department of Food Chemistry, University of Kaiserslautern, Kaiserslautern, Germany;*
[2]*Institute of Pathology, University of Zürich, Switzerland;*
and [3]*Department of Chemistry, University of Missouri, Columbia, MO, USA*

The potent carcinogen N-nitrosodiethanolamine (NDELA) becomes mutagenic to *Salmonella typhimurium* TA98 and TA100 when activated by alcohol dehydrogenase from yeast or horse liver. Metabolic pathways different from α-oxidation might therefore be important for the activation of N-nitroso-2-hydroxyalkylamines such as NDELA. In an in-vitro test system (Namalva cells), neither NDELA nor N-nitrosoethyl-2-hydroxyethylamine was genotoxic, whereas the corresponding metabolites from alcohol dehydrogenase-mediated oxidation, N-nitroso-2-hydroxymorpholine and N-nitroso-ethylethanalamine, induced single-strand breaks even at low doses. An immuno-slot-blot assay was used to study the formation of O^6-2-hydroxyethyldeoxyguanosine in rat liver after oral administration of different N-nitroso-2-hydroxyalkylamines. When given at equimolar doses (0.375 mmol/kg), DNA hydroxyethylation was considerably lower (6.7 μmol/mol deoxyguanosine) with NDELA than with N-nitrosoethyl-2-hydroxyethylamine (48.7 μmol/mol deoxyguanosine) or N-nitrosomethyl-2-hydroxyethylamine (72.1 μmol/mol deoxyguanosine). N-Nitroso-2-hydroxymorpholine did not form detectable levels of O^6-2-hydroxyethyldeoxyguanosine.

The potent carcinogen N-nitrosodiethanolamine (NDELA) has been shown to become mutagenic to *Salmonella typhimurium* TA98 and TA100 when activated by alcohol dehydrogenase from yeast or horse liver. Metabolic pathways other than α-oxidation might be important for the activation of N-nitroso-2-hydroxyalkylamines such as NDELA (Denkel *et al.*, 1986; Sterzel & Eisenbrand, 1986).

The aim of the present work was to obtain further information about the significance of these pathways *in vitro* and *in vivo*. The genotoxic activities of NDELA and N-nitrosoethyl-2-hydroxyethylamine (NEHEA) and of the corresponding metabolites from alcohol dehydrogenase-mediated oxidation, N-nitroso-2-hydroxymorpholine (NHMOR) and N-nitrosoethylethanalamine (NEEALA), were tested in Namalva cells, which is a human lymphoblastoid cell line virtually devoid of activating enzymes (Rabson *et al.*, 1966).

After incubation with the nitrosamines (1 h, 37°C), the cells were washed twice and induction of single-strand breaks was measured using the alkaline filter elution technique (Kohn et al., 1981; Sterzel et al., 1985). Neither NDELA (300 mM) nor NEHEA (100 mM) induced breaks, whereas NHMOR and NEEALA did so at low doses (5 mM; Figs 1 and 2). The potent DNA damaging activity of NHMOR and of the open-chain aldehyde NEEALA in Namalva cells is reflected by similar effects *in vivo* in rat liver. NHMOR and NEEALA are also strongly acting mutagens in *Salmonella* tester strains.

Figure 1. Elution profile of DNA of Namalva cells treated with various doses of *N*-nitroso-2-hydroxymorpholine

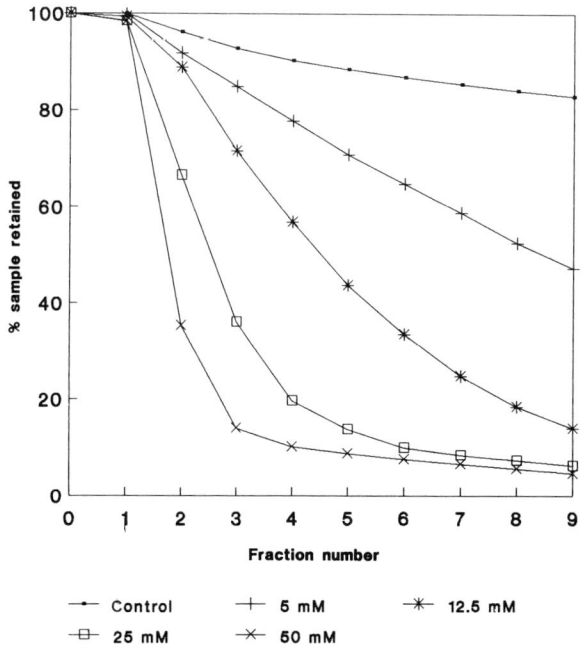

Little is known about the type of DNA lesions that are responsible for the mutagenic and genotoxic potential of NDELA and related compounds. We studied the formation of O^6-(2-hydroxyethyl)-2'-deoxyguanosine (O^6-HEtdGua) in male Wistar rats after oral administration of different *N*-nitroso-2-hydroxyalkylamines. Four hours after application, the animals were killed and the kidneys and livers removed; DNA was isolated by caesium chloride gradient centrifugation. O^6-HEtdGua formation was studied using an immuno-slot-blot technique with a rabbit antibody (Ludeke & Kleihues, 1988). O^6-HEtdGua formation by NDELA was proportional to the dose within a range of 20-200 mg/kg (Fig. 3).

A comparison of the extent of O^6-HEtdGua formation induced by equimolar doses (0.375 mmol/kg) of NDELA, NEHEA, *N*-nitrosomethyl-2-hydroxyethylamine (NMHEA) and NHMOR revealed that NMHEA and NEHEA had a considerably stronger hydroxyethylating potential than NDELA (Table 1). NHMOR did not form detectable levels of O^6-HEtdGua. Since NHMOR is a major metabolic intermediate formed from NDELA by alcohol dehydrogenase-mediated oxidation, this finding suggests that alternative and/or additional metabolic transformations must be operative for O^6-hydroxyethylation effected by NDELA.

Figure 2. Elution profile of DNA of Namalva cells treated with various doses of N-nitrosoethylethanalamine

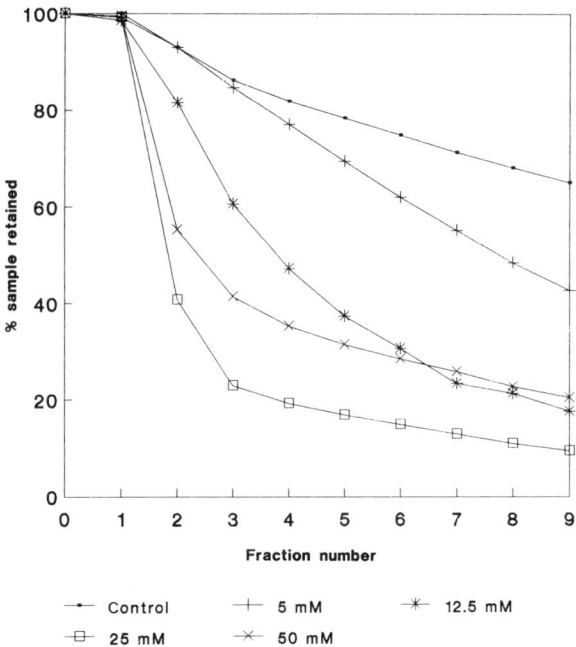

Table 1. Formation of O^6-(2-hydroxyethyl)-2'-deoxyguanosine (O^6-HEtdGua) in liver and kidney after a single administration of 0.375 mmol/kg nitrosamine to rats

Substrate[a]	O^6-HEtdGua(μmol/mol deoxyguanosine)	
	DNA from liver	DNA from kidneys
NMHEA	72.1	6.2
NEHEA	48.7	NT[b]
NDELA	6.7	0
NHMOR	2.0	1.4
Control	1.7	1.2

[a] NMHEA, N-nitrosomethyl-2-hydroxyethylamine; NEHEA, N-nitrosoethyl-2-hydroxyethylamine; NDELA, N-nitrosodiethanolamine; NHMOR, N-nitroso-2-hydroxymorpholine
[b] Not tested

Figure 3. O^6-(2-Hydroxyethyl)-2'-deoxyguanosine (O^6-HEtdGua) in rat liver DNA 4 h after administration of various doses of N-nitrosodiethanolamine (NDELA)

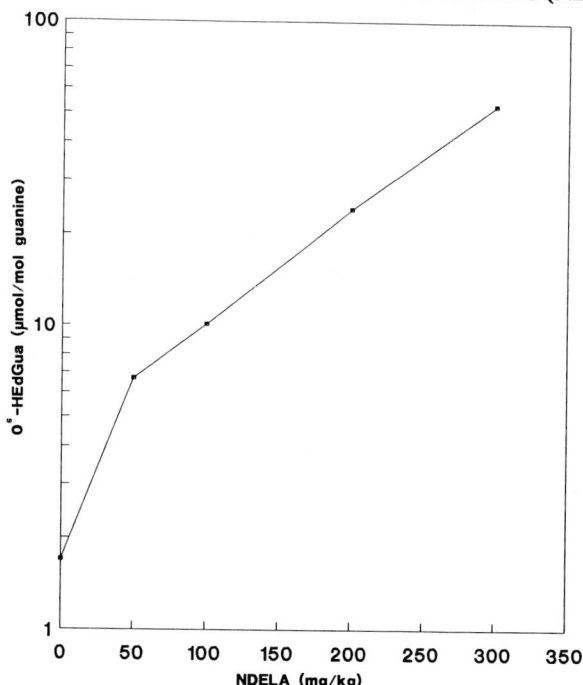

Since α-C-hydroxylation does not appear to play a role in the metabolism of NDELA, a possible alternative might be β-esterification, as proposed by Koepke *et al.* (1988) and Sterzel and Eisenbrand (1986).

References

Denkel, E., Pool, B.L., Schlehofer, J. & Eisenbrand, G. (1986) Biological activity of N-nitrosodiethanolamine and potential metabolites which may arise after activation by alcoholdehydrogenase in Salmonella typhimurium, in mammalian cells and in vivo. *J. Cancer Res. Clin. Oncol.*, 111, 149-153

Kohn, K.W., Ewig, R.A.G., Erickson, L.C. & Zwelling, L.A. (1981) Measurement of strand breaks and cross-links by alkaline elution. In: Friedberg, E.C. & Hanawalt, P.C., eds, *DNA Repair*, Vol. 1, New York, Marcel Dekker, pp. 379-401

Koepke, S.R., Creasia, D.R., Knutsen, G.L. & Michejda, C.J. (1988) Alkylation of DNA in rats by N-nitrosomethyl(2-hydroxyethyl)amine: dose response and persistence of the alkylated lesions in vivo. *Cancer Res.*, 48, 1537-1542

Ludeke, B. & Kleihues, P. (1988) Formation and persistence of O^6-(2-hydroxyethyl)-2'-deoxyguanosine in DNA of various rat tissues following a single dose of N-nitroso-N-(2-hydroxyethyl)urea. An immuno-slot-blot study. *Carcinogenesis*, 9, 147-151

Rabson, A.S., O'Connor, G.T., Baron, S., Whang, J.J. & Legallais, F.Y. (1966) Morphologic, cytogenetic and virologic studies in vitro of a malignant lymphoma from an African child. *Int. J. Cancer*, 1, 89-96

Sterzel, W. & Eisenbrand, G. (1986) N-Nitrosodiethanolamine is activated in the rat to an ultimate genotoxic metabolite by sulfotransferase. *J. Cancer Res. Clin. Oncol.*, 111, 20-24

Sterzel, W., Bedford, P. & Eisenbrand, G. (1985) Automated determination of DNA using the fluorochrome Hoechst 33458. *Anal. Biochem.*, 147, 462-467

Relevance to Human Cancer of *N*-Nitroso Compounds,
Tobacco Smoke and Mycotoxins.
Ed. I.K. O'Neill, J. Chen and H. Bartsch
Lyon, International Agency for Research on Cancer
© IARC, 1991

N-NITROSOBUTYL(4-HYDROXYBUTYL)AMINE α-HYDROXYLATION BY RAT LIVER AND UROTHELIAL CELL HOMOGENATES

L. Airoldi[1], C. Magagnotti, M. Bonfanti, M. Moret & R. Fanelli

*Laboratory of Environmental Pharmacology and Toxicology,
Istituto di Ricerche Farmacologiche Mario Negri, Milan, Italy*

Determination of molecular nitrogen formed as a consequence of nitrosamine α-hydroxylation provides a useful means for studying the extent of activation of these compounds in target and nontarget organs and tissues. α-Hydroxylation in rat liver and urothelial cells was compared using as substrate doubly ^{15}N-labelled *N*-nitrosobutyl(4-hydroxybutyl)amine (^{15}N-NBHBA), a potent bladder carcinogen in rodents. Both enzyme sources metabolized ^{15}N-NBHBA through the α-hydroxylation pathway. $^{15}N_2$ production was dependent on the amount of substrate incubated. V_{max} values for $^{15}N_2$ production by urothelial cells and by liver postmitochondrial supernatant were 4.47 and 3.21 nmol/mg protein per h, respectively.

The initial step in the biotransformation of nitrosamines is believed to be enzymatic hydroxylation at the carbon atom α to the nitroso group. All subsequent steps are nonenzymatic, and the final result is the formation of alkylating species and molecular nitrogen in stoichiometric quantity. As suggested by several authors, the amount of molecular nitrogen formed can be used as an indicator of this metabolic step (Cottrell *et al.*, 1977; Milstein & Guttenplan, 1979). The use of doubly ^{15}N-labelled nitrosamines and the detection of $^{15}N_2$ produced by gas chromatography–mass spectrometry simplify measurement of the extent of α-hydroxylation (Cottrell *et al.*, 1977; Kroeger-Koepke *et al.*, 1981).

In order to establish the importance of the target organ in the activation of bladder carcinogens, we compared α-hydroxylation activities in rat liver and urothelial cells using as substrate doubly ^{15}N-labelled *N*-nitrosobutyl(4-hydroxybutyl)amine (^{15}N-NBHBA), a potent bladder carcinogen in rodents. The $^{15}N_2$ formed was analysed by gas chromatography–selected ion monitoring (manuscript submitted for publication). Neon was used as the internal standard to determine $^{15}N_2$ quantitatively.

Male CD-COBS rats, weighing 200 ± 10 g, were killed after an overnight fast. Their livers were homogenized in four volumes of 0.05M phosphate buffer pH 7.4 containing 0.15M KCl and 0.005M $MgCl_2$, and the homogenate was centrifuged at 9000 *g* for 20 min, in order to obtain the postmitochondrial supernatant (S9). Various amounts of ^{15}N-NBHBA were incubated for 60 min at 37°C with S9 preparation (1–2 mg protein) and the NADPH generating system in a final volume of 0.9 ml.

---

[1]To whom correspondence should be addressed

Urothelial cells were obtained by gently scraping the bladder mucosa with a scalpel. Cells from one bladder were homogenized in minimum essential medium and incubated for 60 min at 37°C with 0.5–5 µmol 15N-NBHBA in the presence of a NADPH generating system in a final volume of 0.9 ml.

Hepatic S9 and urothelial cell samples were placed in sealed vials from which the air was purged and replaced by O_2 containing 5% CO_2 and 0.015% of neon. At the end of incubation, 100 µl of the head-space gas were analysed by gas chromatography–selected ion monitoring.

Figure 1 shows the concentration-dependent rate of production of $^{15}N_2$ by liver S9 preparations. The curve appears to be biphasic, suggesting the existence of different enzyme affinities. The calculated K_m values were 0.54 mM and 5.25 mM; V_{max} values were 3.21 and 12.24 nmol/mg protein per h. With urothelial cell homogenates, the kinetic parameters were K_m 3.27 mM and V_{max} 4.47 nmol/mg protein per h. These observations confirm that bladder carcinogens can also be activated within the target organ, as an alternative to liver metabolism (Airoldi et al., 1987).

Figure 1. Production of labelled molecular nitrogen by liver postmitochondrial preparation

Points are means ± SE of triplicate assays at each substrate concentration indicated; $^{15}N_2$ production is expressed as nmol/mg protein per h.

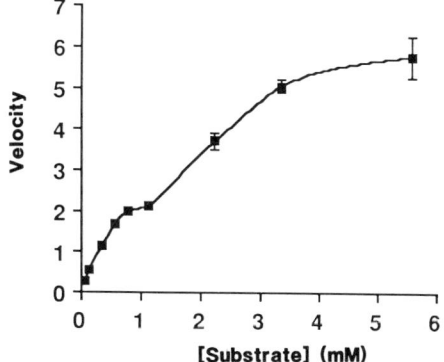

The susceptibility of tissues to chemical carcinogens is determined by the balance between the capacity of the tissue to biotransform chemicals and its ability to repair damaged DNA. Our results indicate that preparations of both liver and bladder metabolize 15N-NBHBA *in vitro* through the α-hydroxylation pathway. The high K_m value observed with urothelial cell homogenates suggests that this tissue has a lower capacity than the liver to activate 15N-NBHBA. The finding that the V_{max} values are similar indicates that, under conditions of enzyme saturation, urothelial cells can activate 15N-NBHBA as efficiently as liver. Since NBHBA is a bladder rather than a liver carcinogen, the mechanisms of repair in the liver may be more effective than those in the bladder.

Acknowledgement

This work was supported by the Italian National Council for Research, Special Project Oncology, Contract No. 88.01115.44.

References

Airoldi, L., Bonfanti, M., Magagnotti, C. & Fanelli, R. (1987) Development of an experimental model for studying bladder carcinogen metabolism using the isolated rat urinary bladder. *Cancer Res.*, **47**, 3697-3700

Cottrell, R.C., Lake, B.G., Phillips, J.C. & Gangolli, S.D. (1977) The hepatic metabolism of ^{15}N labelled dimethylnitrosamine in the rat. *Biochem. Pharmacol.*, **26**, 809-913

Kroeger-Koepke, M.B., Koepke, S.R., McClusky, G.A., Magee, P.N. & Michejda, C.J. (1981) α-Hydroxylation pathway in the in vitro metabolism of carcinogenic nitrosamines: N-nitrosodimethylamine and N-nitroso-N-methylaniline. *Proc. Natl Acad. Sci. USA*, **78**, 6489-6493

Milstein, S. & Guttenplan, J.B. (1979) Near quantitative production of molecular nitrogen from metabolism of dimethylnitrosamine. *Biochem. Biophys. Res. Commun.*, **87**, 337-342

Relevance to Human Cancer of *N*-Nitroso Compounds,
Tobacco Smoke and Mycotoxins.
Ed. I.K. O'Neill, J. Chen and H. Bartsch
Lyon, International Agency for Research on Cancer
© IARC, 1991

MECHANISM OF DNA BINDING BY THE OESOPHAGEAL CARCINOGEN *N*-NITROSO-*N*-METHYLANILINE

S.R. Koepke, M.B. Kroeger-Koepke & C.J. Michejda

Laboratory of Chemical and Physical Carcinogenesis, ABL-Basic Research Program, National Cancer Institute-Frederick Cancer Research and Development Center, Frederick, MD, USA

N-Nitroso-*N*-methylaniline (NMA) is a strong oesophageal carcinogen in rats but exhibits few overt genotoxic effects. Previous work from our laboratory established that NMA is readily metabolized by cytochrome P450-catalysed *N*-demethylation to produce the benzenediazonium ion (BDI), a relatively stable but reactive electrophilic agent. We have also shown that BDI reacts with DNA to form an acid-labile adduct. We have now shown that BDI, generated chemically or by the metabolism of NMA *in vitro*, reacts with DNA to form a triazene coupling product at the *N*6-position of adenine residues. This adduct has also been shown to be produced in the liver DNA of rats treated with NMA. Procedures for the isolation of DNA and for analysis of the adduct are presented.

N-Nitroso-*N*-methylaniline (NMA) has long been known to be an oesophageal carcinogen in rats, with great specificity for that tissue (Druckrey *et al.*, 1967; Napalkov & Pozharisski, 1969; Goodall *et al.*, 1970; Kroeger-Koepke *et al.*, 1983). NMA is an excellent substrate for microsomal mixed-function oxygenases and presumably for one of the isozymes of cytochrome P450, since the reaction is inhibited by carbon monoxide (Kroeger-Koepke & Michejda, 1979; Gold *et al.*, 1987). The principal metabolic reaction is *N*-demethylation, which results in formation of the benzenediazonium ion (BDI). This ion was trapped as *para*-hydroxyazobenzene when 14C-NMA was subjected to metabolism by a metabolic activation system from rat liver in the presence of phenol (Koepke *et al.*, 1987). The yield of the azobenzene was 70%, on the basis of the amount of NMA consumed, which indicates the predominant nature of the demethylation reaction. Studies of the metabolism *in vitro* (Kroeger-Koepke *et al.*, 1981) and *in vivo* (Michejda *et al.*, 1982) of NMA labelled with 15N in both nitrogens revealed that a significant portion of the nitrogen in the molecule was not released in the form of molecular nitrogen, in contrast to the aliphatic nitrosamine, *N*-nitrosodimethylamine (Kroeger-Koepke *et al.*, 1981). This experiment suggested that a substantial portion of the BDI formed during metabolism was captured by nucleophiles, without the release of nitrogen. This conclusion was strengthened further by the finding that BDI formed chemically or by metabolism of NMA reacted with exogenous DNA (Koepke *et al.*, 1987). Another important finding in that study was that BDI was released from the DNA by treatment with mild acid, which indicated that the DNA adduct was acid-labile and reverted back to its starting material.

A number of unsuccessful attempts have been made to detect DNA adducts from NMA. It became clear to us that one possible reason for this lack of success was that the adduct was unstable and did not survive the normal DNA isolation and hydrolysis procedures. Aromatic diazonium ions, such as BDI, in contrast to aliphatic diazonium ions, frequently react with nucleophiles to form azo compounds, in the manner of the aforementioned coupling product with phenol. Thus, our attention became focused on these types of products in the reaction of BDI with DNA. C-Azo adducts were thought to be unlikely, since most of them are stable to mild acid. Our work on triazenes (e.g., Sieh et al., 1980 and subsequent papers) and especially the papers by Stock and co-workers (Chin et al., 1981; Hung & Stock, 1982) suggested to us that the unstable adduct may be a triazene, specifically involving the adenine residues of DNA. The following data indicate that this hypothesis is correct.

Adduct formation in vitro

The hypothesis was that BDI reacted with the adenine residues in DNA to form 6-(1-phenyltriazeno)purine. It had been shown earlier (Chin et al., 1981) that this adduct reverted to starting materials when treated with acid. Thus, it was imperative to devise a method wherein the DNA could be isolated without adduct decomposition; once isolated, the adducted DNA had to be 'fixed', so that further manipulation would not result in loss of the lesions. It was not difficult to reisolate exogenous calf thymus DNA after reactions in vitro. Reactions of benzenediazonium hexafluorophosphate with calf thymus DNA in buffer required repeated precipitation of the DNA with ethanol. Following that, the DNA, dissolved in pH 7.4 phosphate buffer, was treated with an excess of $NaBH_4$. Experiments with authentic 6-(1-phenyltriazeno)purine revealed that reduction of the triazene with borohydride transformed the triazene into 6-hydrazinopurine ($N6$-aminoadenine) in quantitative yield. Following the reduction, the DNA solution was acidified (pH 2) and heated to 100°C for 45 min. The sample was lyophilized, dissolved in methanol and treated with para-(dimethylamino)benzaldehyde. The aldehyde had been shown to react with the hydrazinopurine to form a hydrazone, which was characterized by mass spectrometry and which had the advantage of being fluorescent. The DNA sample was analysed by high-performance liquid chromatography on a C-18 reversed phase column (60% methanol, 40% water, 1 ml/min). The peaks were detected using a spectrofluorimeter, operating at an excitation wavelength of 265 nm and emission of 420 nm. The hydrazone, appearing as an inflection on another peak, was repeatedly collected, concentrated and rechromatographed under identical conditions. The final fraction was collected, concentrated and analysed by high-resolution fast-atom bombardment mass spectrometry and was found to have an identical spectrum to the authentic material. The exact mass of the molecular ion was 282.1465 (calculated: 282.1466). A similar procedure was used to identify the adduct from the reaction of NMA, activated by the rat liver preparations with calf thymus DNA. In that case, however, the presence of other macromolecules in the incubation medium necessitated a short phenol extraction of the DNA. This is generally not the procedure of choice since the adduct appears to be unstable in the presence of phenol (see below); nevertheless, the abbreviated extraction did allow enough adduct to survive so that a positive identification of 6-hydrazinopurine hydrazone could be made.

Identification of the adduct in vivo

Attempts to obtain evidence for the adduct in rat liver DNA following treatment of animals with 225 mg/kg NMA (1 x LD$_{50}$) by gavage in corn oil failed when the modified phenolic extraction procedure was used. A modified procedure was developed. Specifically, nuclei from the liver of a rat treated with 225 mg/kg NMA and killed 4 h later were isolated by the procedure of Bolognesi *et al.* (1981). DNA on membrane filters (4.5 μm) was purified according to the procedure of Leadon and Cerutti (1982). The filter, with the attached DNA, was placed into 25 ml 0.1M phosphate buffer (pH 7.5) and treated with 1 g NaBH$_4$ (a large excess). After 1 h reaction time, the pH was adjusted to 2.0 with 1N HCl, the sample was lyophilized and the residue was suspended in 10 ml methanol and filtered. The solution was then treated with 10 mg 4-(dimethylamino)naphthaldehyde and one drop of 1N HCl overnight. After removal of the solvent *in vacuo*, the residue was triturated three times with diethyl ether to remove most of the unreacted aldehyde. The residue was dissolved in methanol, filtered and analysed by high-performance liquid chromatography using a C-18 reverse phase column (elution conditions: 70% methanol, 30% water at a rate of 1.5 ml/min). The peaks were detected by fluorimetry (excitation, 260 nm; emission, 480 nm). Repeated rechromatography of the area corresponding to the retention time of the hydrazone resulted in a sample with sufficient purity (the major contaminant was the unreacted aldehyde) for high–resolution electron impact mass spectrometry. The spectrum was identical to that of the authentic hydrazone, and the exact mass found was 331.1545 amu (expected: 331.1548 amu). We used the napthaldehyde derivative in this case because it had somewhat better chromatographic characteristics and had a higher fluorescent yield. The reactions are summarized in Figure 1. Quantification of this adduct in DNA is difficult to achieve with the present analytical method. The current detection limit is about 10 μmol adduct/mol adenine. We estimate that the actual yield is closer to 1000 μmol adduct/mol adenine, and may actually be higher.

These data clearly demonstrate that NMA forms a DNA adduct efficiently. This adduct was not detected in the past because standard DNA extraction procedures invariably caused its decomposition.

Acknowledgement

Research sponsored by the National Cancer Institute, DHHS, under contract No. NO1-CO-74101 with BRI. The contents of this publication do not necessarily reflect the views or policies of the Department of Health and Human Services, nor does mention of trade names, commercial products, or organizations imply endorsement by the US Government.

Figure 1. Isolation of adduct of DNA with *N*-nitroso-*N*-methylaniline

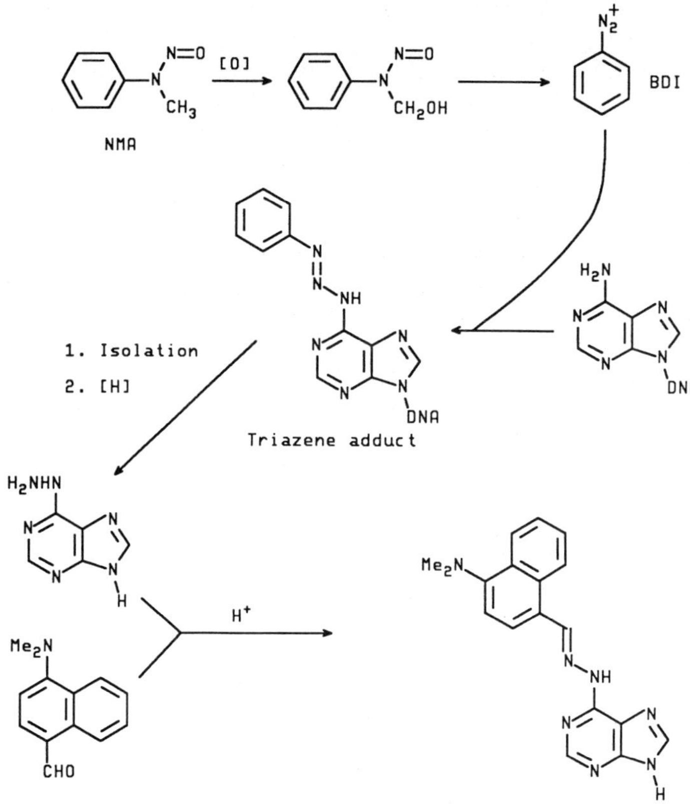

References

Bolognesi, C., Cesarone, C.F. & Santi, L. (1981) Evaluation of DNA damage by alkaline elution technique after *in vivo* treatment with aromatic amines. *Carcinogenesis*, **2**, 265-268

Chin, A., Hung, M.-H. & Stock, L.M. (1981) Reactions of benzene-diazonium ions with adenine and its derivatives. *J. Org. Chem.*, **46**, 2202-2207

Druckrey, H., Preussmann, R., Ivankovic, S. & Schmähl, D. (1967) Organotropic carcinogenicity of 65 different N-nitroso compounds in BD rats (in German). *Z. Krebsforsch.*, **69**, 103-201

Gold, B., Farber, J. & Rogan, E. (1987) An investigation of the metabolism of N-nitroso-N-methylaniline by phenobarbital- and pyrazole-induced Sprague-Dawley rat liver and esophagus derived S9. *Chem.-Biol. Interact.*, **61**, 215-228

Goodall, C.M., Lijinsky, W., Tomatis, L. & Wenyon, C.E.M. (1970) Toxicity and oncogenicity of nitrosomethylaniline and nitrosomethylcyclohexylamine. *Toxicol. Appl. Pharmacol.*, **17**, 426-432

Hung, H.-M. & Stock, L.M. (1982) Reactions of benzenediazonium ions with guanine and its derivatives. *J. Org. Chem.*, **47**, 448-453

Koepke, S.R., Kroeger-Koepke, M.B. & Michejda, C.J. (1987) N-Nitroso-N-methylaniline: possible mode of DNA modification. In: Bartsch, H., O'Neill, I. & Schulte-Hermann, R., eds, *The Relevance of N-Nitroso Compounds to Human Cancer: Exposure and Mechanisms* (IARC Scientific Publications No. 84), Lyon, IARC, pp. 68-70

Kroeger-Koepke, M.B. & Michejda, C.J. (1979) Evidence for several demethylase enzymes in the oxidation of dimethylnitrosamine and phenylmethylnitrosamine by rat liver fractions. *Cancer Res.*, **39**, 1587-1591

Kroeger-Koepke, M.B., Koepke, S.R., McClusky, G.A. Magee, P.N. & Michejda, C.J. (1981) α-Hydroxylation pathway in the *in vitro* metabolism of carcinogenic nitrosamines: N-nitrosodimethylamine and N-nitroso-N-methylaniline. *Proc. Natl Acad. Sci. USA*, **78**, 6489-6493

Kroeger-Koepke, M.B., Reuber, M.D., Iype, P.T., Lijinsky, W. & Michejda, C.J. (1983) The effect of substituents in the aromatic ring on carcinogenicity of N-nitrosomethylaniline in F344 rats. *Carcinogenesis*, **4**, 157-160

Leadon, S.A. & Cerutti, P.A. (1982) A rapid and mild procedure for the isolation of DNA from mammalian cells. *Anal. Biochem.*, **120**, 282-288

Michejda, C.J., Kroeger-Koepke, M.B., Koepke, S.R., Magee, P.N. & Chu, C. (1982) Nitrogen formation during *in vivo* and *in vitro* metabolism of *N*-nitrosamines. In: Magee, P.N., ed., *Nitrosamines and Human Cancer* (Banbury Report 12), Cold Spring Harbor, NY, CSH Press, pp. 69-85

Napalkov, N. & Pozharisski, R.M. (1969) Morphogenesis of experimental tumors of the esophagus. *J. Natl Cancer Inst.*, **42**, 927-931

Sieh, D.H., Wilbur, D.S. & Michejda, C.J. (1980) Preparation of trialkyltriazenes. A comparison of the N-N bond rotation in trialkyltriazenes and aryldialkyltriazenes by variable temperature ^{13}C NMR. *J. Am. Chem. Soc.*, **102**, 3883-3887; and subsequent papers in the series

Relevance to Human Cancer of N-Nitroso Compounds,
Tobacco Smoke and Mycotoxins.
Ed. I.K. O'Neill, J. Chen and H. Bartsch
Lyon, International Agency for Research on Cancer
© IARC, 1991

METABOLIC DENITROSATION OF N-NITROSAMINES: MECHANISM AND BIOLOGICAL CONSEQUENCES

K.E. Appel[1], S. Görsdorf[1], T. Scheper[1], B. Spiegelhalder[2],
M. Wiessler[2], M. Schoepke[1], C. Engeholm[1] & R. Kramer[1]

[1]Max von Pettenkofer Institute, Federal Health Office, Berlin;
and [2]Institute for Toxicology and Chemotherapy, German Cancer
Research Center, Heidelberg, Germany

NADPH-dependent microsomal metabolism of N-nitrosamines results in both oxidative dealkylation and denitrosation of the molecule. For denitrosation, two enzymatic mechanisms have been proposed: (i) cytochrome P450 (P450)-dependent one–electron reduction of the nitrosamine molecule, resulting in the formation of nitric oxide (NO) and secondary and primary amine, and (ii) liberation of NO *via* an oxidative mechanism mediated by a P450-dependent one-electron abstraction. In order to clarify the mechanism of denitrosation, the metabolism and kinetics of N-nitrosodibenzylamine (NDBzA) and its corresponding secondary amine dibenzylamine were studied. The main metabolites of NDBzA are benzaldehyde, the primary amine benzylamine and nitrite. An important finding is that benzaldehyde is generated more rapidly from dibenzylamine than from the parent NDBzA. During reductive denitrosation of NDBzA, the oxygen atom in benzaldehyde is derived from air, while benzaldehyde generated *via* the oxidative mechanism of denitrosation receives its oxygen atom from water due to hydrolysis of the intermediary benzylidenebenzylamine. Microsomal incubation of NDBzA in buffer containing ^{18}O-H_2O resulted in no incorporation of ^{18}O from water into benzaldehyde, which could be related to the formation of the corresponding benzylidenebenzylamine. It is concluded that NDBzA is denitrosated by the proposed reductive mechanism. Current belief is that denitrosation leads to detoxification of the NA molecule; however, toxic effects cannot be excluded if the conversion of NO into NO_2^- and NO_3^- involves intermediary formation of the NO_2 radical. NO, NO_2, NO_2^- and NO_3^- were therefore tested for their ability to induce single-strand breaks in DNA of Chinese hamster (V79) cells using the alkaline elution assay. V79 cells were exposed to NO and NO_2 at 0–500 ppm for 5–30 min. NO, NO_2^- and NO_3^- treatment did not induce detectable DNA damage; but NO_2 led to a dose- and time-dependent increase in the rate of single-strand breaks. The lowest effective concentration was 5 ppm for 20 min. Therefore, NO_2 formed during metabolism of NA within cells may contribute to the toxic effects, in addition to those due to alkylation.

Mechanistic aspects

Denitrosation is catalysed by cytochrome P450; two mechanisms have been proposed. One is P450-dependent one-electron reduction of the nitrosamine molecule, resulting in the formation of NO and the secondary amine, which may then be oxidatively dealkylated (Appel & Graf, 1982; Appel et al., 1980, 1987a). This hypothesis is supported by the finding of secondary amines during metabolism in vitro of N-nitroso-N-methylaminopyridines (Heydt-Zapf et al., 1982), N-nitrosodiphenylamine (Appel et al., 1984, 1987b) and some N-nitrosoureas. The other suggested mechanism is an oxidative mechanism closely related to α-C-hydroxylation (Haussmann & Werringloer, 1987; Wade et al., 1987). A P450-mediated one-electron abstraction of the amine nitrogen generates NO via the formation of an aminium cation radical. Loss of a proton leads to an alkylidenaminoalkane, which hydrolyses via an unstable carbinolamine to the primary amine and the corresponding aldehyde (Fig. 1). This mechanism is deduced from metabolic investigations of N-nitrosodimethylamine. Methylamine has been identified as the main metabolite in this pathway in vitro (Keefer et al., 1987) and in vivo (Heath & Dutton, 1958). Dimethylamine was detected, but in small amounts.

Figure 1. Postulated oxidative and reductive denitrosation of N-nitrosodibenzylamine in buffer containing ^{18}O-H_2O. The resulting benzaldehyde and the ratio between ^{18}O and ^{16}O incorporation was determined by gas chromatography-mass spectrometry.

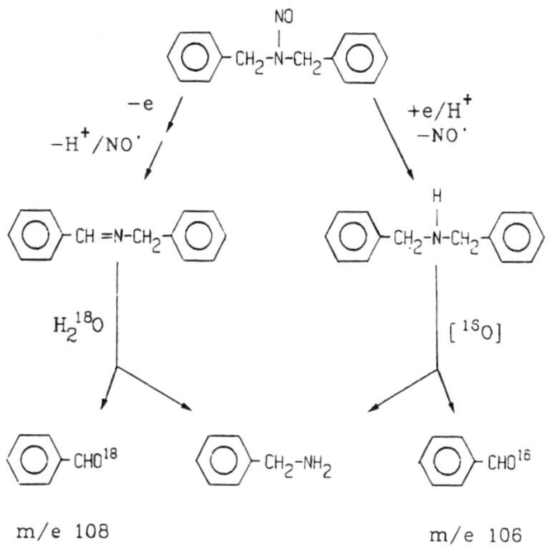

We studied the metabolism and kinetics of N-nitrosodibenzylamine (NDBzA) in order to investigate the mechanism of denitrosation. The main metabolites of NDBzA are benzaldehyde, benzyl alcohol, benzoic acid, benzylamine and nitrite. Dibenzylamine and N-hydroxydibenzylamine were not detected in mouse or rabbit liver microsomes; however, metabolism of dibenzylamine, in rabbit liver microsomes resulted in the formation of benzylamine benzaldehyde, benzyl alcohol and N-hydroxydibenzylamine. The generation of benzylamine from dibenzylamine (determined as V_{max}) is about twice as fast as that from NDBzA (Table 1).

Table 1. Kinetics of the metabolism of *N*-nitrosodibenzylamine/dibenzylamine in liver microsomes from mice pretreated with phenobarbital, as determined by Lineweaver–Burk plots

Substrate	Metabolite	K_m (μM)	V_{max} (nmol/min per mg)
N-Nitrosodibenzylamine	Benzylamine	232	8.3
	Nitrite	237	8.1
Dibenzylamine	Benzylamine	206	14.9
		18 300	83.3
	Benzaldehyde	283	17.5
		19 200	83.3

During reductive denitrosation of NDBzA, the oxygen atom in benzaldehyde is derived from air, while benzaldehyde generated *via* oxidative denitrosation would receive its oxygen atom from water due to hydrolysis of the intermediary benzylidenebenzylamine (Fig. 1). The metabolism of NDBzA was therefore studied in buffer containing ^{18}O-H_2O, and the ratio of ^{18}O-labelled benzaldehyde to ^{16}O-benzaldehyde determined by gas chomatography/mass spectrometry. The results obtained showed no incorporation of ^{18}O from water into benzaldehyde which could be related to formation of the corresponding alkylidenaminoalkane (Fig. 2). These findings and those of others – e.g., the different influences of antibodies against P450 on denitrosation and dealkylation activities (Amelizad et al., 1988) — indicate that denitrosation of NDBzA probably occurs by the reductive mechanism. After cleavage of the N–N bond, the secondary amine may remain bound to the catalytic site of P450, owing to its higher affinity in relation to the nitrosamine, and rapidly undergo oxidative dealkylation to the primary amine in the next reaction cycle of P450.

Biological consequences

We are interested in the possible toxic effects of NO and NO_2 because NO is a metabolite of nitrosamines in mammalian systems. Chemically, the conversion of NO into NO_2^- and NO_3^- is coupled to intermediate formation of NO_2; however, in contrast to NO, the formation of NO_2 has not yet been demonstrated, and it is therefore a putative metabolite in the metabolism of nitrosamines. Our results indicate that these nitrogen oxides may contribute to the genotoxic activity of *N*-nitrosodiphenylamine, which cannot be activated by the usual α-C-hydroxylation pathway (Appel et al., 1987b). We therefore investigated the capacity of NO and NO_2 to generate single-strand breaks in V79 cells.

Cells were cultivated as monolayers in Earle's minimal essential medium supplemented with 10% heat-inactivated fetal calf serum, 100 units/ml penicillin and 10 μg/ml streptomycin, obtained from commercial sources. Cultures were incubated at 37°C in a

100% humidified atmosphere with 5% CO_2. V79 cells were then seeded into 60-mm plastic dishes and washed three times with 10 ml Hanks' balanced salt solution. The dishes were then placed upside down in a gas-exposure chamber with an inner volume of 800 ml. NO and NO_2 (500 ppm in N_2) were rarified with N_2 to obtain the concentrations required. A flow rate of 200 ml/min was regulated by two flow meters. Cells were exposed to the gases at concentrations of 0–500 ppm over periods of 5–30 min. After exposure, the cells were harvested by trypsinization into ice-cold Merchant solution, and cell viability was determined.

Figure 2. Ratio between ^{18}O and ^{16}O incorporation into benzaldehyde after microsomal metabolism of *N*-nitrosodibenzylamine (NDBzA)[a]

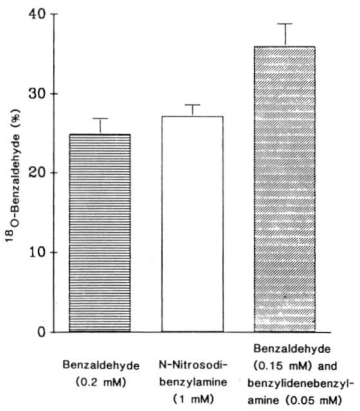

[a]Microsomal incubations were performed with liver microsomes from mice treated with phenobarbital, Tris KCl buffer enriched with ^{18}O-H_2O and an NADPH regenerating system. A concentration of 0.2 mM benzaldehyde resulted in about 25% incorporation of ^{18}O into benzaldehyde during a 3-min incubation period due to a time- and concentration-dependent exchange of ^{16}O with ^{18}O. The concentration tested corresponds to that found after incubation of 1 mM NDBzA under the same incubation conditions, representing both oxidative dealkylation and denitrosation metabolic pathways, irrespective of whether denitrosation occurs by a reductive or an oxidative mechanism. Incubation of 0.15 mM benzaldehyde in combination with 0.05 mM benzylidenebenzylamine resulted in an incorporation rate of nearly 40% ^{18}O. The two concentrations represent the result of denitrosation following an oxidative mechanism. When NDBzA was incubated for 3 min, an ^{18}O-incorporation rate of about 27% was found, corresponding to the first experiment in which ^{16}O-benzaldehyde was incubated. Thus, benzaldehyde formed during denitrosation of NDBzA obtained its oxygen atom from air and not from water. To simulate the metabolic formation of benzaldehyde and benzylidenebenzylamine from NDBzA, the two substances were added continuously to the incubation mixture using a motor-driven microsyringe until the required concentration was achieved. The results are means ± SD of at least four independent experiments.

Exposure of cells to NO at concentrations of 5, 20, 50, 100 and 500 ppm for either 10 or 30 min did not induce detectable DNA damage. In contrast, NO_2 significantly increased the elution rate, depending on the length of the exposure period and the concentration (Figs 3 and 4); the lowest effective concentration was 5 ppm NO_2 for 20 min. As NO_2 is converted into NO_2^- and NO_3^-, we studied whether these ions contribute to the DNA damaging effect of NO_2. For example, after 20-min exposure to 200 ppm NO_2, the amount of nitrite found corresponded to about 300 μM in a 5-ml incubation system; however, incubation of V79 cells with sodium nitrate or sodium nitrite at concentrations up to 1 mM for 2 h induced no single-strand breaks (data not shown). The DNA damaging activity of NO_2 cannot therefore be ascribed to the formation of NO_2^- or NO_3^- during exposure.

Figure 3. Time-dependent induction of DNA single-strand breaks after exposure of V79 cells to 100 or 200 ppm NO_2 for 5-20 min[a]

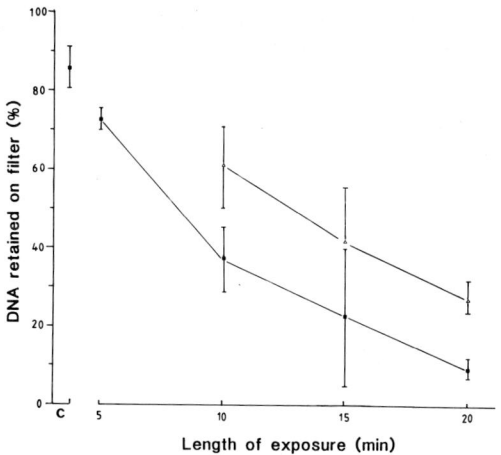

[a]The results are means ± SD of elution runs. The control value (c) represents 12 experiments with 23 elution runs; the values for treated cells represent 2-7 experiments with 4-14 elution runs; △, 100 ppm; ■, 200 ppm. The procedure followed was that of Kohn et al. (1981), with modifications. The amount of DNA eluted and remaining on the filter was determined using the microfluorimetric assay described by Cesarone et al. (1971) using 33258 Hoechst dye. The resulting fluorescence was measured on a model RF-520 Shimadzu differential spectrofluorophotometer at 360 nm (excitation) and 450 nm (emission).

Figure 4. Dose-dependent induction of DNA single-strand breaks after exposure of V79 cells to various concentrations of NO_2 for 20 min[a]

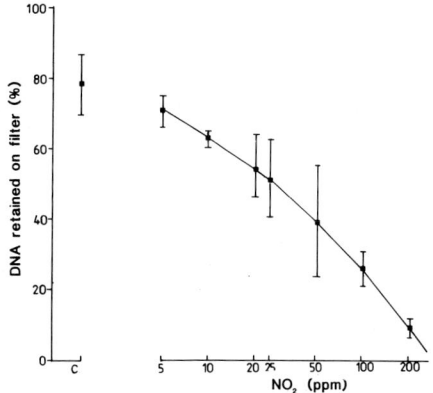

[a]The results are means ± SD of elution runs. The control value (c) represents 12 experiments with 23 elution runs; the values for the treated cells represent 2-6 experiments with 4-12 elution runs. Single-strand breaks were detected as described in the legend to Figure 3.

NO_2 may induce genotoxic effects by reacting directly or indirectly with genetic material. To our knowledge, however, no publication has described the direct interaction of NO_2 radicals with DNA bases *in vivo* or *in vitro*. Indirect mechanisms that lead to single-strand breaks and mutagenic events might include the ability of NO_2 to generate free-radical reactions with proteins and lipids. It is well known that NO_2 can react with alkanes and alkenes to initiate lipid peroxidation, and the biological consequences might include DNA damage and cellular degeneration.

Bartsch and coworkers (Ahotupa *et al.*, 1987a,b; Hietanen *et al.*, 1987) have studied oxidative damage induced by *N*-nitrosodimethylamine alone and in combination with dietary lipids in rats. Administration of the nitrosamine effectively increased lipid peroxidation, as determined by measuring ethane in exhaled air, and ethane exhalation remained elevated for several days after administration of single doses. Similarly, lipid peroxidation in the liver increased rapidly, showing a peak 20 min after treatment. The mechanism by which *N*-nitrosodimethylamine induces lipid peroxidation is unknown. Free-radical species generated during its metabolism might be involved, and, since NO_2 may be formed during denitrosation of the nitrosamine in cells, this free-radical species might be capable of initiating the observed peroxidation processes.

An important question is whether the concentration of NO_2 that is achieved in cells during denitrosation of nitrosamines is sufficient to generate effects such as lipid peroxidation and single-strand breaks. Preliminary results obtained with metabolically active hepatocytes show on the basis of measured NO_2^- concentrations that the amount of intermediary NO_2 formed after denitrosation of *N*-nitrosodiphenylamine might be of the same order of magnitude as the amount of NO_2 absorbed during exposure of V79 cells.

It has been shown in long-term inhalation studies that NO_2 cannot induce tumours in laboratory animals (Henschler & Ross, 1966). Therefore, the relevance *in vivo* of the induction by NO_2 of single-strand breaks in V79 cells is not clear. Certain defence mechanisms may be more effective *in vivo* that in artificial in-vitro systems. Free-radical quenchers, such as thiols, and antioxidants, like tocopherols and β-carotene, have been shown to be anticarcinogenic. Vitamin E has been tested as a prophylactic agent in animals exposed to NO_2, preventing or delaying mortality and alterations to various biochemical parameters.

Our results indicate, however, that NO_2 formed within the cell during metabolism of xenobiotics such as *N*-nitrosamines may contribute to the initiation of lipid peroxidation and to the observed genotoxic effects.

Acknowledgements

This work was supported by Deutsche Forschungsgemeinschaft.

References

Ahotupa, M., Béréziat, J.-C., Bussacchini-Griot, V., Camus, A.-M. & Bartsch, H. (1987a) Lipid peroxidation induced by *N*-nitrosodimethylamine (NDMA) in rats *in vivo* and in isolated hepatocytes. *Free Rad. Res. Commun.*, **3**, 285-291

Ahotupa, M., Bussacchini-Griot, V., Béréziat, J.-C., Camus, A.-M. & Bartsch, H. (1987b) Rapid oxidative stress induced by N-nitrosamines. *Biochem. Biophys. Res. Commun.*, **146**, 1047-1054

Amelizad, Z., Appel, K.E., Oesch, F. & Hildebrandt, A.G. (1988) Effect of antibodies against cytochrome P-450 on demethylation and denitrosation of N-nitrosodimethylamine and N-nitrosomethylaniline. *J. Cancer Res. Clin. Oncol.*, **114**, 380-384

Appel, K.E., Schrenk, D., Schwarz, M., Mahr, B. & Kunz, W. (1980) Denitrosation of N-nitrosomorpholine by liver microsomes; possible role of cytochrome P-450. *Cancer Lett.*, **9**, 13-20

Appel, K.E. & Graf, H. (1982) Metabolic nitrite formation from N-nitrosamines: evidence for a cytochrome P-450-dependent reaction. *Carcinogenesis*, **3**, 293-296

Appel, K.E., Rühl, C.S., Spiegelhalder, B. & Hildebrandt, A.G. (1984) Denitrosation of diphenylnitrosamine in vivo. *Toxicol. Lett.*, **23**, 353-358

Appel, K.E., Schoepke, M., Scheper, T., Görsdorf, S., Bauszus, M., Rühl, C.S., Kramer, R., Ruf, H.H., Spiegelhalder, B., Wiessler, M. & Hildebrandt, A.G. (1987a) Some aspects of cytochrome P-450 dependent denitrosation of N-nitrosamines. In: Bartsch, H., O'Neill, I.K. & Schulte-Hermann, R., eds, *The Relevance of N-Nitroso Compounds to Human Cancer: Exposures and Mechanisms* (IARC Scientific Publications No 84), Lyon, IARC, pp. 117-123

Appel, K.E., Görsdorf, S., Scheper, T., Ruf, H.H., Rühl, C.S. & Hildebrandt, A.G. (1987b) Metabolic denitrosation of diphenylnitrosamine: a possible bioactivation pathway. *J. Cancer Res. Clin. Oncol.*, **113**, 131-136

Cesarone, C.F., Bolognesi, C. & Santi, L. (1979) Improved microfluorometric DNA determination in biological material using 33258 Hoechst. *Anal. Biochem.*, **100**, 188-197

Haussmann, H.-J. & Werringloer, J. (1987) Mechanism and control of denitrosation of N-nitrosodimethylamine. In: Bartsch, H., O'Neill, I.K. & Schulte-Hermann, R., eds, *The Relevance of N-Nitroso Compounds to Human Cancer: Exposures and Mechanisms* (IARC Scientific Publications No. 84), Lyon, IARC, pp. 109-112

Heath, D.F. & Dutton, A. (1958) The detection of metabolic products from dimethylnitrosamine in rats and mice. *Biochem. J.*, **70**, 619-626

Henschler, D. & Ross, W. (1966) Zur Frage einer cancerogenen Wirkung inhalierter Stickstoffoxyde. *Naunyn-Schmiedebergs Arch. Exp. Pathol. Pharmakol.*, **253**, 495-507

Heydt-Zapf, G., Eisenbrand, G. & Preussmann, R. (1982) Metabolism of carcinogenic and non-carcinogenic N-nitroso-N-methylaminopyridines. I. Investigations in vitro. *Carcinogenesis*, **3**, 445-448

Hietanen, E., Ahotupa, M., Béréziat, J.-C., Bussacchini, V., Camus, A.-M. & Bartsch, H. (1987) Elevated lipid peroxidation in rats induced by dietary lipids and N-nitrosodimethylamine and its inhibition by indomethacin monitored via ethane exhalation. *Toxicol. Pathol.*, **15**, 93-96

Keefer, L.K., Anja, T., Wade, D., Woun, T. & Yang, C.H.S. (1987) Concurrent generation of methylamine and nitrite during denitrosation of N-nitrosodimethylamine by rat liver microsomes. *Cancer Res.*, **47**, 447-452

Kohn, K.W., Ewig, R.A.G., Erickson, L.C. & Zwelling, L.A. (1981) Measurement of strand breaks and cross-links by alkaline elution. In: Friedberg, E.C. & Hanawalt, P.C., eds, *DNA Repair*, New York, Marcel Dekker, pp. 379-401

Wade, D., Yang, C.S., Metral, C.J., Roman, J.M., Hrabie, J.A., Riggs, C.W., Anjo, T., Keefer, L.K. & Mico, B.A. (1987) Deuterium isotope effect on denitrosation and demethylation of N-nitrosodimethylamine by rat liver microsomes. *Cancer Res.*, **47**, 3373-3377

SUBCELLULAR FRACTIONS OF RAT ORGANS CONTAIN NITROREDUCTASES WHICH REDUCE *N*-NITRODIMETHYLAMINE TO *N*-NITROSODIMETHYLAMINE

E. Frei, M. Hassel, B. Spiegelhalder & M. Wiessler

*Institute of Toxicology and Chemotherapy,
German Cancer Research Center, Heidelberg,
Germany*

The nasal carcinogen *N*-nitrodimethylamine was reduced to its *N*-nitroso analogue *N*-nitrosodimethylamine *in vivo*. Liver–soluble fraction and brain mitochondria contained enzymes capable of reducing this compound. The activity was only partly inhibited by oxygen; NADPH and NADH served as cosubstrates. *N*-Nitromethylamine is also carcinogenic and was reduced by liver cytoplasm to a protein binding species (methyldiazohydroxide?).

The target organ for the carcinogenicity of *N*-nitrodimethylamine (NTDMA) in rats is the nasal mucosa. The product of NTDMA oxidative metabolism, namely *N*-nitromethylamine (NTMA), induces tumours of the spinal cord and spinal nerves (Hassel *et al*., 1987) of neurogenic origin. The mechanisms of activation to the ultimate carcinogen(s) might include reduction of the nitro to the nitroso compounds.

Metabolic studies in vivo

NTDMA was reduced to *N*-nitrosodimethylamine (NDMA) in the intact animal, and both compounds were excreted in urine and found in exhaled air when the concentrations of NTDMA were above 0.5 mmol/kg (Table 1). Since Spiegelhalder *et al*. (1982) observed enhanced excretion of administered NDMA in urine of ether-treated rats, and Keefer *et al*. (1985) found a decreased elimination of NDMA from rat blood, we exposed NTDMA-treated rats to ether. Excretion of NTDMA in urine was influenced only at low NTDMA doses, but the amount of NDMA in urine and breath increased significantly. In the 23-h experiments, in which urine was fractioned, we found that most of both compounds was excreted during the first 4 h after administration. Since NDMA was also detected after intravenous administration of NTDMA and not only after gavage, we can probably exclude reduction by gut bacteria (Hassel *et al*., 1987 and this paper).

NTDMA reductases

Liver cytoplasm, incubated as described in the legend to Figure 1, showed NTDMA reductase activity. Female rats showed a higher capacity to reduce NTDMA than males, producing 19.1 ± 0.3 nmol NDMA/mg protein × 30 min, *versus* 0.24 ± 0.01 nmol NDMA/mg protein × 30 min at 0.5 mM NTDMA, with NADPH as the cosubstrate. Cytoplasm prepared from nasal mucosa also reduced NTDMA; the activities were similar in animals of each sex and similar to that in liver cytoplasm of females. Depending on the concentration, both NADPH and NADH could serve as reducing substrates, the former being more effective at concentrations below 5 mM NTDMA and the latter at higher concentrations. Figure 1 shows the Lineweaver-Burk plots using NADPH for NTDMA concentrations between 0.01 and 1 mM and NADH for NTDMA concentrations from 0.1

Table 1. Excretion of *N*-nitrodimethylamine (NTDMA) and *N*-nitrosodimethylamine (NDMA) after administration of NTDMA[a]

Dose of NTDMA (mmol/kg bw)	Dose/rat (mg)	Exposed to ether [b]	Urine NTDMA μg	Urine NTDMA % of dose	Urine NDMA	mol NDMA/ mol NTDMA(%)	Breath NTDMA	Breath NDMA
4.5-h								
0.02 (oral)	0.45	–	0.031	0.007	–	–	–	–
0.02 (iv)	0.44	–	0.018	0.004	–	–	–	–
0.11 (oral)	2.23	–	0.067	0.003	–	–	–	–
0.11 (iv)	2.22	–	0.444	0.020	–	–	–	–
0.11 (iv)	2.22	+	4.441	0.200	0.032	0.85	–	–
23-h								
0.5 (oral)	9.00	+	106.1	1.2	0.72	0.85	4.01	–
1.0 (oral)	18.00	–	316.5	1.8	0.76	0.24	2.91	0.008
1.0 (oral)	18.00	+	355.8	2.0	2.24	0.73	2.96	0.036
2.0 (oral)	36.00	+	688.3	1.9	2.08	0.36	5.38	0.190

[a] 200-g male Sprague-Dawley rats were fasted overnight and administered the NTDMA in physiological saline, as indicated. They were then placed in metabolic cages, and urine was collected into a flask cooled in dry ice/acetone, extracted and analysed by gas chromatography-thermal energy analysis (Spiegelhalder *et al.*, 1983).
[b] After administration of NTDMA, animals were exposed to ether until they lost their righting reflex (~1 min) and then immediately transferred to the metabolic cage.
[c] Air was pumped through the metabolic cage and collected in two wash flasks, the first containing 100 ml 0.1M NaOH, the second 200 ml 0.1N NaOH. Flasks were changed three times during the experiment. Wash solutions were extracted and analysed by gas chromatography-thermal energy analysis. Results are expressed as total micrograms.

to 10 mM. The reason for the low V_{max} we observed with both cosubstrates might be suboptimal incubation conditions. Enzyme activities vary in different cytoplasmic preparations; the activity is labile to storage even at –70°C and is inhibited by Mg^{++} but inhibited to only 50% by oxygen. The decrease in NTDMA in many incubations was greater than the increase in NDMA; therefore, a further reduction of NDMA may occur *via* a possible intermediate to hydrazine, i.e., a four- or six-electron reduction instead of a two-electron reduction. Most of the nitroreductases described for aromatic *C*-nitro compounds perform four- or six-electron reductions (e.g., Abou-Khalil *et al.*, 1985;

Holtzmann et al., 1981; Ball & Lewtas, 1985). Xanthinoxidase (EC 1.2.3.2.) and NADH:lipoamide-oxidoreductase (EC 1.6.4.3.) were ineffective in reducing NTDMA.

Köchli et al. (1980) described a nitroreductase in the inner membrane of brain mitochondria. Using the conditions they described (pyrophosphate, 60 mM pH 9; 0.05% Triton X-100; 0.1 mM NADH; 0.5 mM NTDMA), we found that mitochondria from brain and spinal cord could reduce NTDMA: 31.0 ± 1.3 nmol NDMA/mg protein $\times$ 20 min in brain mitochondria and 37.1 ± 0.1 nmol NDMA in spinal cord mitochondria. Mitochondria from liver were inactive. As with the cytoplasmic incubations described above, the decrease in NTDMA was often greater than the formation of NDMA. The enzyme was insensitive to oxygen. Since both types of tumour initiated by NTDMA and NTMA are of neurogenic origin, and spinal cord is the target of NTMA, this reductase could be the key enzyme in activation of these nitramines.

Figure 1. Lineweaver–Burk graphs of reduction of N-nitrodimethylamine by NADPH (A) or NADH (B)

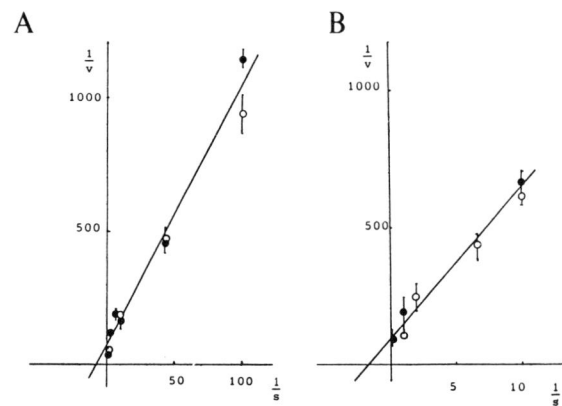

Cytoplasm (100 000 g supernatant) from male Sprague-Dawley rats killed by CO_2 was prepared from livers homogenized in 60 mM phosphate buffer pH 7.4 containing 0.5% KCl. Incubations were performed at 37°C, and contained, in 3 ml: 20 mg protein, 1 mM NADPH (A) or 1 mM NADH (B). Before adding N-nitrodimethylamine at the concentrations indicated, the samples were bubbled with argon for 5 min. Controls were performed with heat-inactivated cytoplasm. Three determinations each of two cyloplasmic preparations are shown (○, ●). A: $K_m = 0.124$ mM; $V_{max} = 0.013$ nmol NDMA/mg protein $\times$ min; $r = 0.99$ B: $K_m = 0.580$ mM; $V_{max} = 0.011$ nmol NDMA/mg protein $\times$ min; $r = 0.98$

No NTDMA reduction was observed with microsomes from liver, nasal mucosa or spinal cord under an argon atmosphere.

Reductive activation of NTMA

The hypothetical product of NTMA reduction would be methyldiazohydroxide, the ultimate carcinogen of NDMA activation. Due to the reactivity of this compound, analysis of its reaction products is difficult. We therefore have only indirect evidence that NTMA is reduced in liver cytoplasm. Using the conditions described for NTDMA, 30% of the radioactivity of ^{14}C-NTMA was associated with protein, precipitated by methanol. In incubations containing heat-inactivated cytoplasm, no such association of 14C-NTMA-derived radioactivity with proteins was observable. No metabolite was detected in high-performance liquid chromatograms of the water/methanol supernatant (reverse phase; tetrabutyl-ammonium formate, pH 7.3).

Conclusions and further studies

To our knowledge, this is the first description of the reduction of an aliphatic N-nitro compound to the corresponding N-nitrosamine. We suggest that reduction to the ultimate carcinogen is the main activation pathway of NTMA. NTDMA may be either oxidized to NTMA and formaldehyde or reduced to the potent carcinogen NDMA. Enzyme systems capable of reducing aliphatic N-nitro compounds would produce highly noxious N-nitroso compounds, which themselves might be transformed to hydrazines as an alternative to cytochrome P450 activation. Future studies will concentrate on elucidating the properties of the enzymes responsible for the observed reduction.

Acknowledgements

The authors would like to acknowledge the expert technical assistance of Andrea Litterer and Gerd Würtele. Part of this work was supported by the Deutsche Forschungsgemeinschaft.

References

Abou-Khalil, S., Abou-Khalil, W.H. & Yunis, A.A. (1985) Identification of a mitochondrial p-dinitrobenzene reductase activity in rat liver. *Pharmacology*, **31**, 301-308

Ball, L.M. & Lewtas, J. (1985) Rat liver subcellular fractions catalyze aerobic binding of 1-nitro(^{14}C)pyrene to DNA. *Environ. Health. Perspect.*, **62**, 193-196

Hassel, M., Frei, E., Scherf, H.R. & Wiessler, M. (1987) Investigations into the pharmacodynamics of the carcinogen N-nitrodimethylamine. In: Bartsch, H., O'Neill, I.K. & Schulte-Hermann, R., eds, *The Relevance of N-Nitroso Compounds to Human Cancer: Exposures and Mechanisms* (IARC Scientific Publications No. 84), Lyon, IARC, pp. 150-152

Holtzman, J.L., Crankshaw, D.L., Peterson, F.J. & Polunszek, C.F. (1981) The kinetics of the aerobic reduction of nitrofurantoin by NADPH-cytochrome P_{450} (c) reductase. *Mol. Pharmacol.*, **20**, 669-673

Keefer, L.K., Garland, W.A., Oldfield, N.F., Swagzdis, J.E. & Mico, B.A. (1985) Inhibition of N-nitrosodimethylamine metabolism in rats by ether anesthesia. *Cancer Res.*, **45**, 5457-5460

Köchli, H.W., Wermuth, B. & von Wartburg, J.-P. (1980) Characterization of a mitochondrial NADH dependent nitro reductase from rat brain. *Biochim. Biophys. Acta*, **616**, 133-142

Spiegelhalder, B., Eisenbrand, G & Preussmann, R. (1982) Urinary excretion of N-nitrosamines in rats and humans. In: Bartsch, H., O'Neill, I.K., Castegnaro, M. & Okada, M., eds, *N-Nitroso Compounds: Occurrence and Biological Effects* (IARC Scientific Publications No. 41), Lyon, IARC, pp. 443-449

Spiegelhalder, B., Eisenbrand, G. & Preussmann, R. (1983) Volatile N-nitrosamines in beer and other beverages by direct extraction using a kieselguhr column. In: Preussmann R., O'Neill, I.K., Eisenbrand, G., Spiegelhalder B. & Bartsch, H., eds, *Environmental Carcinogens: Selected Methods of Analysis*, Vol. 6, *N-Nitroso Compounds* (IARC Scientific Publications No. 45), Lyon, IARC, pp. 135-142

Relevance to Human Cancer of N-Nitroso Compounds,
Tobacco Smoke and Mycotoxins.
Ed. I.K. O'Neill, J. Chen and H. Bartsch
Lyon, International Agency for Research on Cancer
© IARC, 1991

TOXICOKINETIC STUDIES OF N-NITROSAMINE CARCINOGENESIS

A.J. Streeter[1], R.W. Nims & L.K. Keefer

*Chemistry Section, Laboratory of Comparative Carcinogenesis,
National Cancer Institute – Frederick Cancer Research Facility, Frederick, MD, USA*

The toxicokinetics of a single dose of N-nitroso(2-hydroxyethyl)methylamine has been characterized in rats of each sex, and analysis of the data revealed that the previously observed sex difference in the organotropism of this carcinogen may be due in part to the greater capacity of the liver of the female to metabolize the compound.

Among the most important factors involved in the induction of carcinogenesis by chemical agents such as N-nitrosamines are those governing their absorption, distribution, metabolism and elimination, and thus regulating their concentration at the sites at which they exert their carcinogenic activity. Some of the most powerful methods of determining the amounts of a chemical that reach various organs and tissues are those used in toxicokinetics. In this report, the importance of such investigations in elucidating the mechanisms of carcinogen action is illustrated using an N-nitrosamine to which humans are reportedly exposed.

N-Nitroso(2-hydroxyethyl)methylamine

The identification of 7-(2-hydroxyethyl)guanine residues in the hepatic DNA of rats administered N-nitrosoethylmethylamine by von Hofe *et al.* (1986) implicated N-nitroso-(2-hydroxyethyl)methylamine (NHEMA) as an intermediate in the process and, consequently, as a possible metabolite of N-nitrosoethylmethylamine. Humans are at risk of exposure to both carcinogenic nitrosamines, since N-nitrosoethylmethylamine has been found in cigarette smoke (Brunnemann *et al.*, 1980) and NHEMA has been suggested to be a decomposition product of N-nitrosodiethanolamine, which is present in cutting fluids (Loeppky *et al.*, 1979). NHEMA is itself a particularly interesting carcinogen, since it has been observed that the incidence of hepatocellular carcinomas was greater in female rats given the compound while that of squamous-cell carcinoma of the nasal cavity was greater in males (Koepke *et al.*, 1988a). This sex difference was reflected in correspondingly higher amounts of DNA methylation in the livers of female rats (Koepke *et al.*, 1988b). These findings might be due to more extensive hepatic metabolism of oral doses of NHEMA by females, leading to more liver tumour formation, while lower levels of metabolism in males would allow larger amounts of NHEMA to reach the systemic circulation and, consequently, other susceptible tissues such as the nasal squamous epithelium, where more tumours could result.

---

[1]To whom correspondence should be sent c/o PRI, Spring House, PA 19477-0776

In order to test this hypothesis, we characterized the single-dose toxicokinetics in eight-week-old Fischer 344 rats by analysis using high-performance liquid chromatography of approximately 20 serial blood samples from each animal. Graphs of changes in blood concentration of NHEMA with time following intravenous administration to groups of rats of each sex are presented in Figure 1. Non-compartmental toxicokinetic analysis (Riegelman & Collier, 1980; Gibaldi & Perrier, 1982) of the results yielded the parameters listed in Table 1. While there was no sex difference in the apparent steady-state volume of distribution of the compound, which was equivalent to the total body water (Altman & Dittmer, 1974), a more rapid elimination was seen in females, as evidenced by a shorter

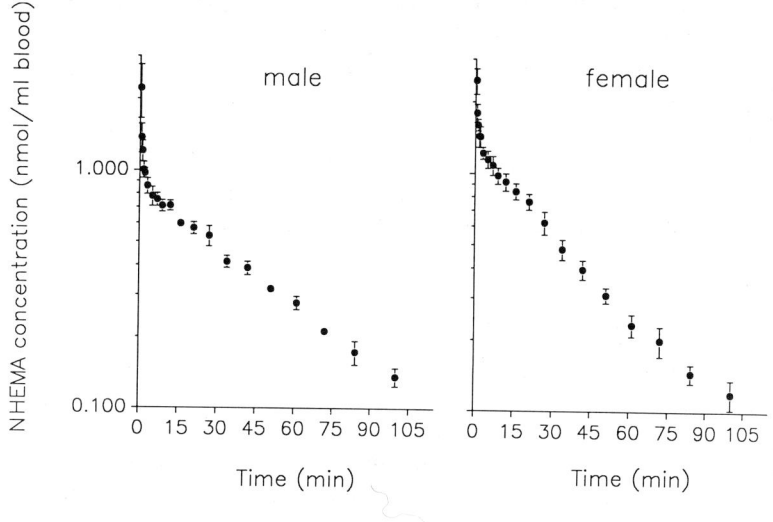

Fig. 1. Blood concentrations of *N*-nitroso(2-hydroxyethyl)methylamine (NHEMA) in groups of male and female rats as a function of time after a single intravenous bolus of 618 ± 34 (n = 4) and 847 ± 67 nmol/kg (n = 5), respectively (mean ± SE at each sampling time)

terminal half-life and mean residence time of the molecules in the systemic circulation. Consequently, the proportionality constant for the rate of elimination from the systemic circulation (systemic blood clearance, CL) for NHEMA was greater in females, a difference that was not completely explained by the slightly greater proportionality constant for the rate of elimination *via* the kidneys (renal blood clearance, CL_R) in rats of that sex. The amount of compound absorbed and reaching the systemic circulation unchanged (systemic bioavailability, F) was determined to be 78% and 69% in separate groups of rats given NHEMA by the intragastric route at doses of approximately 10 µmol/kg in males and females, respectively. Calculations of the bioavailability of total radioactive material from a radiolabelled compound can be used to determine the absorption of that compound (Riegelman *et al.*, 1973), and the use of 14C- labelled NHEMA revealed that approximately 100% of the parent compound was absorbed in animals of each sex. Under these circumstances, if all of the systemic blood clearance that is not due to renal clearance is the result of hepatic metabolism, then the hepatic blood flow (Q) can be calculated from

$$Q = (CL - CL_R)/(1 - F).$$

Substitution into the equation of the appropriate toxicokinetic parameters gives estimations for Q of 56 and 49 ml/min per kg bw for male and female rats, respectively. Since these values are in good agreement with those previously published (Altman & Dittmer, 1974), hepatic metabolism is probably responsible for most of the difference in the systemic blood clearance of NHEMA. Using the assumptions stated above, in addition to our observation that plasma protein binding of NHEMA is negligible, the intrinsic hepatic clearance (CL_I), that is the intrinsic ability of liver tissue to eliminate NHEMA, can be calculated from

$$CL_I = (CL/F - CL_R)/(1 + CL_R/Q).$$

The significantly greater value of CL_I found for females (Table 1) demonstrates a greater capacity for their livers to metabolize NHEMA.

Table 1. Toxicokinetic parameters in rats following a single intravenous dose of *N*-nitroso(2-hydroxyethyl)methylamine

Parameter[a]	Male	Female
Body weight (g)	163 ± 5	126 ± 3[b]
Dose (nmol/kg)	618 ± 34	847 ± 67[c]
t½β (min)	37.4 ± 1.7	27.2 ± 1.2[d]
MRT (min)	52.7 ± 2.6	39.2 ± 2.0[d]
V_{ss} (ml/kg)	685 ± 31	652 ± 23
CL (ml/min/kg)	13.1 ± 0.9	16.9 ± 1.3[c]
CL_R (ml/min/kg)	0.81 ± 0.02	1.45 ± 0.14[d]
CL_I (ml/min/kg)	16.7 ± 2.1	22.5 ± 0.3[c]

[a] t½β, terminal elimination half-life; MRT, mean residence time; V_{ss}, apparent steady-state volume of distribution; CL, systemic blood clearance; CL_R, renal blood clearance, estimated from percentage of dose excreted unchanged in the urine of separate groups of male and female rats administered 10.7 and 14.2 μmol/kg, respectively, intravenously; CL_I intrinsic hepatic clearance
[b] $p < 0.001$, [c] $p < 0.05$, [d] $p < 0.01$, significantly different from the male group (Student's *t* test)

Conclusion

Toxicokinetic experiments can be valuable, either by giving a direct answer to a problem, as in the case of sex differences in NHEMA carcinogenesis, or by indicating the direction for future studies, such as the need to investigate the properties of individual tissues in view of the postulated existence of extrahepatic metabolism of *N*-nitrosodimethylamine in certain species (Gombar *et al.*, 1988).

References

Altman, P.L. & Dittmer, D.S. (1974) *Biology Data Book*, 2nd ed., Vol. III, Bethesda, MD, Federation of American Societies for Experimental Biology

Brunnemann, K.D., Fink, W. & Moser, F. (1980) Analysis of volatile N-nitrosamines in mainstream and sidestream smoke from cigarettes by GLC-TEA. *Oncology*, **37**, 217-222

Gibaldi, M. & Perrier, D. (1982) *Pharmacokinetics*, 2nd ed., New York, Marcel Dekker

Gombar, C.T., Harrington, G.W., Pylypiw, H.M., Bevill, R.F., Thurmon, J.C., Nelson, D.R. & Magee, P.N. (1988) Pharmacokinetics of N-nitrosodimethylamine in swine. *Carcinogenesis*, **9**, 1351-1354

von Hofe, E., Kleihues, P. & Keefer, L.K. (1986) Extent of DNA 2-hydroxyethylation by N-nitrosomethylethylamine and N-nitrosodiethylamine *in vivo*. *Carcinogenesis*, **7**, 1335-1337

Koepke, S.R., Creasia, D.R., Knutsen, G.L. & Michejda, C.J. (1988a) Carcinogenicity of hydroxyalkylnitrosamines in F344 rats: contrasting behavior of β- and γ-hydroxylated nitrosamines. *Cancer Res.*, **48**, 1533-1536

Koepke, S.R., Kroeger-Koepke, M.B., Bosan, W., Thomas, B.J., Alvord, W.G. & Michejda, C.J. (1988b) Alkylation of DNA in rats by N-nitrosomethyl(2-hydroxyethyl)amine: dose response and persistence of the alkylated lesions *in vivo*. *Cancer Res.*, **48**, 1537-1542

Loeppky, R.N., Gnewuch, C.T., Hazlitt, L.G. & McKinley, W.A. (1979) N-Nitrosamine fragmentation and N-nitrosamine transformation. In: Anselme, J.-P., ed, N-*Nitrosamines* (ACS Symposium Series No. 101), Washington DC, American Chemical Society, pp. 109-123

Riegelman, S. & Collier, P. (1980) The application of statistical moment theory to the evaluation of *in vivo* dissolution time and absorption time. *J. Pharmacokinet. Biopharm.*, **8**, 509-534

Riegelman, S., Rowland, M. & Benet, L.Z. (1973) Addendum 1. Use of isotopes in bioavailability testing. *J. Pharmacokinet. Biopharm.*, **1**, 83-87

Relevance to Human Cancer of N-Nitroso Compounds,
Tobacco Smoke and Mycotoxins.
Ed. I.K. O'Neill, J. Chen and H. Bartsch
Lyon, International Agency for Research on Cancer
© IARC, 1991

ENZYME KINETICS OF N-NITROSODIMETHYLAMINE DEMETHYLASE IN RODENTS AND HUMANS

J.S.H. Yoo, H. Ishizaki & C.S. Yang

*College of Pharmacy, Rutgers University,
Piscataway, NJ, USA*

A variety of K_m values have been reported for hepatic microsomal N-nitrosodimethylamine demethylase (NDMAd). We demonstrated previously that the biologically important, high affinity (K_mI) form of microsomal NDMAd is manifested by cytochrome P450IIE1 (also known as P450ac and P450j). The K_mI value of NDMAd was, however, affected greatly by assay conditions: the possible presence of inhibitors and the presence of cytochrome b_5. We re-examined the K_mI value by testing the effect of enzyme concentrations and of different types of enzyme preparations on the K_m. The K_mI value ranged from 15 to 22 µM, as estimated by the direct linear plot, using a microsomal protein concentration in the range of 0.1 to 0.8 mg/ml with correction for substrate utilization. A slight yet significant dependency of microsomal protein concentration on the K_m ($r = 0890$; $p < 0.05$) was seen. When five different microsomal preparations were compared, the K_mI value ranged from 14 to 24 µM (median, 20 µM), as estimated by the direct linear plot. The K_m estimated by the commonly used Eadie-Hofstee plot did not differ from that by the direct linear plot. These K_m values are close to the values obtained in studies with isolated cells and tissue slices. The K_mI form of NDMAd (P450IIE1 and its orthologues) is present in rats, mice, rabbits, hamsters and guinea-pigs. It is responsible for the age-dependent differences between rats and hamsters and for the sex-related differences in mouse kidneys, and for the bioactivation and toxicity of NDMA. This enzyme also exists in human liver microsomes. The human enzyme cross-reacted with anti-rat-P450IIE1 and showed similar kinetic parameters to the rat enzyme in NDMA metabolism. These studies clarify the kinetic behaviours of NDMA metabolism and demonstrate the importance of P450IIE1 (or its orthologues) in the metabolic activation of NDMA in rodents and possibly in humans.

N-Nitrosodimethylamine (NDMA) requires metabolic activation to exert its cytotoxic and carcinogenic actions. The key metabolic activation step is believed to be the oxygenation of α-carbon (which leads to demethylation), catalysed by a cytochrome P450-dependent enzyme system commonly known as NDMA demethylase (NDMAd). The early work of Lai and Arcos (1980) and others resulted in multiple K_m values for hepatic microsomes in this enzyme system: NDMAd I, 0.2-0.3 mM and NDMAd II, 44-51 mM. We have studied the enzymology of NDMA metabolism and observed a very low K_m value of 60–70 µM (Tu & Yang, 1983; Yang *et al.*, 1989), in addition to the K_m values of the previously reported NDMAd I and II (Lai & Arcos, 1980) in rat liver microsomes. We subsequently designated this low K_m as K_mI and the previously used NDMAd I and II as K_mII and K_mIII (Yoo et al., 1987a). Nevertheless, even lower K_m values than the K_mI have been reported for NDMA clearance and metabolism by isolated perfused rat liver (8.3 µM), rat liver slices (25 µM), and isolated guinea-pig liver cells (10–20 µM) (Skipper et al., 1983; Hauber et al., 1984; Swann, 1984). In the study reported here, we re-examined the Michaelis-Menten constant of the K_mI form of microsomal NDMAd by testing the effect of enzyme concentrations and of different types of enzyme preparations on the K_m and by estimating the K_m using computerized direct linear plot analysis.

Experimental methods

Male Sprague–Dawley rats from Taconic, Inc. (Germantown, NY), with body weights of 90–100 g, were subjected to one of the following conditions: no inducer treatment (control), intragastric administration of acetone at a dose of 5 ml/kg bw once 20 h before sacrifice, and fasting for 48 h. Microsomes were prepared by differential centrifugation. NDMAd activity was determined as nanomoles of HCHO formed per minute per milligram of protein using a colorimetric assay (Yoo et al., 1987b). We calculated velocities and mean substrate (NDMA) concentrations in order to correct errors due to varied amounts of substrate utilization during the incubation period (Segal, 1975). We then estimated the K_m using the computerized direct linear plot analysis (Brady & Ishizaki, 1989) and compared it with the K_m obtained by the Eadie–Hofstee plot.

Results and discussion

The K_mI form of NDMAd can be induced by several fold by factors such as acetone, ethanol, isopropanol, isoniazid, fasting and diabetes, and has been purified and characterized as P450IIE1 or P450ac (reviewed by Yang & Yoo, 1988; Yang et al., 1989). During the past several years, studies have been done in our laboratory on the metabolic activation of NDMA, and we were able to demonstrate the key role of NDMAd in the activation of this compound in several different systems (reviewed by Yang et al., 1989). (i) We demonstrated that P450IIE1 was more efficient than other P450 species in activating NDMA to a mutagen in Chinese hamster V79 cells. (ii) Some of the previously reported differences with species and age in NDMA activation were attributed to differences in the quantity of P450IIE1 present in rats and hamsters of different ages. (iii) Pretreatment of rats with ethanol or acetone increased NDMA-induced methylation of DNA in vitro and in vivo and potentiated NDMA-induced hepatotoxicity.

More recently, we reported that the K_mI value of rat liver microsomal NDMAd was lowered to 40–50 µM when a previously used NADPH-generating system consisting of NADP, isocitrate and isocitrate dehydrogenase was replaced by a new system consisting of NADP, glucose 6-phosphate and glucose-6-phosphate dehydrogenase (Yoo et al., 1987b). In the present study, we examined the effects of enzyme concentration, different enzyme sources, and different methods of calculation on the K_m value of NDMAd. We obtained a K_mI value of 15–22 µM for NDMAd by the direct linear plot, using a microsomal protein concentration in the range of 0.1 to 0.8 mg/ml with correction for substrate utilization (Table 1). Furthermore, we observed a slight yet significant dependency of microsomal protein concentration on the K_m ($r = 0.890$; $p < 0.05$). Unlike the case of imipramine metabolism (Gillette, 1963), this dependency remained even when the concentration of aqueous substrate, calculated using the partition coefficient of NDMA, was used to estimate the K_m (data not shown). The direct linear plot was proposed for estimation of the K_m to avoid the risk of using transformed data with extreme data points, as in the commonly used equations, Lineweaver–Burk and Eadie–Hofstee plots (reviewed by Brady & Ishizaki, 1989). Nevertheless, the result obtained by the Eadie–Hofstee plot did not differ from that by the direct linear plot: the K_mI value ranged from 16 to 24 μM with a significant dependency of protein concentration on the K_m ($r = 0.961; p < 0.01$). When velocities and mean substrate concentrations were used to estimate the K_m, reasonably accurate values could be obtained from incubations in which up to 40% of substrate (NDMA) was used (data not shown).

Table 1. Effect of microsomal protein concentration on K_mI value

Microsomal protein (mg/ml)	K_mI (μM)	
	Direct linear	Eadie–Hofstee
0.1	15 ± 3[a]	16 ± 2
0.2	16 ± 3	16 ± 3
0.3	19 ± 1	19 ± 1
0.4	21[b]	21
0.5	21 ± 6	21 ± 6
0.8	22[c]	24
Microsomal protein versus K_mI:	$r = 0.890$ ($p < 0.05$)	$r = 0.961$ ($p < 0.01$)

[a] Mean ± SD of three determinations in duplicate
[b] Average of two determinations in duplicate
[c] A single determination in duplicate

We then selected a microsomal protein concentration of 0.5 mg/ml and compared the K_mI values obtained with five different microsomal preparations. As shown in Table 2, the K_mI estimated by the direct linear plot ranged from 14 to 24 μM (median, 20 μM); these were not significantly different ($p > 0.01$) as analysed by the Newman–Keuls multiple comparison test. The K_mI values estimated by the Eadie–Hofstee plot again did not differ from those obtained by the direct linear plot. In conclusion, our results obtained by two different methods for estimating K_m are comparable, and the K_mI (14–24 μM) is close to the low K_m values obtained in liver slices and hepatocytes (Hauber et al., 1984; Swann, 1984).

Table 2. Comparison of K_mI values in microsomes from different sources

Microsomal source	K_mI (μM)[a]	
	Direct linear	Eadie-Hofstee
Acetone-induced 1	24 ± 3	24 ± 2
Acetone-induced 2	14 ± 3	15 ± 3
Control 1	23 ± 2	22 ± 1
Control 2	20 ± 3	20 ± 3
Fasting (48 h)	16 ± 2	15 ± 2

[a] Mean ± SD of three determinations in duplicate with 0.5 mg microsomal protein/ml incubation

Acknowledgements

This work was supported by NIH Grants GM 38336 and CA 37037.

References

Brady, J.F. & Ishizaki, H. (1989) A BASIC program for the estimation of Michaelis-Menten parameters by the direct linear plot. *Comput. Meth. Prog. Biomed.*, **28**, 271–272

Gillette, J.R. (1963) Metabolism of drugs and other foreign compounds by enzymatic mechanisms. *Prog. Drug Res.*, **6**, 11-73

Hauber, G., Frommberger, R., Remmer, H & Schwenk, M. (1984) Metabolism of low concentrations of N-nitrosodimethylamine in isolated liver cells of the guinea pig. *Cancer Res.*, **44**, 1343-1346

Lai, D.Y. & Arcos, J.C. (1980) Minireview: dialkylnitrosamine bioactivation and carcinogenesis. *Life Sci.*, **27**, 2149-2165

Segel, I.W. (1975) *Enzyme Kinetics*, New York, John Wiley & Sons

Skipper, P.L., Tomera, J.F., Wishnok, J.S., Brunengraber, H. & Tannenbaum, S.R. (1983) Pharmacokinetics model for N-nitrosodimethylamine based on Michaelis-Menten constants determined with isolated perfused rat liver. *Cancer Res.*, **43**, 4786-4790

Swann, P.F. (1984) Effect of ethanol on nitrosamine metabolism and distribution. Implications for the role of nitrosamines in human cancer and for the influence of alcohol consumption on cancer incidence. In: O'Neill, I.K., von Borstel, R.C., Miller, C.T., Long, J. & Bartsch, H., eds, *N-Nitroso Compounds: Occurrence, Biological Effects and Relevance to Human Cancer* (IARC Scientific Publications No. 57), Lyon, IARC, pp. 501-512

Tu, Y.Y. & Yang, C.S. (1983) High-affinity nitrosamine dealkylase system in rat liver microsomes and its induction by fasting. *Cancer Res.*, **43**, 623-629

Yang, C.S. & Yoo, J.S.H. (1988) Dietary effects on drug metabolism by the mixed-function oxidase system. *Pharmacol. Ther.*, **38**, 53-72

Yang, C.S., Yoo, J.S.H., Ishizaki, H. & Hong, J. (1990) Cytochrome P450IIE1: roles in nitrosamine metabolism and mechanisms of regulation. *Drug Metab. Rev.*, **22**, 147–159

Yoo, J.S.H., Ning, S.M., Patten, C.J. & Yang, C.S. (1987a) Metabolism and activation of N-nitrosodimethylamine by hamster and rat microsomes: comparative study with weanling and adult animals. *Cancer Res.*, **47**, 992-998

Yoo, J.S.H., Cheung, R.J., Patten, C.J., Wade, D. & Yang, C.S. (1987b) Nature of N-nitrosodimethylamine demethylase and its inhibitors. *Cancer Res.*, **47**, 3378-3383

Relevance to Human Cancer of *N*-Nitroso Compounds,
Tobacco Smoke and Mycotoxins.
Ed. I.K. O'Neill, J. Chen and H. Bartsch
Lyon, International Agency for Research on Cancer
© IARC, 1991

ACCELERATION OF *N*-NITROSATION REACTIONS BY ELECTROPHILES

B.P. Sullivan[1], T.J. Meyer[1], M.T. Stershic[2] & L.K. Keefer[2]

[1]*Department of Chemistry, The University of North Carolina, Chapel Hill, NC; and* [2]*Chemistry Section, Laboratory of Comparative Carcinogenesis, National Cancer Institute, Frederick Cancer Research Facility, Frederick, MD, USA*

Selected mechanisms by which electrophiles can facilitate *N*-nitrosamine formation are reviewed. Special attention is given to a recently discovered reaction in which nitrogen in its lowest (ammonia) oxidation state is efficiently converted to *N*-nitroso compounds by oxidation in the presence of secondary amines; an electrophilic transition metal centre (E^+) makes this reaction possible by initially *N*-coordinating the ammonia ($E^+ + NH_3 \rightarrow E-NH_3^+$). Other mechanisms considered include: the conversion of nitrite under nonacidic conditions *via* nitro complexes to nitrosatively active transition metal nitrosyl intermediates ($E^+ + NO_2^- \rightarrow E-NO_2^- \rightarrow E-NO^{2+}$); catalysis of *N*-nitrosamine formation in nitrite-amine mixtures by electrophilic carbon centres that initiate the reaction by attack on the amine ($E^+ + R_2NH \rightarrow E-NR_2$); and coordination of nitrite by carbon electrophiles to form activated O-bound species ($E^+ + ONO^- \rightarrow E-O-NO$) capable of performing the required *N*-nitrosation. The findings suggest that acceleration of *N*-nitrosamine-forming reactions by electrophiles may be a critical factor to consider in attempting to rationalize, predict and control the distribution of carcinogenic *N*-nitroso compounds *in vivo* and in the environment.

Nucleophiles such as the thiocyanate ion are generally recognized to be capable of playing a crucial role in the formation of carcinogenic *N*-nitroso compounds to which people are exposed (Ridd, 1961; Fan & Tannenbaum, 1973; Licht & Deen, 1988), but little consideration has thus far been given to the possibility that electrophiles might generally also be important determinants of the extent of physiological and environmental synthesis of *N*-nitrosamines. The purpose of this review is to present a brief summary of several known mechanisms of electrophilic involvement in *N*-nitrosamine-forming reactions, re-emphasizing the potential importance of considering electrophilic as well as nucleophilic catalysts in future attempts to understand how *N*-nitrosamine carcinogens arise environmentally and in the body.

Rate enhancement by electrophiles: coordination effects

Carcinogenic N-nitroso compounds are not normally formed to a measurable extent when nitrite is mixed with amines or amides, unless an electrophilic catalyst such as a hydrogen ion (H^+), or proton, is present (Ridd, 1961). In acidified nitrite-amine mixtures, protons initiate the nitrosation reaction by *coordinating*, or electrophilically attaching themselves to, the nitrite ion. Double protonation of the same oxygen atom converts the otherwise unreactive nitrite anion to the nitrous acidium ion (H_2O-NO^+), which can either attack the amine directly or be converted by nucleophiles such as thiocyanate and nitrite to even better nitrosating agents (NCS–NO and N_2O_3, respectively). Various mechanisms of N-nitrosamine formation known to be initiated by O-protonation of nitrite were reviewed by Ridd (1961).

For a long time, it was believed that the hydrogen ion was the only electrophile capable of catalysing nitrosation, i.e., that nitrite could not be induced to react with amines under nonacidic conditions (Ridd, 1961). More recently, however, analytical surveys of the environment led to the recognition that facile N-nitrosation of amines by nitrite ion could also occur in media containing vanishingly small concentrations of hydrogen ion, such as alkaline metal-working fluids (Fan *et al.*, 1977). These findings suggested that other electrophiles must be able to mimic the proton in its ability to induce N-nitrosamine formation in mixtures of nitrite with amines. Several mechanisms by which this can happen have now been demonstrated experimentally.

Just as the proton attaches itself to the nitrite oxygen, for example, certain other electrophiles can activate nitrite for nitrosative attack on amines by O-coordinating it. Thus, the organic-soluble salt, bis(triphenylphosphine)nitrogen(1+) nitrite (PNP^+ ONO^-), converted pyrrolidine to N-nitrosopyrrolidine in high yield when mixed with dichloromethane, a commonly used solvent for extracting N-nitrosamines from foodstuffs. Formation of N-nitrosopyrrolidine appears to proceed by the mechanism shown in Equation 1, the first step of which is electrophilic attack of solvent on the nitrite oxygen (Fanning & Keefer, 1987).

(a) $Cl-CH_2-Cl + ONO^- \rightarrow Cl-CH_2-O-NO + Cl^-$

(b) $Cl-CH_2-O-NO + R_2NH \rightarrow R_2N-CH_2-O-NO + HCl$ (1)

(c) $R_2N-CH_2-O-NO \rightarrow R_2N-NO + CH_2O$

Electrophiles can also initiate N-nitrosamine formation by N-coordinating the nitrite. One example of this occurs during the rapid formation of N-nitrosamines in alkaline nitrite–amine mixtures under catalysis by pentacyanoiron(II) ion. Detailed kinetic investigations show that the mechanism depicted in Equation 2 governs this reaction (L.K. Keefer, D.G. Williams, K.L. Poff, L.E. Castro, A.M. Dorries, S.-J. Uhm & J.M. Malin, submitted for publication). Note the involvement of both nitro complexes and metal nitrosyl species as critical intermediates in this pathway. Similar mechanisms have been implicated in the enzymatic N-nitrosation of nitrite by nitrite reductase (Weeg-Aerssens *et al.*, 1988).

(a) $Fe(CN)_5(L)^{3-} + ONO^- \rightarrow L$ (water or other ligand) $+ Fe(CN)_5-NO_2^{4-}$

(b) $Fe(CN)_5-NO_2^{4-} + 2H^+ \rightarrow Fe(CN)_5-NO^{2-}$ (nitroprusside) $+ H_2O$ \hfill (2)

(c) $Fe(CN)_5-NO^{2-} + R_2NH \rightarrow [Fe(CN)_5-NONHR_2]^{2-}$

(d) $[Fe(CN)_5-NONHR_2]^{2-} + B$ (a base) $+ L \rightarrow BH^+ + R_2N-NO + Fe(CN)_5(L)^{3-}$

Electrophiles can also initiate *N*-nitrosamine formation in amine-nitrite mixtures by *N*-coordinating the amine. For example, the catalysis of such reactions by formaldehyde in neutral or alkaline solutions is thought to proceed by initial attack of the electrophilic carbonyl carbon on the amine nitrogen to form the iminium ion, as outlined in Equation 3 (Keefer & Roller, 1973). Such a mechanism is presumed to be responsible for a considerable portion of the *N*-nitrosodiethanolamine formed in di- and triethanolamine-containing metal-working lubricants to which machinists are exposed, many of which fluids contain bacteriostatic additives that release formaldehyde (Loeppky et al., 1983).

(a) $CH_2O + R_2NH + H^+ \rightarrow R_2N=CH_2^+ + H_2O$

(b) $R_2N=CH_2^+ + ONO^- \rightarrow R_2N-CH_2-O-NO$ \hfill (3)

(c) $R_2N-CH_2-O-NO \rightarrow R_2N-NO + CH_2O$

Facilitation of nitrosation by electrophiles via *electron transfer*

In addition to the simple coordination effects described above, electrophiles can utilize *redox* processes to stimulate *N*-nitrosamine formation. Electron transfer is obligatory in *N*-nitrosation reactions in which nitric oxide occurs as an intermediate — for example, including those initiated by redox enzymes that produce NO such as nitrate reductase (Leach et al., 1987; Calmels et al., 1988; Ji & Hollocher, 1988; Ralt et al., 1988) and the enzyme(s) responsible for production of nitrate, nitrite and *N*-nitrosamines during the oxidation of arginine by stimulated macrophages (Marletta, 1988).

We have been interested in modelling this type of process nonenzymatically in order to suggest possible mechanisms by which such transformations might occur in the biological milieu. As a starting point, we assumed that the relevant enzyme in arginine oxidation may contain a transition metal centre at its active site that is capable of both coordinating the reduced nitrogen species and removing electrons from it. Using the simpler substrate, ammonia, in place of the arginine and a substitutionally rather stable osmium centre as the coordination site to slow the individual steps of the overall transformation for closer study, we found that *N*-coordination of ammonia followed by electrochemical oxidation in the presence of morpholine at pH 6.8 and at the biologically reasonable applied potential of +0.65 V led to complexes of *N*-nitrosomorpholine (NMOR) with the formula $[(tpy)(bpy)Os(NMOR)][PF_6]_2$, where tpy is 2,2'(6:6'),2''-terpyridine and bpy is 2,2'-bipyridine (Stershic et al., 1988). A suggested mechanism for the transformation is shown in Equation 4.

(a) $[(tpy)(bpy)Os(NH_3)]^{2+} \xrightarrow{-e^-} [(tpy)(bpy)Os(NH_3)]^{3+}$

(b) $[(tpy)(bpy)Os(NH_3)]^{3+} \xrightarrow{-e^-, -2H^+} [(tpy)(bpy)Os(NH)]^{2+}$

(c) $[(tpy)(bpy)Os(NH)]^{2+} \xrightarrow{R_2NH} [(tpy)(bpy)Os(NH_2\text{-}NR_2)]^{2+}$ (4)

(d) $[(tpy)(bpy)Os(NH_2\text{-}NR_2)]^{2+} \xrightarrow{-2e^-, -2H^+} [(tpy)(bpy)Os(NNR_2)]^{2+}$

(e) $[(tpy)(bpy)Os(NNR_2)]^{2+} \xrightarrow[-2H^+]{-2e^-, +H_2O} [(tpy)(bpy)Os(\overset{\underset{\Vert}{O}}{N}NR_2)]^{2+}$.

We believe that this model is noteworthy, for several reasons. For one thing, it demonstrates that an electrophilic centre can simultaneously coordinate and oxidize its ligand, suggesting that an enzyme could also combine the two functions advantageously. Secondly, the nitrosamine product of Equation 4 is isolated as a novel *N*-bound complex that, on *N*-protonation, could serve as a useful, stable model of the transition state involved in other important mechanisms of *N*-nitrosame formation, such as the pentacyanoiron(II)-catalysed reaction dicussed above (see Equation 2(*d*)). Finally, the transformation covers a variety of nitrogen oxidation states in the conversion of -3 nitrogen into an *N*-nitroso group ($+3$ nitrogen), several of which might be used in a chemically similar enzymatic process to intercept an amine molecule nitrosatively. In the mechanism of Equation 4, for example, the electrophilic imido (-1 nitrogen) complex, $[(typ)(bpy)Os(NH)]^{2+}$, is viewed as attacking the amine. Release of nitric oxide ($+2$ nitrogen) as the effective *N*-nitrosating precursor could also occur by such a mechanism if the redox reaction produced a nitrosyl ligand capable of extrusion as •NO.

Conclusion

Electrophiles can initiate *N*-nitrosamine formation in nitrite–amine mixtures by *N*-coordinating the amine as well as by *O*- or *N*-coordinating nitrite. They also can induce redox processes of importance in *N*-nitrosamine formation, as exemplified in the mechanism of Equation 4. Such pathways may contribute significantly to the burden of *N*-nitroso compounds to which humans are exposed. It is recommended that such phenomena be explicitly considered in future attempts to account for and predict the distribution of *N*-nitrosamine carcinogens in the body and in the environment.

References

Calmels, S., Ohshima, H. & Bartsch, H. (1988) Nitrosamine formation by denitrifying and non-denitrifying bacteria: implication of nitrite reductase and nitrate reductase in nitrosation catalysis. *J. Gen. Microbiol.*, **134**, 221-226

Fan, T.Y., Morrison, J., Rounbehler, D.P., Ross, R., Fine, D.H., Miles, W. & Sen, N.P. (1977) *N*-Nitrosodiethanolamine in synthetic cutting fluids: a part-per-hundred impurity. *Science*, **196**, 70-71

Fan, T.-Y. & Tannenbaum, S.R. (1973) Factors influencing the rate of formation of nitrosomorpholine from morpholine and nitrite: acceleration by thiocyanate and other anions. *J. Agric. Food Chem.*, **21**, 237-240

Fanning, J.C. & Keefer, L.K. (1987) Rapid formation of a potent nitrosating agent by solvolysis of ionic nitrite in dichloromethane. *J. Chem. Soc. Chem. Commun.*, 955-956

Ji, X.-B. & Hollocher, T.C. (1988) Mechanism for nitrosation of 2,3-diaminonaphthalene by *Escherichia coli*: enzymatic production of NO followed by O_2-dependent chemical nitrosation. *Appl. Environ. Microbiol.*, **54**, 1791-1794

Keefer, L.K. & Roller, P.P. (1973) N-Nitrosation by nitrite ion in neutral and basic medium. *Science*, **181**, 1245-1247

Leach, S.A., Thompson, M. & Hill, M. (1987) Bacterially catalysed *N*-nitrosation reactions and their relative importance in the human stomach. *Carcinogenesis*, **8**, 1907-1912

Licht, W.R. & Deen, W.M. (1988) Theoretical model for predicting rates of nitrosamine and nitrosamide formation in the human stomach. *Carcinogenesis*, **9**, 2227-2237

Loeppky, R.N., Hansen, T.J. & Keefer, L.K. (1983) Reducing nitrosamine contamination in cutting fluids. *Food Chem. Toxicol.*, **21**, 607-613

Marletta, M.A. (1988) Mammalian synthesis of nitrite, nitrate, nitric oxide, and N-nitrosating agents. *Chem. Res. Toxicol.*, **1**, 249-257

Ralt, D., Wishnok, J.S., Fitts, R. & Tannenbaum, S.R. (1988) Bacterial catalysis of nitrosation: involvement of the *nar* operon of *Escherichia coli*. *J. Bacteriol.*, **170**, 359-364

Ridd, J.H. (1961) Nitrosation, diazotisation, and deamination. *Q. Rev.*, **15**, 418-441

Stershic, M.T., Keefer, L.K., Sullivan, B.P. & Meyer, T.J. (1988) Formation of complexed nitrosamines by oxidation of coordinated ammonia in the presence of secondary amines. *J. Am. Chem. Soc.*, **110**, 6884-6885

Weeg-Aerssens, E., Tiedje, J.M. & Averill, B.A. (1988) Evidence from isotope labelling studies for a sequential mechanism for dissimilatory nitrite reduction. *J. Am. Chem. Soc.*, **110**, 6851-6856

NITROSAMINE ACTIVATION AND DETOXICATION THROUGH FREE RADICALS AND THEIR DERIVED CATIONS

R.N. Loeppky & Y.E. Li

*Department of Chemistry, University of Missouri,
Columbia, MO, USA*

The enzymatic activation of simple dialkylnitrosamines is widely perceived to involve cytochrome P450-mediated α-hydroxylation to generate an unstable α-hydroxynitrosamine. This process can also result in the denitrosation of the nitrosamine, presumably through a common intermediate. We present evidence that the critical intermediate is the alkynitrosaminomethyl free radical, which we generated by thermal decomposition of a nitrosamino acid perester. The radical rapidly loses NO to generate an N-alkylmethyleneimine. The radical is also produced during the Ce(IV) oxidation of β-hydroxynitrosamines after fragmentation, where it not only loses NO but is oxidized further to a cation which reacts with water to form an α-hydroxynitrosamine. These results provide models of the activation and detoxication pathways for β-oxidized nitrosamines.

Metabolic alkyl-chain shortening is a process of importance for dialkyl nitrosamines with chains of three or more carbons (Loeppky et al., 1981). The role of this process in the carcinogenic activation of these compounds is uncertain, although much evidence suggests the importance of N-nitrosodipropylamine and its β-oxidized metabolites (Nagel et al., 1984). It is well known that the α-carbon of these compounds is incorporated into DNA as a chain-shortened methyl group (Krueger, 1971; Leung et al., 1980), but the biochemistry of this process is not well delineated and what is known is often in conflict with results concerning the α-hydroxylation activation of these β-oxidized nitrosamines.

Much of our past and present research has been devoted to exploring chemical and biochemical models for the carcinogenic activation of β-oxidized nitrosamines and the role of chain shortening. We demonstrated that β-hydroxynitrosamines undergo a chemical, retro-aldol-like, base-induced cleavage to methylalkylnitrosamines (Loeppky et al., 1982) and have produced evidence for the existence of a similar biochemical pathway (Loeppky & Outram, 1982). Our results and other data in the literature suggested a possible role of oxidative enzymes in the chain-shortening reaction. We hypothesized that either one-electron oxidation of a β-hydroxynitrosamine or one-electron reduction of a β-ketonitrosamine, with appropriate proton loss or addition, could result in an alkoxy radical at the 2-carbon of the nitrosamine chain, as shown in Scheme 1. The production of such a radical could be mediated by an appropriate metalloenzyme. Alkoxy radicals are well known to fragment to aldehydes or ketones and shorter alkyl radicals. Application of this process to a nitrosamine would generate an α-nitrosaminomethyl radical which could abstract a hydrogen to generate the chain-shortened methylalkylnitrosamine.

Scheme 1

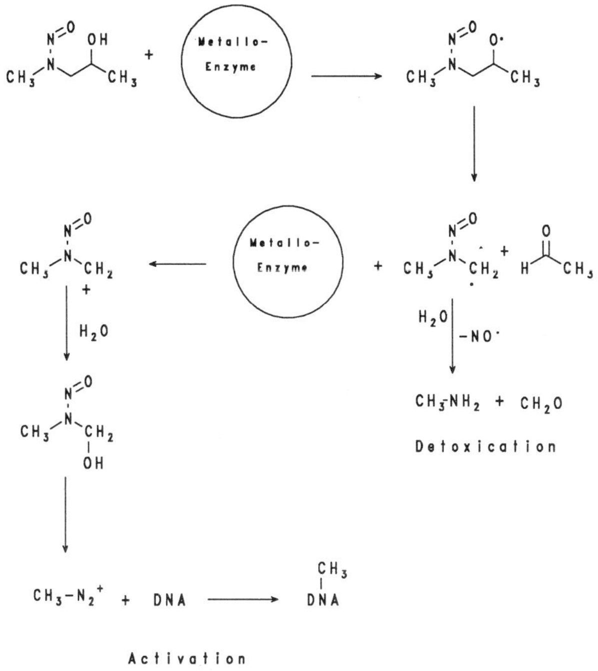

In order to test these hypotheses, we investigated the reaction of Ce(IV), a well known one-electron oxidant of alcohols and our 'surrogate metalloenzyme', with several β-hydroxynitrosamines. This research led to the discovery of a possible activation pathway for these nitrosamines and produced surprising evidence with regard to metabolic detoxication pathways for all nitrosamines (Loeppky & Li, 1988). For most of this research, we used N-nitrosocyclohexylmethanolamine as the model substrate. This compound was chosen because the cyclohexyl group provides an easy means for tracking the chemistry by gas chromatography-mass spectrometry. While this compound has no special carcinogenic or environmental significance, our data suggest that results obtained with it are applicable to other nitrosamines. Ce(IV) oxidation of N-nitrosocyclohexylethanolamine in aqueous acetonitrile gives cyclohexylamine, cyclohexylformamide and a set of products derived from the cyclohexyl diazonium ion. The origin of these substances is depicted in Scheme 2. The production of cyclohexylamine was unexpected, and it is the major product of the reaction, formed in what we call the 'detoxication' pathway. That and chain shortening are the principal subjects of this article.

Scheme 2

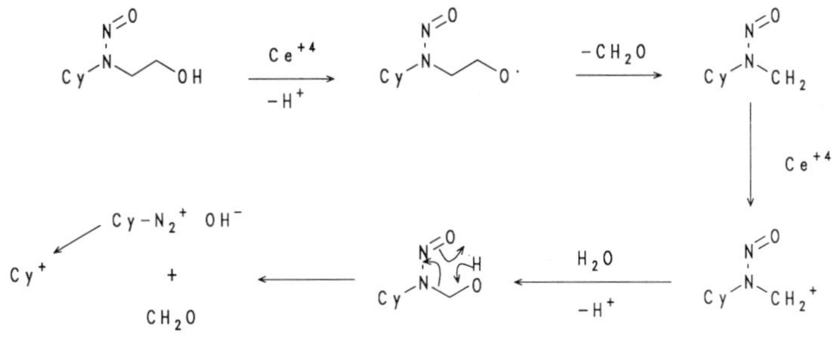

The 'activation' pathway

Alkoxy radicals fragment rapidly by β-scission, in which case the radical product of the reaction is the *N*-nitrosocyclohexylaminomethyl radical. Extensive studies of Ce(IV) alcohol oxidation have shown that the radicals formed by β-scission are oxidized further to carbocations. In this case, the product of transformation is the *N*-nitrosocyclohexylaminomethyl carbocation, which reacts with water to give the corresponding α-hydroxynitrosamine. This species decomposes first to cyclohexyl diazonium ion and formaldehyde by the route proposed some years ago and elucidated experimentally by Mochizuki *et al.* (1980) and others. Carbocations derived from the diazonium ion react readily with nucleophiles in solution.

This model illustrates a process that could generate electrophiles directly from β-oxidized nitrosamines. One-electron oxidation of a β-hydroxynitrosamine and proton loss produces an alkoxy free radical. Rapid β-scission generates an aldehyde and the alkylnitrosaminomethyl radical. A second one-electron metalloenzyme oxidation of this species before it is lost from the complex produces a cation which reacts with water to give the unstable α-hydroxynitrosamine. Hydrogen abstraction by the alkylnitrosaminomethyl free radical glutathione or some other hydrogen donor would generate the chain-shortened methylalkylnitrosamine. Subsequent α-oxidation of this species would generate methyl diazonium ions. Testing of the hydrogen atom abstraction hypothesis is discussed below.

The 'detoxication' pathway

As stated above, cyclohexylamine is the major reaction product of Ce(IV) oxidation of *N*-nitrosocyclohexylethanolamine. We were able to demonstrate that cyclohexylmethyleneimine is the immediate precursor of the amine. Two routes were envisioned for its formation — one from the free radical and one from its derived carbocation. The carbocation is resonance stabilized, as shown below. This stabilization occurs at the expense of the N–NO bond strength, however, and its facile denitrosation by water or another nucleophile to generate cyclohexylmethyleneimine and nitrous acid can be anticipated (Scheme 3).

Although the cyclohexylnitrosaminomethyl free radical is formed by β-scission, extrusion of NO, a stable radical, in yet another β-scission to give cyclohexylmethyleneimine is likely (Scheme 4). In fact, this process has been proposed by several groups with respect to the

N-nitrosomethylaminomethyl radical. Appel *et al.* (1979) and others have shown that nitrosamines undergo metabolic denitrosation in competition with metabolic activation. Mechanistic studies of this detoxication pathway by the groups of Yang *et al.* (1985), Patten *et al.* (1986), Haussmann and Werringloer (1987), Keefer *et al.* (1987) and Wade *et al.* (1987) have provided evidence that activation (α-oxidation) and detoxifying denitrosation are mediated by the same cytochrome P450 isozyme and have a common intermediate, the alkylnitrosaminomethyl radical. Yet the properties of this radical are unknown, and the mediation of the related cation could explain the observations discussed above.

Scheme 3

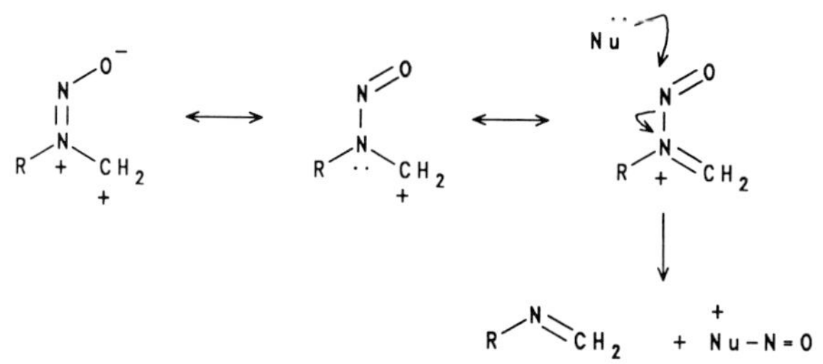

Scheme 4

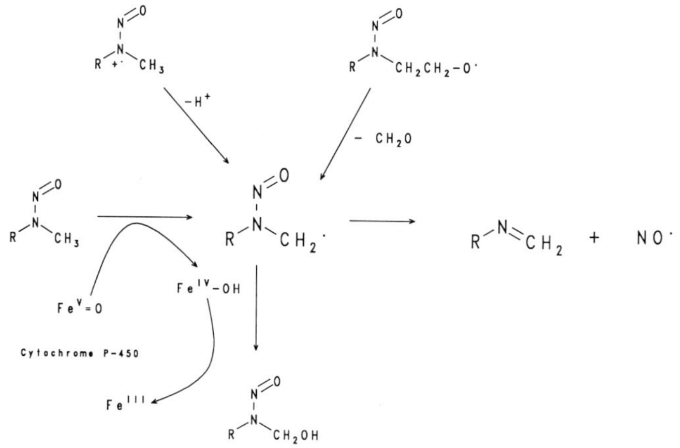

Generation and properties of the alkylnitrosaminomethyl free radical

We have sought an unequivocal method of generating alkylnitrosaminomethyl radicals to test our hypotheses as well as those of others. *tert*-Butyl peresters of alkylcarboxylic acids have proven to be a reliable source of alkyl free radicals because of their facile thermal decomposition (Eq. 1). The desired peresters were synthesized *via* the readily available N-nitroso-N-alkylglycines, although this is not a trivial synthetic undertaking; activation of the carboxyl group of nitrosamino acids results in cyclization to form sydnones, unless extreme care is taken. After numerous unsuccessful attempts, the desired perester was prepared in good yield by the following procedure.

Ethyl bromoacetate was reacted with cyclohexylamine to give ethyl N-cyclohexylglycine. Saponication followed by nitrosation gave N-nitroso-N-cyclohexylglycine, which was carefully purified and dried prior to the critical step. The nitrosamino acid was added to two equivalents of N,N'-carbonyl diimidazole under N_2 in dry tetrahydrofuran at 0–5°C and stirred for 30 min. This was followed by the dropwise addition of 1.5 equivalents of dry (98%) tert-butylhydroperoxide in tetrahydrofuran. After stirring at 0–5°C for 2 h, the tetra-

Equation 1

$$R-\overset{O}{\underset{\|}{C}}-O-O-\underset{CH_3}{\underset{|}{C}}\begin{matrix}CH_3\\-CH_3\end{matrix} \xrightarrow{\Delta} R-\overset{O}{\underset{\|}{C}}-O\cdot + \cdot O-\underset{CH_3}{\overset{CH_3}{C}}-CH_3$$

$$R\cdot + CO_2 \qquad CH_3\cdot + (CH_3)_2CO$$

hydrofuran was removed under vacuum (25°C). Extraction of the product into hexane followed by silica gel chromatography (20% ether in hexane) gave the perester, N-nitroso-tert-butyl-N-cyclohexyl-2-aminoperoxyethanoate (NCGP) in 54% yield as a crystalline solid (melting-point, 68–69°C). Although the spectroscopic and chemical characteristics (thermal lability and iodometric assay) of this compound were entirely consistent with its assigned structure, X-ray crystallographic analysis provided absolute proof (Fig. 1). The tert-butylperester of N-nitrososarcosine was prepared in the same way.

Fig. 1. X-ray crystal structure of N-nitroso-*tert*-butyl-N-cyclohexyl-2-aminoperoxyethanoate

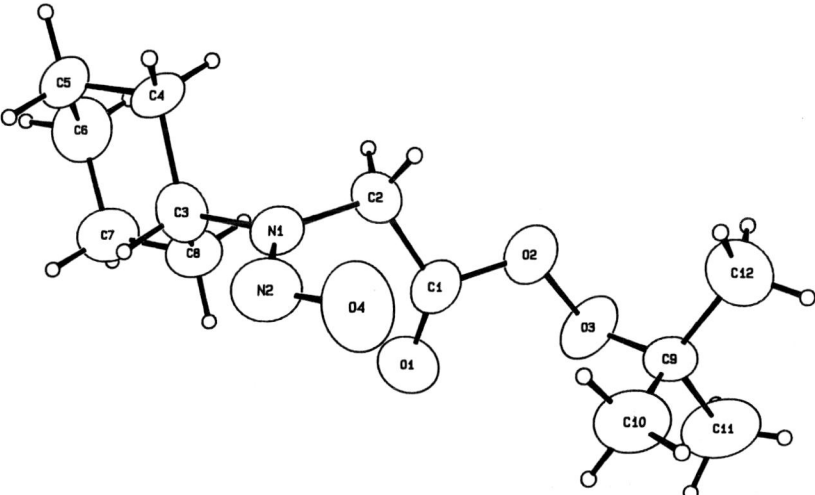

NCGP decomposed smoothly at 60°C in decalin with a half-life of 284 min (k = 2.4 × 10⁻³/ min; ΔH = 30 kcal/mol; ΔS = 13 cal/deg/mol) as determined by high-performance liquid chromatography. Decomposition in benzene at reflux is more suitable for gas chromatography-mass spectrometric analysis of the products, which are given in Scheme 5. The major decomposition product was cyclohexylmethyleneimine. Only trace amounts of *N*-nitrosocyclohexylmethylamine, the hydrogen abstraction product were observed. Other products included acetone, *tert*-butyl alcohol, biphenyl and cyclohexylbenzene. The presence of the last product suggests the formation of cyclohexyl radicals, as well.

Scheme 5

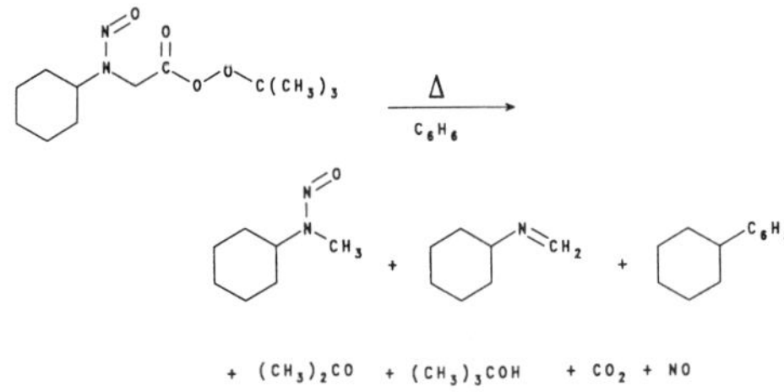

Even though extrusion of NO from this radical represents one of its intrinsic chemical properties, the rate of decomposition of NCGP suggests that the alkylnitrosaminomethyl radical is more stable than a benzyl radical. Rates of perester decomposition are known to be proportional to the stability of the incipient alkyl radical formed from loss of CO_2. NCGP decomposes five to six times more rapidly than the compound that generates the benzyl radical under the same conditions.

These data provide convincing support for the hypothesis that alkylnitrosaminomethyl free radicals easily lose NO to give alkylmethyleneimines. Homolysis of the O-O bond produces the *tert*-butoxy radical and the carboxyl radical, which rapidly loses CO_2 to give the *N*-nitrosocyclohexylmethylamino radical (Scheme 6). The loss of NO must be very rapid. Decomposition of NCGP in cumene, or in the presence of *tert*-butyl thiol, both very effective hydrogen atom donors, did not give rise to a significant increase in the yield of *N*-nitrosocyclohexylmethylamine.

Scheme 6

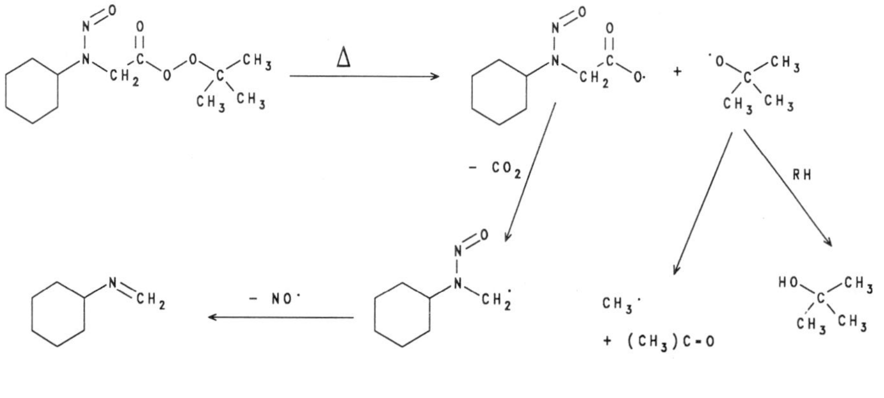

The finding that NO is lost rapidly from the alkylnitrosaminomethyl radical is in good agreement with results on competitive cytochrome P450 α-hydroxylation and denitrosation. The mechanism of cytochrome P450 action (see Scheme 4) suggests that the radical is rapidly scavenged by the $Fe^{IV}OH$ intermediate to give the α-hydroxynitrosamine (rebound). If denitrosation is to compete, it also must be very fast. The data of Wade *et al.* (1987) and Keefer *et al.* (1987b) suggest that with NDMA hydroxylation is four times faster than denitrosation. These relative rates may vary from nitrosamine to nitrosamine and from isoenzyme to isoenzyme.

Acknowledgements

The support of this research by a grant (ES 03593-04) from the National Institute of Environmental Health Sciences, and the helpful comments and suggestions of Dr C.J. Michejda, Dr S.R. Koeppke and Dr L.K. Keefer are gratefully acknowledged.

References

Appel, K.E., Rickart, R., Schwarz, M. & Kunz, H.W. (1979) Influence of drugs on activation and inactivation of hepatocarcinogenic nitrosamines. *Arch. Toxicol.*, **2** (Suppl.), 471-477

Haussmann, H.-J. & Werringloer, J. (1987) Mechanism and control of denitrosation of *N*-nitrosodimethylamine. In: Bartsch, H., O'Neill, I.K. & Schulte-Hermann, R., eds, *The Relevance of N-Nitroso Compounds to Human Cancer: Exposures and Mechanisms* (IARC Scientific Publications No. 84), Lyon, IARC, pp. 109-112

Keefer, L.K., Anjo, T., Heur, Y.-H., Yang, C.S. & Mico, B.A. (1987a) Potential for metabolic deactivation of carcinogenic *N*-nitrosodimethylamine *in vivo*. In: Bartsch, H., O'Neill, I.K. & Schulte-Hermann, R., eds, *The Relevance of N-Nitroso Compounds to Human Cancer: Exposures and Mechanisms* (IARC Scientific Publications No. 84), Lyon, IARC, pp. 113-116

Keefer, L.K., Anjo, T., Wade, D., Wang, T. & Yang, C.S. (1987b) Concurrent generation of methylamine and nitrite during denitrosation of *N*-nitrosodimethylamine by rat liver microsomes. *Cancer Res.*, **47**, 447-452

Krueger, F.W. (1971) Metabolismus von nitrosaminen *in vivo*. I. Uber die β-oxydation aliphatischer Di-n-alkylnitrosamine: Die Bildung von 7-Methylguanin neben 7-Propyl, bzw. 7-Butylguanin nach Applikation von Di-n-propyl- oder Di-n-butylnitrosamin. *Z. Krebsforsch.*, **76**, 147-154

Leung, K.H., Park, K.K. & Archer, M.C. (1980) Methylation of DNA by N-nitroso-2-oxopropylpropylnitrosamine: formation of O^6 and 7-methylguanine and studies on the methylation mechanism. *Toxicol. Appl. Pharmacol.*, **53**, 29-34

Loeppky, R.N. & Li, Y.-N. (1988) Diazonium ion derived products from the Ce(IV) oxidation of (beta)-hydroxy nitrosamines. *Chem. Res. Toxicol.*, **1**, 334-336

Loeppky, R.N. & Outram, J.R. (1982) A biochemical retroaldol cleavage of β-hydroxynitrosamines. In: Bartsch, H., O'Neill, I.K., Castegnaro, M. & Okada, M., eds, *N-Nitroso Compounds: Occurrence and Biological Effects* (IARC Scientific Publications No. 41), Lyon, IARC, pp. 459-472

Loeppky, R.N., Outram, J.R., Tomasik, W. & McKinley, W. (1981) Chemical and biochemical transformations of β-oxidized nitrosamines. In: Scanlan, R.A. & Tannenbaum, S.R., eds, *N-Nitroso Compounds*, Washington DC, American Chemical Society, pp. 21-37

Loeppky, R.N., McKinley, W.A., Hazlitt, L.G. & Outram, J.R. (1982) Base induced fragmentation of β-hydroxy nitrosamines. *J. Org. Chem.*, **47**, 4833-4841

Mochizuki, M., Anjo, T. & Okada, M. (1980) Isolation and characterization of N-alkyl-N-(hydroxymethyl)-nitrosamines from N-alkyl-N-(hydroperoxymethyl)nitrosamines by deoxygenation. *Tetrahedron Lett.*, **21**, 3693-3694

Nagel, D., Helgeson, A.S., Lewis, R. & Lawson, T. (1984) Comparative metabolism of β-oxidized nitrosamines. In: O'Neill, I.K., von Borstel, R.C., Miller, C.T., Long, J. & Bartsch, H., eds, *N-Nitroso Compounds: Occurrence, Biological Effects and Relevance to Human Cancer* (IARC Scientific Publications No. 57), Lyon, IARC, pp. 417-421

Patten, C., Ning, S.M., Lu, A.Y.H. & Yang, C.S. (1986) Acetone-inducible cytochrome P-450: purification, catalytic activity and interaction with cytochrome b5. *Arch. Biochem. Biophys.*, **251**, 629-638

Wade, D., Yang, C.S., Metral, C.J., Roman, J.M., Grabie, J.A., Riggs, C.W., Anjo, T., Keefer, L.K. & Mico, B.A. (1987) Deuterium isotope effect on denitrosation and demethylation of N-nitrosodimethylamine by rat liver microsomes. *Cancer Res.*, **47**, 3373-3377

Yang, C.S., Tu, Y.Y., Koop, D.R. & Coon, M.J. (1985) Metabolism of nitrosamines by purified rabbit liver cytochrome P-450 isozymes. *Cancer Res.*, **45**, 1140-1145

ALKYLATING POTENCY OF NITROSATED AMINO ACIDS AND PEPTIDES

S.E. Shephard, I. Meier & W.K. Lutz

Institute of Toxicology, Swiss Federal Institute of Technology and University of Zurich, Schwerzenbach, Switzerland

The alkylating potency of unstable N-nitrosamino acids and N-nitrosopeptides was investigated *in vitro* using 4-(*para*-nitrobenzyl)pyridine (NBP) as nucleophile. Of the amino acids, Met and those with an aromatic side chain were the most potent. The relative overall alkylating potency was 23:10:5:4:2:1: for Trp, Met, His, Tyr, Phe and Gly, respectively. The homo-dipeptides were much more potent than the amino acids, with relative potencies of 400:110:100:8:3:1, for Trp-Trp, Tyr-Tyr, Met-Met, Asp-Asp, Phe-Phe and Gly, respectively. In the one-phase reaction system (in which NBP is already present during the nitrosation reaction at acidic pH), all amino acids tested showed a second-order reaction for nitrite. In the two-phase system (in which NBP is added only after bringing the nitrosation reaction mixture to neutrality), all amino acids tested except one again showed a second-order reaction for nitrite (Phe, His, Asp and the dipeptide artificial sweetener aspartame); only Met under these conditions had a reaction order of one for nitrite. This could mean that nitrosation of the side chain of Met produces a second N-nitroso product which is relatively stable in acid but reacts with NBP under neutral conditions. In the human stomach, this side-chain nitrosation might become more important than the reactions at the primary amino group, firstly because of the greater stability of the product(s) in acid and secondly because of the first-order reaction rate for nitrite. A decrease in nitrite concentration from the millimolar concentrations of the in-vitro assay to the micromolar concentrations in the stomach reduces the reaction rate by a factor of 1000 for the side-chain nitrosation, whereas a million-fold reduction will be observed for nitrosation of the amino group.

Although unstable primary N-nitroso compounds (NOC) formed endogenously in the gastric lumen are detoxified to a large extent by lumen nucleophiles, there is evidence that these labile NOC could in some cases be sufficiently stable that a fraction might diffuse through the stomach lining and cause local DNA damage (Huber & Lutz, 1984). Using in-vitro screening tests designed to mimic partially the situation in the stomach, a cross-section of dietary NOC precursors (including ureas, guanidines, primary amines, amino acids and dipeptides) was assayed for their activation-independent mutagenic activity and/or alkylating activity immediately following nitrosation (Shephard *et al.*, 1987). As some amino acids and dipeptides showed high reactivity as compared to other precursor classes, more comprehensive and detailed studies on the nitrosation and alkylating characteristics of amino acids and dipeptides have been undertaken.

Overall alkylating potency

The nitrosation and alkylation kinetics were measured by simple colorimetric assays using 4-(*para*-nitrobenzyl)pyridine (NBP) as nucleophile. The tests exploit the fact that alkylated NBP forms a blue chromophore at pH 10. In the so-called one-step NBP test, precursor, nitrite and NBP are present simultaneously at acidic pH. Electrophilic species generated by the breakdown of unstable NOC are trapped immediately by NBP. Experimental details have been described elsewhere (Shephard et al., 1987).

The rates of NBP adduct formation by amino acid and dipeptide precursors in the one-step test are shown in Figure 1. Most striking was the enormous range of reaction rates, from Glu and Gln, which had no detectable activity (less than 1% of the rate of Gly), up to Trp–Trp which reacted 400 times faster than Gly — a span of over four orders of magnitude. Of the amino acids, Met and those with an aromatic side chain were the most reactive. The majority of the aliphatic amino acids were slightly less reactive than Gly. The homo–dipeptides tended to be more reactive than the corresponding amino acids. This was especially pronounced among the most potent amino acid–peptide pairs, where addition of a second residue increased the rate of adduct formation by a factor of 10–30 fold (Trp, 20-fold; Tyr, 29-fold; Met, 10-fold). Striking, too, was the increase in reactivity of aspartame (Asp-Phe-methylester, an artificial sweetener) over that of Asp-Asp (13-fold) and of Asp (180-fold). The reactivity of the peptides was governed mainly by the identity of the *N*-terminal amino acid (e.g., compare Ala-Gly and Gly-Ala, or Met-Gly and Gly-Met with the respective homodipeptides). As with the amino acids, peptides with an aromatic side chain had greater reactivity.

Reaction order for nitrite

Preliminary work showed that the reactivity of both Gly and Gly–Gly increased proportionally to the square of the nitrite concentration, thus following amine- rather than amide-type kinetics (see Mirvish, 1975). Furthermore, *N*-acetylglycine had no reactivity (Shephard *et al.*, 1987). Thus, the alkylating activity being measured appeared to result from nitrosation of the primary amine; however, among the most potent amino acids, additional reactions on the side chains are also possible (Bonnett & Nicolaidou, 1977). The reaction order for nitrite was investigated for these special amino acids.

Using the one-step NBP test, the reaction rate was measured at three or more different nitrite concentrations in the range 10–75 mM (typically, 30, 40, 50 mM). When log(reaction rate) was plotted against log(nitrite concentration), the slope of the resulting line gave the reaction order for nitrite. With each amino acid tested, the production of alkylated NBP was, within experimental error, dependent upon the square of the nitrite concentration (data not shown). This suggested either that the side-chain reactions (i) had the same kinetic order as the reaction at the primary amine and thus were indistinguishable from it or (ii) occurred much more slowly than nitrosation of the primary amine, or that (iii) the product of side-chain nitrosation was stable at acidic pH and therefore undetectable.

To distinguish between these possibilities, more detailed experiments were carried out using the two-step NBP test, which is a closer approximation of the situation *in vivo*. In this system, the nitrosation and alkylation reactions are separated spatially and temporally:

nitrosation is first carried out at acidic pH in the absence of NBP, followed by alkylation of NBP at neutral pH. The experimental details have been described previously (Shephard et al., 1987). The reaction order for nitrite was determined again for Met, His and Phe. Asp and aspartame were also tested, since the latter had shown high reactivity in the one-step test. Because of interfering colour reactions, Trp and Tyr could not be tested in the two-step system. Very short nitrosation times (20, 40 and 60 sec) were used in order to obtain the initial nitrosation reaction rate, before significant breakdown of NOC could occur.

Fig. 1. Overview of 4-(*para*-nitrobenzyl)pyridine (NBP) alkylating activity of A, amino acids and B, peptides following nitrosation in the NBP one-step test

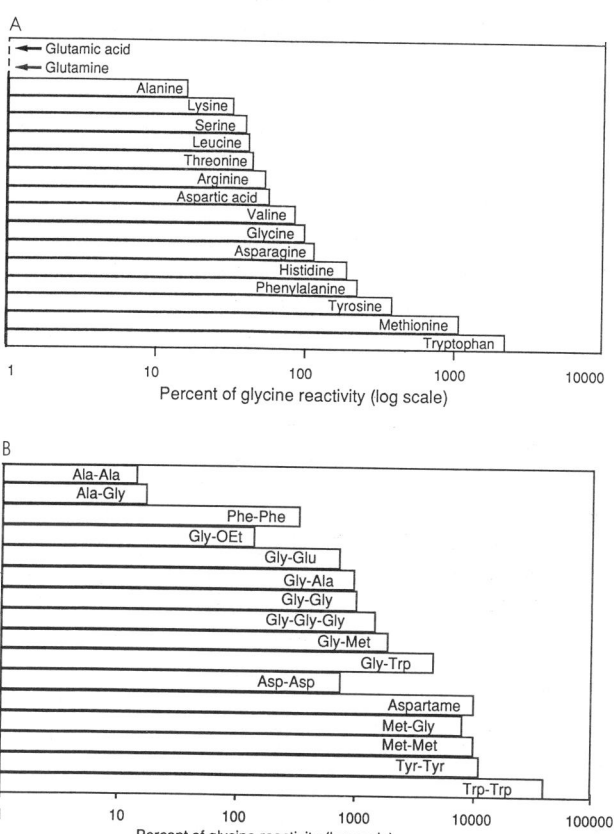

Under these conditions, Phe, His, Asp and aspartame all again showed the expected second-order reaction for nitrite (data not shown). Nitrosation of Met, however, showed a *linear* dependence on nitrite concentration. The difference in the behaviour of Met in the two test systems could be due to the production of a nitrosated product at the side chain. In the human stomach, this side-chain nitrosation might become more important than the reactions at the primary amino group, firstly because of the greater stability of the product(s) in acid and secondly because of the first-order reaction rate for nitrite. A decrease in nitrite concentration from the millimolar concentrations of the in-vitro assay to the micromolar concentrations in the stomach reduces the reaction rate by a factor of 1000 for the side-chain nitrosation, whereas a million-fold reduction will be observed for the nitrosation of the amino group.

Nitrosation reaction profile

The time course of the nitrosation and alkylation reactions was investigated with the two-step test (Fig. 2). Met nitrosated very rapidly — the NOC concentration peaked after only 5 min of nitrosation — then broke down rapidly in acidic solution (half-life, 20 min at pH 2.5). The nitrosated product also reacted extremely rapidly with NBP: the reaction was finished by 5 min (no increase in product was seen after that time). Phe, in contrast,

Fig. 2. Time course of the nitrosation and alkylation reactions of selected amino acids and aspartame in the 4-(*para*-nitrobenzyl)pyridine) (NBP) two-step test

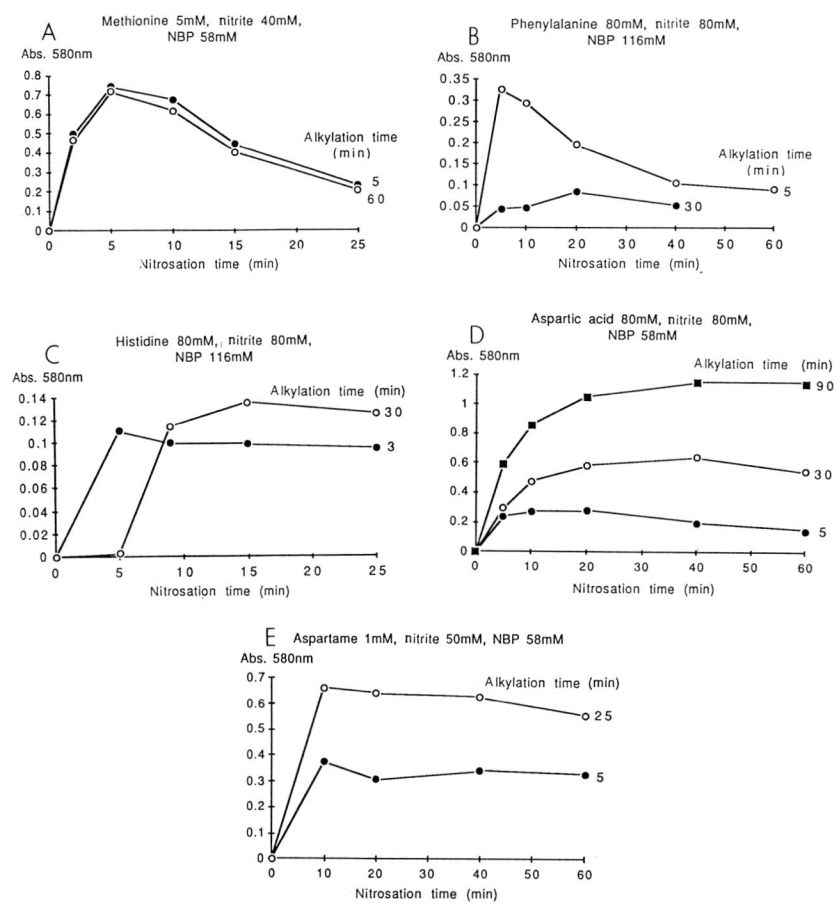

made two nitrosated products. One was nitrosated very rapidly (peak concentration after 5 min) and also alkylated NBP very rapidly; however, the NBP adduct was unstable at pH 7. At longer alkylation times, a second nitrosation peak emerged at 20 min of nitrosation, indicating a product that both nitrosated and alkylated more slowly. The former peak was probably due to the primary amine, the second, weaker reaction to the side chain. His similarly made two products, one with extremely rapid alkylation and a nitrosation peak at 5 min, and the second, again weaker, at longer nitrosation and alkylation times. Asp and aspartame showed yet another pattern of behaviour: they both made nitrosated products that were practically stable at acidic pH (half-life, > 200 min). Alkylation of NBP occurred rather more slowly but reached high levels after 90 min and longer. These nitrosation and alkylation profiles demonstrate the inhomogeneity of the amino acids with respect to nitrosating speed, reaction order for nitrite, alkylating potency and stability of nitroso product — factors that must be taken into account when evaluating the genotoxic risk posed by the in-vivo nitrosation of amino acids and peptides.

Although in-vitro screening tests clearly cannot replace in-vivo assays, they do offer valuable information for the ranking of chemical classes as well as of individual precursors with respect to the formation of alkylating NOC. On the basis of the results of these tests, in-vivo nitrosation experiments are now under way with important precursors.

References

Bonnett, R. & Nicolaidou, P. (1977) Nitrite and the environment. The nitrosation of alpha-amino acid derivatives. *Heterocycles*, 7, 637-659

Huber, K.W. & Lutz, W.K. (1984) Methylation of DNA in stomach and small intestine of rats after oral administration of methylamine and nitrite. *Carcinogenesis*, 5, 1729-1732

Mirvish, S.S. (1975) Formation of N-nitroso-compounds: chemistry, kinetics, and in vivo occurence. *Toxicol. Appl. Pharmacol.*, 31, 325-351

Shephard, S.E., Hegi, M. & Lutz, W.K. (1987) In-vitro assays to detect alkylating and mutagenic activities of dietary components nitrosated *in situ*. In: Bartsch, H., O'Neill, I.K. & Schulte-Hermann, R., eds, *The Relevance of N-Nitroso Compounds to Human Cancer: Exposures and Mechanisms* (IARC Scientific Publications No. 84), Lyon, IARC, pp. 232-236

Relevance to Human Cancer of N-Nitroso Compounds,
Tobacco Smoke and Mycotoxins.
Ed. I.K. O'Neill, J. Chen and H. Bartsch
Lyon, International Agency for Research on Cancer
© IARC, 1991

MACROPHAGES PRODUCE NITRITE, NITRATE AND NITROSAMINES AFTER ADDITION OF CATALASE

M. Miwa[1], M. Watanabe[2], T. Nishida[1] & K. Shinohara[1]

[1]*National Food Research Institute, Tsukuba, Ibaraki; and*
[2]*Department of Agricultural Chemistry, The University of Tokyo, Tokyo, Japan*

Mouse macrophages produced nitrite and N-nitrosomorpholine after incubation with catalase. A macrophage cell line, J774.1 (1×10^6 cells/ml), was incubated with catalase (500 U/ml) and morpholine (5 mM); after 48 h incubation at 37°C, macrophages produced nitrite (100 μM) and N-nitrosomorpholine (1 μM). Stimulation of J774.1 cells with catalase enhanced interleukin-1 production and tumour-killing activity against mastocytoma P815 cells. Flow cytometric analysis showed that catalase was bound to the surface of the macrophages.

Macrophages and their immortalized cell lines, when stimulated with lipopolysaccharide (LPS) and interferony (IFN), produced nitrite, nitrate (Stuehr & Marletta, 1985) and N-nitroso compounds (Miwa *et al.*, 1987, 1989). We investigated the effect of reactive oxygen scavengers and found that when macrophages were incubated with catalase, they produced nitrite, nitrate and N-nitrosomorpholine (NMOR) without LPS/IFN treatment.

Effect of reactive oxygen scavengers

A macrophage-like cell line (J774.1, 1×10^6 cells/ml) was treated with LPS (10 μg/ml), IFN (100 U/ml) and the reactive oxygen scavengers, superoxide dismutase, catalase or mannitol, obtained from commercial sources. After a 48-h incubation, supernatants were collected and analysed for nitrite and NMOR, as reported previously (Miwa *et al.*, 1987). We first investigated whether reactive oxygen intermediates were involved in the production of nitrite, nitrate and NMOR. Table 1 shows that catalase slightly inhibited LPS/IFN-induced nitrite production. To examine whether this inhibition was due to an adverse effect of catalase on the cells or whether catalase acts as a scavenger, macrophages were treated with the scavengers in the absence of LPS/IFN. We found that catalase could still induce significant amounts of nitrite and NMOR. A similar result was observed in thioglycolate-induced mouse peritoneal macrophages. A dose-response study using doses of 0–5000 U/ml catalase showed that nitrite and NMOR production reached a maximum at 500 U/ml.

Analysis of arginine, ornithine and citrulline in culture medium

L-Arginine has been reported to be a precursor of nitrite, nitrate and NMOR (Hibbs *et al.*, 1987; Iyengar *et al.*, 1987). L-Arginine can be metabolized to L-ornithine by the enzyme arginase which is secreted by activated macrophages; L-ornithine is then metabolized to L-citrulline. We determined L-arginine, L-ornithine and L-citrulline in culture medium before and after incubation (Table 2). A dose of 560 μM L-arginine was recovered almost completely as the three amino acids after 48 h of incubation. The pattern of production of the amino acids was different with catalase and LPS/IFN stimulation, although nitrite production was similar, suggesting that the mechanism of nitrite production by catalase is not same as that by LPS/IFN.

Table 1. Effect of radical scavengers on synthesis of nitrite and *N*-nitrosomorpholine (NMOR) by J774.1 macrophages[a]

Stimulus[b]	Scavenger[c]	NO_2^- (nmol/10^6 cells)	NMOR (nmol/10^6 cells)
+	None	74 ± 7	0.9 ± 0.2
	SOD	82 ± 7	1.2 ± 0.1
	Catalase	57 ± 2	0.5 ± 0.0
	Mannitol	80 ± 2	0.8 ± 0.1
−	None	2 ± 3	ND
	SOD	3 ± 2	ND
	Catalase	35 ± 5	0.2 ± 0.0
	Mannitol	3 ± 2	ND

[a]J774.1 macrophages (1 × 10^6 cells/ml) were incubated with the various scavengers for 48 h in the presence or absence of the stimulus, and the supernatants were collected and analysed for nitrite and NMOR.
[b]Lipopolysaccharide (10 μg/ml) and interferon-γ (100 U/ml)
[c]SOD, superoxide dismutase, 300 U/ml; catalase, 2000 U/ml; mannitol, 10 mM

Table 2. Nitrite, arginine, citrulline and ornithine concentrations in culture medium after incubation of 560 μM arginine for 48 h[a]

Stimulus	Nitrite (μM)	Arginine (μM)	Citrulline (μM)	Ornithine (μM)	All three (μM)
None	4 ± 2	421 ± 17	0	158 ± 19	579
Catalase	82 ± 2	76 ± 10	236 ± 15	264 ± 17	576
LPS/IFN	80 ± 6	120 ± 13	298 ± 16	139 ± 17	557

[a]J774.1 macrophages were incubated with catalase (500 U/ml) or lipopolysaccharide and interferon-γ (LPS/IFN), after which the supernatants were collected and analysed for nitrite and amino acids. Other experimental conditions were as described in footnote *a* to Table 1. Mean ± SD of three samples

Interleukin-1 and tumour-killing assay

Macrophages activated by LPS/IFN secrete many bioactive compounds other than nitrite/nitrate. We determined the interleukin-1 (IL-1) and tumour-killing activity of catalase-treated macrophages. IL-1 activity was evaluated by IL-1-dependent growth of a T-helper cell line, D10.G4.1 (Kaye *et al.*, 1984). Tumour-killing activity was evaluated as follows: 3H-thymidine-labelled mastocytoma P815 cells (5 × 10^4 cells/ml) were cultured with macrophages (1 × 10^6 cells/ml) and catalase (500 U/ml); after a 48-h incubation, the surviving P815 cells were harvested and the radioactivity analysed. The amount of IL-1 secreted by catalase-treated macrophages was similar to that secreted after LPS/IFN treatment. Tumour-killing activity was 38% after catalase treatment and 45% after LPS/IFN treatment.

Flow cytometry analysis

Binding of fluorescein isothiocyanate (FITC)-labelled catalase to macrophages was analysed by flow cytometry. Macrophages (1 x 10⁶cells/ml) were treated with FITC-catalase (500 U/ml) at 4°C for 1 h and analysed with a fluorescence-activated cell sorter. Figure 1 shows the staining of macrophages with FITC-catalase and FITC-ovalbumin; the latter, which had no effect on nitrite production, was used as a negative control. Almost all the macrophages stained with FITC-catalase but not with FITC-ovalbumin, indicating that catalase binds to the surface of macrophages through a receptor.

Fig. 1 Flow cytometry analysis of catalase binding to macrophages

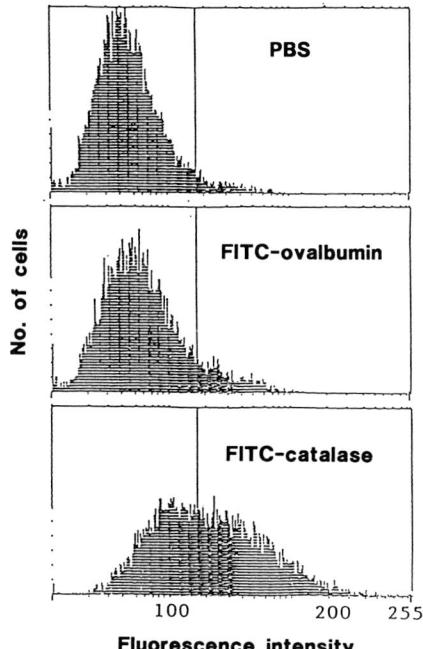

J774.1 macrophages were incubated with fluorescein isothiocyanate (FITC)-labelled catalase or ovalbumin or phospate-buffered saline (PBS) at 4°C for 60 min and washed. Flow cytometry was performed on a fluorescence-activated cell sorter.

We have found that catalase is as strong a stimulant of macrophages as LPS/IFN. We have also found that some food components act as stimulants to produce nitrite/nitrate and N-nitrosamines (data not shown). Our results show that endogenous formation of nitrite/nitrate and N-nitrosamines by macrophages could occur not only in the presence of immunostimulants but also with various other stimulants.

References

Hibbs, J.B., Jr, Taintor, R.R. & Vavrin, Z. (1987) Macrophage cytotoxicity: role for L-arginine deiminase and imino nitrogen oxidation to nitrite. *Science*, **235**, 473-476

Iyengar, R., Stuehr, D.J. & Marletta, M.A. (1987) Macrophage synthesis of nitrite, nitrate, and N-nitrosamines: precursors and role of the respiratory burst. *Proc. Natl Acad. Sci. USA*, **84**, 6369-6373

Kaye, J., Gills, S., Mizel, S.B., Shevach, E.M., Malek, T.R., Dinarello, C.A., Lachman, L.B. & Janeway, C.A., Jr (1984) Growth of a cloned helper T cell line induced by a monoclonal antibody specific for the antigen receptor: interleukin 1 is required for the expression of receptors for interleukin 2. *J. Immunol.*, **133**, 1339-1345

Miwa, M., Stuehr D.J., Marletta, M.A., Wishnok, J.S. & Tannenbaum, S.R. (1987) Nitrosation of amines by stimulated macrophages. *Carcinogenesis*, **8**, 955-958

Miwa, M., Tsuda, M., Kurashima, Y., Hara, H., Tanaka, Y. & Shinohara, K. (1989) Macrophage-mediated N-nitrosation of thioproline and proline. *Biochem. Biophys. Res. Commun.*, **159**, 373-378

Stuehr D.J. & Marletta, M.A. (1985) Mammalian nitrate biosynthesis: mouse macrophages produce nitrite and nitrate in response to *Escherichia coli* lipopolysaccharide. *Proc. Natl Acad. Sci. USA*, **82**, 7738-7742

USE OF MONOCLONAL ANTIBODIES TO IDENTIFY CYTOCHROME P450 ISOZYMES IN RAT LIVER MICROSOMES THAT HYDROXYLATE N-NITROSOMETHYLAMYLAMINE AT EACH OF SIX POSITIONS

S.S. Mirvish[1], C. Ji[1], Q. Huang[1], S. Wang[1], S.S. Park[2] & H.V. Gelboin[2]

[1]*Eppley Institute for Research in Cancer, University of Nebraska Medical Center, Omaha, NE; and* [2]*National Cancer Institute, Washington DC, USA*

Inhibition of enzyme activity by monoclonal antibodies (MAbs) was used to indicate which cytochrome P450 isozymes in Sprague–Dawley rat liver microsomes catalyse hydroxylation of the oesophageal carcinogen N-nitrosomethyl-n-amylamine (NMAA) to give 2- to 5-hydroxy-NMAA (HO-NMAA), formaldehyde and pentaldehyde. Liver microsomes (0.3–0.6 mg protein) were incubated (15 min, 23°C) with 0.4 mg MAb and, after adding NMAA to 6 mM, incubated for 20 min at 37°C. Mixtures were analysed for HO-NMAAs by gas chromatography–thermal energy analysis and for aldehydes by high-performance liquid chromatography of their 2,4-dinitrophenylhydrazones. The percentage inhibition by each MAb indicates the percentage metabolism by the corresponding P450 isozyme(s). These results indicate that the MAb to P450 IIB1 cross-reacts with P450 IIE1 and that the MAb to male-specific constitutive IIC11 cross-reacts with female-specific IIC12. Taking this into account, the main results were as follows. With uninduced male microsomes, 4-hydroxylation was catalysed mainly by IIC11 and demethylation by IIC11 and IIE1. With uninduced female microsomes, P450s reacting with the MAb to IIC11 (probably mainly IIC12) were responsible for most of the 4-hydroxylation and demethylation. With 3-methylcholanthrene-induced male microsomes, most 3-hydroxylation and some depentylation were due to IA1 or IA2. With phenobarbital-induced microsomes, all six reactions, but especially 4-hydroxylation and depentylation, were largely due to IIB1. With Aroclor-induced microsomes, all six reactions were catalysed by IIB1 and IA1 or IA2. The role of P450 IIC11 in 4- (ω-1)-hydroxylation was striking.

N-Nitrosomethyl-n-amylamine (NMAA) is an oesophageal and nasal cavity carcinogen in rats (Bulay & Mirvish, 1979), probably because of organ-specific cytochrome P450 isozymes that activate NMAA by α-hydroxylation. These P450s may include that identified in rat oesophagus (Kleihues *et al.*, 1989) and orthologues of two isolated from rabbit nasal mucosa (Ding & Coon, 1988). NMAA is metabolized by freshly removed rat tissue slices to 2- to 5-hydroxy-NMAA (HO-NMAA). Oesophageal and nasal mucosa produce prominent 2-, 3- and 4-HO-NMAA; liver, mostly 4-HO-NMAA; and lungs, mostly 5-HO-NMAA (Mirvish et al., 1985, 1988).

As with liver slices, liver microsomes from adult male Sprague-Dawley rats produce mainly 4-HO-NMAA (Ji et al., 1989). They also yielded the α-hydroxylation products, formaldehyde and pentaldehyde (i.e., produced demethylation and depentylation), and nitrite. Phenobarbital (PB), 3-methylcholanthrene (3MC) and Aroclor 1254 induced certain of these hydroxylations; K_m values were 0.3–2.1 mM. These results indicate that various P450 isozymes hydroxylate NMAA at different positions. To confirm this view and to identify the isozymes involved, we used monoclonal antibodies (MAbs) that bind to and inactivate specific P450 isozymes in microsomes. Microsomes were incubated with a MAb and then with NMAA, and HO-NMAAs and aldehydes were determined in separate experiments. Most of these results were presented at a previous meeting (Mirvish et al., 1989).

Methods

Microsomes were isolated by differential centrifugation of liver homogenates prepared from six- to eight-week-old Sprague–Dawley rats, including males induced with 3MC, PB or Aroclor 1254 (Ji et al., 1989). They were stored at −80°C in 0.1 M potassium phosphate buffer containing 20% glycerol and 15 mg/l dithiothreitol. Test MAbs (Park et al., 1982, 1984; Ko et al., 1987; Park et al., 1989) are listed in the tables. Results were always compared to a run in which MAb HyHel-9, which does not react with P450s, was included.

Microsomes (0.3–0.6 mg protein) were incubated with 0.4 mg MAb in 20–40 μl buffer for 15 min at 22-24°c with occasional swirling. We then added 100 mM potassium phosphate buffer (pH 8.0), 2 mM NADP, 10 mM glucose-6-phosphate, 1 unit glucose-6-phosphate dehydrogenase, 10 mM $MgCl_2$ and (in aldehyde assays) 5 mM semicarbazide•HCl [pH, 7.4 (HO-NMAA studies or 7.0 (aldehyde studies)]. After 2 min incubation at 37°C, NMAA in 20–40 μl water was added to 6 mM (final volume, 500 μl), and the mixture was incubated for 20 min at 37°C with shaking twice per sec. HO-NMAAs were determined as described previously (Mirvish et al., 1988). To determine formaldehyde and pentaldehyde, their 2,4-dinitrophenylhydrazones were prepared and analysed by high-performance liquid chromatography (HPLC; Farrelly, 1980). The final CH_3CN extract was concentrated to 0.2 ml, and 0.1 ml was injected onto the HPLC column, which was eluted with $CH_3CN:H_2O$ 3:1. Each experiment with HO-NMAA and aldehyde included duplicate tubes with each MAb (HyHel and 2-4 other MAbs) and, in some tests, tubes without any MAb. Aldehyde tests included aldehyde standards and blanks, both with microsomes but without NMAA. A finding that use of a MAb produced X% of the metabolism with HyHel indicated that (100 − X)% of this metabolism was due to the P450 isozyme(s) with which the MAb reacted.

Results

Metabolite yields without MAb were 5–20% less than those with HyHel, indicating a nonspecific enhancement by added protein. Metabolite yields (see HyHel columns of Tables 1 and 2) were similar to those reported by Ji et al., (1989). Uninduced male rat liver microsomes mainly catalysed 4-hydroxylation, with considerable depentylation (Table 1). The MAb to male-specific P450 IIC11 (P450h, see Nebert et al., 1989, for nomenclature) inhibited 63% of 4-hydroxylation and some demethylation and depentylation. Demethylation was also inhibited by the MAb to IIB1 and that to IIE1, with a 'total inhibition' of 171%. Uninduced female liver microsomes produced similar yields of metabolites to those formed by male microsomes; and 4-hydroxylation and demethylation were each inhibited 50–60% by the MAb to IIC11.

3MC-induced male microsomes showed induction of 3-hydroxylation and depentylation and a suppression of 5-hydroxylation (Table 2). Most of the 3-hydroxylation and some of the 4-hydroxylation and depentylation by these microsomes were performed by P450 IA1 or IA2, with about 20% each of 2-, 3-, 4- and 5-hydroxylation catalysed by IIE1. PB induced five of the six reactions, especially depentylation and 4-hydroxylation. With these microsomes, 70% of 4-hydroxylation and depentylation and 50–60% of the remaining reactions were due to the induced P450 IIB1 or IIB2. Aroclor induced five of the six reactions, especially 3-hydroxylation and depentylation, but suppressed 5-hydroxylation; these reactions were due to P450 IIB1 or IIB2 and IA1 or IA2 in varying proportions, except that demethylation and 5-hydroxylation were not catalysed by IA1 or IA2.

Discussion

Cytochrome P450 IIB1 (or IIB2), IIE1 and IIC11 contributed an apparent total of 171% to NMAA demethylation by uninduced microsomes from male rat liver. This is attributed to overlap of MAb specificities. Use of the double immunodiffusion test and inhibition of enzyme activities showed that MAb 2-66-3 to IIB1 and MAb 1-7-1 to IA1 and IA2 did not cross-react (Park et al., 1982, 1984) and that MAbs 1-68-11 to IIC11 and 1-91-3 to IIE1 did not cross-react with IA1, IA2 or IIB1 (Ko et al., 1987; Park et al., 1989). However, cross-reactions between MAbs to IIB1, IIC11 and IIE1 were not tested for.

In PB-induced male microsomes, which are rich in P450 IIB1, MAb 2-66-3 to IIB1 inhibited depentylation of NMAA more than demethylation, and especially inhibited 4- and 5-hydroxylation, whereas, in uninduced male microsomes, 2-66-3 inhibited demethylation by 73% but did not much affect depentylation or HO-NMAA formation; MAb 1-91-3 to IIE1 also inhibited only demethylation. We conclude that, in uninduced male microsomes, 2-66-3 probably does not act mainly on IIB1, the level of which is <1% of total P450 without induction (Guengerich, 1987), but rather acts by inhibiting IIE1. Inhibition of IIC11 by 2-66-3 is unlikely because this would have caused 2-66-3 to inhibit 4-hydroxylation. Therefore, NMAA demethylation in uninduced male microsomes is attributed 46% to IIC11, 35% to IIE1 and, perhaps, 19% to IIB1. It is of interest that three members of the P450 family II may share the ability to demethylate NMAA.

Uninduced female rat liver microsomes contain little P450 IIC11 but contain mostly IIC12 (P450i; Guengerich, 1987). Therefore, the finding that 4-hydroxylation and demethylation by these microsomes were each 50–60% inhibited by MAb 1-68-11 to IIC11 suggests a cross-reaction with IIC12.

Despite these problems of overlap, we conclude that we have identified the P450 isozymes responsible for most or all of NMAA 4-hydroxylation and demethylation by uninduced male microsomes and 3-hydroxylation by 3MC-induced microsomes, and at least

Table 1. Monoclonal antibody (MAb) inhibition of metabolism of *N*-nitrosomethylamylamine (NMAA) by uninduced rat liver microsomes[a]

Metabolite	Male microsomes[b]						Female microsomes				
	HyHel (nmol/min per mg)	Percent HyHel					HyHel (nmol/min per mg)	Percent HyHel			
		1-68-11 (IIC11)	1-7-1 (IA1+IA2)	2-66-3 (IIB1)	1-91-3 (IIE1)			1-68-11 (IIC11)	2-66-3 (IIB1)	1-7-1 (IA1+IA2)	
2-HO-NMAA	0.09 ± 0.01 (9)	99 ± 6 (6)	89 ± 3 (3)	89 ± 6	122 ± 6 (7)		046 ± 0.04 (6)	91 ± 11 (4)	85 ± 0 (2)	89 ± 4 (2)	
3-HO-NMAA	0.26 ± 0.02	94 ± 5	79 ± 3	87 ± 5	122 ± 7		0.19 ± 0.02	82 ± 6	78 ± 6	70 ± 4	
4-HO-NMAA	1.22 ± 0.07	37 ± 4	97 ± 4	83 ± 5	117 ± 6		0.71 ± 0.06	42 ± 3	66 ± 3	88 ± 2	
5-HO-NMAA	0.46 ± 0.04	129 ± 18	101 ± 1	92 ± 4	110 ± 7		0.28 ± 0.05	83 ± 6	79 ± 7	108 ± 2	
Formaldehyde	1.30 ± 0.10	54 ± 5	109 ± 7	27 ± 8	65 ± 14		0.66 ± 0.13	45 ± 1	50 ± 10	94 ± 3	
		(4)	(2)	(4)	(2)		(2)	(2)	(2)	(2)	
Pentaldehyde	1.16 ± 0.06	71 ± 5	89 ± 1	108 ± 1	94 ± 7		0.59 ± 0.15	83 ± 11	114 ± 8	110	

[a] Results are given as mean ± SE; numbers in parentheses are numbers of tests, which are the same for all HO-NMAAs and for both aldehydes
[b] MAbs 1-31-2 (P450 IA1 and IA2), 2-13-1 (IIA1 and IIA2) and 1-98-1 (IIE1) did not affect HO-NMAA or (for 2-13-1) aldehyde formation. Here and in the table headings, parentheses give the P450 isozyme(s) for which the MAb was designed.

Table 2. Monoclonal antibody (MAb) inhibition of *N*-nitrosomethylamylamine (NMAA) metabolism by induced liver microsomes from male rats[a]

Metabolite	3MC microsomes			PB microsomes			Aroclor microsomes		
	HyHel (nmol/min per mg)	Percent HyHel		HyHel (nmol/min per mg)	Percent HyHel		HyHel (nmol/min per mg)	Percent HyHel	
		1-7-1 (IA1 +IA2)	1-91-3 (IIE1)			2-66-3 (IIB1)		1-7-1 (IA1 +IA2)	2-66-3 (IIB1)
2-HO-NMAA	0.11 ± 0.01 (8)	73 ± 3 (6)	80 ± 4 (2)	0.23 ± 0.01 (4)	51 ± 4 (3)		0.24 ± 0.04 (4)	76 ± 10 (3)	73 ± 13 (4)
3-HO-NMAA	1.32 ± 0.11	40 ± 2	79 ± 2	0.77 ± 0.04	45 ± 8		3.68 ± 0.57	35 ± 4	70 ± 11
4-HO-NMAA	0.91 ± 0.60	65 ± 2	74 ± 3	3.81 ± 0.20	28 ± 4		2.75 ± 0.18	55 ± 6	44 ± 4
5-HO-NMAA	0.27 ± 0.01	98 ± 6	83 ± 4	0.46 ± 0.02	49 ± 6		0.25 ± 0.01	86 ± 12	43 ± 5
Formaldehyde	1.14 ± 0.23 (2)	105 ± 10 (2)	-	2.9 (1)	42 ± 7 (2)		1.83 ± 0.01	89 ± 10 (2)	36 ± 10 (2)
Pentaldehyde	2.69 ± 0.48	61 ± 8		6.7	28 ± 0		11.2 ± 0.15	57 ± 6	52 ± 1

[a] Results are given as mean ± SE. Numbers in parentheses are numbers of tests, which are the same for all HO-NMAAs and for both aldehydes. HO-NMAA formation was inhibited < 20% by the following MABs: for 3MC microsomes, 2-66-3 and 1-31-2 (IA1 and IA2); for PB microsomes, 1-91-3 and 1-7-1; for Aroclor microsomes, 1-91-3 and 1-98-1 (IIE1). 3MC, 3-methylcholanthrene; PB, phenobarbital

half of all six reactions in PB- and Aroclor-induced microsomes. This is the first time that P450 IIC11 has been implicated in nitrosamine metabolism. The catalysts of the remaining reactions, e.g., NMAA depentylation in uninduced microsomes, remain unknown. NMAA depentylation is important because this reaction yields methyldiazonium ion, which methylates DNA guanine, a reaction closely linked with carcinogenesis (Von Hofe et al., 1987).

Acknowledgements

This work was supported by NIH grants RO1-CA-35628 and core grant CA-36727 from the National Cancer Institute and core grant ACS-SIG-16 from the American Cancer Society.

References

Bulay, O. & Mirvish, S.S. (1979) Carcinogenesis in rat esophagus by intraperitoneal injection of different doses of methyl-n-amylnitrosamine. *Cancer Res.*, **39**, 3644-3648

Ding, X. & Coon, N.J. (1988) Purification and characterization of two unique forms of cytochrome P-540 from rabbit nasal microsomes. *Biochemistry*, **27**, 8330-8337

Farrelly, J.G. (1980) A new assay for the microsomal metabolism of nitrosamines. *Cancer Res.*, **40**, 3241-3244

Guengerich, F.P. (1987) Cytochrome P-450 enzymes and drug metabolism. *Progr. Drug Metab.*, **10**, 1-54

Ji, C., Mirvish, S.S., Nickols, J., Ishizaki, H., Lee, M.J. & Yang, C.S. (1989) Formation of hydroxy derivatives, aldehydes and nitrite from N-nitrosomethyl-n-amylamine by rat liver microsomes and by purified cytochrome P-450 IIB1. *Cancer Res.*, **49**, 5299-5304

Kleihues, P., Jeltsch, W., Meier, T. & Ludeke, B. (1989) Bioactivation of asymmetric N-nitrosomethyl-alkylamines in rat tissues derived from the ventral endoderm; structure-activity relationships and characterization of extrahepatic P450 isozymes. *Proc. Am. Assoc. Cancer Res.*, **30**, 165

Ko, I.Y., Park, S.S., Song, B.J., Patten, C., Tan, Y., Hah, Y.C., Yang, C.S. & Gelboin, H.V. (1987) Monoclonal antibodies to ethanol-induced rat liver cytochrome P-450 that metabolizes aniline and nitrosamines. *Cancer Res.*, **47**, 3101-3109

Mirvish, S.S., Wang, M.Y., Smith, J.W., Deshpande, A.D., Makary, M.H. & Issenberg, P. (1985) β- to ω-Hydroxylation of the esophageal carcinogen methyl-n-amylnitrosamine by the rat oesophagus and related tissues. *Cancer Res.*, **45**, 577-583

Mirvish, S.S., Ji, C. & Rosinsky, S. (1988) Hydroxy metabolites of methyl-n-amylnitrosamine produced by esophagus, stomach, liver, and other tissues of the neonatal to adult rat and hamster. *Cancer Res.*, **48**, 5663-5668

Mirvish, S.S., Ji, C., Huang, Q., Park S.S. & Gelboin, H.V. (1989) Use of monoclonal antibodies (MAbs) to identify cytochrome P450 isozymes in rat liver microsomes that metabolize methyl-n-amylnitrosamine (MNAN). *Proc. Am. Assoc. Cancer Res.*, **30**, 155

Nebert, D.W., Nelson, D.R., Adesnik, M., Coon, M.J., Estabrook, R.W., Gonzalez, F.J., Guengerich, F.P., Gunsalus, I.C., Johnson, E.F., Kemper, B., Levin, W., Phillips, I.R., Sato. R. & Waterman, M.R. (1989) The P450 superfamily; updated listing of all genes and recommended nomenclature for the chromosomal loci. *DNA.*, **8**, 1-13

Park, S.S., Fujino, T., West, D., Guengerich, F.P. & Gelboin, H.V. (1982) Monoclonal antibodies that inhibit enzyme activity of 3-methylcholanthrene-induced cytochrome P-450. *Cancer Res.*, **42**, 1798-1808

Park, S.S., Fujino, T., Miller, H., Guengerich, F.P. & Gelboin, H.V. (1984) Monoclonal antibodies to phenobarbital-induced rat liver cytochrome P-450. *Biochem. Pharmacol.*, **33**, 2071-2081

Park, S.S., Waxman, D.J., Lapenson, D.P., Schenkman, J.B. & Gelboin H.V. (1989) Monoclonal antibodies to rat liver cytochrome P-450 2c/RIM5 that regiospecifically inhibit steriod metabolism. *Biochem. Pharmacol.*, **38**, 3067-3074

Von Hofe, E., Schmerold, I., Lijinsky, W., Jeltsch, W. & Kleihues, P. (1987) DNA methylation in rat tissues by a series of homologous aliphatic nitrosamines ranging from N-nitrosodimethylamine to N-nitrosomethyldodecylamine. *Carcinogenesis.*, **8**, 1337-1341

Relevance to Human Cancer of *N*-Nitroso Compounds,
Tobacco Smoke and Mycotoxins.
Ed. I.K. O'Neill, J. Chen and H. Bartsch
Lyon, International Agency for Research on Cancer
© IARC, 1991

PARTICIPATION OF PHENOBARBITAL-INDUCIBLE CYTOCHROME P450 IN THE MUTAGENIC ACTIVATION OF *N*-NITROSOPROPYLAMINES BY LIVER AND LUNG 9000 g FRACTIONS FROM FIVE ANIMAL SPECIES AND MAN

Y. Mori[1,3] & Y. Konishi[2]

[1]*Laboratory of Radiochemistry, Gifu Pharmaceutical University, Gifu;*
and [2]*Department of Oncological Pathology, Cancer Center,*
Nara Medical College, Nara, Japan

The mutagenicity of nine carcinogenic *N*-nitrosopropylamines was studied by the Ames preincubation assay using 9000 g supernatant (S9) fractions or alcohol dehydrogenase. Treatment of animals with polychlorinated biphenyls or phenobarbital resulted in a marked increase in the ability of liver S9 to activate *N*-nitrosobis(2-hydroxypropyl)amine, *N*-nitroso(2-hydroxypropyl)(2-oxopropyl)amine, *N*-nitrosobis(2-oxopropyl)amine, *N*-nitrosobis(2-acetoxypropyl)amine, *N*-nitroso-2,6-dimethylmorpholine, *N*-nitroso(2-hydroxypropyl)methylamine, *N*-nitroso(2-oxopropyl)methylamine, *N*-nitroso(2,3-dihydroxypropyl)methylamine and *N*-nitroso(2,3-dihydroxypropyl)(2-hydroxypropyl)amine to mutagens, whereas 3-methylcholanthrene induction was not effective. All reactions required NADP as a cofactor for mutagenic activation, and nitrogen, carbon monoxide, cytochrome c and metyrapone considerably inhibited their mutagenic activities, whereas 7,8-benzoflavone did not. Five propanol derivatives were not mutagenic in the presence of NAD and alcohol dehydrogenase. We conclude that the phenobarbital-inducible major cytochrome P450 in liver S9 from five animal species tested was selectively involved in mutagenic activation. The same cytochrome in human liver S9 and in lung S9 from three rodent species also activated the mutagenicity of *N*-nitroso(2-hydroxypropyl)methylamine.

Several investigators have examined the activation of promutagens by different forms of cytochrome P450, and two have been implicated — phenobarbital (PB)-P450 and 3-methylcholanthrene (MC)-P448 (Tagashira *et al.*, 1985). However, few data are available on the activating enzymes involved in mutagenic activation of *N*-nitrosamines. In previous studies we showed that nine *N*-nitrosopropylamines — *N*-nitrosobis(2-hydroxypropyl)amine (NDHPA), *N*-nitroso(2-hydroxypropyl)(2-oxopropyl)amine (NHPOPA), *N*-nitrosobis(2-oxopropyl)amine (NDOPA), *N*-nitrosobis(2-acetoxypropyl)amine (NDAcPA),

---

[3]To whom correspondence should be addressed

N-nitroso-2,6-dimethylmorpholine (NDMMOR), *N*-nitroso(2-hydroxypropyl)methylamine (NHPMA), *N*-nitroso(2-oxopropyl)methylamine (NOPMA), *N*-nitroso(2,3-dihydroxypropyl)methylamine (NDHPMA) and *N*-nitroso(2,3-dihydroxypropyl)-(2-hydroxypropyl)amine (NDHPHPA) — were mutagenic to *Salmonella typhimurium* strain TA100 in the presence of a 9000 g fraction (S9) of liver or lung from five animal species and from humans (Mori & Konishi, 1987). The present communication deals with the enzyme specificity in these mutagenic activations and the inducibility of cytochrome P450 by NDHPA.

The mutagenicity of *N*-nitrosopropylamines (10 mg) was tested by the preincubation assay in the presence of complete S9, microsomes or cytosol from liver of rats treated with polychlorinated biphenyls (PCB). The compounds were shown to be pure by high-performance liquid chromatography (Mori *et al.*, 1986a). In the presence of cytosol alone, the number of revertants was less than twice the spontaneous rate, except with NHPOPA and NDOPA; these are directly acting mutagens. The activities in the presence of microsomes plus glucose-6-phosphate dehydrogenase were 60–70% of those in the presence of complete S9, and when microsomes and 4 mM NADPH instead of $NADP^+$ were used, the activities were approximately 80% of the control. These results show that the activation system is localized mainly in the microsomal fraction and requires NADPH as a cofactor.

To test further the idea that a microsomal enzyme is responsible for the mutagenic activation, the effects of pretreatment of rats, mice and hamsters with PCB, PB and MC on the activation of the nine *N*-nitrosopropylamines were examined (Table 1). NDHPA and NDHPHPA were weakly mutagenic in the presence of liver S9 from untreated animals, but the other compounds showed clear mutagenicity. Mutagenic activities were induced to the same extent by PCB and PB treatment of animals, but MC pretreatment did not activate mutation even though it induced microsomal cytochrome P448 contents in the livers of the three animal species.

Figure 1 shows the effects of a cofactor, $NADP^+$, and of several inhibitors of P450 on mutagenic activation by liver S9 from PCB-treated animals. All mutagenic activity was completely or considerably (NHPOPA and NDOPA) decreased by removal of $NADP^+$ from the assay system. Preincubation in an atmosphere of CO or N_2 (data not shown) resulted in a marked reduction in mutagenicity. Similarly, activity decreased by 32–93% by addition of cytochrome c or metyrapone to the S9 mixture. Essentially the same results were obtained when PB-induced liver S9 from the three rodent species was used for the incubation, addition of metyrapone resulting in a decrease in mutagenicity. In contrast, 7,8-benzoflavone showed no inhibitory effect on the activities of either PCB- (Fig. 1) or 3-MC-induced liver S9 from the rodents. Six of the *N*-nitrosopropylamines, excluding NDHPA, NDOPA and NDHPHPA, were also activated by liver S9 from untreated rabbits and monkeys; only NHPMA was activated by human liver S9 or by lung S9 from the rodent species treated with PCB. Essentially the same results were obtained with regard to the cofactor requirement and the effect of P450 inhibitors (Fig. 2; Mori *et al.*, 1986b,c).

The effects of NDHPA treatment of rats on the activities of mixed-function oxidase and of hepatic enzymes involved in the mutagenicity of *N*-nitrosopropylamine were examined one, three and six days after a single intraperitoneal injection of NDHPA at a dose of 3 g/kg

Table 1. Effects of inducers on mutagenicity of *N*-nitrosamines and microsomal cytochrome P450 content (results given as revertants of *Salmonella typhimurium* TA100 per plate)[a]

Species	Inducer[b]	*N*-Nitrosamine[c]										P450 content (nmol/ml)
		NDHPA	NHPOPA	NDOPA	NHPMA	NOPMA	NDAcPA	NDMMOR	NDHPHPA	NDHPMA		
Rat	PCB	314 (100)	797 (100)	413 (100)	871 (100)	1126 (100)	1149 (100)	904 (100)	132 (100)	336 (100)		18.75 (100)
	PB	301 (96)	609 (76)	348 (84)	970 (111)	788 (70)	862 (75)	643 (71)	147 (111)	258 (77)		16.73 (89)
	3-MC	58 (19)	348 (44)	240 (58)	428 (49)	105 (9)	245 (21)	357 (39)	53 (40)	171 (51)		8.74 (47)
	None	69 (22)	353 (44)	187 (45)	326 (37)	103 (9)	237 (21)	342 (38)	56 (42)	163 (49)		4.56 (24)
Hamster	PCB	293 (100)	869 (100)	346 (100)	1745 (100)	174 (100)	331 (100)	1486 (100)	131 (100)	345 (100)		20.55 (100)
	PB	235 (80)	787 (90)	341 (99)	1794 (103)	168 (97)	284 (86)	1427 (96)	142 (108)	259 (75)		16.15 (79)
	3-MC	59 (20)	339 (39)	211 (61)	756 (43)	132 (76)	141 (43)	284 (20)	46 (35)	158 (46)		9.34 (45)
	None	55 (19)	315 (36)	194 (56)	734 (42)	123 (70)	134 (40)	267 (18)	40 (31)	171 (50)		4.67 (23)
Mouse	PCB	133 (100)	468 (100)	262 (100)	861 (100)	65 (100)	389 (100)	716 (100)	131 (100)	307 (100)		11.32 (100)
	PB	137 (104)	425 (91)	269 (103)	937 (109)	610 (94)	310 (80)	640 (89)	130 (99)	234 (76)		10.41 (92)
	3-MC	57 (43)	339 (72)	208 (79)	514 (60)	141 (22)	204 (52)	483 (67)	51 (39)	167 (54)		8.60 (76)
	None	17 (13)	350 (75)	185 (71)	530 (62)	154 (24)	215 (55)	453 (63)	57 (43)	156 (51)		6.98 (62)

[a]Assay was carried out with a dose of 10 mg; the number of spontaneous revertants (130) was subtracted. Values in parentheses show the percentage activity in relation to that obtained with liver 9000 *g* (S9) from animals treated with polychlorinated biphenyls
[b]PCB, polychlorinated biphenyls; PB, phenobarbital; 3-MC, 3-methylcholanthrene
[c]NDHPA, *N*-nitrosobis(2-hydroxypropyl)amine; NHPOPA, *N*-nitroso(2-hydroxypropyl) (2-oxopropyl)amine; NDOPA, *N*-nitrosobis (2-oxopropyl)amine; NDAcPA, NDHPA, *N*-nitrosobis(2-acetoxypropyl)amine; NDMMOR, *N*-nitroso-2,6-dimethylmorpholine; NHPMA, *N*-nitroso (2-hydroxypropyl)methylamine; NOPMA, *N*-nitroso-(2-oxopropyl)methyla mine; NDHPMA, *N*-nitroso (2,3-dihydroxypropyl)methylamine; NDHPHPA, *N*-nitroso (2,3-dihydroxypropyl) (2-hydroxypropyl)amine

body weight or after oral administration for six weeks to a calculated total intake of 3.36 g NDHPA per rat. In neither experiment was there a significant difference among control and NDHPA-treated groups with regard to liver weight, microsomal protein or P450 content or aniline hydroxylase activity in the S9 fraction. Futhermore, NDHPA treatment had no effect on the mutagenic activities of any of the nine N-nitrosopropylamines (Mori et al., 1986c).

The mutagenicity of the five β- and γ-hydroxypropyl derivatives, NDHPA, NHPOPA, NHPMA, NDHPMA and NDHPHPA, was also examined at doses of 0.5-20 mg in phosphate buffer (pH 7.4) containing bacteria supplemented with alcohol dehydrogenase (ADH) (441 units/tube) from yeast and NAD, the latter in a five-fold molar excess to the amounts of N-nitrosamine. However, the NAD/ADH system did not activate any of the propanols. On the contrary, the directly acting mutagen NHPOPA was completely inactivated by addition of NAD/ADH to the preincubation buffer.

Fig. 1. Effects of NADP$^+$ and inhibitors on mutagenic activation of N-nitrosopropylamines by liver 9000 g fraction from rats (A), hamsters (B) or mice (C) pretreated with polychlorinated biphenyls. Mutagenic activities of NDHPA (□), NHPOPA (▨), NDOPA (▨), NHPMA (▨), NOPMA (▨), NDAcPA (▨), NDMMOR (▥), NDHPHPA (▨) and NDHPMA (▨)

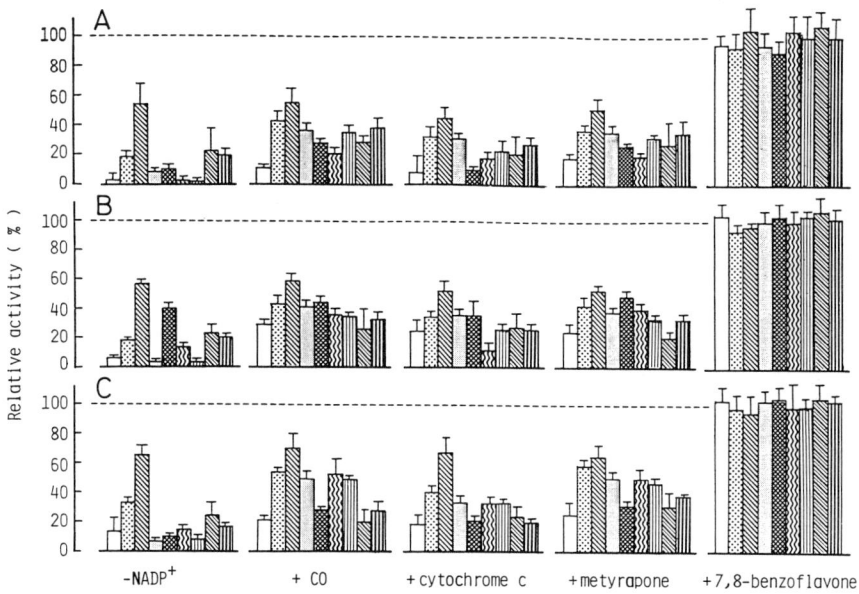

(See footnote c of Table 1 for definitions of abbreviations). The assay was carried out with doses of 10 mg. The results represent mean ± SE (n = 3) expressed as a percentage of controls without inhibition.

These results demonstrate that PB-P450 in liver and lung S9 from various animal species and humans is selectively involved in mutagenic activation of these nine N-nitrosopropylamines, although NDHPA cannot induce microsomal PB-P450 and that

ADH is not involved in the mutagenic activation of five of them. Tagashira et al., (1985) reported that carcinogens can be classified into four groups according to whether their mutagenic activation depends on MC-P448 or PB-P450: (i) activated selectively by MC-P448; (ii) activated predominantly by MC-P448; (iii) activated equally by MC-P448 and PB-P450; and (iv) activated predominantly by PB-P450. They showed that the majority of the 22 carcinogens tested belonged to groups (i) and (ii). We have reported that nine aminoazo compounds were also selectively activated to mutagens by MC-P448 (Mori et al., 1983). The results of this study with nine N-nitrosopropylamines and the finding of Suzuki et al. (1983) with N-nitrosobutylamine clearly indicate that a fifth group of carcinogens, activated selectively by PB-P450, exists. All of these ultimate mutagens may be oxidative metabolites, and α-oxidation by PB-P450 may be relevant mainly for their metabolic activation but induction of PB-P450 may not be necessary for their carcinogenic activation.

Fig. 2. Effects of $NADP^+$ and inhibitors on mutagenic activation of N-nitroso-(2-hydroxypropyl)methylamine by human liver 9000 g fraction. Assay carried out with a dose of 10 mg; in the presence of liver 9000 g from human A (□), B (▨) and C (■), the nitrosamine induced 188, 238 and 174 revertants/plate, respectively, from which the number of spontaneous revertants (130) was subtracted.

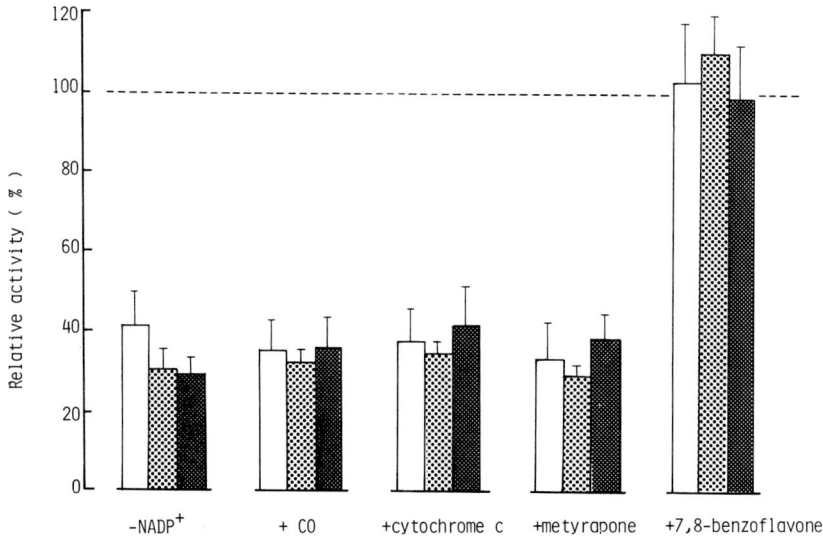

The results represent mean ± SE (n = 6) expressed as a percentage of controls without inhibition.

References

Mori, Y. & Konishi, Y. (1987) A comparative study of the mutagenic activation of carcinogenic N-nitrosopropylamines by various animal species and man. In: Bartsch, H., O'Neill, I.K. & Schulte-Hermann, R., eds, *The Relevance of N-Nitroso Compounds to Human Cancer: Exposures and Mechanisms* (IARC Scientific Publications No. 84), Lyon, IARC, pp. 141-143

Mori, Y., Niwa, T. & Toyoshi, K. (1983) Immunochemical study on the contributions of two molecular species of cytochrome P-450 in mutagenesis by selected aminoazo dyes. *Carcinogenesis*, **4**, 1487-1489

Mori, Y., Toyoshi, K., Denda, A. & Konishi, Y. (1986a) Mutagenic activation of carcinogenic N-nitrosopropylamines by liver S9 fractions from mice, rats and hamsters: evidence for a cytochrome P-450-dependent reaction. *Carcinogenesis*, **7**, 375-379

Mori, Y., Toyoshi, K., Maruyama, H. & Konishi, Y. (1986b) Activation of carcinogenic N-nitrosopropylamines to mutagens by lung and pancreas S9 fractions from various animal species and man. *Mutat. Res.*, **160**, 159-169

Mori, Y., Toyoshi, K., Nagai, H., Koda, A. & Konishi, Y. (1986c) A comparative study of the mutagenic activation of N-nitrosopropylamines by various animnal species and man: evidence for a cytochrome P-450 dependent reaction. *Jpn. J. Cancer Res. (Gann)*, **77**, 107-117

Suzuki, E., Mochizuki, M., Wakabayashi, Y. & Okada, M. (1983) In vitro metabolic activation of N,N-dibutylnitrosamine in mutagenesis. *Gann*, **74**, 51-59

Tagashira, Y., Yonekawa, H., Watanabe, J., Hara, E., Hayashi, J., Gotoh, O. & Kawajiri, K. (1985) Metabolic activation of chemical carcinogens by two molecular species of cytochrome P-450. In: Tagashira, Y. & Omura, S., eds, *P-450 and Chemical Carcinogenesis*, Tokyo, Japan Scientific Society Press, pp. 69-79

BIOLOGICAL AND CHEMICAL PROPERTIES OF ALKANEDIAZOTATES AS ACTIVE SPECIES OF N-NITROSO COMPOUNDS

S. Ukawa & M. Mochizuki

Kyoritsu College of Pharmacy, Tokyo, Japan

The mutagenicity and chemical reactivity of (E)- and (Z)-potassium alkanediazotates, as precursors of corresponding alkanediazohydroxides, were investigated. In three microbial strains, *Salmonella typhimurium* TA1535 and *Escherichia coli* WP2 and WP2hcr^-, the effect of changing the alkyl group on mutagenic potency was similar for (E)- and (Z)-diazotates, N-alkyl-N-nitrosoureas and α-hydroxynitrosamines. The capacity to alkylate nicotinamide, measured in an aqueous phosphate buffer, decreased with increasing alkyl chain length. Specific mutagenicity in *S. typhimurium* TA1535 was linearly related to alkylating activity. These results confirm that alkanediazohydroxides are the active alkylating species of N-nitroso compounds, and that their mutagenicity is determined by their alkylating activity.

α-Hydroxynitrosamines, which are active metabolites of carcinogenic N-nitroso compounds, decompose to alkanediazohydroxides to give the ultimate alkylating species, alkyldiazonium ions. We describe the mutagenicity and alkylating activity of the two geometric isomers of alkanediazohydroxides, which may be common active species of N-nitroso compounds, using their potassium salts, (E)- and (Z)-potassium alkanediazotates.

Mutagenicity of potassium alkanediazotates

(E)- and (Z)-Potassium alkanediazotates (alkyl = Me, Et, Pr and Bu) were prepared by published methods (Thiele, 1910; Müller *et al.*, 1963). Their mutagenic potency in *Salmonella typhimurium* TA1535 and *Escherichia coli* WP2 and WP2hcr^- was linearly related to the concentration of each chemical. (E)-Diazotates were more mutagenic than (Z)-diazotates, and the mutagenic potency of (Z)-diazotates was similar to that of corresponding N-alkyl-N-nitrosoureas, which decompose to alkanediazohydroxides nonenzymatically. Specific mutagenicity per micromole of chemical was defined by the slope of the linear part of the dose-mutagenicity relation. Figure 1 shows the specific mutagenicity of (E)- and (Z)-diazotates and of the corresponding α-hydroxynitrosamines and N-alkyl-N-nitrosoureas in the three microbial strains. The effect of changing the alkyl group on relative mutagenicity was similar for the two series of diazotates and the two series of N-nitroso compounds. These results support the notion that carcinogenic N-nitroso compounds decompose to alkanediazohydroxides, which are the common active alkylating species of N-nitroso compounds.

Fig. 1. Specific mutagenicities of (E)- and (Z)-potassium alkanediazotates, N-alkyl-N-nitrosoureas and N-nitroso-N-hydroxymethylalkylamines[a]

Fig. 2. Relationship between alkylating activity and specific mutagenicity in *Salmonella typhimurium* TA 1535

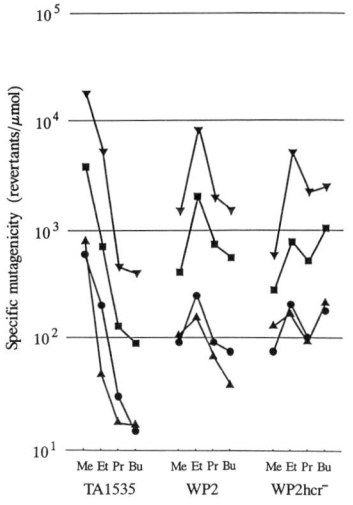

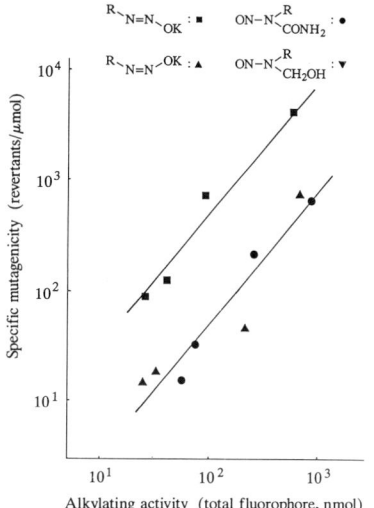

[a] Mutagenicity was assayed in *Salmonella typhimurium* TA1535 and *Escherichia coli* WP2 and WP2hcr⁻. Specific mutagenicity per micromole was calculated from the linear part of the dose-mutagenicity relation.

Alkylating activity of potassium alkanediazotates

The alkylating activity of series of (E)- and (Z)-potassium alkanediazotates and N-alkyl-N-nitrosoureas (alkyl = Me, Et, Pr and Bu) was measured in aqueous phosphate solution. After incubation in 0.1 M phosphate buffer (pH 7.4) containing nicotinamide, alkylated nicotinamide was converted quantitatively to the final fluorophore, 7-alkyl-3-phenyl-2,7-naphthyridin-1[7H]-one, by treatment with acetophenone (Sano *et al.*, 1983). The fluorophore was chromatographed using a Zorbax SCX-300 column with a mixed eluent of acetonitrile-0.04 M monobasic ammonium phosphate. This fluorometric high-performance liquid chromatography method is a slight modification of the method described by Hirayama *et al.* (1985). Table 1 shows the total fluorophore formed when 5 μmol of chemical were reacted with 300 μmol nicotinamide. The alkylating activities of the (E)- and (Z)-diazotates and of N-alkyl-N-nitrosoureas with the same alkyl groups were similar under these conditions, but the activity was decreased as the alkyl chain length increased. The alkylating activity towards nicotinamide in an aqueous solution was linearly related to the specific mutagenicity of (E)- and (Z)-diazotates and N-alkyl-N-nitrosoureas in *S. typhimurium* TA1535 (Fig. 2). This relationship reveals that DNA alkylation is one of the most important factors determining the relative mutagenicity of the compounds.

Acknowledgement

This work was supported in part by a Grant-in-Aid for Cancer Research from the Foundation for Promotion of Cancer Research, Japan.

Table 1. Alkylating activity of (E)- and (Z)-potassium alkanediazotates and N-alkyl-N-nitrosoureas[a]

Compound	Total fluorophore formed (nmol)			
	R = Me	ET	Pr	Bu
R\\N=N\\OK	560	88.3	41.8	25.5
R OK\\ /\\N=N	675	209	33.9	24.1
ON–N(R)(CONH$_2$)	849	247	72.8	54.8

[a] Solutions containing 5 μmol in dimethyl sulfoxide were incubated with 200 μl of 0.1 M phosphate buffer (pH 7.4) containing 300 μmol nicotinamide. Fluorophore was formed after reaction with 15 μmol acetophenone.

References

Hirayama, T., Yoshida, K., Uda, K., Nohara, M. & Fukui, S. (1985) High-performance liquid chromatographic determination of N'-alkylnicotinamide in urine. *Anal. Biochem.*, **147**, 108-113

Müller, E., Hoppe, W., Hagenmaier, H., Haiss, H., Huber, R., Rundel, W. & Suhr, H. (1963) Über Diazoverbindungen. XX. Struktur isomerer Diazotate: das Hantzschsche Methyldiazotat. *Chem. Ber.*, **96**, 1712-1719

Sano, A., Asabe, Y., Suzuki, M. & Takitani, S. (1983) Fluorescence reaction of common epoxides with nicotinamide and acetophenone. *Bunseki Kagaku*, **32**, E90-E100

Thiele, J. (1910) Über Nitrosohydrazine. Isoazotate und Azoverbindungen der Fettreihe. *Justus Liebigs Ann. Chem.*, **376**, 239-268

Relevance to Human Cancer of N-Nitroso Compounds,
Tobacco Smoke and Mycotoxins.
Ed. I.K. O'Neill, J. Chen and H. Bartsch
Lyon, International Agency for Research on Cancer
© IARC, 1991

ACTIVITY OF O^6-ALKYLGUANINE-DNA ALKYL TRANSFERASE IN THE LIVER, KIDNEY AND WHITE BLOOD CELLS OF RATS OF DIFFERENT AGES

A. Likhachev[1], N. Zhukovskaya[1], V. Anisimov[1], J. Hall[2] & N. Napalkov[1]

[1]*N.N. Petrov Research Institute of Oncology, Leningrad, USSR; and*
[2]*International Agency for Research on Cancer, Lyon, France*

Age-related differences in the sensitivity of rats to alkylating carcinogens may be dependent on various factors, including the cellular levels of O^6-alkylguanine-DNA alkyltransferase (AT). In the present study, the levels of AT were measured in protein extracts prepared from liver, kidney and peripheral white blood cells of male outbred rats aged 1, 4, 14, 22 and 36 months. The AT level (expressed as activity per milligram protein) in liver extracts was lower in rats aged 1, 4 or 36 months than in extracts prepared from rats aged 14 or 22 months. This observation of a variation in AT level with age is in agreement with our previous results. The AT levels in kidney and white blood cells did not differ significantly with age, and in all cases the AT levels were lower than those observed in the liver extracts, the kidney extracts having more AT activity than the white blood cell extracts. The total protein content of both liver and kidney tissues, calculated per gram of wet tissue, increased to a maximum at 14 months and subsequently declined, the total protein content being always higher in the liver than in the kidney. In contrast, the DNA content per gram of wet tissue was highest in young animals and subsequently declined to a minimum at 14 months. The implications of this inverse relationship to the levels of AT activity are discussed.

Age-related differences in the sensitivity of rats to alkylating carcinogens may depend on (i) activity of the monooxygenases that convert them into alkylating species (except for directly acting agents); (ii) rate of cell proliferation and DNA synthesis; and/or (iii) efficiency of DNA repair systems, particularly that for the promutagenic lesion O^6-alkylguanine in DNA (see, e.g., Margison, 1985, and references therein). In the present study, the levels of O^6-alkylguanine-DNA alkyltransferase (AT) were assayed as described by Harris *et al.* (1983), using protein extracts prepared from liver, kidney and peripheral white blood cells of male outbred rats from the N.N. Petrov Research Institute, aged 1, 4, 14, 22 and 36 months.

Significant differences were seen in the activities of AT in the livers of rats of different ages (Table 1). The levels were lower in four-month-old rats than in one-month-old animals, and there was an apparent increase in AT activity with increasing age, reaching a maximum at 14-22 months. In older rats (36 months old), the level of liver AT was decreased and was comparable with that in rats of the youngest group (the comparisons are statistically significant at the $p < 0.05$ level). No such difference in AT activity with age was found in either of the other two tissues studied, namely kidney and peripheral white blood cells (Table 1).

In all age groups, AT activity in the liver was 1.5- to five-fold higher than in kidney, and five- to 40-fold higher in white blood cells, the differences being greatest in older rats. The variation in AT activity between different animals within each age group was greatest in the liver. It would appear that AT expression is regulated independently in the different tissues examined here, as there was little correlation between the levels measured in protein extracts prepared from different tissues from the same animal. The results suggest that the level of AT activity in one tissue is not necessarily indicative of that in a different organ within any one animal.

The AT levels measured are expressed as picomoles of $O6$-methylguanine removed per milligram of total protein, the protein being measured by the Lowry method (Lowry et al., 1951). However, published data indicate that cellular protein concentrations may change substantially with age (reviewed by Anisimov, 1987). We found (Table 2) that the total protein content of both liver and kidney, calculated per gram of wet tissue, increased to a maximum at 14 months and subsequently declined. The DNA content (measured according to Gerson et al., 1986) of both tissues showed an inverse pattern: it was highest in young animals and subsequently declined to a minimum at 14 months, the values being similar to those reported by Craddock et al. (1984). The total protein content was always higher in the liver than in the kidney, whereas the total DNA content was higher in the kidney at all ages except in the oldest animals examined. This apparent increase is probably due to the increase in ploidy that occurs in old animals. No such regularity was seen in white blood cells, as shown by the protein:DNA ratios in Table 2. Thus, expression of AT activity relative to the protein or DNA content of a tissue may differ substantially with age, and the DNA and protein content are important factors that must be considered when comparing AT activities in different organs or in the same organs of different species.

A comparison of the relative amounts of DNA and total protein and the activity of AT in liver extracts allows us to speculate that adult (one- to two-year-old rats) should be most resistant to hepatocarcinogenesis after exposure to N-nitroso compounds, as the liver-cell genome of these animals is best protected by AT. In addition, the rate of cell proliferation, which is also a critical determining factor for initiation of carcinogenesis, is known to be very low in the livers of adult animals. This suggestion has been confirmed in experiments on the carcinogenicity of N-nitroso compounds in rats (Anisimov, 1987).

Our findings are in agreement with previous observations obtained on rats of the same stock (Likhachev, 1985). However, Woodhead et al. (1985) found no statistically significant age-associated difference in the AT activity of liver, spleen, brain or kidney in rats aged 1-24 months (males) or 1-37 months (females), even though age-related variations in the levels of this DNA-repairing protein were observed. Like us, they found that AT was more active in the liver than in the kidney in animals of all age groups. The discrepancies in the age-associated pattern of AT found in the above-mentioned studies could result from strain-dependent peculiarities of the rats used in the experiments. In other species, e.g., in

Table 1. Rat tissue O^6-alkylguanine–DNA alkyltransferase activity[a]

Organ	Age (months)			
	1	4	14	22	36
Liver	0.261 ± 0.01 (5)[b]	0.198 ± 0.011 (5)	0.348 ± 0.057 (5)	0.52 ± 0.05 (5)	0.329 ± 0.031 (7)
Kidney	0.135 ± 0.01 (5)	0.128 ± 0.014 (5)	0.126 ± 0.009 (5)	0.102 ± 0.01 (5)	0.091 ± 0.009 (7)
White blood cells	0.038 (1)	0.034 ± 2.9 × 10⁻³ (4)	0.026 ± 2.67 × 10⁻³ (3)	0.013 ± 9.8 × 10⁻³ (5)	0.021 ± 6 × 10⁻³ (4)

[a] Data represent the mean ± SEM of O^6-alkylguanine–DNA alkyltransferase activity measured in protein extracts prepared from independent tissue samples, expressed as pmol O^6-methylguanine removed per mg protein.
[b] Numbers in parentheses indicate the number of independent donors sampled.

mice, AT activity at different ages depends heavily on the strain of animals (Nakatsura et al., 1989).

Acknowledgements

Part of this work was supported by EEC Contract No. EV4V0040-F (CD).

Table 2. Protein and DNA content of liver, kidney and white blood cells from rats of different ages[a]

Age (months)	Substrate	Tissue		
		Liver	Kidney	White blood cells[b]
1	Protein	118.1 ± 12.5	99.9 ± 2.97	
	DNA	3.56 ± 0.27	5.41 ± 0.26	
	Protein:DNA	33.1 ± 2.42	18.7 ± 0.81	31.4 ± 7.66
4	Protein	160.2 ± 7.00	122.9 ± 4.69	
	DNA	2.77 ± 0.23	3.96 ± 0.28	
	Protein:DNA	59.8 ± 4.51	32.8 ± 4.08	17.4 ± 4.97
14	Protein	178.9 ± 6.25	143.3 ± 11.16	
	DNA	2.20 ± 0.22	2.45 ± 0.22	
	Protein:DNA	82.3 ± 4.05	61.5 ± 6.78	93.6 ± 27.0
22	Protein	166.7 ± 5.30	111.5 ± 8.38	
	DNA	2.43 ± 0.23	2.80 ± 0.17	
	Protein:DNA	72.2 ± 6.27	41.7 ± 3.01	59.9 ± 16.0
33	Protein	128.7 ± 7.96	102.1 ± 5.57	
	DNA	3.38 ± 0.17	2.89 ± 0.27	
	Protein:DNA	38.5 ± 2.69	37.4 ± 3.92	107.9 ± 34.9

[a] Data represent the mean ± SEM of protein and DNA content (mg/g wet tissue) in tissue or cell homogenates prepared from independent samples taken from eight rats.

[b] For white blood cells, protein and DNA concentrations were measured on the same sample and the values used to calculate the DNA:protein ratio.

References

Anisimov, V.N. (1987) *Carcinogenesis and Ageing*, Vols 1 & 2, Boca Raton, FL, CRC Press

Craddock, V.M., Henderson, A.R. & Gash, S. (1984) Repair and replication of DNA in rat brain and liver during foetal and post-natal development, in relation to nitroso-alkyl-urea induced carcinogenesis. *J. Cancer Res. Clin. Oncol.*, **108**, 30-35

Gerson, S.L., Trey, J.E., Miller, K. & Berger, N.A. (1986) Comparison of O^6-alkylguanine-DNA alkyltransferase activity based on cellular DNA content in human, rat and mouse tissues. *Carcinogenesis*, **7**, 745-749

Harris, A.L., Karran, P. & Lindahl, T. (1983) O^6-Methylguanine-DNA methyltransferase of human lymphoid cells: structural and kinetic properties and absence in repair deficient cells. *Cancer Res.*, **43**, 3247-3252

Likhachev, A. (1985) Effect of age on DNA repair in carcinogenesis due to alkylating agents. In: Likhachev, A., Anisimov, V. & Montesano, R., eds, *Age-related Factors in Carcinogenesis* (IARC Scientific Publications No. 58), Lyon, IARC, pp. 239-246

Lowry, O.H., Rosebrough, N.J., Farr, A.L. & Randall, R.H.J. (1951) Protein measurement with the Folin phenol reagent. *J. Biol. Chem.*, **193**, 265-275

Margison, G.P. (1985) The effects of age on the metabolism of chemical carcinogens and inducibility of O^6-methylguanine methyltransferase. In: Likhachev, A., Anisimov, V. & Montesano, R., eds, *Age-related Factors in Carcinogenesis* (IARC Scientific Publications No. 58), Lyon, IARC, pp. 225-237

Nakatsura, Y., Aoki, K. & Ishikawa, T. (1989) Age and strain dependence of O^6-methylguanine DNA methyltransferase activity in mice. *Mutat. Res.*, **219**, 51-56

Woodhead, A.D., Merry, B.J., Cao, E.-H., Holehan, A.M., Grist, E. & Carlson, C. (1985) Levels of O^6-methylguanine acceptor protein in tissues of rats and their relationship to carcinogenicity and aging. *J. Natl Cancer Inst.*, **75**, 1141-1145

Relevance to Human Cancer of N-Nitroso Compounds,
Tobacco Smoke and Mycotoxins.
Ed. I.K. O'Neill, J. Chen and H. Bartsch
Lyon, International Agency for Research on Cancer
© IARC, 1991

EXCISION OF IMIDAZOLE RING-OPENED N7-HYDROXY-ETHYLGUANINE FROM CHLOROETHYLNITROSOUREA-TREATED DNA BY *ESCHERICHIA COLI* FORMAMIDOPYRIMIDINE-DNA GLYCOSYLASE

J. Laval[1], F. Lopès[1], J.C. Madelmont[2], D. Godenèche[2], G. Meyniel[2], Y. Habraken[3], T.R. O'Connor[1] & S. Boiteux[1]

[1]*Groupe Réparation des lésions radio et chimioinduites, UA147
Centre National de Recherche Scientifique and U140
Institut National de la Santé et de la Recherche Medicale,
Institut Gustave Roussy, Villejuif;*
[2]*U71 Institut National de la Santé et de la Recherche Médicale,
Clermont Ferrand;*
and [3] *Groupe Radiochimie de l'ADN, U247,
Institut National de la Santé et de la Recherche medicale, Institut Gustave Roussy,
Villejuif, France*

Alkylkation of the N7 of guanine residues in DNA favours the opening of the imidazole ring, yielding a formamidopyrimidine (Fapy). This Fapy residue blocks DNA replication and is actively excised by a DNA glycosylase. We have cloned and sequenced the *Escherichia coli* gene responsible for synthesis of the enzyme, which has also been purified to homogeneity. It was found to have associated apurinic/apyrimidinic (AP) lyase activity, nicking DNA at AP sites. Chloroethylnitrosoureas are used in cancer chemotherapy. The lesions induced in DNA by these compounds, including N7-chloro- and hydroxy-ethylguanine, are excised by *E. coli* 3-methyladenine DNA glycosylase II, and we report that the corresponding imidazole ring-opened forms are repaired by Fapy-DNA glycosylase. Human cells have the counterpart to these enzymes, which could contribute to the repair of these lesions during chemotherapy.

The chloroethylnitrosoureas are effective in the treatment of malignancies such as lymphomas and gliomas (Schein *et al.*, 1984); their antitumour action is related to their ability to modify DNA (Ludlum & Tong, 1985). The major base modification products in chloroethylnitrosourea-treated DNA are N7-alkylguanines, including N7-chloroethylguanine and N7-hydroxyethylguanine (Singer & Grundberger, 1983).

Alkylation at the N7 position of guanine in DNA results in labilization of both the glycosidic bond and the imidazole ring, yielding apurinic sites and 2,6-diamino-4-hydroxy-5N-alkylformamidopyrimidine (Singer & Grundberger, 1983; Boiteux *et al.*, 1984). Formation of imidazole ring-opened (iro) products was observed in the DNA of rats treated with N-methyl-N-nitrosourea (Kadlubar et al., 1984), and a specific DNA glycosylase excises the iro N7-methylguanine in *Escherichia coli* and mammalian cells (Chetsanga & Lindahl, 1979; Margison & Pegg, 1981). The *fpg* gene that codes for the formamidopyrimidine (Fapy)-DNA glycosylase of *E. coli* has been cloned and sequenced and the protein purified to homogeneity (Boiteux et al., 1987).

These observations prompted us to study the excision of iro products from chloroethylnitrosourea-treated DNA. A double-stranded polynucleotide, poly(dG-dC), was reacted with [^{14}C(chloroethyl)]-labelled N'-(2-chloroethyl)-N-[2-(methylsulfinyl)-ethyl-N'-nitrosourea (CMSOEN$_2$; Madelmont et al., 1985), which has shown promising antitumour activity against gliomas and melanomas (Bourut et al., 1986). The CMSOEN$_2$-poly(dG-dC) was further incubated under alkaline conditions (Boiteux et al., 1984) in order to convert N7-alkylguanines into iro products. These modified polynucleotides were used as substrates for the Fapy-DNA glycosylase of *E. coli*. We report here that the iro N7-hydroxyethylguanine is efficiently repaired by Fapy-DNA glycosylase.

Repair of iro N7-*hydroxyethylguanine by* E.coli *Fapy-DNA glycosylase*

Iro N7-hydroxyethylguanine was prepared from N7-hydroxyethylguanosine by alkaline cleavage of the imidazole ring and further elimination of the ribosyl residue by formic acid treatment (Boiteux et al., 1984). The resulting products were separated by high-performance liquid chromatography using a C18 μ Bondapak reversed-phase column isocratically eluted at 1.5 ml/min with 5% methanol (v/v) in 20 mM phosphate buffer pH 4.5. Two major peaks were resolved, eluting at 4.5 and 5.5 min, respectively. Rechromatography of each showed that the two products are slowly interconverted to give the initial elution pattern. By analogy to iro N7-methylguanine (Boiteux et al., 1984), these observations suggest that the two species described are rotameric forms of the same molecule: 2,6-diamino-4-hydroxy-5N-hydroxyethylformamidopyrimidine. Fapy-DNA glycosylase was purified from *E. coli* cells harbouring the pFPG60 plasmid according to the procedure described by Boiteux et al. (1987). The purity of the enzyme was assessed by observation of a single protein band on sodium dodecyl sulfate-polyacrylamide gel.

A double-stranded DNA-like polymer, poly(dG-dC), was reacted with [^{14}C(chloro–ethyl)]-CMSOEN$_2$ to yield CMSOEN$_2$-poly(dG-dC). In order to convert N7-alkylguanines to iro products, the CMSOEN$_2$-poly(dG-dC) was incubated under alkaline conditions to yield iro-CMSOEN$_2$-poly(dG-dC). These modified poly(dG-dC) were then incubated in the presence of increasing amounts of Fapy-DNA glycosylase. Radioactive material was released from the iro-CMSOEN$_2$-poly(dG-dC) by the Fapy-DNA glycosylase but not from CMSOEN$_2$-poly(dG-dC) (Fig. 1).

The nature of the radioactive material liberated by Fapy-DNA glycosylase into the ethanol-soluble fraction was analysed by high-performance liquid chromatography and found to elute almost exclusively with iro N7-hydroxyethylguanine. No radioactivity was recovered at the positions of N7-hydroxyethylguanine and N7-chloroethylguanine (Fig. 2).

Discussion

We have shown that Fapy-DNA glycosylase does not release radioactivity from poly(dG-dC) treated with the antitumour chloroethylnitrosourea, CMSOEN$_2$. This result was not unexpected since N7-hydroxyethylguanine and N7-chloroethylguanine are repaired by the 3-methyladenine-DNA glycosylase II of *E. coli* (Carter *et al.*, 1988). In contrast, alkaline-treated CMSOEN$_2$-poly(dG-dC) is a substrate for the Fapy-DNA glycosylase, the major product of excision being the iro *N*7-hydroxyethylguanine. This result confirms the broad substrate specificity of the Fapy-DNA glycosylase, which might include all iro purines (Boiteux *et al.*, 1989).

The therapeutic action of antitumour drugs is limited by the tumour resistance phenomenon. The mechanism of the resistance to chloroethylnitrosoureas of human glial tumour cells was studied by Bodell *et al.* (1988), who showed that resistant cell lines exhibit lower levels of DNA modification, including cross-links and modified bases. The decrease in the number of cross-links is probably due to the higher level of *O*6-alkyltransferase in resistant cell lines (Bodell *et al.*, 1988); the lower level of base modifications could be due to the presence of increased levels of DNA glycosylases, these enzymes being the mammalian counterparts of 3-methyladenine-DNA glycosylase II (Carter *et al.*, 1988) and Fapy-DNA glycosylase (this report) of *E. coli*.

Fig. 1. Excision of radioactive material from poly(dG-dC) treated with *N'*-(2-chloroethyl)-*N*-[2-(methylsulfinyl)ethyl]-*N'*-nitrosourea; CMSOEN$_2$-poly(dG-dC) and imidazole ring-opened (iro)–CMSOEN$_2$-poly(dG-dC) by *Escherichia coli* formamidopyrimidine (Fapy)-DNA glycosylase

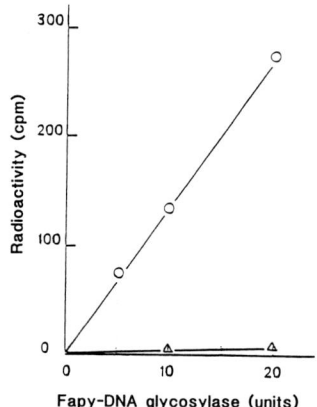

Poly(dG-dC) was reacted with 200 μCi [^{14}C(chloroethyl)]-CMSOEN$_2$(5 mCi/mmol). The specific activity of CMSOEN$_2$-poly(dG-dC) was 800 cpm/μg. Imidazole ring opening was obtained by treatment of CMSOEN$_2$- poly(dG-dC) with 0.2 N NaOH for 30 min at 37°C. Reaction mixtures contained 3000 cpm of modified substrate and increasing amounts of Fapy-DNA glycosylase. The assays were performed as described by Boiteux *et al.* (1987). Enzyme units were defined using iro *N*7-methylguanine as substrate (Boiteux *et al.*, 1984). The radioactivity released in the ethanol-soluble fraction was measured by scintillation spectroscopy. Symbols: (Δ) CMSOEN$_2$- poly(dG-dC); (O) iro-CMSOEN$_2$-poly(dG-dC)

Acknowledgements

We should like to thank Dr D.B. Ludlum (University of Massachussetts Medical School, Worcester, MA, USA) for the kind gift of *N*7-chloroethylguanine and *N*7-hydroxyethylguanine marker molecules. This work was supported by grants from the Institut

National de la Santé et de la Recherche Médicale, the Centre National de Recherche Scientifique and the Association pour la Recherche sur le Cancer.

Fig. 2. High-performance liquid chromatography profile of radioactivity released from imidazole ring-opened (iro) (CMSOEN$_2$)-poly(dG-dC) by *Escherichia coli* formamidopyrimidine (Fapy)-DNA glycosylase

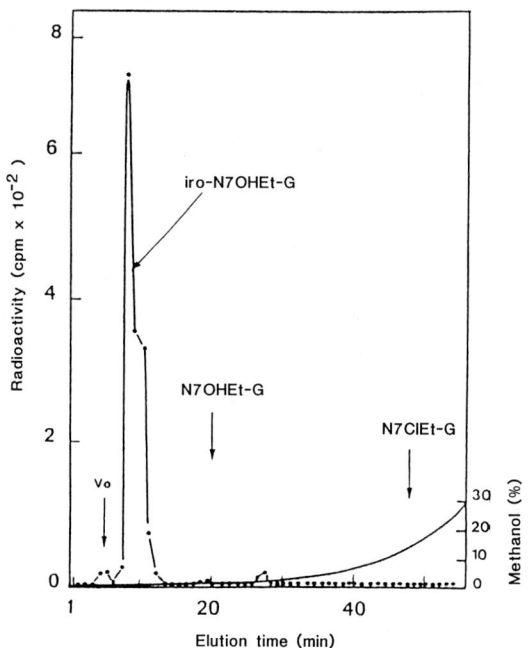

Iro-CMSOEN$_2$-poly(dG-dC) (15 000 cpm) was incubated with 200 units of Fapy-DNA glycosylase. Radioactive material released into the ethanol-soluble fraction was dried, resuspended in water and supplemented with authentic marker molecules. The mixture was analysed by high-performance liquid chromatography using a C18 μ Bondapak column, eluted at 1.5 ml/min with a 90-ml gradient of 0.0–30% methanol (v/v) in 20 mM phosphate buffer pH 4.5. Markers were detected by ultraviolet absorption at 254 nm and the reaction products by scintillation spectroscopy. Iro-N7OHEt-G, imidazole ring-opened N7-hydroxyethylguanine; N7CLEt-G, N7-chloroethylguanine

References

Bodell, W.J., Tōduka, K. & Ludlum, D.B. (1988) Differences in DNA alkylation products formed in the sensitive and resistant human glioma cells treated with N-(2-chloroethyl)-N-nitrosourea. *Cancer Res.*, **48**, 4489-4492

Boiteux, S., Belleney, J., Roques, B.P. & Laval, J. (1984) Two rotameric forms of imidazole ring-opened 7-methylguanine are present in alkylated polynucleotides. *Nucleic Acids Res.*, **12**, 5429-5439

Boiteux, S., O'Connor, T.R. & Laval, J. (1987) Formamidopyrimidine-DNA glycosylase of *E. coli*: cloning and sequencing of the *fpg* structural gene and overproduction of the protein. *EMBO J.*, **6**, 3177-3183

Boiteux, S., Bichara, M., Fuchs, R.P.P. & Laval, J. (1989) Excision of the imidazole ring-opened form of N-2-aminofluorene-C(8)-guanine adduct in poly (dG-dC) by *E. coli* formamidopyrimide DNA glycosylase. *Carcinogenesis*, **10**, 1905-1909

Bourut, C., Chenu, E., Godenèche, D., Madelmont, J.C., Maral, R. & Mathé, R. (1986) Cytostatic action of two nitrosoureas derived from cysteamine. *Br. J. Pharmacol.*, **89**, 539-546

Carter, C.A., Habraken, Y. & Ludlum, D.B. (1988) Release of 7-alkylguanines from haloethylnitrosourea-treated DNA by *E. coli* 3-methyladenine-DNA glycosylase II. *Biochem. Biophys. Res. Commun.*, **155**, 1261-1265

Chetsanga, C.J. & Lindahl, T. (1979) Release of 7-methylguanine whose imidazole has been opened from damaged DNA by a DNA glycosylase from *E. coli*. *Nucleic Acids Res.*, **6**, 3673-3683

Kadlubar, F.F., Beranek, D.T., Weiss, C.C., Evans, F.E., Cox, R. & Irving, C.C. (1984) Characterization of ring-opened 7-methylguanine and its persistence in bladder epithelial DNA after treatment with the carcinogen N-methylnitrosourea. *Carcinogenesis*, **5**, 587-592

Ludlum, D.B. & Tong, W.P. (1985) DNA modifications by the nitrosoureas: chemical nature and cellular repair. In: Muggia, F.M., ed., *Experimental and Clinical Progress in Cancer Chemotherapy*, Boston, Martinus Nijhoff, pp. 141-154

Madelmont, J.C., Godenèche, D., Pury, D., Duprat, J., Chabart, J., Plagne, R., Mathé, G. & Meyniel, G. (1985) New cysteamine 2-chloroethylnitrosoureas synthesis and preliminary antitumour results. *J. Med. Chem.*, **28**, 1346-1350

Margison, G.P. & Pegg, A.E. (1981) Enzymatic release of 7-methylguanine from methylated DNA by rodent liver extracts. *Proc. Natl Acad. Sci. USA*, **78**, 861-865

Schein, P.S., Tew, K.D. & Mathé, G. (1984) Pharmacology of nitrosourea anticancer agents. In: Berkarda, B., Karre, K. & Mathé, E., eds, *Clinical Chemotherapy*, Vol. III, New York, Thieme-Stratton, pp. 264-274

Singer, B. & Grundberger, D., eds (1983) *Molecular Biology of Mutagens and Carcinogens*, New York, Plenum Press, pp. 45-96

RESISTANCE TO N-NITROSO COMPOUNDS IN CELLS TREATED WITH VARIOUS PHYSICAL AND CHEMICAL AGENTS

P. Lefebvre & F. Laval

Groupe Radiochimie de l'ADN, U247 Institut National de la Santé et de la Recherche Médicale, Institut Gustave Roussy, Villejuif, France

N-Nitrosoureas produce various lesions in cellular DNA (Singer & Kusmierek, 1982). In *Escherichia coli*, the lethal lesion N^3-methyladenine is repaired by the sequential action of a specific DNA glycosylase and an apyrimidinic endonuclease (Laval, 1977), whereas O^6-methylguanine, which is mutagenic and probably lethal (Saffhill *et al.*, 1985), is repaired by a specific suicide protein, the O^6-methylguanine–DNA methyltransferase (Lindahl *et al.*, 1982). In *E. coli*, these repair activities may be increased during the adaptive response, which is induced by pretreating the bacteria with low, nontoxic doses of an alkylating agent (reviewed by Walker, 1984). Similar repair activities occur in mammalian cells (reviewed by Yarosh, 1985). As N-nitrosoureas are widely used in cancer therapy, we considered it important to investigate whether repair activity could be enhanced in mammalian cells. We have measured the influence of various pretreatments on cellular N^3-methyladenine glycosylase and transferase activities and on cell sensitivity to N-nitrosoureas.

H4 rat hepatoma cells were grown in Dulbecco's medium supplemented with 5% fetal calf serum and 5% horse serum. Cell survival was measured by plating (Lefebvre & Laval, 1986). N^3-Methyladenine glycosylase activity was determined by incubating cell extracts with DNA treated with ^{3}H-dimethylsulfoxide and measuring the release of N^3-methyladenine by high-performance liquid chromatography (Laval, 1985). Transferase activity was measured by incubating cell extracts with DNA treated with ^{3}H-N-methyl-N-nitrosourea and measuring the remaining amount of O^6-methylguanine in the substrate by high-performance liquid chromatography, as described previously (Boiteux & Laval, 1985).

Repair activities

H4 cells were treated with a single, low, toxic dose of various chemical and physical agents. Table 1 shows the number of transferase molecules in H4 cells 48 h after irradiation with γ-rays or ultraviolet light or treatment for 1 h with various compounds. Treatment with DNA damaging agents increased transferase activity, but treatment with compounds that do not interact with cellular DNA (e.g., metabolic inhibitors) did not (data not shown). The increase was maximal 48 h after treatment, then the transferase activity decreased and reached the control value after 120 h. When the cells were treated repeatedly with the DNA damaging agent (e.g., γ-rays delivered every 48 h), the number of transferase molecules per cell was enhanced for the duration of the treatment.

Table 1. Number of transferase molecules in H4 cells 48 h after treatment with DNA damaging agents

DNA damaging agent[a]	Dose	No. of transferase molecules/cell[b]	Relative number
None	-	54 000	1
γ-Rays	300 rads	285 000	5.27
Ultraviolet light	10 J/m²	135 600	2.50
cis-Pt II	5 μM	207 300	3.83
MNNG	10 μM	134 000	2.48
MMS	1 mM	260 000	4.80
EMS	10 mM	211 200	3.90
N-Methyl-9-hydroxyellipticine	7.5 μM	252 000	4.66

[a] cis-Pt II, cisplatin; MNNG, N-methyl-N'-nitroso-N'-nitrosoguanidine; MMS, methylmethanesulfonate; EMS, ethylmethane sulfonate. Cell survival, determined by plating, was 30% with MNNG, γ-rays, ultraviolet light and ellipticine and 50% with MMS and cis-PtII.
[b] The number of transferase molecules was calculated by measuring the activity of cell extracts on DNA treated with ³H-N-methyl-N-nitrosourea (Boiteux & Laval, 1985).

These treatments also increased N^3-methyladenine glycosylase activity: extracts from treated cells released two to three times more 3-methyladenine residues from DNA treated with 3H-dimethylsulfoxide than did extracts from control cells.

Cytotoxicity

As these two inducible activities are implicated in the repair of DNA lesions induced by N-nitrosoureas, we measured the killing effects of compounds on control cells and on cells pretreated with the various agents. Table 2 shows that the enhanced repair activities are biologically active as pretreated cells become more resistant to the toxic effects of N-nitrosoureas.

Table 2. Toxicity of alkylating compounds to H4 cells with and without pretreatment with DNA damaging agents

Alkylating agent[a]	Pretreatment	D(10)[b]
MMS	None	1.5 mM
MMS	γ-Rays (300 rads)	2.75 mM
MNNG	None	35 M
MNNG	MNNG (10 M)	70 M
BCNU	None	100 M
BCNU	γ-Rays (300 rads)	170 M

[a] MMS, methylmethane sulfonate; MNNG, N-methyl-N'-nitro-N'-nitrosoguanidine; BCNU, N-bischoroethyl-N-nitrosourea. Cells were incubated for 1 h with the alkylating drug 48h after the pretreatment. Survival was determined by plating.
[b] Concentration that reduced survival to 10% of the control

Therefore, two mechanisms for the repair of alkylated bases can be enhanced in mammalian cells by DNA damaging agents. This induction seems to occur preferentially in tumour cell lines (data not shown). It is well known in cancer therapy that some tumours develop resistance to radio- or chemotherapy during the course of treatment. Increased repair of alkylated bases might be one of the processes that modify the sensitivity of tumour cells to treatment with N-nitrosoureas.

Acknowledgements

This work was supported by grants from the Institut National de la Santé et de la Recherche Médicale and from the Association pour la Recherche sur le Cancer (Villejuif).

References

Boiteux, S. & Laval, F. (1985) Repair of O^6-methylguanine, by mammalian cell extracts, in alkylated DNA and poly (dG-m^5dC).(poly dG-m^5dC) in B and Z forms. *Carcinogenesis*, **6**, 805-807

Laval, J. (1977) Two enzymes are required for strand incision in repair of alkylated DNA. *Nature*, **269**, 829-831

Laval, F. (1985) Repair of methylated bases in mammalian cells during adaptive response to alkylating agents. *Biochimie*, **67**, 361-364

Lefebvre, P. & Laval, F. (1986) Enhancement of O^6-methylguanine-DNA-methyltransferase activity induced by various treatments in mammalian cells. *Cancer Res.*, **46**, 5701-5705

Lindahl, T., Demple, B. & Robins, P. (1982) Suicide inactivation of the E. coli O^6-methylguanine-DNA-methyltransferase. *EMBO J.*, **1**, 1359-1363

Saffhill, R., Margison, G.P. & O'Connor, P.J. (1985) Mechanisms of carcinogenesis induced by alkylating agents. *Biochim. Biophys. Acta*, **823**, 111-140

Singer, B. & Kusmierek, J.T. (1982) Chemical mutagenesis. *Ann. Rev. Biochem.*, **51**, 655-693

Walker, G.C. (1984) Mutagenesis and inducible responses to deoxyribonucleic acid damage in Escherichia coli. *Microbiol. Rev.*, **48**, 60-93

Yarosh, D.B. (1985) The role of O^6-methylguanine-DNA-methyltransferase in cell survival, mutagenesis and carcinogenesis. *Mutat. Res.*, **145**, 1-16

EFFECT OF LONG-TERM FEEDING OF NIVALENOL ON AFLATOXIN B_1-INITIATED HEPATOCARCINOGENESIS IN MICE

Y. Ueno[1], T. Kobayashi[1], H. Yamamura[1], T. Kato[1],
F. Tashiro[2], K. Nakamura[3] & K. Ohtsubo[3]

[1]*Department of Toxicology and Microbial Chemistry, Faculty of Pharmaceutical Sciences, Science University of Tokyo, Ichigaya, Tokyo;* [2]*Department of Biological Science and Technology, Science University of Tokyo, Noda; and* [3]*Department of Clinical Pathology, Tokyo Metropolitan Institute of Gerontology, Itabashi, Tokyo, Japan*

Nivalenol, a trichothecene, occurs widely in cereals and foods; our current two-year feeding trial has revealed no tumorigenic activity in female mice. To investigate whether dietary nivalenol modulates the development of aflatoxin B_1 (AFB_1)-initiated hepatocarcinogenesis, one-week old C57Bl/6 × C3H F_1 mice were injected intraperitoneally with 6 mg/kg bw AFB_1 and six weeks later fed diets containing 0, 6 or 12 ppm nivalenol for one year. Male mice in all three groups developed hepatocellular carcinomas and adenomas, while the incidences in females were 31% in those given AFB_1 alone and 20% and 0 in those given AFB_1 with 6 and 12 ppm nivalenol, respectively. These findings indicate that dietary nivalenol suppresses AFB_1-initiated hepatocarcinogenesis in female mice, presumably by acting on the promotion step.

Carcinogenesis is now considered to be a progressive process which may involve a sequence of cellular and tissue changes consequent to initiation and promotion (Farber & Cameron, 1980). Aflatoxin B_1 (AFB_1) is a well-known fungal carcinogen which selectively induces hepatocellular carcinoma in experimental animals. Covalent binding to DNA (Essigman *et al.*, 1977; Ueno *et al.*, 1983) and subsequent activation of oncogenes (McMahon *et al.*, 1986; Tashiro *et al.*, 1986) are involved in the process of AFB_1-induced hepatocarcinogenesis. A single intraperitoneal injection of AFB_1 to infant mice selectively induced hepatic tumours one year later (Vesselinovitch *et al.*, 1972). This evidence suggests that this mycotoxin is a potent initiator in the hepatocarcinogenic process.

Trichothecenes such as nivalenol, deoxynivalenol and others, produced by *Fusarium* species, are often detected in cereals and foodstuffs (Ueno, 1987). In previous experiments, we found that nivalenol was not tumorigenic in mice (Ohtsubo *et al.*, 1989)

AFB_1 and nivalenol are produced by different pathogenic plant fungi. Although the host-parasite relationships of the two compounds are different, simultaneous exposure to these two mycotoxins is likely to occur; however, the contribution of the combination to human and animal diseases, particularly to hepatocarcinogenesis, is unknown.

On the basis of the evidence that (i) AFB_1 is a potent initiator of hepatocarcinogenesis, (ii) that newborn mice are a good model for investigating hepatocarcinogenicity and (iii) that nivalenol is not tumorigenic but occurs widely in foodstuffs, we used a single-dose administration of AFB_1 in infant mice to evaluate whether dietary nivalenol can modulate AFB_1-initiated hepatocarcinogenicity.

Materials and methods

Groups of one-week-old C57Bl/6 × C3H F_1 mice of each sex were injected intraperitoneally with 6 mg/kg bw AFB_1; seven weeks later, they were fed pellets containing 0, 6 or 12 ppm nivalenol, prepared by mixing the basal diet with mouldy rice containing *Fusarium nivale* Fn 2B, for one year. Gas chromatographic analysis at three-month intervals showed that nivalenol was distributed uniformly in the pellets, and its content did not change over one year. The numbers of mice employed were 56 in the control group, 51 given AFB_1 alone, 41 given AFB_1 and 6 ppm nivalenol and 38 given AFB_1 and 12 ppm nivalenol. All animals were weaned at four weeks of age. Changes in body weight were monitored once weekly; at 71 weeks, all mice were sacrificed for pathological examination.

Incidence and characteristics of liver tumours

Several mice died before the age of 60 weeks, but no tumorous change was observed in either control or treated groups. The incidences of liver tumours in animals of each sex sacrificed at the age of 71 weeks are summarized in Table 1. Most of the hepatocellular carcinomas were well differentiated and showed a trabecular structure. Mitotic figures were numerous and abnormal in the carcinomas in males of all treated groups. The incidence and microscopic morphology of the AFB_1-initiated hepatomas were not altered by feeding with nivalenol.

Liver tumours were observed at a much lower frequency in AFB_1-treated females than in males, and most occurred as a single nodule. The total incidence of liver tumours was only 31%, and their degree of malignancy was far lower than that in males. The notable finding in this experiment is that in females fed 6 ppm nivalenol only one hepatocellular adenoma, one hepatocellular carcinoma and one haemangioma were seen in 15 animals, reducing the total incidence of liver tumours to 20%, and that with 12 ppm nivalenol no liver tumour occurred.

Future problems

We have demonstrated that nivalenol represses AFB_1-induced hepatocarcinogenesis in female mice but not in males. As in many other reports (Peers *et al.*, 1976), the incidence of hepatocellular carcinomas was higher in males than in females. The initiating dose of AFB_1 that induced 100% tumours in males appears to have been too high to reveal any tumour suppressing effect of nivalenol, but it was adequate to induce 30% incidence in females. As a result, the incidences of hepatic tumours in the females were lowered by increasing concentrations of dietary nivalenol. It is conceivable that the 'weak' suppressive effect of nivalenol on AFB_1-induced hepatocarcinogenesis is due to interference in the promotion and/or selection step. Since nivalenol and related trichothecene mycotoxins inhibit protein synthesis at the ribosomal level in mammalian cells (Ueno, 1987) and their activity is not influenced by the sex of animals or by sex hormones (data not shown), the

Table 1. Incidences of liver tumours in infant mice treated with aflatoxin B_1 (AFB_1) and different levels of nivalenol

Group	Effective no.		No. of mice with tumours						Total (%)	
			Hepatocellular adenoma		Hepatocellular carcinoma		Other			
	Male	Female	Male	Female	Male	Female	Male	Female	Male	Female
Control	25	21	0	0	1	0	0	0	1 (4)	0
AFB_1 alone	19	26	2	6	17	1	0	1a	19 (100)**	8 (31)**
AFB_1 plus 6 ppm nivalenol	20	15	1	1	19	1	0	1b	20 (100)**	3 (20)*
AFB_1 plus 12 ppm nivalenol	18	19	3	0	15	0	0	0	18 (100)**	0***

[a] Kupffer-cell sarcoma
[b] Haemangioma
*$p < 0.05$, **$p < 0.01$ in comparison with controls
***$p < 0.01$ in comparison with AFB_1 alone

different susceptibilities of the two sexes to AFB_1 (Gurtoo & Motycka, 1976; Chemesky et al., 1982; Monroe & Eaton, 1988) may have played the principal role.

AFB_1 is the most potent carcinogen ever known and has been suspected to be involved in human liver cancer. The ubiquitous occurrence of nivalenol and deoxynivalenol in marketed cereals and foodstuffs raised the question whether simultaneous contamination with these two kinds of mycotoxins might occur. We know practically nothing about the long-term effects of trichothecenes on humans at the parts per billion level. The results of our recent study (Ohtsubo et al., 1989) suggest that mice fed nivalenol live longer than controls. Even if contamination by both AFB_1 and nivalenol can prevent hepatic tumours, as shown by our results, the effects of continuous exposure to these mycotoxins are not known. On the basis of the results of this and previous (Ohtsubo et al., 1989) experiments, we consider that the permissible levels of trichothecenes in food and feeds should be re-evaluated.

References

Chemesky, P., Jayaraj, A. & Richardson, A. (1982) The effect of testosterone treatment on the ability of female rat liver S9 to activate aflatoxin B_1 and aminofluorene in the Ames/*Salmonella* system. *Mutat. Res.*, **103**, 267-273

Essigman, J.M., Croy, R.G., Nadzan, A.M., Busby, W.F., Jr, Reihold, V.N., Buchi, G. & Wogan, G.N. (1977) Structural identification of the DNA adduct formed by aflatoxin B_1 *in vitro*. *Proc. Natl Acad. Sci. USA*, **74**, 1870-1874

Farber, E. & Cameron, R. (1980) The sequential analysis of cancer development. *Adv. Cancer Res.*, **31**, 125-220

Gurtoo, H.L. & Motycka, L. (1976) Effects of sex differences on the *in vitro* and *in vivo* metabolism of aflatoxin B_1 in the rat. *Cancer Res.*, **36**, 4663-4671

McMahon, G., Hanson, L., Lee, J.L. & Wogan, G.N. (1986) Identification of an activated c-Ki-ras oncogenes in rat liver tumours induced by aflatoxin B_1. *Proc. Natl Acad. Sci. USA*, **83**, 9411-9422

Monroe, D.H. & Eaton, D.L. (1988) Effects of modulation of hepatic glutathione to biotransformation and covalent binding of aflatoxin B_1 to DNA in the mouse. *Toxicol. Appl. Pharmacol.*, **94**, 118-129

Ohtsubo, K., Ryu, J.C., Nakamura, K., Izumiyama, N., Tanaka, T., Yamamura, H., Kobayashi, T. & Ueno, Y. (1989) Chronic toxicity of nivalenol in mice: 2-years feeding trial with *Fusarium nivale* Fn 2B-moulded rice. *Food Chem. Toxicol.*, **27**, 591-593

Peers, F.G., Gilman, G.A. & Linsell, C.A. (1976) Dietary aflatoxins in human liver cancer. A study in Swaziland. *Int. J. Cancer*, **17**, 167-176

Tashiro, F., Morimura, S., Hayashi, K., Makino, R., Kawamura, H., Horikoshi, N., Nemoto, K., Ohtsubo, K., Sugimura, T. & Ueno, Y. (1986) Expression of the c-Ha-ras and c-myc genes in aflatoxin B_1-induced hepatocellular carcinomas. *Biochem. Biophys. Res. Commun.*, **138**, 858-864

Ueno, Y. (1987) *Mycotoxins in Food*, London, Academic Press, p. 123

Ueno, Y., Ishii, K., Omata, Y., Kamataki, T. & Kato, R. (1983) Specificity of hepatic cytochrome P-450 isozymes from PCB-treated rats and participation of cytochrome b_5 in the activation of aflatoxin B_1. *Carcinogenesis*, **4**, 1071-1073

Vesselinovitch, S.D., Mihailovich, N., Wogan, G.N., Lombard, L.S. & Rao, K.V.N. (1972) Aflatoxin B_1, a hepatocarcinogen in the infant mouse. *Cancer Res.*, **32**, 2289-2291

DISTRIBUTION AND EXCRETION OF ³H-STERIGMATOCYSTIN IN RATS

D.S. Wang[1], H.L. Sun[1], F.Y. Xiao[2],
X.H. Ji[1], Y.X. Liang[1] & F.G. Han[2]

[1]*Beijing Institute for Cancer Research and,*
[2]*The 309th Military Hospital, Beijing, China*

In order to investigate the distribution and excretion of sterigmatocystin, a carcinogenic mycotoxin, radioactively labelled compound was studied in rats. The highest concentration of radioactivity in serum appeared 3 h after administration of 0.5 µCi/g bw. The half-lives of distribution and excretion were 0.51 h and 43.9 h, respectively. The radioactivity was concentrated mainly in liver, stomach, kidney, duodenum and lung and to a lesser extent in fat, muscle, testis, rectum and bone. By 48 h, 56.4% had been excreted in faeces and 20.1% in urine. Biliary excretion may be the major route of excretion of sterigmatocystin.

Sterigmatocystin is a carcinogenic mycotoxin. We have studied its distribution and excretion in rats using radiolabelled compound in order to obtain theoretical evidence to further research on the relationship between sterigmatocystin and the etiology of cancers such as those of the stomach and liver.

Materials and methods

Sterigmatocystin labelled with tritium, at a specific activity of 20 mCi/mg and a radiochemical purity of over 95%, was diluted with dimethyl sulfoxide to 1.0 mCi/ml for storage and further diluted to 100.0 µCi/ml for administration. A group of 30 male Wistar rats was given ³H-sterigmatocystin orally at a dose of 0.5 µCi/g bw. The animals were then placed randomly in pairs in metabolic cages. Blood was drawn from the femoral vein, and two samples of serum (100 µl), faecal matter (50 mg) and urine (100 µl) were obtained at observation times and every 24 h. Animals were killed at various intervals after treatment and dissected. Two control animals were given 0.5 ml dimethyl sulfoxide.

Two tissue samples weighing about 50 mg each were taken from lung, liver, fat, testis, kidney, muscle, bone, rectum, duodenum and stomach. Each sample was digested with 800 µl methanoic acid and 200 µl hydrogen peroxide at 70°C for 1 h.

For radioactivity measurements, 5.0 ml scintillating solution were added to 100 µl digested solution and counts per minute were measured with a Packard 4460 liquid scintillometer.

Biological half-life of ^{3}H-sterigmatocystin in serum

The half-life for the distribution of radioactive compound in serum was 0.51 h and that for excretion, 43.9 h respectively (Fig. 1). By 7.3 days, 93.8% of the total had been excreted, indicating prolonged internal exposure to the compound.

Fig. 1. Distribution and extretion of 3H-sterigmatocystin in serum of rats

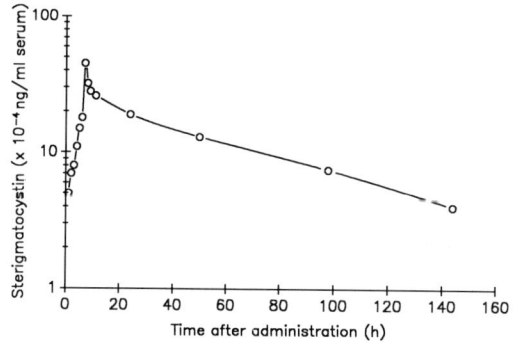

Distribution of ^{3}H-sterigmatocystin in tissues

Radioactivity was concentrated mainly in liver, stomach, kidney, duodenum and lung and to a lesser extent in fat, muscle, testis, rectum and bone (Fig. 2); these differences were significant ($p < 0.05$, rank test). Radioactivity was highest in liver and decreased slowly, indicating that metabolic detoxification of sterigmatocystin might be carried out in that organ. It is possible therefore that the hepatocarcinogenicity of sterigmatocystin is related to its high concentration in liver (Purchase et al., 1970). Similarly, the high radioactivity and persistence of 3H-sterigmatocystin in kidney may correspond to its toxic effects on that organ (Purchase et al., 1969; van der Watt et al., 1970) The initially high radioactivity in the gastrointestinal tract decreased markedly after two days, due not only to absorbtion but perhaps also to attachment to gastrointestinal mucous membrane. Although there has been no report of an association between treatment with sterigmatocystin and stomach cancer in animals, acanthomatous changes were found in stomach (Purchase et al., 1970).

Excretion of ^{3}H-sterigmatocystin

Six hours after administration, excretion of 3H-sterigmatocystin in faeces increased quickly to a total of 45.2% of the administered dose at 24 h; at that time, only 15.9% had been excreted in urine (Fig. 3). By 48, 96 and 144 h, 56.1%, 62.4% and 64.4% had been excreted in faeces and 20.1%, 21.0% and 23.8% in urine, respectively, indicating that the biliary excretion *via* the faeces is the primary route of excretion for sterigmatocystin.

References

Purchase, I.F.H. & van der Watt, J.J. (1969) Acute toxicity of sterigmatocystin to rats. *Food Cosmet. Toxical.*, **7**, 135-139

Purchase, I.F.H. & van der Watt, J.J. (1970) Carcinogenicity of sterigmatocystin. *Food Cosmet. Toxicol.*, **8**, 289-295

van der Watt, J.J. & Purchase, I.F.H. (1970) The acute toxicity of retrororsine, aflatoxin and sterigmatocystin in vervet monkeys. *Br. J. Exp. Pathol.*, **51**, 183-190

Fig. 2. Distribution of 3H-sterigmatocystin in rat tissues

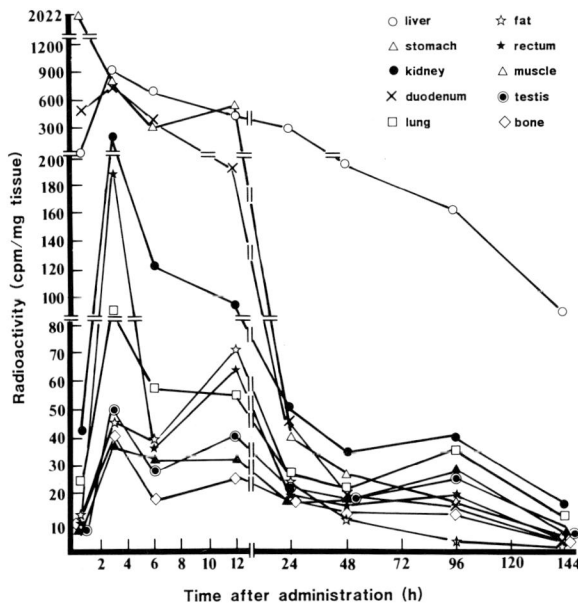

Fig. 3. Excretion of 3H-sterigmatocystin in urine (○) and (●) faeces of rats

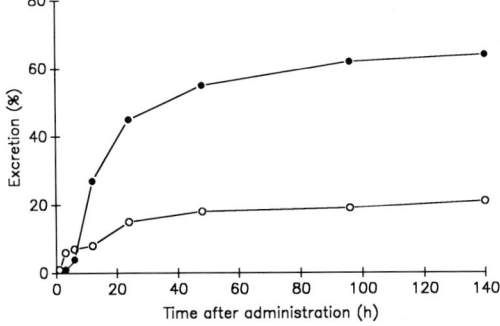

AFLATOXIN B_1-DNA BINDING AND AFLATOXIN B_1-GLUTATHIONE CONJUGATION WITH ISOLATED HEPATOCYTES FROM RATS AND HAMSTERS

L.L. Ho, E.C. Jhee[1], P. Gopalan, K. Tsuji & P.D. Lotlikar[2]

*Fels Institute for Cancer Research and Molecular Biology,
Temple University School of Medicine, Philadelphia, PA, USA*

Binding of aflatoxin B_1 (AFB_1) to DNA and AFB_1-glutathione conjugation during the metabolism of AFB_1 have been examined with freshly isolated hepatocytes from male Fischer rats and Syrian hamsters. Even though there was no significant difference in cytochrome P450 and glutathione contents, there were marked differences in the metabolism of AFB_1 (33 nM) in hepatocytes from these two species. Thus, AFB_1-DNA binding was six-fold higher in the rat than in hamster hepatocytes, whereas AFB_1-glutathione conjugation was 12-fold higher in hamster than in rat hepatocytes. The addition of 0.5 mM diethylmaleate had no significant effect in rats, whereas its presence produced a nine-fold increase in AFB_1-DNA binding with 85% inhibition of thiol conjugation in hamster hepatocytes. Styrene oxide (1 mM) produced 50% and 25-fold increases in AFB_1-DNA binding in rat and hamster hepatocytes, respectively, with corresponding decreases in thiol conjugation. Triethyltin bromide (50 μM) inhibited both processes by 50% in rat hepatocytes, whereas it produced a nine-fold increase in AFB_1-DNA binding with a concomitant decrease in thiol conjugation in hamster hepatocytes. These results suggest that glutathione S-transferases play a more significant role in modulating AFB_1-DNA binding in hamster than in rat hepatocytes.

The hepatocarcinogen aflatoxin B_1 (AFB_1) undergoes metabolic activation *via* epoxidation, and the interaction of this epoxide with DNA is believed to be responsible for initiation of carcinogenesis (Miller & Miller, 1977). The carcinogenicity of AFB_1 and its hepatic binding to DNA *in vivo* in rats (susceptible) and hamsters (resistant) do not correlate well with the microsome-mediated activation of AFB_1 in these two species (Herrold, 1969; Garner & Wright, 1975; Swenson *et al.*, 1977; Moore *et al.*, 1982; Lotlikar *et al.*, 1984). Subsequent subcellular experiments have shown that cytosolic glutathione S-transferases play an important role in modulating AFB_1-DNA binding (Raj *et al.*, 1984). In the present study, we used freshly isolated intact hepatocytes from rats and hamsters to examine both AFB_1-DNA binding and AFB_1-thiol conjugation during the metabolism of AFB_1 and the effects of diethylmaleate, styrene oxide and triethyltin bromide on these reactions.

---

[1]Present address: Department of Biochemistry, Chonbuk National University, College of Dentistry, Chonju, Republic of Korea
[2]To whom requests for reprints should be addressed

Hepatocytes were isolated by the collagenase method from male Fischer rats and Syrian golden hamsters, as described previously (Lotlikar et al., 1989). Hepatic glutathione and cytochrome P450 levels were determined by the methods of Ellman (1959) and Omura and Sato (1964), respectively.

Duplicate aliquots of freshly isolated hepatocytes (2×10^7 cells/5 ml) were incubated in the presence of modified Hank's balanced salt solution containing 2 mM $CaCl_2$, 25 mM HEPES buffer (pH 7.4), 0.5% serum albumin, 33 nM [3H]AFB_1 containing 1 µCi [3H]AFB_1 and various chemicals (as indicated) in 95% O_2–5% CO_2 at 37°C for 1 h. AFB_1, diethylmaleate, styrene oxide and triethyltin bromide were all dissolved in dimethyl sulfoxide. After incubation, cells were homogenized and nuclei were sedimented at 600 g for 10 min. Nuclear supernatants were used for extraction, separation and quantification of free hydroxy metabolites and AFB_1-glutathione conjugate; and isolation of DNA from the sedimented nuclei and quantification of AFB_1–DNA binding were performed as described previously (Lotlikar et al., 1989).

There was no significant difference between rats and hamsters in either cytochrome P450 (19.6 ± 3.3 (SE) versus 23.8 ± 1.4 nmol/10^8 cells) or glutathione contents (2.1 ± 0.2 versus 1.9 ± 0.3 µmol/10^8 cells) of isolated hepatocytes.

Effects of diethylmaleate, styrene oxide and triethyltin bromide on AFB_1 metabolism in isolated hepatocytes

There were marked differences in AFB_1 metabolism in hepatocytes from rats and hamsters (Table 1). Thus, AFB_1–DNA binding was six-fold higher in rat than in hamster hepatocytes, whereas AFB_1-glutathione conjugation was 12-fold higher in hamster than in rat hepatocytes. The ratio of AFB_1-glutathione conjugation to AFB_1–DNA binding was 60-fold higher in hamster than in rat cells, indicating a preponderance of inactivation of the AFB_1 epoxide in the former species. These data are in agreement with results of subcellular experiments which indicated that epoxidation alone cannot account for the differences in AFB_1–DNA binding in these two species (Lin et al., 1978; Lotlikar et al., 1984); cytosolic glutathione S-transferases were shown to play an important role in modulating AFB_1–DNA binding (Raj et al., 1984).

Even though incubation of hepatocytes from both species with 0.5 mM diethylmaleate for 5 min lowered glutathione levels to the same extent as in the absence of diethylmaleate (60%; data not shown), this did not have much effect on AFB_1–DNA binding in rat hepatocytes. However, its presence produced a nine-fold increase in AFB_1–DNA binding in hamsters, with a concomitant inhibition of thiol conjugation (Table 1). One explanation for these findings may be differences between the two species in the K_m for glutathione of transferases, hamster hepatocytes having a higher K_m.

Styrene oxide (1 mM) produced 50% and 25-fold increases in AFB_1–DNA binding in rat and hamster hepatocytes, respectively, with concomitant decreases in thiol conjugation (Table 1). In earlier studies, we examined the role of styrene oxide in glutathione S-transferase-dependent modulation of AFB_1 binding (Lotlikar et al., 1984; Raj et al., 1984), since other studies had indicated that epoxide hydrase was not a modulator of AFB_1–DNA binding (Gurtoo & Bejba, 1974; Lin et al., 1978; Lotlikar et al., 1984). The increased AFB_1–DNA binding and inhibition of AFB_1-thiol conjugation in the presence of styrene oxide observed in the present study indicate that the effect of the oxide is due predominantly to inhibition of glutathione S-transferases. Similar effects have been observed with styrene oxide in hepatocytes from rats pretreated with phenobarbital (Lotlikar *et al.*, 1989) or butylated hydroxyanisole (Jhee *et al.*, 1989).

Triethyltin bromide (50 µM) produced a nine-fold increase in DNA binding, with a concomitant drop in thiol conugation in hamster hepatocytes, whereas it inhibited both process in rat hepatocytes (Table 1). This compound has been shown to be an effective inhibitor of several purified rat liver glutathione S-transferases (Tipping *et al.*, 1979; Alin *et al.*, 1985), but no study has been performed with purified transferases from hamsters.

Table 1. Binding of aflatoxin B_1 (AFB_1) to DNA and conjugation to glutathione during AFB_1 metabolism in isolated hepatocytes in pmol/10^8 cells per h[a]

Addition[b]	Rat		Hamster	
	AFB_1-DNA	AFB_1-thiol	AFB_1-DNA	AFB_1-glutathione
-	5.5 ± 0.8	10.4 ± 1.2	0.9 ± 0.2*	120 ± 15*
Diethylmaleate	6.8 ± 1.7	8.0 ± 1.3	8.1 ± 1.1*	17 ± 4*
Styrene oxide	8.2 ± 1.2*	4.5 ± 0.7*	25.4 ± 3.8*	10 ± 3*
Triethyltin bromide	2.9 ± 0.5*	4.0 ± 1.1*	9.1 ± 1.3*	18 ± 6*

[a] Results are means of three separate analyses on individual livers with variations as SE.
[b] Hepatocytes were preincubated with 0.5 mM diethylmaleate for 5 min before incubation with 33 nM [^{3}H]AFB_1 or with 0.5 mM styrene oxide or 50 µM triethyltin bromide during incubation with [^{3}H]AFB_1 for 1 h.
*Data statisitically significant at $p < 0.05$ when compared with respective controls

The amounts of free hydroxylated metabolites of AFB_1 generated during AFB_1 metabolism were similar in hepatocytes from rats and hamsters: AFQ_1, 13.0 ± 2.3 (SE) *versus* 14.8 ± 1.4; AFM_1, 27.5 ± 5.2 *versus* 33.6 ± 2.5; AFP_1, 21.0 ± 1.7 *versus* 21.8 ± 2.6, for rats *versus* hamsters, respectively. The amounts of conjugated hydroxy metabolites of AFB_1 released after β-glucuronidase and sulfatase treatment were much smaller than those of free hydroxylated metabolites (data not shown).

Our results suggest that glutathione S-transferases play a more significant role in modulating AFB_1–DNA binding in hamster than in rat hepatocytes. These data are compatible with those of studies of AFB_1–DNA binding *in vivo* and of hepatocarcingenicity in these two species.

Acknowledgements

This work was supported in part by grants CA-31641 and CA-12227 and a training grant CA-09214 (P.G., a trainee on this grant) from the National Cancer Institute, Department of Health and Human Services, by grant SIG-6 from the American Cancer Society and Samuel S. Fels Fund of Philadelphia.

References

Alin, P., Jensson, H., Guthenberg, C., Danielson, U.H., Tahir, M.K. & Mannervik, B. (1985) Purification of major basic glutathione transferase isoenzymes from rat liver by use of affinity chromatography and fast protein liquid chromatofocusing. *Anal. Biochem.*, **146**, 313-320

Ellman, G.L. (1959) Tissue sulfhydryl groups. *Arch. Biochem. Biophys.*, **82**, 70-77

Garner, R.C. & Wright, C.M. (1975) Binding of [^{14}C]aflatoxin B_1 to cellular macromolecules in the rat and hamster, *Chem.-Biol. Interactions*, **11**, 123-131

Gurtoo, H.L. & Bejba, N. (1974) Hepatic microsomal mixed function oxygenase: enzyme multiplicity for the metabolism of carcinogens to DNA-binding metabolites. *Biochem. Biophys. Res. Commun.*, **61**, 735-742

Herrold, K.M. (1969) Aflatoxin induced lesions in Syrian hamsters. *Br. J. Cancer*, **23**, 655-660

Jhee, E.C., Ho, L.L., Tsuji, K., Gopalan, P. & Lotlikar, P.D. (1989) Effect of butylated hydroxyanisole pretreatment on aflatoxin B_1–DNA binding and alfatoxin B_1-glutathione conjugation in isolated hepatocytes from rats. *Cancer Res.*, **49**, 1357-1360

Lin, J.K., Kennan, K.A., Miller, E.C. & Miller, J.A. (1978) Reduced nicotinamide adenine dinucleotide phosphate-dependent formation of 2,3-dihydro-2,3-dihydroxyaflatoxin B_1 by hepatic microsomes. *Cancer Res.*, **38**, 2424-2428

Lotlikar, P.D., Jhee, E.C., Insetta, S.M. & Clearfield. M.S. (1984) Modulation of microsome-mediated aflatoxin B_1 binding to exogenous and endogenous DNA by cytosolic glutathione S-transferases in rat and hamster livers. *Carcinogenesis*, **5**, 269-276

Lotlikar, P.D., Raj, H.G., Bohm, L.S., Ho, L.L., Jhee, E.C., Tsuji, K. & Gopalan, P. (1989) A mechanism of inhibition of aflatoxin B_1–DNA binding in the liver by phenobarbital pretreatment of rats. *Cancer Res.*, **49**, 951-957

Miller, J.A. & Miller, E.C. (1977) Ultimate chemical carcinogens as reactive mutagenic electrophiles. In: Hiatt, H.H., Watson, J.D. & Winsten, J.A., eds, *Origins of Human Cancer*, Cold Spring Harbor, NY, CSH Press, pp. 605-627

Moore, M.R., Pitot, H.C., Miller, E.C. & Miller, J.A. (1982) Cholangiocellular carcinomas induced in Syrian golden hamsters administered aflatoxin B_1 in large doses. *J. Natl Cancer Inst.*, **68**, 271-278

Omura, T. & Sato, R (1964) The carbon monoxide-binding pigment of liver microsomes. I. Evidence for its hemoprotein. *J. Biol. Chem.*, **239**, 2370-2378

Raj, H.G., Clearfield, M.S. & Lotlikar, P.D. (1984) Comparative kinetic studies on aflatoxin B_1–DNA binding and aflatoxin B_1-glutathione conjugation with rat and hamster livers *in vitro*. *Carcinogenesis*, **5**, 879-884

Swenson, D.H., Lin, J.K., Miller, E.C. & Miller, J.A. (1977) Aflatoxin B_1-2,3-oxide as a probable intermediate in the covalent binding of aflatoxin B_1 and B_2 to rat liver DNA and ribosomal RNA *in vivo*. *Cancer Res.*, **37**, 172-181

Tipping, E., Ketterer, B., Christodoulides, L., Elliott, B.M., Aldridge, W.N. & Bridges, J.W. (1979) The interactions of triethyltin with rat glutathione S-transferases A, B and C. Enzyme-inhibition and equilibrium dialysis studies. *Chem.-biol. Interactions*, **24**, 317-327

RELIABILITY OF A SHORT-TERM TEST FOR HEPATO-CARCINOGENESIS INDUCED BY AFLATOXIN B_1

Y. Li, R.Q. Yan, G.Z. Qin, L.L. Qin & X.X. Duan

*Department of Pathology, Guangxi Cancer Institute,
Nanning, Guangxi, China*

The reliability of a short-term test for hepatocarcinogenesis induced by aflatoxin B_1 (AFB_1) was tested by comparing the early appearance of γ-glutamyl transpeptidase (GGT)-positive foci with the occurrence of primary liver cancer at a later stage. All rats received a basic short-term treatment with AFB_1 intraperitoneally, during which three experimental groups received Chinese green tea or 2000 or 5000 ppm butylated hydroxyanisole in the diet and a control group received basic diet. Some of the rats in each group were sacrificed at the end of the short-term procedure, and the remainder were observed up to 92 weeks. The livers of all animals were examined for GGT-positive foci or primary liver tumours. The GGT-positive foci were most numerous and largest and the incidence of liver tumours was highest in the control group. These findings suggest that GGT-positive foci are a valuable preneoplastic marker for AFB_1-induced hepatocarcinogenesis, that the short-term model is fairly reliable, and that both Chinese green tea and butylated hydroxyanisole inhibit AFB_1-induced hepatocarcinogenesis.

In order to test the reliability of γ-glutamyl transpeptidase (GGT)-positive hepatocytic foci as a preneoplastic marker for hepatocarcinogenesis induced by aflatoxin B_1 (AFB_1), rats were observed for a long period after short-term exposure to AFB_1, in a model established in our laboratory to screen the effects of inhibitors of AFB_1 (Yan et al., 1987).

The experimental procedure is outlined in Figure 1. Briefly, 200 inbred six-week-old male Wistar rats were divided into four groups: one received only the short-term test procedure; the three others received Chinese green tea or 2000 or 5000 ppm butylated hydroxyanisole (BHA) from ten days before receiving AFB_1 to day 3 after withdrawal of the carcinogen. The control group was fed basic diet only. At the end of the short-term test, 12–13 rats in each group were sacrificed, and liver tissues were stained for GGT. The number and size of GGT foci were measured with a micrometer and calculated by computer. Other rats were fed continuously with basic diet and were observed up to week 92 of the experiment, when they were killed and liver tissues examined for primary liver cancer.

Both the number and size of GGT-positive foci in the control group were larger than in the groups that received Chinese green tea, 2000 ppm BHA or 5000 ppm BHA (Table 1), and these differences were statistically significant. Likewise, the incidence of primary liver cancer was highest in controls and lower in the three experimental groups (Table 2). In addition, the control rats had the shortest survival, the earliest occurrence of liver tumours, more cancer nodules and the highest incidence of pulmonary metastasis. All of the tumours were positive for GGT, except one poorly differentiated one.

Fig. 1. Experimental procedure for induction of hepatocarcinogenesis in rats

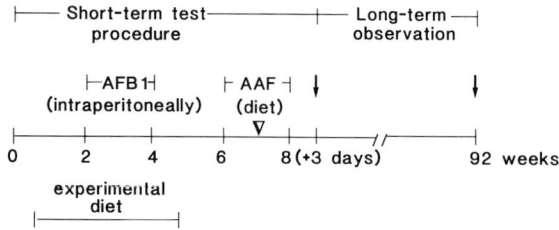

↓, sacrifice; ∇, partial hepatectomy; AFB_1, aflatoxin B_1 (400 µg/kg bw); AAF, 2-acetylaminofluorene (0.015%)

Table 1. γ-Glutamyl transpeptidase (GGT)-positive hepatocytic foci induced in rats in a short-term procedure

Treatment[a]	No. of rats	GGT-positive foci	
		No./cm	mm/cm
Control	12	10.53 ± 4.84	0.125 ± 0.162
Green tea	13	5.58 ± 3.10 **	0.081 ± 0.077 ***
2000 ppm BHA	13	2.22 ± 1.75 ***	0.059 ± 0.052 ***
5000 ppm BHA	13	1.64 ± 0.93 ***	0.059 ± 0.050 ***

[a]BHA, butylated hydroxyanisole
**, $p < 0.01$
***, $p < 0.001$

Table 2. Primary liver cancer in rats after long-term observation following induction of hepatocarcinogenesis in a short-term model

Treatment[a]	No. of rats	No. of rats with tumours	Incidence (%)	Week in which first tumour occurred	No. of tumours/rat (mean ± SD)	No. of rats with lung metastases
Control	23	4	17.39	54	1.75 ± 0.96	2
Green tea	22	2	9.10	60	1.00 ± 0	0
2000 ppm BHA	21	1	4.76	55	1	1
5000 ppm BHA	25	1	4.00	78	1	0

[a]BHA, butylated hydroxyanisole

These findings demonstrate that: (i) there is a good correlation between the induction of GGT-positive foci at an early stage and the later development of primary liver cancer in AFB_1-induced hepatocarcinogenesis, and (ii), as we reported previously (Yan et al., 1986; Qing et al., 1987; Li et al., 1987, 1989), the common natural beverage, Chinese green tea, and the food additive BHA can effectively inhibit AFB_1-induced hepatocarcinogenesis.

References

Li, Y., Yan, R.Q., Qing, G.Z., Qing, L.L. & Duan, X.X. (1987) The comparison of inhibitory effects of green tea and butylated hydroxyanisole on aflatoxin B_1 induced hepatocarcinogenesis in rats (in Chinese). *J. Guangxi Med. Coll.*, **3**, 10-14

Li, Y., Yan, R.Q., Qing, G.Z., Qing, L.L. & Duan, X.X. (1989) Time-related inhibitory effect of green tea diet on AFB_1-induced hepatocarcinogenesis in rats (in Chinese). *J. Guangxi Med. Coll.*, **1**, 10-15

Qing, G.Z., Yan, R.Q., Li, Y., Qing, L.L. & Duan, X.X. (1987) Effect of green tea and infused tea residue on aflatoxin B_1 induced liver cancer in rats. In: *Proceedings of International Tea-Quality-Human Health Symposium (Hangzhou, China)*, pp. 282-285

Yan, R.Q., Chen, Z.Y., Qing, G.Z., Li, Y. & Qing, L.L. (1986) Inhibitory effect of Chinese green tea on aflatoxin B1 (AFB1)-induced hepatocarcinogenesis in rats. *Hepatology*, **16**, 25-26

Yan, R.Q., Chen, Z.Y., Qing, G.Z., Li, Y. & Qing, L.L. (1987) Inhibitory effect of Chinese traditional herbs on aflatoxin B1 (AFB1)-induced hepatocarcinogenesis in rats. *Hepatology*, **18**, 9-10

CHEMICAL TRANSFORMATION OF HUMAN EMBRYONIC NASOPHARYNGEAL EPITHELIAL CELLS *IN VITRO*

Z.C. Chen, S.C. Pan & K.T. Yao[1]

*Cancer Research Institute, Hunan Medical University
Changsha, Hunan, China*

The association of Epstein-Barr virus (EBV) with poorly differentiated carcinoma of the nasopharynx is well known; however, certain environmental factors, such as nitrosamines, are also important for the development of this cancer (Ho, 1975). N,N'-Dinitrosopiperazine (DNPZ) can induce nasopharyngeal carcinoma in rats and increased sister chromatid exchange frequency in human embryonic nasopharyngeal epthelial (HENE) cells. We have now demonstrated the transformation by DNPZ of HENE cells, which had a prolonged life span, anchorage-independent growth, chromosomal aberrations, tumorigenicity and morphological and ultrastructural alterations. These transformed cells might derive from the columnar epithelium of the nasopharynx, as indicated by the positive histochemical reaction with CAM 5.2 antikeratin antibody. Negative results in an immunofluorescence test for EBV nuclear antigen and Southern hybridization for EBV DNA rule out the participation of this virus in the neoplastic transformation of HENE cells by DNPZ.

The role of chemical carcinogens and especially nitrosamines in the pathogenesis of nasopharyngeal carcinoma (NPC) cannot be ignored (Ho, 1975). We have successfully induced NPC in rats with N,N'-dinitrosopiperazine (DNPZ; Yao *et al.*, 1981). In this paper, we present data on the in-vitro transformation of human embryonic nasopharyngeal epithelial (HENE) cells by DNPZ.

Cultivation of normal HENE cells in vitro

Epithelial cells from human embryonic nasopharynx were cultivated according to the method developed in our laboratory (Zhu *et al.*, 1984). Untreated HENE cells were used as controls; these usually reached senescence and died after one or two passages, with an average survival time of 40 days.

Exposure of cultured cells to DNPZ

Data obtained in our laboratory (Li *et al.*, 1988) indicate that HENE cells can metabolize DNPZ directly, without the aid of liver microsomal enzymes. The optimal dose of DNPZ for transformation was determined to be 250 µg/ml. Primary and secondary cell cultures received two exposures to DNPZ at an interval of two days. Most of the exposed cells died; the small portion that survived grew into 'island-like' foci, and then into a monolayer after the eighth passage. These cells have been subcultured over 100 passages (Fig. 1).

---

[1]To whom correspondence should be sent

Fig. 1. Transformed human embryonic nasopharyngeal epithelial cells with morphological alterations. × 120

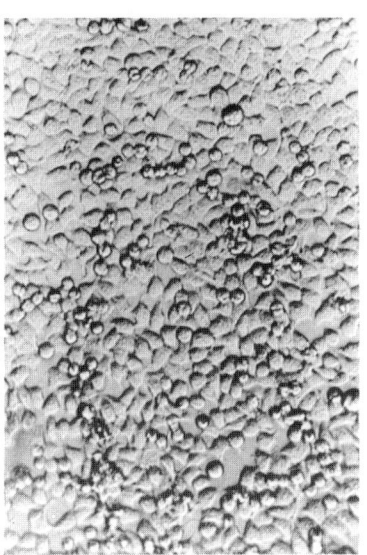

Characteristics of transformed HENE cells

The transformed cells showed polymorphism under the light microscope, being round, oval and polygonal with large nuclei and frequent multiple nucleoli so that the nucleus:cytoplasm ratio was inverted. Electron microscopy revealed that the cells were covered with numerous microvilli and their nuclear membranes were irregular and highly uneven. Desmosomes could be found between the cells. The transformed cells also showed a positive histochemical reaction to CAM 5.2 antikeratin antibody, characteristic of columnar epithelial cells.

The population doubling time of transformed HENE cells was 49 h — much shorter than that of normal HENE cells (Cao et al., 1984). The cells also showed decreased serum requirement (survived in 1% serum-supplemented medium) and anchorage-independent growth in soft agar (colony forming efficiency, 0.2%), but did not grow into multilayers.

At passage 80, the transformed cells were tested for tumorigenicity by subcutaneous inoculation into the axillar region of BALB/c-nu/nu nude mice. Solid tumours occurred at the site of inoculation seven days later; histological examination showed one tumour to be a poorly differentiated squamous-cell carcinoma (Fig. 2).

Numerical and structural chromosomal aberrations were found in the transformed HENE cells by G-banding analysis (Fig. 3). The modal number was 74, and many markers were observed. Sister chromatid exchange frequency was significantly increased (9.76 ± 0.51 (SE)) over that in normal cells (5.20 ± 0.79).

Fig. 2. Histological appearance of the tumours in nude mice after inoculation of transformed human embryonic nasopharyngeal epithelial cells — a poorly differentiated squamous-cell carcinoma. Haematoxylin and eosin; × 400

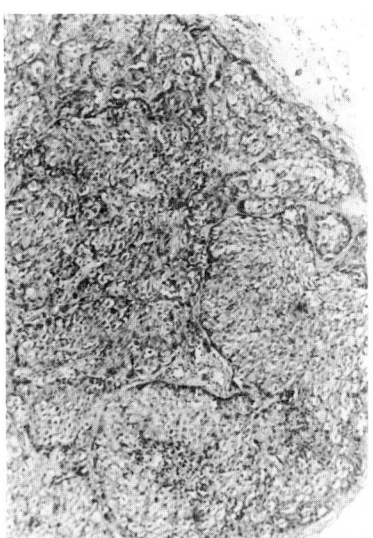

In order to rule out the participation of Epstein-Barr virus (EBV) in the chemical transformation of HENE cells, we carried out an immunofluorescence test for EBV nuclear antigen and Southern hybridization of EcoR1 digested DNA extracted from the transformed cells (at passages 30, 50 and 80) with a ^{32}P-labelled W-fragment of EBV DNA. All of the results were negative (Chen et al., 1990).

Relevance to the etiology of human NPC

EBV is closely associated with NPC, but evidence for the direct transformation of human nasopharyngeal epthelial cells by EBV is still lacking. Consumption of salted fish, a source of nitrosamines, has been suspected to be a risk factor. Carcinomas in the nasal and paranasal regions, but not NPC, were induced in Wistar rats by feeding them with steamed salted marine fish. *N*-Nitrosodimethylamine is the predominant volatile nitrosamine in salted fish. Our study is the first report of direct transformation of HENE cells with DNPZ, and the tumour that developed from inoculated transformed HENE cells in nude mice was a poorly differentiated squamous-cell carcinoma. Our findings provide a clue to the environmental factors, which vary from place to place, involved in the pathogenesis of NPC, and indicate the need to study the relationship between EBV and chemical carcinogens in inducing this cancer.

Fig. 3. Karyotype of transformed human embryonic nasopharyngeal epithelial cells. Giemsa stain; × 1000

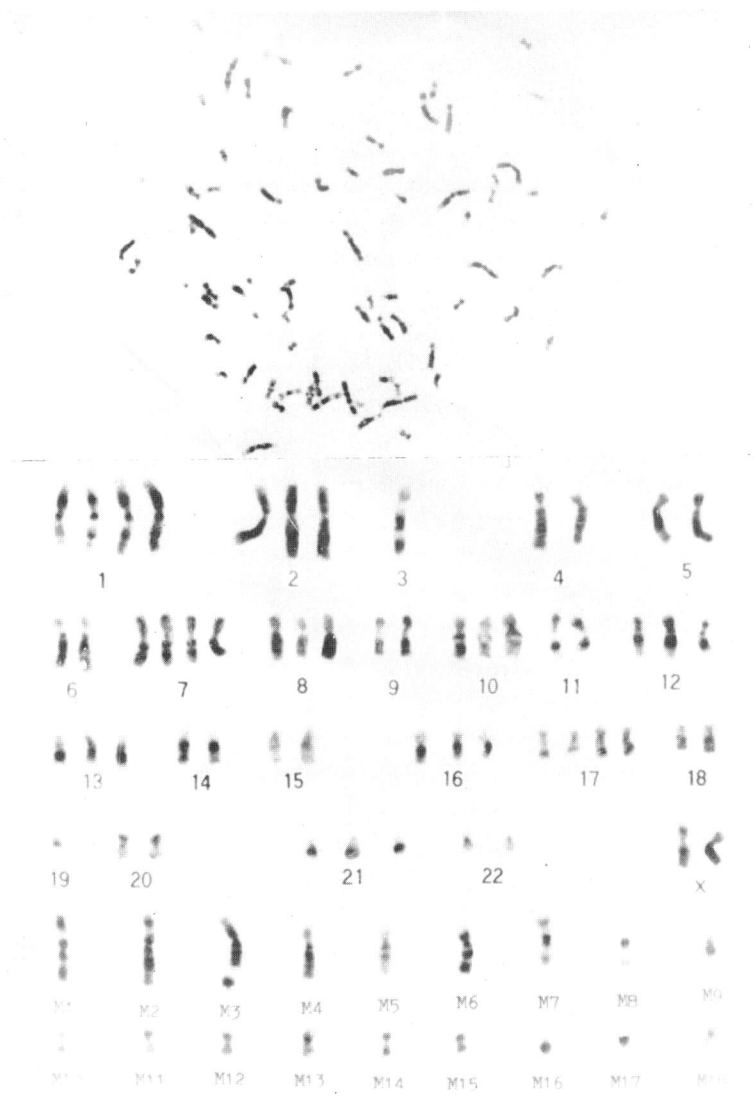

Acknowledgements

We are sincerely grateful to Professor B.C. Feng for supplying us with antikeratin antibody. We thank Dr F.X. Wang for aid in electron microscopy, Mr H.C. Zhu and Mr Z.H. Chen for expert technical assistance, and Ms S. Yu for analysing the karyotypes of the transformed cells. This work was supported by a grant (65)36-2-7 awarded by the National Committee of Science and Technology.

References

Cao, Y., Chen, Z.C. & Yao, K.T. (1984) Biological characteristic observation of explant culture of human embryonic nasopharyngeal epithelium. V. The growth kinetics (in Chinese). *Bull. Hunan Med. Coll.*, **9**, 329-334

Chen, Z.C., Pan, S.C. & Yao, K.T. (1990) Study of histological characteristics of DNPZ-transformed HENE cells (submitted)

Ho, H.C. (1975) Epidemiology of nasopharyngeal carcinoma. *J. R. Coll. Surg.*, **20**, 223-235

Li, Y.P., Yao, K.T., Hu, Y.Q. & Pan, S.C. (1988) Dinitrosopiperazine-induced sister chromatid exchanges in human nasopharyngeal epithelial cells (in Chinese). *Aizheng*, **7**, 167-170

Yao, K.T., Pan, S.C., Huang, J.L. & Wen, D.S. (1981) Further investigation of experimental induction of nasopharyngeal carcinoma in rats by dinitrosopiperazine (in Chinese). *Bull. Hunan Med. Coll.*, **6**, 1-6

Zhu, H.C., Cao, Y., Luo, M.Q. & Yao, K.T. (1984) Biological characteristic observation of explant culture of human embryonic nasopharyngeal epithelium. I. The culture of human embryonic nasopharyngeal epithelium in vitro (in Chinese). *Bull. Hunan Med. Coll.*, **9**, 105-108

Relevance to Human Cancer of *N*-Nitroso Compounds,
Tobacco Smoke and Mycotoxins.
Ed. I.K. O'Neill, J. Chen and H. Bartsch
Lyon, International Agency for Research on Cancer
© IARC, 1991

CONTROL OVER THE SEQUENCE SPECIFICITY OF DNA ALKYLATION: SYNTHESES AND REACTIONS WITH ^{32}P-END-LABELLED DNA OF *N*-ALKYL-*N*-NITROSOUREAS LINKED TO MINOR GROOVE BINDING LEXITROPSINS

B. Gold, K.M. Church, R.L. Wurdeman, Y. Zhang & F.X. Chen

*Eppley Institute for Research in Cancer and Allied Diseases
and Department of Pharmaceutical Sciences, University of Nebraska
Medical Center, Omaha, NE, USA*

The syntheses of *N*-2-chloroethyl-*N*-nitrosoureas (Cl-ENU) that are covalently linked to a series of minor groove binding lexitropsins related to distamycin A are reported. The lexitropsins of 2-Cl-ENU show a sequence specificity for alkylating an adenine toward the ends of its DNA affinity binding domains. The reaction of DNA with 1-(2-chloroethyl)-3-cyclohexyl-1-nitrosourea does not yield these products. Therefore, the linking of the 2-Cl-ENU to the minor groove binder qualitatively and quantitatively alters the DNA observed.

N-2-Chloroethyl-*N*'-alkyl-*N*-nitrosoureas (Cl-ENU) are clinically useful antineoplastic agents employed in the treatment of human brain malignancies. The major site for DNA adduction by Cl-ENU is at *N*7-guanine, with modifications also found at O^6-guanine, *N*3-cytosine, *N*3-thymidine and the phosphate backbone. To gain insight into the factors responsible for DNA modification and to alter and/or increase the overall yield of DNA adducts, a series of compounds was synthesized in which a Cl-ENU moiety is bridged to a minor groove recognizing lexitropsins (lex).

Syntheses

The nitrosoureas linked to lex dipeptides (see Fig. 1) were prepared using methods based on work by Lown and Krowicki (1985). The nitrosourea was attached in the final stage of the synthesis using acyl transfer chemistry (Martinez *et al.*, 1982).

DNA footprinting

Because of the hydrolytic instability of Cl-ENU (t $\sim$ 25 min), it was not feasible to determine their binding specificity; however, the *N*-methyl- and *N*-ethylnitrosourea analogues are sufficiently stable and were used in DNA footprinting studies with methidiumpropyl EDTA Fe(II) (Van Dyke & Dervan, 1983). The recognition sites of the lex are at 5'-T$_{222}$TTAA, A$_{205}$TATTAAA and T$_{186}$TTAT (Fig. 2). The nitrosourea moiety has little qualitative effect on the sites protected from methidiumpropyl Fe(II)-mediated strand scission, although, on the basis of the quantity of compound required to protect the DNA, the nitrosourea functionality diminishes the affinity binding. As with distamycin A, all compounds showed specificity for A-T-rich regions.

Fig. 1. Structures of *N*-2-chloroethyl-*N'*-alkyl-*N*-nitrosoureas (Cl-ENU) bridged to lexitropsins (lex)

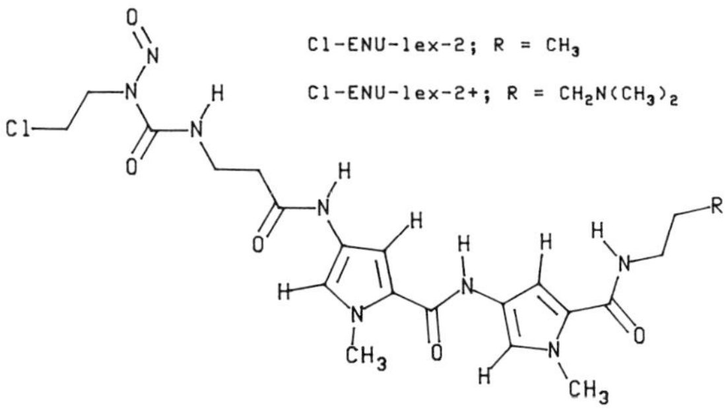

Cl-ENU-lex-2; R = CH_3
Cl-ENU-lex-2+; R = $CH_2N(CH_3)_2$

Fig. 2. Sequence of 85 base-pair restriction fragment and location of footprinted affinity binding sites (underlined regions)

```
      230       220       210       200       190       180       170       160
5'-ATAACCATGTTTAACTAACCATGCACATATTAAAGGCCATTCTTCTTTATCAACCAAACCAAAGTCTCCTGG

3'-TATTGGTACAAATTGATTGGTACGTGTATAATTTCCGGTAAGAAGAAATAGTTGGTTTGGTTTCAGAGGACC
```

Reactions of Cl-ENU-lex with ^{32}P-end-labelled DNA fragments

The restriction fragment, prepared (Maxam & Gilbert, 1980) from a plasmid containing the promotor region for the canine parvovirus coat protein (Rhode, 1985), and calf thymus DNA (~ 100 µM) were dissolved in pH 8.0 buffer containing the desired concentration of NaCl or cationic DNA affinity binder. This DNA solution was incubated with the nitrosourea at 37°C for 2 h. The reactions were terminated and the DNA washed with ethanol and dried *in vacuo*. The DNA was heated at 90°C for 15 min to depurinate/depyrimidinate any thermally labile sites and then treated with piperidine to convert these abasic sites into single-strand breaks. The reacted DNA was suspended in loading buffer and denatured at 90°C for 2 min and then cooled in ice. The DNA cleavage fragments were analysed on a 12% polyacrylamide (7.8 M urea) denaturing gel, which was run at 65 W. The standard Maxam–Gilbert G and/or G + A reaction lanes were included as sequence markers.

The cleavage of DNA by 1-(2-chloroethyl)-3-cyclohexyl-1-nitrosourea (CCNU), Cl-ENU-lex-2 and Cl-ENU-lex-2+, after sequential neutral thermal hydrolysis and piperidine-mediated cleavage at apurinic/apyrimidinic sites, is compared in Figure 3.

CCNU shows an alkylation pattern restricted to G sites (Hartley et al., 1986); both the cationic and neutral derivatives of Cl-ENU-lex-2 also induced fragmentations similar to CCNU at G residues, but additional bands were observed near Cl-ENU-lex DNA affinity binding sites. These cleavage sites are not observed using G-lane chemistry. The co-migration of all of the bands generated by Cl-ENU-lex with the Maxam–Gilbert 5′- and 3′-phosphate markers suggests that the nitrosoureas also generate phosphate terminuses. In the TTTAA sequence, the internal A is a strong cleavage site. Weak cleavage at terminal A is seen only at the highest concentration. At the A-T-rich stretch between T_{203} and A_{195}, the most intense band is at the 5′-A in the A_3 run, with a gradual decrease in intensity proceeding from the central to the 3′-A. There is an additional cleavage at A_{179} in the TTTAT sequence. All of these cleavages are qualitatively inhibited by distamycin A and significantly reduced with 100 mM NaCl.

Fig. 3. DNA alkylation: dose-response[a]

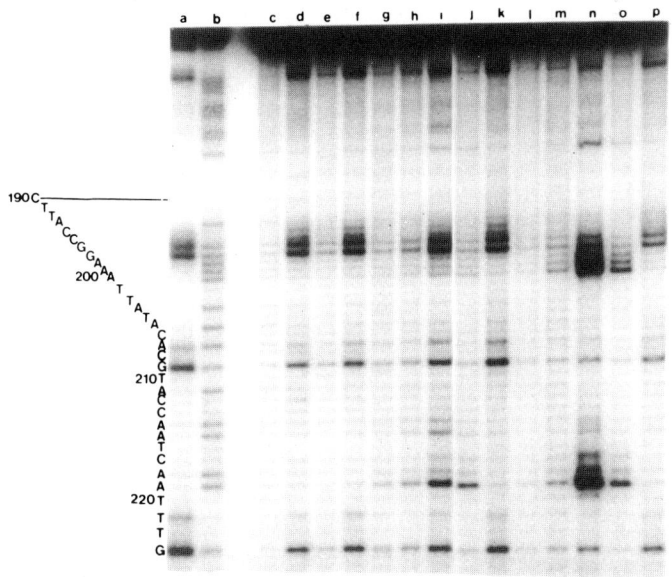

[a] 86-base-pair fragment (Crick strand) using neutral thermal treatment followed by piperidine: lane A: Maxam–Gilbert G; lane B, Maxam–Gilbert G + A; lane C, control; lane D, 500 µM 1-(2-chloroethyl)-3-cyclohexyl-1-nitrosourea (CCNU); lane E, 500 µM CCNU + 100 mM NaCl; lane F, 500 µM CCNU + 100 µM distamycin A; lanes G–I, 50, 100, 500 µM Cl-ENU-lex-2+, respectively; lane J, 500 µM Cl-ENU-lex-2+ + 100 mM NaCl; lane K, 500 µM Cl-ENU-lex-2+ + 100 µM distamycin A; lanes L–N, 50, 100, 500 µM Cl-ENU-lex-2, respectively; lane O, 500 µM Cl-ENU-lex-2 + 100 mM NaCl; lane P, 500 µM Cl-ENU-lex-2 + 100 µM distamycin A

Acknowledgements

The parvovirus clone was made available by S. Rhode. We thank P. Dervan for the sample of methidiumpropyl EDTA used in the DNA footprinting experiments. Supported by NIH grant CA29088 and NCI Center grant CA36727 and ACS Core grant SIG-16.

References

Hartley, J.A., Gibson, N.W., Kohn, K.W. & Mattes, W.B. (1986) DNA sequence selectivity of guanine-N7 alkylation by three antitumor chloroethylating agents. *Cancer Res.*, **46**, 1943-1947

Lown, J.W. & Krowicki, K. (1985) Efficient total syntheses of the oligopeptide antibiotics netropsin and distamycin. *J. Org. Chem.*, **50**, 3774-3779

Martinez, J., Oiry, J., Imbach, J.L. & Winternitz, F. (1982) Activated N-nitrosocarbamates for regioselective synthesis of N-nitrosoureas. *J. Med. Chem.*, **25**, 178-182

Rhode, S.L., III (1985) Nucleotide sequence of the coat protein gene of canine parvovirus. *J. Virol.*, **54**, 630-633

Van Dyke, M.W. & Dervan, B.P. (1983) Chromomycin, mithramycin, and olivomycin binding sites on heterogeneous deoxyribonucleic acid. Footprinting with (methidiumpropyl-EDTA)iron(II). *Biochemistry*, **22**, 2373-2377

Relevance to Human Cancer of N-Nitroso Compounds,
Tobacco Smoke and Mycotoxins.
Ed. I.K. O'Neill, J. Chen and H. Bartsch
Lyon, International Agency for Research on Cancer
© IARC, 1991

NITROTYROSINE AS A NEW MARKER FOR ENDOGENOUS NITROSATION AND NITRATION

H. Ohshima, I. Brouet, M. Friesen & H. Bartsch

International Agency for Research on Cancer, Lyon, France

A sensitive and selective method has been developed for analysing 3-nitrotyrosine (NTTYR), an exposure marker for exogenous and endogenous nitrosating or nitrating agents, in tissue and blood proteins by gas chromatography–thermal energy analysis. Using this method, a number of kinetic studies were carried out. Free and protein-bound tyrosine were reacted easily to yield NTTYR. The method was also applied to the study of NTTYR formation *in vivo*; a dose-dependent increase in NTTYR was seen in both plasma proteins and haemoglobin obtained from rats 24 h after intraperitoneal injection of various doses (0.5–2.5 μmol/rat) of tetranitromethane. Major urinary metabolites of NTTYR, given orally to rats, were isolated and identified as 3-nitro-4-hydroxyphenylacetic acid (NHPA) and 3-nitro-4-hydroxyphenyllactic acid. About 44% and 5% of the oral dose of NTTYR (100 μg/rat), respectively, was excreted as these metabolites. Some human urine samples were analysed for NHPA by gas chromatography–thermal energy analysis after ethyl acetate extraction and high-performance liquid chromatography purification; 2.8 ± 2.3 (mean ± SD; n = 11) μg/24 h, ranging from 0–7.9 μg/24 h, were detected (detection limit, 0.2 μg/l). In conclusion, NTTYR in proteins or its metabolites in urine could be readily analysed by gas chromatography–thermal energy analysis as a new, additional marker for endogenous nitrosation and nitration.

In order to study nitrosation by nitrogen oxides in the lungs and by activated macrophages in inflamed tissues, we are developing a sensitive, specific method for monitoring nitrosation *in situ*. We have chosen 3-nitrotyrosine (NTTYR) in protein as an exposure marker, because it has been reported to be readily formed in peptides and proteins by reactions with nitrite or nitrogen dioxide generated by radiolysis (Knowles *et al.*, 1974; Cheng & Lin, 1979; Prutz *et al.*, 1985; Natake & Ueda, 1986). We describe here a new method for the analysis of NTTYR and a number of kinetic studies on its formation. This novel monitoring method is now being validated in studies in animals *in vitro* and *in vivo*.

Analysis of NTTYR by a gas chromatograph coupled with a thermal energy analyser

In order to analyse NTTYR by gas chromatography (GC), we have examined various derivatization methods that have been applied to GC analyses of amino acids.

tert-Butyldimethylsilyl derivatization is the most suitable for NTTYR analysis, because the derivative is easily formed and stable. Derivatization with fluorinated anhydrides, such as heptafluorobutyric acid anhydride and pentafluorobenzoyl bromide, resulted in denitration, and no peak corresponding to NTTYR could be detected by GC. Thus, conditions for the *tert*-butyldimethylsilyl derivatization and GC-thermal energy analysis (TEA) were optimized as described in the footnote to Table 1. A TEA model 543 was used as detector, with a pyrolysing temperature of 700°C. Under these conditions, nitroaromatic compounds could be detected selectively (Lafleur & Mills, 1981), the detection limit for NTTYR being as low as 0.5 ng/injection.

Kinetics of NTTYR formation

Using this new method, the optimal pH for NTTYR formation from tyrosine and nitrite was found to be around 2.5 (Fig. 1). When 2.5 mmol/l each of tyrosine and nitrite were reacted at pH 2.5 and 37°C for 1 h, the yield of NTTYR was 53 µmol/l (2.1%), which was about 50 times greater than that of *N*-nitrosoproline (NPRO) formed from proline and nitrite under identical conditions. NTTYR formation was proportional to both the tyrosine and nitrite concentrations (Fig. 2). Ascorbic acid inhibited NTTYR formation; but potassium thiocyanate, a well-known catalyst for nitrosation of secondary amines, did not catalyse the reaction.

Figure 1. pH profiles for formation of 3-nitrotyrosine (NTTYR) and *N*-nitrosoproline (NPRO)[a]

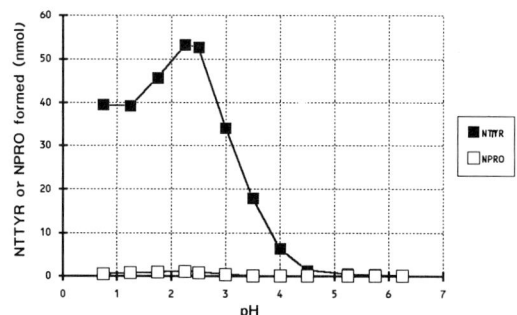

[a] Experiments were carried out at 37°C in 100 mmol/l aqueous citrate-citric acid buffer in a pH range of 1.0–6.0. The reaction was initiated by adding a nitrite solution to the buffer solution containing an amino acid (L-tyrosine or L-proline) to give a final volume of 2 ml; the final concentration of nitrite and amino acid was 2.5 mmol/l. The reaction was stopped after 1 h by adding 1 ml of 20% ammonium sulfamate in 3.6N H_2SO_4. NTTYR was purified on a C_{18} Sep-pak cartridge and analysed by gas chromatography–thermal energy analysis as described in the footnote to Table 1.

NTTYR was also easily formed when protein (bovine serum albumin or gelatine) was incubated with nitrite under both neutral and acidic conditions (Table 1). For comparison, formation of NPRO in nitrosated gelatine was determined after enzymatic hydrolysis of protein, followed by GC–TEA analysis (Dunn & Stich, 1984). About 100 times more NTTYR than NPRO was formed under acidic conditions, indicating that tyrosine in protein reacted easily with nitrite to form NTTYR, but only *N*-terminal proline residues in protein were nitrosated to form NPRO. NTTYR was also easily formed in proteins after reaction in vitro with tetranitromethane (TNM), a well-known nitrosating and nitrating agent (Riordan et al., 1966) and a lung carcinogen in rats and mice (Stowers et al., 1987).

NTTYR formation in blood proteins

The same methods were used to study NTTYR formation in blood proteins. Rat blood (5 ml) was incubated in vitro with 0.1 mmol/l nitrite or TNM at 37°C for 3 h. The plasma proteins were precipitated with $(NH_4)_2SO_4$ (80% saturation), dialysed against distilled water and hydrolysed in 6N HCl at 110°C for 24 h. Globin was prepared from haemoglobin according to Carmella and Hecht (1987) and hydrolysed in 6 N HCl, as above. NTTYR in these hydrolysates was further purified on a C_{18} Sep-pak cartridge and derivatized for GC–TEA. Under these conditions, appreciable amounts (19.5 and 7.9 nmol/100 mg protein) of NTTYR were detected in plasma proteins of rat blood incubated with nitrite and TNM, respectively; NTTYR was not found in globin. The plasma proteins were also hydrolysed enzymatically with proteinase and analysed for the protein-bound form of NPRO, but none could be detected.

Figure 2. Effects of concentrations of tyrosine (A) and nitrite (B) on formation of 3-nitrotyrosine (NTTYR)[a]

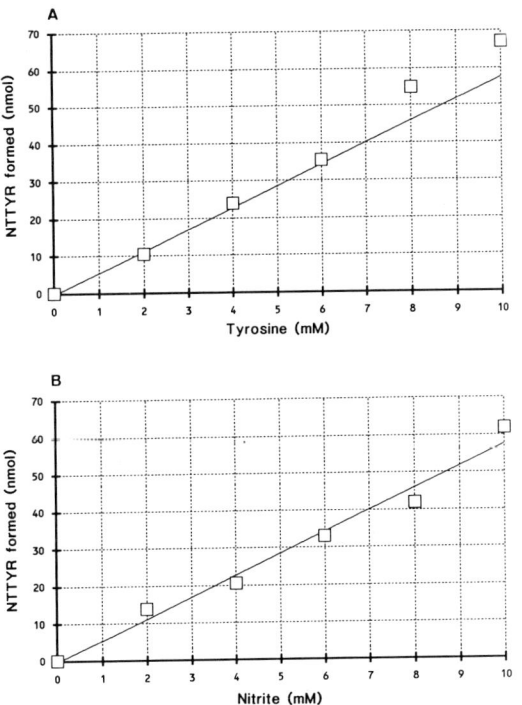

[a]Reactions were carried out at 37°C for 1 h in 100 mmol/l citrate buffer, pH 3.5, containing (A) tyrosine (0–10 mmol/l) and $NaNO_2$ (constant at 5 mmol/l) or (B) tyrosine (constant at 5 mmol/l) and nitrite (0–10 mmol/l).

On the basis of these results, experiments in animals were carried out. Male BDVI rats (150 g) received intraperitoneal injections of various doses of TNM (0–5 µmol/rat) in 1 ml of 40% ethanol. Blood obtained from rats 24 h later was analysed for NTTYR as described above. The concentration of NTTYR increased in a dose-dependent fashion in both plasma proteins and globin (Figure 3).

Table 1. Formation of 3-nitrotyrosine (NTTYR) and N-nitrosoproline (NPRO) in proteins incubated with $NaNO_2$ at pH 7.2 and 3.5[a]

pH	Time (h)	$NaNO_2$ (mM)	NTTYR/100 mg protein (nmol)		NPRO/100 mg protein (nmol) (gelatine)
			BSA	Gelatine	
7.2	24	0.0	0.38	0.04	0.16
7.2	24	0.1	0.19	0.04	0.44
7.2	24	1.0	2.19	0.19	0.54
3.5	24	0.0	0.43	0.06	0.18
3.5	2	0.1	16.2	45	0.70
3.5	24	0.1	505	818	3.60
3.5	2	1.0	–	224	3.50
3.5	24	1.0	2150	848	8.31

[a] 20 ml of 0.5% bovine serum albumin (BSA) or gelatine, the pH of which had been adjusted to 7.2 and 3.5 with NaOH or acetic acid, were incubated at 37°C with $NaNO_2$ at a final concentration of either 0.1 mmol/l or 1.0 mmol/l. The reaction was stopped by adding 5 ml of 25% trichloroacetic acid solution containing 10% $(NH_4)_2SO_4$. Proteins were further precipitated and washed with ethanol:ether mixture (1:1). About 50 mg protein were then hydrolysed *in vacuo* in 2 ml of 6 N HCl at 110°C for 24 h. The hydrolysate, to which ^{14}C-NTTYR (~ 5000 dpm) had been added as an internal standard, was applied to a C_{18} Sep-pak cartridge. The Sep-pak was first washed with 5 ml distilled water, and NTTYR was then eluted with 3 ml methanol, which was evaporated to dryness *in vacuo*. To the residue, 20 µl pyridine and 80 µl N-methyl-N-(*tert*-butyldimethylsilyl)trifluoroacetamide (Regis, Morton Grove, IL, USA) were added and derivatized at 70°C for 20 min. The derivatized sample (10 µl) was injected onto a gas chromatograph equipped with a 2 m × 3 mm i.d. glass column containing 3% OV-17 on Chromosorb W (80–100 mesh). The temperatures of the column oven and injection port were 250 and 300°C, respectively. The carrier gas, argon, was used at a flow rate of 40 ml/min. A Thermal Energy Analyzer 543 was used as detector, with pyrolysing temperature at 700°C.

Identification of major urinary metabolites of NTTYR

Analysis of urine samples from rats that had received NTTYR by gavage showed the presence of three additional peaks on GC–TEA chromatograms after formation of *tert*-butyldimethylsilyl derivatives. Two of the metabolites were isolated and identified as 3-nitro-4-hydroxyphenylacetic acid (NHPA) and 3-nitro-4-hydroxyphenyllactic acid (NHPL) (Figure 4). Identification was based on identical chromatographic and mass spectral data for the purified urine samples and the synthesized authentic compounds. When 100 µg NTTYR in 1 ml of 0.9% saline were given orally to five male BD VI rats (150 ± 10 g), 43.7 ± 4.4 and 5.3 ± 1.2% of the dose were excreted as NHPA and NHPL, respectively, and 2.2 ± 0.7% as an unidentified metabolite (calculated as NHPA).

Figure 3. Presence of 3-nitrotyrosine (NTTYR) in plasma proteins and globin 24 h after intraperitoneal treatment of rats with various doses of tetranitromethane (TNM)[a]

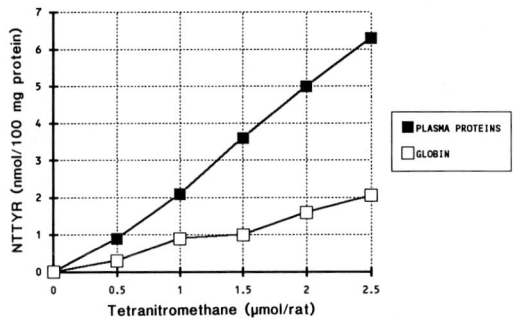

[a]Male BD VI rats (~ 150 g) were given intraperitoneal injections of 1 ml 40% ethanol containing TNM (0–5 μmol). Blood was obtained from ether-anaesthetized rats 24 h after injection of TNM, and plasma proteins and globin were prepared and analysed as described in the text and in the footnote to Table 1.

Figure 4. Structures of 3-nitrotyrosine and its major urinary metabolites

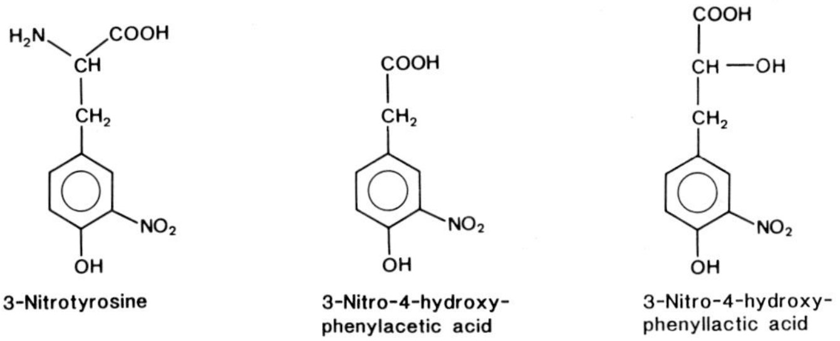

3-Nitrotyrosine

3-Nitro-4-hydroxy-phenylacetic acid

3-Nitro-4-hydroxy-phenyllactic acid

Some human urine samples were also analysed for NHPA. Urine samples (15 ml) to which 14C-NHPA had been added as an internal standard were acidified with 1 ml of 20% ammonium sulfamate in 3.6N Na_2SO_4 and extracted three times with 30 ml ethyl acetate. The concentrated ethyl acetate extracts were then purified by high-performance liquid chromatography using a semipreparative silica gel column with a dichloromethane–methanol gradient. The fraction containing NHPA was derivatized to form the *tert*-butyldimethylsilyl derivative and analysed by GC–TEA. In this preliminary study, nonsmokers (n = 5) excreted from a nondetectable level to 5.8 μg/day (mean ± SD, 2.7 ± 2.1) and smokers (n = 6) from nondetectable to 7.9 μg/day (2.9 ± 2.7). Further studies will be needed to investigate the origin of NHPA in human urine.

During these analyses, human urine was found to contain many TEA-responsive compounds which could be detected by GC–TEA after derivatization, suggesting that they may be *C*-nitro compounds. Some of the major compounds present in human urine are now being isolated and their structures characterized.

Conclusion

Both free and protein-bound tyrosine have been shown to react very easily with nitrosating and nitrating agents to form NTTYR *in vitro* and *in vivo*. Thus, NTTYR in proteins and its metabolites in urine could be analysed as new or additional markers for endogenous nitrosation and nitration. However, further studies are needed to investigate whether NTTYR can be formed *in situ* by nitrogen oxides in the lung or by activated macrophages in inflamed tissues. For these purposes, rabbit antisera against NTTYR–protein conjugate have been raised using the method of Helman and Givol (1971). These are now being used for the immunohistochemical detection of NTTYR-containing proteins as a marker for nitrosation *in situ*.

Acknowledgements

We wish to thank Mrs E. Bayle for secretarial work and Mr J.-C. Béréziat for help with the animal experiments.

References

Carmella, S.G. & Hecht, S.S. (1987) Formation of hemoglobin adducts upon treatment of F 344 rats with the tobacco specific nitrosamines 4-(methylnitrosamino)-1-(3-pyridyl)-1-butanone and N'-nitrosonornicotine. *Cancer Res.*, 47, 2626-2630

Cheng, C.-C. & Lin, J.-K. (1979) Estimation of nitrotyrosine derivatives in nitrosated foods. *J. Formosan Med. Assoc.*, 78, 1027-1036

Dunn, B.P. & Stich, H.F. (1984) Determination of free and protein-bound *N*-nitrosoproline in nitrite-cured meat products. *Food Chem. Toxicol.*, 22, 609-613

Helman, M. & Givol, D. (1971) Isolation of nitrotyrosine-containing peptides by using an insoluble-antibody column. *Biochem. J.*, 125, 971-974

Knowles, M.E., McWeeny, D.J., Couchman, L. & Thorogood, M. (1974) Interaction of nitrite with proteins at gastric pH. *Nature*, 247, 288-289

Lafleur, A.L. & Mills, K.M. (1981) Trace level determination of selected nitroaromatic compounds by gas chromatography with pyrolysis/chemiluminescent detection. *Anal. Chem.*, 53, 1202-1205

Natake, M. & Ueda, M. (1986) Changes in food proteins reacted with nitrite at gastric pH. *Nutr. Cancer*, 8, 41-45

Prutz, W.A., Monig, H., Butler, J. & Land, E.J. (1985) Reactions of nitrogen dioxide in aqueous model systems: oxidation of tyrosine units in peptides and proteins. *Arch. Biochem. Biophys.*, 243, 125-134

Riordan J.F., Sokolovsky, M. & Vallee, B.L. (1966) Tetranitromethane. A reagent for the nitration of tyrosine and tyrosyl residues of proteins. *J. Am. Chem. Soc.*, 88, 4104-4105

Stowers, S.J., Glover, P.L., Reynolds, S.H., Boone, L.R., Maronpot, R.R. & Anderson, M.W. (1987) Activation of the K-*ras* proto-oncogene in lung tumors from rats and mice chronically exposed to tetranitromethane. *Cancer Res.*, 47, 3212-3219

TOBACCO-RELATED CANCER

LUNG CANCER AND THE CHANGING CIGARETTE

D. Hoffmann, I. Hoffmann & E.L. Wynder

*Naylor Dana Institute for Disease Prevention,
American Health Foundation, Valhalla, NY, USA*

Epidemiological studies have shown that the long-term smoker of low-yield cigarettes has a 20-50% lower risk of lung cancer than the smoker of high-yield cigarettes. This risk reduction is attributed to changes in the make-up of cigarettes and especially to the introduction of filter tips. Other changes relate to the use of tobaccos that produce lower smoke yields, including reconstituted and expanded tobaccos, as well as utilization of porous cigarette paper and perforated filter tips. New developments in the make-up of commercial cigarettes must be monitored in order to prevent unfavourable introductions. Although a smoke-free society should be the major public health goal, recent consumer statistics do not support this goal. Thus, a strong social case is made for further developments in the low-yield cigarette.

In 1961, Wynder and Day offered three postulates for the causation of noncommunicable diseases, including cancer:

(1) The greater and the more prolonged the exposure to the factor, the greater the risk of a population involved.

(2) The epidemiological pattern should be consistent with the distribution of the factor.

(3) Removal or reduction of the risk factors for a given population group should be followed by a reduction in the incidence of disease.

More than 100 epidemiological studies from various countries have demonstrated a dose-response relationship between the number of cigarettes smoked and the risk for cancer of the lung (IARC, 1986; US Department of Health and Human Services, 1989). Laboratory studies have substantiated these findings by documenting the dose-response relationship between exposure to cigarette smoke and tumours of the upper respiratory tract in hamsters (Dontenwill, 1974) and that between application of smoke condensate to the skin of mice and rabbits and tumour yield at the site of application (Wynder & Hoffmann, 1967). These findings clearly satisfy postulate (1) for a causative association between cigarette smoking and lung cancer.

Postulate (2) is sustained by the correlation between age at onset of cigarette smoking and depth of inhalation with risk for lung cancer. Further support is gleaned from the observation that, by comparison to cigarette smokers, primary cigar and pipe smokers have a lower risk for cancer of the lung but the same risk for cancer of the oral cavity (US Department of Health and Human Services, 1989).

The observed decrease in lung cancer risk upon cessation of cigarette smoking supports postulate (3) for cancer causation. More importantly, it also has major public health implications. Independent of the length of smoking history, the risk for tobacco-related diseases diminishes further as the period of abstention from smoking increases (Table 1). Thus, the demand for a smoke-free society by the year 2000 (Koop, 1986) deserves the full support of the medical and scientific community. Unfortunately, the recent statistics on tobacco use do not support this goal. Between 1971-75 and 1979-81, the percentage increase in cigarette consumption in all parts of the world, except Europe, exceeded the increase in population (Fig. 1).

Figure 1. Change in apparent cigarette consumption and adult population size, by region; 1971-75 to 1979-81[a]

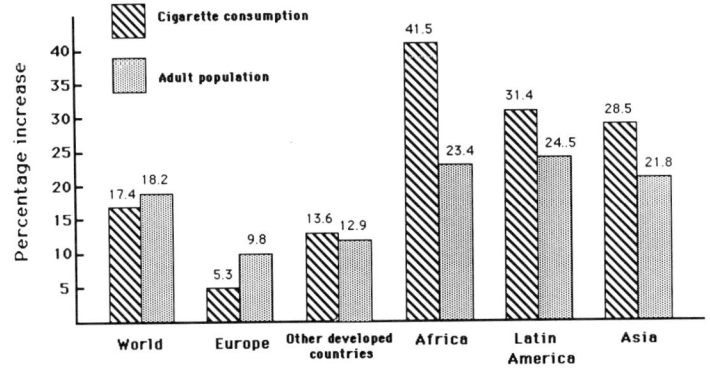

[a] From World Health Organization (1986)

These data on cigarette consumption reinforce the need for health education about the harmful effects of cigarette smoking and for the widest availability of smoking cessation programmes. However, these figures also show that a strong case can be made for less toxic cigarettes, since far too many people seemingly cannot or will not give up the smoking habit.

The changing cigarette — epidemiological observations

Since the early epidemiological studies on the association of cigarette smoking and lung cancer clearly showed a dose-response relationship, product modification was seen as one possible approach to reducing exposure (Wynder & Hoffmann, 1962). Changes in the design of cigarettes, most notably use of filter tips but also alterations of the composition of cigarettes, were implemented in several countries in the late 1950s and early 1960s. Since then, the yields of smoke condensate ('tar') and of nicotine in the smoke stream have fallen drastically. For example, between 1960 and 1984, sales-weighted average 'tar' delivery of cigarettes in the USA, the UK and the Federal Republic of Germany decreased from 26, 31 and 25 mg to 14.2, 13.7 and 12.8 mg, respectively, that is, a reduction of 45-50%. During the same period, nicotine yields decreased by about 38-44% (Adlkofer et al., 1988). As stated above, the introduction of filter tips was the first and most evident change in cigarette design, and more than 50% of smokers in North America and Europe were using filter cigarettes by 1965 (Weber, 1972). Most of the epidemiological assessments of lung cancer risk have therefore involved comparisons of statistics on smokers of plain cigarettes with those on long-term smokers of filter cigarettes (> 10 years).

The relative risks for lung cancer of smokers who switched to filter cigarettes in comparison with the risk for lung cancer of smokers of plain cigarettes are listed in Table 2. The reduction in risk in these studies varies from 20 to 50% depending on the type of cigarette smoked for the longest time, thus also supporting postulate (3). This observation appears to pertain primarily to a reduction in the Kreyberg I lung cancer type (including squamous-cell, large-cell and oat-cell lung carcinoma); reduction in Kreyberg II type of lung cancer appears to be limited to male smokers (Wynder & Kabat, 1988).

Table 1. Lung cancer mortality ratios in ex-cigarette smokers, by number of years since stopping smoking[a]

Study population	Years since stopping smoking	Mortality ratio	
British physicians	1-4	16.0	
	5-9	5.9	
	10-14	5.3	
	≥15	2.0	
	Current smokers	14.0	
US veterans	1-4	18.83	
	5-9	7.73	
	10-14	4.71	
	15-19	4.81	
	≥ 20	2.10	
	Current smokers	11.28	
Japanese males	1-4	4.65	
	5-9	2.50	
	≥ 10	1.35	
	Current smokers	3.79	
		Number of cigarettes smoked per day	
		1-19	≥ 20
US males aged 50-69	< 1	7.20	29.13
	1-4	4.60	12.00
	5-9	1.00	7.20
	≥ 10	0.40	1.06
	Current smokers	6.47	13.67

[a] From US Department of Health and Human Services (1982)

On the basis of the extent of sales-weighted 'tar' reduction in the smoke of filter cigarettes, an even greater decline in the risk for lung cancer might have been predicted. Two factors may explain why the reality differs from the expectation: (i) sales-weighted 'tar' and nicotine values are based on standardized machine-smoking of cigarettes, and (ii) most smokers of low-yield cigarettes compensate for the reduced nicotine delivery by smoking more, by drawing puffs more frequently and more intensely, and by inhaling more deeply (Herning *et al.*, 1981; Haley *et al.*, 1985; Augustine *et al.*, 1989).

Epidemiological studies have also reported a reduction in the risks for cancers of the larynx and urinary bladder for long-term smokers of filter cigarettes (Wynder & Stellman, 1979; Vineis *et al.*, 1984; Wynder *et al.*, 1988).

In addition to comparing the cancer risks of smokers of plain and filter cigarettes, epidemiological studies have been concerned with the relative risk of smoking black cigarettes *versus* blended or bright cigarettes. It was found that smokers of cigarettes made exclusively of black tobaccos, such as those that are commonly smoked in France, North Africa and Cuba, are at higher risk for cancers of the lung, larynx and urinary bladder than are smokers of blended or bright cigarettes (Joly *et al.*, 1983; Vineis *et al.*, 1984, 1985; Benhamou *et al.*, 1985, 1987; DeStefani *et al.*, 1987).

The changing cigarettes — technical developments

The introduction of cigarettes with filter tips in North America and Europe around 1940 represents the first major change in the make-up of machine-made cigarettes. However, it was not until the first reports on smoking and lung cancer in the early 1950s that consumers demanded cigarettes with reduced 'tar'. In response, cigarettes with cellulose acetate filter tips were offered. These filter tips are usually 17-30-mm long and contain a few per cent of plasticizers such as glyceryl triacetate. In 1988, 96% of the cigarettes sold in the USA had filter tips; about half of them were 85-mm long and about one-third were 100-mm long (US Department of Agriculture, 1989). Similar developments were seen in other American countries, in Japan and in some parts of Europe, while in 1987 filter-tipped cigarettes in China accounted for only 26% and in the USSR, 27% (Anon., 1988).

The degree to which 'tar' and nicotine can be retained by conventional filter tips is limited to about 50%, primarily due to the fact that cigarette smokers will usually not accept a draw resistance above 140 mm (Kuhn & Klus, 1976). Apart from removing 'tar' and nicotine from the smoke, cellulose acetate can selectively remove up to 80% of certain hydrophilic, volatile smoke components, such as phenols and volatile *N*-nitrosamines (Wynder & Hoffmann, 1967; Brunnemann *et al.*, 1977).

The introduction of perforated filter tips around 1968 had a profound impact on the cigarette market (Norman, 1982). In general, such filter tips have one or more rows of perforation in the wrapper at the half-way point of the filter column. As a result, the smoke is diluted by air entering through the holes during puff-drawing without causing an increase in draw resistance. Enhancing the air dilution of the smoke stream by optimal perforation of the filter tip diminishes the velocity of the air that enters the burning cone. This, in turn, reduces the oxygen deficiency of certain zones in the burning cone, which results in the selective reduction of CO, nitrogen oxides and volatile aldehydes in these cigarettes (Newsome & Keith, 1965; Baker, 1984).

Even though cigarette length has increased from 70 mm to 85 mm, and even to 100 mm and 120 mm, the average weight of the tobacco in the typical US cigarette has decreased from 1300 mg in the 1940s to 750 mg in the 1980s (Norman, 1982). This is due mainly to the utilization in present-day cigarettes of tobacco material with higher filling power, namely reconstituted tobacco, tobacco ribs and expanded tobacco. Reconstituted tobacco sheets are made from tobacco dust and fines and from opened ribs. They can be prepared by slurry or paper-like processes. Expanded tobacco is obtained by a freeze-drying process (Wynder & Hoffmann, 1967; Perfetti, 1987).

The composition of cigarette smoke and its toxicity and carcinogenic potency are also very much affected by the tobacco type(s) used. The composition of the tobacco filler in cigarettes in different countries is primarily the result of traditional usage and availability of tobacco types. In the USA, the Federal Republic of Germany and Scandinavia, blends of the four major tobacco types are used; in China, flue-cured tobacco is the dominant type for cigarettes; while cigarettes made exclusively from flue-cured tobaccos are preferred in the UK and Finland. A large segment of cigarette smokers in France, Italy, North Africa and several Central and South American countries prefer black tobaccos.

Bright tobaccos contain generally high levels of reducing sugars (15-20% of dry weight) and carbohydrates but relatively lower amounts of di- and tricarboxylic acids ($<4\%$) and nitrate ($<0.1\%$), while burley and black tobaccos are low in reducing sugars ($<2\%$), relatively high in di- and tricarboxylic acids ($>10\%$) and high in nitrate ($>1\%$); this is especially true for the ribs. Oriental and Maryland tobaccos are minor components of cigarette blends because of their specific smoke flavours. Their leaves contain moderate amounts of reducing sugars (10-15%) and of di- and tricarboxylic acids ($<10\%$), while their nitrate content is low (Neurath & Ehmke, 1964; Tso, 1972). The nicotine concentration of the leaves depends on the variety chosen; however, bright, oriental and Maryland tobaccos contain less nicotine ($<2\%$) than burley and black tobaccos (Tso, 1972).

The composition of a typical blend currently formulated for US cigarettes is shown in Table 3 (the composition of expanded tobaccos is described by Perfetti, 1987). Other changes in the make-up of commercial cigarettes include the use of cigarette paper of higher porosity as well as reduction of the circumference of some cigarettes (DeBardeleben *et al.*, 1978; Owens, 1978). Together, all of these changes have led to a gradual decrease in the sales-weighted 'tar' and nicotine yields in the smoke of commercial US cigarettes (Figure 2).

The changing cigarette — effect on tumorigenicity of smoke

As discussed earlier, smokers of low-yield cigarettes tend to compensate for the reduced delivery of nicotine and perhaps other smoke components; however, they do not compensate fully. The 20-50% reduction in risk for lung cancer of long-term cigarette smokers supports the concept that low-yield cigarettes have reduced carcinogenic potential. The IARC monograph on tobacco smoking states: 'no substantial cause (or cofactor) has so far been identified that offers a plausible explanation for the observed magnitude of the reduction of risk for lung cancer, other than changes in the cigarette design which include reduction in tar content.' (IARC, 1986).

Whether other factors are indeed responsible for the reduced carcinogenic potential of the changed cigarette can at this time be evaluated only in laboratory studies. Concepts of 'less harmful' cigarettes have frequently been devised by investigators outside the industry

Table 2. Relative risk for lung cancer by type of cigarette smoked (filter *versus* nonfilter) in men

Reference	Type of study	Relative risk
Hawthorne & Fry (1978)	Cohort	0.8
Rimington (1981)	Cohort	0.7
Bross & Gibson (1968)	Case-control	0.6
Wynder et al. (1970)	Case-control	0.6
Dean et al. (1977)	Case-control	0.5
Wynder & Stellman (1979)	Case-control	0.6-0.9[a]
Lubin (1984)	Case-control	0.6[b]

[a] Depending on number of cigarettes smoked daily
[b] Men who smoked only filter cigarettes

Figure 2. US sales-weighted average tar and nicotine yields (adapted from Norman, 1982)[a]

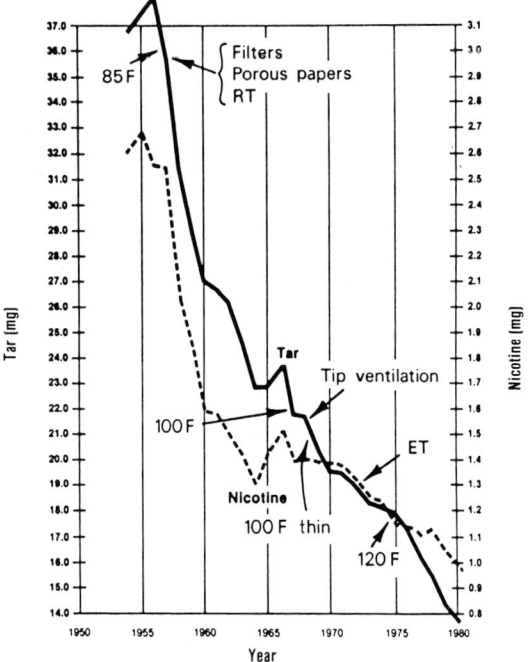

[a]RT, reconstituted tobacco; ET, expanded tobacco; F, cigarettes with filter tips; numbers, lengths of filter cigarettes in millimetres. Arrows denote years in which specific changes were first introduced

Table 3. Composition of a typical US blend for cigarettes[a]

Component	%
Flue-cured leaf	32
Burley leaf	20
Maryland leaf	2
Oriental leaf	10
Cut, rolled stems	6
Reconstituted sheet	22
Dip casing	4
Flavours/humectants	4

[a] From Perfetti (1987)

(Wynder & Hoffmann, 1962; Hoffmann et al., 1980). Various processes have been tested for their efficacy by initially analysing the smoke of the modified cigarettes for 'tar', nicotine, CO, benzo[a]pyrene and tobacco-specific N-nitrosamines (TSNA). When this analytical profile showed a reduction in the smoke yields, determinations of volatile aldehydes, volatile N-nitrosamines, cyanide, phenols and catechol were made. Experimental cigarettes with significantly changed analytical smoke profiles were then designated for evaluation of the toxicity and tumorigenic potential of their smoke. Such bioassays were done with 'tars' as well as with smoke itself (Wynder & Hoffmann, 1967; Dontenwill, 1974), as summarized in Table 4. Several of the modifications described here have been incorporated in the design of present-day low-yield cigarettes. Modifications in cigarette make-up have also led to the selective reduction of certain smoke constituents and have specifically contributed towards lowering the toxic and tumorigenic potential of cigarette smoke. The decline of benzo[a]pyrene levels in the smoke of a leading US non-filter cigarette as measured over 1958-79 is one indicator of selective reduction of a carcinogen; while 'tar' and nicotine in the smoke of this cigarette were reduced by 31% and 39%, the reduction in benzo[a]pyrene was from 36 to 16 ng/cigarette, i.e., 56% (Hoffmann et al., 1980).

In evaluating preventive strategies that involve product modification, it is important to monitor not only specific indicators but also the overall chemical composition of the smoke of commercial cigarettes. Use of nitrate-rich tobaccos and ribs increases the potential for higher smoke yields of TSNA (Brunnemann et al., 1983). Consequently, low-yield cigarettes can deliver higher amounts of carcinogenic TSNA than some plain cigarettes, as was recently shown for some cigarettes in the Federal Republic of Germany by Fischer et al. (1989). Another concern is the addition of flavouring agents to the tobacco of low-yield cigarettes (LaVoie et al., 1985). These few citations underscore how important it is that nonindustrial scientists monitor new developments in the make-up of cigarettes.

Outlook

Future research and development will probably bring about new features in the design of commercial cigarettes. Plans have been announced for the marketing of filter cigarettes made with tobaccos from which the bulk of the nicotine has been removed by supercritical extraction with CO_2 (Grubbs & Howell, 1987). This development is potentially beneficial

Table 4. Reductions in the biological activity of smoke from experimental cigarettes[a]

Method	Smoke constituents			Selective reduction in biological activity[b]		Remarks[c]
	'Tar'	Nicotine	Benzo[a]pyrene	Carcinogenicity	Tumour promotion	
Agricultural						
Tobacco type (bright-burley)[d]	+	+	+	+	+	Increase in TSNA
New cultivars	+	+	+	+	?	
Fertilization (nitrate)	+	+	+	+	?	Increase in TSNA
Tobacco processing						
Cut	±	±	±	±?	?	
Use of tobacco midribs	+	+	+	++	++	
Reconstituted tobacco sheets (RTS)[e]	+	+	+	++	±	Some RTS give high CO
RTS/paper process	++	+	+	++	±±	
Expanded tobacco laminae	+	++	+	±?	±±	
Expanded tobacco midribs	+	++	+	++	?	
Cigarette production						
Paper porosity	+	+	+	±	?	
Cellulose acetate filters	+	+	+	±±	±±	
Charcoal filters[f]	+	+	+	±±	±±	
Perforated filters	++	++	++	±	±±	Smoker's compensation

[a] Adapted from Wynder & Hoffmann (1982). Methods known to be applied to commercial US cigarettes. Reductions: ++, > 50%; +, significant; ±?, questionable; ?, unknown
[b] Comparison of gram-to-gram 'tar' on mouse skin tests and/or smoke inhalation with hamsters
[c] TSNA, tobacco-specific N-nitrosamines
[d] Replacing bright with burley tobaccos
[e] Data given for RTS relate to those not made by the paper process
[f] Reductions in 'tar', nicotine, benzo[a]pyrene (and other nonvolatiles) and volatile N-nitrosamines are generally greater with cellulose acetate filters than with charcoal filters.

in that it promises a selective reduction in the major habituating agent of tobacco and a precursor for carcinogenic TSNA. A recent attempt to market a modified cigarette that 'heats rather than burns tobacco' has not been accepted by consumers (R.J. Reynolds Tobacco Co., 1988).

Because large numbers of people continue to smoke cigarettes, development of less harmful cigarettes should not be rejected *per se*. One may agree with a recent editorial in the *New York Times* (Anon., 1989), which reads in part, 'Obviously, no smoking is better than smoking, but the best should not be the enemy of the good. There is a strong social case for encouraging manufacturers to develop safer cigarettes that will sell.'

Acknowledgements

We are most grateful to many colleagues at the American Health Foundation who have made significant contributions to our research programme in tobacco carcinogenesis. We thank Mrs Bertha Stadler for editorial assistance. Our studies are supported by research grants No. CA-17613, CA-29580 and CA-32617 from the US National Cancer Institute.

References

Adlkofer, F., Scherer, G. & Thurau, K. (1988) Rauchen und Gesundheit: Chancew durch Produktmodifikation. In: the 'Deutsche Lebensmittelchemikertag 1988', Bremen, FRG, Sept. 14-16, 1988, p. 22

Anon. (1988) Percentage share of filter cigarettes compared with total production. *Tob. J. Int.*, **5**, 362

Anon. (1989) Safer cigarettes. *New York Times*, 3 March, p. A38

Augustine, A., Harris, R.E. & Wynder, E.L. (1989) Compensation as a risk factor for lung cancer in smokers who switch from nonfilter to filter cigarettes. *Am. J. Public Health*, **79**, 188-191

Baker, R. (1984) The effect of ventilation on cigarette combustion mechanics. *Recent Adv. Tob. Sci.*, **10**, 88-150

Benhamou, S., Benhamou, E., Tirmarche, M. & Flamant, R. (1985) Lung cancer and use of cigarettes: a French case-control study. *J. Natl Cancer Inst.*, **74**, 1169-1175

Benhamou, E., Benhamou, S. & Flamant, R. (1987) Lung cancer and women: results of a French case-control study. *Br. J. Cancer*, **55**, 91-95

Bross, I.D.J. & Gibson, R. (1968) Risk of lung cancer in smokers who switch to filter cigarettes. *Am. J. Public Health*, **58**, 1396-1403

Brunnemann, K.D., Yu, L. & Hoffmann, D. (1977) Assessment of carcinogenic volatile N-nitrosamines in tobacco and in mainstream and sidestream smoke of cigarettes. *Cancer Res.*, **37**, 3218-3222

Brunnemann, K.D., Masaryk, J. & Hoffmann, D. (1983) The role of tobacco stems in the formation of N-nitrosamines in tobacco and cigarette mainstream and sidestream smoke. *J. Agric. Food Chem.*, **31**, 1221-1224

Dean, G., Lee, P.N., Todd, G.F. & Wicken, A.J. (1977) *Report on a Second Retrospective Study in Northeast England. Part I. Factors Related to Mortality from Lung Cancer, Bronchitis, Heart Disease and Stroke in Cleveland County with Particular Emphasis on the Relative Risks Associated with Smoking Filter and Plain Cigarettes* (Research Paper 14), London, Tobacco Research Council

DeBardeleben, M.Z., Chaflin, W.E. & Gannon, W.F. (1978) Role of cigarette physical characteristics on smoke composition. *Recent Adv. Tob. Sci.*, **4**, 85-111

DeStefani, E., Correa, P., Oreggia, F., Leiva, J., Rivero, S., Fernandez, G., Deneo-Pellegrini, H., Zavala, D. & Fontham, E. (1987) Risk factors for laryngeal cancer. *Cancer*, **60**, 3087-3091

Dontenwill, W.P. (1974) Tumorigenic effect of chronic cigarette smoke inhalation on Syrian golden hamsters. In: Karbe, E. & Park, J.F., eds, *Experimental Lung Cancer. Carcinogenesis and Bioassays*, New York, Springer, pp. 331-359

Fischer, S., Spiegelhalder, B. & Preussmann, R. (1989) Tobacco-specific nitrosamines in mainstream smoke of West German cigarettes – tar is not a sufficient index for the carcinogenic potential of cigarette smoke. *Carcinogenesis*, **10**, 169-173

Grubbs, H.J. & Howell, T.M. (1987) *Process for the Removal of Basic Materials from Plant Products* (European Patent Office Publ. No. 0 280 817)

Haley, N.J., Sepkovic, D.W., Hoffmann, D. & Wynder, E.L. (1985) Compensation with nicotine availability as a single variable. *Clin. Pharmacol. Ther.*, **38**, 164-170

Hawthorne, V.M. & Fry, J.S. (1978) Smoking and health: the association between smoking behavior, total mortality and cardiorespiratory disease in West Central Scotland. *J. Epidemiol. Commun. Health*, **32**, 260-266

Herning, R.I., Jones, R.T., Bachman, J. & Mines, A.H. (1981) Puff volume increases when low-nicotine cigarettes are smoked. *Br. Med. J.*, **283**, 187-189

Hoffmann, D., Tso, T.C. & Gori, G.B. (1980) The less harmful cigarette. *Prev. Med.*, **9**, 287-296

IARC (1986) *IARC Monographs on the Evaluation of the Carcinogenic Risk of Chemicals to Humans*, Vol. 38, *Tobacco Smoking*, Lyon

Joly, O.G., Lubin, J.H. & Caraballoso, M. (1983) Dark tobacco and lung cancer in Cuba. *J. Natl Cancer Inst.*, **70**, 1033-1039

Koop, C.E. (1986) The quest for a smoke-free young America by the year 2000. *J. School Health*, **56**, 8-9

Kuhn, H. & Klus, H. (1976) Possibilities for the reduction of nicotine in cigarette smoke. In: Wynder, E.L., Hoffmann, D. & Gori, G.B., eds, *Proceedings of the Third World Conference on Smoking and Health*, Vol. 1, *Modifying the Risk for the Smoker* (DHEW Publ. No. (NIH) 76-1221), Washington DC, US Department of Health, Education, and Welfare, pp. 463-494

LaVoie, E.J., Shigematsu, A., Tucciarone, P.L., Adams, J.D. & Hoffmann, D. (1985) Comparison of steam-volatile components in commercial cigarette, pipe, and chewing tobacco. *J. Agric. Food Chem.*, **33**, 876-879

Lubin, J.H. (1984) Modifying risk of developing lung cancer by changing habits of cigarette smoking. *Br. Med. J.*, **289**, 92

Neurath, G. & Ehmke, H. (1964) Study of nitrate content of tobacco (in German). *Beitr. Tabakforsch.*, **2**, 333-334

Newsome, J.R. & Keith, C.H. (1965) Variation of the gas phase composition within a burning cigarette. *Tob. Sci.*, **9**, 65-69

Norman, V. (1982) Changes in smoke chemistry of modern day cigarettes. *Recent Adv. Tob. Sci.*, **8**, 141-177

Owens, W.F., Jr (1978) Effect of cigarette paper on smoke yield and composition. *Recent Adv. Tob. Sci.*, **4**, 3-24

Perfetti, T.A. (1987) Measurement of the quality of cigarette products. *Recent Adv. Tob. Sci.*, **13**, 119-161

R.J. Reynolds Tobacco Co. (1988) *New Cigarette Prototypes that Heat Instead of Burn Tobacco*, Winston-Salem, NC

Rimington, J. (1981) The effect of filters on the incidence of lung cancer in cigarette smokers. *Environ. Res.*, **24**, 162-166

Tso, T.C. (1972) *Physiology and Biochemistry of Tobacco Plants*, Stroudsburg, PA, Dowden, Hutchinson & Ross, p. 393

US Department of Agriculture (1989) *Tobacco Products (Tobacco — Situation Outlook Rep., TS-206)*, Washington DC, Economic Research Service, pp. 4-8

US Department of Health and Human Services (1982) *The Health Consequences of Smoking. Cancer. A Report of the Surgeon-General* (DHHS Publ. NO. (PHS) 82-50179), Bethesda, MD

US Department of Health and Human Services (1989) *Reducing the Health Consequences of Smoking. 25 Years of Progress. A Report of the Surgeon-General* (DHHS Publ. No. (CDC) 89-8411), Bethesda, MD

Vineis, P., Estève, J. & Terracini, B. (1984) Bladder cancer and smoking in males: types of cigarettes, age at start, effect of stopping and interaction with occupation. *Int. J. Cancer*, **34**, 165-170

Vineis, P., Ciccone, G. & Ghisetti, V. (1985) Cigarette smoking and bladder cancer in females. *Cancer Lett.*, **26**, 61-66

Weber, K.H. (1972) Tar and nicotine in German cigarettes. *Prev. Med.*, **1**, 446-450

World Health Organization (1986) Change in apparent cigarette consumption and adult population size, by regions, 1971-75 to 1979-81. In: *WHO Health Statistics*, Geneva, pp. 16-19

Wynder, E.L. & Day, E. (1961) Some thoughts on the causation of chronic disease. *J. Am. Med. Assoc.*, **175**, 997-999

Wynder, E.L. & Hoffmann, D. (1962) Present status of laboratory studies on tobacco carcinogenesis. *Acta Pathol. Microbiol. Scand.*, **52**, 119-132

Wynder, E.L. & Hoffmann, D. (1967) *Tobacco and Tobacco Smoke. Studies in Experimental Carcinogenesis*, New York, Academic Press

Wynder, E.L. & Hoffmann, D. (1982) Tobacco. In: Schottenfeld, D. & Fraumeni, J.F., Jr, eds, *Cancer Epidemiology and Prevention*, Philadelphia, Saunders, pp. 277-292

Wynder, E.L. & Kabat, C. (1988) The effect of low-yield cigarette smoking on lung cancer risk. *Cancer*, **62**, 1223-1230

Wynder, E.L. & Stellman, S.D. (1979) The impact of long-term filter cigarette usage on lung and larynx cancer risk: a case control study. *J. Natl Cancer Inst.*, **62**, 471-477

Wynder, E.L., Mabuchi, K. & Beattie, E.J., Jr (1970) The epidemiology of lung cancer. Recent trends. *J. Am. Med. Assoc.*, **213**, 2221-2228

Wynder, E.L., Augustine, A., Kabat, G.C. & Hebert, J.R. (1988) Effect of the type of cigarette smoked on bladder cancer risk. *Cancer*, **61**, 622-627

Relevance to Human Cancer of N-Nitroso Compounds,
Tobacco Smoke and Mycotoxins.
Ed. I.K. O'Neill, J. Chen and H. Bartsch
Lyon, International Agency for Research on Cancer
© IARC, 1991

ENVIRONMENTAL DETERMINANTS OF LUNG CANCER IN SHENYANG, CHINA

Z.Y. Xu[1], W.J. Blot[2,5], G. Li[1], J.F. Fraumeni, Jr[2], D.Z. Zhao[1], B.J. Stone[2], Q. Yin[1], A. Wu[3], B.E. Henderson[3] & B.P. Guan[4]

[1]*Liaoning Public Health and Anti-epidemic Station, Shenyang, Liaoning, China;* [2]*National Cancer Institute, Bethesda, MD, USA;* [3]*University of Southern California, Los Angeles, CA, USA; and* [4]*Liaoning Cancer Institute, Shenyang, Liaoning, China*

To investigate determinants of the high rates of lung cancer in Shenyang, an industrial city in north-eastern China, a case-control study was conducted. Interviews with 1249 lung cancer patients and 1345 population-based controls revealed that cigarette smoking was the main cause of lung cancer. Smoking accounted for 55% of the lung tumours in men and 37% in women. In addition, air pollution from coal-burning heating and cooking devices was significantly linked to lung cancer, with risks rising in proportion to duration of exposure to indoor pollutants. Measurement of benzo[a]pyrene revealed average wintertime levels in air that were nearly 60 times the recommended upper limit for US cities, with even higher concentrations indoors in traditional single-storey homes using coal-burning *kang* (stoves). Occupational factors were also involved, the risk being elevated by three fold among smelter workers. Soil levels of arsenic and other metals rose with increasing proximity to the Shenyang copper smelter, and elevated risks of lung cancer were found among men, but not women, living within 1 km of its central stacks. Prior nonmalignant lung disease was common and was reported more often among the lung cancer patients than among controls. The findings suggest that cigarette smoking and environmental pollutants combine to account for most of the excess risk of lung cancer in this population.

Mortality from lung cancer varies substantially in China, with the highest rates generally found in the north-eastern provinces (National Cancer Control Office, 1980; Xu *et al.*, 1986). The north-south gradient is particularly prominent among females. Since the incidence of lung cancer is high in general among Chinese women — even among nonsmokers — some of the world's highest rates of lung cancer in females are found in northern China (Waterhouse *et al.*, 1982; Gao *et al.*, 1987). To investigate environmental determinants, we initiated a case-control study in Shenyang, the largest city and capital of Liaoning province in north-eastern China. Herein, we review the principal findings, some of which have been presented in more detail elsewhere (Xu *et al.*, 1989).

---

[5]To whom correspondence should be addressed

Selection of cases and controls

All cases of primary lung cancer newly diagnosed between September 1985 and September 1987 among residents of urban Shenyang in persons aged 30-69 years were sought for inclusion in the study. In total, 1249 lung cancer cases (729 males, 520 females) were enrolled after exclusion of 60 who had died or were too ill and nine who could not be located or who refused to participate. Review of available diagnostic material revealed that, among the histologically confirmed cases (83% of all male cases, 73% of all female cases), the percentages of squamous-cell carcinomas, adenocarcinomas, oat-cell carcinomas and other cell types, respectively, were 51%, 27%, 15% and 8% among men and 32%, 38%, 17% and 13% among women.

Controls were randomly selected from population rosters available for each of the nearly 1500 neighbourhood committees in Shenyang. A total of 1345 controls (788 men, 577 women) were enrolled. The age and sex distribution of the controls closely matched those of the cases. The median ages among the men and women in this study were 59 and 56, respectively.

Interviews

All subjects were interviewed by trained personnel using a structured, precoded questionnaire. Information was obtained on lifetime histories of smoking, occupation and residence. Characteristics of each house lived in for three or more years were obtained, including presence of *kang* (coal-burning stoves, often underneath beds, with exposed pipes leading to vents in the side walls of the house) and other heating devices used in north-eastern China. Radon levels were measured in homes of female cases and controls. In addition, information was gathered on prior medical conditions, familial cancer, diet, occupational exposure, passive smoking and other variables.

Effects of tobacco smoking

Most of the lung cancer patients (86% of men, 55% of women) were tobacco smokers, whereas the percentages of smokers among controls were significantly lower (70% and 35%). Overall, smoking was associated with a 2.7-fold (95% confidence interval (CI), 2.1-3.5) increased risk of lung cancer among men and a 2.6-fold (95% CI, 2.0-3.3) increase among women. Relative risks (RR) increased sharply with both duration and intensity of smoking and were higher for squamous- and oat-cell cancers than for adenocarcinoma. Among the heaviest smokers (men smoking at least 30 cigarettes/day and women smoking at least 20 cigarettes/day for 40 or more years), the risks of squamous- and oat-cell cancers were more than 20 times greater than those of nonsmokers.

Calculations of attributable risk indicated that 55% of the lung cancers in men and 37% in women could be attributed to smoking. These figures are lower than those in western populations, since fewer cigarettes are smoked in Shenyang: the median numbers of cigarettes smoked per day by male and female cases were 19 and 7, respectively. The prevalence of smoking among female controls in Shenyang, however, was nearly twice the Chinese national average (Weng *et al.*, 1987). Smoking thus accounts for part (although less than half) of the excess lung cancer mortality in Shenyang, in comparison with national rates. The higher prevalence of smoking was limited to women aged 50 and over, while few

Indoor air pollution

Risks of lung cancer were elevated among persons who used *kang* in which coal was burned directly under their beds. The RR reached about 2.0 among men and women who had used these burning *kang* for 20 or more years. Risks also increased in proportion to several other exposures to burning coal and declined with use of central (primarily gas) heating. Table 1 shows the RR associated with an index of exposure to indoor air pollution from home heating and cooking. Significant elevations in risk with increasing exposure were observed for both men and women. The index was derived from objective information provided on the types of heating and cooking devices used in each home in which the subject had lived. Evaluation of subjective measures of exposure yielded similar results, as cases (particularly females) more often than controls perceived that their homes were 'smokey' during winter heating.

Table 1. Relative risks (RR, with 95% confidence interval, CI) of lung cancer according to an index of exposure to indoor air pollution[a]

Air pollution index[b]	Male		Female	
	RR	95% CI	RR	95% CI
I (low)	1.0	-	1.0	-
II	1.1	0.8-1.4	1.2	0.9-7.8
III	1.2	0.9-1.6	1.3	0.9-1.9
IV (high)	1.6	1.2-2.3	1.5	1.0-2.4

[a] All risks adjusted for age, education and smoking
[b] The index takes into account duration of exposures across all residences to sources of heating and cooking.

We measured wintertime benzo[*a*]pyrene concentrations indoors and outdoors at ten traditional single-storey houses (where *kang* were used), ten two-storey houses (usually heated by coal stoves) and ten taller ($\geq$3-storey) buildings (with central heating systems). Outdoor concentrations were similar (about 60 ng/m3, a level 60 times the recommended upper limit for urban air in the USA (Shy & Struba, 1982)), but indoor levels varied considerably. Average levels in single-storey homes heated by *kang* reached 75 ng/m3. Levels were somewhat lower in two-storey houses but declined to about 30 ng/m3 in taller apartment buildings.

Our findings are qualitatively similar to those from Yunnan province in southern China, where living in houses without chimneys, with heavy pollution from smoke from soft local coals, accounted for the clustering of excess lung cancer among both men and women (Mumford *et al.*, 1987). In Yunnan, such exposures predominated over cigarette smoking as risk factors for lung cancer, whereas smoking accounted for more cases in Shenyang.

Outdoor air pollution

The lung cancer cases more often reported that they lived in neighbourhoods where the outdoor air was smokey (Table 2). Cases also reported that their residence was more often within 200 m of a factory, although the excess was significant only for residences near nonferrous metallurgical factories. We plotted distances of residences from the most prominent factory, the Shenyang copper smelter. The RR associated with living within 1 km of the smelter, after controlling for smoking and for working in the smelter, was 3.0 (95% CI, 1.5–6.0) among males, then declined to expected levels beyond this distance. No concomitant excess risk with close proximity was seen among females, however. We sampled undisturbed soil in concentric circles around the smelter and found highest levels of arsenic, lead, copper and cadmium and other metals within 1 km, with sharp declines as the distance increased, indicating that emissions had reached the highly populated surrounding neighbourhood. The finding among males is consistent with reports of excess lung cancer following residential exposure to smelter emissions in the USA and Sweden (Brown et al., 1984; Pershagen, 1985), but the absence of an effect among women precludes a causal interpretation. Occupation in the smelter was also associated with increased risk of lung cancer (RR, 3.6; 95% CI, 1.6–8.2), in agreement with reports of excess lung cancer among arsenic-exposed smelter workers throughout the world (Lubin et al., 1981).

Table 2. Relative risks (RR, with 95% confidence intervals, CI) of lung cancer according to perceived neighbourhood smokiness

Outdoor environment	Males		Females	
	RR	95% CI	RR	95% CI
Not smokey	1.0	–	1.0	–
Somewhat/slightly smokey	1.5	1.2-2.0	1.4	1.1-2.0
Smokey	2.3	1.7-2.9	2.5	1.8-3.5

Chronic lung disease

Prior chronic lung disease was reported commonly — particularly chronic bronchitis (22% of cases, 12% of controls) and tuberculosis (18% of cases, 14% of controls). The excess among cases persisted after excluding diagnoses of these conditions within three years of the lung cancer diagnosis. The association with tuberculosis was stronger when it had been diagnosed within 4–10 years of lung cancer diagnosis, while the association with chronic bronchitis was stronger when it had occurred more than 20 years prior to diagnosis of lung cancer. The link with tuberculosis was seen for all cell types, while the association with chronic bronchitis was seen primarily for squamous- and oat-cell cancers. These findings add to growing evidence that suggests that both recent and past lung diseases may increase susceptibility to lung cancer (Zheng et al., 1987; Wu et al., 1988).

Other factors

Lung cancer in family members was reported more often by cases than controls, although the percentages with affected first-degree relatives were small. Dietary histories were taken, but the frequency of consumption of various foods did not differ markedly between cases and controls. In particular, no protective effect of carotene-containing foods was seen. Passive smoking effects could not be demonstrated, perhaps because exposures to environmental tobacco smoke were dominated by high indoor air pollution from heating and cooking sources. Although there were more smelter workers among the cases, the distributions by major occupational category were generally similar among participants. The relative stability of the population (median duration in the subjects' current residence was 24 years) and the year-long alpha-track measurements yielded estimates of radon exposure that probably reflect exposures relevant to lung cancer risk. No trend of rising risk with increasing radon level was observed, however.

In summary, this large, population-based study, conducted in an area with some of the highest rates of lung cancer in China (and in the world among females) and heavy pollution from residential and industrial sources, provided an opportunity to evaluate a variety of risk factors for lung cancer. Smoking was found to be the major cause of the tumours, a finding not surprising to western observers but unexpected by many people in Liaoning, who had considered that other factors (including air pollution) were more important. The study shows, however, that use of *kang* and other coal-burning stoves contributes significantly to the lung cancer burden in north-eastern China and provides incentive for further evaluation of air pollutants in the etiology of lung cancer.

References

Brown, L.M., Pottern, L.M. & Blot, W.J. (1984) Lung cancer in relation to environmental pollutants emitted from industrial sources. *Environ. Res.*, 34, 250-261

Gao, Y.T., Blot, W.J., Zheng, W., Ershow, A.G., Hsu, C.W., Levin, L.F., Zhang, R. & Fraumeni, J.F. (1987) Lung cancer among Chinese women. *Int. J. Cancer*, 40, 604-609

Lubin, J.H., Pottern, L.M., Blot, W.J., Tokudome, S., Stone, B.J. & Fraumeni, J.F. (1981) Respiratory cancer among copper smelter workers: recent mortality statistics. *J. Occup. Med.*, 23, 779-784

Mumford, J.L., He, X., Chapman, R.S., Cao, S., Harris, D., Li, X., Xian, Y., Jiang, W., Xu, C., Chuang, J., Wilson, R. & Cooke, M. (1987) Lung cancer and indoor air pollution in Xuan Wei, China. *Science*, 235, 217-220

National Cancer Control Office (1980) *Atlas of Cancer Mortality in the People's Republic of China*, Beijing, China Map Press

Pershagen, G. (1985) Lung cancer mortality among men living near an arsenic-emitting smelter. *Am. J. Epidemiol.*, 122, 684-694

Shy, C.M. & Struba, R.J. (1982) Air and water pollution. In: Schottenfeld, D. & Fraumeni, J., eds, *Cancer Epidemiology and Prevention*, Philadelphia, Saunders, pp. 336-363

Waterhouse, J., Muir, C., Shanmugaratnam, K. & Powell, J., eds (1982) *Cancer Incidence in Five Continents, Vol. IV* (IARC Scientific Publications No. 42), Lyon, IARC

Weng, X., Hong, Z. & Chen, D. (1987) Smoking prevalence in Chinese aged 15 and above. *Chin. J. Med.*, 100, 886-892

Wu, A.H., Yu, M.C., Thomas, D.C., Pike, M.C. & Henderson, B.E. (1988) Personal and family history of lung disease as risk factors for adenocarcinoma of the lung. *Cancer Res.*, 48, 7279-7284

Xu, Z.Y., Blot, W.J. & Fraumeni, J.F. (1986) Geographic variation in female lung cancer in China. *Am. J. Public Health*, 76, 1249-1250

Xu, Z.Y., Blot, W.J., Xiao, H.P., Wu, A., Feng, Y.P., Stone, B.J., Sun, J., Ershow, A.G., Henderson, B.E. & Fraumeni, J.F. (1989) Smoking, air pollution and the high rates of lung cancer in Shenyang, China. *J. Natl Cancer Inst.*, **81**, 1800-1806

Zheng, W., Blot, W.J., Liao, M.L., Wang, Z., Levin, L.J., Zhang, J., Fraumeni, J.F. & Gao, Y.T. (1987) Lung cancer and prior tuberculosis infection in Shanghai. *Br. J. Cancer*, **56**, 501-504

Relevance to Human Cancer of *N*-Nitroso Compounds,
Tobacco Smoke and Mycotoxins.
Ed. I.K. O'Neill, J. Chen and H. Bartsch
Lyon, International Agency for Research on Cancer
© IARC, 1991

BETEL QUID AND ORAL CANCER: PROSPECTS FOR PREVENTION

P.C. Gupta

*Basic Dental Research Unit, WHO Collaborating Center for
Oral Cancer Prevention, Tata Institute of Fundamental
Research, Bombay, India*

Betel-quid chewing is an ancient and socially accepted practice. The introduction of tobacco reinforced this practice, and now almost all habitual chewers of betel quids include tobacco. It is well established that chewing of betel quid with tobacco causes oral cancer and is largely responsible for the high incidence of oral cancer in several South Asian countries. The feasibility of primary prevention of oral cancer was studied in a population-based prospective intervention study. A cohort of 12 212 betel-quid chewers and smokers was exposed to a programme of health education for stopping chewing and smoking and subjected to annual examinations for detection of oral precancerous lesions. Evaluations after one, five and eight years showed that primary prevention of oral cancer is feasible and practicable. Early detection of oral cancer is an important control measure. In a secondary prevention study, 53 basic health workers were trained in the detection and referral of lesions suspected of being oral cancer. Over one year, they examined more than 39 000 high-risk individuals, resulting in the detection of 20 cases of oral cancer. The sensitivity and specificity of their diagnoses was assessed through a re-examination of a 5% sample: we concluded that it was possible to incorporate a secondary prevention programme into the existing health care system.

The chewing of betel quid is a very ancient practice in India and many other South Asian countries. References to this habit in stone and other inscriptions are about a millenium old, and literature references are about two millennia old. Betel-quid chewing has been part of religious and cultural rituals and enjoys complete social acceptance.

Basically, the betel quid consists of betel leaf, areca nut, lime and condiments, and sweetening and flavouring agents, which depend upon individual and local preferences. Tobacco was introduced into India, as everywhere else, from the New World in the sixteenth century by Europeans. It soon became an ingredient of the betel quid and because of this association enjoyed social acceptance.

The relationship between betel-quid chewing and oral cancer was postulated in the late nineteenth and early twentieth centuries by British surgeons, who noted that oral cancer was rare in Great Britain but common in India. Subsequently, several studies have been made of this association. The terms used to describe the habit have not been consistent, however, and the same terms have been used differently in different studies. This has resulted in a considerable amount of difficulty in interpretation. A major source of confusion was the description of the habit as 'betel-nut' chewing, with the occasional addition of tobacco. This description led to the impression that the addition of tobacco to the chewing quid was of little or marginal prevalence. This impression was corrected comparatively recently by population-based house-to-house surveys of chewing habits in six different areas of India. Chewing habits were widespread in three of the six areas, and in these three areas 92-98% of chewers who did not smoke included tobacco in their quid (Mehta et al., 1969, 1972). These findings demonstrate that the betel-quid chewing habit is usually the habit of chewing betel quid with tobacco.

'Chewing habit' is another loose term, which may include areca-nut chewing, tobacco-lime chewing and betel-quid chewing. Unless otherwise specified, the term betel-quid chewing implies the chewing of betel quid containing tobacco. Although the prevalence of tobacco chewing has been declining over the last few decades, it is estimated that there are still at least 40 million regular tobacco chewers in India.

Betel quid and oral cancer

The relationship between betel-quid chewing and oral cancer has been reviewed extensively (IARC, 1985). The association has been demonstrated in numerous case-control and cohort studies, and evidence from experimental studies points in the same direction. The overall evaluation of the IARC working group was that the evidence is sufficient for the carcinogenicity of betel quid with tobacco, meaning that chewing of betel quid with tobacco causes oral cancer. Neither the epidemiological nor the experimental evidence for the carcinogenicity of betel quid without tobacco was sufficient (Gupta et al., 1982). This does not mean, however, that the chewing of betel quid without tobacco is an innocuous habit. Areca nut in betel quid causes oral submucous fibrosis, which is a debilitating disease with no known cure (Bhonsle et al., 1987); it is also a precancerous lesion (Murti et al., 1985). A person with submucous fibrosis who is exposed to carcinogens in tobacco in the form of chewing or smoking has a higher risk of developing oral cancer than persons without the disease.

It has been estimated that about 30% of oral cancer can be attributed to the habit of chewing betel quid with tobacco (World Health Organization, 1984). Betel-quid chewing and tobacco smoking have a synergistic effect on the risk for oral cancer, and the combined habits of chewing and smoking are quite common. Thus, an additional 50% of oral cancers can be attributed to the combination. It is therefore clear that betel-quid chewing is the single most important factor responsible for the high incidence of oral cancer in India and other South Asian countries.

Betel quid and oral precancer

Betel-quid chewing is strongly associated with white lesions of the oral cavity, the most important of which is leukoplakia. In several hospital-based studies in India, leukoplakia was found to be highly prevalent among people who chewed betel quid and rare among those who did not chew betel quid and did not smoke. These findings were confirmed in population-based cross-sectional studies (Mehta et al., 1969, 1972). The most compelling evidence came from a ten-year prospective study of a random sample of 10 287 individuals in Ernakulam district with annual follow-ups (Gupta et al., 1980). Not a single new leukoplakia was diagnosed among individuals who did not chew or smoke, although there were 30 523 person-years of observation in this category. The annual incidence of leukoplakia among betel-quid chewers, however, was 2.5 per thousand among men and 3 per thousand among women. A total of 13 new cases of oral cancer were detected, all of them were preceded by a precancerous lesion or condition. All 13 individuals were betel-quid chewers, although four of them smoked as well. Nine oral cancers developed from a pre-existing leukoplakia.

Primary prevention

Since betel-quid chewing is a causal factor for oral cancer, it can be hypothesized that stopping betel-quid chewing will lead to a decrease in the risk for oral cancer. Supportive evidence for this hypothesis comes from a population-based intervention study in Ernakulam district. A cohort of 12 212 individuals which included betel-quid chewers and smokers was interviewed about betel-quid chewing habits and was examined by dentists for the presence of oral precancerous lesions, first in a baseline survey and annually thereafter. This cohort was then exposed to a health education programme for stopping betel-quid chewing and tobacco smoking through the use of documentary films, posters, newspaper articles, radio messages and other means, in a carefully controlled manner with regular monitoring. Health education was also provided through personal communication by dentists and social scientists. After one year of follow-up, it was reported that although only a small percentage of individuals had stopped their tobacco habits, the regression of leukoplakia was significantly higher among those who had stopped or reduced their chewing and smoking habits (Mehta et al., 1982).

After five years of follow-up, health education was shown to have been very helpful in stopping tobacco habits and had been especially helpful to betel-quid chewers (Gupta et al., 1986a). The age-adjusted incidence rates of leukoplakia were found to be much lower in the intervention cohort than in a control cohort provided by an earlier ten-year follow-up study in the same area but among different individuals. This study was similar to the intervention study in all respects except that there was no concentrated programme of health education on tobacco habits. For betel-quid chewers, the incidence in the intervention cohort was five times lower than that in the control cohort for men and two times lower among women (Gupta et al., 1986b).

Similar results were reported after eight years of follow-up (Gupta et al., 1989a). The percentage of individuals in the intervention cohort who stopped their tobacco habits continued to increase, although at a slow pace. The incidence of leukoplakia continued to be substantially and significantly lower in the intervention cohort than in the control cohort. No reduction in the incidence of oral cancer could be demonstrated directly, however. Since most cases of oral cancer in this population develop from oral precancerous lesions (Gupta et al., 1989b), a decrease in the incidence of oral precancerous lesions was construed as indicative of a decrease in risk for future oral cancers.

It has thus been demonstrated that an effective health education programme on betel-quid chewing and tobacco smoking would lead to primary prevention of oral cancer.

Secondary prevention

Most cases of oral cancer in India are detected at an advanced stage, resulting in disfigurement if treatment is successful or high mortality. This finding has been persistent, in spite of the fact that the oral cavity is an easily accessible site for examination and oral cancer can be detected at an early stage through careful examination. In India, the dentist:population ratio is very low, especially in rural areas (e.g., 1:631 000 in Maharashtra), and it is not practical to recommend regular oral examination by dentists as a measure for the control of oral cancer.

In the government health care delivery system in rural areas, the health providers on the lowest rung of the ladder are basic health workers (also known as multipurpose health workers). They are attached to a primary health care centre and visit households in their jurisdiction once every three months. In a feasibility study of secondary prevention of oral cancer, 53 basic health workers from two administrative blocks of Ernakulam district were trained in identifying high-risk individuals (betel-quid chewer or tobacco smoker aged 35 years or over) and examining the mouth for early detection of oral cancer through identification of precancerous lesions. Over a period of one year, along with their routine work, these basic health workers examined 39 331 high-risk individuals; they referred 523 (1.3%) individuals with suspicious lesions to an oral cancer detection centre. Some 377 (72%) individuals arrived at the detection centre, and this resulted in the early detection of 20 cases of oral cancer. The sensitivity and specificity of examination by basic health workers was assessed through a re-examination of a 5% sample by dentists. The conclusion was drawn that it is feasible and practicable to implement secondary preventive measures as part of the rural health care delivery system in India (Mehta et al., 1986). A somewhat similar study was reported from Sri Lanka, with almost identical results (Warnakulasuriya et al., 1984).

It should be pointed out that the detection of precancerous lesions and their proper management may lead to primary prevention of oral cancer as well.

Future studies

The studies carried out so far have demonstrated that primary and secondary prevention of oral cancer among betel-quid chewers and smokers is feasible and practicable. Future studies are planned to assess the effectiveness of teaching mouth self-examination to high-risk individuals. We also propose to train basic health workers in imparting health education on tobacco use and to study its effectiveness.

Acknowledgements

Supported by National Institutes of Health, USA, under US-India Fund research agreement No. N-406-560.

References

Bhonsle, R.B., Murti, P.R., Gupta, P.C., Mehta, F.S., Sinor, P.N., Irani, R.R. & Pindborg, J.J. (1987) Regional variations in oral submucous fibrosis in India. *Commun. Dent. Oral Epidemiol.*, **15**, 225-229

Gupta, P.C., Mehta, F.S., Daftary, D.K., Pindborg, J.J., Bhonsle, R.B., Jalnawalla, P.N., Sinor, P.N., Pitkar, V.K., Murti, P.R., Irani, R.R., Shah, H.T., Kadam, P.M., Iyer, K.S.S., Iyer, H.M., Hegde, A.K., Chandrashekhar, G.K., Shroff, B.C., Sahiar, B.E. & Mehta, M.N. (1980) Incidence of oral cancer and natural history of oral precancerous lesions in a 10-year follow-up study of Indian villagers. *Commun. Dent. Oral Epidemiol.*, **8**, 287-333

Gupta, P.C., Pindborg, J.J. & Mehta, F.S. (1982) Comparison of carcinogenicity of betel quid with and without tobacco: an epidemiological review. *Ecol. Dis.*, **1**, 213-219

Gupta, P.C., Aghi, M.B., Bhonsle, R.B., Murti, P.R., Mehta, F.S., Mehta, C.R. & Pindborg, J.J. (1986a) Intervention study of chewing and smoking habits for primary prevention of oral cancer among 12 212 Indian villagers. In: Zaridze, D.G. & Peto, R., eds, *Tobacco: A Major International Health Hazard* (IARC Scientific Publications No. 74), Lyon, IARC, pp. 307-318

Gupta, P.C., Mehta, F.S., Pindborg, J.J., Aghi, M.B., Bhonsle, R.B., Murti, P.R., Daftary, D.K., Shah, H.T. & Sinor, P.N. (1986b) Intervention study for primary prevention of oral cancer among 36,000 Indian tobacco users. *Lancet*, **i**, 1235-1238

Gupta, P.C., Mehta, F.S., Pindborg, J.J., Daftary, D.K., Aghi, M.B., Bhonsle, R.B. & Murti, P.R. (1989a) A primary prevention study of oral cancer among Indian villagers. Eight-year follow-up results. In: Hakama, M., Beral, V., Cullen, J. & Parkin, M., eds, *Evaluating Effectiveness of Primary Prevention of Cancer* (IARC Scientific Publications No. 103), Lyon, IARC, pp. 149-156

Gupta, P.C., Bhonsle, R.B., Murti, P.R., Daftary, D.K., Mehta, F.S. & Pindborg, J.J. (1989b) An epidemiologic assessment of cancer risk in oral precancerous lesions in India with special reference to nodular leukoplakia. *Cancer*, **63**, 2247-2252

IARC (1985) *IARC Monographs on the Evaluation of the Carcinogenic Risk of Chemicals to Humans*, Vol. 37, *Tobacco Habits other than Smoking; Betel-quid and Areca-nut Chewing; and Some Related Nitrosamines*, Lyon

Mehta, F.S., Pindborg, J.J., Gupta, P.C. & Daftary, D.K. (1969) Epidemiologic and histologic study of oral cancer and leukoplakia among 50,915 villagers in India. *Cancer*, **24**, 832-849

Mehta, F.S., Gupta, P.C., Daftary, D.K., Pindborg, J.J. & Choksi, S.K. (1972) An epidemiologic study of oral cancer and precancerous conditions among 101,761 villagers in Maharashtra, India. *Int. J. Cancer*, **10**, 134-141

Mehta, F.S., Aghi, M.B., Gupta, P.C., Pindborg, J.J., Bhonsle, R.B., Jalnawalla, P.N. & Sinor, P. (1982) An intervention study of oral cancer and precancer in rural Indian populations: a preliminary report. *Bull. World Health Organ.*, **60**, 441-446

Mehta, F.S., Gupta, P.C., Bhonsle, R.B., Murti, P.R., Daftary, D.K. & Pindborg, J.J. (1986) Detection of oral cancer using basic health workers in an area of high oral cancer incidence in India. *Cancer Detect. Prev.*, **9**, 219-225

Murti, P.R., Bhonsle, R.B., Pindborg, J.J., Daftary, D.K., Gupta, P.C. & Mehta, F.S. (1985) Malignant transformation rate in oral submucous fibrosis over a 17-year period. *Commun. Dent. Oral Epidemiol.*, **13**, 340-341

Warnakulasuriya, K.A.A.S. (1984) Utilization of primary health care workers for early detection of oral cancer and precancer cases in Sri Lanka. *Bull. World Health Organ.*, **62**, 243-250

World Health Organization (1984) Control of oral cancer in developing countries. Report of a WHO meeting. *Bull. World Health Organ.*, **62**, 817-830

Relevance to Human Cancer of N-Nitroso Compounds,
Tobacco Smoke and Mycotoxins.
Ed. I.K. O'Neill, J. Chen and H. Bartsch
Lyon, International Agency for Research on Cancer
© IARC, 1991

LUNG CANCER: POLITICAL MEASURES

J.M. Mackay

Asian Consultancy on Tobacco Control, Hong Kong

In countries where prolonged smoking of manufactured cigarettes is a widely established habit, it is responsible for about 90% of lung cancer. As lung cancer is usually incurable, even with expensive technology, the key to its control lies in prevention. World experience has shown the crucial need for government commitment, funding and action in controlling the epidemic of tobacco-related disease. It is recommended that each country establish a national council of 'tobacco or health' to coordinate a comprehensive tobacco control programme. This programme should incorporate data collection, including evaluation of specific anti-tobacco measures; legislative measures, including strong, rotating health warnings, limits on harmful substances, establishment of smoke-free areas, bans on any new forms of tobacco use, and a total ban on all direct or indirect promotion of tobacco products; health education campaigns; and taxation and price policies. The support and involvement of the medical profession is vital. Obstacles to success include the effect of advertising revenue in silencing the media, the inertia of governments and the medical profession, but most importantly the tobacco industry — the largest, wealthiest, most determined and strongest opposition to tobacco control worldwide.

In countries where tobacco smoking is an established habit, it is responsible for about 90% of lung cancer. As lung cancer is rarely curable, even with expensive technology, the key to reducing its frequency lies in prevention. While the prevalence of cigarette consumption is decreasing in industrialized countries, it is increasing in nonindustrialized countries which have fewer legislative controls and other measures that, in industrialized countries, limit the use of tobacco. Different countries are at very different points in the five-stage process of lung cancer prevention: epidemiology, social science, action, decrease in smoking prevalence and, finally, reduction in lung cancer mortality.

Science and epidemiology

Conclusive data on the hazards of tobacco exist on which preventive public health action can be based now, without waiting for further research. Continuing surveys on tobacco prevalence, mortality and morbidity related to tobacco use, attitudinal surveys and the economic impact of tobacco remain necessary, however, in order to assess the scope of the problem in each country, to illustrate the necessity of preventive measures and to evaluate the most effective of these measures.

Preventive measures

World experience has shown the crucial need for government commitment, funding and action in establishing a national programme to reduce the tobacco epidemic. It is important that all countries establish a national focal point to stimulate, support and coordinate anti-tobacco activities. The experience of such bodies is that it is not possible to cooperate with the tobacco industry, which is committed to the opposite goal — of increasing tobacco sales.

Ban on all tobacco promotion, advertising and sponsorship

Advertising conveys the message that smoking is associated with success, pleasure, relaxation, machoism, sports, freedom, beauty in nature, slimness, sophistication and sexuality. Castleden (1983), Charlton (1986) and Aitken *et al.* (1988) have shown that children are aware of and are influenced by tobacco advertising. Roemer (1986) reported that in 20 countries there was a total ban on advertising, in 17 there were strong partial bans, and in 21 there were moderate, partial bans. Partial bans have only partial effects and are frequently circumvented by ingenious, indirect advertising and sponsorship. A total ban allows children to grow up free from all commercial pressure to smoke and, as Bjartveit and Lund (1987) have shown in Norway, this leads to a decrease in the numbers of children who smoke.

Ban on sales to youth

Banning sales to young people may be good in principle, but in practice it is difficult to enforce. The countereffect is that, by suggesting that smoking is an adult activity, some youths, who wish to appear adult or daring, may even be encouraged to smoke. Saito (1987) reported that, in Japan, where persons under 20 years of age are not allowed to smoke, the adult rates of smoking are nevertheless established in the teenage years, indicating the ineffectiveness of that law.

Effective, rotating health warnings

Roemer (1986) noted that in 53 countries health warnings are required. Ramström (1980) in Sweden has shown that smokers read and are aware of informative, rotating warnings. In many countries, governments are moving from single, mild warnings, such as 'Smoking may harm your health', to tougher, rotating health warnings, such as: 'Smoking kills,' 'Smoking causes lung cancer,' 'Tobacco is addictive,' 'Smoking in pregnancy harms your baby,' 'Smoking causes heart disease,' and 'Quit smoking and feel healthier.'

Limits on harmful substances

Lowering the levels of tar in cigarettes can prevent about one-third of cases of lung cancer, although quitting can prevent more. A ceiling of about 10–15 mg tar per cigarette is a reasonable current aim. Smokers have an exaggerated perception of the benefits of low-tar cigarettes, however, so the tobacco industry should never be allowed to suggest that a lower tar cigarette is a 'safe' cigarette — especially since cigarettes kill even more people by other diseases than by lung cancer, and such evidence as is available on these other diseases suggests that differences in machine-measured tar delivery may involve little difference in risk.

Ban on new forms of tobacco

New forms of manufactured tobacco products are constantly being launched. Pre-emptive bans are true preventive health measures in that they avoid an additional range of cancers.

Smoke-free areas

Smoking is not only unpleasant to nonsmokers but may also give them cancer (IARC, 1986). In many countries, smoking has been banned in public areas, public transport, places of work (especially health premises), schools and government offices. Most people are nonsmokers; thus, the freedom of the majority to breathe clean air is an important consideration. Bans also help smokers: the creation of smoke-free areas encourages smokers to cut down or quit, thus helping them to make a decision in the best interest of their own health.

Price policy

Increasing taxes on cigarettes is a very effective way of reducing smoking without loss of revenue to governments. Most smokers give cost and health as the two main reasons for quitting. Lewit *et al.* (1981) and Warner (1984) reported that in the USA, for example, for every 10% increase in tax there is a 4% decrease in the number of smokers and a 14% decrease in the number of teenage smokers. The World Health Organization (1984) noted that 'Millions of lives could be saved if steep taxes were imposed on tobacco.' Grossman (1983) concluded that increasing taxes has a particularly beneficial effect upon young people and the poor, who have less money to spend and are therefore more likely to quit.

Health information and education

Health information and education form an important part of a comprehensive anti-tobacco programme, by educating both decision makers and the population to understand and accept legislative and other anti-tobacco measures. In contrast to the attractive 'Come and join us' images used by the tobacco industry, many health educators have traditionally used depressing, boring health statistics and finger-wagging 'Don't smoke' messages, which may encourage adults to quit but seem to have little effect in preventing young people from starting to smoke. Health education — especially that geared to youth — is now moving towards positive, healthy images.

Overcoming the obstacles

The media, the medical profession and politicians must be persuaded that smoking is harmful and that action is needed; yet these groups are often obstacles to tobacco control.

The media

Warner (1985) stated that 'Studies dating back to the 1930s provide evidence that the media's dependence on revenue from cigarette advertising has repeatedly led to suppression of smoking and health matters,' concluding 'It seems likely that there are more people who smoke today than there would be in an environment of responsible media coverage. The result is an avoidable excess burden of suffering and premature death. As long as cigarette advertising remains legal and widespread, its influence on editorial coverage of smoking and health is likely to persist.'

As well as avoiding financial pressures that distort media coverage, it is important to avoid the pressure to regard tobacco-related issues as stale news. Simply saying that smoking causes cancer does not produce front-page headlines. Doctors involved with anti-tobacco programmes must learn to become public relations experts, to write press releases, to nurture journalists, to highlight newsworthy items and, perhaps most difficult of all, to drop their professional jargon and find lively ways of presenting complicated medical statistics accurately — roles for which we have received little training.

The health profession

Doctors are more often involved with cure than with prevention. Medical societies usually have little money in comparison with the wealth of the tobacco industry. Both individual doctors and medical societies are often reluctant to be involved in political issues or public confrontations with the well-groomed representatives of the tobacco industry. Tobacco is a health problem, but the resolution of this problem is political. Decisions on nationwide containment of tobacco use, for example by legislation, lie with governments not with hospitals. If doctors are in the business of prevention, however, they also are in the business of politics. Doctors who do not address the political dimensions of the tobacco epidemic cannot hope to contribute substantially to reducing the scope of this epidemic.

The support of medical organizations to national anti-tobacco efforts is essential: in giving health information to the public, in advising governments and in participating in anti-tobacco actions. Statements from international and national health organizations can have an influential effect by indicating solidarity on this issue. In Hong Kong, all 65 medical societies have agreed in writing that tobacco is harmful to health — a powerful statement of support. Medical meetings can be declared 'smoke-free meetings'. Individual doctors can set an example by not smoking, making their offices smoke-free, displaying posters, giving out pamphlets on how to quit, participating in anti-smoking efforts, refusing tobacco money and not buying shares in the tobacco industry.

Walking a more lively path, one finds organizations like 'BUGA-UP' (Billboard Utilising Graffitists Against Unhealthy Promotions), spawned in Australia, and 'DOC' (Doctors Ought to Care) in the USA. These groups comprise health professionals and others who have become weary of the inactivity of governments on this public health issue. Staging eye-catching 're-facing' of billboards, using catchy slogans, cheerfully disrupting tobacco-sponsored events and promotional displays with alternative messages and appearing dressed in skeleton suits, these groups certainly focus media and public attention on tobacco.

Governments and politicians

Governments are often worried about 'losing' immediate tax revenue, and politicians may have pressure put upon them by large tobacco companies. In fact, tobacco use drains the economy by medical and health costs, lost productivity, welfare costs, costs of fires and cost of the use of land that could be used to grow food.

The tobacco industry

The international tobacco industry is the largest, most determined and strongest opposition to tobacco control. It is organized globally, commands considerable political influence and continues to deny the main evidence about the health effects of tobacco. It has vast amounts of money to sponsor sport, the arts, academic institutions and many other organizations, and sponsorship of institutions makes them less likely to speak out against tobacco.

Nowadays, a country in which effective legislation is attempted can expect a coordinated and intensive confrontation with the international tobacco industry, and use of double standards. The international tobacco companies, although based in countries with long-established bans, strenuously fought Hong Kong's ban on tobacco advertising on television. The same companies are breaking the stated regulations of China by advertising cigarettes.

Sections of the US government have threatened trade sanctions against Hong Kong, Japan, the Republic of Korea, Taiwan and Thailand if those countries do not open their markets to the sale or advertising of US cigarettes. McNeil (1988) reported a comment of the Philip Morris tobacco company: 'The suspension of the tariffs in Japan and the recent opening of the market in Taiwan are the direct result of effective negotiations by the Office of the US Trade Representative [of the US government].' The political coercion of trace sanctions is noteworthy by an industry that so often speaks of freedom. The tobacco companies make no apology for this. Chan (1988) reported that a representative of the R.J. Reynolds company even said 'We expect such support [from the US government]. That's why we vote them in.'

A class action suit has been brought in the Philippines against US tobacco companies for failure to provide the same level of protection for Filipino children as that provided for children in their country of origin. If successful, this would result in a ban on television advertising, health warnings and enforcement of the same tar levels as in the same cigarettes sold in the USA. This case has implications for many developing countries.

Litigation is a more recent and interesting development in the 'tobacco war'. Ramström (1986) reported that compensation had been awarded for the harmful effects of passive smoking in causing lung cancer in Sweden; there has recently been a similar case in Australia; and headline-catching cases are now in progress in the USA (Eichenwald, 1988), which reverberate on tobacco shares on Wall Street. The number of such cases will undoubtedly increase over the next few decades and could possibly bring the tobacco industry to its knees financially.

A note of optimism

Is it possible to be optimistic about tobacco control? Worldwide, the health statistics give little comfort, and death and disability from tobacco will certainly increase over the next few decades. The behaviour of the tobacco industry gives no comfort either. But optimism stems from several points: data on health effects are slowly being collated — a crucial first step. While some countries have still taken virtually no action against tobacco, health concerns are mobilizing in many others. Remarkably similar battles are being fought all over the world. There is now sharing of international expertise in countering the tobacco industry at the political and legislative level. Roemer (1986) reported that, by mid-1986, over 70 countries had enacted legislation to control smoking and many had strengthened existing legislation — a continuing process. An ancient Chinese saying goes: 'A journey of 10 000 miles begins with a single step.' The first steps in cancer reduction have already been taken, but there are many miles ahead.

References

Aitken, P.P., Leather, D.S., Scott, A.L. & Squair, S.I. (1988) Cigarette brand preferences of teenagers and adults. *Health Promotion*, **2**, 219-226

Bjartveit, K. & Lund, K.E. (1987) *Smoking Control in Norway*, Oslo, National Council on Smoking and Health, pp. 1-4

Castleden, W.M. (1983) Advertising, cigarettes and young smokers. *Med. J. Aust.*, **1**, 196-197

Chan, G. (1988) Tobacco giants and COSH trade fiery words over smoking. *South China Morning Post*, October 16, p. 4

Charlton, A. (1986) Children's advertisement-awareness related to their views on smoking. *Health Educ. J.*, **45**, 75-78

Eichenwald, K. (1988) Tobacco case verdict hailed. Report on Rose Cipollone case from *New York Times*. *Hong Kong Standard*, June 16

Grossman, M. (1984) Taxation and cigarette smoking in the United States. In: *Proceedings of the 5th World Conference on Smoking and Health (Winnipeg, 10-15 July, 1983)*, Ottawa, Canadian Council on Smoking and Health, Vol. 1, pp. 483-487

IARC (1986) *IARC Monographs on the Evaluation of the Carcinogenic Risk of Chemicals to Humans*, Vol. 38, *Tobacco Smoking*, Lyon

Lewit, E.M., Coate, D. & Grossman, M. (1981) The effects of government regulation on teenage smoking. *J. Law Econ.*, **24**, 545-569

McNeil, M. (1988) Cigarette firms look overseas as the market dries up in the US. *South China Morning Post*, January 26, p. 4

Ramström, L.M. (1980) The Swedish programme for smoking control – current progress and future plans. In: Ramström, L.M., ed., *The Smoking Epidemic, A Matter of Worldwide Concern*, Stockholm, Almqvist & Wiksell

Ramström, L.M. (1986) A case of lung cancer classified as occupational injury due to passive smoking at the workplace – a ruling by the Swedish Insurance Court of Appeal. Press Release, National Smoking and Health Association, Sweden; January 13

Roemer, R. (1986) *Recent Developments in Legislation to Combat the World Smoking Epidemic* (WHO/SMO/HLE/86.1), Geneva, World Health Organization

Saito, R. (1987) Smoking among young women in Japan. In: Aoki, M., Hisamichi, S. & Tominaga, S., eds, *Smoking and Health 1987 (Excerpta Medica International Congress Series 780)*, Amsterdam, Elsevier, 517-519

Wald, N.J., Nanchahal, K., Thomson, S.G. & Cuckle, H.S. (1986) Does breathing other people's tobacco smoke cause lung cancer? *Br. Med. J.*, **293**, 1217-1222

Warner, K.E. (1984) Cigarette taxation; doing good by doing well. *J. Public Health Policy*, **5**, 312-319

Warner, K.E. (1985) Cigarette advertising and media coverage of smoking and health. *New Engl. J. Med.*, **312**, 384-388

World Health Organization (1984) World tobacco tax could help save millions of lives. *WHO Tobacco Alert Series 1*, Vol. 2, No. 4:1

Relevance to Human Cancer of N-Nitroso Compounds,
Tobacco Smoke and Mycotoxins.
Ed. I.K. O'Neill, J. Chen and H. Bartsch
Lyon, International Agency for Research on Cancer
© IARC, 1991

ANALYSIS AND PYROLYSIS OF SOME N-NITROSAMINO ACIDS IN TOBACCO AND TOBACCO SMOKE

K.D. Brunnemann, M.V. Djordjevic, R. Feng & D. Hoffmann

*Naylor Dana Institute for Disease Prevention,
American Health Foundation, Valhalla, NY, USA*

A new tobacco-specific nitrosamine, 4-(N-nitrosomethylamino)-4-(3-pyridyl)butyric acid (iso-NNAC), has been identified in tobacco, and its structure was confirmed by gas chromatography–mass spectrometry following enrichment of a tobacco extract. The levels of iso-NNAC ranged from 0.01 to 0.95 ppm. It does not induce DNA repair in primary rat hepatocytes and is inactive as a tumorigenic agent in strain A mice. In order to study the fate of nitrosamino acids during smoking, we spiked cigarettes with the following N-nitrosamino acids: iso-NNAC, 3-(nitrosomethylamino)propionic acid (NMPA), 4-(nitrosomethylamino)butyric acid (NMBA), N-nitrososarcosine (NSAR) and N-nitrosoproline (NPRO). NMPA and NMBA were partially transferred, unchanged, during smoking and partially formed the corresponding methyl esters, while pyrolysis of NSAR and NPRO resulted mainly in their decarboxylating products. This is the first time that the pyrosynthesis of methyl esters has been observed during smoking.

We have isolated and identified a new tobacco-specific nitrosamine (TSNA) and consider it a potential biomarker for exposure to TSNA. In addition, we studied the fate of certain nitrosamino acids during the burning of a cigarette in order to determine how they contribute to the formation of volatile nitrosamines and to the carcinogenic potential of cigarette smoke.

Tobacco-specific nitrosamines

Nitrosation of nicotine with sodium nitrite (Figure 1) *in vitro* gives rise to three TSNA: N'-nitrosonornicotine (NNN), 4-(nitrosomethylamino)-1-(3-pyridyl)-1-butanone (NNK) and 4-(nitrosomethylamino)-4-(3-pyridyl)butanol (NNA; Hecht *et al.*, 1978). Although NNA was found in neither tobacco nor tobacco smoke, its reduction product, 4-(nitrosomethylamino)-4-(3-pyridyl)-1-butanol (iso-NNAL), has been identified in dry and moist snuff together with the reduction product of NNK, 4-(nitrosomethylamino)-1-(3-pyridyl)-1-butanol (NNAL; Brunnemann *et al.*, 1987). The identification of iso-NNAL led to the hypothesis that 4-(N-nitrosomethylamino)-4-(3-pyridyl)butyric acid (iso-NNAC), an oxidation product of NNA, would be present in tobacco.

We developed an analytical procedure for the determination of iso-NNAC in tobacco, consisting of an aqueous extraction of tobacco and subsequent solvent partition with ethyl acetate at pH 2 (nitrosamino acid fraction), pH 9 (TSNA fraction) and pH 4 (iso-NNAC fraction). The final fractions were methylated and analysed by gas chromatography-thermal energy analyser (GC-TEA; Djordjevic *et al.*, 1989). Table 1 presents the levels of iso-NNAC and other nitrosamino acids in different tobaccos. 3-(Nitrosomethylamino)propionic acid (NMPA) and *N*-nitrosoproline (NPRO) were the most abundant nitrosamino acids in all tobaccos analysed; 4-(nitrosomethylamino)butyric acid (NMBA) was present at lower levels, and iso-NNAC at concentrations of 0.01–0.95 ppm. The ranked order of abundance of the nitrosamino acids in tobacco was: NPRO > NMPA > NMBA > iso-NNAC.

Figure 1. Formation of tobacco-specific *N*-nitrosamines

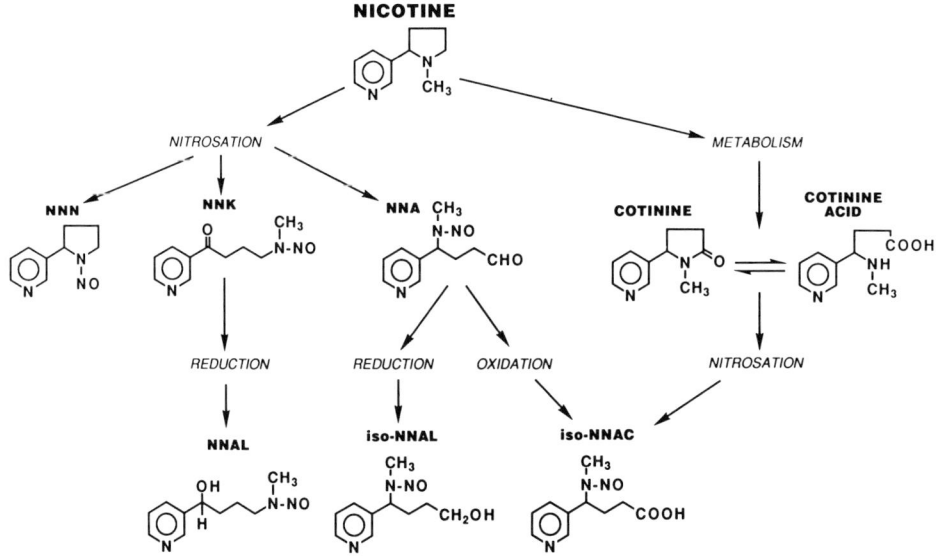

In order to verify the structure of iso-NNAC, we analysed the enriched pH 4 fraction by capillary gas chromatography-mass spectrometry (GC–MS). The compound isolated from tobacco was confirmed as iso-NNAC. In order to study the biological activity of iso-NNAC, we employed the rat hepatocyte primary culture/DNA repair test (Djordjevic *et al.*, 1989; Williams *et al.*, 1989); in addition, iso-NNAC was bioassayed for lung adenomas in strain A/J female mice (Rivenson *et al.*, 1989). Iso-NNAC did not induce DNA repair in primary rat hepatocytes and, at 200 µmol/mouse, did not increase the rate of lung adenomas.

Pyrolysis studies

Carcinogenic volatile *N*-nitrosamines, such as *N*-nitrosodimethylamine and *N*-nitrosopyrrolidine, have been detected in tobacco and tobacco smoke. We postulated that they may be formed by decarboxylation of nitrosamino acids occurring in tobacco. We therefore selected NMPA, NMBA, *N*-nitrososarcosine (NSAR) and NPRO and applied 5 mg of each dissolved in 50 µl water to the tobacco column of 20 cigarettes. The cigarettes were then smoked under standard conditions and the smoke analysed by GC–TEA.

Table 1. Contents of nitrosamino acids in different tobacco products

Product type	Sample	Nitrosamino acids (μg/g dry weight)[a]				
		NMPA	NMBA	NPRO	iso-NNAC	Total
Chewing tobacco	KY 1S1	1.0	0.05	0.7	0.03	1.8
	A	0.6	0.03	0.2	0.02	0.8
Moist snuff	KY 1S3	4.6	0.40	6.6	0.13	11.8
	A	3.2	0.26	7.1	0.05	10.6
	E	11.0	0.12	4.5	0.21	15.8
Dry snuff	KY 1S2	13.1	1.54	15.4	0.95	31.0
	A	1.2	0.14	3.0	0.05	4.4
	B	4.5	0.46	8.1	0.21	13.3
Cigarette tobacco	KY 1R1	0.16	ND	0.59	0.01	0.76
	KY 1R4F	0.63	ND	0.57	0.01	1.21
	A	3.10	0.17	2.62	0.05	5.94

[a] Abbreviations: NMPA, 3-(nitrosomethylamino)propionic acid; NMBA, 4-(nitrosomethylamino)butyric acid; NPRO, N-nitrosoproline; ND, not detected

NMPA did not yield its decarboxylation product but partially formed its methyl ester (Figure 2). This fraction was further enriched by rotary thin-layer chromatography, and subsequent GC-MS analysis confirmed the structural identity of the methyl ester of NMPA (Figure 3). In addition, we observed that some NMPA was carried over intact into the mainstream smoke. Pyrolysis of NMBA on a tobacco column also resulted in formation of the methyl ester, as confirmed by GC-MS. NMBA was also decarboxylated and was carried over into the smoke to some extent. NSAR was primarily decarboxylated to form N-nitrosodimethylamine. NPRO was decarboxylated to a large extent to form N-nitrosopyrrolidine, although some methylation and transfer into the smoke occurred. This is the first time that pyrosynthesis of methylesters has been observed during smoking.

Interestingly, when iso-NNAC was pyrolysed on a tobacco column, it did not undergo decarboxylation but its condensation product was formed (0.07%); 0.9% of iso-NNAC was transferred unchanged into the mainstream smoke.

In order to study the mechanism of formation of these esters, we spiked cigarettes with 3-(N-nitrosoethylamino)propionic acid, which resulted in the formation of both the methyl and ethyl esters at a ratio of 2:1. Additional mechanistic studies with deuterated 3-(N-nitrosoethylamino)propionic acid are under way.

Future studies

Future studies will focus on the conditions leading to the endogenous formation of iso-NNAC and its analysis in urine. Iso-NNAC may serve as a biomarker for measuring exposure to TSNA and/or the extent of endogenous formation.

Figure 2. Gas chromatography–thermal energy analyser (TEA) traces of volatile N-nitrosamine reference mixture (top) and of pyrolysis fraction of 3-(N-nitrosomethylamino)propionic acid (bottom)[a]

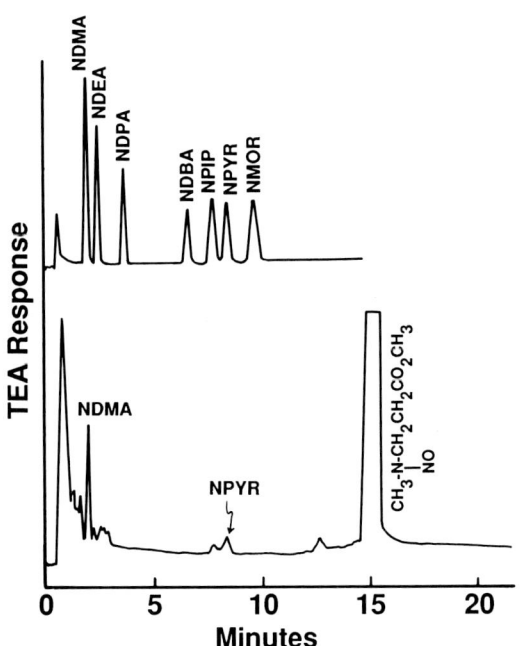

[a] Abbreviations: NDMA, N-nitrosodimethylamine; NDEA, N-nitrosodiethylamine; NDPA, N-nitrosodipropylamine; NDBA, N-nitrosodibutylamine; NPIP, N-nitrosopiperidine; NPYR, N-nitrosopyrrolidine; NMOR, N-nitrosomorpholine

Acknowledgement

This study was supported by Grant CA-29580 from the US National Cancer Institute.

References

Brunnemann, K.D., Genoble, L. & Hoffmann, D. (1987) Identification and analysis of a new tobacco-specific N-nitrosamine, 4-(methylnitrosamino)-4-(3-pyridyl)-1-butanol. *Carcinogenesis*, **8**, 465-469

Djordjevic, M.V., Brunnemann, K.D. & Hoffmann, D. (1989) Identification and analysis of a nicotine-derived N-nitrosamino acid and other nitrosamino acids in tobacco. *Carcinogenesis*, **10**, 1725-1731

Hecht, S.S., Chen, C.B., Ornaf, R.M., Jacobs, E., Adams, J.D. & Hoffmann, D. (1978) Reaction of nicotine and sodium nitrite: formation of nitrosamines and fragmentation of the pyrrolidine ring. *J. Org. Chem.*, **43**, 72-76

Rivenson, A., Djordjevic, M.V., Amin, S. & Hoffmann, D. (1989) A study of tobacco carcinogenesis. XLIV. Bioassay in A/J mice of some N-nitrosamines. *Cancer Lett.*, **47**, 111–114

Williams, G.M., Mori, H. & McQueen, C.A. (1989) Structure–activity relationships in the rat hepatocyte DNA-repair test for 300 chemicals. *Mutat. Res.*, **221**, 263-286

Figure 3. Mass spectra of the methyl ester of 3-(nitrosomethylamino)propionic acid: A, reference; B, isolated from tobacco smoke

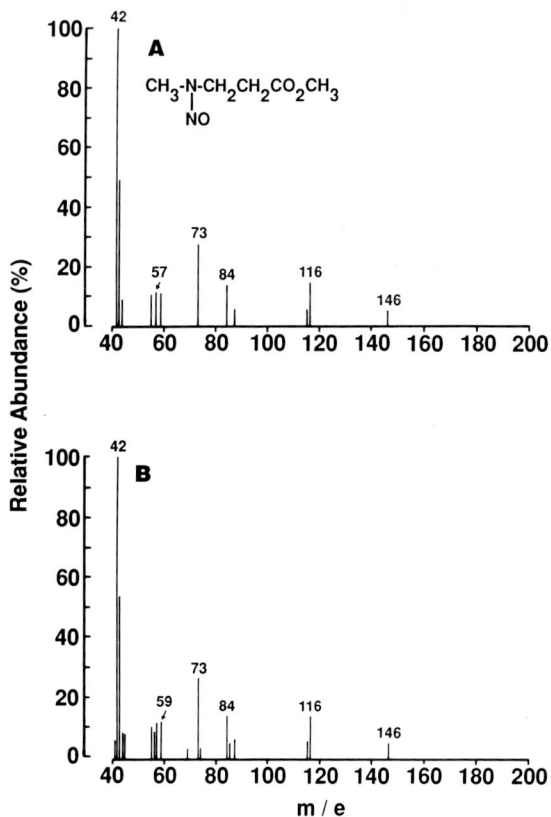

STUDIES IN TOBACCO CARCINOGENESIS

D. Hoffmann, A.A. Melikian & K.D. Brunnemann

*Naylor Dana Institute for Disease Prevention,
American Health Foundation, Valhalla, NY, USA*

The vapour phase of freshly generated cigarette mainstream smoke, of sidestream smoke and of environmental tobacco smoke was analysed for such tumorigenic agents as benzene, 1,3-butadiene and acrolein with a newly developed, highly sensitive gas chromatography–mass selective detection method. The major carcinogen in tobacco smoke, catechol, was studied in regard to its specific action on the metabolism of benzo[*a*]pyrene in mouse lung and mouse skin. The major tobacco-specific carcinogens in tobacco and its smoke are the nicotine-derived *N*-nitrosamines, *N*'-nitrosonornicotine and 4-(nitrosomethylamino)-1-(3-pyridyl)-1-butanone. A third nitrosamine that can be formed *in vitro* by nitrosation of nicotine is 1-(nitrosomethylamino)-1-(3-pyridyl)butylaldehyde. This aldehyde is not present in tobacco products, but its noncarcinogenic oxidation product, 4-(nitrosomethylamino)-1-(3-pyridyl)butyric acid, was found in tobacco and can be formed from the major nicotine metabolite, cotinine. It is also likely that this acid can be formed by endogenous reactions.

The objective of our studies in tobacco carcinogenesis lies in the elucidation of mechanisms of carcinogenesis and the determination of bioactive constituents in tobacco and tobacco smoke. This paper describes some of the chemical–analytical, biochemical and biological evaluations in progress.

Analysis of the vapour phase of tobacco smoke

The vapour phase of cigarette mainstream smoke and of sidestream smoke contains known tumorigenic agents (IARC, 1986). An analytical procedure was developed for the determination of selected gas-phase components in unaged smoke, utilizing cryofocusing capillary gas chromatography and mass selective detection. The latter was used in the selective ion monitoring mode, allowing us to scan for three selective ions, typical for each compound, during specific 'time windows'. Mainstream smoke was analysed *via* a ten-port gas sampling valve on a puff-by-puff basis. The concentrations of benzene and 1,3-butadiene increased only slightly with progressive numbers of puffs, except for cigarettes with charcoal-containing filter tips when volatiles that have been selectively retained in the filter from early puffs are released into the mainstream of later puffs. The mainstream smoke of one cigarette contains 6–73 µg benzene, 5–88 µg toluene, 16–70 µg 1,3-butadiene, 90–1060 µg isoprene and 8–260 µg acrolein. The sidestream smoke of one cigarette contains 350–650 µg benzene and 200–360 µg 1,3-butadiene. In the ambient air of rooms polluted with tobacco smoke we found 7-36 µg/m3 benzene and 0.8-4.5 µg/m3 1,3-butadiene.

Effect of catechol on the metabolism of benzo[a]pyrene

Bioassays have revealed that catechol (1,2-dihydroxybenzene), a major phenolic constituent of tobacco smoke, is a potent carcinogen and a weak co-initiator with benzo[a]pyrene (BP). The mechanism underlying these biological phenomena is not known. In vivo, catechol alters the penetration of BP into mouse skin as well as its metabolism. Specifically, catechol suppresses secondary steps in the metabolism of BP, namely epoxidation of 7,8-dihydroxy-7,8-dihydro-BP (BP-7,8-diol) to 7,8-dihydroxy-9,10-epoxy-7,8,9,10-tetrahydro-BPs (BPDEs). Co-application of catechol with BP or with BP-7,8-diol also decreases the speed with which the hydrocarbons penetrate skin. When applied to mouse skin with the racemic BP-7,8-diol, catechol is as potent a cocarcinogen as it is with BP itself. However, we found differences in the effects of catechol on the extent of metabolic activation of racemic and enantiomeric BP-7,8-diols and also with regard to DNA binding. Catechol suppresses epoxidation of the moderately carcinogenic (+)-BP-7,8-diol to a greater extent than that of the more active (−) enantiomer. In the presence of catechol, less of the major adduct is formed between metabolites of the (+)-BP-7,8-diol and DNA, while the presence of the phenolic compound has no significant impact on formation of the major DNA adduct derived from the (−) enantiomer. Consequently, catechol affects the proportion of major adducts derived from (+)- and (−)-BP-7,8-diols in mouse skin.

Tobacco-specific N-nitrosamines

N-Nitrosation of nicotine in vitro leads to three tobacco-specific N-nitrosamines, N'-nitrosonornicotine (NNN), 4-(nitrosomethylamino)-1-(3-pyridyl)-1-butanone (NNK) and 4-(nitrosomethylamino)-4-(3-pyridyl)butylaldehyde (NNA; Figure 1). While tobacco and tobacco smoke contain significant amounts of the highly carcinogenic NNN and NNK, NNA has not been identified in any tobacco product. However, both the reduction product of NNA, 4-(nitrosomethylamino)-1-(3-pyridyl)-1-butanol (iso-NNAL; 66–2500 ng/g) and the oxidation product of NNA, 4-(nitrosomethylamino)-4-(3-pyridyl)butyric acid (iso-NNAC; 10–950 ng/g) have been identified in processed tobacco. This acid is not genotoxic, nor does it induce tumours in mice. Iso-NNAC is also formed in vitro from the major nicotine metabolite, cotinine. Biochemical studies in progress indicate that iso-NNAC may also be formed in vivo after N-nitrosation of cotinine and cotinine acid.

Acknowledgement

Our studies in tobacco carcinogenesis are supported by grants No. CA-29580 and CA-43910 from the US National Cancer Institute.

Reference

IARC (1986) *IARC Monographs on the Evaluation of the Carcinogenic Risk of Chemicals to Humans*, Vol. 38, *Tobacco Smoking*, Lyon

Figure 1. *N*-Nitrosation products of nicotine[a]

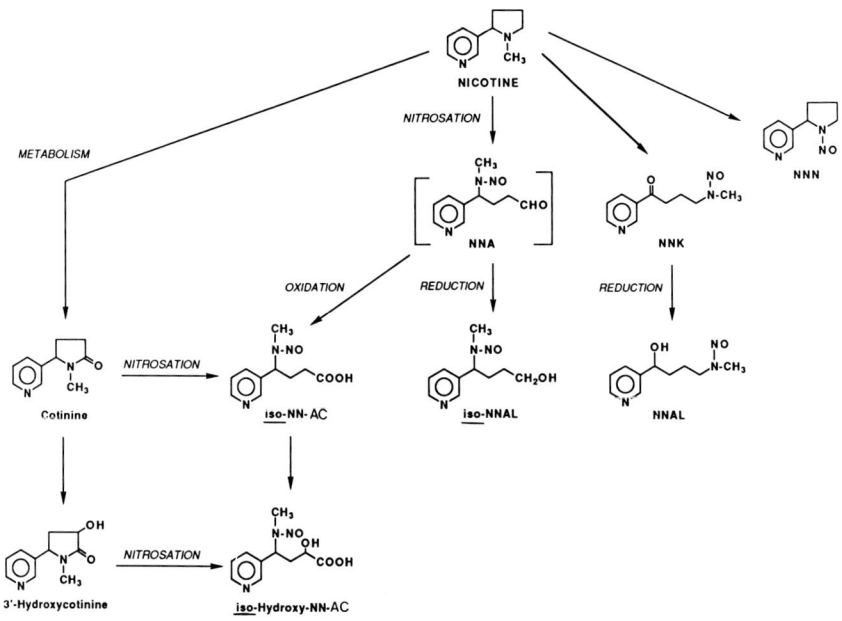

[a] NNN, *N'*-nitrosonornicotine; NNA, 4-(nitrosomethylamino)-4-(3-pyridyl)butylaldehyde; NNK, 4-(nitrosomethylamino)-1-(3-pyridyl)-1-butanone; NNAC, 4-(nitrosomethylamino)-4-(3-pyridyl)-butyric acid; NNAL, 4-(nitrosomethylamino)-1-(3-pyridyl)-1-butanol

Relevance to Human Cancer of N-Nitroso Compounds,
Tobacco Smoke and Mycotoxins.
Ed. I.K. O'Neill, J. Chen and H. Bartsch
Lyon, International Agency for Research on Cancer
© IARC, 1991

CARCINOGENIC SUBSTANCES IN SOVIET TOBACCO PRODUCTS

D.G. Zaridze[1], R.D. Safaev[1], G.A. Belitsky[1],
K.D. Brunnemann[2] & D. Hoffmann[2]

[1]*Department of Epidemiology and Prevention, All-Union Cancer Research Centre, the USSR Academy of Medical Sciences, Moscow, USSR; and* [2]*Naylor Dana Institute for Disease Prevention, American Health Foundation, Valhalla, NY, USA*

Chemical carcinogens were determined in mainstream smoke from nonfilter cigarettes produced and consumed in the USSR and in *nass*, a mixture of tobacco, lime, ash and cotton oil. Cigarettes contained high levels of tar (23–25 mg/cigarette) and nicotine (1.5–1.9 mg/cigarette) and, generally, a high content of polycyclic aromatic hydrocarbons, which are major epithelial carcinogens, N-nitrosamines, which are organ-specific carcinogens, and some carcinogenic metals, such as arsenic and chromium. *Nass* contained the tobacco-specific N-nitroso compounds, N'-nitrosonornicotine, N'-nitrosoanatabine, N'-nitrosoanabasine and 4-(N-nitrosomethylamino)-1-(3-pyridyl)-1-butanone, as well as volatile N-nitrosamines, but at levels lower than in other types of chewing tobacco and snuff. The low levels in *nass* are due to the short ageing process used, in contrast to commercially produced chewing tobacco and fine-cut snuff, which are highly processed products requiring long ageing and fermentation.

In the USSR, the proportion of filter cigarettes smoked increased from 0.3% in 1963 to 21.3% in 1982; however, 90% of the cigarettes consumed in this country fall within the category of 'high-tar' cigarettes (Zaridze *et al.*, 1986a). In some areas of the USSR, particularly in the central Asian republics, a smokeless tobacco product, *nass*, is widely used. Use of *nass* has been associated with a high incidence of oral cancer and a high prevalence of leukoplakia (Zaridze *et al.*, 1986b).

Carcinogens in cigarette smoke

Two types of nonfilter cigarettes were smoked according to ISO requirements (International Standards Organization, 1986) in a 20-port smoking machine (Borgwald). The concentration of polycyclic aromatic hydrocarbons was determined by spectroluminescence (Khesina *et al.*, 1983), metals by atomic emission spectroscopy (Westcott & Spincer, 1974; Jenkins, 1986), and carcinogenic nitrosamines by gas chromatography-thermal energy analysis (Brunnemann *et al.*, 1977; Adams *et al.*, 1983).

The cigarettes contained high levels of tar (23–25 mg/cigarette) and nicotine (1.5–1.9 mg/cigarette), and the smoke contained very high concentrations of some carcinogenic polycyclic aromatic hydrocarbons (Table 1). The concentrations of metals in Soviet cigarette smoke, particularly those of chromium and arsenic, which are known human carcinogens, were higher than those reported previously (IARC, 1986; Table 2).

Table 1. Polycyclic aromatic hydrocarbons (PAH) in the smoke of Soviet cigarettes

PAH	Carcinogenic activity in animals	Concentration range (ng/cigarette)	Concentration range reported by IARC (1986)
Benzo[a]pyrene	+++	101–108	10–50
Dibenz[a,h]anthracene	++	27–76	4–40
Benzo[b]fluoranthene	++	38–48	4–30
Benzo[k]fluoranthene	++	6–25	6–12
Dibenzo[a,i]pyrene	++	3–9	1.7–3.2
Dibenzo[a,h]pyrene	++	5–9.5	
Dibenzo[a,c]anthracene	+	60–76	
Benz[a]anthracene	+	69–88	40–70
Chrysene	+	47	40–60
Coronene	+	2–6	
Pyrene	Cocarcinogen	96–203	50–200
Fluoranthene	Cocarcinogen	340–350	100–260
Benz[ghi]perylene	Cocarcinogen	9–54	60
Benzo[e]pyrene	Promoter	10–44	5–40

Table 2. Content of metals in Soviet cigarette smoke

Metal	Concentration range in Soviet cigarettes (µg/cigarette)	Reported by IARC (1986) (µg/cigarette)
Na	0.4-0.42	1.3
Mg	2.0-2.37	0.07
Cr	0.3-0.5	0.004-0.069
Mn	0.075-0.175	0.003
Fe	0.45	0.042
Co	0.005-0.1	0.0002
Ni	0.1-0.3	0.0-0.51
Cu	0.1-0.25	0.19
Zn	2.2-5.5	0.012-0.022
Cd	0.11-0.31	0.007-0.35
Al	0.55-1.05	0.22
Pb	0.3-0.55	0.017-0.98
As	0.2-0.25	0.012-0.022
Se	0.05	0.001-0.063
Ca	11.5-13.6	
Sr	0.09-13.6	
Be	ND	
Mo	0.5	
Ti	ND	

ND, not detected

The concentrations of carcinogenic N-nitrosamines in Soviet cigarettes were in the upper range of values reported for nonfilter cigarettes or exceeded it, except for 4-(N-nitrosomethylamino)-1-(3-pyridyl)-1-butanone (Table 3).

Table 3. Content of N-nitrosamines[a] in Soviet cigarette smoke

N-Nitrosamine	Concentration range in Soviet cigarettes (ng/cigarette)	Concentration range reported by IARC (1986) (ng/cigarette)
NDMA	20-35	2-20
NPYR	66-82	1.5-30
NNN	580-840	310
NAT	320-540	370
NNK	57-97	150

[a] NDMA, N-nitrosodimethylamine; NDEA, N-nitrosodiethylamine; NPYR, N-nitrosopyrrolidine; NNN, N'-nitrosonornicotine; NAT, N'-nitrosoanatabine; NNK, 4-(N-nitrosomethylamino)-1-(3-pyridyl)-1-butanone

The cigarettes tested thus had elevated contents of the major epithelial contact carcinogens and cocarcinogens identified in the smoke of nonfilter cigarettes, of the major organ-specific carcinogens, N-nitrosamines, which are by far most powerful carcinogens in tobacco smoke, as well as of some carcinogenic metals.

Carcinogens in nass

Nass is a mixture of tobacco, lime, ash and cottonseed oil. The results of chemical analyses of four samples are presented in Table 4. The samples contained both tobacco-specific N-nitroso compounds and volatile N-nitrosamines, but at levels lower than those found in other types of chewing tobacco and snuff.

Table 4. Content of N-nitrosamines[a] in *nass*

N-Nitrosamine	Concentrations in samples of *nass* (ng/g)				US snuff tobacco[b] (ng/g)
	A	B	C	D	
NPIP	9.0	7.7	8.0	6.0	Not detected
NPYR	8.8	1.8	1.7	4.3	0-360
NNN	519	143	119	516	2200-33 000
NAT	289	39	39	167	1700-40 000
NAB	34	3.0	4.0	17	100-1900
NNK	108	16	29	126	600-8300

[a] NPIP, N-nitrosopiperidine; NPYR, N-nitrosopyrrolidine; NNN, N'-nitrosonornicotine; NAT, N'-nitrosoanatabine; NAB, N'-nitrosoanabasine; NNK, 4-(N-nitrosomethylamino)-1-(3-pyridyl)-1-butanone
[b] Ranges in four leading commercial brands (Brunnemann et al., 1982)

The finding of low levels of *N*-nitrosamines in *nass* is not surprising, because, contrary to commercially produced chewing tobacco and fine-cut snuff tobacco which are highly processed products requiring long ageing and fermentation, *nass* production requires only a short ageing process. The relatively low levels of these compounds in *nass* suggest that the relatively strong genotoxic activity of this product is due primarily to other chemicals — possibly oxidized phenolics (Zaridze *et al.*, 1986b). The absence of local carcinogenic effects of *nass* in experimental animals (Kiseleva *et al.*, 1976) suggests that the increased risk for cancer of the mouth associated with *nass* chewing is due either to the endogenous formation of carcinogens from components contained in *nass* or to a combination of such components with other endogenous or exogenous factors (Zaridze *et al.*, 1986b).

The age-standardized incidence rates of oral cancer in areas with a high prevalence of *nass* use, although high (6.0 and 3.0 per 100 000 population for males and females, respectively) in comparison with other regions of the USSR, are much lower than the highest or even intermediate rates worldwide (Muir *et al.*, 1987). This might be related to the relatively low content of tobacco-specific *N*-nitroso compounds.

References

Adams, J.D., Brunnemann, K.D. & Hoffmann, D. (1983) Rapid method for the analysis of tobacco-specific *N*-nitrosamines by gas-liquid chromatography with a thermal energy analyzer. *J. Chromatogr.*, 256, 347-351

Brunnemann, K.D., Yu, L. & Hoffmann, D. (1977) Assessment of volatile carcinogenic *N*-nitrosamines in tobacco and in mainstream and sidestream smoke from cigarettes. *Cancer Res.*, 37, 3218-3222

Brunnemann, K.D., Scott, J.C. & Hoffmann, D. (1982) *N*-Nitrosomorpholine and other volatile *N*-nitrosamines in snuff tobacco. *Carcinogenesis*, 3, 693-696

IARC (1986) *IARC Monographs on the Evaluation of the Carcinogenic Risk of Chemicals to Humans*, Vol. 38, *Tobacco Smoking*, Lyon

International Standards Organization (1986) *International Standard ISO 3308. Cigarettes. Routine Analytical Cigarette-smoking Machine. Definitions and Standard Conditions*, Geneva

Jenkins, R.A. (1986) Occurrence of selected metals in cigarette tobaccos and smoke. In: O'Neill, I.K., Schuller, P. & Fishbein, L., eds, *Enviromental Carcinogens: Selected Methods of Analysis*, Vol. 8, *Some Metals: As, Be, Cd, Cr, Ni, Pb, Se, Zn* (IARC Scientific Publications No. 71), Lyon, IARC, pp. 129-138

Khesina, A. Ya., Khitrovo, I.A. & Gevorkyan, B.Z. (1983) The quantitative detection of PAH in the environmental pollutants by spectroluminiscence (in Russian). *J. Appl. Spectrosc.*, 38, 928-934

Kiseleva, N.S., Milievskaja, I.L. & Chaklin, A.V. (1976) Development of tumours in Syrian hamsters during prolonged experimental exposure to nass. *Bull. World Health Organ.*, 54, 591-605

Muir, C., Waterhouse, J., Mack, T., Powell, J. & Whelan, S., eds (1987) *Cancer Incidence in Five Continents, Vol. V* (IARC Scientific Publications No. 88), Lyon, IARC

Westcott, D. & Spincer, D. (1974) The cadmium, nickel and lead content of tobacco and cigarette smoke. *Beitr. Tabakforsch.*, 7, 217-221

Zaridze, D.G., Matiakin, E.G., Stich, H.F., Hoffmann, D., Blettner, M., Poljakov, B.P., Rosin, M.P. & Brunnemann, K.D. (1986a) The effect of nass use and smoking on the risk of oral leukoplakia. *Cancer Detect. Prev.*, 9, 435-440

Zaridze, D.G., Dvoirin, V.V. & Kobljakov, V.A. (1986b) Smoking patterns in the USSR. In: Zaridze, D. & Peto, R., eds, *Tobacco: A Major International Health Hazard* (IARC Scientific Publications No. 74), Lyon, IARC, pp. 75-86

Relevance to Human Cancer of N-Nitroso Compounds,
Tobacco Smoke and Mycotoxins.
Ed. I.K. O'Neill, J. Chen and H. Bartsch
Lyon, International Agency for Research on Cancer
© IARC, 1991

TOBACCO-SPECIFIC NITROSAMINES IN COMMERCIAL CIGARETTES: POSSIBILITIES FOR REDUCING EXPOSURE

S. Fischer, B. Spiegelhalder & R. Preussmann

Institute of Toxicology and Chemotherapy, German Cancer Research Center, Heidelberg, Germany

Tobacco-specific nitrosamines (TSNA) are powerful carcinogens found in tobacco and tobacco smoke in relatively high concentrations. Tar delivery, which is generally accepted as an index for the carcinogenic potential of cigarette smoke, must be declared in most European countries. In this investigation of more than 170 types of commercial cigarettes from several European countries and the USA, no correlation was observed between tar delivery and mainstream smoke concentration of N'-nitrosonornicotine (NNN) and 4-(N-nitrosomethylamino)-1-(3-pyridyl)-1-butanone (NNK). Therefore, although crucial, tar delivery alone is not a sufficient index for the carcinogenic potential of cigarette smoke. It is proposed that TSNA concentrations be determined for characterization of the carcinogenic potential of cigarettes with low and ultra-low tar yields and that these be declared by an additional and adequate parameter. The mainstream smoke concentrations of NNN and NNK are given by the amounts of preformed compounds in tobacco, which is dependent on the nitrate content of the tobacco and the tobacco type. A further important determinant of the exposure of smokers to TSNA is the total volume drawn through a cigarette while smoking, which is dependent on puff volume and puff frequency and which directly influences TSNA transfer. Smokers inhale higher volumes when smoking low-nicotine cigarettes, so that low NNN:nicotine and NNK:nicotine ratios result in decreased exposure to TSNA. Reduction of exposure to TSNA can be achieved by selecting tobaccos with low levels of preformed TSNA (low nitrate content, small amounts of burley tobaccos and stems) and by manufacturing cigarettes with low NNN:nicotine and NNK:nicotine ratios.

Tobacco-specific nitrosamines (TSNA), which are powerful carcinogens, have been found in tobacco and tobacco smoke in relatively high concentrations (Hoffmann & Hecht, 1985). In order to reduce smokers' exposure to TSNA, we have investigated the main factors that influence their concentrations in mainstream smoke. Since tar delivery is generally accepted as an index of the carcinogenic potential of cigarette smoke, and must be declared in most European countries, this measure was compared with the amounts of TSNA, and especially N'-nitrosonornicotine (NNN) and 4-(N-nitrosomethylamino)-1-(3-pyridyl)-1-butanone (NNK), in cigarettes.

More than 170 types of commercial cigarettes from Austria, Belgium, Germany, France, Italy, Poland, Switzerland, the United Kingdom, the USA and the USSR were analysed for

TSNA in tobacco and in mainstream smoke as well as for nitrate in tobacco, by the methods of Spiegelhalder *et al.* (1989) and Fischer and Spiegelhalder (1989). The declared values of tar and nicotine were used. The results are presented in Table 1.

Tar delivery, which is widely considered to reflect the carcinogenic potential of cigarette smoke, did not correlate with the amounts of the strong carcinogens NNN and NNK in mainstream smoke (NNN: $r^2 = 0.18$; NNK: $r^2 = 0.14$). Furthermore, there was no correlation between nicotine and TSNA deliveries (NNN: $r^2 = 0.13$; NNK: $r^2 = 0.10$). Thus, the concentration of TSNA should also be determined and declared by an additional and adequate parameter (Fischer *et al.*, 1989a).

The mainstream smoke concentrations of NNN and NNK strongly depend on the amounts of preformed NNN and NNK in tobacco (Fischer *et al.*, 1989b). We saw a constant ratio between the two concentrations, which was not dependent on the level of nitrate in tobacco, except for NNK in nitrate-rich, dark tobacco cigarettes, nor on the nicotine level. Spiking the cigarettes with the nitrosamine precursors nicotine (at 10 mg/cigarette) and nitrate (at 4–20 mg/cigarette) prior to smoking did not significantly change the mainstream smoke concentrations of NNN and NNK (Fischer *et al.*, 1989b). These data indicate that NNN is not formed during smoking, and synthesis of NNK is very unlikely, at least for tobaccos with low nitrate levels. Thus, the NNN and NNK found in mainstream smoke is derived from preformed nitrosamines in the tobacco (Fischer *et al.*, 1989b).

The amount of preformed TSNA in tobacco is determined mainly by the nitrate level in the tobacco (Fischer *et al.*, 1989c), and this depends on the tobacco type, especially for NNK. The lowest TSNA concentrations were observed in oriental-type cigarettes, which contain little nitrate. In low-nitrate, Virginia-type cigarettes, the TSNA concentrations were also low, but NNK was present at the same or much higher concentrations than NNN, whereas in other cigarettes NNN levels exceeded NNK levels. The highest TSNA concentrations were found in cigarettes made of dark tobaccos, which are high in nitrate. In blended cigarettes, both high and low TSNA concentrations were found, correlating with the nitrate level of the tobacco composition.

A further important determinant of TSNA concentration in mainstream smoke is the total volume drawn through a cigarette while smoking, which is dependent on the puff volume and the puff frequency and which directly influences TSNA transfer (Fig. 1; Fischer *et al.*, 1989d). Since the smoker's interest is to maintain an adequate nicotine intake, smokers inhale higher volumes from low-nicotine cigarettes, resulting in higher intakes of TSNA and tar in relation to standard smoking conditions. Thus, with low ratios of NNN:nicotine and NNK:nicotine, the smoker's exposure to TSNA is decreased.

A reduction in the exposure of smokers to TSNA could be achieved by selecting cigarettes made of tobaccos with low concentrations of preformed TSNA, i.e., those with a low nitrate content, and containing little burley tobaccos and stems, and by manufacturing cigarettes with low ratios of NNN:nicotine and NNK:nicotine.

Table 1. Ranges of tar, nicotine, N'-nitrosonornicotine (NNN) and 4-(N-nitrosomethylamino)-1-(3-pyridyl)-1-butanone (NNK) deliveries in mainstream smoke (MS) as well as concentrations of preformed nitrosamines and nitrate in tobacco of commercial cigarettes from several European countries and the USA

Country	No. of samples	Tar (mg/cig min–max)	Nicotine (mg/cig min–max) MS	NNN (ng/cig) MS (min–max)	NNN Tobacco (min–max)	NNK (ng/cig) MS (min–max)	NNK Tobacco (min–max)	Nitrate (mg/cig; min–max)
Austria	5	9–15	0.7–0.9	306–1122	42–172	12–100	92–310	4.2–8.0
Belgium	7	13–16	1.0–1.3	504–1939	38–203	20–150	219–594	1.8–10.8
France	20	6–44	0.3–2.7	120–6019	11–1000	19–498	57–990	1.5–19.4
Germany	55	1–28	0.1–2.0	50–5316	5–625	ND–432	ND–1120	0.6–14.4
Italy	10	NA	NA	632–12 454	21–1353	8–1749	153–10 745	6.2–13.3
Poland	6	NA	NA	870–2760	121–347	38–105	140–450	4.4–12.8
Switzerland	3	12–15	0.9–1.2	1280–2208	121–226	69–124	450–554	6.4–7.8
United Kingdom	12	LT–MT	NA	140–1218	17–123	18–103	92–433	1.4–8.0
USA	20	NA	NA	993–1947	54–197	41–145	433–733	6.2–13.5
USSR	9	NA	NA	60–850	23–312	4–40	ND–150	1.7–9.1

NA, not available; ND, not detected (MS, <4 ng/cigarette; tobacco, <50 ng/cigarette); LT, low tar; MT, middle tar

Figure 1. Dependence of concentration of N'-nitrosonornicotine (NNN; ◇) and 4-(N-nitrosomethylamino)-1-(3-pyridyl)-1-butanone (NNK, ⊙) in mainstream smoke on the total volume drawn through a low-tar, blended filter cigarette

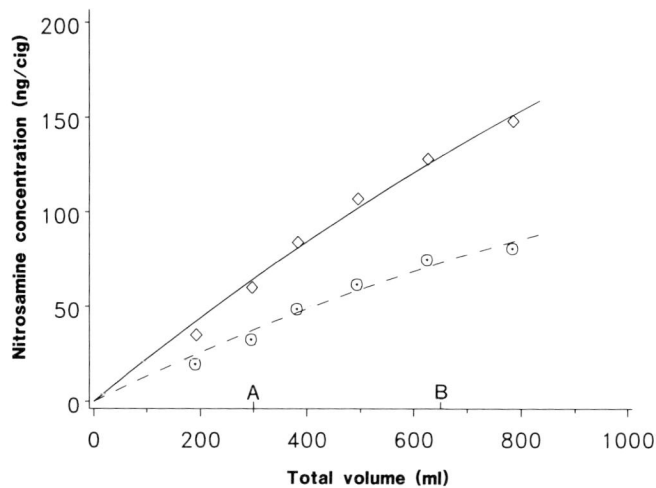

A, approximate total volume for standard smoking conditions; B, approximate average inhalation volume for low-tar, low-nicotine cigarettes

References

Fischer, S. & Spiegelhalder, B. (1989) Improved method for the determination of tobacco-specific nitrosamines in tobacco smoke. *Beitr. Tabakforsch.*, **14**, 145-153

Fischer, S., Spiegelhalder, B. & Preussmann, R. (1989a) Tobacco-specific nitrosamines in West German cigarettes – tar alone is not a sufficient index for the carcinogenic potential of cigarette smoke. *Carcinogenesis*, **14**, 169-173

Fischer, S., Spiegelhalder, B. & Preussmann, R. (1989b) Investigations on the origin of tobacco-specific nitrosamines in mainstream smoke of cigarettes. *Carcinogenesis* (submitted for publication)

Fischer, S., Spiegelhalder, B. & Preussmann, R. (1989c) Preformed tobacco-specific nitrosamines in tobacco - role of nitrate and influence of tobacco type. *Carcinogenesis*, **10**, 1511-1517

Fischer, S., Spiegelhalder, B. & Preussmann, R. (1989d) Influence of smoking parameters on the delivery of tobacco-specific nitrosamines in cigarette smoke - a contribution to relative risk evaluation. *Carcinogenesis*, **10**, 1059-1066

Hoffmann, D. & Hecht, S.S. (1985) Nicotine derived N-nitrosamines and tobacco related cancer: current status and future directions. *Cancer Res.*, **45**, 935-944

Spiegelhalder, B., Kubacki, S.J. & Fischer, S. (1989) A method for the determination of tobacco-specific nitrosamines, nitrate and nitrite in tobacco leaves and processed tobacco. *Beitr. Tabakforsch.*, **14**, 135-144

OCCURRENCE OF AND EXPOSURE TO N-NITROSO COMPOUNDS IN TOBACCO

A.R. Tricker & R. Preussmann

Institute of Toxicology and Chemotherapy, German Cancer Research Center, Heidelberg, Federal Republic of Germany

The concentrations of 21 N-nitroso compounds in smokeless tobaccos are presented. Tobacco-specific nitrosamines accounted for 70–90% of the total identified N-nitroso compounds. Daily exposure of smokeless tobacco users to preformed N-nitroso compounds may exceed 200 µg/day in certain populations.

We have identified 21 volatile, nonvolatile and tobacco-specific N-nitrosamines (TSNA) in smokeless tobacco products available commercially in 1987–88 (Tricker & Preussmann, 1988, 1989). Selected representative data are presented in Table 1. Currently, N-nitrosodimethylamine, N-nitrosoethylmethylamine, N-nitrosopiperidine and N-nitrosopyrrolidine are the most commonly found volatile nitrosamines; traces of N-nitrosodipropylamine and N-nitrosodibutylamine are present in heavily cured and/or fermented tobacco products. Trace levels of N-nitrosomorpholine are still found in some UK and Swedish oral tobacco products packed in waxed containers. The most commonly found nonvolatile N-nitrosamino acids and their derivatives are N-nitrosoproline, N-nitrosopipecolic acid, N-nitrosohydroxyproline, N-nitrososarcosine, 3-(N-nitroso-N-methylamino)propionic acid and 4-(N-nitroso-N-methylamino)butyric acid. Some heavily cured/fermented tobaccos, in particular *zarda* and some American moist snuffs (data not presented), were found to contain N-nitrosoazetidine 4-carboxylic acid (< 200 µg/kg) and N-nitrosothiazolidine 4-carboxylic acid (< 280 µg/kg). The presence of other nonvolatile N-nitrosamines, amenable to gas chromatography–thermal energy analysis only after methylation or silylation, were also found. TSNA are by far the most abundant N-nitroso compounds present in tobacco and usually account for 70–90% of the total. 4-(N-Nitrosomethylamino)-4-(3-pyridyl)-1-butanol (< 8100 µg/kg) has been detected in about 60% of all tobacco samples analysed (n = > 120), including various forms of smoking tobacco, indicating far greater exposure to this compound than previously reported (Hecht & Hoffmann, 1988).

In order to make a rough estimate of potential exposure to carcinogenic N-nitroso compounds, and in particular to 4-(N-nitrosomethylamino)-1-(3-pyridyl)-1-butanol and N'-nitrosonornicotine during use of smokeless tobacco, mean daily exposure to the identified N-nitroso compounds was estimated by multiplying the concentrations of individual compounds by the average amount of tobacco consumed by the tobacco-using population (Table 2). This method of calculating exposure is based on the assumption that

Table 1. N-Nitroso compounds in smokeless tobaccos (μg/kg)

N-Nitrosamine[a]	UK oral tobacco (5 samples)		Swedish moist snuff (5 samples)		Indian zarda (11 samples)		European nasal snuff (10 samples)	
	Mean	Range	Mean	Range	Mean	Range	Mean	Range
NDMA	37	6.0–82	1.5	1.0–2.5	11	2.0–31	20	2.0–82
NEMA	1.0	ND–2.	ND		1.0	ND–2.0	1.6	ND–8.5
NPIP	20	2.5–40	ND		0.3	ND–2.0	3.8	ND–17
NPYR	120	64–190	5.0	4.5–6.0	100	6.0–69	44	1.5–13
NMOR	0.5	ND–1.	1.0	ND–1.0	ND		ND	
NDELA	105	ND–220	19	8–31	9.5	ND–54	12	ND–42
NSAR	120	29–240	19	8–31	49	ND–350	21	ND–85
NPMA	3980	1360–8300	1340	1040–1820	2050	22–18 000	1620	490–4260
NBMA	640	62–1470	70	53–94	170	ND–2040	160	76–410
NAzCA	ND		ND		18	ND–140	ND	
NPRO	2260	330–4950	1100	630–1820	2850	280–18 000	3950	770–873
NPIC	540	83–1760	36	ND–130	260	ND–2040	80	ND–310
NTCA	6	ND–28	21	ND–69	48	ND–280	7.4	ND–46
NHPRO	330	92–610	140	ND–230	69	ND–190	98	46–27
iso-NNAl	65	ND–150	27	ND–80	1420	120–8100	360	ND–1590
NAB/NAT	3520	1980–4800	2640	1650–3250	16 030	780–99 100	3120	1030–7830
NNN	3960	1090–7630	3360	2100–4800	13 420	400–79 000	7840	2390–18 750
NNK	2800	400–8250	790	400–1040	4030	220–24 100	2430	580–6430
Total[b]	19 770	6130–36 450	9570	6220–12 500	40 530	1670–241 000	19 780	7190–42 530

[a] NDMA, N-nitrosodimethylamine; NEMA, N-nitrosoethylmethylamine; NPIP, N-nitrosopiperidine; NPYR, N-nitrosopyrrolidine; NMOR, N-nitrosomorpholine; NDELA, N-nitrosodiethanolamine; NSAR, N-nitrososarcosine; NPMA, N-nitrosopropylamine; NBMA, N-nitrosobutylmethylamine; NAzCA, N-nitrosoazetidine 4-carboxylic acid; NPRO, N-nitrosoproline; NPIC, N-nitrosopipecolic acid; NTCA, N-nitrosothiazolidine 4-carboxylic acid; NHPRO, N-nitrosohydroxyproline; iso-NNAl, 4-(N-nitrosomethylamino)-4-(3-pyridyl)-1-butanol; NAB/NAT, N′-nitrosoanabasine/N′-nitrosoanatabine; NNN, N′-nitrosonornicotine; NNK, 4-(N-nitrosomethylamino)-1-(3-pyridyl)-1-butanone
[b] Total identified N-nitroso compounds
ND, not detected; limits of detection: volatile nitrosamines, 0.1 μg/kg; nonvolatile nitrosamines, 1.0 μg/kg; tobacco-specific nitrosamines, 1.0 μg/kg tobacco

100% of TSNA are extracted from saliva. Österdahl and Slorach (1988) have shown that this is not true in Swedish snuff dippers, whose saliva contains higher levels of TSNA than can be accounted for by levels of preformed nitrosamines in tobacco. Thus, endogenous formation of TSNA in saliva (Österdahl & Slorach, 1988) and under simulated gastric conditions (Tricker et al., 1988) probably results in higher exposure to TSNA than indicated in Table 2.

Table 2. Exposure to *N*-nitroso compounds from use of oral tobacco

Tobacco-using population	Use (g/day)	Daily nitrosamine exposure (μg/day)	
		Total	NNN/NNK[a]
Dry snuff (Europe)	2.0	39.6	20.5
Oral tobacco (UK)	4.5	89.0	30.4
Moist snuff (Sweden)	15.9	152.1	66.0
Zarda (India)	5.0	203.0	87.5

[a]NNN, *N'*-nitrosonornicotine; NNK, 4-(*N*-nitrosomethylamino)-1-(3-pyridyl)-1-butanone

We conclude that high levels of *N*-nitroso compounds present in smokeless tobaccos expose the consumer to a considerable exogenic burden of potentially carcinogenic compounds, and in particular TSNA.

References

Hecht, S.S. & Hoffmann, D. (1988) Tobacco-specific nitrosamines, an important group of carcinogens in tobacco and tobacco smoke. *Carcinogenesis*, **9**, 875-884

Österdahl, B.-G. & Slorach, S. (1988) Tobacco-specific *N*-nitrosamines in saliva of habitual male snuff dippers. *Food Addit. Contam.*, **5**, 581-586

Tricker, A.R. & Preussmann, R. (1988) The occurrence of *N*-nitroso compounds in zarda tobacco. *Cancer Lett.*, **42**, 113-118

Tricker, A.R. & Preussmann, R. (1989) Preformed nitrosamines in smokeless tobacco. In: Maskins, A.P., ed., *Tobacco and Cancer, Perspectives in Prevention Research*, Amsterdam, Elsevier, pp. 35-47

Tricker, A.R., Haubner, R., Spiegelhalder, B. & Preussmann, R. (1988) The occurrence of tobacco-specific nitrosamines in oral tobacco products and their potential formation under simulated gastric conditions. *Food Chem. Toxicol.*, **26**, 861-865

LOCALIZATION OF DNA ADDUCTS FORMED IN THE NASAL CAVITY OF THE RAT BY THE TOBACCO-SPECIFIC NITROSAMINE 4-(N-NITROSOMETHYLAMINO)-1-(3-PYRIDYL)-1-BUTANONE (NNK)

J. Van Benthem[1], J.W.G.M. Wilmer[2,3], L. Den Engelse[1] & E. Scherer[1]

[1]*The Netherlands Cancer Institute (Antoni van Leeuwenhoek Huis), Division of Chemical Carcinogenesis, Amsterdam; and* [2]*TNO-CIVO Toxicology and Nutrition Institute, Department of Biological Toxicology, Zeist, The Netherlands*

The tissue localization of the DNA adducts O^6- and 7-methylguanine induced in the nasal cavity by the nicotine-derived carcinogen 4-(N-nitrosomethylamino)-1-(3-pyridyl)-1-butanone (NNK, 30 mg/kg intraperitoneally) has been investigated immunocytochemically in male Sprague-Dawley rats. Adduct-specific nuclear staining, indicative of the metabolic activation of NNK to a methylating compound, was observed in both respiratory and olfactory mucosa. In the respiratory epithelium, strong staining was generally observed in areas devoid of goblet cells. Less intense staining was observed both in the serous gland cells and their efferent ducts in the respiratory submucosa, whereas the mucous gland cells were unstained. In the olfactory mucosa, the sustentacular and basal cells of the olfactory epithelium were moderately stained; staining varied substantially from site to site. No DNA adduct was detected in the olfactory cells. Strong nuclear staining, similar to that in the respiratory mucosa, was observed in the cells of the Bowman glands of the olfactory submucosa. A similar distribution of methylated DNA bases in nasal tissues has been observed in rats after exposure to other N-nitrosamines and in Syrian hamsters after exposure to NNK. This finding may indicate that in man the same cell types undergo DNA adduct formation after exposure to NNK and other N-nitrosamines.

Immunocytochemical visualization of carcinogen-DNA adducts provides a rapid, simple approach to assessing carcinogen metabolism and the tissue localization of DNA modifications (Scherer *et al.*, 1988, 1989). In the present study, we used highly specific polyclonal antisera to investigate the tissue-specific induction of O^6-methylguanine (O^6-MeG) and 7-methylguanine (7-MeG) by the tobacco-specific nitrosamine 4-(N-nitrosomethylamino)-1-(3-pyridyl)-1-butanone (NNK) in the nasal cavity of rats. NNK is a potent animal carcinogen, causing tumours in the nasal cavity, lung, trachea, liver and pancreas of rats (Hecht *et al.*, 1980; Hoffmann *et al.*, 1981; Rivenson *et al.*, 1988).

---

[3]Present address: Dow Europe SA, Bachtobelstr. 3, CH-8810 Horchen, Switzerland

Methods

Rabbit antisera raised against an O^6-MeG-haemocyanin conjugate and an imidazole ring-opened 7-MeG-haemocyanin conjugate were obtained from Dr P. Kleihues (Lüdeke & Kleihues, 1988) and Dr R. Montesano (Degan *et al.*, 1988), respectively. Both antisera were used at a dilution of 1:15000.

Adult male Sprague-Dawley rats (400-450 g) from the breeding colony of the Netherlands Cancer Institute were kept on softwood bedding, two per polycarbonate cage, and fed standard food pellets (Hope Farms, Woerden, The Netherlands) *ad libitum*. Groups of two rats received a single intraperitoneal injection of either NNK (30 mg/kg body weight in saline) or solvent only and were sacrificed after 6 h. The nasal cavity was gently flushed through the posterior opening of the ductus pharyngeus with freshly prepared modified Carnoy's solution (chloroform:methanol:acetic acid, 6:3:1). The samples were fixed for 2 h under continued movement, kept overnight in 70% ethanol and cut with a water-cooled diamond-band saw into 2-mm transverse slices at the levels indicated by Woutersen and Feron (1989).

After dehydration in 100% ethanol and acetone for 30 min each, the slices were infiltrated overnight at room temperature with a mixture of methylmethacrylate and softener (8:2 by volume; K-plast™ from Medim, Giessen, FRG). After addition of 4% initiator, the slices were agitated for 10 min; polymerization was started by raising the temperature to 37°C.

Sections of 2 μm were stained by a capillary block protocol (Scherer & Van Benthem, 1990). In order to differentiate between submucosal glandular structures and to recognize goblet cells, selected sections were counterstained (after the complete immunocytochemical staining procedure) by the periodic acid-Schiff (PAS) reaction. Levels three (mainly respiratory mucosa) and five (mainly olfactory mucosa) were evaluated.

Nuclear staining patterns

Immunocytochemical visualization of DNA adducts could be greatly improved by the use of plastic embedded tissues instead of cryostat sections, in particular for bony organs like the nose. O^6-MeG- and 7-MeG-specific staining was restricted to the nuclei; the staining intensity was not decreased by the plastic embedding and removal. The clearer morphology in the plastic-embedded sections facilitated discrimination of different cell types. Loss of DNA during the staining procedure was negligible, as judged from the intensity of Feulgen staining performed after the complete immunocytochemical procedure but prior to staining with diaminobenzidine.

The nuclear staining patterns observed for O^6-MeG and 7-MeG were similar in the different tissues of the nasal cavity. In the stratified squamous mucosa of the nasal vestibulum and the ventral meatus, nuclear staining was slight and observed only in the epithelial cells. Strong nuclear staining was observed in the respiratory and olfactory mucosa.

In the respiratory epithelium, O^6-MeG and 7-MeG were found predominantly in areas lacking goblet cells (Figures 1A, 2A). The goblet cells themselves appeared to be unstained (Figure 1B). At the level 3, the epithelium was relatively flat and the nuclei were round (Figure 1A), whereas at level 5 the epithelial cells were more columnar and had elongated nuclei (Figure 2A). This observation contradicts preliminary indications obtained from cryostat sections of dissected turbinates (Van Benthem *et al.*, 1989a) not counterstained by PAS. Figures 1A and 2A show that not all nuclei were stained.

In the respiratory submucosa, strong nuclear immunostaining was observed in the PAS-negative serous gland cells (Figures 1C, 1D) and especially in their efferent ducts (Figure 1C). The PAS-positive mucous glands did not show DNA adduct-specific staining (Figure 1B).

Figure 1. Immunocytochemical staining for the DNA adducts indicated in rat nasal cavity (level 3) 6 h after a single intraperitoneal injection of 30 mg/kg 4-(*N*-nitrosomethyl-amino)-1-(3-pyridyl)-1-butanone

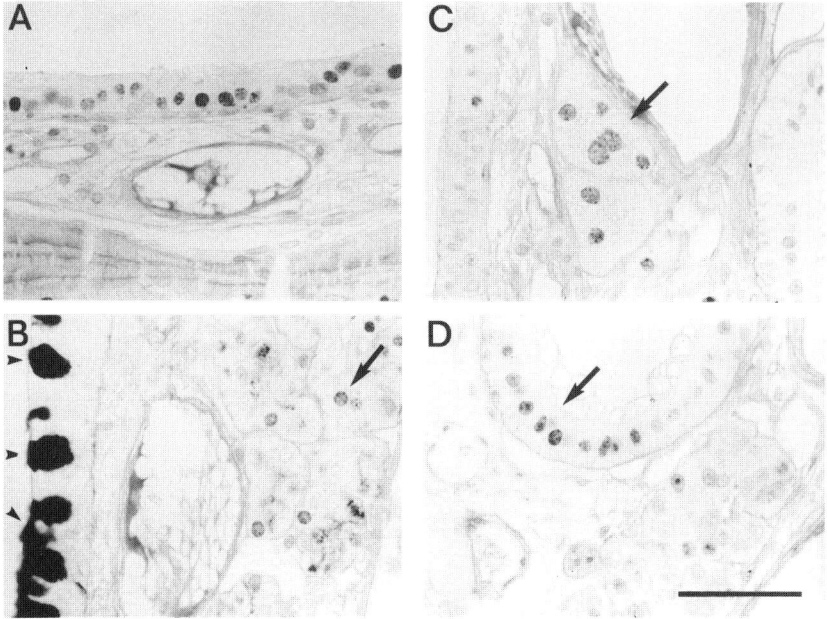

All sections are counterstained by the periodic acid-Schiff (PAS) reaction and diluted 1:20 in haematoxylin. A, Strongly stained cuboidal respiratory epithelium; the thin submucosa contains no glandular structures and is unstained (O^6-methylguanine); B, Respiratory epithelium (lower septum) containing many goblet cells (dark PAS staining, arrow heads) is unmodified (O^6-methylguanine); nuclei of submucosal PAS-negative glands (arrow) are modified. C, Strong nuclear staining in serous gland cells (arrow; 7-methylguanine). D, Efferent duct cells of serous glands (arrow) are strongly stained (7-methylguanine). Bar, 50 μm

In the olfactory mucosa (in the upper part of level 3 and the ethmoid turbinates of level 5), DNA modification by NNK was most prominent in the submucosal Bowman glands (Figures 2C, 2D). These glands were characterized by strongly positive PAS-staining of cytoplasm and of the ductal lumen. The basal cells (Figure 2C) and the sustentacular cells (Figure 2D) of the olfactory epithelium exhibited DNA modification only in certain areas. The olfactory cells remained unstained.

Discussion

In the respiratory epithelium, six cell types can be distinguished by electron microscopy (Monteiro-Riviere & Popp, 1984). Which cell types are stained cannot presently be specified; immunocytochemistry at the electron microscope level will be needed to answer this question. One of these cell types, the nonciliated columnar cell, resembles the bronchiolar Clara cell with regard to the abundance of apical smooth endoplasmic reticulum. This indicates a substantial oxidative metabolic capacity. Strong DNA modification in Clara cells by NNK has been reported (Belinski et al., 1987; Van Benthem et al., 1989b), suggesting that the nonciliated columnar cell of the respiratory epithelium will be among the strongly stained cell types.

Figure 2. Immunocytochemical staining for the DNA adducts indicated in rat nasal cavity (level 5) 6 h after a single intraperitoneal injection of 30 mg/kg 4-(N-nitrosomethylamino)-1-(3-pyridyl)-1-butanone

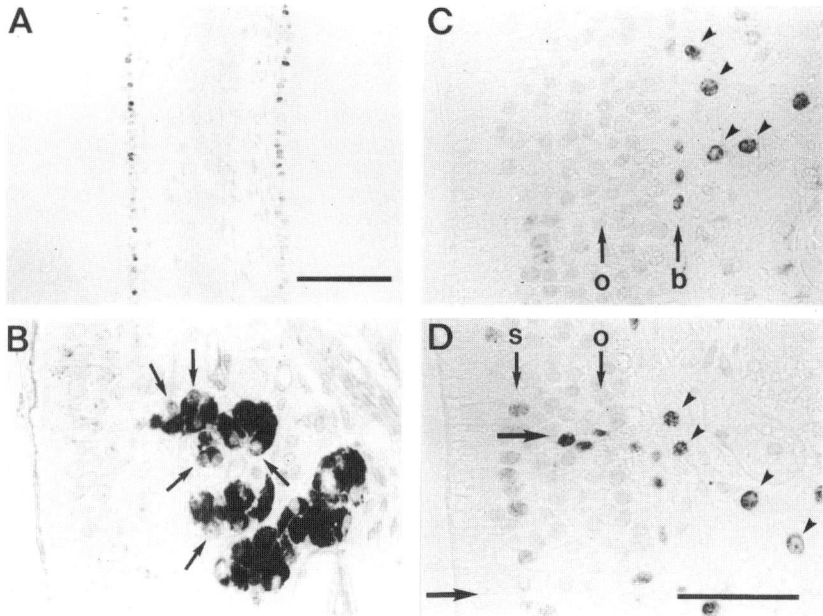

A, C, D, are slightly counterstained with 1:20 diluted haematoxylin and B with PAS and diluted haematoxylin. A, Strongly stained columnar respiratory epithelium (7-methylguanine). Bar, 100 μm. B, Olfactory mucosa: the contents of the Bowman glands are strongly PAS-positive; the nuclei (arrows) are strongly stained (7-methylguanine). In some parts of the olfactory mucosa (C), nuclear staining (7-methylguanine) is observed in nuclei of the basal layer (b); in other parts of the olfactory mucosa (D), staining is seen in nuclei of the sustentacular layer (s), whereas nuclei of the olfactory layer (o) are always unstained. Nuclear staining is also present in Bowman gland cells (arrow heads). Two efferent ducts (D, heavy arrows) of Bowman glands are visible. Bar for B, C, and D, 50 μm

We showed recently that silver-enhanced immunogold staining of carcinogen-DNA adducts was feasible with 2-μm plastic sections (J. Van Benthem & E. Scherer, unpublished results), yielding the same staining patterns as obtained with the present immunoperoxidase technique. A similar distribution of methylated DNA bases in nasal tissues has been observed in rats after exposure to other N-nitrosamines and in Syrian hamsters after exposure to NNK (data not shown). This finding may indicate that the same cell types in man undergo DNA adduct formation after exposure to NNK and other N-nitrosamines.

We have shown in this immunocytochemical study that DNA modification by NNK in the nasal cavity of the rat is highly cell specific. Most of the modified cell types could be identified by light microscopy, with the exception of the stained cells of the respiratory epithelium, for which additional studies using immunostaining at the electron microscope level will have to be performed.

Acknowledgements

This study was supported in part by grant no. NKI 88-19 from the Dutch Cancer Society, and by the Scientific Advisory Committee on Smoking and Health. The technical assistance of W.R. Leeman, C. Schippers-Gillissen, E. Vermeulen and H.H.K. Winterwerp is greatly appreciated. Thanks are also due to Dr P. Kleihues and Dr R. Montesano for their generous gift of the antisera against O^6- and 7-methylguanine, respectively.

References

Belinski, S.A., White, C.M., Devereux, T.R., Swenberg, J.A. & Anderson, M.W. (1987) Cell selective alkylation of DNA in rat lung following low dose exposure to the tobacco specific carcinogen 4-(N-methyl-N-nitrosamino)-1-(3-pyridyl)-1-butanone. *Cancer Res.*, **47**, 1143-1148

Degan, P., Montesano, R. & Wild, C.P. (1988) Antibodies against 7-methyldeoxyguanosine: its detection in rat peripheral blood lymphocyte DNA and potential applications to molecular epidemiology. *Cancer Res.*, **48**, 5065-5070

Hecht, S.S., Chen, C.B., Ohmori, T. & Hoffmann, D. (1980) Comparative carcinogenicity in F344 rats of the tobacco-specific nitrosamines, N'-nitrosonornicotine and 4-(N-methyl-N-nitrosamino)-1-(3-pyridyl)-1-butanone. Cancer Res., **40**, 298-302

Hoffmann, D., Castonguay, A., Rivenson, A. & Hecht, S.S. (1981) Comparative carcinogenicity and metabolism of 4-(*N*-methylnitrosamino)-1-(3-pyridyl)-1-butanone and N'-nitrosonornicotine in Syrian golden hamsters. *Cancer Res.*, **41**, 2386-2393

Lüdeke, B.I. & Kleihues, P. (1988) Formation and persistence of O^6-(2-hydroxyethyl)-2'-deoxyguanosine in DNA of various rat tissues following a single dose of N-nitroso-N-(2-hydroxyethyl)urea. An immuno-slot-blot study. *Carcinogenesis*, **9**, 147-151

Monteiro-Riviere, N.A. & Popp, J.A. (1984) Ultrastructural characterization of the nasal respiratory epithelium in the rat. *Amer. J. Anat.*, **169**, 31-43

Rivenson, A., Hoffmann, D., Prokopczyk, B., Amin, S. & Hecht, S.S. (1988) Induction of lung and exocrine pancreas tumors in F344 rats by tobacco-specific and Areca-derived N-nitrosamines. *Cancer Res.*, **48**, 6912-6917

Scherer, E. & Van Benthem, J. (1990) Improved immunocytochemical staining of carcinogen-DNA adducts by a capillary slot block system. *J. Histochem. Cytochem.*, **38**, 433-436

Scherer, E., Van Benthem, J., Terheggen, P.M.A.B., Vermeulen, E., Winterwerp, H.H.K. & Den Engelse, L. (1988) Immunocytochemical analysis of DNA adducts at the single cell level: a novel tool for experimental carcinogenesis, chemotherapy and molecular epidemiology. In: Bartsch, H., Hemminki, K. & O'Neill, I.K., eds, *Methods for Detecting DNA Damaging Agents in Humans: Applications in Cancer Epidemiology and Prevention* (IARC Scientific Publications No. 89), Lyon, IARC, pp. 286-291

Scherer, E., Van Den Berg, T., Vermeulen, E., Winterwerp, H.H.K. & Den Engelse, L. (1989) Immunocytochemical analysis of O^6-alkylguanine shows tissue specific formation in and removal from esophgeal and liver DNA in rats treated with methylbenzylnitrosamine, dimethylnitrosamine, diethylnitrosamine and ethylnitrosourea. *Cancer Lett.*, **46**, 21-29

Van Benthem, J., Wilmer, J.W.G.M., Winterwerp, H.H.K., Leeman, W.R., Scherer, E. & Den Engelse, L. (1989a) Immunocytochemical study of formation and repair of NNK-induced DNA adducts in nasal epithelium. In: Feron, V.J. & Bosland, M.C., eds, *Nasal Carcinogenesis in Rodents: Relevance to Human Health Risk*, Wageningen, Pudoc, pp. 152-158

Van Benthem, J., Wilmer, J.W.G.M., Leeman, W.R., Winterwerp, H.H.K., Den Engelse, L. & Scherer, E. (1989b) Immunocytochemical study of DNA adduct formation by the tobacco-specific nitrosamine 4-(N-methyl-N-nitrosamino)-1-(3-pyridyl)-1-butanone (NNK) in rat and hamster. In: Maskens, A.P., Molimard, R., Preussmann, R. & Wilmer, J.W., eds, *Tobacco and Cancer. Perspectives in Preventive Research*, Amsterdam, Elsevier, pp. 49-58

Woutersen, R.A. & Feron, V.J. (1989) Localization of nasal tumours in rats exposed to acetaldehyde or formaldehyde. In: Feron, V.J. & Bosland, M.C., eds, *Nasal Carcinogenesis in Rodents: Relevance to Human Health Risk*, Wageningen, Pudoc, pp. 70-75

Relevance to Human Cancer of N-Nitroso Compounds,
Tobacco Smoke and Mycotoxins.
Ed. I.K. O'Neill, J. Chen and H. Bartsch
Lyon, International Agency for Research on Cancer
© IARC, 1991

FEASIBILITY OF A PROSPECTIVE STUDY OF SMOKING AND MORTALITY IN QIDONG, CHINA

J.G. Chen[1], R. Peto[2], Z.T. Sun[3] & Y.R. Zhu[1]

[1]*Qidong Liver Cancer Institute, Jiangsu, China;* [2] *ICRF Cancer Studies Unit, University of Oxford, UK; and* [3]*Cancer Institute, Chinese Academy of Medical Science, Beijing, China*

Qidong is a rural county in eastern China with particularly good facilities for epidemiological research: cigarette use by adult males is widespread (70% now smoke), male lung cancer rates already appear to be rising, the population is stable and well served by a county-wide network of health care facilities, and systematic county-wide registration of all deaths has existed since the mid-1970s, causes currently being assigned according to the 9th International Classification of Diseases.

To help assess the feasibility of a prospective study of smoking and tobacco-related disease in rural China, we describe the current prevalence of smoking, the history of tobacco use, the age-specific mortality rates from lung cancer and the overall mortality rates from various broad groups of disease in one particular county, Qidong. Qidong is a low-lying, densely populated agricultural county located at the mouth of the Yangtze River, in Jiangsu province, just north of Shanghai municipality. There is little turnover of population: almost all the adults were born in the county and will probably continue living there. A central registry, with age, sex and ICD coding of the causes of all deaths in Qidong, is maintained by the Qidong Liver Cancer Institute. Direct information on tobacco consumption before 1949 is not available, although retrospective questioning of old people about their previous habits has been attempted. Since 1959, manufactured cigarette sales to the county as a whole have been recorded. The Qidong Statistics Bureau reported that the population of Qidong has about doubled, from 620 000 in 1950 to 1 130 000 in 1986; and that annual cigarette sales in 1950, 1960, 1970 and 1986 were 195, 378, 638 and 1294 million pieces, respectively (Table 1). The mean daily sales of manufactured cigarettes per person reached 3.15 in 1986, as against only 0.86 in 1950. Almost all of the cigarettes are consumed by men, so although no exactly representative survey of age- and sex-specific smoking habits is available, the consumption per adult male is probably well over 10/man per day.

All causes of death coded in ICD9 for 1987

A population-based registry system of all causes of death has been established in Qidong since 1974, although registration may have been seriously incomplete in the mid-1970s. In

1987, the causes were classified according to the 9th International Classification of Diseases (ICD 9), as recommended by the Health Ministry of China. The leading causes of death in Qidong in 1987 are given in Table 2. In classifying causes of death, particular attention is given to cancers, most of which are diagnosed at the county hospital, which is adjacent to the Qidong Liver Cancer Institute that houses the death registry.

Table 1. Trends in manufactured cigarette sales in Qidong

Year	Total sales of manufactured cigarettes (millions)	Annual sales per person	Daily sales per person
1949	97	160	0.44
1950	195	314	0.86
1951	243	382	1.05
1952	304	466	1.28
1953	338	492	1.35
1954	359	508	1.39
1955	376	528	1.45
1956	463	615	1.69
1957	392	509	1.40
1958	382	493	1.35
1959	381	480	1.32
1960	378	475	1.30
1961	153	191	0.52
1962	76	90	0.25
1963	195	224	0.61
1964	361	405	1.11
1965	611	668	1.83
1966	342	365	1.00
1967	388	407	1.11
1968	425	437	1.20
1969	607	608	1.67
1970	638	627	1.72
1971	656	637	1.74
1972	631	608	1.67
1973	664	634	1.74
1974	599	566	1.55
1975	747	700	1.92
1976	730	678	1.86
1977	708	654	1.79
1978	721	661	1.81
1979	680	621	1.70
1980	688	627	1.72
1981	778	706	1.93
1982	729	656	1.80
1983	1009	905	2.48
1984	920	822	2.25
1985	1092	973	2.67
1986	1294	1150	3.15

Table 2. Causes of death in 1987 in Qidong county, China, as recorded by the Qidong Liver Cancer Institute registry

Cause of death 9th ICD category	Age 0–34 M	Age 0–34 F	Age 35–69 M	Age 35–69 F	Age ≥ 70 M	Age ≥ 70 F
011: Pulmonary tuberculosis	11	5	73	32	42	16
Rest of 001–138 (infectious and parasitic diseases	27	19	20	9	13	18
150: Cancer of oesophagus	0	0	32	12	36	21
151: Cancer of stomach	4	2	112	61	103	45
155: Cancer of liver	67	19	358	102	28	13
162: Cancer of lung	4	2	102	38	65	36
Rest of 140–230-4 (malignant neoplasms)	21	14	127	121	59	92
430–438: Stroke	4	3	120	111	319	417
410–414: Ischaemic heart disease	6	1	38	33	89	110
Rest of 390–439 (vascular diseases)	2	9	30	26	101	131
466, 490–3, 496: Bronchitis, emphysema, asthma	7	7	230	92	642	523
Rest of 460–519 (other respiratory diseases)	60	45	8	10	44	61
571: Cirrhosis	20	2	156	72	47	37
Rest of 520–579 (other digestive diseases)	13	12	48	36	77	88
Rest of 001–799 (other medical causes)	138	93	66	77	198	302
800–999: External causes	192	109	100	54	47	62
All causes	538	352	1620	886	1910	1972
(Rate per 100 000)	(147)	(97)	(943)	(501)	(9338)	(6036)

Age-specific mortality from lung cancer

During 1972–86, a total of 2572 deaths were attributed to lung cancer (although there is inevitably some under-registration, and, conversely, there is some possibility of confusion between primary and secondary disease, since many cases were diagnosed only by radiology and symptoms). During this period, the crude mortality rate attributed to lung cancer was 16/100 000 (22/100 000 males and 10/100 000 females).

The age-specific lung cancer mortality rates recorded in 1972–86 among females do not appear to be increasing. The age-specific rates for males were already higher than those for females in 1972-74, and the male rates increased significantly between 1972-74 and 1984–86.

Prevalence of smoking among male adults

A sampling survey on smoking prevalence was completed in 1987 among male peasants aged over 30 years: 70% are current smokers, 5% being heavy smokers (over 23 pieces/day) and 32% being light smokers (1–7 pieces/day); 69% of current and ex-smokers started to smoke before 24 years of age, the median age of starting being 22.

Manufactured cigarettes are the main type of tobacco used in Qidong. In 1987, the prevalence of cigarette smoking was 69% among adult males, while that of water-pipe smoking was 11%. Most water-pipe smokers, however, also smoke cigarettes.

Table 3, which is based only on retrospective recall of previous habits by adult males in 1987, suggests that the prevalence of smoking may have approximately doubled over the past few decades, but the limitations of such retrospective enquiries are obvious.

Table 3. Retrospective estimation of the prevalence of smoking in different periods in Qidong county, China

Year	No. of males interviewed at age over 15 years [a]	No. then smoking		
		Cigarette	Water pipe	Pipe
1987	694[b]	476 (69%)	74	0
1976	694[c]	447 (64%)	69	3
1966	566	420 (74%)	63	4
1958	426	245 (58%)	40	6
1949	293	154 (53%)	28	6
(1939)	(161)	(50) (31%)	(4)	(1)

[a] Estimated from those aged over 30 years at present
[b] Over age 30 in 1987
[c] Over age 19 in 1976 (and hence over age 30 in 1987)

Potential epidemiological enquiries

It has been estimated (Peto, 1987) that current Chinese smoking habits may eventually cause a total of about two million deaths a year from various neoplastic, vascular and respiratory diseases during the second quarter of the next century. This study indicates that lung cancer is already a common cause of death in Qidong (Table 2), and there has recently been a substantial further increase in cigarette consumption (Table 1). Male lung cancer death rates are increasing, and, even though liver cancer is particularly common in Qidong, lung cancer may eventually overtake it as a cause of death.

Qidong is in many ways a particularly suitable area for prospective, or other, field research on smoking and disease in rural China: (i) a county-wide network of health care has been in place since the early 1970s, and information on all causes of death is now collected routinely (Chen, 1987); (ii) these causes are already registered in such a way they could routinely be linked with personal identifiers; and (iii) there is a high prevalence of smoking, particularly of cigarettes. Qidong is, moreover, one of the 65 Chinese counties fo which diet, life style and mortality rates were described extensively in a recent monograph (Chen et al., 1990), and comparison of the characteristics of Qidong with those of other counties may be of further assistance in planning epidemiological studies.

Acknowledgement

This work was supported by National Natural Science Foundation of China.

References

Chen, J.G. (1987) Experience with the method of classifying death-causes that has been used in Qidong for the past 12 years (in Chinese). *Chin. J. Health Stat.*, **4**, 52-54

Chen, J.S. Campbell, T.C., Li, J.Y. & Peto, R. (1990) *Diet, Life-style and Mortality in China,* Oxford, Oxford University Press
Peto, R. (1987) Tobacco-related deaths in China. *Lancet,* **ii**, 211

LACK OF PROMOTING ABILITY OF SNUFF IN RATS INITIATED WITH 4-NITROQUINOLINE-*N*-OXIDE

S.L. Johansson[1], J.M. Hirsch[2], P.-A. Larsson[3], J. Saidi[1] & B.-G. Österdahl[4]

[1]*Department of Pathology and Microbiology, University of Nebraska Medical Center, Eppley Institute for Research on Cancer and Allied Diseases, Omaha, NE, USA;* [2]*Department of Oral Surgery, University of Göteborg, Göteborg, Sweden;* [3]*Department of Clinical Virology, University of Göteborg, Göteborg, Sweden; and* [4]*Nutrition Laboratory Swedish National Food Administration, Uppsala, Sweden*

In an experiment to evaluate the carcinogenicity and promoting capacity of snuff, a reservoir was created in the lower lip of male Sprague–Dawley rats. Groups of 30 rats were treated with snuff only (twice a day on five days a week), propylene glycol (solvent) three times weekly for four weeks, painting of the hard palate with 4-nitroquinoline-*N*-oxide (4-NQO) three times weekly for four weeks followed by snuff, 4-NQO only for four weeks, or cotton pellets only (twice a day on five days a week). The experiment was continued up to 108 weeks. High levels of tobacco-specific nitrosamines were found in the snuff (a commercial US brand). Rats treated with snuff only, 4-NQO followed by snuff and 4-NQO only had a significantly higher number of squamous-cell tumours and hyperplastic squamous lesions of the lip, oral and nasal cavity and forestomach than solvent or untreated controls. The total number of neoplasms was significantly higher in rats treated with snuff only and with 4-NQO followed by snuff in comparison to the other groups. Thus, snuff and 4-NQO by themselves can induce benign and malignant tumours. Snuff appears to have a general tumorigenic effect but lacked promoting ability after initiation with 4-NQO.

Epidemiological studies have demonstrated that snuff dipping is associated with an increased risk for developing oral cancer (Winn *et al.*, 1981). However, in contrast to the human situation, sufficient evidence for the carcinogenicity of snuff in animals is lacking (IARC, 1985). Creation of a surgical pouch in the lower lip (Hirsch & Thilander, 1981) allows long-term administration of snuff, which has resulted in induction of some oral tumours (Hirsch & Johansson, 1983; Hecht *et al.*, 1986). The aim of the present investigation was to evaluate the tumour promoting effects of snuff in rats initiated with a subcarcinogenic dose of 4-nitroquinoline-*N*-oxide (4-NQO) as well as to determine the effects of long-term administration of snuff. Details of this work have been published previously (Johansson *et al.*, 1989).

Material, methods and experimental design

Five groups of male Sprague–Dawley rats underwent surgical creation of a canal in the lower lip. After two weeks of healing, they were divided into the following groups: group

I, 30 rats received snuff application twice a day starting at week 4 and continuing until week 108 of the experiment; group II, 29 rats received propylene glycol on the palate three times weekly for four weeks, with no further treatment during the remaining 104 weeks; group III, 29 rats received 4-NQO dissolved in propylene glycol, as described for group II; group IV, 30 rats received 4-NQO as for group III for the first four weeks and then snuff for up to 104 weeks; and group V, 29 rats received a cotton pellet dipped in physiological saline twice a day on five days a week for the entire 108 weeks. The animals were killed when moribund or at 108 weeks of the experiment. Complete autopsies were performed, with histological examination of the lip, test canal, palate, oral and nasal cavities, lungs, heart, liver, oesophagus, forestomach, glandular stomach, kidneys, urinary bladder and grossly abnormal tissues.

The snuff used was a commercially available US brand. 4-NQO (Sigma Chemical Corp., St Louis, MO, USA) was dissolved in propylene glycol to a concentration of 0.5%. Tobacco-specific nitrosamines were determined by isothermal gas-liquid chromatography interfaced with a thermal energy analyser. The detection limits of the method were 0.01–0.02 mg tobacco-specific nitrosamines/kg wet weight of snuff. The mean contents were found to be 5.14 ± 1.39 mg/kg N'-nitrosonornicotine, 5.10 ± 1.02 mg/kg N-nitrosoanatabine and 0.89 ± 0.12 mg/kg 4-(N-nitrosomethylamino)-1-(3-pyridyl)-1-butanone

Results and conclusion

Table 1 gives the incidence and location of tumours in the different groups. The incidences of hyperplastic and dysplastic lesions of the lip canal, hard palate and forestomach were significantly higher in groups I and IV than in the other groups; the total numbers of tumours were also significantly higher in groups I and IV than in groups II, III and V. Of special interest was the occurrence of two lip sarcomas in group I and three in group IV and of two hepatomas in group I and one in group IV. Thus, snuff by itself appears to have independent carcinogenic activity towards the lip, oral cavity and nasal cavity, supporting the results of epidemiolgoical studies in humans. It also appeared to have a general tumorigenic effect but did not seem to display any significant promoting effect.

Future studies will involve initiation with 4-NQO and 7,12-dimethylbenz[a]anthracene in the lip canal followed by snuff administration, with evaluation of the presence of DNA adducts of tobacco-specific nitrosamines in oral epithelial cells.

Acknowledgements

Supported by funds from the Department of Pathology and Microbiology, University of Nebraska Medical Center, NIH Research Grant No. CA 36727 and a grant from the Smokeless Tobacco Research Council No. 0144-02.

References

Hecht, S.S., Rivenson, A., Braley, J., Dibello, J., Adams, J.D. & Hoffmann, D. (1986) Induction of oral cavity tumors in F344 rats by tobacco specific nitrosamines and snuff. *Cancer Res.*, **46**, 4162-4166

Hirsch, J.M. & Johansson, S.L. (1983) Effect of long-term application of snuff on the oral mucosa – an experimental study in the rat. *J. Oral Pathol.*, **12**, 187-198

Hirsch, J.M. & Thilander, H. (1981) Snuff-induced lesions of the oral mucosa – an experimental model in the rat. *J. Oral Pathol.*, **10**, 342-353

IARC (1985) *IARC Monographs on the Evaluation of the Carcinogenic Risk of Chemicals to Humans*, Vol. 37, *Tobacco Habits other than Smoking; Betel-quid and Areca-nut Chewing; and Some Related Nitrosamines*, Lyon, pp. 116-148

Johansson, S.L., Hirsch, J.M., Larsson, P.A., Saidi, J. & Österdahl, B.-G. (1989) Snuff-induced carcinogenesis: effect of snuff in rats initiated with 4-nitroquinoline N-oxide. *Cancer Res.*, **49**, 3063-3069

Winn, D.M., Blot, W.J., Shy, C.M., Pickle, L.W., Toledo, A. & Fraumeni, J.F. (1981) Snuff dipping and oral cancer among women in the southern United States. *New Engl. J. Med.*, **304**, 745-749

Table 1. Incidence and distribution of neoplasms in groups of rats treated with snuff (I), propylene glycol (II), 4-nitroquinoline-N-oxide (4-NQO) (III), 4-NQO plus snuff (IV) or saline (V)

Neoplasm	Group[a]				
	I (29)	II (28)	III (29)	IV (28)	V (29)
Lip					
Squamous-cell papilloma	1				
Squamous-cell carcinoma *in situ*				1	
Squamous-cell carcinoma	1				
Undifferentiated sarcoma	2			3	
Hard palate					
Squamous-cell papilloma	1				
Squamous-cell carcinoma *in situ*	1				
Squamous-cell carcinoma	2		2	4	
Tongue					
Squamous-cell papilloma			2	1	
Squamous-cell carcinoma			2	1	
Nasal cavity					
Squamous-cell papilloma	1				
Squamous-cell carcinoma	1			1	
Oesophagus, squamous-cell carcinoma			1		
Forestomach, squamous-cell carcinoma	1		2	2	
Hepatoma	2			1	
Mammary adenocarcinoma				1	
Renal pelvic tumour		1			
Wilms' tumour				1	
Leydig-cell tumour				1	1
Pituitary adenoma	1				
Malignant lymphoma	4		1	1	
Subcutis					
Malignant fibrous histiocytoma	2	1		1	
Neurofibrosarcoma		1	1		
Neurofibroma		2			
Skin					
Undifferentiated sarcoma			2		1
Fibroma	1			1	
Total	23	4	13	18	3

[a] In parentheses, effective number of animals

Relevance to Human Cancer of N-Nitroso Compounds,
Tobacco Smoke and Mycotoxins.
Ed. I.K. O'Neill, J.Chen and H. Bartsch
Lyon, International Agency for Research on Cancer
© IARC, 1991

CHARACTERIZATION OF ACTIVATION AND DEACTIVATION PATHWAYS OF 4-(N-NITROSOMETHYLAMINO)-1-(3-PYRIDYL)-1-BUTANONE (NNK) IN RAT HEPATOCYTES

L. Liu, M.A. Alaoui-Jamali, N. El Alami & A. Castonguay

*Laboratory of Cancer Etiology and Chemoprevention,
School of Pharmacy, Laval University, Quebec City, Canada*

We have characterized the metabolism of 4-(N-nitrosomethylamino)-1-(3-pyridyl)-1-butanone (NNK) in cultured rat hepatocytes and have established the relationship between various metabolic pathways and single-strand breaks (SSB) in DNA. Metabolism of [5-3H]-NNK by carbonyl reduction, α-carbon hydroxylation and pyridine N-oxidation was linear from 0.5 to 6 h with 0.25–2 × 10⁶ hepatocytes. Using an alkaline elution assay, we observed that NNK induces SSB in DNA in a dose- and time-dependent manner. SSB induced by NNK were rejoined partially within 2 h and totally by 12 h after exposure. NNK-N-oxide produced a smaller number of SSB than NNK, suggesting that pyridine N-oxidation of NNK is a deactivation pathway. Carbonyl reduction of NNK led to 4-(N-nitrosomethylamino)-1-(3-pyridyl)-1-butan-1-ol (NNAl). Reaction of NNK with methyl magnesium iodide gave 4-(N-nitrosomethylamino)-1-(methyl)-1-(3-pyridyl)butan-1-ol (1-MeNNAl) 82% yield. NNAl, but not 1-MeNNAl, can be reoxidized to NNK. Doses of 5 mM NNAl and 1-MeNNAl both induced SSB, indicating that NNAl does not require reconversion to NNK to be activated to DNA damaging intermediates. α-Methylene hydroxylation resulted in the formation of 4-oxo-4-(3- pyridyl)butanal. At equimolar concentration (5 mM), the aldehyde was more damaging than NNK to hepatocyte DNA. The results of this study demonstrate that NNK is activated by rat hepatocytes and that metabolites formed by α-carbon hydroxylation induce SSB.

Unlike N'-nitrosonornicotine (NNN), 4-(N-nitrosomethylamino)-1-(3-pyridyl)-1-butanone (NNK) and N-nitrosodimethylamine (NDMA) are hepatocarcinogenic in Fischer 344 rats. Reactive electrophilic intermediates formed by activation of N-nitrosamines can damage DNA. Such damage, as quantified by the alkaline elution/rat hepatocyte assay (Bradley *et al.*, 1982), correlates well with carcinogenic and mutagenic activities and has therefore been proposed as a good predictor of the carcinogenic potential of N-nitroso compounds in humans.

Metabolism

Hepatocytes were isolated by two-step collagenase perfusion (Bradley & Sina, 1984) and cultured (2 × 10⁶/dish) in minimum essential medium containing [5-3H]-NNK (5 μM).

Figure 1. Metabolic pathways of 4-(N-nitrosomethylamino)-1-(3-pyridyl)-1-butanone (NNK) in primary cultures of rat hepatocytes[a]

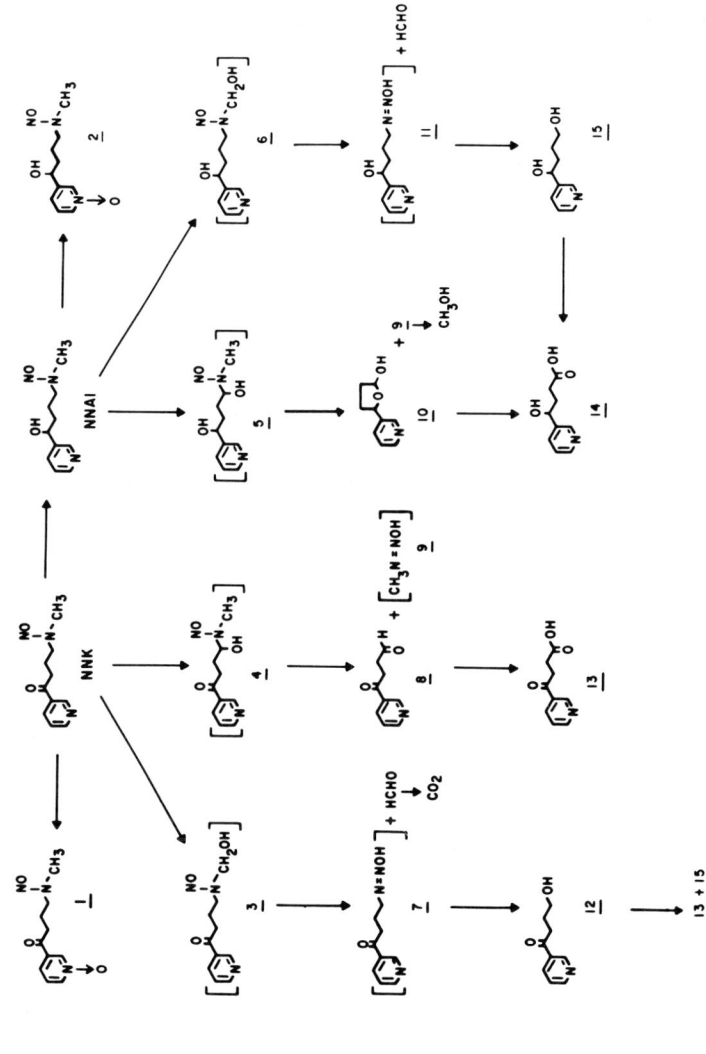

[a] Structures in brackets are hypothetical intermediates; NNAl, 4-(N-nitrosomethylamino)-1-(3-pyridyl)butan-1-ol

NNK metabolites were assayed by reverse-phase high-performance liquid chromatography, as described previously (Castonguay et al., 1983). The three major metabolic pathways of NNK are illustrated in Figure 1. Carbonyl reduction of NNK yields the N-nitroso alcohol, 4-(N-nitrosomethylamino)-1-(3-pyridyl)-1-butan-1-ol (NNAl), at a rate of 0.54 nmol/h, which was linear up to 6 h. Pyridine N-oxidation of NNK and NNAl yields NNK-N-oxide (*1*) and NNAl-N-oxide (*2*), respectively; and α-carbon hydroxylation of NNK and NNAl gives metabolites *12*, *13*, *14* and *15*. The rate of α-carbon hydroxylation (1248 pmol/10⁶ cells) was greater than that of pyridine N-oxidation (146 pmol/10⁶ cells), and both were linear up to 6 h of culture. Of the four metabolites generated by α-carbon hydroxylation, the keto acid *13* was the most abundant, and its formation must involve the keto aldehyde *8* along with the methylating species *9* as intermediates.

DNA single-strand breaks

The kinetics of formation of DNA single-strand breaks (SSB) was studied in cells treated with NNK (5 mM) and by elution of DNA at pH 12.3. Figure 2 shows that NNK induced SSB in a dose-dependent manner. More interestingly, lower, nontoxic concentrations such as 1 and 5 mM also induced significant numbers of SSB: the percentages of DNA eluted from the filter after 6 h were 43, 69 and 81% with 1, 5 and 10 mM NNK. Plotting the elution rates *versus* the concentrations (1–10 mM) of NNK gives a straight line with a slope of 0.008 and a correlation coefficient equal to 0.98 (data not shown). At equimolar concentrations (5 mM), NNN induced fewer DNA SSB than NNK (Figure 3): the percentages of DNA eluted from the filters after 6 h elution were 12% and 70%, and the elution rates were 0.014 and 0.095 for NNN and NNK, respectively. The rate of DNA elution obtained with NNN, however, was similar to those observed with NNK-N-oxide (0.017) and NNN-N-oxide (0.013), which are generated by N-oxidation of the parent compounds. Both metabolites induced a low frequency of DNA SSB, and this finding supports the hypothesis that pyridine N-oxidation is a pathway for deactivation of NNK (Castonguay et al., 1983). These results suggest either that NNN-N-oxide and NNK-N-oxide are not good substrates for NNK activation enzymes or that the alkylating species derived from NNN-N-oxide and NNK-N-oxide do not damage DNA. The NNAl analogue 4-(N-nitrosomethylamino)-1-(methyl)-1-(3-pyridyl)butan-1-ol (1-MeNNAl), which cannot be reconverted to NNK, was synthesized by reacting NNK with methyl magnesium iodide. The rates of elution of DNA damaged with 5 mM NNAl (0.073) and 5 mM 1-MeNNAl (0.054) were slightly lower than those obtained with 5 mM NNK (0.095; Figure 3): the percentages of DNA eluted from the filter after 6 h were 42 and 48% for 1-MeNNAl and NNAl, respectively. These results suggest that NNAl is itself activated to intermediates that induce SSB and that NNAl does not necessarily require reconversion to NNK to damage DNA.

While the role of DNA methylation in the carcinogenesis of NNK is well documented (Hecht et al., 1986), the contribution of aldehyde *8* to the carcinogenesis process has not been investigated. We observed that aldehyde *8* was a strong DNA damaging agent; and at equimolar concentrations (5 mM) was more damaging than NNK to hepatocyte DNA (Figure 4). Our study underlines the importance of aldehyde *8* in NNK-induced DNA damage.

Activation and deactivation pathways of NNK 513

Figure 3. DNA single-strand breaks induced by 4-(N-nitrosomethylamino)-1-(3-pyridyl)-1-butanone (NNK) and its analogues

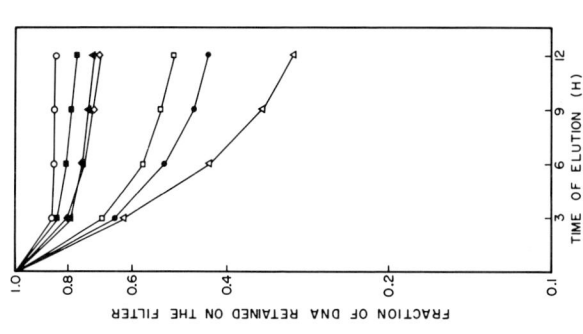

Concentrations, 5 mM; length of exposure, 4 h. ○, control (0.5% dimethyl sulfoxide); ■, N'-nitrosonornicotine (NNN)-N-oxide; ▲, NNK-N-oxide; ◇, NNN; □, 1-Me4-(N-nitrosomethylamino)-1-(3-pyridyl)-butan-1-ol (NNAl); ●, NNAl; △, NNK. Each point represents the mean of four determinations.

Figure 2. DNA single-strand breaks induced by 4-(N-nitrosomethylamino)-1-(3-pyridyl)-1-butanone at various concentrations

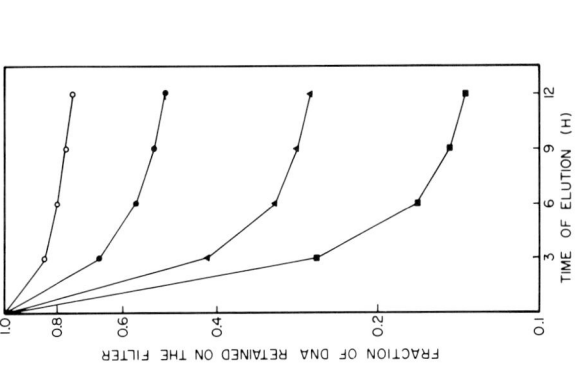

Length of exposure, 4 h. Each point represents the mean of two determinations. ○, control (0.5% dimethyl sulfoxide); ●, 1 mM; ▲, 5 mM; ■, 10 mM

Figure 4. DNA single-strand breaks induced by 4-(N-nitrosomethylamino)-1-(3-pyridyl)-1-butanone (NNK) and the keto aldehyde 8 (see Figure 1)

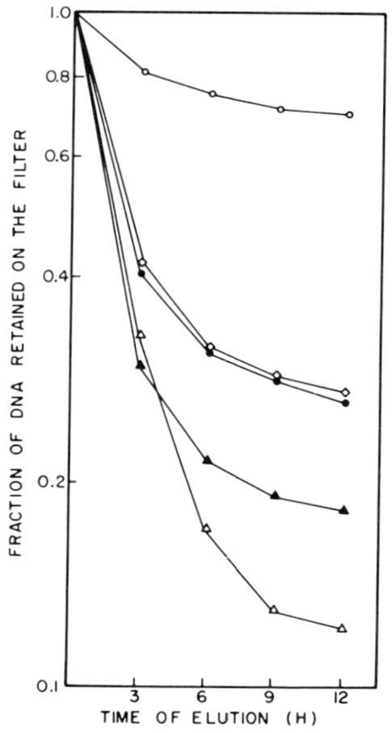

Length of exposure, 4 h. ○, control (0.5% dimethyl sulfoxide); □, 5 mM NNK; ●, 1 mM keto aldehyde 8; ▲, 5 mM keto aldehyde 8; △, 10 mM keto aldehyde 8

Figure 5. DNA single-strand breaks induced by 4-(N-nitrosomethylamino)-1-(3-pyridyl)-1-butanone (NNK) and N-nitrosodimethylamine (NDMA)

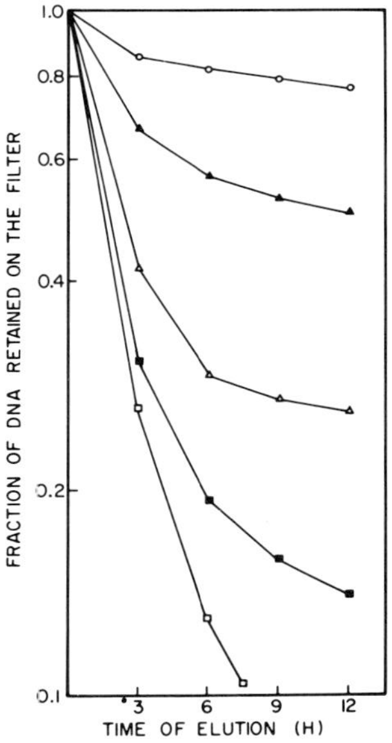

Length of exposure, 4 h. ○, control (0.5% dimethyl sulfoxide); ▲, 1 mM NNK; △, 5 mM NNK; ■, 1 mM NDMA; □, 5 mM NDMA

We also compared NNK with two related N-nitrosamines, NNN and NDMA. NDMA induced more SSB than NNK (Figure 5): the percentages of DNA eluted from the filters after 6 h elution were 43 and 70% with 1 and 5 mM NNK, respectively, and 79 and up to 87% with 1 and 5 mM NDMA, respectively. All three nitrosamines can thus induce SSB, but fewer SSB were induced by NNN than by NNK or NDMA. These results correlate with the hepatocarcinogenicity of these three N-nitrosamines.

In order to assess repair of DNA SSB induced by NNK, hepatocytes were treated for 1 h with 5 and 10 mM NNK and then incubated in NNK-free medium for various periods. The rate of repair was dependent on the time of culture in the absence of NNK: 63% of the SSB induced by NNK were rejoined partially by 2 h and totally by 12 h after exposure. Even after exposure to 10 mM NNK, repair was complete within 12 h (data not shown).

Future studies

The rat hepatocyte/alkaline elution assay could be used to study modulation of NNK hepatocarcinogenicity by chemopreventive agents. Experiments in progress will define the role of the keto aldehyde 8 in the carcinogenicity of NNK.

Acknowledgement

This study was supported by grant MA-9959 from the Medical Research Council of Canada.

References

Bradley, M.O. & Sina, J.S. (1984) Methods for detecting carcinogens and mutagens with the alkaline elution/rat hepatocyte assay. In: Kilbey, B.J., Legator, M., Nichols, W. & Ramel, C., eds, *Handbook of Mutagenicity Test Procedures*, Amsterdam, Elsevier, pp. 71-82

Bradley, M.O., Dysart, G., Fitzsimmons, K., Harbach, P., Lewin, J. & Wolf, S. (1982) Measurements by filter elution of DNA single- and double-strand breaks in rat hepatocytes: effects of nitrosamines and γ-irradiation. *Cancer Res.*, **42**, 2592-2597

Castonguay, A., Lin, D., Stoner, G.D., Radok, P., Furuya, K., Hecht, S.S., Schut, H.A.J. & Klaunig, J.E. (1983) Comparative carcinogenicity in A/J mice and metabolism by cultured mouse peripheral lung of N'-nitrosonornicotine, 4-(N-methylnitrosamino)-1-(3-pyridyl)-1-butanone and their analogues. *Cancer Res.*, **43**, 1223-1229

Hecht, S.S., Trushin, N., Castonguay, A. & Rivenson, A. (1986) Comparative tumorigenicity and DNA methylation in F344 rats by 4-(N-methylnitrosamino)-1-(3-pyridyl)-1-butanone and N-nitrosodimethylamine. *Cancer Res.*, **46**, 498-502

Relevance to Human Cancer of N-Nitroso Compounds,
Tobacco Smoke and Mycotoxins.
Ed. I.K. O'Neill, J. Chen and H. Bartsch
Lyon, International Agency for Research on Cancer
© IARC, 1991

ACTIVATION OF N'-NITROSONORNICOTINE BY HYDROGEN PEROXIDE *IN VITRO*

J. Nair[1], U.J. Nair[1], A.J. Amonkar[2] & S.V. Bhide[1]

[1]*Carcinogenesis Division and* [2]*Bio-organic Unit,*
Cancer Research Institute, Tata Memorial Centre,
Parel, Bombay, India

Betel-quid ingredients were found to produce reactive oxygen species, such as superoxide anion and hydrogen peroxide, *in vitro*. We demonstrated that N'-nitrosonornicotine (NNN) can be converted to its active metabolite, hydrogen peroxide, nonenzymatically in the presence of ferrous ions and ethylenediaminetetracetic acid (EDTA) at pH 7.2. Three ultimate metabolites of NNN — NNN-1-N-oxide, 4-hydroxy-4-(3- pyridyl)butyric acid and 4-oxo-4-(3-pyridyl)butyric acid — and nornicotine were detected by high-performance liquid chromatography. 3H-NNN and 14C-NNN interact with calf thymus DNA in the presence of hydrogen peroxide, ferrous ion and EDTA. The results suggest that formation of reactive oxygen species in the presence of NNN may be a key factor in the initiation of oral tumours in tobacco and betel-quid chewers.

The saliva of chewers of betel quid with tobacco contains several N-nitroso compounds, including tobacco-specific nitrosamines (Wenke *et al.*, 1984; Nair *et al.*, 1985). We reported earlier the formation of reactive oxygen species, such as superoxide anion and hydrogen peroxide (H_2O_2) by catechu and areca-nut extract *in vitro* (Nair *et al.*, 1987). The possible co-existence of N-nitroso compounds and reactive oxygen species during the chewing of betel quid with tobacco led us to investigate the nonenzymatic activation of N'-nitrosonornicotine (NNN), a carcinogenic, tobacco-specific nitrosamine.

Materials and methods

NNN, NNN-1-N-oxide, 4-hydroxy-4-(3-pyridyl)butyric acid (hydroxy acid) and 4-oxo-4-(3-pyridyl)butyric acid (keto acid) were synthesized according to the reported methods (McKennis *et al.*, 1964; Hu *et al*, 1974; Hecht *et al.*, 1980). 3H(G)-NNN (79.4 mCi/mmol) was prepared at Bhabha Atomic Research Centre, Bombay, India, by tritium exchange and was purified by silica gel column chromatography. N'-Pyrrolidine-2-14C-NNN (51.7 mCi/mmol) was obtained from NEN and calf thymus DNA from Sigma. High-performance liquid chromatography (HPLC) was carried out on a Waters system fitted with U6K injector, 6000A solvent delivery system, 410 ultraviolet detector (254 nm) and a μ Bondapak C_{18} column. NNN and its metabolites were separated with 5 mM acetic acid (pH 4.0) containing 5% acetonitrile and 0.2% tetrahydrofuran, by isocratic solvent elution at a rate of 1 ml/min.

The 110-μl reaction system contained 10 μl NNN (100 mM in normal saline), 60 μl bicarbonate buffer pH 7.2 (25 mM containing 100 mM NaCl), 10 μl ethylenediaminetetraacetic acid (EDTA) (20 mM), 10 µl FeSO$_4$ (5 mM) and 10 µl H$_2$O$_2$ (8.8 mM), which were incubated at 37°C for 90 min and 10 µl 0.2 M HCl which were added before the reaction mixture was frozen. Suitable aliquots of the reaction mixture were injected for the HPLC, and the metabolites of NNN (NNN-1-N-oxide, hydroxy and keto acids) and nornicotine were identified and quantified using authentic standards.

To study the interaction of radiolabelled NNN with DNA, 900 µl of reaction system, containing 250 µl DNA (2 mg/ml), 350 µl bicarbonate buffer pH 7.2, 3H-NNN (2 µCi) or 14C-NNN (0.2 µCi), 100 µl EDTA, 100 µl FeSO$_4$ and 100 µl H$_2$O$_2$ (8.8 or 44.0 mM) were incubated at 37°C for 120 min. The DNA was precipitated by adding 5 M sodium acetate and cold ethanol, washed five times with cold ethanol:water (1:1) and ethanol, and finally with ether, and dried. The DNA was redissolved in distilled water, and suitable aliquots were used to determine the radioactivity on a scintillation counter.

Results

Figure 1 shows a typical chromatograph of NNN and its metabolites obtained after reaction. From 1 µmol NNN, 3.8 ± 3.0 nmol hydroxy acid, 1.31 ± 0.3 nmol NNN-1-N-oxide, 2.4 ± 0.3 nmol keto acid and 3.4 ± 1.0 nmol nornicotine were formed (values are means of three different experiments ± SD). Reaction of NNN with only ferrous ions and EDTA resulted in the formation of nornicotine; after reaction with H$_2$O$_2$ alone, NNN-1-N-oxide was detected.

Figure 1. High-performance liquid chromatography profile of N'-nitrosonornicotine (NNN) and its metabolites after chemical activation with H$_2$O$_2$

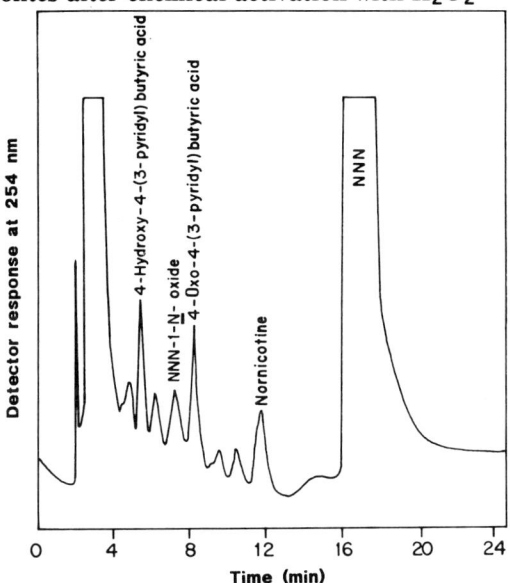

The results of interactions between labelled NNN and calf thymus DNA are given in Table 1. With 3H-NNN, more radioactivity was incorporated into DNA with the complete reaction system than in the absence of H$_2$O$_2$, EDTA and ferrous ion, and more radioactivity was incorporated into DNA when larger amounts of H_2O_2 were used in the reaction system. Since radioactivity was incorporated into DNA in control experiments with 3H-NNN, due perhaps to nonspecific exchange of tritium from NNN to DNA, we carried out a similar experiment with 14C-NNN. Radioactivity was found in DNA only with the complete reaction mixture, which confirmed the results of 3H-NNN incorporation.

Table 1. Interaction of N'-nitrosonornicotine (NNN) with calf thymus DNA with H_2O_2 *in vitro*

Reaction system	Incorporation of radioactivity (dpm $\times$ 10^{-2}/mg DNA)[a]	
	^{3}H	^{14}C
Complete reaction system with 0.88 μmol H_2O_2	3124, 2796	46
Without H_2O_2 and Fe^{2+}	842, 792	5
Without Fe^{2+}	1652, 1205	4
Complete reaction system with 4.4 μmol H_2O_2	5366, 5107	205

[a] 3H(G)-NNN or pyrrolidine-2-14C-NNN

Discussion

Oxidation of N-nitrosodimethylamine and N-nitrosopiperidine by nonenzymatic hydroxylation with molecular oxygen has been reported (Hsieh *et al.*, 1976). The products were mutagenic in the Ames assay (*Salmonella typhimurium* TA1535) only after activation with rat liver microsomal preparations. Malling (1966) and Mayer (1971) demonstrated the mutagenicity of N-nitrosomethylamine and N-nitrosodiethylamine in *Saccharomyces cerevisiae* and *Neurospora crassa*, respectively, with a chemical hydroxylation system. One of the mechanisms of chemical oxidation by molecular oxygen is believed to be through the formation of H_2O_2 (Udenfriend *et al.*, 1954). During the chewing of betel quid with tobacco, reactive oxygen species and carcinogenic N-nitroso compounds may co-exist in the oral cavity of the chewers. Our results indicate that if this situation arose, NNN might be converted to its reactive metabolites, which might interact with macromolecules. The results of an experiment in which 4-(N-nitrosomethylamino)-4-(3-pyridyl)-1-butanone and H_2O_2 were applied simultaneously in the cheek pouch of hamsters and 28% developed oral papillomas, with none in controls (S.V. Bhide, unpublished data), support this observation. Further work is in progress to explore the capacity of betel-quid ingredients to activate carcinogenic molecules nonenzymatically and to estimate the load of reactive oxygen species during the chewing of betel quid.

Acknowledgements

The authors would like to thank Mrs Evelyn Bayle for secretarial assistance.

References

Hecht, S.S., Chen, C.B. & Hoffmann, D. (1980) A study of chemical carcinogenesis. 29. Metabolic β-hydroxylation and N-oxidation of N'-nitrosonornicotine. *J. Med. Chem.*, **23**, 1175-1178

Hsieh, S., Kraft, P.L., Archer, M.C. & Tannenbaum, S.R. (1976) Reaction of nitrosamines in the Udenfriend system: principal products and biological activity. *Mutat. Res.*, **35**, 23-28

Hu, M.W., Bondinell, W.E. & Hoffmann, D. (1974) Synthesis of carbon-14-labelled myosmine, nornicotine and N'-nitrosonornicotine. *J. Labelled Compd.*, **10**, 79-88

Malling, H.V. (1966) Mutagenicity of two potent carcinogens, dimethyl nitrosamines and diethyl nitrosamines in *Neurospora crassa*. *Mutat. Res.*, **3**, 537-540

Mayer, V.W. (1971) Mutagenicity of dimethyl nitrosamine and diethyl nitrosamine for *Saccharomyces* in an *in vitro* hydroxylating system. *Mol. Gen. Genet.*, **112**, 289-294

McKennis, J., Jr, Schwartz, S.L., Turnbull, L.B., Tamaki, E. & Bowman, E.R. (1964) The metabolic formation of γ-(3-pyridyl)-γ-hydroxy butyric acid and its possible intermediary role in the mammalian metabolism of nicotine. *J. Biol. Chem.*, **239**, 3981-3989

Nair, J., Ohshima, H., Friesen, M., Croisy, A., Bhide, S.V. & Bartsch, H. (1985) Tobacco specific and betel nut specific N-nitroso compounds: occurrence in saliva and urine of betel quid chewers and formation *in vitro* by nitrosation of betel quid. *Carcinogenesis*, **6**, 295-303

Nair, U.J., Floyd, R.A., Nair, J., Bussachini V., Friesen, M. & Bartsch, H. (1987) Formation of reactive oxygen species and 8-hydroxy deoxyguanosine in DNA *in vitro* with betel quid ingredients. *Chem.-Biol. Interactions*, **63**, 157-169

Udenfriend, S., Clark, C.T., Axelrod, J. & Brodie, B.B. (1954) Ascorbic acid in aromatic hydroxylation I. A model system for aromatic hydroxylation. *J. Biol. Chem.*, **208**, 731-739

Wenke, G., Brunnemann, K.D., Hoffmann, D. & Bhide, S.V. (1984) A study of betel quid carcinogenesis. IV. Analysis of the saliva of betel chewers. A preliminary report. *J. Cancer Res. Clin. Oncol.*, **108**, 110-113

ANTIMUTAGENIC AND ANTICARCINOGENIC EFFECTS OF BETEL LEAF EXTRACT AGAINST THE TOBACCO-SPECIFIC NITROSAMINE 4-(N-NITROSOMETHYLAMINO)-1-(3-PYRIDYL)-1-BUTANONE (NNK)

S.V. Bhide[1], P.R. Padma[1] & A.J. Amonkar[2]

[1]*Carcinogenesis Division and* [2]*Bio-organic Unit,*
Cancer Research Institute, Tata Memorial Centre,
Parel, Bombay, India

Earlier studies showed that betel leaf inhibits the mutagenic action of standard mutagens like benzo[a]pyrene and dimethylbenz[a]anthracene. Since tobacco-specific nitrosamines are the major carcinogens present in unburnt forms of tobacco, we studied the effect of an extract of betel leaf on the mutagenic and carcinogenic actions of one of the most potent, 4-(N-nitrosomethylamino)-1-(3-pyridyl)-1-butanone (NNK). Betel-leaf extract and hydroxychavicol suppressed the mutagenicity of NNK in both the Ames and the micronucleus test. In studies in mice, betel-leaf extract reduced the tumorigenic effects of NNK by 25%. Concurrent treatment with the extract also inhibited the decreases in levels of vitamin A in liver and plasma induced by NNK. Betel leaf thus has protective effects against the mutagenic, carcinogenic and adverse metabolic effects of NNK in mice.

The habit of chewing tobacco has been shown to be associated causally with oral cancer (Sanghvi, 1981; Bhide *et al.*, 1984; Winn, 1984). In addition, people who chew tobacco with lime have a higher risk for oral cancer than those who chew tobacco in a betel quid (Khanolkar, 1944, 1950), suggesting that betel quid contains some protective factor. We showed earlier that betel-leaf extract (BLE) inhibits the mutagenicity of benzo[a]pyrene and dimethylbenz[a]anthracene (Nagabhushan *et al.*, 1987). Since tobacco-specific N-nitrosamines are the only carcinogens present in unburnt tobacco products like chewing tobacco (IARC, 1985), we tested the effect of BLE on the mutagenic and carcinogenic effects of 4-(N-nitrosomethylamino)-1-(3-pyridyl)-1-butanone (NNK), which is a strong tobacco-specific carcinogen. We also tested the effects of hydroxychavicol, a phenolic antimutagen present in betel leaf (Amonkar *et al.*, 1986, 1989), against the effects of NNK.

NNK was synthesized according to the method of Hecht *et al.* (1983). BLE and hydroxychavicol were prepared according to procedures described earlier (Amonkar *et al.*, 1989; Padma *et al.*, 1989).

Antimutagenicity studies

The antimutagenic effects of BLE and hydroxychavicol against NNK were tested in the Ames test (Maron & Ames, 1983) and in the micronucleus test (Schmid, 1975). At a dose of 200 μg/plate (Padma et al., 1989), both BLE and hydroxychavicol suppressed the mutagenic action in *Salmonella typhimurium* strain TA100 of all three concentrations of NNK tested (Table 1) in the presence of an exogenous metabolic activation system.

Table 1. Effects of betel leaf extract (BLE) and hydroxychavicol (HC) on the concentration-dependent mutagenicity of 4-(*N*-nitrosomethylamino)-1-(3-pyridyl)-1-butanone (NNK) in *Salmonella typhimurium* strain TA100 with postmitochondrial rat liver fraction

NNK (μg/plate)	No. of his^+ revertants/plate		
	NNK alone	+ BLE (200 μg/ml)	+ HC (200 μg/ml)
0	123 ± 1	124 ± 7	89 ± 4
250	320 ± 9	284 ± 8	124 ± 14
500	499 ± 41	301 ± 10	222 ± 17
1000	824 ± 26	471 ± 4	374 ± 19

Values are means ± SE of eight plates from two independent experiments

In the micronucleus test, NNK at 500 mg/kg body weight induced 1.3 ± 0.05 micronuclei per 100 polychromatic erythrocytes as compared to 0.3 ± 0.02 in controls given distilled water. In the presence of BLE and hydroxychavicol, the numbers of micronuclei were 0.5 ± 0.04 and 0.9 ± 0.05, respectively, which were significantly different from that in NNK-treated animals. Animals treated with only BLE and hydroxychavicol showed 0.4 ± 0.025 and 0.42 ± 0.025 micronuclei, respectively.

Anticarcinogenic effects

The anticarcinogenic effect of BLE against NNK was tested in long-term studies on inbred male Swiss mice. NNK (1 mg/day three times a week; total dose, 22 mg) was administered on the tongues of mice following a 3–4 h administration of atropine (1% in drinking-water) to decrease salivation and facilitate retention of the nitrosamine in the oral cavity, as is the case with chewers. BLE was supplied in drinking-water (2.5 mg/animal per day) for the duration of NNK treatment. After treatment, animals were killed periodically to monitor tumour incidence or when moribund. All dead animals were autopsied and tissues fixed in 10% formalin for histological analysis.

In the group treated with NNK alone, 19/29 (65.5%) animals had tumours (17 lung adenomas, three forestomach papillomas and two hepatomas) while in the group treated with NNK and BLE, 11/27 (41%) animals had tumours (eight lung adenomas, three forestomach papillomas and one hepatoma). The per cent decrease in tumour incidence is not statistically significant.

Effect of BLE on vitamin A status in liver and plasma

The levels of vitamin A in both liver and plasma have been shown to be depleted by treatment with NNK (Padma, 1988). We studied the effect of BLE at 12–14 and 20–22 months on vitamin A levels, using the method described by Neeld and Pearson (1963). The results are shown in Table 2. Animals treated with BLE alone had significantly higher levels of liver vitamin A than untreated animals at both time points, while the levels in plasma were elevated only during the earlier period.

Table 2. Levels of vitamin A in the liver and plasma of Swiss male mice treated with 4-(N-nitrosomethylamino)-1-(3-pyridyl)-1-butanone (NNK) alone or in combination with betel-leaf extract (BLE)

Treatment	Vitamin A levels at	
	12–14 months	20–22 months
Liver		
Untreated controls	208.2 ± 10.8	196.1 ± 6.4
NNK alone	32.7 ± 1.4	57.3 ± 4.0
BLE alone	261.1 ± 13.2	245.3 ± 8.8
NNK + BLE	148.8 ± 2.6	135.9 ± 3.9
Plasma		
Untreated controls	91.1 ± 1.3	100.4 ± 3.0
NNK alone	34.1 ± 1.3	59.2 ± 6.2
BLE alone	216.6 ± 12.2	105.2 ± 5.5
NNK + BLE	93.0 ± 1.9	110.1 ± 5.9

Results are means for six animals ± SE

The levels of vitamin A in the livers of the animals treated with NNK and BLE were significantly lower than those in untreated controls at both times, but the levels were significantly higher than those in animals treated with NNK alone. The levels of circulating vitamin A in animals treated with NNK and BLE were comparable to the control levels at both intervals but were significantly higher than those in animals treated with NNK alone.

Potential use of BLE as an anticarcinogen

Our study shows that BLE is nonmutagenic in both the Ames and micronucleus tests. These results are consistent with previous reports in which betel leaf was found to be nonmutagenic in the Ames test (Shirname *et al.*, 1983; Nagabhushan *et al.*, 1987) and in the V79 and human lymphoblastoid cell lines (Umezawa *et al.*, 1981). Our observation that betel leaves are not carcinogenic is also consistent with the results of previous studies in mice (Bhide *et al.*, 1979) and rats (Mori *et al.*, 1979).

BLE has been shown to suppress the mutagenicity of polycyclic aromatic hydrocarbons (Nagabhushan *et al.*, 1987), as well as that of NNK and N'-nitrosonornicotine (Padma, 1988), and to be anticarcinogenic against benzo[a]pyrene in the hamster cheek pouch model (Rao, 1984) and in the forestomach tumour model in mice (Padma, 1988), against dimethylbenz[a]anthracene in the rat mammary tumour model (Rao *et al.*, 1985) and against N'-nitrosonornicotine in Swiss mice (Padma, 1988). The antimutagenic and anticarcinogenic action of BLE against this wide variety of agents may be attributed to the presence of compounds like chlorophyll, phenolics like eugenol and hydroxychavicol (Amonkar *et al.*, 1986) and vitamins like A, ascorbic acid (Aykroyd, 1963) and vitamin E (unpublished data) in betel leaves.

We found that vitamin A levels were elevated in both liver and plasma of BLE-treated animals. It has also been observed that BLE induces a significant increase in liver ascorbic acid levels (Padma, 1988). Vitamin A has been shown to exert a protective action against carcinogens (McCormick & Moon, 1982; Goodwin et al., 1986), and ascorbic acid has been shown to prevent the initiation of skin tumours following the application of a promoter (Slaga & Bracken, 1977). The protective effect may therefore be mediated partly by vitamin A and ascorbic acid.

Thus, the inclusion of betel leaf may reduce the carcinogenic risk of tobacco chewers, and these results support the hypothesis that betel-quid chewers may have some protection against cancer (Khanolkar, 1944, 1950).

References

Amonkar, A.J., Nagabhushan, M., D'Souza, A.V. & Bhide, S.V. (1986) Hydroxychavicol: a new phenolic antimutagen from betel leaf. *Food Chem. Toxicol.*, **24**, 1321-1324

Amonkar, A.J., Padma, P.R. & Bhide, S.V. (1989) Protective effect of hydroxychavicol, a phenolic component of betel leaf, against the tobacco specific carcinogens. *Mutat. Res.*, **210**, 249-253

Aykroyd, W.R. (1963) Table of food values. III. Vitamins. In: Gopalan, C. & Balsubramanian, S.C., eds, *The Nutritive Value of Indian Foods and the Planning of Satisfactory Diets*, New Delhi, Indian Council of Medical Research, pp. 114-133

Bhide, S.V., Shivapurkar, N.M., Gothoskar, S.V. & Ranadive, K.J. (1979) Carcinogenicity of betel quid ingredients: feeding mice with aqueous extract and the polyphenol fraction of betel nut. *Br. J. Cancer*, **40**, 922-926

Bhide, S.V., Shah, A.S., Nair, J. & Nagaraj Rao, D. (1984) Epidemiological and experimental studies on tobacco-related oral cancer in India. In: O'Neill, I.K., von Borstel, R.C., Miller, C.T., Long, J. & Bartsch, H., eds, *N-Nitroso Compounds: Occurrence, Biological Effects and Relevance to Human Cancer* (IARC Scientific Publications No. 57), Lyon, IARC, pp. 851-857

Goodwin, W.J., Jr, Bordash, G.D., Huijing, F. & Altman, N. (1986) Inhibition of hamster tongue carcinogenesis by selenium and retinoic acid. *Ann. Otol. Rhinol. Laryngol.*, **95**, 162-166

Hecht, S.S., Lin, D. & Castonguay, A. (1983) Effects of α-deuterium substitution on the mutagenicity of 4-(methylnitrosamino)-1-(3-pyridyl)-1-butanone (NNK). *Carcinogenesis*, **4**, 305-310

Khanolkar, V.R. (1944) Oral cancer in Bombay, India: a review of 1000 consecutive cases. *Cancer Res.*, **4**, 313-319

Khanolkar, V.R. (1950) Cancer in India. *Acta Unio Int. Contra Cancrum*, **6**, 881-890

Maron, D.M. & Ames, B.N. (1983) Revised methods for the Salmonella mutagenicity test. *Mutat. Res.*, **113**, 173-215

McCormick, D.L. & Moon, R.C. (1982) Influence of delayed administration of retinyl acetate on mammary carcinogenesis. *Cancer Res.*, **42**, 2639-2643

Mori, H., Matsubara, N., Ushimaru, Y. & Hirino, I. (1979) Carcinogenicity examination of betle nuts and *Piper betle* leaves. *Experientia*, **35**, 384-385

Nagabhushan, M., Amonkar, A.J., D'Souza, A.V. & Bhide, S.V. (1987) Nonmutagenicity of betel leaf and its antimutagenic action against environmental mutagens. *Neoplasm*, **34**, 159-167

Neeld, J.B. & Pearson, W.N. (1963) Macro- and micro-methods for the determination of serum vitamin A trifluoroacetic acid. *J. Nutr.*, **79**, 454-462

Padma, P.R. (1988) *Mutagenicity and Carcinogenicity of N'-Nitrosonornicotine and 4-(Methylnitrosamino)-1-(3-pyridyl)-1-butanone, and their Mechanism of Action*. PhD Thesis submitted to the University of Bombay, India

Padma, P.R., Amonkar, A.J. & Bhide, S.V. (1989) Antimutagnic effects of betel leaf extract against the mutagenicity of two tobacco-specific N-nitrosamines. *Mutagenesis*, **4**, 154-156

Rao, A.R. (1984) Modifying influences of betel quid ingredients on B(a)P induced carcinogenesis on the buccal pouch of hamster. *Int. J. Cancer*, **33**, 581-586

Rao, A.R., Sinha, A. & Selvan, R.S. (1985) Inhibitory action of *Piper betle* leaf on initiation of 7,12-dimethylbenzanthracene induced mammary carcinogenesis. *Cancer Lett.*, **26**, 207-214

Sanghvi, L.D. (1981) Cancer epidemiology: the Indian scene. *J. Cancer Res. Clin. Oncol.*, **99**, 1-14
Schmid, W. (1975) The micronucleus test. *Mutat. Res.*, **31**, 9-15
Shirname, L.P., Menon, M.M., Nair, J. & Bhide, S.V. (1983) Correlation of mutagenicity and tumorigenicity of betel quid and its ingredients. *Nutr. Cancer*, **5**, 87-91
Slaga, T.J. & Bracken, W.M. (1977) The effects of antioxidants on skin tumor initiation and aryl hydrocarbon hydroxylase. *Cancer Res.*, **37**, 1631-1635
Umezawa, K., Fujie, S., Matsushima, T., Katoh, Y., Tanaka, M. & Takayama, S. (1981) Morphological transformation, sister chromatid exchange and mutagenesis assay of betel constituents. *Toxicol. Lett.*, **8**, 17
Winn, D.M. (1984) Tobacco chewing and snuff dipping: an association with human cancer. In: O'Neill, I.K., von Borstel, R.C., Miller, C.T., Long, J. & Bartsch, H., eds, N-*Nitroso Compounds: Occurrence, Biological Effects and Relevance to Human Cancer* (IARC Scientific Publications No. 57), Lyon, IARC, pp. 837-849

Relevance to Human Cancer of N-Nitroso Compounds,
Tobacco Smoke and Mycotoxins.
Ed. I.K. O'Neill, J. Chen and H. Bartsch
Lyon, International Agency for Research on Cancer
© IARC, 1991

EFFECT OF VITAMIN A STATUS OF RATS ON METABOLIZING ENZYMES AFTER EXPOSURE TO TOBACCO EXTRACT OR N'-NITROSONORNICOTINE

U.J. Nair, N. Ammigan, M. Nagabhushan, A.J. Amonkar & S.V. Bhide

Cancer Research Institute, Tata Memorial Centre, Parel, Bombay, India

The effects of N'-nitrosonornicotine (NNN) and tobacco extract on hepatic and pulmonary biotransformation enzymes were studied in rats fed vitamin A-sufficient or -deficient semisynthetic diets. Basal levels of cytochrome P450, benzo[a]pyrene hydroxylase, benzphetamine demethylase, glutathione S-transferase and glutathione were lower in the group on the deficient diet. Treatment with tobacco extract or NNN significantly increased the levels of these enzymes in the sufficient diet group. However, in the deficient group, phase I enzymes were significantly increased, but glutathione and glutathione S-transferase levels were drastically reduced. Urine from animals on the deficient diet and treated with tobacco extract or NNN were mutagenic in the Ames *Salmonella*/microsome test. The results suggest that altered metabolism resulting from a vitamin A-deficient diet may be an important factor in susceptibility to carcinogens.

Tobacco use is a global phenomenon, but in India and other Afro-Asian countries exposure to tobacco is often concurrent with poor nutrition. Vitamin A and B complex deficiencies and protein-calorie malnutrition are widespread. We present here results on the effect of vitamin A status on levels of biotransformation enzymes in rats treated with tobacco extract (TE) or N'-nitrosonornicotine (NNN).

Animals, diets and treatment

Male weanling Sprague-Dawley rats were fed a semisynthetic diet composed of 20% vitamin-free casein, 39.8% starch, 30% dextrose, 5% refined peanut oil, 4% salt mixture, 0.2% choline chloride and 1% vitamin mixture; the diets were either vitamin A-free or vitamin A-sufficient containing 20 000 IU retinyl acetate/kg diet. After three months, 75% of the LD_{50} dose of TE and of NNN was administered in three intraperitoneal injections at an interval of 24 h, and animals were sacrificed 24 h after the last injection. Urine was collected at intervals from 0 to 72 h in metabolic cages, pooled, concentrated and tested for mutagenicity in the Ames *Salmonella*/microsome assay (Yamasaki & Ames, 1977). Liver and lung microsomes were prepared (Santhanam *et al.*, 1988); and determinations were made of cytochrome P450 (Omura & Sato, 1964), benzo[a]pyrene hydroxylase (Dehnen *et al.*, 1973), benzphetamine-N-demethylase (Farell & Correia, 1980), cytosolic glutathione

(Maron et al., 1979) and glutathione-*S*-transferase (Habig et al., 1974). Protein (Lowry et al., 1951) and hepatic vitamin A levels (Neeld & Pearson, 1963) were also determined. TE was prepared by extracting a chewing variety of *Nicotiana tabacum* in dichloromethane, flash evaporation and redissolving it in dimethyl sulfoxide. NNN was synthesized according to the method of Hu et al. (1974). Statistical analyses were done using Student's *t* test.

Results and discussion

After three months, the rate of body weight gain in animals fed the vitamin A-deficient diet was lower than that in the supplemented group, although growth did not cease completely. Analysis of the livers showed that vitamin A was completely depleted; the animals were therefore considered to be vitamin A-deficient. Hepatic vitamin A levels in vitamin A-sufficient rats were decreased by treatment with TE/NNN by $\simeq 45\%$. The mean vitamin A levels were 417 ± 12, 212 ± 14 and 225 ± 13 µg/g liver in control, TE- and NNN-treated groups, respectively.

In agreement with the results of Colby et al. (1975), we found that animals fed the vitamin A-deficient diet had depressed drug metabolizing capabilities, as indicated by decreased levels of metabolizing enzymes in both lung (Table 1) and liver (Table 2). All TE/NNN treated animals showed higher levels of activating enzymes; however, the level of induction was higher in groups fed the vitamin A-deficient diet than in the supplemented group. Azais-Braesco et al. (1989) have also reported increased induction of cytochrome P450 by livers of DDT-treated, vitamin A-deficient Wistar rats. While the protective components glutathione and glutathione-*S*-transferase were also induced after treatment of vitamin A-sufficient rats, a drastic depletion was observed in the deficient groups. In tests for mutagenicity, only the urine of treated, deficient animals was mutagenic in the Ames assay, urine from TE-treated animals showing mutagenicity to strain TA98 and that from NNN-treated animals to TA100, after metabolic activation (Table 3).

Our results clearly demonstrate that vitamin A deficiency concurrent with tobacco exposure may lead to increased formation of reactive species, which, in the absence of efficient deactivation by the glutathione-*S*-transferase system, could be an important factor in increasing the risk for cancer after exposure to tobacco.

Acknowledgement

The authors thank Mrs E. Bayle for secretarial assistance.

References

Azais-Braesco, V., Pascal, G., Maseschi, J.P., Fayet, Y., Degiuli, A., Letoublon, R., Got, R. & Frot-Coutaz, J. (1989) Influence of vitamin A status and DDT on vitamin-A dependent protein mannosylation in rat liver. *Chem.-Biol. Interact.*, **69**, 259-267

Colby, D., Kramer, R.E., Gseiner, J.W., Robinson, D.A., Krause, K. & Canaday J.W. (1975) Hepatic drug metabolism in retinol deficient rats. *Biochem. Pharmacol.*, **24**, 1644-1646

Dehnen, W., Tomingas, R. & Ross, J.E. (1973) A modified method for the analysis of benzo(a)pyrene hydroxylase. *Anal. Biochem.*, **53**, 373

Farell, G.G. & Correia, M.A. (1980) Structural and functional reconstitution of hepatic cytochrome P–450 in vivo – reversal of allylisopropyl acetamide-mediated destruction of the hemoprotein by exogenous heme. *J. Biol. Chem.*, **255**, 10128-10133

Habig, W.H., Pabst, M.J. & Jakoby, W.B. (1974) Glutathione-S-transferase, the first enzymatic step in mercapturic acid formation. *J. Biol. Chem.*, **249**, 7130-7139

Table 1. Effect of tobacco extract (TE) and N'-nitrosonornicotine (NNN) on pulmonary levels of metabolizing enzymes in Sprague–Dawley rats given vitamin A-sufficient and -deficient diets[a]

Diet and treatment (mg/kg bw)	Cytochrome P450 (nmol/mg protein)	Benzo[a]pyrene (BP) hydroxylase (pmol hydroxy-BP/min per mg protein)	Benzphetamine-N-demethylase (nmol formaldehyde/min per mg protein)	Glutathione-S transferase (nmol 1-chlorodinitrobenzene conjugated/min per mg protein)	Glutathione (nmol/g tissue)
Vitamin A-sufficient					
Control	0.37 ± 0.06	210 ± 10	0.37 ± 0.03	165 ± 2	1.80 ± 0.10
TE (244)	0.55 ± 0.06* (+ 48%)	262 ± 15* (+ 25%)	0.42 ± 0.02	225 ± 6* (+ 36%)	2.43 ± 0.12* (+ 35%)
NNN (206)	0.51 ± 0.02* (+ 38%)	237 ± 12 (+ 13%)	0.40 ± 0.04	197 ± 3* (+ 19%)	2.18 ± 0.10* (+ 21%)
Vitamin A-deficient					
Control	0.31 ± 0.03	175 ± 10	0.34 ± 0.04	84 ± 5	1.60 ± 0.10
TE (169)	0.68 ± 0.06* (+ 119%)	271 ± 60* (+ 55%)	0.44 ± 0.02* (+ 29%)	40 ± 3* (− 52%)	1.01 ± 0.06* (− 38%)
NNN (150)	0.54 ± 0.07* (+ 74%)	305 ± 12* (+ 74%)	0.40 ± 0.03	50 ± 1* (− 40%)	1.17 ± 0.03* (− 28%)

[a] Means ± SE; figures in parantheses indicate percent change from control
* $p < 0.005$ (vs corresponding control)

Table 2. Effect of tobacco extract (TE) and N'-nitrosonornicotine (NNN) on hepatic levels of metabolizing enzymes in Sprague–Dawley rats given vitamin A-sufficient and -deficient diets[a]

Diet and treatment (mg/kg bw)	Cytochrome P450 (nmol/mg protein)	Benzo[a]pyrene (BP) hydroxylase (pmol hydroxy-BP/min per mg protein)	Benzphetamine-N-demethylase (nmol formaldehyde/min per mg protein)	Glutathione-S transferase (nmol 1-chlorodinitrobenzene conjugated/min per mg protein)	Glutathione (nmol/g tissue)
Vitamin A-sufficient					
Control	0.78 ± 0.02	398 ± 10	0.61 ± 0.04	715 ± 10	6.51 ± 0.20
TE (244)	1.56 ± 0.02* (+ 109%)	498 ± 14* (+ 25%)	0.83 ± 0.02* (+ 36%)	852 ± 12* (+ 19%)	9.30 ± 0.10* (+ 43%)
NNN (206)	1.20 ± 0.02* (+ 54%)	438 ± 12* (+ 10%)	0.74 ± 0.04* (+ 21%)	825 ± 15* (+ 15%)	8.91 ± 0.2* (+ 37%)
Vitamin A-deficient					
Control	0.65 ± 0.02	305 ± 10	0.52 ± 0.02	505 ± 19	5.37 ± 0.60
TE (169)	1.65 ± 0.20* (+ 154%)	550 ± 12* (+ 80%)	1.04 ± 0.04* (+ 100%)	312 ± 06* (− 38%)	2.63 ± 0.14* (− 51%)
NNN (150)	1.21 ± 0.14* (+ 86%)	530 ± 10* (+ 74%)	0.90 ± 0.02* (+ 73%)	348 ± 04* (− 31%)	3.22 ± 0.12* (− 40%)

[a] Means ± SE; figures in parentheses indicate percent change from control
* $p < 0.005$ (vs corresponding control)

Table 3. Effect of dietary vitamin A status on urinary excretion of mutagens detected in the Ames *Salmonella*/microsome assay after metabolic activation

Strain	Treatment[a]	No. of revertants/plate[b]	
		Vitamin A-sufficient	Vitamin A-deficient
TA98	None	32 ± 1	28 ± 4
(SR = 30)	DMS0	32 ± 1	64 ± 4
	TE	39 ± 6	165 ± 11*
TA100	None	137 ± 8	122 ± 7
(SR = 100)	DMSO	134 ± 15	120 ± 2
	NNN	146 ± 14	459 ± 16*

[a] DMSO, dimethyl sulfoxide; TE, tobacco extract; NNN, N'-nitrosonornicotine
[b] Mean $\pm$ SD of eight plates from two independent experiments; spontaneous revertants (SR) were not subtracted; viable cell count was 2×10^8 cells/ml
* $p < 0.005$ as compared to DMSO controls

References (contd)

Hu, M.W., Bondinell, W.E. & Hoffmann, D. (1974) Chemical studies on tobacco smoke. XXIII. Synthesis of carbon-14 labelled myosmine, nornicotine and N'-nitrosonornicotine. *J. Labelled Compd Radiopharm.*, **10**, 79-88

Lowry, O.H., Rosebrough, R.J., Farr, A.L. & Randall, R.J. (1951) Protein mesurement with a folin phenol reagent. *J. Biol. Chem.*, **193**, 265-275

Maron, M.S., De Piere, J.W. & Mannerwich, B. (1979) Levels of glutathione, glutathione reductase and glutathione-S-transferase activities in rat lung and liver. *Biochem. Biophys. Acta*, **582**, 67-78

Neeld, J.B., Jr & Pearson, W.N. (1963) Macro- and micromethods for the determination of serum vitamin A using trifluoroacetic acid. *J. Nutr.*, **79**, 454-462

Omura, J. & Sato, R. (1964) The carbon monoxide binding pigment of liver microsomes. I. Evidence for its hemoprotein nature. *J. Biol. Chem.*, **239**, 2370-2378

Santhanam, U., Nair, U.J. & Bhide, S.V. (1988) Effect of vitamin A deficiency on induction of enzymes metabolizing different carcinogens. *Indian J. Exp. Biol.*, **26**, 337-340

Yamasaki, E. & Ames, B.N. (1977) Concentration of mutagens from urine by adsorption with the nonpolar resin XAD-2: cigarette smokers have mutagenic urine. *Proc. Natl Acad. Sci. USA*, **74**, 3555-3559

INHIBITION OF TOBACCO-SPECIFIC NITROSAMINE 4-(N-NITROSOMETHYLAMINO)-1-(3-PYRIDYL)-1-BUTANONE (NNK) TUMORIGENESIS WITH AROMATIC ISOTHIOCYANATES

M.A. Morse, K.I. Eklind, S.S. Hecht & F.L. Chung[1]

Section of Nucleic Acid Chemistry, Division of Chemical Carcinogenesis, American Health Foundation, Valhalla, NY, USA

4-(N-Nitrosomethylamino)-1-(3-pyridyl)-1-butanone (NNK) is a potent tobacco-specific carcinogenic nitrosamine. At low doses, it induces primarily lung tumours in mice, hamsters and rats, regardless of the route of administration. Its unique organ specificity and potency suggest its possible role in the high incidence of lung cancer in smokers. The goal of this study was to find agents that would potentially prevent NNK tumorigenesis. Previous results led us to test phenethyl isothiocyanate (PEITC) on NNK tumorigenesis in a two-year bioassay in Fischer 344 rats. The NNK-treated group developed 80% lung tumour incidence, whereas NNK-treated rats fed PEITC diets had only 40% lung tumour incidence. Incidences in other organs were not affected by this treatment. We also tested PEITC in a 16-week, short-term bioassay against NNK-induced lung adenomas in A/J mice. Pretreatment of mice with PEITC by gavage at four daily doses of 5 µmol or 25 µmol reduced the formation of NNK-induced lung adenomas by 70% or 100%, respectively. Interestingly, benzyl isothiocyanate and phenyl isothiocyanate, the lower homologues of PEITC, were inactive in this bioassay. Using a protocol similar to that used in the bioassays, PEITC was shown to decrease DNA methylation by NNK in the lungs of rats and mice and suppress the metabolism of NNK by mouse lung microsomes. These results are consistent with the previous data, suggesting that the inhibition of NNK-induced lung tumour formation by PEITC is a consequence of reduced DNA methylation caused by inhibition of NNK metabolism. As an extension of the structure-activity study, we also tested phenylpropyl and phenylbutyl isothiocyanate in the A/J mouse bioassay. These isothiocyanates were remarkably potent inhibitors of NNK-induced lung adenoma: at doses of 5 µmol, they completely inhibited lung adenoma formation caused by NNK treatment. These results provide a basis for future chemoprevention studies on lung cancer induction associated with exposure to NNK in tobacco smoke.

---

[1]To whom correspondence should be addressed

4-(N-Nitrosomethylamino)-1-(3-pyridyl)-1-butanone (NNK; Figure 1) is the most potent carcinogenic N-nitrosamine so far found in tobacco and tobacco smoke (Hecht & Hoffmann, 1988). The organ-specific effect of NNK in the induction of lung tumours in all species tested indicates its possible role in the development of lung cancer among smokers (IARC, 1985). Therefore, it would be of great importance to find compounds, either synthetic or dietary-related, that can counteract the carcinogenic action of NNK.

Our previous studies demonstrated that pretreatment of rats with a diet containing phenyl isothiocyanate (PITC), benzyl isothiocyanate (BITC) or phenethyl isothiocyanate (PEITC; Figure 1) resulted in reduced metabolic demethylation of NNK in hepatic microsomes as well as a decrease in hepatic DNA methylation by NNK (Chung et al., 1985), suggesting that these aromatic isothiocyanates can potentially inhibit NNK carcinogenesis. In this study, we examined the effects of PEITC on NNK tumorigenesis in Fischer 344 rats and of PITC, BITC and PEITC on NNK tumorigenesis in A/J mice. We compared these results to their effects on NNK-induced DNA methylation in liver, lung and nasal cavity of rats and in lung of mice, all target tissues of NNK tumorigenesis. In a separate bioassay, we also evaluated phenylpropyl- and phenylbutylisothiocyanates (PPITC and PBITC), higher homologues of PEITC, and oxopyridylbutylisothiocyanate (OPBITC), an isothiocyanate related to NNK, for their effects on NNK tumorigenesis in A/J mice. Both PBITC and OPBITC are newly synthesized arylalkyl isothiocyanates.

Figure 1. Structures of isothiocyanates and of 4-(N-nitrosomethylamino)-1-(3-pyridyl)-1-butanone (NNK)[a]

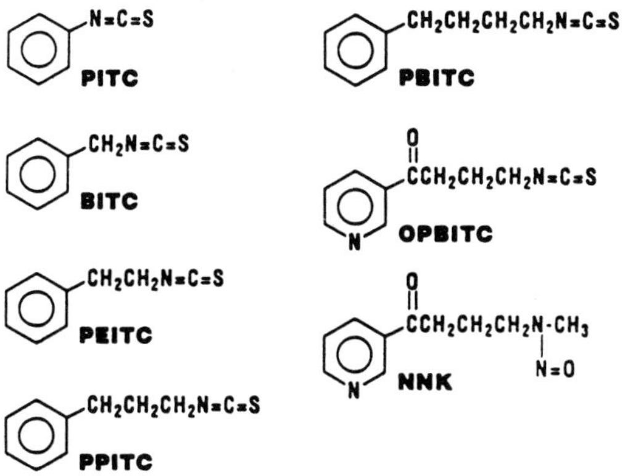

[a] Abbreviations: PITC, phenyl isothiocyanate; BITC, benzyl isothiocyanate; PEITC, phenethyl isothiocyanate; PPITC, phenylpropyl isothiocyanate; PBITC, phenylbutyl isothiocyanate; OPBITC, oxopyridylbutylisothiocyanate

Table 1 shows the tumour incidences in lung, liver and nasal cavity of Fischer 344 rats after treatment with NNK, NNK plus PEITC or PEITC. The tumour incidences induced by NNK alone were within the range expected on the basis of the results of previous bioassays. Only 43% of rats fed a diet containing PEITC before and during NNK treatment developed lung tumours, whereas 80% did so in the group fed the control diet. The PEITC diet did not alter the incidences of tumours in the liver or nasal cavity induced by NNK.

Table 1. Incidences of lung, liver and nasal cavity tumours after treatment with NNK, NNK plus PEITC or PEITC[a]

Group no.	Treatment	No. of rats	Lung			Liver			Nasal cavity		
			Adenoma	Carcinoma	Total	Adenoma	Carcinoma	Total	Benign[b]	Malignant[c]	Total
1	NNK	40	8	24	32 (80)	12	3	15 (38)	8	3	11 (28)
2	NNK plus PEITC	40	5	12[d]	17 (43)[e]	9	5	14 (35)	6	1	7 (18)
3	PEITC	20	0	0	0	4	2	6 (28)	0	1	1 (5)
4	Control	20	1	0	1 (15)	3	1	4 (20)	0	1	1 (5)

[a] NNK, 4-(N-nitrosomethylamino)-1-(3-pyridyl)-1-butanone; PEITC, phenethyl isothiacyanate. Male Fischer 344 rats, 8 weeks of age, were randomized into four groups; groups 2 and 3 were fed PEITC in the diet (3 µmol/g diet) ad libitum for 21 weeks, while groups 1 and 4 were given only NIH-07 diets. After the first week of feeding, NNK (1.76 mg/kg body weight) was administered to groups 1 and 2 by subcutaneous injection three times weekly for 20 weeks. The experiment was terminated after 104 weeks. Gross lesions and representative samples of all major organs were processed for microscopic examination.
[b] Squamous-cell papillomas, transitional-cell papillomas, polyps
[c] Squamous-cell carcinoma
[d] One animal had a squamous-cell carcinoma and 11 had adenocarcinomas
[e] $p < 0.05$ compared to NNK group

To determine the effects of PEITC on the formation of DNA adducts by NNK, we used experimental conditions analogous to those used in the bioassay. Table 2 shows the effects of two weeks' feeding of PEITC on DNA methylation by NNK in liver, lung and nasal mucosa of rats. The levels of 7-methylguanine in DNA of the liver and nasal mucosa of rats were not affected; in lung, however, the levels were reduced from 10.4 to 5.9 µmol/mol guanine, a reduction of nearly 50%.

In the A/J mouse bioassay, we examined the effects of PEITC and its homologues PITC and BITC on NNK-induced lung adenomas (Table 3). A single intraperitoneal administration of NNK at a dose of 10 µmol/mouse resulted in a 100% incidence of pul-

Table 2. DNA methylation in rats treated with 4-(N-nitrosomethylamino)-1-(3-pyridyl)-1-butanone (NNK) and fed control or phenethyl isothiocyanate (PEITC) diets[a]

Diet	7-Methylguanine (µmol/mol guanine)		
	Lung	Liver	Nasal mucosa
Control	10.4 ± 1.3[b]	20.6 ± 0.9[b]	21.5[c]
PEITC	5.9 ± 0.6[d]	22.8 ± 0.7	31.5

[a] Groups of six male Fischer 344 rats fed control or test diets containing 3 µmol/g of diet PEITC for 2 weeks. Beginning on day 11 of feeding, [^{3}H-CH$_3$]NNK was administered subcutaneously daily at a dose of 0.6 mg/kg body weight for four consecutive days. Four hours after the last NNK dosing, rats were sacrificed and tissue DNA was isolated for analysis of 7-methylguanine
[b] Mean ± SE for six rats
[c] Mean of two pooled preparations (two to three rats/pool)
[d] $p < 0.05$ compared to values in control group

Table 3. Effects of isothiocyanates on adenomas induced by 4-(N-nitrosomethylamino)-1-(3-pyridyl)-1-butanone (NNK) and on formation of of O^6-methylguanine (O^6-MeG) in lungs of A/J mice[a]

Group no.	Pretreatment[b]	Daily dose (µmol)	No. of mice	Mice with tumours (%)	No. of tumours/mouse[c]	O^6-MeG (µmol/mol guanine)[c]
1	None	–	30	100	10.7[1] ± 0.8	30.9[1] ± 5.9
2	PEITC	5	18	89	2.6[2] ± 0.4	3.9[2] ± 1.2
3	PEITC	25	20	30[d]	0.3[3] ± 0.1	ND[e]
4	BITC	5	20	100	7.6[1] ± 0.5	26.1[1] ± 6.7
5	PITC	5	20	100	9.5[1] ± 1.2	29.7[1] ± 4.4

[a] Groups of 20-30 female A/J mice, maintained on AIN-76A diet, were administered corn oil or isothiocyanates by gavage daily for four consecutive days. Two hours after the final gavaging, a single dose of NNK (10 µmol/mouse) was administered intraperitoneally. Sixteen weeks after NNK administration, mice were sacrificed and pulmonary adenomas were counted. For assay of O^6-MeG, groups of five mice were administered corn oil or isothiocyanates by gavage for four consecutive days. Two hours after the final gavaging, NNK was administered intraperitoneally at a dose of 10 µmol/mouse. Mice were sacrificed 6 h after NNK administration. DNA was isolated from lung and hydrolysed in 0.1 N HCl for 60 min. O^6-MeG was analysed by strong cation exchange high-performance liquid chromatography and fluorescence detection.
[b] PEITC, phenethyl isothiocyanate; BITC, benzyl isothiocyanate; PITC, phenyl isothiocyanate
[c] Mean ± SE; mean bearing different superscripts under each column are statistically different ($p < 0.05$) from one another as determined by analysis of variance followed by Newman-Keul's ranges test.
[d] Significantly ($p < 0.01$) less than that of group 1 as determined by the chi-square test.
[e] Not detected

monary adenomas, with a multiplicity of 10.7 tumours/mouse in only 16 weeks. The 5-μmol daily dose (20 μmol total) of PEITC did not significantly reduce the proportion of mice that developed pulmonary adenomas, but resulted in an approximately 70% reduction in tumour multiplicity. The 25-μmol daily dose (100 μmol total) of PEITC resulted in a 70% reduction in the percentage of mice that developed tumours and nearly complete inhibition of tumour multiplicity. However, pretreatment with BITC or PITC at 5 μmol/day resulted in no significant change in the percentage of mice that developed tumours or in tumour multiplicity. Both BITC and PITC proved too toxic to be tested at a daily dose of 25 μmol.

The effects of these isothiocyanates on NNK-induced O^6-methylguanine (O^6-MeG) in A/J mouse lung DNA were also investigated. The same dosing regimen as employed in the pulmonary adenoma assays was used. Six hours after NNK administration, the 5-μmol daily dose of PEITC had resulted in an 87% reduction in O^6-MeG levels, while the 25-μmol daily dose gave undetectable levels. Neither BITC nor PITC pretreatment resulted in a significant reduction in O^6-MeG levels. The effects of the isothiocyanates on NNK-induced O^6-MeG formation are thus in good agreement with their effects on NNK lung tumorigenicity.

Our results indicate an upward trends in inhibitory potency as the alkyl chain length increases. To test this hypothesis, we assayed NNK using PPITC, PBITC, OPBITC, PITC, BITC and PEITC as inhibitors (Table 4). PPITC and PBITC exhibited remarkable inhibitory activities; both were considerably more potent than PEITC, and at a daily dose of 5 μmol for four days, they completely prevented the development of lung tumours by NNK. OPBITC was inactive. As in previous bioassays, PITC and BITC were inactive, whereas PEITC reduced lung adenoma multiplicity by 60-70%. The multiplicity and tumour incidences in the groups treated only with isothiocyanates were similar to those in a corn oil control group.

DNA methylation by NNK was reduced in the lungs of rats and mice administered PEITC; however, it was not affected in the liver and nasal cavity of rats fed PEITC or in the lungs of mice fed PITC and BITC. These effects are consistent with the effects of these isothiocyanates on tumorigenicity induced by NNK in the two species and clearly suggest that the protective effect of PEITC against NNK-induced lung tumours is due to its ability to inhibit DNA methylation by NNK. Since there is a parallel relation between NNK tumorigenicity and DNA methylation, ability to inhibit DNA methylation could be used as a means of screening potential inhibitors of NNK carcinogenicity.

These data represent the first demonstration of inhibition of NNK tumorigenesis by any compound (Morse et al., 1989a,b). PEITC is a product of the hydrolysis of gluconasturtiin, which is commonly found in turnips and rutabagas (Tookey et al., 1980). PPITC and PBITC are both synthetic compounds. The goals of our future studies are to develop inhibitors of higher efficacy by structure-activity studies and, ultimately, to test them in high-risk groups such as heavy smokers.

Table 4. Effects of isothiocyanates on pulmonary adenoma induction by 4-(N-nitrosomethylamino)-1-(3-pyridyl)-1-butanone (NNK) in A/J mice[a]

Treatment[b]	No. of mice	Weight (g; mean ± SE)	No. of tumours/ mouse (mean ± SE)[c]	Mice with tumours (%)
Corn oil + saline	29	24.2 ± 0.6	0.3[1] ± 0.1	31
Corn oil + NNK	39	21.9 ± 0.3	9.2[2] ± 0.5	100
PITC + NNK	30	23.4 ± 0.4	9.8[2] ± 0.9	100
BITC + NNK	29	22.1 ± 0.4	10.4[2] ± 0.7	100
PEITC + NNK	28	22.4 ± 0.3	3.3[3] ± 0.4	93
PPITC + NNK	30	22.0 ± 0.3	0.4[1] ± 0.1	37[d]
PBITC + NNK	28	22.2 ± 0.5	0.4[1] ± 0.1	32
OPBITC + NNK	28	23.1 ± 0.4	7.9[2] ± 1.0	96

[a] Groups of 20-40 A/J mice were administered corn oil or isothiocyanate (5 μmol/mouse per day) by gavage daily for four consecutive days. Two hours after the final dose of corn oil or inhibitor, a single dose of saline or NNK (10 μmol/mouse) in saline was administered intraperitoneally. Sixteen weeks after NNK administration, mice were sacrificed and pulmonary adenomas were counted.
[b] PITC, phenyl isothiocyanate; BITC, benzyl isothiocyanate; PEITC, phenethyl isothiocyanate; PPITC, phenylpropyl isothiocyanate; PBITC, phenylbutyl isothiocyanate; OPBITC, oxypyridylbutyl isothiocyanate
[c] Means in this column that bear different superscipts are significantly different from one another as determined by analysis of variance followed by Newman-Keul's ranges test. Saline-treated groups and NNK-treated groups were tested separately.
[d] Significantly different from the appropriate control group as determined by the chi-square test

References

Chung, F.L., Wang, M. & Hecht, S.S. (1985) Effects of dietary indoles and isothiocyanates on N-nitrosodimethylamine and 4-(methylnitrosamino)-1-(3-pyridyl)-1-butanone α-hydroxylation and DNA methylation in rat liver. *Carcinogenesis*, 6, 539-543

Hecht, S.S. & Hoffmann, D. (1988) Tobacco-specific nitrosamines: an important group of carcinogens in tobacco and tobacco smoke. *Carcinogenesis*, 9, 875-884

IARC (1985) *IARC Monographs on the Evaluation of the Carcinogenic Risk of Chemicals to Humans*, Vol. 37, *Tobacco Habits Other than Smoking: Betel-quid and Areca-nut Chewing; and Some Related Nitrosamines*, Lyon, pp. 209-224

Morse, M.A., Wang, C.X., Stoner, G.D., Mandal, S., Conran, P.B., Amin, S.G., Hecht, S.S. & Chung, F.L. (1989a) Inhibition of 4-(methylnitrosamino)-1-(3-pyridyl)-1-butanone-induced DNA adduct formation and tumorigenicity in the lung of F344 rats by dietary phenethyl isothiocyanate. *Cancer Res.*, 49, 549-553

Morse, M.A., Hecht, S.S. & Chung, F.L. (1989b) Effects of aromatic isothiocyanates on tumorigenicity, O^6-methylguanine formation, and metabolism of the tobacco-specific nitrosamine 4-(methylnitrosamino)-1-(3-pyridyl)-1-butanone in A/J mouse lung. *Cancer Res.*, 49, 2894-2897

Tookey, H.L., VanEtten, C.H. & Daxenbichler, M.E. (1980) Glucosinolates. In: Liener, I.E., ed., *Toxic Constituents of Plant Foodstuffs*, 2nd ed., Ch. 4, New York, Academic Press, pp. 103-142

Relevance to Human Cancer of N-Nitroso Compounds,
Tobacco Smoke and Mycotoxins.
Ed. I.K. O'Neill, J. Chen and H. Bartsch
Lyon, International Agency for Research on Cancer
© IARC, 1991

MODULATION OF GENOTOXIC ACTIVITY OF TOBACCO SMOKE

R.M. Balansky[1], P.M. Blagoeva & Z.I. Mircheva

*Laboratory of Chemical Mutagenesis and Carcinogenesis,
National Centre of Oncology, Sofia, Bulgaria*

Tobacco smoke (TS) caused a three- to nine-fold increase in the frequency of his^+ revertants in *Salmonella typhimurium* TA98 but not in TA97a, TA100 or TA102. Activation by a post-mitochondrial fraction obtained from the liver of rats pretreated with Aroclor-1254 or methylcholanthrene was required; fractions from phenobarbital-pretreated or untreated rats had no effect. Vitamins A and E, but not ascorbic acid, inhibited the TS-induced mutagenesis by up to 63%, whereas glutathione and cysteine increased it slightly. Na_2SeO_3, but neither $CoCl_2$ nor caffeine, inhibited the mutagenic effect of TS by 46-56%. In Chinese hamster ovary cells, both Na_2SeO_3 and caffeine strongly potentiated the number of chromosomal aberrations induced by TS, while theophilline slightly reduced its clastogenic effect. Treatment of mice with TS for 60 min/day increased the frequency of micronuclei in polychromatic erythrocytes in bone marrow and in fetal liver and the number of NCE micronuclei in peripheral blood by four to five fold. Simultaneous treatment of mice with TS and Na_2SeO_3 reduced the clastogenic effect of TS. Ascorbic acid had no effect on clastogenicity but reduced toxicity as measured by body weight loss. Both Na_2SeO_3 and ascorbic acid suppressed the induction of TS-induced hyperplastic and metaplastic changes in bronchial mucosa but had no effect on the number of urethane-induced lung adenomas. Vitamins A and E and ascorbic acid may have a protective effect against the toxic and genotoxic activities of TS.

Chemoprevention of genotoxicity due to tobacco smoke (TS) might contribute to reducing the cancer risk linked to tobacco smoking. Thus, a better understanding of the role of modifiers in this process is needed.

Modulation in vitro *of TS-induced mutagenesis in bacteria*

We have seen previously a three- to nine-fold increase in mutation rate in *Salmonella typhimurium* TA98 but not in TA97a, TA100 or TA102 treated with TS (Balansky *et al.*, 1987, 1988). The addition of an exogenous metabolic activation system, obtained from the livers of Aroclor-1254 or methylcholanthrene pretreated rats was required; no activation was produced with liver from phenobarbital-pretreated or untreated rats. A dose-dependent

[1] To whom correspondence should be addressed

inhibition of TS-induced mutagenesis up to 57-63% ($p < 0.05$) was seen in *S. typhimurium* TA98 treated with vitamins A or E (0.08-0.8 mg/plate) (Figure 1). Inhibition of 46-56% ($p < 0.05$) was observed when Na_2SeO_3 (25-100 μg/plate) was added to the top agar (Figure 1). Neither ascorbic acid (0.2-2.0 mg/plate), $CoCl_2$ (0.2-1.0 mg/plate) nor caffeine (0.2-0.8 mg/plate) influenced this process. However, the addition of glutathione (0.6-2.4 mg/plate) or cysteine (0.12-0.48 mg/plate) increased TS-induced mutagenesis slightly.

Figure 1. Dose-dependent inhibition of mutagenesis induced by tobacco smoke (TS; 240 cm³ in a 16-l glass chamber) in *Salmonella typhimurium* TA98 by Na_2SeO_3 and vitamins A and E; spontaneous mutation rate, 33-41 his^+ revertants per plate

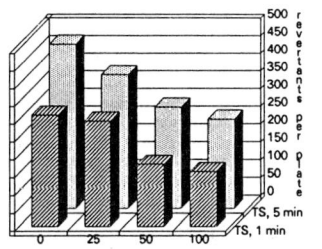
Na_2SeO_3 (μg/plate)

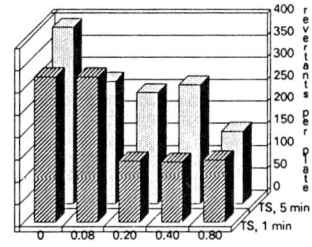
Vitamin A (mg/plate)

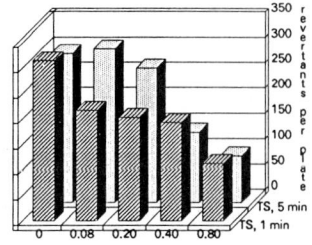
Vitamin E (mg/plate)

Modulation in vitro *of TS-induced clastogenicity in Chinese hamster ovary cells*

Direct treatment of Chinese hamster ovary (CHO) cells with TS enhanced the number of metaphases that had chromosomal aberrations (breaks and exchanges). Treatment of CHO cells with Na_2SeO_3 (2.0-5.0 μg/ml) 2-4 h before or 18 h after exposure to TS increased the chromosome damaging effect of TS by 80.9-92.3% ($p < 0.05$). Similar treatments with caffeine (0.1-0.2 mg/ml) potentiated TS-induced clastogenicity by 97-141% ($p < 0.01$). In contrast, treatment of cells with theophilline (0.05-0.2 mg/ml) 4-15 h before exposure to TS caused a 31-53% ($p < 0.01$) decline in the frequency of TS-induced chromosomal aberrations.

Modulation in vivo *of TS-induced clastogenicity in bone marrow, peripheral blood and fetal liver of mice*

Single or multiple treatment of BDF_1 (C57Bl x DBA_2) mice with TS (60 min/day) caused a dose-dependent, four- to five-fold increase in the number of polychromatic and NCE micronuclei in mouse bone marrow and peripheral blood, respectively (Balansky *et al.*, 1988). In addition, TS was shown to have transplacental clastogenic activity in fetal liver after single or multiple treatment of pregnant mice during the last third of gestation. Four months' treatment of mice with TS (60 min/day) did not change significantly the number of lung adenomas induced by urethane (1.0 g/kg). Ascorbic acid (0.04-0.3%) and Na_2SeO_3 (5-10 ppm) did not influence adenoma formation but suppressed TS-induced hyperplastic and metaplastic changes in bronchial mucosa. Simultaneous treatment of mice with

Na$_2$SeO$_3$ and TS decreased the number of NCE micronuclei as compared with that in mice given TS alone; ascorbic acid did not have a similar effect (Figure 2). Ascorbic acid slightly reduced the toxicity of TS, as measured by body weight loss.

Supplementation of the diet with vitamins A, and E and ascorbic acid might have a protective effect against the toxic and genotoxic activities of TS.

Figure 2. Numbers of micronuclei in the peripheral blood of mice treated with tobacco smoke (TS; 600 cm^3 in a 14-L glass chamber, four exposures of 15 min each with 1-min intervals during which a total air change was made); Na$_2$SeO$_3$ and ascorbic acid were added to drinking-water starting ten days before the first TS exposure and continuing up to the end of the experiment.

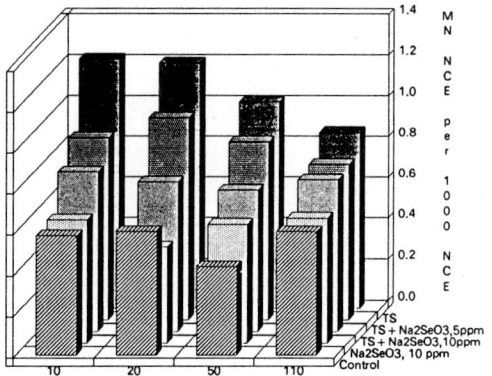
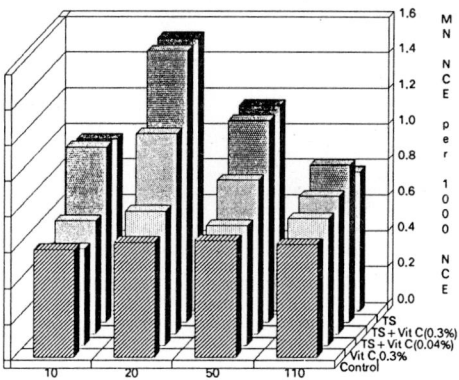

TS treatment (days)   TS treatment (days)

Acknowledgements

Supported in part by the Ministry of Culture, Science and Education, grant 317. Part of this investigation was carried out under the tenure of a Research Fellowship awarded by the International Agency for Research on Cancer, Lyon, France, and spent at the British Columbia Cancer Research Centre, Vancouver, BC, Canada.

References

Balansky, R.M., Blagoeva, P.M. & Mircheva, Z.I. (1987) Investigation of the mutagenic activity of tobacco smoke. *Mutat. Res.*, **188**, 13-19

Balansky, R.M., Blagoeva, P.M. & Mircheva, Z.I. (1988) The mutagenic and clastogenic activity of tobacco smoke. *Mutat. Res.*, **208**, 237-241

PREVENTION OF EXPOSURE TO
N-NITROSO COMPOUNDS

CHINESE TEA INHIBITS THE OCCURRENCE OF OESOPHAGEAL TUMOURS INDUCED BY N-NITROSOMETHYLBENZYLAMINE AND BLOCKS ITS FORMATION IN RATS

C. Han & Y. Xu

Institute of Nutrition and Food Hygiene, Chinese Academy of Preventive Medicine, Beijing, China

Chinese tea can inhibit the occurrence of oesophageal tumours not only by blocking the formation of N-nitrosomethylbenzylamine (NMBzA) *in vivo*, but also by inhibiting the carcinogenesis of preformed NMBzA. The incidence of oesophageal tumours in rats intubated with preformed NMBzA (5 mg/kg bw per week) and with the precursors of NMBzA (methylbenzylamine, 1 mM/kg bw; nitrite, 0.5 mM/kg bw) without Chinese tea was higher than in tea-treated groups; the same tendency was found with regard to the severity of the oesophageal lesions. The anticarcinogenic effects of five varieties of Chinese tea were different from one another.

The possible anticarcinogenicity of tea has been suggested, since it inhibited the mutagenesis caused by some chemical mutagens and the growth of tumour cells (Hara, 1986; Wang *et al.*, 1988), the occurrence of precancerous lesions in rat liver caused by aflatoxin B_1 (Chen *et al.*, 1987) and the occurrence of skin tumours in mice (Yoshizawa *et al.*, 1987). A series of studies by our research group confirmed that Chinese tea can strongly block the formation of N-nitroso compounds both *in vitro* and *in vivo* and showed a clear dose–response relationship (Wu *et al.*, 1988; Wang & Wu, this volume). In the present study, we examined the effect of Chinese tea *in vivo* on oesophageal tumours induced by preformed N-nitrosomethylbenzylamine (NMBzA) and by precursors of NMBzA.

Methods

Five varieties of Chinese tea were selected according to their ability to inhibit N-nitroso compound formation *in vitro* and to their concentrations of polyphenols. These were jasmine tea and Oolong tea from Fujian, green tea and black tea from Hainan and green tea from Hangzhou. All of the leaves were collected freshly in the same year. In the first experiment, a water infusion of tea (1 g tea/50 ml boiling water, kept at room temperature for 30 min) was used as drinking-water for the rats, and made freshly daily. In the second experiment, these water infusions were concentrated under vacuum at 60–70°C and 300–400 mm Hg and kept at –20°C.

Wistar rats, 75-110 g bw, were maintained at 21 ± 4° C on a 12-h light cycle, housed at two rats per cage, and allowed free access to standard rodent chow and water (or tea). For the first experiment, 280 rats were divided into seven groups of 40 (20 males, 20 females); five groups received one of the teas plus NMBzA (twice weekly intubations to give 5 mg/kg bw per week), one received NMBzA only, and one was given no treatment. In the second experiment, 140 male rats were divided into seven groups of 20; five groups received one of the teas plus methylbenzylamine (1.0 mM/kg bw freshly prepared and mixed with the tea) plus $NaNO_3$ (0.5 mM/kg bw freshly prepared before intubation); one group received methylbenzylamine and $NaNO_3$, and one group received only $NaNO_2$. Half of the rats were sacrificed at six weeks and half at 12 weeks, and gross and histopathology examination was carried out. The whole oesophagus was fixed in Bouin's solution and embedded in wax. Slides were stained with haematoxylin and eosin and examined by light microscopy.

Effects of water infusions of tea on the tumorigenicity of preformed NMBzA

No significant difference in body weight or in consumption of tea was seen among the seven groups of animals. After six weeks of NMBzA administration, the oesophageal mucosa was smooth and normal by gross examination in all animals; however, histopathological examination showed abnormal basal cells in the mucosal epithelium and proliferation and atypical proliferation in oesophageal mucosa in all animals receiving NMBzA (Table 1). After 12 weeks of NMBzA administration, the main lesions in the oesophagus included various degrees of proliferation in nodes, thickening of the mucosa, mucosal ulcers and papillomas. The number of animals with papillomas, the number of lesions and the severity of the lesions, reflected by the size of the tumours and the number of tumours in each tumour-bearing animal, were all greater in animals that received only NMBzA than in the five tea-treated groups (Table 2). Histopathological findings (proliferation, papilloma, atypical proliferation and cancer) were similar to those observed grossly.

Table 1. Effects of Chinese teas on induction of histopathological lesions in rat oesophagus by N-nitrosomethylbenzylamine (NMBzA) after six weeks of intubation

Treatment	No. of rats	Abnormal changes		Epithelial changes		
		No.	%	Abnormal basal cells	Proliferation	Atypical proliferation
None	20	0	0	0	0	0
Hainan black tea + NMBzA	20	5	25	1	3	3
Fujian jasmin tea + NMBzA	19	3	16	0	3	3
Fujian Oolong tea + NMBzA	20	4	20	0	1	3
Hainan green tea + NMBzA	17	10	59	1	6	6
Hangzhou green tea + NMBzA	19	5	26	0	6	2[a]
NMBzA	17	17	100*	7	26	4

[a] Including a papilloma
* Significantly different from other groups, χ^2-test, $p < 0.01$

Table 2. Effects of Chinese teas on induction of oesophageal papillomas by N-nitrosomethylbenzylamine (NMBzA) in rats observed grossly after 12 weeks

Treatment	No. of rats	Tumour incidence		Size of papilloma (cm)				No. of tumours/ tumour-bearing rat
		No.	%	> 0.1	> 0.2	> 0.4	> 0.8	
None	20	0	0	0	0	0	0	0.0
Hainan black tea + NMBzA	18	12	67	13	7	6	0	2.2
Fujian jasmin tea + NMBzA	18	8	44**	8	13	0	0	2.6
Fujian Oolong tea + NMBzA	19	8	42**	12	9	0	0	2.6
Hainan green tea + NMBzA	20	13	65	19	16	4	0	3.0
Hangzhou green tea + NMBzA	19	10	53*	13	10	0	0	2.3
NMBzA	20	18	90	60	22	10	2	5.2

* Compared with NMBzA alone, χ^2 test, $p < 0.05$
** Compared with NMBzA alone, χ^2 test, $p < 0.01$

Effects of concentrated tea on the tumorigenicity of precursors of NMBzA

Rats given the two precursors of NMBzA showed decreased body weight gain during the experiment, reflecting the toxic effect of NMBzA formed *in vivo*. After 12 weeks' treatment, the effects of Chinese teas were similar to those seen in the first experiment (Table 3). The results are consistent with our studies on the inhibitory effect of Chinese tea on *N*-nitrosamine formation *in vitro* and in humans (Wu *et al.*, 1988; Wang & Wu, this volume). The present study supports the suggestion that Chinese tea can inhibit *N*-nitrosamine carcinogenesis by blocking the formation of *N*-nitroso compounds *in vivo*.

Acknowledgement

We thank Mr K.X. Xue and Mr B.G. Li of the Institute of Basic Medicine, Chinese Academy of Medical Sciences (Beijing) for technical assistance in preparing pathology specimens, and Professor J. Gao for reviewing the pathological data.

References

Chen, Z.Y., Yan, R.Q., Qin, G.Z. & Qin, L.L. (1987) Effect of six edible plants on the development of AFB1-induced gamma-glutamyltranspeptidase-positive hepatocyte foci in rats. *Chin. J. Oncol.*, **9**, 109-111

Hara, Y. (1986) *Antimutagenesis and Anticarcinogenesis Mechanisms (Basic Life Science 39)*, New York, Plenum, p. 556

Wang, Z.Y., Das, M., Bickers, D.R. & Mukhtar, H. (1988) Interaction of epicalechins derived from green tea with rat hepatic cytochrome P-450. *Drug Metab. Dispos.*, **16**, 98-103

Wu, Y.N., Wang, H.Z., Xu, D.G., Li, F.M., Xu, Y., Han, C. & Lu, S.X. (1988) The influence of Chinese teas on the *in vivo* formation of *N*-nitroso compound in human (in Chinese). *J. Hyg. Res.*, **17**, 37-40

Yoshizawa, S., Horiuchi, T., Fujiki, H., Yoshida, T., Okuda, T. & Sugimura, T. (1987) Antitumor promoting activity of (-)-epigallocatechin gallate, the main constituent of "tannin" in green tea. *Phytother. Res.*, **1**, 44-47

Table 3. Effects of Chinese teas on induction of oesophageal tumours in rats after administration of precursors of
N-nitrosomethylbenzylamine (NMBzA)

Treatment	No. of rats	Tumour incidence		Size of papillomas (mm)			No. of tumours/ tumour-bearing rat
		No.	%	≤ 0.1	> 0.1	> 0.3	
None	20	0	0	0	0	0	0.0
Hainan black tea + precursors	21	1	5	1	0	0	1.0
Fujian jasmine tea + precursors	19	3	16	3	0	0	1.0[a]
Fujian Oolong tea + precursors	17	2	12	2	0	0	1.0[b]
Hainan green tea + precursors	21	4	19	5	0	0	1.3[b]
Hangzhou green tea + precursors	20	1	5	1	0	0	1.0
Precursors	20	9	95[c]	209	95	7	16.4

[a] $p < 0.05$ in comparison with precursor group
[b] $p < 0.01$ in comparison with precursor group
[c] $p < 0.01$ in comparison with other groups

INHIBITORY EFFECT OF CHINESE TEA ON N-NITROSATION IN VITRO AND IN VIVO

H. Wang & Y. Wu

Institute of Nutrition and Food Hygiene, Chinese Academy of Preventive Medicine, Beijing, China

The inhibitory effect of 145 samples of Chinese tea on the formation of N-nitrosomorpholine was studied *in vitro*. The rates of inhibition by green tea, crush, tear, curl (CTC) black tea, brick tea, jasmine tea, Oolong tea, sun-dried tea and black tea were positively correlated with their polyphenol contents. An inhibitory effect of green and black tea on endogenous N-nitrosation was also confirmed in humans. Drinking tea after a meal had a greater effect than drinking it before a meal.

Tea is the most popular beverage in China. Since it contains polyphenolic compounds and ascorbic acid, it may be able to modify N-nitrosation. The objective of this study was to evaluate the possible role of Chinese tea as an inhibitor of N-nitrosation and to compare the relative inhibitory potencies of different teas *in vitro* and *in vivo*.

Inhibition of N-*nitrosation* in vitro

Samples were collected from tea factories in different parts of China and stored at room temperature. The effects of 145 varieties on the formation of N-nitrosomorpholine (NMOR) were studied. Five grams of tea were soaked in 50 ml boiling water for 5 min, and the filtrate was used as the test solution. In an experiment to test for a possible dose-response relationship, 0.5-4.0 ml of test solution were added to test tubes containing 4.0 ml of pH 3.0 buffer and 20 µmol morpholine; then, 20 µmol $NaNO_2$ were added to each tube. The volume was made up to 10.0 ml with distilled water and the reaction mixtures were incubated at 37°C for 30 min in a water bath. At the end of this time, 2.0 ml of 20% ammonium sulfamate in 3.6 N H_2SO_4 were added to each tube to stop the nitrosation reaction. The NMOR formed was determined by gas chromatography-thermal energy analysis. Low concentrations of tea extract promoted NMOR formation, while higher concentrations inhibited it (Figure 1). In order to compare the relative inhibitory potency of various kinds of tea, the volume of tea test solution added was fixed at 2.0 ml. The inhibitory effect was expressed as the blocking rate (BR):

$$BR = \left(1 - \frac{\text{NMOR formed in the presence of tea}}{\text{NMOR formed in the absence of tea}}\right) \times 100\%.$$

The results are given in Table 1. After one year's storage at room temperature, the BRs of the teas were decreased slightly (Figure 2).

The polyphenol contents of the tea samples were inversely correlated with the amounts of NMOR formed in vitro ($r = 0.73, p < 0.01$; data not shown). This implies that the BR of tea is based on its polyphenol content.

Figure 1. Formation of *N*-nitrosomorpholine (NMOR) from precursors in the presence of varying concentrations of three teas[a]

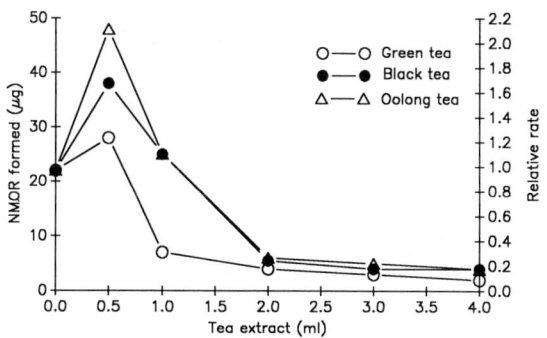

[a]Relative rate: NMOR formed in the presence of tea extract/NMOR formed in the absence of tea extract

Table 1. Blocking rates of different Chinese teas on *N*-nitrosomorpholine (NMOR) formation *in vitro*

Tea	No. of samples	NMOR (μg; mean ± SD)	Blocking rate[a] (%)
Positive control	16	21.99 ± 1.71	-
Green tea	60	2.41 ± 0.61	89.04
CTC black tea	13	2.86 ± 1.00	86.99
Brick tea	7	3.21 ± 0.79	85.40
Jasmine tea	21	3.30 ± 1.85	84.99
Oolong tea	9	3.88 ± 0.84	82.37
Sun-dried tea	13	8.38 ± 0.48	61.89
Chinese black tea	22	9.90 ± 3.95	54.98

[a]Blocking rate = $\left(1 - \dfrac{\text{NMOR formed in the presence of tea}}{\text{NMOR formed in the absence of tea}}\right) \times 100\%$

Inhibition of *N-nitrosation* in vivo

In order to evaluate the effect of tea *in vivo*, 14 healthy young men (nonsmokers) were divided into two groups, one that drank tea before breakfast and one that drank tea after breakfast, and were tested using the *N*-nitrosoproline (NPRO) test (Ohshima & Bartsch, 1981). All subjects received a controlled diet throughout the experimental period. On day 3, subjects ingested 300 mg NaNO$_3$ 1 h before breakfast (for those that drank tea before breakfast) or 30 min after breakfast (for the second group). Thirty minutes later, 300 mg proline were administered. On day 5, subjects were given 300 mg NaNO$_3$ and green tea (5 g tea soaked in 2 × 250 ml boiling water) and 30 min later, 300 mg proline. On day 7, subjects were treated in the same way as on day 5 except that 5 g black tea were substituted for green tea. On each experimental day, after NaNO$_3$ was given, 24-h urine samples were collected in 2-l plastic bottles containing 10 g NaOH. The volumes of the urine samples were recorded. After thorough mixing, 100-ml aliquots of urine were transferred into plastic bottles and stored immediately at -20°C prior to analysis. Both green and black tea strongly inhibited N-nitrosation, and drinking tea after the meal was more inhibitory than drinking tea before the meal (Table 2).

Figure 2. Change in inhibitory effect of tea on formation of N-nitrosomorpholine (NMOR) after one-year storage at room temperature

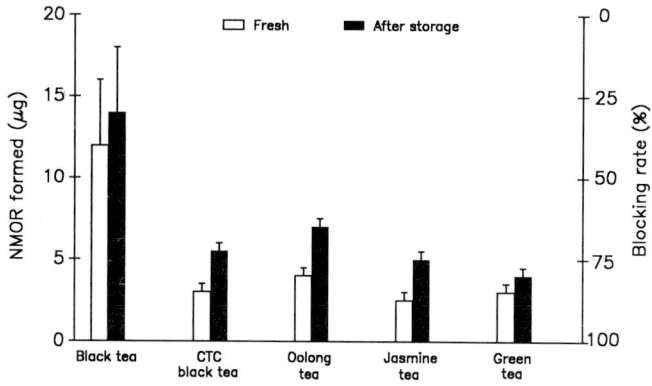

Table 2. Inhibiton of N-nitrosation in human volunteers by tea

Treatment	Tea before meal		Tea after meal	
	NPRO (μg)[a]	Relative excretion[b]	NPRO (μg)	Relative excretion
Control diet	3.83 ± 1.83	0.36	6.67 ± 5.32	0.36
Nitrate plus proline	10.20 ± 2.53	1.00	18.30 ± 5.97	1.00
Nitrate, proline and green tea (5 g)	5.20 ± 2.67	0.49	5.49 ± 1.97	0.30
Nitrate, proline and black tea (5 g)	6.24 ± 1.42	0.58	7.90 ± 4.42	0.43

[a] 24-h urinary excretion of N-nitrosoproline (NPRO), mean ± SD
[b] Taking the amount of NPRO excreted without tea as 1.00

In another experiment, four men and eight women were given 1, 3 and 5 g of green tea in different periods using the same experimental design as described above. The 24-h urinary excretions of NPRO (mean ± SD) was 0.72 ± 0.84 µg (basal excretion), 5.05 ± 5.16 µg (nitrate plus proline), 1.64 ± 1.56 µg (plus 1 g tea), 0.31 ± 0.50 µg (plus 3 g tea) and 0.21 ± 0.37 µg (plus 5 g tea). Thus, ingestion of 3-5 g tea per day can effectively block the endogenous synthesis of NPRO and reduce the carcinogenic burden in human beings.

Reference

Ohshima, H. & Bartsch, H. (1981) Quantitative estimation of endogenous nitrosation in human by monitoring N-nitrosoproline excreted in the urine. *Cancer Res.*, **42**, 3658-3662

INHIBITORY EFFECT OF ELLAGIC ACID ON GENOTOXICITY INDUCED BY AFLATOXINS B_1 AND G_1

T. Górski, E. Górska, J. Odlanicki & M. Sikora

Department for Cancer Prophylaxis and Education, Sanitary Epidemiological Station, Lodz, Poland

Wood *et al.* (1982) reported that ellagic acid has exceptional antimutagenic activity. We have studied the effect of this phenolic plant compound on reversion of *Salmonella typhimurium* and DNA damage induced by aflatoxins B_1 and G_1. We used the Ames plate method with *S. typhimurium* strains TA97, TA98 and TA100 in the presence of metabolic activation by the microsomal fraction of rat liver induced by Aroclor 1254. Ellagic acid was added at 1000 μg/plate before treatment with aflatoxins. DNA damage was measured by the alkaline elution method, using DNA from rat liver. Rats were treated *in vivo* by intraperitoneal injection of aflatoxins 4 h before killing; 100 mg/kg bw ellagic acid were injected intraperitoneally 0.5 h before aflatoxin treatment. The amount of DNA eluted from hepatocytes was determined, and the percentage of DNA that had undergone fragmentation and thus passed through the filter was used as the basis for evaluating DNA damage. The results are given as averages of two to five experiments.

Ellagic acid inhibited the genotoxic effects of aflatoxins B_1 and G_1 in *S. typhimurium* (Table 1), and reduced the bacteriotoxic action of higher doses. In the alkaline elution test, the percentage of DNA passing through the filter after treatment with 1.0 mg/kg bw aflatoxin B_1 was reduced from 26.9% to 24.9% in the presence of ellagic acid; after treatment with 25 mg/kg bw, a reduction from 37.5% to 26.9% was seen with ellagic acid. Ellagic acid may inhibit the covalent binding of carcinogenic metabolites to DNA (Teel, 1986).

References

Teel, R.W. (1986) Ellagic acid binding to DNA as a possible mechanism for its antimutagenic and anticarcinogenic action. *Cancer Lett.*, 30, 324–336

Wood, A.W., Huang, M.T., Chang, R.L., Newmark, H.L., Lehr, R.E., Yagi, H., Sayer, J.M., Jerina, D.M. & Conney, A.H. (1982) Inhibition of the mutagenicity of bay-region diol epoxides of polycyclic aromatic hydrocarbons by naturally occurring plant phenols: exceptional activity of ellagic acid. *Proc. Natl Acad. Sci. USA*, 79, 5513–5517

Table 1. Effect of ellagic acid on reversion of *S. typhimurium* induced by aflatoxins B_1 (AFB) and G_1 (AFG)

Treatment	Ellagic acid	No. revertants per plate					
		TA97		TA98		TA100	
		AFB	AFG	AFB	AFG	AFB	AFG
Aflatoxin (µg)							
0.1	−	662	573	239	283	735	1058
	+	139	139	20	22	186	154
1.0	−	562	754	972	715	1307	2500
	+	158	113	28	164	164	191
10.0	−	326	475	260	205	1253	1642
	+	161	225	36	77	193	337
50.0	−	toxic	toxic	toxic	toxic	607	toxic
	+	271	356	46	261	246	511
100.0	−	toxic	−	toxic	−	503	−
	+	209	−	70	−	241	−
Spontaneous revertants		122	105	28	32	181	187
Dimethylsulfoxide (control)		130	90	30	30	175	180

SOME APPROACHES TO PREVENTION OF ENDOGENOUS FORMATION OF N-NITROSAMINES IN HUMANS

P.P. Dikun, V.B. Ermilov & I.A. Shendrikova

N.N. Petrov Research Institute of Oncology of the USSR Ministry of Health, Leningrad, USSR

In a series of experiments on formation of N-nitrosamines from precursors in the gastric juice, we investigated individual nitrosation capacity and the influence of inhibitors on this process. In the presence of a molar ratio of 2:1 ascorbic acid:$NaNO_2$, we found increased nitrosamine formation in 34% of gastric juice samples. Ferulic acid and caffeic acid generally suppressed nitrosation at high concentrations and stimulated it at low concentrations. The concentration at which the transition from a stimulating to an inhibitory effect occurs depends on individual differences in the gastric juice.

The results of a number of studies (Bartsch *et al.*, 1983; Lu *et al.*, 1984; Ohshima *et al.*, 1985; Wagner *et al.*, 1985) provide convincing evidence that ascorbic acid can suppress the synthesis of N-nitrosamines in human organisms. This conclusion was reached on the basis of statistical analyses of data obtained from two groups of people: a control group with no additional ascorbic acid intake and an experimental group with ascorbic acid added to their diet. Under these circumstances, however, it is impossible to exclude the possibility of an enhancing (anomalous) effect of the ascorbic acid.

The only study in which such an effect could be seen is that of Wagner *et al.* (1985), in which endogenous synthesis of N-nitrosamines with and without additional ascorbic acid and α-tocopherol was compared in the same organism. Although few individuals were studied and the doses of vitamins were high, a slight increase was seen in the urinary excretion of N-nitrosamines in two of six people given ascorbic acid and in two of four given tocopherol. An effect of ascorbic acid was also reported from a study by Nair *et al.* (1986).

The aim of our study was to investigate the effect of individual differences on the activity of inhibitors of N-nitrosamine formation (Ermilov *et al.*, 1986; Dikun *et al.*, 1987, 1988; Ermilov *et al.*, 1989a,b,c), with a view to preventing the endogenous formation of N-nitrosamines.

Effects of individual differences on N-nitrosamine formation

The main method used was measurement of nitrosation of amines by $NaNO_2$ (molar ratio, 1:2) in individual samples of human gastric juice. The assumption was that gastric juice taken from an empty stomach contains the main indicators of the individual properties of the organism that influence N-nitrosamine formation. Qualitative and quantitative determination of N-nitrosamines was carried out with a gas chomatograph and a thermal energy analyser (TEA-502).

The first step was to study the ability of 157 samples of human gastric juice to nitrosate dimethylamine and the influence of ascorbic acid on this process. The results obtained in the absence of ascorbic acid (Fig. 1) clearly demonstrate that the different samples of gastric juice are capable of both enhancing and suppressing nitrosation as compared to the buffer medium. Hence, inhibitors such as ascorbic acid act on a background of other modifying factors.

Fig. 1. Synthesis of *N*-nitrosodimethylamine from dimethylamine and $NaNO_2$ in samples of gastric juice from humans with normal mucosa (●), chronic gastritis (Δ), gastric ulcer (○), polyps (▲) and cancer of the stomach (□). The line shows the results in buffer in different solutions.

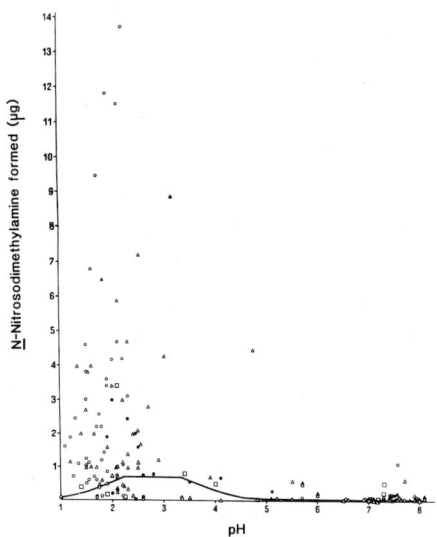

To study the effect of ascorbic acid, a ratio of 2:1 ascorbic acid:$NaNO_2$ was chosen, since it has been reported (Mirvish, 1981) that under these conditions *N*-nitrosamine formation is completely suppressed. In only 51% of the samples did ascorbic acid have a clear inhibitory effect (average, 56%); in 15% of cases, there was no pronounced (>20%) effect of ascorbic acid; and in the remaining 34% of samples an anomalous effect was seen in which the presence of ascorbic acid increased *N*-nitrosodimethylamine (NDMA) formation, by an average of five fold. Similar effects were seen using aminopyrine and morpholine as the amine, and with all three amines using α-tocopherol as the inhibitor.

The next phase of the study was an investigation of the correlation between the efficacy of the inhibitors and their concentration. Three samples were used, each comprising a mixture of ten individual samples of gastric juice, with pH values of 6.1, 3.2 and 1.7. We used two amines, dimethylamine and aminopyrine, for nitrosation and three inhibitors, ascorbic acid, ferulic acid and caffeic acid, known for their high efficacy in suppressing nitrosation (Newmark & Mergens, 1981). All of the inhibitors we used, depending on their concentration in the reaction medium, can act as either inhibitors or stimulators of N-nitrosamine formation in gastric juice (see, e.g., Fig. 2).

Figure 3 makes it possible to compare the effects of all three antioxidants on the same amine (dimethylamine) in two different samples of gastric juice, and Figure 4 the effects on different amines in the same sample of gastric juice. It can be seen that all three antioxidants can manifest an anomalous effect at low ratios of antioxidant to nitrosating agent.

Fig. 2. Gas chromatographic-thermal energy analyser chromatograms of N-nitrosodimethylamine (NDMA) synthesis in gastric juice (a) without added inhibitor, (b) with caffeic acid added in a ratio of 0.005 to $NaNO_2$, and (c) with caffeic acid added in a ratio of 1.6 to $NaNO_2$

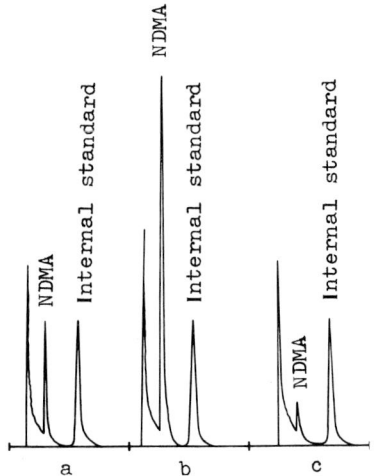

An important aspect of this process is the concentration at which a stimulating effect of antioxidants stops and inhibition begins. Our experiments show that the range of the concentrations at which an anomalous effect is manifested is dependent on other properties of the gastric juice than the pH value. The effect of the antioxidants on nitrosation also depended on the antioxidant and the amine in question and on the ratio of antioxidant:nitrite (Table 1).

Discussion

Individual differences in nitrosation activity play a significant role in the endogenous nitrosation of amines, and individual differences in gastric juice can modulate the influence of antioxidants on N-nitrosamine synthesis. On the basis of such individual properties, recommendations might be given for inhibiting endogenous synthesis of nitrosamines. Our method for measuring the individual nitrosation activity of gastric juice and its influence on the efficacy of inhibitors, together with the N-nitrosoproline test, could be used.

Fig. 3. Influence of changing concentrations of ascorbic acid (▲), ferulic acid (●) and caffeic acid (X) on synthesis of *N*-nitrosodimethylamine (NDMA) from dimethylamine and NaNO₂ in gastric juice (a) pH 6.1 and (b) pH 3.2

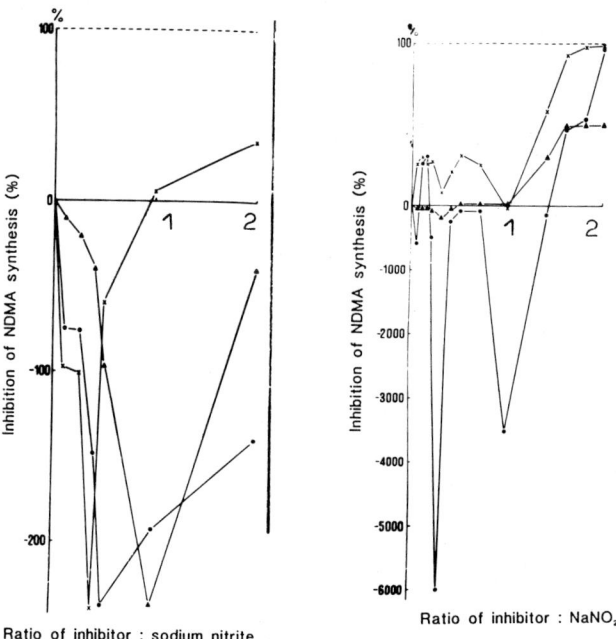

Fig. 4. Influence of changing concentrations of ascorbic acid (▲), ferulic acid (●) and caffeic acid (X) on synthesis of *N*-nitrosodimethylamine (NDMA) from aminopyrine and NaNO₂ in gastric juice pH 3.2

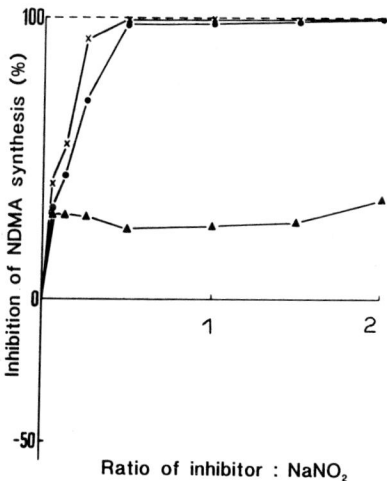

Table 1. Molar ratios of antioxidants and nitrites at which their effect on nitrosation changes from stimulation to inhibition

pH of gastric juice	Ascorbic acid		Ferulic acid		Caffeic acid	
	Dimethyl-amine	Amino-pyrine	Dimethyl-amine	Amino-pyrine	Dimethyl-amine	Amino-pyrine
6.1	1.4	-	1.4	-	0.1	-
3.2	8.0	0.1	-	0.1	1.0	0.1
1.7	1.0	0.4	0.2	0.6	0.2	0.2

-, not determined

Large doses of ascorbic acid can effectively inhibit the nitrosation of amines, but these doses may be unphysiological for some individuals. In these situations, our method could also be used to search for other inhibitors.

References

Bartsch, H., Ohshima, H., Muñoz, N., Crespi, M. & Lu, S.H. (1983) Measurement of endogenous nitrosation in humans: potential applications of a new method and initial results. In: Harris, C.C. & Autrup, H.N., eds, *Human Carcinogenesis*, New York, Academic Press, pp. 833-856

Dikun, P.P., Pavlov, K.A., Ermilov, V.B., Stephanenko, Y.F., Volkov, D.P. & Chernomordikov, V.G. (1987) N-Nitrosocompound assay carried out in gastric juice to assess individual nitrosating ability of human body. *Vopr. Onkol.*, **33**(12), 61-67

Dikun, P.P., Ermilov, V.B. & Shendrikova, I.A. (1988) Efficacy of inhibitors of amine nitrosation in human gastric juice medium and in organism of animals. In: *Proceedings of II Meeting of Oncologists, Radiologists and Roentgenologists of Kazakhstan, Alma-Ata 1988*, pp. 289-291

Ermilov, V.B., Shendrikova, I.A., Volkov, D.P., Stephanenko, Y.F., Chernomordikov, V.G., Pavlov, K.A. & Dikun, P.P. (1986) Anomalous action of nitrosation inhibitors in human gastric juice. *Vopr. Onkol.*, **32**(10), 58-64

Ermilov, V.B., Shendrikova, I.A. & Dikun, P.P. (1989a) On the effect of inhibitors of amine nitrosation in human gastric juice and animal body. *Vopr. Onkol.*, **35**(1), 38-44

Ermilov, V.B., Sultanov, V.S., Shendrikova, I.A. & Dikun, P.P. (1989b) The influence of lignins and some of their model structures on nitrosamine synthesis in the human gastric juice samples and in mice. *Vopr. Onkol.*, **35**(6), 693-698

Ermilov, V.B., Volkov, D.P., Stephanenko, Y.F., Chernomordikov, V.G., Pavlov, K.A. & Dikun, P.P. (1989c) The effect of human gastric juice on dimethylamine nitrosation by sodium nitrite. *Vopr. Onkol.*, **35**(5), 603-607

Lu, S.H., Ohshima, H. & Bartsch, H. (1984) Recent studies on N-nitroso compounds as possible etiological factors in oesophageal cancer. In: O'Neill, I.K., von Borstel, R.C., Miller, C.T., Long, J. & Bartsch, H., eds, *N-Nitroso Compounds: Occurrence, Biological Effects and Relevance to Human Cancer* (IARC Scientific Publications No. 57), Lyon, IARC pp. 947-953

Mirvish, S. (1981) Ascorbic acid inhibition of N-nitroso compound formation in chemicals, food and biological systems. In: Zedek, M. & Lipkin, M., eds, *Inhibition of Tumor Induction and Development*, New York, Plenum Press, pp. 101-126

Nair, J., Ohshima, H., Pignatelli, B., Friesen, M., Malaveille, C., Calmels, S. & Bartsch, H. (1986) Modifiers of endogenous carcinogen formation: studies on in-vivo nitrosation in tobacco users. In: Hoffmann, D., ed., *New Aspects of Tobacco Carcinogenesis* (Banbury Report No. 23), Cold Spring Harbor, NY, CSH Press

Newmark, H. & Mergens, W. (1981) α-Tocopherol (vitamin E) and its relationship to tumor induction and development. In: Zedek, M. & Lipkin, M., eds, *Inhibition of Tumor Induction and Development*, New York, Plenum Press, pp. 127-168

Ohshima, H., Muñoz, N., Nair, J., Calmels, S., Pignatelli, B., Crespi, M., Leclerc, H., Lu, S.H., Bhide, S., Vincent, P., Gounot, A.M. & Bartsch, H. (1985) Monitoring for endogenous formation of N-nitroso compounds by measurement of urinary N-nitrosamino acids. Its application to epidemiological and clinical studies. *Ann. Biol. Clin.*, **43**, 463-474

Wagner, D.A., Shuker, D.E.G., Bilmazes, C., Obiedzinski, M., Baker, I., Young, V.R. & Tannenbaum, S.R. (1985) Effect of vitamins C and E on endogenous sythesis of N-nitrosamino acids in humans: precursor-product studies with (^{15}N) nitrate. *Cancer Res.*, **45**, 6519-6522

INHIBITION BY FATTY ACIDS OF DIRECT MUTAGENICITY OF N-NITROSO COMPOUNDS

K. Takeda, S. Ukawa & M. Mochizuki

Kyoritsu College of Pharmacy, Minato-ku, Tokyo, Japan

Fatty acids inhibited the direct mutagenicity of N-nitroso compounds in *Salmonella typhimurium* TA1535, *Escherichia coli* WP2 and WP*hcr*⁻, and *E. coli* H/r30R (wild) and Hs30R (uvrA). This inhibitory activity was dependent on the concentration of fatty acids, and fatty acids with longer alkyl chain were more potent. Of the N-nitroso compounds tested, α-hydroxy nitrosamines underwent the strongest inhibitory effect. The rate of decomposition was not changed by addition of fatty acids. The partitioning property of the mutagens was altered but not to such a degree as to explain the amount of inhibition. No significant difference in alkylating activity of the N-nitroso compounds was observed in phosphate and acetate buffers. A stronger inhibition of mutagenicity by a butylating mutagen was detected in *E. coli* WP2 than in WP2*hcr*⁻ and in *E. coli* H/r30R than in Hs30R, suggesting that excision repair was a possible mechanism of inhibition. The mutagenicity and cytotoxicity of α-hydroxy nitrosamines in Chinese hamster V79 cells were also inhibited by acetate.

N-Nitroso compounds are suspected to be one of the causes of human cancer; therefore, the identification of agents that inhibit their mutagenicity would be important in preventing cancer. Negishi and Hayatsu (1984) reported that fatty acids inhibit mutagenicity by inhibiting metabolic activation of N-nitroso compounds. N-Nitrosodialkylamines are metabolically activated through α-hydroxylation, and the direct mutagenicity of α-hydroxy nitrosamines has been reported previously (Mochizuki *et al.*, 1982). We therefore investigated whether fatty acids inhibit the direct mutagenicity of N-nitroso compounds, and especially of α-hydroxy nitrosamines.

Inhibitory activity of fatty acids

The mutagenicity of N-nitroso compounds was assayed in *Salmonella typhimurium* TA1535, *Escherichia coli* WP2 and WP2*hcr*⁻, and *E. coli* H/r30R (wild) and Hs30R (*uvr*A). Fatty acids were used as 0.1 M aqueous solutions (pH 5.0 or 6.0), instead of in phosphate buffer.

The inhibitory effects of fatty acids on the mutagenicity or N-nitroso-2-(hydroxymethyl)methylamine in TA1535, WP2 and WP2*hcr*⁻ strains are shown in Figure 1. Less than 1% of the control level of revertants was seen in TA1535 and less than 15% of the control in WP2*hcr*⁻ strain with 50 μmol/plate of acetic acid. A series of fatty acids, from formic acid (HCOOH) to caproic acid ($CH_3(CH_2)_4COOH$), inhibited the mutagenicity of α-hydroxy nitrosamines. The activity was dependent on the concentration of fatty acids (Figure 2) and increased with the length of the alkyl chain. By measuring viable cells (survival), we confirmed that the inhibition was independent of the cytotoxicity of the fatty acids, since the effect was similar in phosphate buffer. Fatty acids also inhibited the mutagenicity of α-hydroperoxy nitrosamines and N-nitroso-N-alkylureas, but not that of α-acetoxy- or α-phosphonooxynitrosamines (Table 1).

Figure 1. Mutagenicity of N-nitroso(2-hydroxymethyl)methylamine in phosphate buffer and in fatty acid solutions in *Salmonella typhimurium* TA1535, *Escherichia coli* WP2 and WP2*hcr*⁻ [a]

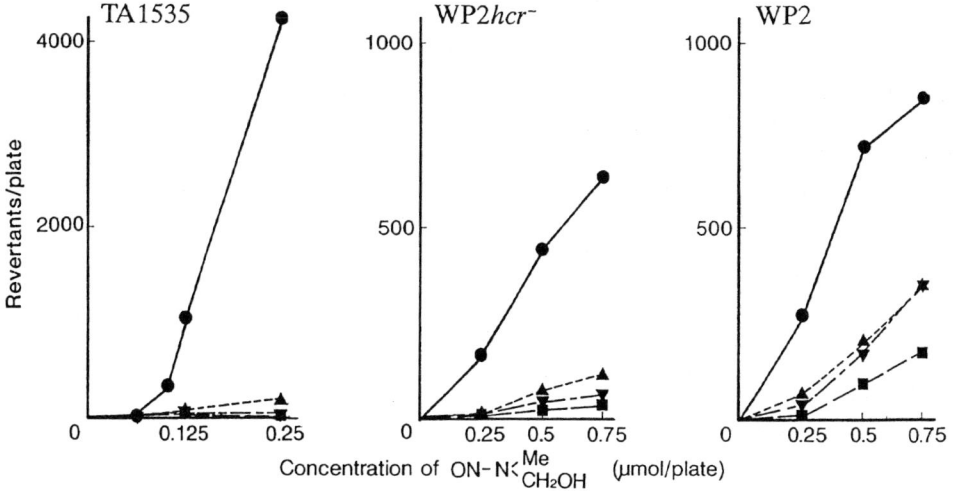

[a] Each fatty acid was used as 0.1 M aqueous solution (pH 5.0); phosphate (●), formate (▲), acetate (▼), propionate (■)

Mechanism of inhibition

The half-lives of α-hydroxy nitrosamines and N-nitroso-N-alkylureas were similar in phosphate buffer and in aqueous fatty acid solutions at the same pH (Table 2), indicating that the mutagens did not react directly with fatty acids. Fatty acids altered the partitioning of N-methyl-N-nitrosourea between octanol and water by less than 3% when used as aqueous solutions, and the degree of the alterations was not enough to explain the amount of the inhibition.

The mutagenicity of a series of N-alkyl-N-nitrosoureas in *S. typhimurium* TA1535 strain was linearly related to their alkylating activity towards nicotinamide (Ukawa & Mochizuki, this volume); in this study, however, no significant difference in the alkylating activity of N-methyl-N-nitrosourea in phosphate and in acetate buffer was observed. The degree of inhibition by acetate of mutagenicity in TA1535 and WP2*hcr*⁻ due to α-hydroxy nitrosamines with different alkyl groups was greatest in compounds with a methyl group and decreased with increasing length of alkyl group. In the WP2 strain, however, the greatest inhibition was seen for a compound with a butyl group. The mutagenicity of N-nitroso-2-(hydroxymethyl)butylamine was inhibited more strongly in *E. coli* WP2 and H/r30R strains than in the corresponding DNA excision repair-deficient WP2*hcr*- and Hs30R strains (Figure 3). These results indicate that the DNA excision repair system may be involved in the inhibition.

Figure 2. Inhibition by fatty acids of N-nitroso-2-(hydroxymethyl)butylamine-induced mutagenicity in *Salmonella typhimurium* TA1535[a]

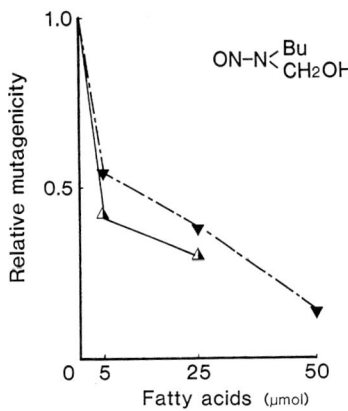

[a]Acetic acid (▼) or caproic acid (△) was used as a 0.1 M aqueous solution (pH 5.0); relative mutagenicity = revertants (acetate buffer)/revertants (phosphate buffer)

Table 1. Inhibitory effect of acetate on the mutagenicity of N-nitroso compounds in *Salmonella typhimurium* TA1535 and *Escherichia coli* WP2*hcr*– [a]

N-Nitroso compound ON-N<R,X		Dose (µmol)	Revertants/plate			
			TA1535		WP2*hcr*–	
R	X		Phosphate	Acetate	Phosphate	Acetate
Me	CH$_2$OH	0.25	4337	16		
		0.5			652	62
Bu	CH$_2$OH	1.0	494	99	2190	660
Me	CH$_2$OOH	0.05	1069	694	218	68
Me	CONH$_2$	2.0	1162	665		
		10.0			1246	535
Bu	CH$_2$OAc	0.75	951	910		
		7.5			467	511
Et	CHCH$_3$ | OPO$_3^{2-}$	0.1	744	726		

[a] Phosphate and acetate were used as 0.1 M buffer solutions at pH 5.0 for α-hydroxy nitrosmaines, and at pH 6.0 for other N-nitroso compounds.

Figure 3. Effect of fatty acids on N-nitroso(2-hydroxymethyl)butylamine-induced mutagenesis in *Escherichia coli* WP2 and WP2*hcr*⁻ and *E. coli* H/r30R and Hs30R[a]

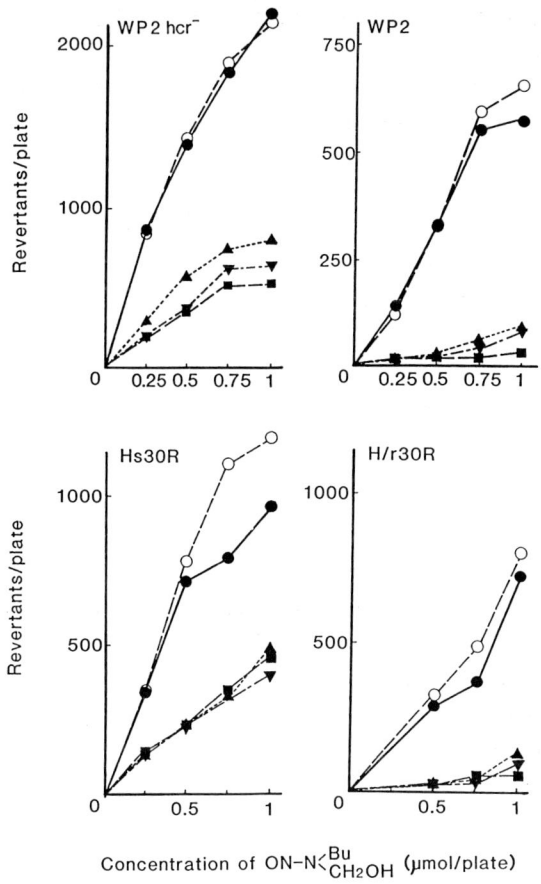

Concentration of ON-N<$^{Bu}_{CH_2OH}$ (μmol/plate)

[a]Fatty acids were used as 0.1 M aqueous solutions (pH 5.0); phosphate (●), citrate (○), formate (▲), acetate (▼) and propionate (□)

Inhibitory effect on mutagenicity in Chinese hamster V79 cells

The mutagenicity and cytotoxicity of α-hydroxynitrosamines in Chinese hamster V79 cells were reported by Mochizuki *et al.* (1984). To investigate the inhibitory effect of fatty acids on mutagenicity in V79 cells, Hanks' solution containing 0.15 M acetate (pH 5.0) was used. Acetate inhibited the mutation frequency and decreased the cytotoxicity of α-hydroxy nitrosamines (Fig. 4).

Table 2. Stability of *N*-nitroso compounds in aqueous solutions

Compound	Buffer[a]	Half-life
ON-N(Bu)(CH$_2$OH)	Phosphate	4.9 min
	Formate	5.0 min
	Acetate	5.1 min
	Butyrate	5.2 min
ON-N(Me)(CH$_2$OH)	Phosphate	5.2 min
	Acetate	5.0 min
ON-N(Et)(CONH$_2$)	Phosphate	11.4 h
	Acetate	11.6 h

[a] Phosphate and fatty acids were used as 0.1 M buffer solutions at pH 5.0 for α-hydroxy nitrosamines, and at pH 6.0 for *N*-ethyl-*N*-nitrosourea.

Acknowledgements

This work was supported in part by a Grant-in-Aid for Cancer Research from the Ministry of Education, Science and Culture. We acknowledge the gift of Chinese hamster V79 cells from Professor T. Kuroki, Institute of Medical Science, University of Tokyo.

References

Mochizuki, M., Anjo, T., Takeda, K., Suzuki, E., Sekiguchi, N., Huang, G.-F. & Okada, M. (1982) Chemistry and mutagenicity of α-hydroxy nitrosamines. In: Bartsch, H., O'Neill, I.K., Castegnaro, M. & Okada, M., eds, N-*Nitroso Compounds: Occurrence and Biological Effects* (IARC Scientific Publications No. 41), Lyon, IARC, pp. 553-559

Mochizuki, M., Osabe, M., Anjo, T., Takeda, K., Suzuki, E., & Okada, M. (1984) Mutagenicity of α-hydroxy nitrosamines in V79 Chinese hamster cells. In: O'Neill, I.K., von Borstel, R.C., Miller, C.T., Long, J. & Bartsch, H., eds, N-*Nitroso Compounds: Occurrence, Biological Effects and Relevance to Human Cancer* (IARC Scientific Publications No. 57), Lyon, IARC, pp. 715-719

Negishi, T. & Hayatsu, H. (1984) Inhibitory effect of saturated fatty acids on the mutagenicity of *N*-nitrosodialkylamine. *Mutat. Res.*, **135**, 87-96

Figure 4. Suppressive activity of acetate on *N*-nitroso(2-hydroxymethyl)methylamine-induced mutagenicity in Chinese hamster V79 cells[a]

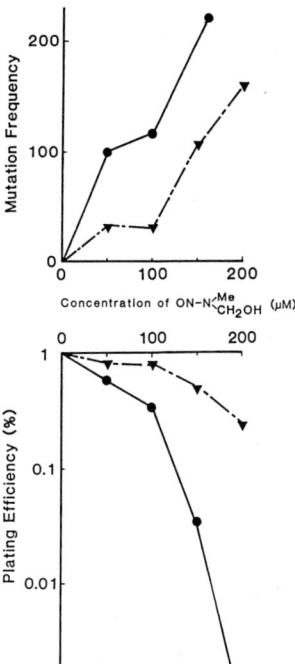

[a]Hanks' solution containing 0.15 M acetate pH 5.0 (▼); Hanks' solution (●)

POTENT INHIBITION OF OESOPHAGEAL METABOLISM OF *N*-NITROSOMETHYLBENZYLAMINE, AN OESOPHAGEAL CARCINOGEN, BY HIGHER ALCOHOLS PRESENT IN ALCOHOLIC BEVERAGES

V.M. Craddock & A.R. Henderson

MRC Toxicology Unit, Carshalton, Surrey, UK

The main cause of oesophageal cancer in western countries is consumption of alcoholic beverages, the degree of risk being much greater for certain spirits than for wine or beer. Risk shows a striking correlation with the content of higher alcohols in the drinks, although the alcohols *per se* have not been shown to be carcinogenic in experimental animals. To test the concept that higher alcohols modulate the oesophageal carcinogenicity of nitrosamines by altering their metabolism, we studied the effect of certain higher alcohols on the metabolism of *N*-nitrosomethylbenzylamine by rat oesophageal mucosal and liver microsomes. In oesophagus, the alcohols were 1000 times more inhibitory than ethanol, and in liver 100 times. This suggests that enhancement of carcinogenesis may not result from an effect on nitrosamine metabolism. Higher alcohols could act by increasing the rate of replication of cells already initiated for malignancy by previous exposure to nitrosamines. Intubation of 2-methylbutanol produces a very much greater increase in oesophageal basal-cell proliferation than does ethanol.

It has been suggested that the association between consumption of alcoholic beverages and oesophageal cancer is due to enhancement by ethanol of the carcinogenicity of the nitrosamines to which man is exposed; however, for consumption of the same amount of ethanol, the degree of risk is much greater for certain spirits than for wine, and greater for wine than for beer (Tuyns *et al.*, 1979; Yu *et al.*, 1988 and others), the risk showing a striking correlation with the concentration of long-branched-chain higher alcohols in the drinks (Postel *et al.*, 1981). As previous work on the action of alcohol on the metabolism of nitrosamines has been largely restricted to the effect of ethanol on the metabolism of *N*-nitrosodimethylamine by liver microsomes, a study was carried out on the effect of certain commonly occurring higher alcohols, mainly 2-methylbutanol, on the metabolism of a nitrosamine that is carcinogenic for the oesophagus, *N*-nitrosomethylbenzylamine (NMBzA), by microsomes of rat oesophageal mucosa and liver.

The method of Labuc and Archer (1982) was used to measure the conversion of NMBzA to benzaldehyde, the metabolite generally assumed to be formed concomitantly with an alkylating intermediate which is the ultimate carcinogen. As 2-methylbutanol is insoluble in aqueous solutions, it was dissolved in a concentration of ethanol that was well below the level found to inhibit NMBzA metabolism. The procedure was checked by adding 2-methylbutanol directly to the microsomal suspension in experiments in which the volume needed was large enough to be practical. The oesophageal mucosae from 12 animals gave sufficient microsomes for one control and one experimental incubation and for determination of protein concentration. Metabolism of NMBzA (5 mM), calculated as nanomoles of benzaldehyde formed per milligram protein per minute, was 0.68 ± 0.08 SD (n = 11) for normal oesophageal microsomes. With liver, there was a wider range of activity between different groups of animals (0.74 ± 0.28, n = 9), but control experiments carried out in each case showed that inclusion of alcohols in the incubation medium gave consistent levels of inhibition.

In the oesophagus, ethanol, even at 500 mM, did not inhibit NMBzA metabolism appreciably until the concentration of nitrosamine was lowered to 0.1 mM (Table 1). In contrast, 2-methylbutanol at 500 mM inhibited the metabolism of NMBzA (5 mM) by 100%, and was still effective at 5.6 mM. The levels of NMBzA and 2-methylbutanol were

Table 1. Effects of ethanol, 2-methylbutanol, 3-methylbutanol and β-phenethyl alcohol on the metabolism of *N*-nitrosomethylbenzylamine (NMBzA) in rat oesophageal mucosal microsomes

Concentration (mM)		Inhibition (%)
NMBzA	Ethanol	
5.0	500	0
1.0	500	8, 6
0.5	500	5
0.1	500	23
0.05	500	48
	2-Methylbutanol	
5.0	500	100
5.0	45	82, 80
5.0	22.5	63
5.0	11.3	55
5.0	5.6	49
5.0	2.8	7
3.0	2.8	45
1.0	1.4	33
0.5	0.7	28
0.1	0.35	28
0.05	0.35	40
	3-Methylbutanol	
5.0	500	100
5.0	11.3	39
	β-Phenylethyl alcohol	
5.0	11.3	55
0.05	0.35	100
0.05	0.18	95

then lowered concomitantly to determine the lowest effective concentration of 2-methylbutanol that could be detected. With 0.05 mM NMBzA, a similar inhibition was produced by 0.35 mM 2-methylbutanol as by 500 mM ethanol. This implies that 2-methylbutanol is approximately 1000 times more inhibitory than ethanol. 3-Methylbutanol and β-phenethylalcohol gave similar results (Table 1). Therefore, although the concentration of higher alcohols in calvados is 100 times lower than that of ethanol (Postel et al., 1981), these alcohols could be ten times more effective than ethanol.

Ethanol inhibited NMBzA metabolism more effectively with liver mircosomes than with oesophageal microsomes (Table 2), but, again, 2-methylbutanol was much more effective than ethanol. With 5 mM NMBzA, 5.6 mM 2-methylbutanol produced a similar inhibition to that of 500 mM ethanol, i.e., 2-methylbutanol was approximately 100 times more inhibitory. The quantitative difference from oesophagus is probably due to the fact that different cytochrome P450 isozymes are involved in the two organs. In liver, the main isozyme responsible for the metabolism of ethanol also metabolizes N-nitrosodimethylamine (Yang et al., 1985). As oesophagus does not metabolize that nitrosamine, the isozyme is probably absent from this organ.

Table 2. Effects of ethanol and 2-methylbutanol on the metabolism of N-nitrosomethylbenzylamine in rat liver microsomes

Concentration (mM)		Inhibition (%)
NMBzA	*Ethanol*	
5.0	750	56
5.0	500	40, 48
5.0	250	19
5.0	50	0
3.0	750	87
3.0	500	97, 90
3.0	250	65
3.0	100	60
	2-Methylbutanol	
5.0	500	100
5.0	100	100
5.0	50	86, 88
5.0	11.25	70, 71
5.0	5.6	45
5.0	2.8	3
3.0	5.6	60
3.0	2.8	39

Duplicate results from repeat experiments using different groups of animals

In contrast to certain other xenobiotics (Wargovich et al., 1988), higher alcohols inhibit the metabolism of nitrosamines concomitantly in liver and oesophagus. This suggests that their association with oesophageal cancer is not mediated by an effect on nitrosamine metabolism. The alcohols could act by increasing the rate of replication of cells already initiated for malignancy by previous exposure to nitrosamines. Intubation of 2-methylbutanol produces a dramatic increase in basal-cell proliferation (V.M. Craddock & R. Edwards, unpublished). Our findings of inhibition of nitrosamine metabolism and mitotic stimulation are in close agreement with those of Mufti et al. (1989).

If compelling evidence is produced that implicates higher alcohols in oesophageal cancer, it should be possible to reduce their concentrations in especially hazardous drinks without detracting from the organoleptic properties of the beverages.

References

Labuc, G.E. & Archer, M.C. (1982) Esophageal and hepatic microsomal metabolism of N-nitroso-N-methylbenzylamine and dimethylnitrosamine in the rat. *Cancer Res.*, **42**, 3181–3186

Mufti, S.I., Becker, G. & Sipes, I.G. (1989) Effect of chronic dietary ethanol consumption on the initiation and promotion of chemically-induced esophageal carcinogenesis in experimental rats. *Carcinogenesis*, **10**, 303–309

Postel, Von W., Adam, L. & Jager, K.H. (1981) Herstellung und Zusammensetzung von Calvados. *Branntweinwirtschaft*, 162–167

Tuyns, A.J., Péquignot, G. & Abbatucci, J.S. (1979) Oesophageal cancer and alcohol consumption: importance of type of beverage. *Int. J. Cancer*, **23**, 443–447

Wargovich, M.J., Woods, C., Eng, V.W.S., Stephens, L.C. & Gray, K. (1988) Chemoprevention of N-nitrosomethylbenzylamine-induced esophgeal cancer in rats by the naturally occurring thioether, diallyl sulfide. *Cancer Res.*, **48**, 6872–6875

Yang, C.S., Tu, Y.T., Koop, D.R. & Coon, M.J. (1985) Metabolism of nitrosamines by purified rabbit liver cytochrome P-450 isozymes. *Cancer Res.*, **45**, 1140–1145

Yu, M.C., Garabrant, D.H., Peters, J.M. & Mack, T.M. (1988) Tobacco, alcohol, diet, occupation, and carcinoma of the oesophagus. *Cancer Res.*, **48**, 3843–3848

PREVENTION OF TUMOUR PRODUCTION IN RATS FED AMINOPYRINE PLUS NITRITE BY SEA BUCKTHORN JUICE

Y. Li & H. Liu

Department of Nutrition and Food Hygiene, Shanxi Medical College, Taiyuan, Shanxi, China

Three groups of Wistar rats were fed a diet containing aminopyrine (0.2%) and $NaNO_2$ (0.2%) and either tap-water (controls), sea buckthorn juice or ascorbic acid solution *ad libitum* for 38 weeks. All 17 rats given tap-water developed tumours in the liver, and six developed tumours in the lungs and four in the kidneys; in rats given sea buckthorn juice, 15 had tumours of the liver, 11 of the lungs and two of the kidneys; the incidences in rats given ascorbic acid were 18/18, 6/18 and 4/18, respectively. The average life span of the group given sea buckthorn juice was 270 days, which was significantly longer ($p < 0.01$) than those of rats given tap-water (195 days) or ascorbic acid (220 days). Microscopic examination of the livers of rats receiving sea buckthorn juice showed fewer foci of carcinogenesis than those of the control and the ascorbic acid groups. The results suggest that sea buckthorn juice can block the endogenous formation of *N*-nitroso compounds more effectively than ascorbic acid and thereby prevent tumour production.

Sea buckthorn (*Hippophae rhamnoids* L.) juice is rich in ascorbic acid and can block the in-vitro nitrosation reaction more effectively than pure ascorbic acid (Li *et al.*, 1987). The objective of the present study was to examine inhibition of endogenous nitrosation by sea buckthorn juice and to explore its possible anticancer properties.

Three groups of eight-week-old Wistar rats were fed a diet containing aminopyrine (0.2%) and $NaNO_2$ (0.2%) and drank either tap-water, ascorbic acid solution (2.5 g/l) or sea buckthorn juice *ad libitum* for 38 weeks. Both juice and the ascorbic acid solution were prepared daily at pH 4. The daily diet and liquid consumption of the three groups were basically the same for the first 15 weeks. All surviving animals were killed at 50 weeks, and complete autopsies and histological examination were carried out.

The rats drinking tap-water developed distended abdomens after ten weeks, and death occurred earlier than in the other groups. The rats given ascorbic acid lived significantly longer ($p < 0.05$) than the tap-water controls but less long ($p < 0.01$) than those in the sea buckthorn group (Table 1).

There was no substantial difference in tumour incidence between rats receiving tap-water and ascorbic acid. The numbers of tumours of the liver and kidney in the sea buckthorn juice group were smaller than those in the other two groups, although the lung tumour incidence in this group was higher (Table 2). A reasonable interpretation might be that the rats given sea buckthorn juice lived longer and metastatic tumours had more time

Table 1. Survival and average life span of rats given aminopyrine and NaNO$_2$ with different liquids

Treatment	No.	Survivors at weeks after initiation of treatment										Average life span (days)
		0–4	5–9	10–14	15–19	20–24	25–29	30–34	35–39	40–44	45–50	
Tap-water	9 M	9	9	9	8	8	6	2	0	0	0	195
	9 F	9	9	9	9	6	1	0	0	0	0	
Ascorbic acid solution	9 M	9	9	9	9	9	8	2	1	1	1	220*
	9 F	9	9	9	9	9	4	2	0	0	0	
Sea buckthorn juice	9 M	9	9	9	9	9	9	7	6	3	2	270**
	8 F	8	8	8	8	8	8	4	3	1	0	

* $p < 0.05$
** $p < 0.01$

to develop in lungs. Chang and Fong (1977) reported that ascorbic acid completely protected rats against hepatic carcinogenesis but not against lung and kidney tumours.

Table 2. Tumour distribution among rats given aminopyrine and $NaNO_2$ and different liquids

Treatment	No. of rats examined	No. of rats bearing tumours in				
		Liver	Lung	Kidney	Stomach	Other organs
Tap-water	17	17	6	4	0	0
Ascorbic acid solution	18	18	6	4	0	1[a]
Sea buckthorn juice	17	15	11	2	0	0

[a] Cystic tumour

The liver tumours in the groups receiving tap-water or ascorbic acid presented with similar evidence of malignancy. However, the livers of the group receiving sea buckthorn juice showed fewer foci of carcinogenesis, which were also less malignant.

Acknowledgements

Supported in part by a grant from the Shanxi Provincial Government. We thank Dr L.H. Jin of the Department of Pathology at Shanxi Medical College for his help in pathological examination.

References

Chan, W.C. & Fong, Y.Y. (1977) Ascorbic acid prevents liver tumour production by aminopyrine and nitrite in the rat. *Int. J. Cancer*, **20**, 268-270

Li, Y., Liu, H. & Song, P.J. (1987) Blocking effect of sea buckthorn juice on the nitrosation of morpholine in vitro (in Chinese). *Chin. J. Prev. Med.*, **21**, 199-201

Relevance to Human Cancer of *N*-Nitroso Compounds,
Tobacco Smoke and Mycotoxins.
Ed. I.K. O'Neill, J. Chen and H. Bartsch
Lyon, International Agency for Research on Cancer
© IARC, 1991

INHIBITION OF BACTERIALLY MEDIATED *N*-NITROSATION BY ASCORBATE: THERAPEUTIC AND MECHANISTIC CONSIDERATIONS

S.A. Leach, C.W. Mackerness, M.J. Hill & M.H. Thompson

*Pathology Division, Public Health Laboratory Service
Centre for Applied Microbiology and Research,
Porton Down, Salisbury, Wiltshire, UK*

Ascorbate is known to inhibit the acid-catalysed *N*-nitrosation reactions of nitrite in the normally acid stomach, suggesting a useful therapeutic application of this compound to reduce exposure to the carcinogenic products of such reactions. However, in the achlorhydric stomach, which is particularly predisposed to cancer, increased exposure to endogenous *N*-nitroso compounds may result from bacterially catalysed reactions. The mechanism of these bacterially mediated reactions is only just beginning to be understood, and, indeed, more than one such mechanism may exist. Despite its usual lack of reactivity towards nitrite at neutral pH, ascorbate proved to be a potent inhibitor of the bacterially mediated (*Pseudomonas aeruginosa*) nitrosation of morpholine, competing with morpholine for the nitrosating agent elaborated by the bacteria from nitrite (the kinetics of the inhibition were classically competitive). This and other data, particularly with regard to the dependence of the bacterially mediated reaction on amine pK_a, are discussed in relation to the potential mechanisms of these bacterially mediated reactions.

Ascorbate reacts more rapidly than secondary amino compounds with the nitrosating agents derived from nitrite at acidic pH (Archer, 1984). The mechanism of this apparent scavenging of nitrite is well known, and its suggested use in the inhibition of endogenous formation of *N*-nitroso compounds in the acidic stomach is well documented (Mirvish *et al.*, 1972; Mirvish, 1981; Ohshima & Bartsch, 1981). *N*-Nitroso compound formation itself is ordinarily an acid-catalysed process, but, under conditions of neutral pH, bacterially mediated reactions have been demonstrated to be of potential importance. [Much of the literature has been reviewed recently (Leach, 1988).] Ordinarily, ascorbate does not react significantly with nitrite at neutral pH (Archer, 1984), and it is under these conditions *in vivo*, for example in the achlorhydric stomach, that there appears to be a particular risk from endogenous formation of *N*-nitroso compounds.

In this paper, therefore, we address two questions: (i) does ascorbate inhibit the bacterially catalysed *N*-nitrosation reaction (at neutral pH), and, if so, (ii) do the characteristics of the inhibitory process cast some light on the mechanism of the *N*-nitrosation reaction itself?

Inhibition of bacterial N-nitrosation by ascorbate

The effect of ascorbate on the bacterially mediated N-nitrosation of morpholine was determined using *Pseudomonas aeruginosa* (BM1030) as the nitrosating bacterium and procedures previously described (Leach *et al.*, 1987). Unexpectedly, considering its lack of reactivity towards nitrite at neutral pH, ascorbate proved to be a potent inhibitor of the bacterially mediated nitrosation of morpholine, the rate of N-nitrosomorpholine formation decreasing in a nonlinear fashion as the ascorbate concentration was increased, with molar ascorbate:nitrite ratios of 0.04 and 0.2, giving 85% and 98% inhibition, respectively (Figure 1). Thus, in this in-vitro system, concentrations of ascorbate that are low in relation to those of nitrite very effectively inhibit bacterially mediated N-nitrosation.

Figure 1. Inhibition of bacterial N-nitrosation by ascorbate

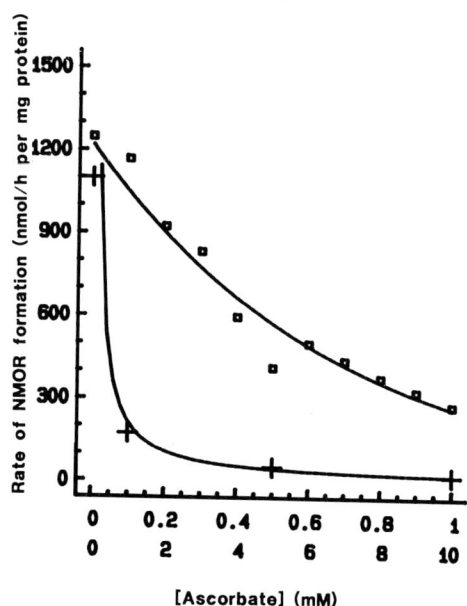

Nitrosation reactions were carried out in 0.1 M Na_2PO_4 buffer, 25 mM $NaNO_2$, 16 mM morpholine, ~ 0.5 mg/ml bacterial protein and 0–1 mM ascorbate (□) or 0–10 mM ascorbate (+) in two separate series of experiments; the final pH of all the incubation mixtures was 8 and incubation was at 37°C for 30 min. Control incubations in the absence of the bacterium (*P. aeruginosa* BM1030) generated negligible amounts of N-nitrosomorpholine (NMOR).

Mechanistic considerations

The molar ascorbate:nitrite ratios shown above further suggest that the effective concentration of 'nitrosating agent' derived from the addition of nitrite to these bacterial cells is at least an order of magnitude lower than the nitrite concentration itself, and it is with this nitrosating agent that the ascorbate and secondary amine compete. It is likely that the discrepancy between the concentration of nitrite and of the apparent 'nitrosating agent' in these systems derives in part from the metabolism by the bacteria of nitrite to products other than N-nitroso compounds. In particular, the reduction of nitrite to nitrogen oxides (NO? and N₂O) and ultimately nitrogen gas is a very important fate in this bacterium, involving the several distinct enzymes of the denitrification sequence (Stouthamer et al., 1982). Indeed, the enzyme-bound nitrite-derived species of this sequence, or even their liberated products, may prove to be the effective bacterially derived 'nitrosating agent' (Figure 2; Garber & Hollocher, 1982; Calmels et al., 1987; Ji & Hollocher, 1988; Leach, 1988; Ralt et al., 1988). Elucidation of a single mechanism whereby bacteria elaborate potential nitrosating agent(s) from nitrite may not be possible, since certain characteristics of bacterially mediated nitrosation differ significantly between species. Thus, the reactions catalysed by denitrifying species (e.g., *P. aeruginosa*) may be distinct from those catalysed by non-denitrifying species (e.g., *Escherichia coli*), particularly in relation to the enzymes thought to be most directly involved in the reaction (Calmels et al., 1988).

Figure 2. Schematic representation of the denitrification sequence with some postulated bacterially derived 'nitrosating agents' depicted in boxes

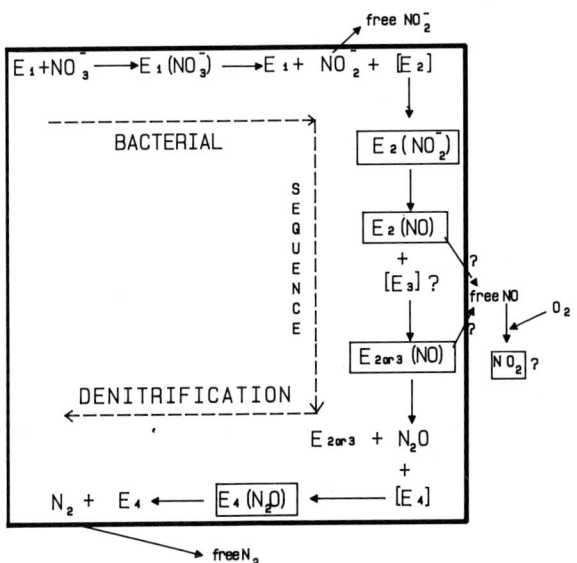

E_{1-4} represent the distinct reductase enzymes.

It is nevertheless conceivable that the bacterially derived nitrosating agent(s) effective at neutral pH are free N_2O_4 and N_2O_3 derived from NO liberated by the bacterial cells and subsequently oxidized by molecular oxygen. Recent work (Ji & Hollocher, 1988) has suggested this to be the case in relation to the reaction between 2,3-diaminonaphthalene and nitrite mediated by *E. coli* (Ralt et al., 1988), an important point in this study being the introduction of molecular oxygen into the system during sample work-up and its subsequent reaction with bacterially derived NO, yielding the nitrosating agents N_2O_3 and N_2O_4. Such involvement of these nitrogen oxides in bacterially mediated reactions would account for the observed inhibition by ascorbate, since ascorbate is known to react with these oxides of nitrogen at neutral pH.

In relation to denitrifying bacteria, however, it is not clear whether organisms such as *P. aeruginosa* ever release free NO. Work employing heavy-isotope labelled (18O, 15N) substrates suggests that, even among denitrifying bacteria, species differ with regard to the precise details of their denitrifying sequence, some liberating free NO and others (including *P. aeruginosa*) not (St John & Hollocher, 1977; Garber & Hollocher, 1981; Hollocher, 1982). Indeed, we have carried out experiments to determine whether free NO ($+O_2 \rightarrow NO_2$, etc) is responsible for the nitrosation reactions mediated by *P. aeruginosa*, as appears to be the case for *E. coli*, as described above (Ji & Hollocher, 1988). In our experiments, a second amine was included in the organic solvent used to extract the *N*-nitroso products from the aqueous incubation mixtures. In all cases, the only significant concentrations recovered were of the *N*-nitroso derivative of the amine that had been present in the aqueous phase. Since the *N*-nitroso products in this case were clearly not elaborated on the introduction of molecular oxygen into the system during the work-up of the samples for analysis, these findings are in contrast to those of experiments using *E. coli*. However, in our nitrosation experiments and those of Calmels *et al.*, strict anaerobiosis was not employed during the bacterial incubations themselves, and it is conceivable that traces of O_2 partitioning into the incubation mixtures were sufficient to oxidize any bacterially generated NO. At the same time, the dissolved oxygen concentration would have had to be sufficiently low not to inhibit bacterial nitrite reduction, which is an enzyme system very sensitive to oxygen tension.

Further evidence that free (NO $\rightarrow N_2O_3$, N_2O_4) evolved during the bacterial metabolism of nitrite is the bacterial nitrosating agent is derived from a comparison of the relative reaction rates of amines of different pK_a by bacterially and chemically mediated mechanisms. It was not clear from the literature how the reactivities of amines to gaseous oxides of nitrogen would relate to amine pK_a. It had been reported (Challis & Kyrtopoulos, 1973; Shuker, 1988) that nitrosation by such agents would show a more random than regular dependence on amine pK_a. These observations, however, were largely based on reactions in organic solvents and in aqueous 0.1 M NaOH. Since the reactive species appears to be the unprotonated amine, the reactions of these nitrogen oxides should show a stronger dependence on amine pK_a in aqueous media at neutral pH. This is demonstrated by the results of experiments carried out in 0.1 M phosphate buffer (pH 7.7) employing standardized aliquots of mixtures of oxides of nitrogen in the gas phase above amine solutions (Figure 3). Comparison of the relative reaction rates reveals a pattern of reactivity that closely mirrors that observed with the bacterially mediated nitrosations both of *E. coli* and particularly of *P. aeruginosa* (Figure 3). This suggests that in both cases the nitrosating agent(s) may derive from bacterially generated free NO (i.e., N_2O_3 and/or N_2O_4) or an enzyme complex (Figure 2) with very similar chemical properties.

If NO is a free intermediate in the denitrification sequence of *P. aeruginosa* and subsequent nitrosation reactions depend upon its aerial oxidation, several observations now require careful consideration. (i) The trapping experiments employing heavy isotope-labelled substrates previously referred to failed to show NO to be a free intermediate of denitrification, despite the fact that (ii) this bacterium has been shown to be capable of nitrosation reactions by a mechanism reported to be inconsistent with mediation by free oxides of nitrogen (Garber & Hollocher, 1982); further, (iii) the more recently determined kinetics of bacterially mediated nitrosation conform to classical enzyme-mediated reactions (Michaelis-Menten kinetics; Calmels et al., 1985; Leach et al., 1985, 1987); and (iv) the inhibitory action of ascorbate is also that of a classical competitive inhibitor (Figure 4).

Figure 3. Dependence of reaction rate (bacterially and acid-catalysed aqueous nitrosations) on amine pK_a

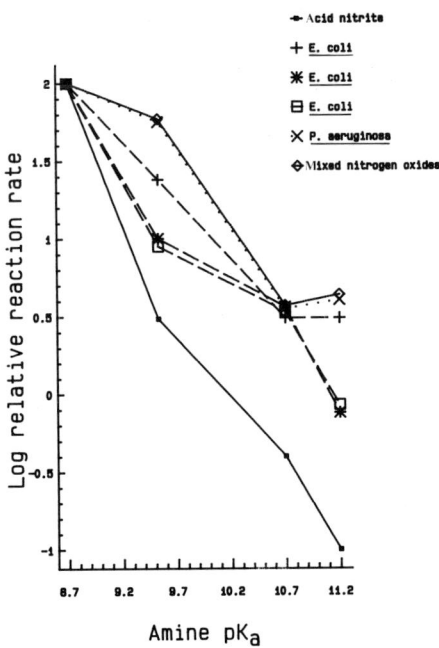

Data from four sources, shown for comparative purposes, using relative reaction rates for each nitrosation mechanism to permit comparison. Acid-catalysed reaction data from Challis and Challis (1982); data for the three strains of *Escherichia coli* from Calmels *et al.* (1985); data for *Pseudomonas aeruginosa* (BM1030) from this study; incubation conditions same as described in legend to Figure 1 but without ascorbate (mean of three determinations); data for oxides of nitrogen from this study (amine [32 mM] in 0.1 M phosphate buffer (final pH, 7.7) in sealed vials exposed with vigorous shaking to standardized aliquots of mixed oxides of nitrogen introduced by gas syringe; reaction time, 5 min; 20 °C (mean of three determinations)

Figure 4. Lineweaver-Burk plots showing the kinetics of the effect of four different concentrations of ascorbate on the bacterially catalysed nitrosation of morpholine

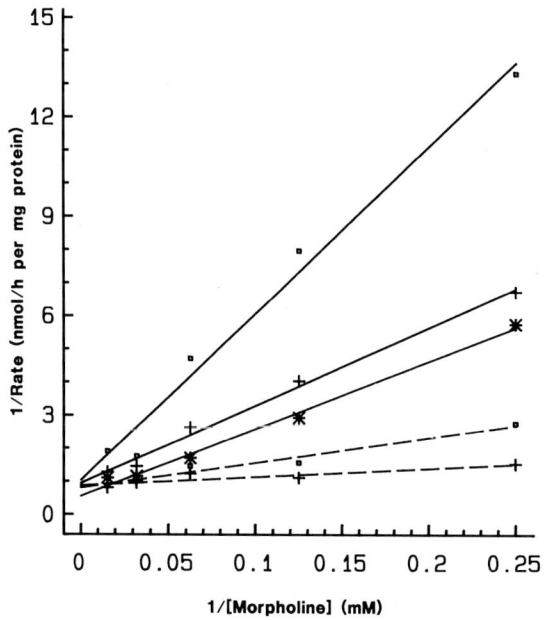

Nitrosation reactions were carried out in 0.1 M phosphate buffer, 25 mM NaNO$_2$, ~ 0.5 mg/ml bacterial protein; ...+..., 0 mM ascorbate; ...□..., 0.2 mM ascorbate; —*— 0.4 mM ascorbate; —+— 0.6 mM ascorbate; —□—, 1.0 mM ascorbate; at 37°C for 30 min at pH 8

Of these observations, the latter two may be most easily explicable in terms of reactions involving nitrogen oxides, since (i) there is an enzymic component in the generation of NO from nitrite and (ii) the use of higher concentrations of substrates which may be toxic could result in the apparent 'saturation' kinetics usually more typical of enzyme-mediated reactions. The first two observations are, however, more difficult to reconcile with the notion that free oxides of nitrogen are the bacterial nitrosating agent. However, some insight may be provided by the denitrification scheme proposed by Averill and Tiedje (1982), in which some enzyme intermediates are proposed that could give rise to free NO; however, the extent of this reaction is determined by bacterial strain and species and environmental influences, particularly the flux of nitrite and reductants through the sequence. Thus, the results of the different studies may critically depend upon the precise conditions to which the bacteria are exposed. In particular, as already discussed, the presence of traces of oxygen in the reaction mixtures in all of the more recent bacterial nitrosation studies may have significantly affected the balance of products of denitrification in favour of NO formation. In contrast, in those studies carried out under more strict anaerobiosis (St John & Hollocher, 1977), free NO was shown not to be a significant free intermediate. It remains to clarify these important points experimentally.

In conclusion, ascorbate is an effective inhibitor of bacterially mediated N-nitrosation in the in-vitro system employed in this study and thus may have relevance in reducing N-nitroso compounds formation in the achlorhydric stomach. Closer examination of the kinetics of this inhibitory process, in conjunction with other supportive data, suggest that the bacterially derived nitrosating agent(s) may be free NO $\longrightarrow$ N_2O_3, N_2O_4 or an enzyme-bound nitrite-derived intermediate of very similar chemical properties.

Acknowledgements

We should like to thank the Cancer Research Campaign for their financial support of this work (CWM).

References

Archer, M.C. (1984) Catalysis and inhibition of N-nitrosation reactions. In: O'Neill, I.K., von Borstel, R.C., Miller, C.T., Long, J. & Bartsch, H., eds, N-*Nitroso Compounds: Occurrence, Biological Effects and Relevance to Human Cancer* (IARC Scientific Publications No. 57), Lyon, IARC, pp. 263-274

Averill, B.A. & Tiedje, J.M. (1982) The chemical mechanism of microbial denitrification. *FEBS Lett.*, 138, 8-12

Calmels, S., Ohshima, H., Vincent, P., Gounot, A.-M. & Bartsch, H. (1985) Screening of microorganisms for nitrosation catalysis at pH 7 and kinetic studies on nitrosamine formation for secondary amines by *E. coli* strains. *Carcinogenesis*, 6, 911-915

Calmels, S., Ohshima, H., Rosenkranz, H., McCoy, E. & Bartsch, H. (1987) Biochemical studies on the catalysis of nitrosation by bacteria. *Carcinogenesis*, 8, 1085-1088

Calmels, S., Ohshima, H. & Bartsch, H. (1988) Nitrosamine formation by denitrifying and non-denitrifying bacteria: implication of nitrite reductase and nitrate reductase in nitrosation catalysis. *J. Gen. Microbiol.*, 134, 221-226

Challis, B.C. & Challis, J.A. (1982) N-Nitrosmaines and N-nitrosimines. In: Patai, S., ed., *The Chemistry of Amino, Nitroso and Nitro Compounds and their Derivatives*, New York, Wiley, pp. 1151-1223

Challis, B.C. & Kyrtopoulos, S.A. (1973) The chemistry of nitroso compounds. Part 11. Nitrosation of amines by the two phase interaction of amines in solution with gaseous oxides of nitrogen. *J. Chem. Soc., Perkin I*, 299-304

Garber, E.A.E. & Hollocher, T.C. (1981) ^{15}N Tracer studies on the role of NO in denitrification. *J. Biol. Chem.*, 256, 5459-5465

Garber, E.A.E. & Hollocher, T.C. (1982) ^{15}N, ^{18}O Tracer studies on the activation of nitrite by denitrifying bacteria. *J. Biol. Chem.*, 257, 8091-8097

Hollocher, T.C. (1982) The pathway of nitrogen and reductive enzymes of denitrification. *Antonie van Leeuwenhoek*, 48, 531-544

Ji, X.B. & Hollocher, T.C. (1988) Mechanism for nitrosation of 2,3-diaminonaphthalene by *Escherichia coli*: enzymatic production of NO followed by O_2-dependent chemical nitrosation. *Appl. Environ. Microbiol.*, 54, 1791-1794

Leach, S.A. (1988) Mechanisms of endogenous N-nitrosation. In: Hill, M.J., ed., *Nitrosamines: Toxicology and Microbiology*, Chichester, Ellis Horwood, pp. 69-87

Leach, S.A., Challis, B.C., Cook, A.R., Hill, M.J. & Thompson, M.H. (1985) Bacterial catalysis of the N-nitrosation of secondary amines. *Biochem. Soc. Trans.*, 13, 381-382

Leach, S.A., Thompson, M. & Hill, M. (1987) Bacterially catalysed N-nitrosation reactions and their relative importance in the human stomach. *Carcinogenesis*, 8, 1907-1912

Mirvish, S.S. (1981) Ascorbic acid inhibition of N-nitroso compound formation in chemical, food and biological systems. In: Zedeck, M.S. & Lipkin, M., eds, *Inhibition of Tumor Induction and Development*, New York, Plenum, pp. 101-126

Mirvish, S.S., Wallcave, L., Eagen, M., & Shubik, P. (1972) Ascorbate-nitrite reaction: possible means of blocking the formation of carcinogenic N-nitroso compounds. *Science*, 177, 65-68

Ohshima, H. & Bartsch, H. (1981) Quantitative estimation of endogenous nitrosation in humans by monitoring N-nitrosoproline excreted in the urine. *Cancer Res.*, 41, 3658-3662

Ralt, D., Wishnok, J.S., Fitts, R. & Tannenbaum, S.R. (1988) Bacterial catalysis of nitrosation: involvement of the *nar* operon of *Escherichia coli*. *J. Bacteriol.*, **170**, 359-364

Shuker, D.E.G. (1988) The chemistry of *N*-nitrosation. In: Hill, M.J., ed., *Nitrosamines: Toxicology and Microbiology*, Chichester, Ellis Horwood, pp. 48-68

St John, R.T. & Hollocher, T.C. (1977) Nitrogen 15 tracer studies on the pathway of denitrification in *Pseudomonas aeruginosa*. *J. Biol. Chem.*, **252**, 212-218

Stouthamer, A.H., Boogerd, F.C. & van Verseveld, H.W. (1982) The bioenergetics of denitrification. *Antonie van Leeuwenhoek*, **48**, 543-553

RETINOIDS PREVENT EPITHELIAL CARCINOGENESIS INDUCED BY N-NITROSO COMPOUNDS

H.Y. Cai

Cancer Institute, Sun Yat-Sen University of Medical Sciences, Guangzhou, China

Two new retinoic acid esters and retinamides synthesized in China, N-(4-ethoxycarbophenyl)retinamide (RI) and N-(4-carboxyphenyl)retinamide (RII), significantly inhibited carcinogenesis induced in the epithelium of the forestomach of mice by N-nitrososarcosine ethyl ester. RI also markedly inhibited carcinogenesis induced in the epithelium of the oesophagus and forestomach in rats by this ester. No sign of hypervitaminosis was noticed with doses as high as six times the therapeutic dose. RI also inhibited precancerous and cancerous lesions in the nasal cavity and nasopharynx and oesophagus of rats induced by dinitrosopiperazine. In a malignant oesophageal epithelial cell line from rats, RE25-3, established in our laboratory, RI and RII inhibited mitosis, proliferation rate, chromosomal aberrations and incorporation of 3H-thymidine into DNA. The ability to form colonies on agar plates was also inhibited by these two compounds.

Interest in the role of retinoids in preventing cancer (Sporn, 1983) has increased over the past 15 years. Retinoids have been found to be effective chemopreventive agents, particularly during the precancerous stage, when they can be used to arrest or reverse the progression to cancer; however, many of them were toxic and caused side-effects. Research has recently been directed to the synthesis of more potent but less toxic retinoids. Retinoids are composed of three components, a ring, a side chain and a polar terminal group; the undesirable toxic effects may be associated with the presence of a terminal carboxyl group. A series of carboxylic esters and amines was synthesized in China (Xu *et al.*, 1981), with lower toxicity but potential effects. We report here on the main compounds which prevent the epithelial carcinogenesis induced by N-nitroso compounds.

Effect on carcinogenesis in the forestomach of mice

Twenty-seven retinoic acid esters and retinamides were prepared by treating the corresponding amine and alcohol with retinyl chloride, which was obtained from retinoic acid and PCl_3 (Xu *et al.*, 1981). Nine retinoic acid esters and retinamides were used to inhibit carcinogenesis in the forestomach epithelium induced by N-nitrososarcosine ethyl ester in mice. Although N-(4-ethoxycarbophenyl)retinamide (RI), and N-(carboxyphenyl)retinamide (RII), RIII, RVI and RVIII (see Table 1) showed significant effects, RI appeared to be most effective, with reproducible results. No toxic effect was observed with a therapeutic dose of RI (6.6×10^{-4} M/kg bw given 11 times), as measured by effect on body weight, whereas retinoic acid caused a significant ($p < 0.001$) decrease in body weight. No sign of hypervitaminosis A was seen with a dose of RI as high as six times the therapeutic dose, although hypervitaminosis was induced with retinoic acid at the same dose. Autoradiography using 3H-thymidine showed a marked inhibitory effect of RII on DNA synthesis in the forestomach epithelium of mice (labelling index ± SE: 10.8 ± 0.85 in eight treated mice and 16.0 ± 1.10 in ten controls; $p < 0.01$) (Cai et al., 1981).

Table 1. Effects of new retinoids on carcinogenesis of the forestomach in mice induced by N-nitrososarcosine ethyl ester[a]

Group[b]	Duration of treatment (days)	Dose of retinoid (mg/kg bw × times)	No. of mice	Inhibition of carcinogenesis (%)
Control	70-85	-	38	
RI	70-85	50 × 21	35	73.0
	82	50 × 20	22	63.6
RII	82	50 × 20	22	54.6
RIII	82	50 × 20	22	54.6
RIV	82	50 × 20	20	20.0
RV	82	45 × 19	19	16.0
RVI	82	45 × 20	22	45.8
RVII	85	50 × 20	19	16.5
RVIII	75	50 × 19	29	56.2
RX	75	50 × 20	24	7.6

[a] 20% sarcosine ethyl ester and 3% $NaNO_2$ mixture fed to mice
[b] RI, N-(4-ethoxycarbophenyl)retinamide; RII, N-(4-carboxyphenyl)retinamide; RIII, N-(5-ethoxycarbophenyl)-retinamide; RIV, N-(5-carboxyphenyl)retinamide; RV, N-(6-ethoxycarbophenyl)retinamide; RVI, N-(3-iodo-6-carboxyphenyl)-retinamide; RVII, 4H(3,1-benzoxazin-4-one), 2-[2,6-dimethyl-8-(2,6,6-trimethyl-1-cyclohexen-1-yl)-1,3,5,7-octatetraenyl]retinamide; RVIII, 4H(3,1,-benzoxazin-4-one),2-[2,6-dimethyl-8-(2,6,6-trimethyl-1-cyclohexen-1-yl)-1,3,5,7-octatetraenyl]- 7-iodoretinamide; RX, N-(5-hydroxyphenyl)retinamide

Effect of RI on carcinogenesis

Female Wistar rats weighing 60-90 g were fed precursors of N-nitrososarcosine ethyl ester (20% sarcosine ethyl ester and 3% $NaNO_2$; 1:1 mixture) to induce carcinogenesis. One group also received 1.1×10^{-4} M/kg bw RI. Dysplasia and carcinomas that developed in both the oesophagus and forestomach were markedly inhibited by RI (Table 2), to a statistically significant degree ($p < 0.01$; Cai et al., 1983).

Dinitropiperazine (DNP) was used to induce dysplasia and cancer in the nasal cavity and nasopharynx of rats. Dysplasia occurred in all animals at doses of 20 or 40 mg/kg bw. In studies in which RI was given after the last dose of DNP, the percentage of precancerous lesions that developed into cancer decreased from 100% in the control group to 61.5% and 76.5% in the RI-treated groups. In studies in which RI was given before DNP, the incidence decreased from 85.7% in the control group to 38.4% in the RI-treated group (Table 3; Cai et al., 1988a). RI also inhibited epithelial carcinogenesis in the oesophagus of rats treated with DNP (data not shown).

Table 2. Inhibitory effects of N-(4-ethoxycarbophenyl)retinamide (RI) on carcinogenesis in the oesophagus and forestomach of rats

Group	Tumour site	No. of rats	Lesions			
			Dysplasia		Carcinoma	
			No.	%	No.	%
Control	Oesophagus	11	9	81.8	7	63.6
	Forestomach	11	5	45.4	2	18.1
	Oesophagus + forestomach				9	81.8
RI	Oesophagus	17	12	70.5	2	11.6
	Forestomach	17	3	17.6	0	
	Oesophagus + forestomach				2	11.6

Table 3. Effect of N-(4-ethoxycarbophenyl)retinamide (RI) on carcinogenesis in the nasal cavity and nasopharynx of rats

Group	No. of rats	Dose (mg/kg bw × times)	Time (days)	Precancerous lesions[a]		Tumours	
				No.	%	No.	%
Therapy[b]							
Control	8	-	160	8	100.0	8	100.0
RI	13	50 × 46	160	13	100.0	8	61.5
Control	10	-	225	10	100.0	10	100.0
RI	13	50 × 46	225	13	100.0	10	76.9
Prevention[c]							
Control	14	-	273-302	14	100.0	12	85.7
RI	13	50 × 64	304	13	100.0	5	38.4

[a] Including basal-cell dysplasia, glandular dysplasia, atypical metaplasia and papilloma
[b] RI given after last dose of dinitrosopiperazine
[c] RI given before dinitrosopiperazine

Effect of RI, RII and retinoic acid on an oesophageal carcinoma cell line in vitro

A malignant transformed cell line (RE25-3) from rat oesophagus was established using N-nitrososarcosine ethyl ester in an in-vivo-in-vitro method in our laboratory (Cai et al., 1988b). RI, RII and retinoic acid all inhibited mitosis (Figure 1), proliferation rate and the incorporation of 3H-thymidine into DNA (Table 4). All three retinoids decreased the number of aberrant chromosomes, and the diploid cells that had been lost during transformation of the RE25-3 cell line were restored (Figure 2). The ability to form colonies on an agar plate was also inhibited by RII (72%), RI (54%) and retinoic acid (48%) (Ho

& Cai, 1988), indicating that RI might be more effective *in vivo* and RII more effective *in vitro*.

Figure 1. Effects of retinoids on mitosis in a malignant transformed cell line from rat oesophagus (RE25-3)

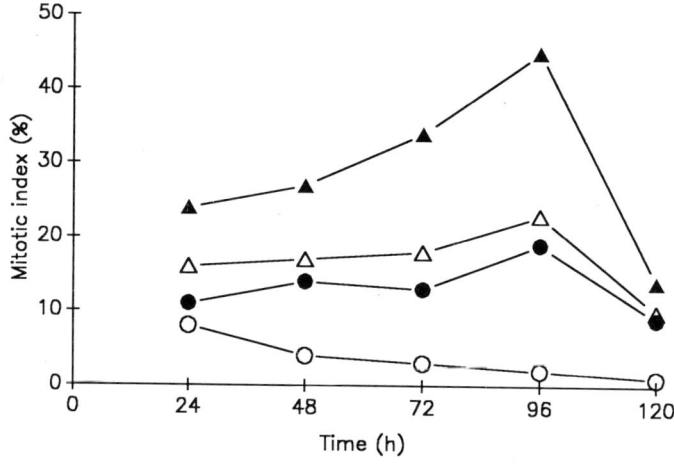

▲, control; ●, retinoic acid; △, N-(4-ethoxycarbophenyl) retinamide (RI); ○, N-(4-carboxyphenyl) retinamide (RII)

Figure 2. Effect of retinoids on chromosomal aberrations in a malignant transformed cell line from rat oesophagus (RE25-3)

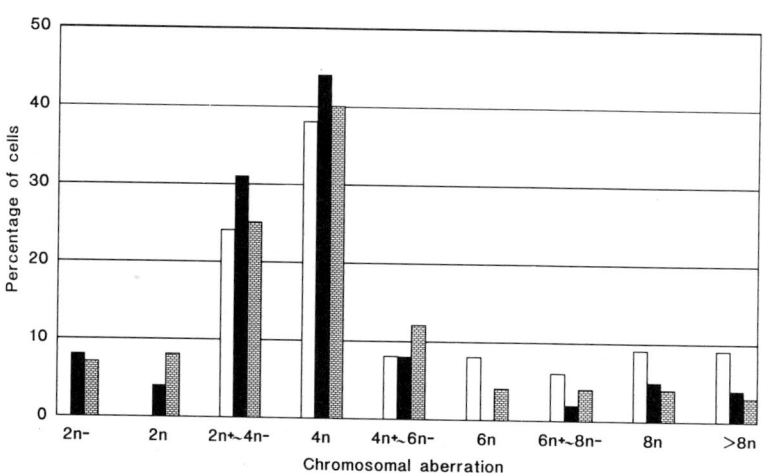

□, control; ■, retinoic acid; □, N-(4-ethoxycarbophenyl) retinamide (RI)

Table 4. Effect of retinoids on DNA synthesis in 5×10^5 RE25-3 cells

Time (h)	Counts per minute ± SE			
	Control	Retinoic acid	RI[a]	RII[b]
12	39 159.0 ± 10 232.8	36 400.0 ± 1716.7	31 843.0 ± 3016.8	19 815.5 ± 1892.9
17	485 42.7 ± 6026.7	42 717.0 ± 3324.5	32 155.0 ± 2960.5	25 045.7 ± 2781.5
24	47 541.0 ± 8080.8	39 420.0 ± 6384.8	34 333.3 ± 5238.9	22 505.0 ± 1974.7

[a] RI, N-(4-ethoxycarbophenyl)retinamide; $p < 0.05$
[b] RII, N-(4-carboxyphenyl)retinamide; $p < 0.01$

Wang et al. (1983) reported that RI and RII inhibited the proliferation of ECA-109C3 human oesophageal cell line *in vitro*, and these compounds also inhibited the expression of Epstein-Barr virus early antigen in Raji cells (Zeng et al., 1983). Our experiments show that they might inhibit cell proliferation and mitosis and control cell differentiation, thus preventing the progression of the precancerous lesions to cancer. Other mechanisms may be involved, however, and further studies should be carried out. Nevertheless, our data indicate that these compounds, and especially RI, could be administered to populations at high risk for oesophageal and nasopharyngeal carcinoma to decrease the cancer incidence rate.

Acknowledgements

Part of this research was supported by the Fund of National Natural Sciences and National Fund for Cancer Research.

References

Cai, H.Y., Zhang, J.S. & Cui, X.X. (1981) Inhibitory effect of some new retinoids on carcinogenesis (in Chinese). *Acta Pharm. Sin.*, **16**, 648-653

Cai, H.Y., Li, B.S. & Lu, S.X. (1988a) Induction of precancerous lesion and cancer of nasopharynx and nasocavity in rats and the inhibitory effect of RI on carcinogenesis (in Chinese). *Chin. J. Pathol.*, **17**, 98-101

Cai, H.Y., Li, Q.X., Ye, Y.L. & Lu, S.X. (1988b) Transformation of rat esophageal precancerous epithelial cells and its biologic characteristics (in Chinese). *Chin. J. Oncol.*, **10**, 256-259

Cai, H.Y., Zhang, J.S., Zheng, G.L. & Cui, X.X. (1983) Inhibitory effect of new retinoic acid analogue on the carcinogenesis of esophagus and forestomach in rats (in Chinese). *Chin. J. Cancer*, **2**, 210-214

Ho, W.M. & Cai, H.Y. (1988) Inhibitory effect of retinoids on the malignancy of transformed esophageal cell line RE25-3 (in Chinese). *Acad. J. Sun Yat-sen Univ. Med. Sci.*, **9**, 23-26

Sporn, M.B. (1983) Retinoids and cancer. *Cancer Surv.*, **2**, 221-222

Wang, R.Z., Pan, Q.Q., Xu, S.P. & Huang, L. (1983) Action of 24 new synthetic retinoids on the proliferation of ECA-109C3 cells in culture (in Chinese). *Chin. J. Oncol.*, **51**, 243-248

Xu, S.P., Guo, Z.R., Yuan, Z.L., Li, L.M. & Huang, L. (1981) Studies on compounds of tumor prevention – synthesis of derivatives of retinoic acid (in Chinese). *Acta Pharm. Sin.*, **16**, 678-686

Zeng, Y., Zhou, H.M. & Xu, J.P. (1983) Inhibitory effect of retinoids on Epstein-Barr virus induction in Raji cells (in Chinese). *Acta Acad. Med. Sin.*, **4**, 251-253

STABILITY OF MUTAGENIC NITROSATED PRODUCTS OF INDOLE COMPOUNDS OCCURRING IN VEGETABLES

H.G.M. Tiedink[1,2], J.A.R. Davies[2], W.M.F. Jongen[3] & L.W. van Broekhoven[2]

[1]*Agricultural University, Department of Toxicology;*
[2]*Centre for Agrobiological Research and* [3]*Agrotechnological Research Institute, Wageningen, The Netherlands*

Levels of indolylglucosinolates in *Brassica* vegetables correlated significantly with the amounts of *N*-nitroso compounds formed in these vegetables after nitrite treatment. Nitrosation of indole-3-carbinol, indole-3-acetonitrile and indole, hydrolysis products of an indolylglucosinolate, resulted in formation of nitrosated products, which were directly mutagenic to *Salmonella typhimurium* TA100. The nitrosated products were unstable at pH 2 but stable at pH 8. Experiments to elucidate the mechanisms behind these differences in stability showed an equilibrium between the nitrosated indole compound and the free compound plus nitrite.

Indole compounds were investigated as potential precursors of *N*-nitroso compounds in *Brassica* vegetables. Special attention was paid to the stability of their nitrosated products.

N-Nitroso compounds formed in Brassica *vegetables*

Thirty-one vegetables were screened for their potential to form directly mutagenic *N*-nitroso compounds by treating aqueous extracts with nitrite (40 mmol/l) at pH 2, 1 h, 37°C; Tiedink *et al.*, 1988). Brassica vegetables showed the highest mutagenicity to *Salmonella typhimurium* TA100 in the absence of metabolic activation. Moreover, the total amount of *N*-nitroso compounds formed in these vegetables was significantly correlated ($p < 0.01$) with their indolylglucosinolate content (Figure 1).

Nitrosation of indole compounds

In the *Brassica* vegetables, glucobrassicin contributed an average of 67% to the indolylglucosinolate levels. This compound formed mainly indole-3-carbinol and, in lesser amounts, indole-3-acetonitrile and indole (Figure 2) after hydrolysis with myrosinase. After treatment with nitrite (final concentration, 40 mmol/l; pH 2, 15 min; 37°C), the nitrosation reaction being stopped by adding an excess of sulfamate compared to nitrite, all of these compounds formed directly mutagenic nitrosated products (Figure 3). About 29% of indole-3-acetonitrile, 18% of indole-3-carbinol and 13% of indole was nitrosated. The number of revertants was doubled at about 100 nmol indole-3-carbinol and indole and at about 200 nmol indole-3-acetonitrile. These results indicate that nitrosated indole,

indole-3-carbinol and indole-3-acetonitrile were mutagenic from about 13 nmol, 18 nmol and 58 nmol, respectively.

Figure 1. Relationship between indolylglucosinolate content of *Brassica* vegetables and the amount of *N*-nitroso compounds formed after nitrite treatment of extracts of the vegetables[a]

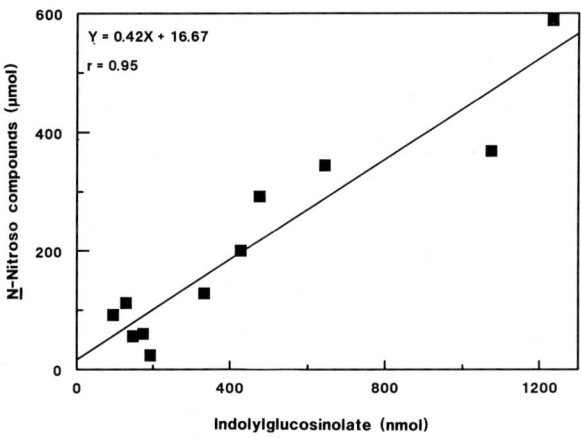

[a]Data expressed per gram fresh material

Figure 2. Chemical structures of indole, indole-3-carbinol and indole-3-acetonitrile

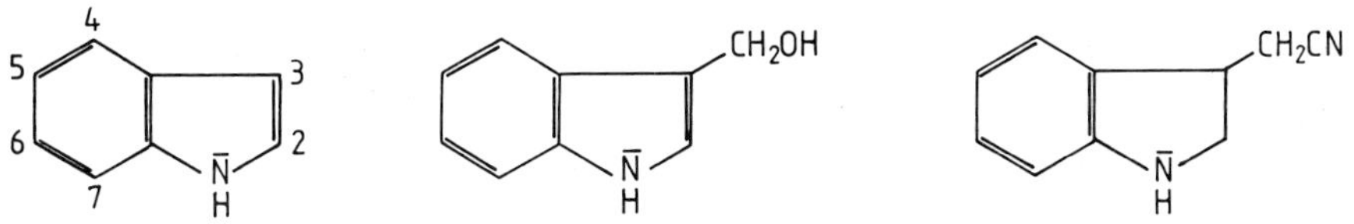

Stability of nitrosated indole compounds

The N-nitroso compounds in some *Brassica* vegetables were more stable at pH 8 than at pH 2. We investigated the stability of the nitrosated indole compounds just after termination of the nitrosation reaction, without changing the pH (~ 2) and after raising the pH to 8. Like the vegetables, the nitrosated indole compounds were unstable at pH 2 and more stable at pH 8 (Figure 4). High-performance liquid chromatography in combination with photohydrolysis detection (Shuker & Tannenbaum, 1983) revealed that nitrosated indole solutions at pH 8 contained nitrite, and the excess of sulfamate was able to scavenge all nitrite at pH 2 but not at pH 8. We concluded that the stability of the nitrosated indole compounds at pH 8 was due to the presence of nitrite, in other words that an equilibrium exists between the nitrosated indole compound and the free compound plus nitrite. This hypothesis was examined, first, by stopping the nitrosation reaction with less than equimolar amounts of sulfamate compared to nitrite. This should result in the presence of nitrite even in pH 2 solutions and therefore in the maintenance of the equilibrium, which should be manifested by the stability of the nitrosated products. We examined the hypothesis, secondly, by stopping the nitrosation reaction with an excess of ascorbic acid, which can scavenge nitrite both at pH 2 and higher. This should result in a shift in the equilibrium to the free indole compound and should be manifested by instability of the nitrosated products at both pH 2 and higher. Figure 4 shows the result of these experiments performed with nitrosated indole-3-acetonitrile in the mutagenicity assay. The results confirmed the existence of an equilibrium.

Figure 3. A, Amounts of nitrosated products formed, and B, dose-response curves in *Salmonella typhimurium* assay (using strain TA100 in the absence of metabolic activation) of indole (●), indole-3-carbinol (■) and indole-3-acetonitrile (▲) after nitrite treatment

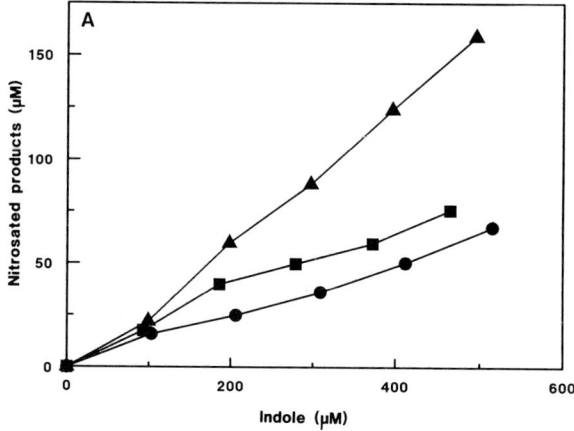

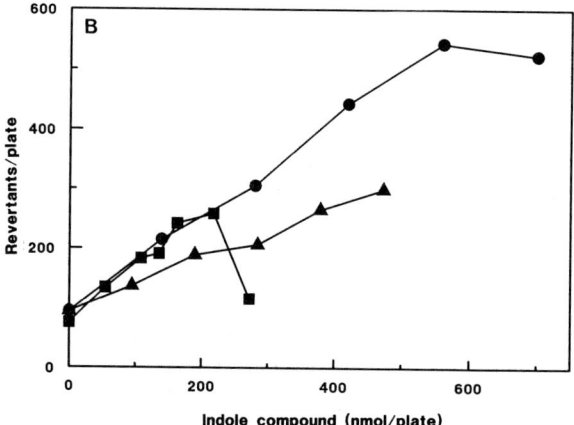

Conclusions

The results indicate that the hydrolysis products of indolylglycosinolates are precursors of the directly mutagenic N-nitroso compounds in *Brassica* vegetables. After nitrosation, the hydrolysis products of glucobrassicin were mutagenic at levels similar to those at which they occur in a normal portion of *Brassica* vegetables. Because the nitrite concentration used in our experiments was high, we also treated indole with low, physiologically feasible nitrite concentrations. This resulted in bacterial mutagenicity, showing that the indole compounds could be nitrosated endogenously. If nitrosated indole compounds are formed in the stomach, they could be stabilized by the presence of nitrite.

Acknowledgements

This work was supported by the Directorate for Nutrition and Quality Affairs of the Ministry of Agriculture and Fisheries.

Figure 4. Mutagenic activity[a] of nitrosated indole-3-acetonitrile (640 nmol/plate) at pH 2 (open symbols) and 8 (filled symbols) after termination of nitrosation reaction with sulfamate 2× (circles) and 0.75× (triangles) or ascorbic acid 2× (squares)[b] the concentration of nitrite

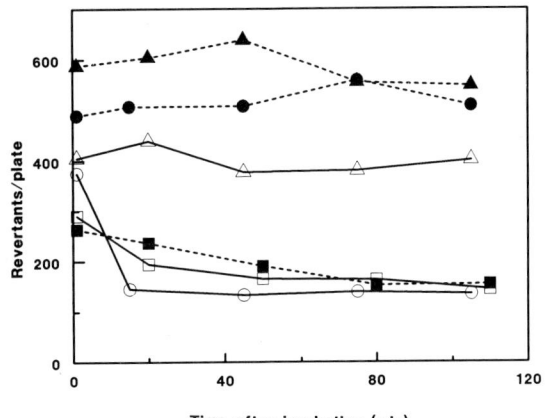

[a] *Salmonella typhimurium* TA100; no metabolic activation; blank (water); 107 revertants/plate
[b] Within ~ 20 min, the pH had dropped to 6.

References

Shuker, D.E.G. & Tannenbaum, S.R. (1983) Determination of nonvolatile N-nitroso compounds in biological fluids by liquid chromatography with postcolumn photohydrolysis detection. *Anal. Chem.*, **55**, 2152-2155

Tiedink, H.G.M., Davies, J.A.R., van Broekhoven, L.W., van der Kamp, H.J. & Jongen, W.M.F. (1988) Formation of mutagenic N-nitroso compounds in vegetable extracts upon nitrite treatment; a comparison with the glucosinolate content. *Food Chem. Toxicol.*, **26**, 947-954

Relevance to Human Cancer of *N*-Nitroso Compounds,
Tobacco Smoke and Mycotoxins.
Ed. I.K. O'Neill, J. Chen and H. Bartsch
Lyon, International Agency for Research on Cancer
© IARC, 1991

NEWLY SYNTHESIZED DITHIOCARBAMATES INHIBIT THE METABOLISM AND TOXICITY OF *N*-NITROSODIMETHYLAMINE

N. Frank, B. Bertram, H.R. Scherf & M. Wiessler

German Cancer Research Centre, Institute of Toxicology and Chemotherapy, Heidelberg, Federal Republic of Germany

Several dithiocarbamates (DTC) of secondary amines and secondary amino acids were tested for their stability in aqueous solution and for their effect on nitrosamine metabolizing enzymes and on the acute toxicity of *N*-nitrosodimethylamine (NDMA) in rats. The following results were found: (i) All DTC tested were stable in alkaline solution; in acidic milieu, only DTC derived from secondary amino acids were moderately stable. (ii) The activity of NDMA-demethylase in rat liver microsomes was inhibited completely by all DTC tested. (iii) The excretion of unmetabolized NDMA in rat urine over 24 h increased from 0.1% without pretreatment to 3.6% of the given NDMA dose when combined with a single dose of DTC. (iv) The acute toxicity of NDMA was reduced by dihydroxyethyldithiocarbamate; when the sulfur compound was administered simultaneously and 24 h after the nitrosamine, lethality was almost completely inhibited. (v) The stability of a compound in aqueous solutions did not affect its activity in the enzyme tests.

Some sulfur compounds are known to affect nitrosamine-induced carcinogenesis (Schmähl *et al.*, 1976; Irving *et al.*, 1979; Hadjiolov *et al.*, 1988). The proposed mechanisms are inhibition of activating enzymes, induction of detoxifying systems and scavenging effects on ultimate nucleophilic carcinogens (see review by Bertram, 1988). Since we have found that the $-CS_2$ function may play a crucial role in inhibiting the metabolism and carcinogenesis of nitrosamines (Frank *et al.*, 1988), we studied the effects of several dithiocarbamates (DTC) on the metabolism of *N*-nitrosodimethylamine (NDMA), on the activity of microsomal nitrosamine-metabolizing enzymes, and on the acute toxicity of NDMA.

Synthesis and stability of DTC

The DTC were synthesized by reacting amines with carbon disulfide in alkaline medium. The compounds and their stabilities are listed in Table 1. Methylhydroxyethyl-, dihydroxyethyl- and diethyldithiocarbamate showed increasing stability with increasing pH and were unstable at pH 1 and pH 3. The amino acid-derived compounds were much more stable at low pH than the amino-derived compounds. It is probable that cyclization took place to form 2-thio-3-alkylthiazolid-5-ones (see Figure 1) under acidic conditions.

Effect of DTC on nitrosamine metabolizing enzymes

In order to elucidate the mechanisms of inhibition of NDMA metabolism by DTC, the activity of NDMA-demethylase was measured with and without DTC pretreatment. Rats were fed DTC (0.5 mmol/kg × day) over a period of ten days, then a liver microsomal fraction was isolated and incubated (60 min) with NDMA (2 mM). Demethylation was measured by monitoring the formaldehyde developed in the assay. Demethylation was found to be completely inhibited.

Effect of DTC on urinary excretion of NDMA

Since the nitrosamine metabolizing enzymes are inhibited, excretion of the unchanged parent compound should be changed. We therefore measured the amount of NDMA in 24-h rat urine after feeding with 100 mg/kg bw DTC and intraperitoneal injection with 100 mg/kg bw NDMA. Urine was extracted with dichloromethane and analysed by gas chromatography-thermal energy analysis. NDMA excretion increased from 0.12% (0.08-0.40; n = 6) in the controls to 3.5% (3.1-5.0; n = 6) after pretreatment with methylhydroxyethyldithiocarbamate (100 mg/kg i.p.) and to 3.7% (2.3-6.7; n = 6) after pretreatment with prolinedithiocarbamate (100 mg/kg i.p.).

Figure 1. Cyclization of amino acid-derived dithiocarbamates
Prolinedithiocarbamate: R; R' = $-CH_2-CH_2-CH_2-$
Sarcosinedithiocarbamate: R = CH_3; R' = H

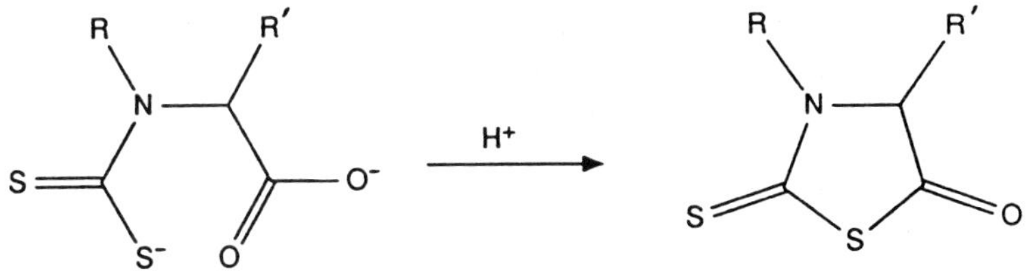

Effect of DTC on acute toxicity of NDMA

The finding that DTC inhibited dealkylation of NDMA and caused a subsequent increase in urinary excretion suggested that toxicity had been reduced. When NDMA was administered orally to rats and dihydroxyethyldithiocarbamate given intraperitoneally (200 mg/kg bw) 1 h prior to the carcinogen, the LD_{50} for NDMA was increased from 40 to 60 mg/kg bw. In another experiment, the influence of the time of DTC administration relative to that of the carcinogen was studied: groups that received the DTC 1 h before plus simultaneously or simultaneously plus 2 h after the carcinogen showed a distinct increase in survival time. The best result was achieved by DTC treatment simultaneously plus 24 h after NDMA, when 9/10 animals survived a dose of 74 mg/kg bw NDMA.

Table 1. Half-lives of dithiocarbamates in relation to pH

pH	Methylhydroxyethyl-dithiocarbamate	Dihydroxyethyl-dithiocarbamate	Sarcosinedithiocarbamate	Prolinedithiocarbamate	Diethyldithiocarbamate
1	< 10 sec	< 10 sec	40.6 (39.6–41.9) min	1.0 (0.0–1.1) min	< 10 sec
3	< 10 sec	< 10 sec	34.2 (33.7–34.9) min	2.1 (2.0–2.1) min	< 10 sec
5	2.1 (2.1–2.3) min	1.1 (1.1–1.2) min	1.5 (1.2–1.6) min	48.1 (47.0–49.1) min	33.6 (32.4–34.2) min
7	3.9 (3.8–4.0) min	1.4 (1.35–1.4) h	3.5 (3.5–3.5) h	Stable[a]	2.4 (2.4–2.5) h
9	Stable	21.4 (20.4–22.6) h	Stable	Stable	Stable

Values are weighted means of three determinations with 95% confidence intervals; degradation was followed by ultraviolet spectroscopy with two maxima at about 280 and 254 nm.
[a] Stable, half-life > 24 h

DTC can therefore inhibit the metabolism and toxicity of NDMA. The advantage of DTC derived from secondary amino acids may be that the amines that are formed during metabolism of the compounds occur naturally. Further, any endogenous transnitrosation to, e.g., N-nitrosoproline, will not be toxic, since N-nitrosoproline is nearly completely excreted in the urine (Ohshima & Bartsch, 1981). Further investigations will clarify if DTC are also suitable agents for use against the toxic effects of other N-nitrosamines and if they inhibit chemical carcinogenesis.

References

Bertram, B. (1988) The modifying effect of disulfiram on the carcinogenicity of some N-nitrosamines: mechanistic investigations. In: Schmähl, D., ed., *Combination Effects in Chemical Carcinogenesis*, Weinheim, Verlag Chemie, pp. 193-229

Frank, N., Bertram, B., Frei, E., Hadjiolov, D. & Wiessler, M. (1988) Influence of thiocompounds on the metabolism of N-nitrosodiethylamine. *Carcinogenesis*, **9**, 1303-1306

Hadjiolov, D., Frank, N., Hadjiolov, N. & Yanev, S. (1988) Effects of potassium ethylxanthogenate on nitrosodiethylamine-induced DNA damage and on liver carcinogenesis. *Arch. Geschwulstforsch.*, **53**, 313-318

Irving, C.C., Tice, A.J. & Murphy, W.M. (1979) Inhibition of N-n-butyl-n-(4-hydroxybutyl)nitrosamine induced urinary bladder cancer in rats by administration of disulfiram in the diet. *Cancer Res.*, **39**, 3030-3034

Ohshima, H. & Bartsch, H. (1981) Quantitative estimation of endogenous nitrosation in humans by monitoring N-nitrosoproline excreted in the urine. *Cancer Res.*, **41**, 3658-3662

Schmähl, D., Kruger, F.W., Habs, M. & Diehl, B. (1976) Influence of disulfiran on the organotropy of the carcinogenic effect of dimethylnitrosamine and diethylnitrosamine in rats. *Z. Krebsforsch.*, **85**, 271-276

NEW SULFENAMIDE ACCELERATORS DERIVED FROM 'SAFE' AMINES FOR THE RUBBER AND TYRE INDUSTRY

C.-D. Wacker, B. Spiegelhalder & R. Preussmann

German Cancer Research Centre, Institute of Toxicology and Chemotherapy, Heidelberg, Germany

A reduction of the high exposures to *N*-nitrosamines in the rubber and tyre industry is possible using the concept of 'safe' amines, in which vulcanization accelerators contain amine moieties that are both difficult to nitrosate and, on nitrosation, yield noncarcinogenic *N*-nitroso compounds. The toxicological and technological properties of more than 50 benzothiazole sulfenamides derived from 'safe' amines have been evaluated. Some of the new compounds show excellent vulcanization properties and seem suitable as replacements for traditional accelerators in this class of compounds.

The occurrence of volatile, carcinogenic *N*-nitrosamines in industrial environments is still quite high. Recent monitoring of workroom air in rubber factories in Germany showed *N*-nitrosamine concentrations up to 41 µg/m3. A reduction of this exposure is indispensable, since new regulations have been introduced into the German rubber industry allowing an upper concentration of total *N*-nitrosamines of only 2.5 µg/m3 (German Ministry of Labour, 1988).

The major source of the carcinogenic *N*-nitrosamines is the use of certain vulcanization accelerators (amino derivatives). The *N*-nitrosamine concentrations cannot be reduced below 2.5 µg/m3 if these accelerators continue to be used. Our concept of 'safe' amines can provide a solution to this problem (Spiegelhalder & Preussmann, 1982). 'Safe' amines are characterized by low nitrosability and/or formation of noncarcinogenic *N*-nitrosamines. Thus, substitution of 'safe' amines for the traditional amino components in different types of accelerators should yield compounds that are still active as accelerators but unable to produce carcinogenic *N*-nitrosamines. We now present data on the first accelerators of the benzothiazole sulfenamide class derived from 'safe' amines, which show vulcanization characteristics similar to those of standard industrial accelerators.

Syntheses and technological properties

Over 50 benzothiazole sulfenamides were prepared by reaction of 2-mercaptobenzothiazole (method A, X = H; Figure 1) with the appropriate amine in the presence of iodine or sodium hypochlorite. Greatly hindered amines (e.g., 2,2,6,6-tetramethylpiperidine) gave poor yields and were better synthesized from amine and benzothiazole sulfenylchloride (method B, X=Cl, without oxidant; Figure 1).

Detailed studies of the technological features of the new compounds revealed substantial differences in their behaviour. Samples of 'safe' benzothiazole sulfenamides were tested for vulcanization characteristics together with a traditional accelerator in a standard rubber formulation. Table 1 summarizes some of the data obtained for the accelerators derived from 'safe' amines in comparison to the industrial standard, morpholino-2-benzothiazolesulfenamide (MBS, 3). It should be noted that good sulfenamide accelerators have short cure times and a high degree of cross-linking.

Table 1. Comparison of vulcanization characteristics of 'safe' (1, 2) and traditional benzothiazole sulfenamide accelerators (3) in a standard rubber mixture; curing temperature, 150°C

Compound no.	Accelerator (benzothiazole sulfenamide)	Scorch time t_2 (min)	Cure time t_{90} min	Cross-linking (daNm)
1	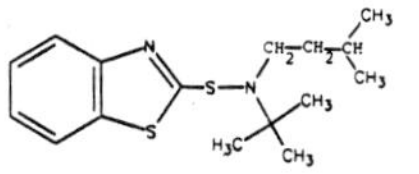	11.8	23.0	13.9
2	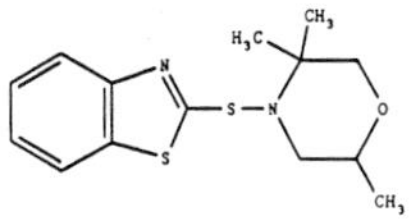	16.4	32.6	12.8
3	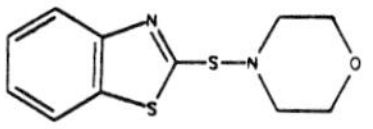	12.4	18.8	14.7

Fig. 1. Methods for the synthesis of benzothiazole sulfenamides: method A, X = H; method B, X = Cl

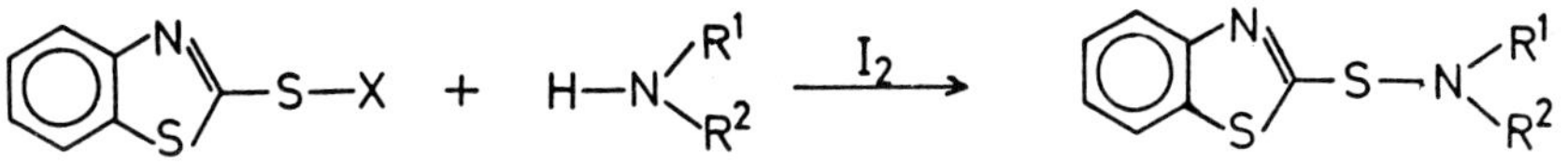

Data for N-(3-methylbutyl)-N-tert-butyl-2-benzothiazolesulfenamide (1) are in the same range as those for MBS, while the morpholine derivative 2, a higher methylated homologue of MBS, gave unsatisfactory results despite its structural relationship to MBS. The similarity between 'safe' accelerator 1 and MBS was further confirmed in two natural rubber mixtures for truck and passenger car tyres. The vulcanizates from 1 showed almost identical properties to those obtained from MBS. Another 'safe' amine-derived sulfenamide accelerator, N-methyl-N-tert-butyl-2-benzothiazolesulfenamide, was shown to have very similar vulcanization characteristics to those of the industrial standard, N-cyclohexyl-2-benzothiazolesulfenamide.

Toxicological investigation

The nitrosability of the new sulfenamides was determined by nitrosation with aqueous $NaNO_2$ at different temperatures and pH values under standardized conditions (Wacker et al., 1987). In general, 'safe' benzothiazole sulfenamides resulted in five to ten times lower concentrations of N-nitrosamines than the industrial standard, MBS (data not shown).

Table 2. Properties of rubber samples obtained with a 'safe' (1) and a traditional benzothiazole sulfenamide accelerator (3)

Compound no.	Accelerator (benzothiazole sulfenamide)	Tensile strength (MPa)	Elongation at break (%)	Rebound elasticity (%)
1	[structure]	26.4	470	54
3	[structure]	23.4	490	56

The mutagenic potential of the corresponding N-nitrosamines was investigated in Ames tests, as described previously (Wacker et al., 1987). No or weak mutagenicity was observed in all cases. Further tests for mutagenicity and carcinogenicity are in progress.

Acknowledgements

This work was supported by the German Ministry of Research and Technology.

References

German Ministry of Labour (1988) *Technische Regeln für Gefahrstoffe* (TRGS 552), Bonn

Spiegelhalder, B. & Preussmann, R. (1982) Nitrosamines and rubber. In: Bartsch, H., O'Neill, I.K., Castegnaro, M. & Okada, M., eds, N-*Nitroso Compounds: Occurrence and Biological Effects* (IARC Scientific Publications No. 41), Lyon, IARC, pp. 231-243

Wacker, C.-D., Spiegelhalder, B., Börzsönyi, M., Brune, G. & Preussmann, R. (1987) Prevention of exposure to N-nitrosamines in the rubber industry: new vulcanization accelerators based on 'safe' amines. In: Bartsch, H., O'Neill, I.K. & Schulte-Hermann, R., eds, *The Relevance of* N-*Nitroso Compounds to Human Cancer: Exposures and Mechanisms* (IARC Scientific Publications No. 84), Lyon, IARC, pp. 370-374

LIST OF PARTICIPANTS

Dr L. Airoldi
Istituto di Ricerche Farmacologiche 'Mario Negri'
Via Eritrea 62
20157 Milan
Italy

Dr K.E. Appel
Max von Pettenkofer Institute
Federal Health Office
Thieallee
1000 Berlin 33
Germany

Dr M.C. Archer
Ontario Cancer Institute
500 Sherbourne Street
Toronto, Ontario
Canada M4X 1K9

Professor G.S. Bailey
Oregon State University
Department of Food Science
Corvallis, OR 97331
USA

Dr H. Bartsch
International Agency for Research on Cancer
150 cours Albert Thomas
69372 Lyon Cédex 08
France

Dr M.R. Berger
Institute of Toxicology and Chemotherapy
German Cancer Research Center
Im Neuenheimer Feld 280
6900 Heidelberg
Germany

Dr S.V. Bhide
Head, Carcinogenesis Division
Cancer Research Institute
Tata Memorial Centre
Parel, Bombay 400012
India

Dr W.J. Blot
Head, Analytical Study Section
Environmental Epidemiology Branch
National Cancer Institute
Landow Building, Room 3C09
Bethesda, MD 20205
USA

Dr F.X. Bosch
Unit of Field and Intervention Studies
International Agency for Research on Cancer
150 cours Albert Thomas
69372 Lyon Cédex 08
France

Dr G. Bouvier
Unit of Environmental Carcinogens and Host Factors
International Agency for Research on Cancer
150 cours Albert Thomas
69372 Lyon Cédex 08
France

Dr S. Boyse
British-American Tobacco Company Limited
PO Box 482
Westminster House
7 Millbank
London SW1P 3JE
UK

Dr K.D. Brunnemann
American Health Foundation
1 Dana Road
Valhalla NY 10595
USA

Dr S. Calmels
Unit of Environmental Carcinogens and Host Factors
International Agency for Research on Cancer
150 cours Albert Thomas
69372 Lyon Cédex 08
France

Dr C.P.J. Caygill
PHLS Communicable Disease Surveillance Centre
61 Colindale Avenue
London NW9 5HT
UK

List of participants

Dr J. Chang–Claude
Institute of Epidemiology and Biometry
German Cancer Research Centre
Im Neuenheimer Feld 280
6900 Heidelberg
Germany

Dr B. Chen
International Programme on Chemical Safety
Division of Environmental Health
World Health Organization
1211 Geneva 27
Switzerland

Dr C.S. Chen
Beijing Institute for Cancer Research
Da-Hong-Luo-Chang Street
Western District
100034 Beijing
China

Dr J. Chen
Chinese Academy of Preventive Medicine
Institute of Nutrition and Food Hygiene
20 Nan Wei Road
Beijing
China

Dr F.-L. Chung
American Health Foundation
1 Dana Road
Valhalla, NY 10595
USA

Dr F. Colby
90 Park Avenue
New York, NY 10016
USA

Dr V.M. Craddock
MRC Toxicology Unit
Woodmansterne Road
Carshalton, Surrey SM5 4EF
UK

Dr G. Eisenbrand
Institute of Food Chemistry & Environmental
 Toxicology
University of Kaiserslautern
PO Box 3049
Erwin-Schroedinger-Str.
6750 Kaiserslautern
Germany

Dr P.B. Farmer
MRC Toxicology Unit
Woodmansterne Road
Carshalton, Surrey SMJ 4EF
UK

Dr S. Fischer
German Cancer Research Centre
Institute of Toxicology and Chemotherapy
Im Neuenheimer Feld 280
6900 Heidelberg
Germany

Dr D. Forman
Imperial Cancer Research Fund
Gibson Laboratories
The Radcliffe Infirmary
Oxford OX2 6HE
UK

Dr N. Frank
Institute of Toxicology
German Cancer Research Centre
Im Neuenheimer Feld 280
6900 Heidelberg
Germany

Dr E. Frei
Institute of Toxicology and Chemotherapy
German Cancer Research Centre
Im Neuenheimer Feld 280
6900 Heidelberg
Germany

Dr H. Furukawa
Laboratory of Biological Science
Meijo University
Yagoto-urayama 15
Tempaku-cho
Tempaku-ku, Nagoya 468
Japan

Dr Y.T. Gao
Director
Shanghai Cancer Institute
2200 Xie Tu Road
Shanghai 200032
China

Dr U. Goff
Manager, Analytical Services and Instruments
Thermedics Inc.
470 Wildwood Street
PO Box 2999
Woburn, MA 01888-1799
USA

List of participants

Dr B. Gold
Eppley Institute for Research on Cancer
University of Nebraska Medical Center
42 and Dewey Avenue
Omaha NE 68105
USA

Dr T. Górski
Sanitary-Epidemiological Station
Department for Cancer Prophylaxis and Education
Wodna str. 40
90–046 Lodz
Poland

Dr R.C. Grafström
Department of Toxicology
Karolinska Institute
Box 60 400
104 01 Stockholm
Sweden

Dr P.C. Gupta
Basic Dental Research Unit
Tata Institute of Fundamental Research
Homi Bhabha Road
Bombay 400 005
India

Dr C. Han
c/o Dr J. Chen
Chinese Academy of Preventive Medicine
Institute of Nutrition and Food Hygiene
20 Nan Wei Road
Beijing
China

Dr S.S. Hecht
Chief, Division of Chemical Carcinogenesis
Naylor Dana Institute for Disease Prevention
American Health Foundation
Dana Road
Valhalla, NY 10595
USA

Professor J.M. Hirsch
Department of Oral Surgery
University of Göteborg
Box 33070
400 33 Göteborg
Sweden

Dr D. Hoffmann
American Health Foundation
1 Dana Road
Valhalla NY 10595
USA

Mrs I. Hoffmann
American Health Foundation
1 Dana Road
Valhalla, NY 10595
USA

Dr J.H. Hotchkiss
Institute of Food Science
Department of Food Science
Stocking Hall
Cornell University
Ithaca NY 14853
USA

Dr C. Janzowski
Lebensmittelchemie & Umweittoxicologie
Fachbereich Chemie
Universitat Kaiserlautern
Erwin-Schrodinger-Strasse
6750 Kaiserslautern
Germany

Dr Y.Z. Jiang
Department of Chemical Etiology
Cancer Institute
Chinese Academy of Medical Sciences
100021 Beijing
China

Dr T. Kawabata
1-43-51, Hiyoshi–cho
Kokubunji-shi
Tokyo 185
Japan

Dr L.K. Keefer
Chief, Chemistry Section
Laboratory of Comparative Carcinogenesis
NCI-FRCF
Bldg 538
Frederick, MD 21701-1013
USA

Dr P. Kleihues
Department of Neuropathology
Institute of Pathology
University Hospital
Schmelzbergstr. 12
8091 Zürich
Switzerland

Dr R. G. Klein
Institute of Toxicology and Chemotherapy
German Cancer Research Centre
Im Neuenheimer Feld 280
6900 Heidelberg
Germany

Dr T. Knight
3 Digby Road
Menston, W. Yorks. LS29 6JB
UK

Ms Y. Kurashima
Biochemistry Division
National Cancer Center Research Institute
5-1-1 Tsukijii
Chuo-Ku
Tokyo 104
Japan

Dr S.A. Kyrtopoulos
The National Hellenic Research Foundation
Biological Research Centre
48, Vassileos Constantinou Avenue
Athens 116 35
Greece

Dr P.-A. Larsson
Department of Clinical Virology
University of Göteborg
Guldhedsgatan 10B
413 46 Göteborg
Sweden

Dr F. Laval
Radiochimie de l'ADN (PR1)
Institut Gustave Roussy
94805 Villejuif Cédex
France

Dr J. Laval
Groupe 'Réparation des lésions radio-chimio induites'
Institut Gustave Roussy
94805 Villejuif Cédex
France

Mr S.A. Leach
Public Health Laboratory Service Centre for Applied Microbiology & Research
Bacterial Metabolism Research Laboratory
Porton Down
Salisbury
Wilts. SP4 0JG
UK

Dr J.F. Lechner
Laboratory of Human Carcinogenesis
National Cancer Institute
Building 37, Room 2C25
Bethesda, MD 20892
USA

Dr W. Lijinsky
NCI-Frederick Cancer Research Facility
BRI Basic Research Program
PO Box B
Frederick, MD 21701
USA

Dr A. Likhachev
N.N. Petrov Research Institute of Oncology
68, Leningradskaya St.
Pesochny 2
189646 Leningrad
USSR

Dr L. Liu
Laboratory of Cancer Etiology and Chemoprevention
School of Pharmacy
Laval University
Quebec City
Canada G1K 7P4

Dr Y. Liu
Department of Toxicology
Karolinska Institute
10401 Stockholm
Sweden

Dr R.N. Loeppky
Department of Chemistry
123 Chemistry Building
University of Missouri
Columbia, MO 65211
USA

Dr P.D. Lotlikar
Fels Research Institute
Temple University Medical School
Philadelphia, PA 19140
USA

Dr B. Ludeke
Department of Pathology
University Hospital
8091 Zurich
Switzerland

Dr P. N. Magee
Fels Research Institute
Temple University School of Medicine
Philadelphia, PA 19140
USA

Dr I. Meier
Institute for Toxicology
Schorenstr. 16
8603 Schwerzenbach
Switzerland

List of participants

Dr P. Mende
Institute for Toxicology and Chemotherapy
German Cancer Research Centre
Im Neuenheimer Feld 280
6900 Heidelberg
Germany

Dr C.J. Michejda
NCI-Frederick Cancer Research Facility
PO Box B
Frederick, MD 21701
USA

Dr S.S. Mirvish
Eppley Institute for Research in Cancer
University of Nebraska Medical Center
42nd and Dewey Avenue
Omaha NE 68105
USA

Dr M. Miwa
National Food Research Institute
2-1-2 Kannondai, Tsukuba
Ibaraki 305
Japan

Professor M. Mochizuki
Kyoritsu College of Pharmacy
Shibakoen 1-5-30
Minato-ku, Tokyo 105
Japan

Dr H. Møller
Institute of Cancer Epidemiology
Danish Cancer Society
Danish Cancer Registry
Rosenvaengets Hovedvej 35
Box 839
2100 København
Denmark

Dr Y. Mori
Laboratory of Radiochemistry
Gifu Pharmaceutical University
5-6-1 Mitahora-higashi
Gifu 502
Japan

Dr J. Nair
Carcinogenesis Division
Cancer Research Institute
Tata Memorial Centre
Parel
Bombay 400012
India

Dr U. J. Nair
Carcinogenesis Division
Cancer Research Institute
Tata Memorial Centre
Parel
Bombay 400012
India

Dr P.J. O'Connor
Paterson Institute for Cancer Research
Christie Hospital and Holt Radium Institute
Manchester M20 9BX
UK

Dr I.K. O'Neill
Unit of Environmental Carcinogens and Host Factors
International Agency for Research on Cancer
150 cours Albert Thomas
69372 Lyon Cédex 08
France

Dr H. Ohshima
International Agency for Research on Cancer
150 cours Albert Thomas
69372 Lyon Cédex 08
France

Dr B.-G. Österdahl
Nutrition Laboratory
Swedish National Food Administration
Box 622
751 26 Uppsala
Sweden

Mr R. Peto
MRC Cancer Studies Unit
Radcliffe Infirmary
Oxford OX2 6HE
UK

Dr B. Pignatelli
International Agency for Research on Cancer
150 cours Albert Thomas
69372 Lyon Cédex 08
France

Dr P.I. Reed
Gastrointestinal Unit
Wexham Park Hospital
Slough SL2 4HL
UK

Professor H. Remmer
Institut für Toxikologie
University of Tübingen
Wilhelmstr. 56
7400 Tübingen
Germany

List of participants

Dr E.B. Sansone
National Cancer Institute
Frederick Cancer Research Facility
PO Box B
Frederick, MD 21701
USA

Dr R.A. Scanlan
Research Office Ads 312A
Oregon State University
Corvallis, Oregon 97331
USA

Dr E. Scherer
The Netherlands Cancer Institute
Plesmanlaan 121
1066 CX Amsterdam
The Netherlands

Dr Gertrud Scherer
Lebensmittelchemie & Umwelttoxifologie
Fachbereich Chemie
Universität Kaiserslautern
6750 Kaiserslautern
Germany

Dr Gerhard Scherer
Analytisch-biologisches Forschunglabor
 Prof. Adlkofer
Goethestrasse 20
8000 Munich 20
Germany

Dr N.P. Sen
Food Research Division
Health Protection Branch
Sir F. Banting Research Centre
Ottawa, Ontario
Canada K1A OL2

Dr Y.M. Shao
c/o Biomedical Research Unit
Global Programme on AIDS
World Health Organization
1211 Geneva 27
Switzerland

Dr D.E.G. Shuker
Unit of Environmental Carcinogens and Host
 Factors
International Agency for Research on Cancer
150 cours Albert Thomas
69372 Lyon Cédex 08
France

Dr M. A. Siddiqi
Department of Biochemistry
University of Kashmir
Srinagar-190006, J&K
India

Dr R. Sierra
Instituto de Investigaciones en Salud
Universidad de Costa Rica
San Jose
Costa Rica

Dr J.M. Sontag
Epidemiology and Biostatistics Program
Division of Cancer Etiology
National Cancer Institute
Executive Plaza North, Room 543
Bethesda, MD 10892
USA

Professor B.W. Stewart
Children's Leukaemia and Cancer Research Unit
Prince of Wales Hospital
Randwick
Sydney, NSW 2031
Australia

Ms K. Sundqvist
Department of Toxicology
Karolinska Institute
Box 60400
Stockholm 104 01
Sweden

Ms K. Takeda
Kyoritsu College of Pharmacy
Shibakoen 1-5-30
Minato-ku
Tokyo 105
Japan

Dr H.G.M. Tiedink
Department of Toxicology
Agricultural University
Bomenweg 2
6703 HD Wageningen
The Netherlands

Dr L. Tomatis
International Agency for Research on Cancer
150 Cours Albert Thomas
69372 Lyon Cédex 08
France

Dr H. Tozawa
3-23-21 Kugenuma Matsugaoka
Fujisawa-shi
Kanagawa-ken 251
Japan

List of participants

Dr A.R. Tricker
Institute of Toxicology and Chemotherapy
German Cancer Research Center
Im Neuenheimer Feld 280
6900 Heidelberg
Germany

Dr M. Tsuda
Biochemistry Division
National Cancer Center Research Institute
1-1 Tsukiji 5-chome
Chuo-ku
Tokyo 104
Japan

Professor Y. Ueno
Department of Toxicology and Microbiol Chemistry
Faculty of Pharmaceutical Sciences
Science University of Tokyo
12 Ichigaya
Funagawara-Machi, Shinjuku-ku
Tokyo
Japan

Ms S. Ukawa
Kyoritsu College of Pharmacy
Shibakoen 1-5-30
Minato-ku
Tokyo 105
Japan

Dr J. Van Benthem
The Netherlands Cancer Institute
Anthoni Van Leeuwenhoekhuis
Division of Chemical Carcinogenesis
Plesmanlaan 121
1066 CX Amsterdam
The Netherlands

Dr C.-D. Wacker
Institute of Toxicology and Chemotherapy
German Cancer Research Center
Im Neuenheimer Feld 280
6900 Heidelberg
Germany

Dr H. Wang
c/o Dr J. Chen
Chinese Academy of Preventive Medicine
Institute of Nutrition and Food Hygiene
20 Nan Wei Road
Beijing, China

Dr A.H. Warfield
Philip Morris Research Center
PO Box 26683
Richmond, Virginia 23261
USA

Mr F. Welsch
9208 Villa Drive
Bethesda, MD 20817
USA

Dr C. P. Wild
Unit of Mechanisms of Carcinogenesis
International Agency for Research on Cancer
150 cours Albert Thomas
69327 Lyon Cédex 08
France

Dr J.S. Wishnok
Massachusetts Institute of Technology
Room 56–313
77 Massachusetts Avenue
Cambridge, MA 02139
USA

Dr C.S. Yang
Department of Chemical Biology and
  Pharmacognosy
College of Pharmacy
PO Box 789
Rutgers University
Piscataway, NJ 08855–0789
USA

Dr M. C. Yu
Norris Cancer Hospital
1441 Eastlake Avenue
Los Angeles, CA 90333
USA

Dr L.S. Zahn
13 Lincoln Road
Great Neck, NY 11021
USA

Dr D.G. Zaridze
Head, Department of Cancer Epidemiology and
  Prevention
All-Union Cancer Research Centre
Academy of Medical Sciences of the USSR
6 Kashirskoye shosse
Moscow
USSR

NOMENCLATURE AND ABBREVIATIONS

Following the system of nomenclature for *N*-nitroso compounds proposed in the Proceedings of the Fifth Meeting in this series, which was based on the IUPAC system of nomenclature, additional proposals are made for systematization of nomenclature and abbreviations of these compounds.

N-NITROSAMINES

1. As in existing recommendations, the N–NO radical is always stated first (abbreviations commence with 'N'); the parts joined to the amine nitrogen follow; and, where appropriate, the names terminate with 'amine' and the abbreviation with 'A'.

2. The parts joined to the amine nitrogen are placed in the following order, both in nomenclature and in abbreviation:

(i) aliphatic and alicyclic radicals
(ii) aromatic radicals
(iii) non-aromatic heterocyclic radicals
(iv) oxidized radicals
(v) alkene radicals
(vi) other types or derivatives

When there are two radicals of the same type, the larger one is given first (by number of carbon atom, then mass).

3. Unless otherwise specified, alkane radicals are normal and unbranched. Branched alkane radicals are denoted by placing *i (iso)*, *s (sec)* and *t (tert)* before the radical name and before the radical abbreviation. The position of substituents on these chains is specified, giving the carbon position before the derivative. Note that an α-keto function turns the amine into an amide, for which a variation in nomenclature and abbreviation is proposed to reflect the significant alteration in chemical properties (see below).

4. The following standard abbreviations are reserved:

D = di or bis (i.e., two radicals of same type attached to the amine nitrogen, as in NDMA and NDHPA).

M, E, P, B, Ph, Bz are reserved for methyl, ethyl, propyl, butyl, phenyl, benzyl, respectively.

PIP, PYR, MOR, PZ, SAR, PRO, THZ and AZ are reserved for piperidine, pyrrolidine, morpholine, piperazine, sarcosine, proline, thiazolidine and azetidine, respectively.

5. Derivation of radicals by hydroxy, keto or acetoxy groups is covered by placing H, O or Ac in front of the respective radical abbreviation.

6. The abbreviation NDELA is retained, due to its widespread usage.

N-NITROSAMIDES

1. Instead of using ammonia as the root for nomenclature, an amide is taken, so that, for example, an α-keto propylamine part of a molecule is a propionamide radical. This complete radical is then placed at the end of the nomenclature and abbreviation, e.g., '-propionamide' and 'PAd'. The initial 'N-nitroso' and 'N' are retained as with N-nitrosamines.
2. As only one more radical can be attached to the nitrogen, it is suggested that this be inserted between the 'N-nitroso' and '-amide' parts, irrespective of its nature. In the case of nitrosoureas, however, the nomenclature ends with -N-nitrosourea and the abbreviations with -NU. For nitrosourethanes, use -NUT.
3. In all other respects the same terms are used as for N-nitrosamines.

N-NITRAMINES

It is proposed to use the same systematic nomenclature as for N-nitrosamines but to represent N-nitro as NT at the beginning, e.g., N-nitrodimethylamine = NTDMA.

EXAMPLES

N-nitrosodimethylamine	NDMA
N-nitrosodi-n-butylamine	NDBA
N-nitrosodi-isobutylamine	NDi-BA
N-nitrosoethylmethylamine	NEMA
N-nitrosopyrrolidine	NPYR
N-nitrosomorpholine	NMOR
N-nitrosohydroxyproline	NHPRO
N'-nitrosonornicotine	NNN
N-nitrosodiethanolamine	NDELA
N-nitrosopropyl(2-hydroxypropyl)amine	NPHPA
N-nitrosomethyl(2-oxobutyl)amine	NMOBA
N-nitrosomethylbutyramide [= N-nitrosomethyl-(1-oxobutyl)amine]	NMBAd
N-nitrosoethylvinylamine	NEVA
N-nitroso(2-hydroxypropyl)(2-oxopropyl)amine	NEHPOPA
N-nitrosobis(2-hydroxypropyl)amine	NDHPA
N-butyl-N-nitrosourea	BNU

AUTHOR INDEX

Airoldi, L., 343
Alaoui-Jamali, M.A., 510
Ambatzi, P., 78
Amelung, F., 322
Ammigan, N., 525
Amonkar, A.J., 516, 520, 525
Anisimov, V., 407
Appel, K.E., 351
Archer, M.C., 329
Autrup, H., 168
Bailey, E., 71
Bailey, G.S., 275
Balansky, R.M., 535
Bao, Y., 244
Barbour, J.F., 242
Bartsch, H., 1, 88, 162, 172, 187, 204, 281, 443
Belitsky, G.A., 485
Béréziat, J.-C., 187
Berger, F., 172
Berger, M.R., 311
Bertram, B., 588
Bhide, S.V., 516, 520, 525
Blagoeva, P., 535
Blankart, M., 238
Blot, W.J., 33, 460
Boiteux, S., 412
Bonfanti, M., 343
Bornkamm, G.W., 204
Bosch, F.X., 48
Bousiotou, V., 78
Bouvier, G., 204
van Broekhoven, L.W., 584
Brouet, I., 88, 443
Brunnemann, K.D., 477, 482, 485
Butler, W.H., 119
Cai, H.Y., 579
Calmels, S., 158, 187
Carmella, S.G., 113
Castonguay, A., 510
Caygill, C.P.J., 137
Chang, Y.S., 33
Chang-Claude, J., 192
Charrière, M., 158
Chen, C., 162, 172
Chen, F.X., 439
Chen, J., 18, 219
Chen, J.G., 502
Chen, Y., 152
Chen, Z.C., 434
Cheng, W.F., 143
Chinnock, A., 162

Chui, S.X., 11
Chung, F.L., 529
Church, K.M., 439
Cocco, G., 146
Correa, P., 83, 192
Craddock, V.M., 564
Crespi, M., 192
Cushnir, J.R., 107
Dashwood, R.H., 275
Davaris, G., 78
Davies, J.A.R., 584
De Montclos, H., 158, 172
Den Engelse, L., 496
Deng, D.J., 152
Dikun, P.P., 552
Djordjevic, M.V., 477
Dong, Z.M., 258
Duan, X.X., 431
Dubreuil, C., 158
Duncan, S., 137
Edler, L., 311
Eimoto, H., 318
Eisenbrand, G., 238, 332, 339
Eklind, K.I., 529
El Alami, N., 510
El Ghissassi, F., 162
Engelholm, C., 351
Erhardt, P., 281
Ermilov, V.B., 552
Fan, C.Y., 119
Fanelli, R., 343
Farmer, P.B., 71, 107
Fazili, Z., 210
Feng, R., 477
Fernandez, F., 137
Fischer, S., 489
Fong, A.T., 275
Forman, D., 22, 129, 146
Frank, N., 588
Fraumeni, J.F., Jr, 33, 460
Frei, E., 358
Friesen, M.D., 102, 443
Fubini, S.L., 182
Gamboa, C., 162
Gao, J., 219
Gao, R.N., 62
Gao, Y.T., 62
Garren, L., 102
Gelboin, H.V., 392
Glogowski, J., 83
Godenèche, D., 412
Gold, B., 439

Gopalan, P., 427
Górska, E., 550
Górski, T., 550
Görsdorf, S., 351
Grafström, R.C., 281
Green, J.A., 71
Guan, B.P., 460
Guo, L.P., 11
Gupta, P.C., 466
Habraken, Y., 412
Hall, J., 407
Hall, R., 137
Hamdi Chérif, M., 158
Han, C., 541
Han, F.G., 424
Haritopoulos, G., 78
Harris, C.C., 294
Hassel, M., 358
Hastings, R., 244
Hautefeuille, A., 162, 172
Hecht, S.S., 54, 113, 529
Heller, D., 244
Henderson, A.R., 564
Henderson, B.E., 460
Hendricks, J.D., 275
Henn, I., 332
Hermann, R., 322
Hill, M.J., 137, 139, 571
Hirsch, J.M., 507
Ho, L.L., 427
Hoffmann, D., 54, 449, 477, 482, 485
Hoffmann, I., 449
Hong, J.Y., 265
Hotchkiss, J.H., 182, 219
Hu, X.N., 11
Huang, Q., 392
Iman, D., 294
Ishizaki, H., 265, 366
Jacob, D., 332
Janowsky, I., 322
Janzowski, C., 332
Jensen, O.M., 168
Jensen, P., 168
Jhee, E.C., 427
Ji, C., 392
Ji, X.H., 424
Ji, Y.S., 143
Jiang, Y.Z., 96
Jin, F., 62
Jin, Z.L., 230
Johansson, S.L., 507
Johnston, B.J., 139

Author Index

Jongen, W.M.F., 584
Kagan, M., 113
Kagan, S.S., 113
Kato, T., 420
Kawabata, T., 214
Keefer, L.K., 33, 362, 370
Keimig, S.D., 226
Khlat, M., 88
Kleihues, P., 286, 339
Klein, R.G., 322
Kneller, R., 33
Knight, T., 146
Kobayashi, T., 420
Koepke, S.R., 346
Konishi, Y., 318, 398
Kramer, R., 351
Kroeger-Koepke, M.B., 346
Kumar, R., 210
Kurashima, Y., 123
Kyrtopoulos, S.A., 78
Lamb, J.H.,107
Lambert, R., 172
Landt, J., 168
Larsson, P.A., 507
Laval, F., 417
Laval, J., 412
Leach, S.A., 137, 146, 571
Lefebvre, P., 417
Leung, C.S., 71
Li, F.M., 11
Li, G., 460
Li, Y., 431, 568
Li, Y.E., 375
Liang, Y.X., 424
Lijinsky, W., 305
Likhachev, A., 407
Liu G.T., 253, 258
Liu, H., 568
Liu, L., 510
Liu, Y., 281
Loeppky, R.N., 244, 339, 375
Lopès, F., 412
Lotlikar, P.D., 427
Lu, S.H., 11
Lu, S.J., 329
Ludeke, B., 286, 339
Lutz, W.K., 383
Ma, Y.F., 253
Mack, W., 197
Mackay, J.M., 471
Mackerness, C.W., 571
Madelmont, J.C., 412
Magagnotti, C., 343
Makarananda, K., 96
Malaveille, C., 162, 172, 204

Manson, M.M., 71
Mao, D.J., 143
Maragos, C.M., 182
Massey, R., 137
McMenamin, M., 294
Meier, I., 383
Meier, T., 286
Melikian, A.A., 482
Mende, P., 223
Meyer, T.J., 370
Meyniel, G., 412
Miao, J., 253, 258
Michejda, C.J., 346
Milligan, J.R., 329
Mircheva, Z.I., 535
Mirvish, S.S., 392
Miwa, M., 388
Mochizuki, M., 404, 558
Møller, H., 168
Moret, M., 343
Mori, Y., 318, 398
Morse, M.A., 529
Mostafa, M.H., 178
Moulinier, B., 172
Muñoz, N., 162, 172, 192
Nagabhushan, M., 525
Nagel, D., 244
Nair, J., 281, 516
Nair, U.J., 516, 525
Nakamura, K., 420
Napalkov, N., 407
Neal, G.E., 96
Nii, H., 318
Nims, R.W., 362
Nishida, T., 388
O'Connor, P.J., 119
O'Connor, T.R., 412
Odlanicki, J., 550
Ohshima, H., 88, 162, 172, 187, 204, 443
Ohtsubo, K., 420
Österdahl, B.G., 235, 507
Packer, P., 146
Padma, P.R., 520
Palli, D., 146
Pan, S.C., 434
Pangalis, G., 78
Park, S.S., 392
Parkin, M., 88
Parry, A., 107
Pedersen, E., 168
Peers, F., 48
Peña, A.S., 162
Peto, R., 502
Pfeifer, A., 294

Pignatelli, B., 162, 172
Pirastu, R., 146
Poirier, S., 158, 204
Poizat, R., 158
Polack, A., 204
Pool-Zobel, B.L., 322, 332
Preston-Martin, S., 197
Preussmann, R., 178, 210, 223, 489, 493, 592
Prévost, V., 102
Qian, Y.Z., 258
Qin, G.Z., 431
Qin, L.L., 431
Qiu, S.L., 192
Raedsch, R., 192
Reddel, R., 294
Reed, P.I., 139
Ruan, L.R., 253
Safaev, R.D., 485
Saidi, J., 507
Sandbothe, J., 244
Sansone, E.B., 226
Scanlan, R.A., 242, 275
Scheper, T., 351
Scherer, E., 496
Scherer, G., 339
Scherf, H.R., 588
Schmähl, D., 311
Schmezer, P., 322
Schoepke, M., 351
Sen, N.P., 232
Shao, Y.M., 204
Shen, X.H., 230
Shendrikova, I.A., 552
Shephard, S.E., 383
Shi, K.X., 143
Shi, Z.Y., 258
Shinohara, K., 388
Shuker, D.E.G., 102, 129
Siddiqi, M.A., 210
Sierra, R., 162
Sikora, M., 550
Skotnicki, S., 329
Smith, T., 265
Sommer, H., 238
Souliotis, V.L., 78
Spiegelhalder, B., 178, 223, 322, 351, 358, 489, 592
Srivatanakul, P., 88
Stershic, M.T., 370
Stillwell, W.G., 83
Stone, B.J., 460
Streeter, A.J., 362
Su, T., 253
Sugimura, M., 318

Author Index

Sukaryodhin, S., 88
Sullivan, B.P., 370
Sun, H.L., 424
Sun, Z.T., 502
Sundqvist, K., 281
Takeda, K., 558
Tannenbaum, S.R., 83
Tashiro, F., 420
Teuchmann, S., 162
de-Thé, G., 158, 204
Thompson, M.H., 571
Thuillier, P., 172
Thurnham, D.I., 192
Tian, J.F., 253
Tiedink, H.G.M., 584
Tozawa, H., 214
Tricker, A.R., 178, 210, 493
Trump, B.F. 294
Tsuda, M., 123
Tsuji, K., 427
Tsutsumi, M., 318
Ueno, Y., 420
Ukawa, S., 404, 558
Van Benthem, J., 496
Wacker, C.-D., 592
Wahrendorf, J., 192
Walters, C.L., 139

Wang, D.S., 424
Wang, H., 546
Wang, S., 392
Wang, X.L., 253
Watanabe, M., 388
Weber, B., 238
Weston, A., 294
Wiessler, M., 332, 351, 358, 588
Wild, C.P., 96
Williams, D.E., 275
Wilmer, J.W.G.M., 496
Wishnok, J.S., 83
Wu, A., 460
Wu, H.Y., 152
Wu, Y., 546
Wurdeman, R.L., 439
Wynder, E.L., 449
Xiao, F.Y., 424
Xing Y.D., 253
Xu, G.W., 33
Xu, H.X., 83, 230
Xu, L.Z., 143
Xu, X.B., 230
Xu, Y., 541
Xu, Y.M., 253, 258
Xu, Z.Y., 460
Yamamoto, K., 318

Yamamura, H., 420
Yan, R.Q., 431
Yang, C.S., 265, 366
Yang, G.R., 192
Yang, S.L., 253
Yang, W.X., 11
Yang, Z.T., 33
Yao, K.T., 434
Yin, Q., 460
Yoo, J.S.H., 366
You, W.C., 33
Yu, M.C., 39
Zankl, H., 332
Zaridze, D.G., 485
Zeller, W.J., 322
Zeng, Y., 204
Zhang, L., 33
Zhang, P., 258
Zhang, R.F., 152
Zhang, Y., 439
Zhao, D.Z., 460
Zhen, Y.Z., 253, 258
Zheng, Q.L., 253
Zheng, W., 62
Zhou, G., 253
Zhu, Y.R., 502
Zhukovskaya, N., 407

SUBJECT INDEX

A

Adduct
    to DNA
        formation of, 339–42, 347–8, 496–501
        measurement of, 71–7
    to haemoglobin
        measurement of, 71–7
Aflatoxin
    and human cancer, 48–53
    binding of to DNA, 427–30
    –glutathione conjugation, 427–30
    in diet, 48–51
    inhibition of genotoxicity of, 550–1
    measurement of, 131–2
    modulation of carcinogenicity of, 275–80, 420–3
    monitoring of exposure to, 96–101
    short-term test for carcinogenicity of, 431–3
Alcohol, as inhibitor of nitrosamine metabolism, 564–7
Aldehyde, in tobacco, 54–5
Alkanediazotate, 404–6
Alkylating agent, measurement of exposure to, 71–7
Alkylation, DNA (*see also* Methylation and *individual alkylated bases*)
    by nitrosated amino acids and peptides, 383–7
    by N-nitroso compounds, 305–10, 412–6
    sequence specificity of, 439–42
O^6-Alkylguanine
    –DNA alkyl transferase, 407–11
    measurement of, 78–82
N-Alkyl-N-nitrosoureas (*see also individual compounds*)
    DNA alkylation by, 439–42
Alternaria alternata
    and oesophageal cancer, 258–62
    contamination of grain by, 253–7
Amine, tertiary aromatic, nitrosation of, 244–52
Anticarcinogenicity, 520–4, 529–34
Antimutagenicity, 520–4, 535–7
Antioxidant status, and cancer mortality, 20
Arecaidine, 281–4
Areca nut, toxicity of, 281–5
Arecoline, 281–4
Ascorbic acid (*see also* Vitamin)
    and brain cancer, 201–2
    and endogenous formation of nitrosamines, 1, 7, 553–5
    and oesophageal cancer, 21
    and toxicity of tobacco smoke, 536–7
    effect of on nitrosamine formation, 126–7, 571–7
    effect of on intragastric environment, 139–42
Aspergillus, measurement of, 132
Azoxymethane, methylation by, 305–10

B

Bacteria
    and catalysis of nitrosation, 1, 2, 3, 6, 160,163,164–5, 187–91, 571–7
    and nasopharyngeal carcinoma, 158–61
    denitrifying, 6
Beer, nitrosamines in, 230–1, 242–3
Benzo[*a*]pyrene, in tobacco, 54, 483
Betel leaf, antimutagenic and anticarcinogenic effects of, 520–4
Betel quid (*see also* Areca nut), and oral cancer, 1, 3, 56, 466–70
Bilharzia (*see* Schistosomiasis)
Biliary tract, cancer of, and nitrosation, 137–8
Bladder (*see* Urinary bladder)
Brain, cancer of, and exposure to N-nitroso compounds, 197–203
Breast, cancer of, and diet, 18–21
Buccal cavity (*see* Oral cavity)

C

Campylobacter pylori (*see also Helicobacter pylori*), 163–6
β-Carotene, and gastric cancer, 23
Catalase, 388–91
Catechol, 483
Chloroethylnitrosourea, 412–6
Cholangiocarcinoma, 88–95
Chronic atrophic gastritis, 5
Cigarette smoking (*see also* Tobacco)
    and coronary heart disease, 66
    and digestive cancer, 65–6
    and gastric cancer, 34–5
    and lung cancer, 56–7, 64–5, 449–65, 471–6
    and oesophageal cancer, 57
    and oral cancer, 55–6

Cigarette smoking (contd)
    and pancreatic cancer, 57–8
    and urinary cancer, 65–6
    prevalence of in China, 63–4
Colon-rectum, cancer of, and diet, 18–21
Cosmetics, N-nitrosoalkanolamines in, 238–41
Cysteamine, 215–8
Cytochrome P450 system
    in metabolism of nitrosamines, 265–74, 351–3, 366–7
    in mutagenic activation, 398–403
    in rat liver microsomes, 392–7

D
Denitrosation, 375–82
    consequences of, 351–57
N-Dialkylnitrosamine, activation of, 286–93, 375–82
Diet (see also Food and individual dietary items)
    and cancer mortality in China, 18–21
    and gastric cancer, 22–8, 33–38
    nitrosamines in, 11–2, 210–3
N,N'-Dinitrosopiperazine, 434–8
Dithiocarbamate, effect on nitrosamine carcinogenicity, 587–90
Dosimetry, for determining human exposure to N-nitroso compounds, mycotoxins and tobacco, 71–134

E
Electrophile, acceleration of N-nitrosation by, 370–4
Ellagic acid, 550–1
Epstein-Barr virus
    activators in food, 204–9
    and formation of nitrosamines, 3, 44
    and transformation of nasopharyngeal epithelium, 437
2-Ethylhexyl 4-N-methyl-N-nitrosaminobenzoate, 244–52

F
Fatty acid, and inhibition of mutagenicity of N-nitroso compounds, 558–63
Fish
    meal, N-nitrosamines in, 214–8
    salted
        and gastric cancer, 26–7
        and nasopharyngeal cancer, 42–5
        N-nitrosamines in, 44
Food (see also Diet)
    fermented
        and cholangiocarcinoma, 91, 95
        and gastric cancer, 34–5
        and nasopharyngeal cancer, 44

Food (contd)
    nitrosamines in, 221
    mouldy, mutagenicity of, 253–7
    nitrosamides in, 152–7
    nitrosamines in, 91, 165–6, 219–22, 230–7
    preserved
        and gastric cancer, 4, 26–8
        and nasopharyngeal cancer, 204–9
        and oesophageal cancer, 212
        nitrosamines in, 221, 232–4
Formamidopyrimidine, 412–6
Free radical, role in nitrosamine activation and detoxication, 375–82
Fruit
    and gastric cancer, 23–5
    and nasopharyngeal cancer, 45
Fusarium, measurement of, 132

G
Gastric cancer (see Stomach, cancer of)
Gastritis
    chronic atrophic (CAG), 35–6, 172–6
    prior, and gastric cancer, 34–5
Genital organs, formation of nitrosamines in, 3
Genotoxicity (see Mutagenicity)
Glycosylase, 412–6
Guvacine, 281–4
Guvacoline, 281–4

H
Haemoglobin, adducts with, 71–7, 113–8
Helicobacter pylori (see also *Campylobacter pylori*), 28–9
Hepatitis B virus, and primary liver cancer, 48–52
Hydrogen peroxide, 516–9
Hydroxylation, 343–5, 392–7
$N7$-Hydroxyethylguanine, 412–6
4-Hydroxy-1-(3-pyridyl)-1-butanone, 113–8

I
Indole, effect of on mycotoxin and nitrosamine carcinogenesis, 275–80, 583–6
Isothiocyanate, 529–34

L
Lexitropsin, 439–42
Liver
    alkylation in, 120
    cancer, 6–7, 18–21
        and aflatoxins, 48–53
    formation of nitrosamines in, 3

Liver fluke
    and endogenous formation of nitrosamines, 1, 3, 4, 6–7, 88–95
    as risk factor for cholangiocarcinoma, 88–95
    effect of infection with on monitoring for aflatoxins, 96–101
Lung, cancer of
    and diet, 18–21
    and cigarette smoking, 56–7, 449–59, 504–5
    and oncogenes, 294–304
    and pollution, 462–3
    and tumour suppressor genes, 294–304
    environmental determinants of, 460–5
    prevention of by political measures, 471–6

M

Macrophage, and mediation of nitrosation, 2, 388–91
Metaplasia, intestinal, and nitrosation, 35–6
3-Methyladenine
    determination of in urine, 102–6
    in urine of patients at risk for gastric cancer, 83–7
Methylation, by N-nitroso compounds, 305–10, 329–31
N7-Methylguanine, in urine of patients at risk for gastric cancer, 83–7
O^6-Methylguanine
    and oesophageal cancer, 7
    formation of by N-nitroso compounds, 305–10
    location of, 119–22
N-Methyl-N-nitrosoethylurea, alkylation by, 305–10
N-Methyl-N-nitrosourea, alkylation by, 305–10, 329–31
Microflora, of nasopharynx, 158–61
Mutagenicity
    inhibition of, 550–1, 558–63
    of *Alternaria alternata*, 253–62
    of areca-nut constituents, 283
    of beans, 165–6
    of gastric juice, 172–7
    of salted fish, 44
    of nitrosated fish sauce, 154
    of N-nitroso compounds, 339, 398–403, 558–63
    of *Penicillium cyclopium*, 253–7
    of preserved foods, 204–9
    of sunscreen ingredients, 251
    modulation of, 535–7
Mycotoxins (*see also individual mycotoxins*)
    measurements of, 131–2
    modulation of carcinogenicity of, 274–80
    mutagenicity of, 253–7

N

Nasal cavity, formation of nitrosamines in, 3, 496–501
Nasopharynx
    cancer of
        and diet, 18–21, 39–47, 204–9
        and microflora, 158–61
        geographical variation in incidence of, 39–40
        racial/ethnic variation in incidence of, 40–1
    epithelium, transformation of, 434–8
Nass, 485–8
Nitrate
    conversion of into nitrite, 5
    excretion, 147
    formed by immunostimulation, 30
    in body fluids, 1, 83–7, 173–4, 178–81
    production of by macrophages, 388–91
    reduction of by bacteria, 160–1
    -rich foods and gastric cancer, 4
Nitrite
    in body fluids, 1, 173–4, 178–81
    production of by macrophages, 388–91
    reaction of, 384–6
    -rich foods and gastric cancer, 4
    trapping by thioproline, 123–8
N-Nitrodimethylamine, 358–61
Nitroreductase, in rat organs, 358–61
Nitrosamides, role of in gastric cancer, 152–7
Nitrosamines (*see also individual compounds*)
    as risk factors for cholangiocarcinoma, 88–95
    inhibition of formation of, 552–7, 568–70
    metabolism of, 265–74, 333
    production of by macrophages, 388–91
    toxicokinetic studies of, 362–5
    volatile, in foods, 204–22, 230–7
N-Nitrosamino acids, in tobacco and tobacco smoke, 477–81
Nitrosation
    endogenous, 1, 2, 3
        and cancer of the biliary tract, 137–8
    exogenous, 1, 160
    inhibition of, 4, 546–9
    in sunscreen ingredients, 244–52
    marker for, 443–8
    measurement of, 3–8, 130–1
    modulation of carcinogenicity of, 275–80
N-Nitroso(acetoxymethyl)methylamine, 329–31
N-Nitrosoalkanolamines, in cosmetics, 238–41

N'-Nitrosoanabasine, 57
  in smokeless tobacco products, 493–4
N'-Nitrosoanatabine, in smokeless tobacco products, 493–4
N-Nitrosoazetidine 4-carboxylic acid, in smokeless tobacco products, 493–4
N-Nitrosobis(2-acetoxypropyl)amine, 398–403
N-Nitrosobis(2-hydroxypropyl)amine
  activation of, 398–403
  formation of, 318–21
  in cosmetics, 238–41
N-Nitrosobis(2-hydroxypropyl)(2-oxopropyl)amine, 398–403
N-Nitrosobis(2-oxopropyl)amine, 305–10, 398–403
N-Nitrosobutyl-3-carboxypropylamine, 332–8
N-Nitrosobutyl(4-hydroxybutyl)amine, 343–5
N-Nitrosobutylmethylamine, 267
  in smokeless tobacco products, 493–4
N-Nitroso compounds (*see also individual compounds*)
  in diet, see Diet
  in tobacco products (*see also* Tobacco-specific nitrosamines), 54–61, 477–81, 493–5
  measurement of
    in gastric juice, 34–5
    in urine, 35–6
  occupational exposures to, 201
  prevention of exposure to, 541–93
  resistance to, 417–9
  total, measurement of, 5
N-Nitrosodibenzylamine, 232–4, 351–57
N-Nitrosodi-n-butylamine
  carcinogenicity of, 332–8
  in diet, 44, 232–4
N-Nitrosodiethanolamine
  effects of low doses of, 311–7
  in cosmetics, 238–41
  in smokeless tobacco products, 493–4
  mutagenicity of, 339–42
N-Nitrosodiethylamine
  effects of low doses of, 311–7
  in cigarette smoke, 52
  in diet, 12–3, 44, 220–1
  modulation of carcinogenicity of, 275–80
  occupational exposure to, 226–9
N-Nitroso(2,3-dihydroxypropyl)(2-hydroxypropyl)amine, 398–403
N-Nitrosodimethylamine
  demethylase, 366–9

N-Nitrosodimethylamine (contd)
  detection of cells for metabolism of, 119–22
  DNA single-strand breaks induced by, 514
  formation of, 358–61
  in diet, 12–3, 44, 208, 211, 220–1, 230–1, 235–6, 242–3
  inhalation of, 322–8
  inhibition of formation of, 552–7
  inhibition of metabolism and toxicity of, 587–90
  in smokeless tobacco products, 493–4
  metabolism of, 265–74
  methylation by, 305–10
N-Nitrosodimethylmorpholine
  activation of, 398–403
  metabolism of, 267
N-Nitrosodi-n-propylamine, in diet, 44, 220–1
N-Nitrosoethylmethylamine, 57, 267, 362
  in smokeless tobacco products, 493–4
  methylation by, 305–10
N-Nitrosoguvacine, 281–4
N-Nitrosoguvacoline, 281–4
N-Nitroso-2-hydroxyalkylamine, effects of, 339–42
N-Nitroso(2-hydroxyethyl)methylamine, 362–4
N-Nitroso(2-hydroxymethyl)butylamine, 560–1
N-Nitroso(2-hydroxymethyl)methylamine, 558–9, 563
N-Nitroso(2-hydroxymethyl)thiazolidine, 232–4
N-Nitroso(2-hydroxypropyl)methylamine, 398–403
N-Nitrosohydroxyproline, 493–4
N-Nitrosomethyl(acetoxybenzyl)amine, 329–31
N-Nitrosomethylamine, 242–3
3-(N-Nitrosomethylamino)butyric acid, 477–81
3-(N-Nitrosomethylamino)propionaldehyde, 281–4
3-(N-Nitrosomethylamino)propionic acid, 477–81
3-(N-Nitrosomethylamino)propionitrile, 281–4
4-(N-Nitrosomethylamino)-1-(3-pyridyl)-1-butanol (NNAl), 57, 268, 483
4-(N-Nitrosomethylamino)-4-(3-pyridyl)-1-butanol, in smokeless tobacco products, 493–4
4-(N-Nitrosomethylamino)-1-(3-pyridyl)-1-butanone (NNK)
  activation and deactivation of, 510–5
  and DNA adducts, 496–501
  in induction of cancer, 54–8
  in smokeless tobacco products, 493–5
  metabolism of, 270–1
  protection against carcinogenicity of, 520–4, 529–34
  quantification of, 113–8, 483, 489–92
4-(N-Nitrosomethylamino)-4-(3-pyridyl)butylaldehyde, 483

4-(*N*-Nitrosomethylamino)-4-(3-pyridyl)butyric acid, 477-81, 483
N-Nitrosomethyl-*n*-amylamine, 392-7
N-Nitroso-*N*-methylaniline, 232-4
    DNA binding by, 346-50
    metabolism of, 267
N-Nitroso-*N*-methylbenzylamine
    in diet, 11-6
    induction of oesophageal tumours by, 541-4
    metabolism of, 267
        inhibition of, 564-7
N-Nitroso-2-methylthiazolidine 4-carboxylic acid (NMTCA), in urine of patients with gastric dysplasia, 36
N-Nitrosomorpholine
    formation of in stomach, 187-91
    in diet, 44
    inhibition of formation of, 546-9, 571-7
    in smokeless tobacco products, 493-4
N'-Nitrosonornicotine
    activation of by hydrogen peroxide, 516-9
    in induction of cancer, 54-8
    in smokeless tobacco products, 493-5
    protection against, 525-8
    quantification of, 113-8, 483, 489-92
N-Nitroso(2-oxopropyl)methylamine, 398-403
N-Nitrosopipecolic acid, in smokeless tobacco products, 493-4
N-Nitrosopiperidine
    in diet, 44, 208, 211
    in smokeless tobacco products, 493-4
N-Nitrosoproline (NPRO)
    test, 1, 30
        and risk for gastric cancer, 36, 83-7
        evaluation of, 130-1
        in relation to vegetable consumption, 168-71
        to measure cancer risk, 146-51
    fate of, 477-81
    in foods, 220
    in smokeless tobacco products, 493-4
N-Nitrosopropylamine
    activation of, 398-403
    in smokeless tobacco products, 493-4
N-Nitrosopyrrolidine
    effects of low doses of, 311-7
    in diet, 44, 208, 211, 235-6
    in smokeless tobacco products, 493-4
    metabolism of, 267

N-Nitrososarcosine, 477-81
    in smokeless tobacco products, 493-4
N-Nitrosothiazolidine
    formation of in fish meal, 214-8
    in foods, 232-4
N-Nitrosothazolidine 4-carboxylic acid (NTCA)
    formation of, 188, 217-8
    in foods, 220, 232-4
    in smokeless tobacco products, 493-4
    measurement of, 123-8
N-Nitrosourea (*see also individual compounds*), analysis of, 223-5
3-Nitrotyrosine, 443-8
Nivalenol
    effect of, 420-3
    measurement of, 132

O

Ochratoxin, measurement of, 132
Oesophagus
    cancer of
        and cigarette smoking, 57
        and diet, 18-21, 210-3, 253-62
        and endogenous nitrosamines, 4, 7
        and 3-methyladenine in urine, 104-5
        inhibition of, 541-4
        precursor lesions of, 192-6
    carcinogen, 346-50, 564-7
Oncogene
    activation of by nitrosamines, 12-5, 57
    in lung cancer, 294-304
Opisthorchis viverrini (*see also* Liver fluke), and cholangiocarcinoma, 89-91, 96
Oral cavity
    cancer of, 54-6, 466-70
    epithelial cells, 281-5
    precancer of, 467-8

P

Pancreas, cancer of, and cigarette smoking, 57-8
Penicillium
    cyclopium, contamination of grain by, 253-7
    measurement of, 132
Polynuclear aromatic hydrocarbons (PAH)
    determination of, 71
Precursor
    lesions of gastric cancer, 33-8
    lesions of oesophageal cancer, 192-6
    of *N*-nitroso compounds, 172-7

Index

R
Retinoids, 578–82
Retinol, 20–1
Rubber
    accelerators, 591–3
    pacifiers, 232–4
    teats, 232–4

S
Salt (*see also* Fish, salted), and gastric cancer, 26–8, 34–5
Schistosoma haematobium, 178–81
Schistosomiasis, and endogenous formation of nitrosamines, 6, 178–81
Sea buckthorn juice, 568–70
Snuff
    as tumour promoter, 507–9
    haemoglobin adducts in users of, 115–8
    nitrosamines in, 235–6
Sterigmatocystin, 424–6
Stomach, cancer of, 1, 4–6
    and ascorbic acid, 139–42
    and cigarette smoking, 35–6
    and diet, 18–21
    formation of *N*-nitroso compounds in, 182–6, 187–91, 553
    in China, 18–21, 33–8, 152–7
    in Costa Rica, 1, 4, 5, 162–7
    precancerous lesions of, 33–8, 172–7
    risk factors for, 22–32, 83–7, 162–7
Sulfenamide accelerators, 591–3
Sunscreen ingredients, nitrosation of amines in, 244–52

T
Tea, and inhibition of nitrosamine carcinogenesis, 541–9
Thiazolidine, 215–8
Thiocyanate, effect of on nitrosamine formation, 126–7
Thioproline, nitrite-trapping capacity of, 123–8
Tobacco
    and cancer, 449–538
    black, nitrosamines in, 56
    carcinogens in, 485–8

Tobacco (contd)
    N-nitrosamino acids in, 477–81
    smoke
        analysis of, 482–4
        measurement of exposure to, 132–4
        modulation of genotoxicity of, 535–7
    smoking of
        and effects on health in China, 62–7, 502–6
        and endogenous formation of nitrosamines, 1
        and gastric cancer, 28–9
    –specific nitrosamines (TSNA; *see also individual compounds*), 54–8
        exposure to, 113–8
        in cigarettes, 489–92
        in *nass*, 487–8
        in snuff, 235–6
        in tobacco smoke, 483, 485–7
α-Tocopherol, 21
N-Trimethyl-*N*-nitrosourea, 182–6
Tumour suppressor gene, and lung cancer, 294–304

U
Urinary bladder
    cancer of, 6
        and schistosomiasis, 178–81
            nitrate, nitrite and nitroso compounds in, 178–81
    carcinogen, 332–8
    infections and endogenous formation of nitrosamines, 1, 3, 4

V
Vegetables
    and excretion of NPRO, 168–71
    and gastric cancer, 23–6, 33–5
    and nasopharyngeal cancer, 45
    nitrosamines in, 95, 583–6
Vitamin, protective effect of against brain cancer, 201–2
Vitamin A
    and inhibition of mutagenesis, 536
    effect of on nitrosamine metabolism, 525–8
Vitamin E, and inhibition of mutagenesis, 536

PUBLICATIONS OF THE INTERNATIONAL AGENCY FOR RESEARCH ON CANCER
Scientific Publications Series

(Available from Oxford University Press through local bookshops)

No. 1 Liver Cancer
1971; 176 pages (*out of print*)

No. 2 Oncogenesis and Herpesviruses
Edited by P.M. Biggs, G. de-Thé and L.N. Payne
1972; 515 pages (*out of print*)

No. 3 N-Nitroso Compounds: Analysis and Formation
Edited by P. Bogovski, R. Preussman and E.A. Walker
1972; 140 pages (*out of print*)

No. 4 Transplacental Carcinogenesis
Edited by L. Tomatis and U. Mohr
1973; 181 pages (*out of print*)

No. 5/6 Pathology of Tumours in Laboratory Animals, Volume 1, Tumours of the Rat
Edited by V.S. Turusov
1973/1976; 533 pages; £50.00

No. 7 Host Environment Interactions in the Etiology of Cancer in Man
Edited by R. Doll and I. Vodopija
1973; 464 pages; £32.50

No. 8 Biological Effects of Asbestos
Edited by P. Bogovski, J.C. Gilson, V. Timbrell and J.C. Wagner
1973; 346 pages (*out of print*)

No. 9 N-Nitroso Compounds in the Environment
Edited by P. Bogovski and E.A. Walker
1974; 243 pages; £21.00

No. 10 Chemical Carcinogenesis Essays
Edited by R. Montesano and L. Tomatis
1974; 230 pages (*out of print*)

No. 11 Oncogenesis and Herpesviruses II
Edited by G. de-Thé, M.A. Epstein and H. zur Hausen
1975; Part I: 511 pages
Part II: 403 pages; £65.00

No. 12 Screening Tests in Chemical Carcinogenesis
Edited by R. Montesano, H. Bartsch and L. Tomatis
1976; 666 pages; £45.00

No. 13 Environmental Pollution and Carcinogenic Risks
Edited by C. Rosenfeld and W. Davis
1975; 441 pages (*out of print*)

No. 14 Environmental N-Nitroso Compounds. Analysis and Formation
Edited by E.A. Walker, P. Bogovski and L. Griciute
1976; 512 pages; £37.50

No. 15 Cancer Incidence in Five Continents, Volume III
Edited by J.A.H. Waterhouse, C. Muir, P. Correa and J. Powell
1976; 584 pages; (*out of print*)

No. 16 Air Pollution and Cancer in Man
Edited by U. Mohr, D. Schmähl and L. Tomatis
1977; 328 pages (*out of print*)

No. 17 Directory of On-going Research in Cancer Epidemiology 1977
Edited by C.S. Muir and G. Wagner
1977; 599 pages (*out of print*)

No. 18 Environmental Carcinogens. Selected Methods of Analysis. Volume 1: Analysis of Volatile Nitrosamines in Food
Editor-in-Chief: H. Egan
1978; 212 pages (*out of print*)

No. 19 Environmental Aspects of N-Nitroso Compounds
Edited by E.A. Walker, M. Castegnaro, L. Griciute and R.E. Lyle
1978; 561 pages (*out of print*)

No. 20 Nasopharyngeal Carcinoma: Etiology and Control
Edited by G. de-Thé and Y. Ito
1978; 606 pages (*out of print*)

No. 21 Cancer Registration and its Techniques
Edited by R. MacLennan, C. Muir, R. Steinitz and A. Winkler
1978; 235 pages; £35.00

No. 22 Environmental Carcinogens. Selected Methods of Analysis. Volume 2: Methods for the Measurement of Vinyl Chloride in Poly(vinyl chloride), Air, Water and Foodstuffs
Editor-in-Chief: H. Egan
1978; 142 pages (*out of print*)

No. 23 Pathology of Tumours in Laboratory Animals. Volume II: Tumours of the Mouse
Editor-in-Chief: V.S. Turusov
1979; 669 pages (*out of print*)

No. 24 Oncogenesis and Herpesviruses III
Edited by G. de-Thé, W. Henle and F. Rapp
1978; Part I: 580 pages, Part II: 512 pages (*out of print*)

Prices, valid for January 1990, are subject to change without notice

List of IARC Publications

No. 25 Carcinogenic Risk. Strategies for Intervention
Edited by W. Davis and C. Rosenfeld
1979; 280 pages (*out of print*)

No. 26 Directory of On-going Research in Cancer Epidemiology 1978
Edited by C.S. Muir and G. Wagner
1978; 550 pages (*out of print*)

No. 27 Molecular and Cellular Aspects of Carcinogen Screening Tests
Edited by R. Montesano, H. Bartsch and L. Tomatis
1980; 372 pages; £29.00

No. 28 Directory of On-going Research in Cancer Epidemiology 1979
Edited by C.S. Muir and G. Wagner
1979; 672 pages (*out of print*)

No. 29 Environmental Carcinogens. Selected Methods of Analysis. Volume 3: Analysis of Polycyclic Aromatic Hydrocarbons in Environmental Samples
Editor-in-Chief: H. Egan
1979; 240 pages (*out of print*)

No. 30 Biological Effects of Mineral Fibres
Editor-in-Chief: J.C. Wagner
1980; **Volume 1:** 494 pages; **Volume 2:** 513 pages; £65.00

No. 31 *N*-Nitroso Compounds: Analysis, Formation and Occurrence
Edited by E.A. Walker, L. Griciute, M. Castegnaro and M. Börzsönyi
1980; 835 pages (*out of print*)

No. 32 Statistical Methods in Cancer Research. Volume 1. The Analysis of Case-control Studies
By N.E. Breslow and N.E. Day
1980; 338 pages; £20.00

No. 33 Handling Chemical Carcinogens in the Laboratory
Edited by R. Montesano *et al.*
1979; 32 pages (*out of print*)

No. 34 Pathology of Tumours in Laboratory Animals. Volume III. Tumours of the Hamster
Editor-in-Chief: V.S. Turusov
1982; 461 pages; £39.00

No. 35 Directory of On-going Research in Cancer Epidemiology 1980
Edited by C.S. Muir and G. Wagner
1980; 660 pages (*out of print*)

No. 36 Cancer Mortality by Occupation and Social Class 1851-1971
Edited by W.P.D. Logan
1982; 253 pages; £22.50

No. 37 Laboratory Decontamination and Destruction of Aflatoxins B_1, B_2, G_1, G_2 in Laboratory Wastes
Edited by M. Castegnaro *et al.*
1980; 56 pages; £6.50

No. 38 Directory of On-going Research in Cancer Epidemiology 1981
Edited by C.S. Muir and G. Wagner
1981; 696 pages (*out of print*)

No. 39 Host Factors in Human Carcinogenesis
Edited by H. Bartsch and B. Armstrong
1982; 583 pages; £46.00

No. 40 Environmental Carcinogens. Selected Methods of Analysis. Volume 4: Some Aromatic Amines and Azo Dyes in the General and Industrial Environment
Edited by L. Fishbein, M. Castegnaro, I.K. O'Neill and H. Bartsch
1981; 347 pages; £29.00

No. 41 *N*-Nitroso Compounds: Occurrence and Biological Effects
Edited by H. Bartsch, I.K. O'Neill, M. Castegnaro and M. Okada
1982; 755 pages; £48.00

No. 42 Cancer Incidence in Five Continents, Volume IV
Edited by J. Waterhouse, C. Muir, K. Shanmugaratnam and J. Powell
1982; 811 pages (*out of print*)

No. 43 Laboratory Decontamination and Destruction of Carcinogens in Laboratory Wastes: Some *N*-Nitrosamines
Edited by M. Castegnaro *et al.*
1982; 73 pages; £7.50

No. 44 Environmental Carcinogens. Selected Methods of Analysis. Volume 5: Some Mycotoxins
Edited by L. Stoloff, M. Castegnaro, P. Scott, I.K. O'Neill and H. Bartsch
1983; 455 pages; £29.00

No. 45 Environmental Carcinogens. Selected Methods of Analysis. Volume 6: *N*-Nitroso Compounds
Edited by R. Preussmann, I.K. O'Neill, G. Eisenbrand, B. Spiegelhalder and H. Bartsch
1983; 508 pages; £29.00

No. 46 Directory of On-going Research in Cancer Epidemiology 1982
Edited by C.S. Muir and G. Wagner
1982; 722 pages (*out of print*)

No. 47 Cancer Incidence in Singapore 1968-1977
Edited by K. Shanmugaratnam, H.P. Lee and N.E. Day
1983; 171 pages (*out of print*)

No. 48 Cancer Incidence in the USSR (2nd Revised Edition)
Edited by N.P. Napalkov, G.F. Tserkovny, V.M. Merabishvili, D.M. Parkin, M. Smans and C.S. Muir
1983; 75 pages; £12.00

No. 49 Laboratory Decontamination and Destruction of Carcinogens in Laboratory Wastes: Some Polycyclic Aromatic Hydrocarbons
Edited by M. Castegnaro, *et al.*
1983; 87 pages; £9.00

No. 50 Directory of On-going Research in Cancer Epidemiology 1983
Edited by C.S. Muir and G. Wagner
1983; 731 pages (*out of print*)

No. 51 Modulators of Experimental Carcinogenesis
Edited by V. Turusov and R. Montesano
1983; 307 pages; £22.50

List of IARC Publications

No. 52 Second Cancers in Relation to Radiation Treatment for Cervical Cancer: Results of a Cancer Registry Collaboration
Edited by N.E. Day and J.C. Boice, Jr
1984; 207 pages; £20.00

No. 53 Nickel in the Human Environment
Editor-in-Chief: F.W. Sunderman, Jr
1984; 529 pages; £41.00

No. 54 Laboratory Decontamination and Destruction of Carcinogens in Laboratory Wastes: Some Hydrazines
Edited by M. Castegnaro, et al.
1983; 87 pages; £9.00

No. 55 Laboratory Decontamination and Destruction of Carcinogens in Laboratory Wastes: Some N-Nitrosamides
Edited by M. Castegnaro et al.
1984; 66 pages; £7.50

No. 56 Models, Mechanisms and Etiology of Tumour Promotion
Edited by M. Börzsönyi, N.E. Day, K. Lapis and H. Yamasaki
1984; 532 pages; £42.00

No. 57 N-Nitroso Compounds: Occurrence, Biological Effects and Relevance to Human Cancer
Edited by I.K. O'Neill, R.C. von Borstel, C.T. Miller, J. Long and H. Bartsch
1984; 1013 pages; £80.00

No. 58 Age-related Factors in Carcinogenesis
Edited by A. Likhachev, V. Anisimov and R. Montesano
1985; 288 pages; £20.00

No. 59 Monitoring Human Exposure to Carcinogenic and Mutagenic Agents
Edited by A. Berlin, M. Draper, K. Hemminki and H. Vainio
1984; 457 pages; £27.50

No. 60 Burkitt's Lymphoma: A Human Cancer Model
Edited by G. Lenoir, G. O'Conor and C.L.M. Olweny
1985; 484 pages; £29.00

No. 61 Laboratory Decontamination and Destruction of Carcinogens in Laboratory Wastes: Some Haloethers
Edited by M. Castegnaro et al.
1985; 55 pages; £7.50

No. 62 Directory of On-going Research in Cancer Epidemiology 1984
Edited by C.S. Muir and G. Wagner
1984; 717 pages (*out of print*)

No. 63 Virus-associated Cancers in Africa
Edited by A.O. Williams, G.T. O'Conor, G.B. de-Thé and C.A. Johnson
1984; 773 pages; £22.00

No. 64 Laboratory Decontamination and Destruction of Carcinogens in Laboratory Wastes: Some Aromatic Amines and 4-Nitrobiphenyl
Edited by M. Castegnaro et al.
1985; 84 pages; £6.95

No. 65 Interpretation of Negative Epidemiological Evidence for Carcinogenicity
Edited by N.J. Wald and R. Doll
1985; 232 pages; £20.00

No. 66 The Role of the Registry in Cancer Control
Edited by D.M. Parkin, G. Wagner and C.S. Muir
1985; 152 pages; £10.00

No. 67 Transformation Assay of Established Cell Lines: Mechanisms and Application
Edited by T. Kakunaga and H. Yamasaki
1985; 225 pages; £20.00

No. 68 Environmental Carcinogens. Selected Methods of Analysis. Volume 7. Some Volatile Halogenated Hydrocarbons
Edited by L. Fishbein and I.K. O'Neill
1985; 479 pages; £42.00

No. 69 Directory of On-going Research in Cancer Epidemiology 1985
Edited by C.S. Muir and G. Wagner
1985; 745 pages; £22.00

No. 70 The Role of Cyclic Nucleic Acid Adducts in Carcinogenesis and Mutagenesis
Edited by B. Singer and H. Bartsch
1986; 467 pages; £40.00

No. 71 Environmental Carcinogens. Selected Methods of Analysis. Volume 8: Some Metals: As, Be, Cd, Cr, Ni, Pb, Se Zn
Edited by I.K. O'Neill, P. Schuller and L. Fishbein
1986; 485 pages; £42.00

No. 72 Atlas of Cancer in Scotland, 1975–1980. Incidence and Epidemiological Perspective
Edited by I. Kemp, P. Boyle, M. Smans and C.S. Muir
1985; 285 pages; £35.00

No. 73 Laboratory Decontamination and Destruction of Carcinogens in Laboratory Wastes: Some Antineoplastic Agents
Edited by M. Castegnaro et al.
1985; 163 pages; £10.00

No. 74 Tobacco: A Major International Health Hazard
Edited by D. Zaridze and R. Peto
1986; 324 pages; £20.00

No. 75 Cancer Occurrence in Developing Countries
Edited by D.M. Parkin
1986; 339 pages; £20.00

No. 76 Screening for Cancer of the Uterine Cervix
Edited by M. Hakama, A.B. Miller and N.E. Day
1986; 315 pages; £25.00

No. 77 Hexachlorobenzene: Proceedings of an International Symposium
Edited by C.R. Morris and J.R.P. Cabral
1986; 668 pages; £50.00

List of IARC Publications

No. 78 Carcinogenicity of Alkylating Cytostatic Drugs
Edited by D. Schmähl and
J.M. Kaldor
1986; 337 pages; £25.00

No. 79 Statistical Methods in Cancer Research. Volume III: The Design and Analysis of Long-term Animal Experiments
By J.J. Gart, D. Krewski, P.N. Lee, R.E. Tarone and J. Wahrendorf
1986; 213 pages; £20.00

No. 80 Directory of On-going Research in Cancer Epidemiology 1986
Edited by C.S. Muir and G. Wagner
1986; 805 pages; £22.00

No. 81 Environmental Carcinogens: Methods of Analysis and Exposure Measurement. Volume 9: Passive Smoking
Edited by I.K. O'Neill,
K.D. Brunnemann, B. Dodet and
D. Hoffmann
1987; 383 pages; £35.00

No. 82 Statistical Methods in Cancer Research. Volume II: The Design and Analysis of Cohort Studies
By N.E. Breslow and N.E. Day
1987; 404 pages; £30.00

No. 83 Long-term and Short-term Assays for Carcinogens: A Critical Appraisal
Edited by R. Montesano,
H. Bartsch, H. Vainio, J. Wilbourn and H. Yamasaki
1986; 575 pages; £48.00

No. 84 The Relevance of N-Nitroso Compounds to Human Cancer: Exposure and Mechanisms
Edited by H. Bartsch, I.K. O'Neill and R. Schulte-Hermann
1987; 671 pages; £50.00

No. 85 Environmental Carcinogens: Methods of Analysis and Exposure Measurement. Volume 10: Benzene and Alkylated Benzenes
Edited by L. Fishbein and
I.K. O'Neill
1988; 327 pages; £35.00

No. 86 Directory of On-going Research in Cancer Epidemiology 1987
Edited by D.M. Parkin and
J. Wahrendorf
1987; 676 pages; £22.00

No. 87 International Incidence of Childhood Cancer
Edited by D.M. Parkin, C.A. Stiller, C.A. Bieber, G.J. Draper,
B. Terracini and J.L. Young
1988; 401 pages; £35.00

No. 88 Cancer Incidence in Five Continents Volume V
Edited by C. Muir, J. Waterhouse, T. Mack, J. Powell and S. Whelan
1987; 1004 pages; £50.00

No. 89 Method for Detecting DNA Damaging Agents in Humans: Applications in Cancer Epidemiology and Prevention
Edited by H. Bartsch, K. Hemminki and I.K. O'Neill
1988; 518 pages; £45.00

No. 90 Non-occupational Exposure to Mineral Fibres
Edited by J. Bignon, J. Peto and
R. Saracci
1989; 500 pages; £45.00

No. 91 Trends in Cancer Incidence in Singapore 1968–1982
Edited by H.P. Lee, N.E. Day and K. Shanmugaratnam
1988; 160 pages; £25.00

No. 92 Cell Differentiation, Genes and Cancer
Edited by T. Kakunaga,
T. Sugimura, L. Tomatis and
H. Yamasaki
1988; 204 pages; £25.00

No. 93 Directory of On-going Research in Cancer Epidemiology 1988
Edited by M. Coleman and
J. Wahrendorf
1988; 662 pages (*out of print*)

No. 94 Human Papillomavirus and Cervical Cancer
Edited by N. Muñoz, F.X. Bosch and O.M. Jensen
1989; 154 pages; £19.00

No. 95 Cancer Registration: Principles and Methods
Edited by O.M. Jensen,
D.M. Parkin, R. MacLennan,
C.S. Muir and R. Skeet
Publ. due 1991; approx. 300 pages £28.00

No. 96 Perinatal and Multigeneration Carcinogenesis
Edited by N.P. Napalkov,
J.M. Rice, L. Tomatis and
H. Yamasaki
1989; 436 pages; £48.00

No. 97 Occupational Exposure to Silica and Cancer Risk
Edited by L. Simonato,
A.C. Fletcher, R. Saracci and
T. Thomas
1990; 124 pages; £19.00

No. 98 Cancer Incidence in Jewish Migrants to Israel, 1961–1981
Edited by R. Steinitz, D.M. Parkin, J.L. Young, C.A. Bieber and
L. Katz
1989; 320 pages; £30.00

No. 99 Pathology of Tumours in Laboratory Animals, Second Edition, Volume 1, Tumours of the Rat
Edited by V.S. Turusov and
U. Mohr
740 pages; £85.00

No. 100 Cancer: Causes, Occurrence and Control
Editor-in-Chief L. Tomatis
1990; 352 pages; £24.00

List of IARC Publications

No. 101 Directory of On-going Research in Cancer Epidemiology 1989/90
Edited by M. Coleman and J. Wahrendorf
1989; 818 pages; £36.00

No. 102 Patterns of Cancer in Five Continents
Edited by S.L. Whelan and D.M. Parkin
1990; 162 pages; £25.00

No. 103 Evaluating Effectiveness of Primary Prevention of Cancer
Edited by M. Hakama, V. Beral, J.W. Cullen and D.M. Parkin
1990; 250 pages; £32.00

No. 104 Complex Mixtures and Cancer Risk
Edited by H. Vainio, M. Sorsa and A.J. McMichael
1990; 442 pages; £38.00

No. 105 Relevance to Human Cancer of N-Nitroso Compounds, Tobacco Smoke and Mycotoxins
Edited by I.K. O'Neill, J. Chen and H. Bartsch
Publ. due 1991; approx. 600 pages £70.00

No. 108 Environmental Carcinogens: Methods of Analysis and Exposure Measurement. Volume 11: Polychlorinated Dioxins and Dibenzofurans
Edited by C. Rappe, H.R. Buser, B. Dodet and I.K. O'Neill
Publ. due 1991; approx. 400 pages; £45.00

No. 109 Environmental Carcinogens: Methods of Analysis and Exposure Measurement. Volume 12: Indoor Air Contaminants
Edited by B. Seifert, B. Dodet and I.K. O'Neill
Publ. due 1991; approx. 400 pages

No. 110 Directory of On-going Research in Cancer Epidemiology 1991
Edited by M. Coleman and J. Wahrendorf
1991; approx. 720 pages; £38.00

No. 112 Autopsy in Epidemiology and Medical Research
Edited by E. Riboli and m. Delendi
1991; approx 250 pages; £25.00

List of IARC Publications

IARC MONOGRAPHS ON THE EVALUATION OF CARCINOGENIC RISKS TO HUMANS

(Available from booksellers through the network of WHO Sales Agents*)

Volume 1 **Some Inorganic Substances, Chlorinated Hydrocarbons, Aromatic Amines, *N*-Nitroso Compounds, and Natural Products**
1972; 184 pages (*out of print*)

Volume 2 **Some Inorganic and Organometallic Compounds**
1973; 181 pages (out of print)

Volume 3 **Certain Polycyclic Aromatic Hydrocarbons and Heterocyclic Compounds**
1973; 271 pages (*out of print*)

Volume 4 **Some Aromatic Amines, Hydrazine and Related Substances, *N*-Nitroso Compounds and Miscellaneous Alkylating Agents**
1974; 286 pages;
Sw. fr. 18.-/US $14.40

Volume 5 **Some Organochlorine Pesticides**
1974; 241 pages (*out of print*)

Volume 6 **Sex Hormones** 1974;
243 pages (*out of print*)

Volume 7 **Some Anti-Thyroid and Related Substances, Nitrofurans and Industrial Chemicals**
1974; 326 pages (*out of print*)

Volume 8 **Some Aromatic Azo Compounds**
1975; 375 pages;
Sw. fr. 36.-/US $28.80

Volume 9 **Some Aziridines, *N*-, *S*- and *O*-Mustards and Selenium**
1975; 268 pages;
Sw.fr. 27.-/US $21.60

Volume 10 **Some Naturally Occurring Substances**
1976; 353 pages (*out of print*)

Volume 11 **Cadmium, Nickel, Some Epoxides, Miscellaneous Industrial Chemicals and General Considerations on Volatile Anaesthetics**
1976; 306 pages (*out of print*)

Volume 12 **Some Carbamates, Thiocarbamates and Carbazides**
1976; 282 pages;
Sw. fr. 34.-/US $27.20

Volume 13 **Some Miscellaneous Pharmaceutical Substances** 1977;
255 pages;
Sw. fr. 30.-/US$ 24.00

Volume 14 **Asbestos**
1977; 106 pages (*out of print*)

Volume 15 **Some Fumigants, The Herbicides 2,4-D and 2,4,5-T, Chlorinated Dibenzodioxins and Miscellaneous Industrial Chemicals**
1977; 354 pages;
Sw. fr. 50.-/US $40.00

Volume 16 **Some Aromatic Amines and Related Nitro Compounds - Hair Dyes, Colouring Agents and Miscellaneous Industrial Chemicals**
1978; 400 pages;
Sw. fr. 50.-/US $40.00

Volume 17 **Some *N*-Nitroso Compounds**
1987; 365 pages;
Sw. fr. 50.-/US $40.00

Volume 18 **Polychlorinated Biphenyls and Polybrominated Biphenyls**
1978; 140 pages;
Sw. fr. 20.-/US $16.00

Volume 19 **Some Monomers, Plastics and Synthetic Elastomers, and Acrolein**
1979; 513 pages;
Sw. fr. 60.-/US $48.00

Volume 20 **Some Halogenated Hydrocarbons**
1979; 609 pages (*out of print*)

Volume 21 **Sex Hormones (II)**
1979; 583 pages;
Sw. fr. 60.-/US $48.00

Volume 22 **Some Non-Nutritive Sweetening Agents**
1980; 208 pages;
Sw. fr. 25.-/US $20.00

Volume 23 **Some Metals and Metallic Compounds**
1980; 438 pages (*out of print*)

Volume 24 **Some Pharmaceutical Drugs**
1980; 337 pages;
Sw. fr. 40.-/US $32.00

Volume 25 **Wood, Leather and Some Associated Industries**
1981; 412 pages;
Sw. fr. 60.-/US $48.00

Volume 26 **Some Antineoplastic and Immunosuppressive Agents**
1981; 411 pages;
Sw. fr. 62.-/US $49.60

Volume 27 **Some Aromatic Amines, Anthraquinones and Nitroso Compounds, and Inorganic Fluorides Used in Drinking Water and Dental Preparations**
1982; 341 pages;
Sw. fr. 40.-/US $32.00

Volume 28 **The Rubber Industry**
1982; 486 pages;
Sw. fr. 70.-/US $56.00

Volume 29 **Some Industrial Chemicals and Dyestuffs**
1982; 416 pages;
Sw. fr. 60.-/US $48.00

Volume 30 **Miscellaneous Pesticides**
1983; 424 pages;
Sw. fr. 60.-/US $48.00

Volume 31 **Some Food Additives, Feed Additives and Naturally Occurring Substances**
1983; 314 pages;
Sw. fr. 60-/US $48.00

List of IARC Publications

Volume 32 Polynuclear Aromatic Compounds, Part 1: Chemical, Environmental and Experimental Data
1984; 477 pages;
Sw. fr. 60.-/US $48.00

Volume 33 Polynuclear Aromatic Compounds, Part 2: Carbon Blacks, Mineral Oils and Some Nitroarenes
1984; 245 pages;
Sw. fr. 50.-/US $40.00

Volume 34 Polynuclear Aromatic Compounds, Part 3: Industrial Exposures in Aluminium Production, Coal Gasification, Coke Production, and Iron and Steel Founding
1984; 219 pages;
Sw. fr. 48.-/US $38.40

Volume 35 Polynuclear Aromatic Compounds, Part 4: Bitumens, Coal-tars and Derived Products, Shale-oils and Soots
1985; 271 pages;
Sw. fr. 70.-/US $56.00

Volume 37 Tobacco Habits Other than Smoking: Betel-quid and Areca-nut Chewing; and some Related Nitrosamines
1985; 291 pages;
Sw. fr. 70.-/US $56.00

Volume 38 Tobacco Smoking
1986; 421 pages;
Sw. fr. 75.-/US $60.00

Volume 39 Some Chemicals Used in Plastics and Elastomers
1986; 403 pages;
Sw. fr. 60.-/US $48.00

Volume 40 Some Naturally Occurring and Synthetic Food Components, Furocoumarins and Ultraviolet Radiation
1986; 444 pages;
Sw. fr. 65.-/US $52.00

Volume 41 Some Halogenated Hydrocarbons and Pesticide Exposures
1986; 434 pages;
Sw. fr. 65.-/US $52.00

Volume 42 Silica and Some Silicates
1987; 289 pages;
Sw. fr. 65.-/US $52.00

Volume 43 Man-Made Mineral Fibres and Radon
1988; 300 pages;
Sw. fr. 65.-/US $52.00

Volume 44 Alcohol Drinking
1988; 416 pages;
Sw. fr. 65.-/US $52.00

Volume 45 Occupational Exposures in Petroleum Refining; Crude Oil and Major Petroleum Fuels
1989; 322 pages;
Sw. fr. 65.-/US $52.00

Volume 46 Diesel and Gasoline Engine Exhausts and Some Nitroarenes
1989; 458 pages;
Sw. fr. 65.-/US $52.00

Volume 47 Some Organic Solvents, Resin Monomers and Related Compounds, Pigments and Occupational Exposures in Paint Manufacture and Painting
1990; 536 pages;
Sw. fr. 85.-/US $68.00

Volume 48 Some Flame Retardants and Textile Chemicals, and Exposures in the Textile Manufacturing Industry
1990; 345 pages;
Sw. fr. 65.-/US $52.00

Volume 49 Chromium, Nickel and Welding
1990; 677 pages;
Sw. fr. 95.–/US$76.00

Volume 50 Pharmaceutical Drugs
1990; 415 pages;
Sw. fr. 65.–/US$52.00

Volume 51 Coffee, Tea, Mate, Methylxanthines and Methylglyoxal
1991; 513 pages;
Sw. fr. 80.–/US$64.00

Supplement No. 1
Chemicals and Industrial Processes Associated with Cancer in Humans (IARC Monographs, Volumes 1 to 20)
1979; 71 pages; (*out of print*)

Supplement No. 2
Long-term and Short-term Screening Assays for Carcinogens: A Critical Appraisal
1980; 426 pages;
Sw. fr. 40.-/US $32.00

Supplement No. 3
Cross Index of Synonyms and Trade Names in Volumes 1 to 26
1982; 199 pages (*out of print*)

Supplement No. 4
Chemicals, Industrial Processes and Industries Associated with Cancer in Humans (IARC Monographs, Volumes 1 to 29)
1982; 292 pages (*out of print*)

Supplement No. 5
Cross Index of Synonyms and Trade Names in Volumes 1 to 36
1985; 259 pages;
Sw. fr. 46.-/US $36.80

Supplement No. 6
Genetic and Related Effects: An Updating of Selected IARC Monographs from Volumes 1 to 42
1987; 729 pages;
Sw. fr. 80.-/US $64.00

Supplement No. 7
Overall Evaluations of Carcinogenicity: An Updating of IARC Monographs Volumes 1-42
1987; 434 pages;
Sw. fr. 65.-/US $52.00

Supplement No. 8
Cross Index of Synonyms and Trade Names in Volumes 1 to 46 of the IARC Monographs
1990; 260 pages;
Sw. fr. 60.-/US $48.00

List of IARC Publications

IARC TECHNICAL REPORTS*

No. 1 Cancer in Costa Rica
Edited by R. Sierra,
R. Barrantes, G. Muñoz Leiva,
D.M. Parkin, C.A. Bieber and
N. Muñoz Calero
1988; 124 pages;
Sw. fr. 30.-/US $24.00

No. 2 SEARCH: A Computer Package to Assist the Statistical Analysis of Case-control Studies
Edited by G.J. Macfarlane,
P. Boyle and P. Maisonneuve (in press)

No. 3 Cancer Registration in the European Economic Community
Edited by M.P. Coleman and
E. Démaret
1988; 188 pages;
Sw. fr. 30.-/US $24.00

No. 4 Diet, Hormones and Cancer: Methodological Issues for Prospective Studies
Edited by E. Riboli and
R. Saracci
1988; 156 pages;
Sw. fr. 30.-/US $24.00

No. 5 Cancer in the Philippines
Edited by A.V. Laudico,
D. Esteban and D.M. Parkin
1989; 186 pages;
Sw. fr. 30.-/US $24.00

No. 6 La genèse du Centre International de Recherche sur le Cancer
Par R. Sohier et A.G.B. Sutherland
1990; 104 pages
Sw. fr. 30.-/US $24.00

No. 7 Epidémiologie du cancer dans les pays de langue latine
1990; 310 pages
Sw. fr. 30.-/US $24.00

No. 8 Comparative Study of Anti-smoking Legislation in Countries of the European Economic Community
Edited by A. Sasco
1990; c. 80 pages
Sw. fr. 30.-/US $24.00
(English and French editions available) (in press)

DIRECTORY OF AGENTS BEING TESTED FOR CARCINOGENICITY (Until Vol. 13 Information Bulletin on the Survey of Chemicals Being Tested for Carcinogenicity)*

No. 8 Edited by M.-J. Ghess,
H. Bartsch and L. Tomatis
1979; 604 pages; Sw. fr. 40.-

No. 9 Edited by M.-J. Ghess,
J.D. Wilbourn, H. Bartsch and
L. Tomatis
1981; 294 pages; Sw. fr. 41.-

No. 10 Edited by M.-J. Ghess,
J.D. Wilbourn and H. Bartsch
1982; 362 pages; Sw. fr. 42.-

No. 11 Edited by M.-J. Ghess,
J.D. Wilbourn, H. Vainio and
H. Bartsch
1984; 362 pages; Sw. fr. 50.-

No. 12 Edited by M.-J. Ghess,
J.D. Wilbourn, A. Tossavainen and
H. Vainio
1986; 385 pages; Sw. fr. 50.-

No. 13 Edited by M.-J. Ghess,
J.D. Wilbourn and A. Aitio 1988;
404 pages; Sw. fr. 43.-

No. 14 Edited by M.-J. Ghess,
J.D. Wilbourn and H. Vainio
1990; c. 370 pages; Sw. fr. 45.-

NON-SERIAL PUBLICATIONS †

Alcool et Cancer
By A. Tuyns (in French only)
1978; 42 pages; Fr. fr. 35.-

Cancer Morbidity and Causes of Death Among Danish Brewery Workers
By O.M. Jensen 1980;
143 pages; Fr. fr. 75.-

Directory of Computer Systems Used in Cancer Registries
By H.R. Menck and D.M. Parkin
1986; 236 pages;
Fr. fr. 50.-

---

* Available from booksellers through the network of WHO sales agents.

† Available directly from IARC